T0338181

5G Physical Layer Technologies

5G Physical Layer Technologies

Mosa Ali Abu-Rgheff
Centre for Security, Communications and Network Research
University of Plymouth
United Kingdom

WILEY

This edition first published 2020

Registered Offices
John Wiley & Sons, Inc., 111 River Street, Hoboken, NJ 07030, USA
John Wiley & Sons Ltd, The Atrium, Southern Gate, Chichester, West Sussex, PO19 8SQ, UK

Editorial Office
The Atrium, Southern Gate, Chichester, West Sussex, PO19 8SQ, UK

For details of our global editorial offices, customer services, and more information about Wiley products visit us at www.wiley.com.

Wiley also publishes its books in a variety of electronic formats and by print-on-demand. Some content that appears in standard print versions of this book may not be available in other formats.

Library of Congress Cataloging-in-Publication Data

Names: Abu-Rgheff, Mosa Ali, author.
Title: 5G physical layer technologies / Mosa Ali Abu-Rgheff, University of Plymouth, UK.
Description: Hoboken, NJ : John Wiley & Sons, Inc., 2020. | Includes bibliographical references and index. |
Identifiers: LCCN 2019014360 (print) | LCCN 2019017528 (ebook) | ISBN 9781119525523 (Adobe PDF) | ISBN 9781119525493 (ePub) | ISBN 9781119525516 (hardback)
Subjects: LCSH: 5G mobile communication systems–Equipment and supplies.
Classification: LCC TK5103.25 (ebook) | LCC TK5103.25 .A29 2019 (print) | DDC 621.39/81–dc23
LC record available at https://lccn.loc.gov/2019014360

Cover Design: Wiley
Cover Image: © jamesteohart/Shutterstock

Set in 10/12pt WarnockPro by SPi Global, Chennai, India
Printed and bound in Singapore by Markono Print Media Pte Ltd

10 9 8 7 6 5 4 3 2 1

Contents

Preface

The cellular wireless generation (G) commonly implies a change in the characteristics of the system such as data speed, access or transmission technology, and operating frequency bands. Each generation has particular standards, capabilities, and technologies, with new architectures that differentiate it from the previous one. Out of the four generations of cellular wireless communication, the 2G network has been the most profitable, with the longest life cycle. 3G network has possibly been less successful than 4G networks.

Not counting the analogue 1G network, new wireless generations are designed to offer a higher speed and amount of data transportation, along with new features and services. The amount of data transported by wireless networks is expected to approach 500 EBs by 2020, compared to 3 exabytes (EBs) in 2010, (1 EB = *1 billion gigabytes).* The growth of data is mainly driven by devices such as smart phones, tablets, and video streaming. In addition, the number of connected devices and data rates will continue to increase exponentially. Consequently, it was clear to researchers and industry that an incremental approach to system improvements to these challenges would not achieve its targets by 2020. There were enormous calls on academia and industry to meet these challenges through innovative new technologies that are practical, smart, and efficient.

The 3GPP has led the campaign for long-term evolution (LTE) of wireless communications, and 3GPP recommendations are arranged in Releases. Release 14 was published in 2017; Release 15 came out at the end of 2018, and Release 16 is due at the end of 2019. Releases 14 and 15 concur with 5G trial service tests and 5G commercial service enrolled in 2020.

5G wireless communications technology comprises economic growth and industrial revolution. 5G will be the backbone of the Internet of Things (IoT), connecting devices, machines, and vehicles. Contemporary wireless networks must be upgraded to the new architectures to implement new technologies. New hardware must be employed to make this revolution possible. Public administrations have to engage with their citizens to build new smart communities and cities. Mobile service providers are required to upgrade to give their subscribers the new experience. 5G also enables the Internet of Medical Things (IoMT), like monitoring devices and wearable equipment for patients suffering from serious health issues, collaborative cancer cloud platforms to fine-tune cancer therapies, and remote surgery. 5G enables smart grids linking all the national power generation, transmission, and distribution resources and management systems. All these applications and many more need knowledge and understanding of the elements and functioning of 5G.

Therefore, design engineers and service providers should update their knowledge about the new wireless systems to make this change happen. Universities are obliged to update their engineering courses to provide trained engineers. Knowledge updating through training to provide skills and well-written technology books would help in this quest. As a matter of fact, many books could be written to explain in depth various 5G technologies, and there are many excellent books on the market now. However, there are also gaps where technologies are either not explained in depth or not discussed at all. The book attempts to fill in some of these gaps and to develop these technologies in depth for the readers' benefit.

The book is aimed at engineers working in industrial design and research and development (R & D), private–public research communities, university academics involved in research, and postgraduate courses (PhD and MSc), including senior engineering degree courses, technical managers, and service providers. Readers should have a degree in electronics or communications.

A key prerequisite in all written engineering books is mathematics. Support on advanced mathematics used is provided in the book as appendices. The proposed book deals with 5G advanced technologies that are key to anyone involved with the physical (PHY) layer topics, such as enabling technologies, full-duplex communications, full-dimension beamforming, massive MIMO channel estimation, multi-user precoding, and many more.

Acknowledgements

It gives me great pleasure to acknowledge the many people who have influenced my thinking and contributed to the presentation of this book. In particular, I am grateful to the staff at the Centre for Security, Communications and Network Research (CSCAN), Plymouth, UK, for their support and encouragement and for sponsoring my use of the University Library. I am obliged to the following professors: Professor David Gesbert at the Mobile Communications Laboratory of EURECOM; Professor Carlo Capsoni at the Politecnico di Milano; and Professor Mao Wang at Southeast University China; for their indispensable advice and discussions. My special thanks to Dr. Timothy Reis at the University of Greenwich, UK, for his invaluable help on mathematical problems. Thanks are also due to the many researchers whose contributions are cited in the book, and apologies to those who were left out for lack of space.

Last, but foremost, I wish to dedicate the book to my late parents and to my family as an acknowledgement of their unfailing love and support throughout the preparation and completion of the book.

Plymouth Devon,
UK

Mosa Ali Abu-Rgheff

List of Mathematical Notation

$\mathbb{C}\mathcal{N}$	Complex normal random distribution
$\mathbb{R}$	A set of real numbers generally over single direction (dimension)
$\mathbb{R}^d$	d-dimensional real vector space
$\mathbf{X}$	Matrix $\mathbf{X}$
$\mathbf{x}$	Vector $\mathbf{x}$
$[\mathbf{X}]_{i,j}$	The element on row i and column j of $\mathbf{X}$
$\mathbf{X}^*$	Conjugate of $\mathbf{X}$
$\mathbf{X}^{\mathrm{T}}$	Transpose of $\mathbf{X}$
$\mathbf{X}^{\mathrm{H}}$	Conjugate and transpose of $\mathbf{X}$
$\det(\mathbf{X})$	Determinant of $\mathbf{X}$
$\mathrm{tr}(\mathbf{X})$	Trace of $\mathbf{X}$
$\|\mathbf{X}\|_F$	Frobenius norm of $\mathbf{X}$
$\lambda(\mathbf{X})$	Eigenvalues of $\mathbf{X}$
$Vec(\mathbf{X})$	Vectorization operator stacks the columns of $\mathbf{X}$ on top of each other
$\mathbf{\Lambda}_{\mathrm{X}}$	Diagonal matrix of the eigenvalues of $\mathbf{X}$
$\mathbf{\Sigma}_{\mathrm{X}}$	Diagonal matrix of the singular values of $\mathbf{X}$
$\mathbf{U}_{\mathrm{X}}$	Eigen or singular vectors of $\mathbf{X}$
σ_{sh}^2	Shadow fading variance
(r, θ, ϕ)	Spherical coordinates: radial distance r, polar angle θ, and azimuthal angle ϕ
Z_0	Reference impedance normally 50 $\mathbf{\Omega}$
$\mathrm{h}(\boldsymbol{x})$	Differential entropy of $\mathbf{x}$
$\boldsymbol{I}(\boldsymbol{y}; \boldsymbol{x})$	Mutual information of $\boldsymbol{x}$ of $\boldsymbol{y}$
λ	Wavelength
j	$\sqrt{-1}$
(a , b]	$\{x \in \mathbb{R} : a < x \leq b\}$
$\rho\,(\lambda_1, \lambda_2)$	Densification gain
σ^2	Noise variance

List of Wireless Network Symbols

B	BS signal bandwidth
M	Number of transmit antennas
N	Number of receive antennas
M_B	Number of BS receive antennas
N_c	Number of cells in the network
K	Number of users
$\mathbf{H}$	MIMO flat-fading channel matrix
$\mathbf{n}$	Additive Gaussian vector
$\mathbf{h}_{ikl}$	Vector channel between the ith BS and the kth user of the lth cell
$\mathbf{D}$	$K \times K$ Large-scale (attenuation and shadow) diagonal matrix whose elements are β
β_{ikl}	Large scale fading for the channel between the ith BS and the kth user of the lth cell
g_{ikl}	Small scale fading vector
α_{ikl}	Log-normal random variable
r_{ikl}	Distance between k^{th} user in the lth cell and the ith BS
$\mathbf{G}$	Uplink channel model
$\boldsymbol{G}_{il}$	Small scale fading matrix between the K users in the lth cell and the M_t antennas of the ith BS
$[\boldsymbol{G}_{il}]_{mk}$	Small scale fading vector between the kth user and the mth antenna
T_C	Channel coherence interval
f_{Max}	Maximum Doppler spread
$\varkappa$	Attenuation power exponent
$\mathbf{R}$	MIMO channel covariance
$\mathbf{R}_t$	Spatial covariance matrices at the transmitter
$\mathbf{R}_r$	Spatial covariance matrices at the receiver
$\boldsymbol{S}_r$	Received spatial matrix
$\mathbf{F}$	Multi-user channel precoding matrix
L_{glass}	Effective glass loss
$L_{concrete}$	Effective concrete loss
L_{angle}	Angular wall loss
L_{i-wall}	Indoor wall loss
L_{body}	Average body loss

List of Abbreviations

3 GPP	Third Generation Partnership Project
6LoWPAN	low-power wireless personal area network using IPv6
E^3F	energy-efficiency evaluation framework
vEPC	virtual evolved packet core
$v$switch	virtual switching function
vRAN	virtual radio access network
vBBU	virtual base band unit
AF_{CA}	antenna factor of the circular array
AF_{LA}	antenna factor of linear array
AF_{RA}	antenna factor of rectangular array
ACK	acknowledgement
ADC	analogue-to-digital conversion
AF	array factor
AGC	automatic gain control
A-GPS	assisted global positioning system
AOA	angle of arrival
AOD	angle of departure
AP	access point
a.s.	almost surely
AS	azimuth spread
AWGN	additive white Gaussian noise
BB	baseband
BBU	baseband unit
BC	broadcast channel
BD	block diagonalization
BF	beamforming
bps, b/s	bits per second
BS	base station
BW	bandwidth
CA	carrier aggregation
CAPEX	capital expenditure
CDIR	channel distribution information at the receiver
CDIT	channel distribution information at the transmitter
CE-SCN	cache-enabled small cell network
CI	channel inversion
CIP	channel inversion precoding
CMI	channel mean information

CoMP	coordinated multi-point
CR	core network
C-RAN	cloud-based radio access network
CRC	cyclic redundancy check
CRE	cell range extension
CSI	channel state information
CSIR	channel state information at the receiver
CSIT	channel state information at the transmitter
D2D	device-to-device
DAS	distributed antenna system
dBi	decibels-isotropic
DL	downlink
DPC	dirty paper coding
DPD	digital predistortion
DRX	discontinuous reception
E-CID	enhanced cell ID
EDGE	Enhanced Data rates for GSM Evolution
eICIC	enhanced inter-cell interference coordination
EM	electromagnetic
eMBMS	evolved Multimedia Broadcast Multicast Services
eNB	evolved Node B
EPC	evolved packet core; electronic product code
EU	European Union
EVD	eigenvalue decomposition
FD	full duplex
FDD	frequency-division duplex
FD-MIMO	full-dimension MIMO
FFR	fractional frequency reuse
F-Post-Cal	full post-precoding calibration
F-Pre-Cal	full pre-precoding calibration
G	generation; giga as in gigahertz GHz
GPRS	general packet radio service
GWCN	gateway core network
HD	half duplex
HSPA	high-speed packet access
i.i.d	independent, identically distributed
IC	interference channel; integrated circuit; interference cancellation
ICT	information and communication technology
IF	intermediate frequency
iff	if and only if
InP	infrastructure provider
IoT	Internet of Things
IP	internet protocol
ITU-R	International Telecommunication Union-Radio
IWC	ice water content
LHS	left-hand side
LOS	line-of-sight
LSPs	large-scale parameters
LTE	long-term evolution

LTE-A	long-term evolution-Advanced
LWC	Liquid water content
M2M	machine-to-machine
MAC	multiple access channel
MBS	macrocell base station
MF	matched filtering
MGF	moment-generating function
MIMO	multiple-input, multiple-output
MME	mobile management entity
MMSE	minimum mean square error
mmWave	millimetre wavelength
MOCN	multi-operator core network
MPI	Moore-Penrose inverse
MRT	maximum ratio transmission
MS	mobile station
MSE	mean squared error
MT	mobile terminal
MTC	machine type communication
MUBF	multi-user beamforming
MU-MIMO	multi-user multiple-input, multiple-output
MVNO	mobile virtual network operator
NFV	network functions virtualization
NFVI	network functions virtualisation infrastructure
NLOS	non-line-of-sight
NPSS	narrowband IoT primary synchronization signal
NRS	narrowband IoT reference signal
NSA	non-standalone
NSSS	narrowband IoT secondary synchronization signal
OPEX	operational expenditure
OTDOA	observed time difference of arrival
P2P	peer to peer
PA	power amplifier
PAS	power azimuth spectrum
PBS	picocell base station
PDF	probability density function
PHY	physical layer
P-Post-Cal	partial post-precoding calibration
PPP	Poisson point process
P-Pre-Cal	partial pre-precoding calibration
PRB	physical resource block
PU	primary user
QoS	quality of service
RAN	radio access network
RBs	resource blocks
RX	receiver
RF	radio frequency
RFAC	RF anechoic chamber
RHS	right-hand side
RL	reinforcement learning

RMa	rural macrocell
RMS	root mean square
ROP	random object process
RRH	remote radio head
RRU	remote radio unit
RS	reference signal
RST	random shape theory
RTS/CTS	request to send / clear to send
RTT	radio time transmission
RZF	regularised zero forcing
SCM	spatial channel model
SIC	self-interference cancellation
SINR	signal power to interference power plus noise power ratio
SNE	squared norm estimation
SNR	signal power to noise power ratio
SON	self-organised network
SP	service provider
SU	secondary user
SUI	Stanford University Interim
SVD	singular value decomposition
TDD	time-division duplex
TIP	telecom infra project
TPE	truncated polynomial expansion
TX	transmitter
UE	user equipment
UL	uplink
UMa	urban macrocell
UMi	urban microcell
VNA	vector network analyzer
WCDMA	wideband code division multiple access
WET	wireless energy transfer
WF	Wiener filtering
WINNER	World Wireless Initiative New Radio
WPC	wireless powered communication
ZF	zero forcing

Structure of the Book

Chapter 1: Introduction

The chapter begins with a brief review of the contemporary cellular networks, examining their pros and cons in terms of future demands for mobile services. The 3GPP-led long-term evolution of wireless communications signifies the beginning for a new, complete system of technologies from 4G to 5G. Since technologies dominate the generations of wireless networks, it was compelling to investigate how these enchanting, ingenious technologies could be implemented in the physical layer of multiple-input, multiple-output (MIMO) wireless systems working in non-ideal environments. In fact, 5G is more than just a mobile communications network. It will have an astounding impact on our society, economy, and almost every aspect of daily life. The massive MIMO system, high speed of data delivery, and increased coverage expected from the 5G are not the only advantages over the 4G. Low latency also provides the possibility of nearly immediate wireless connection in real time. The 5G low latency can provide connections to enormous numbers of IoT smart objects. The chapter provides an overview of the important merits of the contemporary networks, including the capacity regions of the MIMO BC and MIMO MAC systems, the characteristics of the fading wireless channels and the multicell MIMO channels. The intensive concerns for the environment and the impacts of global warming on the climate necessitate appropriate green communications for the twenty-first century and are examined in the chapter. A model for the wireless network power consumption is provided, and it turns out that most power consumption is due to the base stations (BSs). The green cellular mobile communication scenarios are presented in detail. The chapter concludes with a short summary, a list of selected up-to-date references, and a tutorial on the theory and techniques of optimization mathematics.

Chapter 2: 5G Enabling Technologies: Small Cells, Full-Duplex Communications, and Full-Dimension MIMO Technologies

The chapter comprises three important technologies that have tremendous dominance in 5G. The large number of small cells employed in a 5G network was identified as network densification. The analysis and evaluation of network densifications and HetNet model are addressed comprehensively. The traditional RAN is too expensive to deploy in future wireless generations. The cloud-based RAN architecture includes: resource management between macrocells and small cells and mobile small cells is analysed in detail. The cache-enabled small cell is examined, in particular, for file delivery performance and outage probability. The full-duplex (FD) mobile communications implied that a transceiver unit can transmit and receive data simultaneously. The FD technology for 5G is analysed and evaluated, and a complete FD infrastructure design is presented. The full-dimension MIMO technology exploits the 2D active antenna arrays and considers the energy propagation in both vertical and horizontal directions and they are thoroughly analysed and evaluated. Key design parameters are identified and debated elaborately. Additionally, the chapter overviews the current and future reference signals, antenna ports, and physical communications channels. The chapter concludes with a short summary, a list of selected novel references, and appendices.

Chapter 3: 5G Enabling Technologies: Network Virtualization and Wireless Energy Harvesting

The chapter presents two important technologies: network virtualization and wireless energy harvesting. Virtualization is a duplication of a hardware platform using software in such a way that all functionalities are reproduced as 'virtual instance' and are operating similar to the traditional hardware elements. In practice, network virtualization has produced overlay networks on top of the physical hardware. Virtualized infrastructure of wireless network reduces the CAPEX and the OPEX. Virtualization is extended beyond the physical layer to include layer 2 and layer 3 (switching and routing) and includes a transport layer as well. Network functions virtualization (NFV) aspires solutions to the NFV infrastructure to build up many network equipment types onto industry-standard virtual devices. Virtual technology devices comprise *v*RAN, EPC, *v*switch, etc. The wireless powered communication is key to the future generation. It uses a harvested energy to transmit/decode information to/from other devices and currently has many applications in IoT systems, large-scale wireless sensor networks (WSNs) for environment monitoring, smart power grid, and wireless powered mobile devices. mmWave networks have two major attractions for wireless energy harvesting, namely: massive number of antennas and highly dense BSs deployments. The chapter ends with a short summary and a list of selected up-to-date references.

Chapter 4: 5G Enabling Technologies: Narrowband Internet of Things and Smart Cities

Internet of Things (IoT) is a networking technology of a large number of conventional physical objects turned into smart devices with limited bandwidth (BW), power, and processing capabilities. IoT devices intelligently communicate using IP and are operated without the help of human beings. IoT devices exist in buildings, vehicles, and the environment. The IoT architecture moved on to the narrowband IoT (NB-IoT) which is a category of IoT that is introduced by 3GPP employing a small portion (180 kHz for DL and UL) of LTE BW to provide connectivity to the IoT devices. The analysis and evaluation of NB-IoT is presented in the chapter.

Smart city is an urban community that uses modern information and communication technology to collect information to manage available assets and resources efficiently. The analysed information can be used, for example, to monitor and manage services and to provide a better life for its citizens.

The EU smart city model is used to deal with basic requirements for smart cities. The EU model is based on the city performance in *six key fields of urban development*: *smart economy*; *smart mobility*; *smart environment*; *smart people*; *smart living*; and *smart governance*. It is important to note that currently there are no global standards to qualify a city to be smart. The chapter concludes with a short summary, a list of selected up-to-date references, and an appendix.

Chapter 5: Millimeter Wave Massive MIMO technology

The chapter deals with the mmWave massive MIMO, which can be the key 5G technology. The main issues of concern are broadcast channel BC estimation and the mmWave beamforming systems. The first is investigated by the reciprocity technique in a time-division duplex (TDD) network operated protocol. The concept of channel reciprocity implies that the downlink (DL) can be the reciprocal of the estimated uplink (UL). While propagation UL and DL are reciprocal, a practical reciprocity model is much more complicated since the transceivers at both BS and the terminals may not provide reciprocal UL and DL. The chapter analyses this problem in great depth. The beamforming issue is presented, including the concept of hybrid digital and analogue beamforming for mm wave antenna arrays. In addition, the chapter examines the mmWave market and the choice of technologies that can be used and the massive MIMO hardware requirement. The chapter addresses the pros and cons of the contemporary technologies and the pros and cons of the conventional MIMO beamforming schemes. The chapter concludes with a short summary, a list of selected up-to-date references, and appendices.

Chapter 6: mmWave Propagation Modelling: Atmospheric Gaseous and Rain Losses

The chapter deals with two important topics related to mmWave propagation, namely: atmospheric gaseous losses and rain losses. The atmospheric gases include two main gases that account for 99% of the atmospheric gases, namely: oxygen 21% and nitrogen 78%. The remaining (1%) includes water vapour, carbon dioxide (0.037%), and organ (0.9%) and other rare gases. The issue here is how much attenuation these gases inflect on the mmWave transmission. As it appears, the attenuation is caused primarily by oxygen and water vapour molecules. The chapter continues this subject through published research and inclusion of the ITU recommended model. The chapter deals with rain attenuation at mmWave bands with a brief survey of the research development in the field and focuses on the physical rain Capasoni model and the ITU recommended rainfall rate conversion model. Furthermore, the losses by snow and hail are investigated by assessing the physical parameters of an ice slab and by applying the strong fluctuation theory. The chapter concludes with a short summary, a list of selected novel references, and an appendix.

Chapter 7: mmWave Propagation Modelling – Weather, Vegetation, and Building Material Losses

The chapter deals with three topics: weather; vegetation, and building materials, and examines their losses contribution at mmWave propagation. The chapter investigates the clouds and fogs and derives a microphysical model for the representation of rain droplets size distribution, taking in the Rayleigh and Mie scattering distributions and the loss calculation based on the water content. The ITU model for clouds and fog attenuation calculation is described in depth. The attenuation of propagated radio waves in vegetation can be important for low transmit power mobile communications devices, such as those proposed for use in 5G. Attention is given to attenuation in vegetation due to diffraction (top diffraction, side diffraction, ground reflection, and the scattering components). The propagation losses due to various building materials for indoors and the exterior of the house are analysed and evaluated in detail. The chapter closes with a short summary, a list of selected up-to-date references, and appendices.

Chapter 8: Wireless Channel Modelling and Array Mutual Coupling

Massive MIMO channel models considered with correlation inspired models are of two types: models correlated at one end of the link and are appropriate when dealing with single antenna mobile terminals known as Kronecker channel models. The other type of models consider the correlation at both ends of the link jointly and are suitable when dealing with multi antennas mobile terminals known as Weichselberger channel models. The chapter considers both types of correlation inspired models in considerable detail. In addition, considerations are given to the virtual channel representation and the i.i.d Rayleigh channel model. Furthermore, the chapter deals with the mutual coupling in massive MIMO antenna arrays at the BS when arrays operating in transmit and receive modes and the mutually coupled arrays channel capacity. The chapter also provides an overview of contemporary wireless channel fading and statistics. The chapter concludes with a brief summary, a list of novel references, and appendices.

Chapter 9: Massive Array Configurations and 3D Channel Modelling

The chapter deals with two important topics: future massive MIMO array configurations and the 3D channel modelling. The mmWave massive array configuration is likely to be spherical arrays comprised of a large number of radiating elements placed on a spherical surface. For a given beam direction, a sector of the array is excited and the spherical arrays are highly symmetrical devices. The symmetry is also applied to the array elements. mmWave array package with beam-steering capability for the 5G mobile terminal applications was proposed in the literature. Depending on the operating frequency, microstrip patch antennas can also be considered. The chapter considers the array configurations thoroughly. The 3D channel models recommended in 3GPP Release 14 included generating the 3D channel model coefficients based on measured channel was used to model buildings and presented in detail. The chapter concludes with a brief summary, a list of selected novel references, and appendices.

Chapter 10: Massive MIMO Channel Estimation Schemes

Massive MIMO channels estimation in noncooperative TDD networks with time-shifted pilot scheme and channel estimation using coordinated cells in MIMO system using Bayesian estimation method are analysed and evaluated in the chapter. The chapter also examines arbitrary correlated Rician fading channel estimating, using MMSE approach in great detail. The massive MIMO channel calibration is presented in the chapter through two methods, namely the Argos method, which works on a sample basis, and the mutual coupling method, which uses the coupling as a basis to derive the calibration approach. The pre- or post-precoding channel calibration is debated in this chapter as well. The chapter concludes with a short summary, a list of selected up-to-date references, and appendices.

Chapter 11: Linear Precoding Strategies for Multi-user Massive MIMO Systems

SU-MIMO precoding based on The SVD of the channel matrix **H** is considered in the chapter. The channel inversion precoding at the transmitter for BC is analysed and evaluated. Linear ZF precoding for the BC with transmit power constraint based on the Wiesel et al. method is examined thoroughly in the chapter. An important issue related to the outage probability of the channel is considered in depth. The precoding for MIMO channel using transmit filter matrices at the transmit and receive end of the link based on Joham et al. method is considered in the chapter. The ZF, MF, WF are derived in the chapter. An unequal power allocation to the user terminals is derived for ZF precoding scheme. The advantages and the design of regularized ZF design are examined in the chapter. Multi-user BD precoding suitable for users' terminal, each with multiple antennas, is analysed extensively. Massive matrix inversion in linear precoding design is a high complexity process. The chapter explores the replacement of the large matrix inversion with polynomial expansion based on the Cayley-Hamilton theorem. Furthermore, it is possible to truncate the polynomial to limited numbers of terms. The chapter deals with the truncated polynomial expansion precoding providing analysis and evaluation of the TPE for multi-user applications. The chapter ends with a short summary, a list of selected up-to-date references, and appendices.

1

Introduction

1.1 Motivations

Remarkable driving forces motivated scientists and engineers to move their ideas beyond the contemporary conceptual wireless generations to long-term evolution (LTE) of mobile communications towards 5G with power of imagination to create a system that seems at the time to be far-off from possible. After a self-reflection on my experience and overview of the literature, it was compelling to investigate how these enchanting technologies could be implemented to the physical layer of multiple-input multiple-output (MIMO) wireless systems working in a non-ideal environment. Pre-LTE, cellular wireless generation advances mostly implies a partial change in the characteristics of the system to enhance data speed, and/or power efficiencies. Each generation has particular standards, capabilities, limitations, and technologies with modified architectures to differentiate it from the previous one. Operators will find 5G infrastructures are not depending on previous generations but will be based on generic roll-out enterprise, so they have to upgrade their networks infrastructure. 4G mobile services are still developing in both coverage and data speed capabilities. So why do we need 5G?

The key benefits of 5G over 4G will not only be the speed of data delivery, which is expected to be between 10 and 100 Gb/s, but with latency. Low latency provides the possibility of connecting to immediately send and receive data in real-time. In the first quarter of 2018, the average latency for 4G was 75 ms and for LTE–Advanced was 45 ms, both much lower than latency in 3G, which was 135 ms. However, LTE-Advanced latency is still not enough to support real-time response.

5Gs will be ultra-low-latency with a possible range between less than 1 and 10 ms. 5G capacity and performance enhancements attribute to a massive MIMO system and are also important for coping with rapid growth of the Internet of things (IoT). IoT will transform the traditional physical objects into smart objects with the capabilities to see, hear, think, and perform jobs by communicating to each other and sharing information to coordinate their decisions. Indeed, IoT services support our economy to grow and also improve our lives. The 3rd Generation Partnership Project (3GPP) standardized IoT as the narrowband-IoT. IoT needs 5G high capability to cope with the vast number of connected devices. So, the bid for bandwidth (BW) will continue to grow even though IoT is a narrowband system. It is predicted that by 2020, there will be at least 20 billion devices connected to the internet. There is another problem between 4G and IoT and it's the latency. Low latency is a key enabler for coping with the massive number of connected devices.

5G will provide connections to such colossal numbers of Internet-connected IoT smart objects. The amount of data transported by wireless mobile networks is expected to increase rapidly with 5G. As an example, consider the time it takes to download an 800 MB movie. On average, it takes 26 minutes and 40 seconds for 3G (speed 4 kb/s) and 4 minutes and 16 seconds

5G Physical Layer Technologies, First Edition. Mosa Ali Abu-Rgheff.

for 4G (speed 25 Mb/s). This will be reduced to a fraction of a second (0.3 second) with 5G (speed 20 Gb/s).

The current growth of data is mainly driven by devices such as smart phones, tablets, and video streaming. In addition, the number of connected devices and data rates will continue to increase exponentially. Consequently, it was clear to researchers and industry that the incremental approach to system improvements to these challenges would not achieve its targets by 2020. There are enormous demands on academia and industry to meet these challenges through innovative new technologies that are practical, smart, and efficient.

During the last five years, there have been a number of collaborative research projects between academia and industry to develop and test technologies that can enable 5G to deliver its expected performance. These technologies will be part of the 5G networks according to the general consensus of the research community.

5G is more than a mobile network generation. It will advance our economy by generating a substantial contribution to the growth in gross domestic product (GDP) and will enhance our life and society. 5G offers these benefits through a set of technologies connecting people to people and people to everything. The full economic benefits encompass education, automotive industry, transport, building, health care, and entertainment, to name a few. Additionally, 5G Gigabit wireless connectivity and computing capacity help the development of smart cities and smart municipalities across our country, thus providing a cleaner, safer, and healthier environment for citizens both today and tomorrow. Furthermore, industries such as automotive, health care, and IoT will notice the effects of the changeover to 5G. Indeed, IoT will change the way we think of the Internet as a human-to-human interface to a more machine-to-machine platform. In particular, automotive, health care, and the IoT will bring about exciting benefits to our daily lives.

5G will not be an incremental advance on 4G. Indeed, it will be a paradigm shift with massive bandwidth, high base-station (BS) and device densities, and a very large number of antennas. 5G's greatest difficulty of providing a uniform service to a large number of users is mitigated using the formation of heterogeneous (HetNet) networking or small-cell scenarios. Small cell (discussed in Chapter 2) is a key radio cell used in HetNet of coverage 20–400 m and transmitting power of 20 mW–2 W. In addition, 5G applications require a cloud-based architectural elements. Accordingly, 5G radio access has to be flexible, scalable, robust, reliable, and efficient in spectrum and energy.

The outstanding expectations for 5G systems are in *data rate, latency, energy,* and *cost*. The expected aggregate data rate (i.e. network area capacity) in bits/s is expected to increase by 1000X compared to the 4G current optimal data rate, while the *worst data rate* that a user can expect to receive (known as edge rate, or 5% rate) ranges from 100 Mb/s to as much as 1 Gb/s. 5G is expected to increase links data rate between 10X and 100X compared to the 4G data rate. 5G low latency enriches applications such as gaming, tactile internet, wearable devices. 5G is likely to reduce (or at least not increase) energy consumption, and for this reason it is expected to reduce the running cost of the system.

With the purpose of achieving such high provisions, 5G is enabled by a combination of new and well-known technologies. These sets of enabling technologies aim to provide unlimited access to information and the ability to share data anywhere, anytime, by anyone for the comfort of people and to the advantage of business and society.

Half-duplex transceivers (TRXs) are commonly implemented in LTE devices, which means users can either transmit or receive but not both operations simultaneously. This is because of the self-interference generated at the user's end when the user is transmitting and receiving at the same time. 5G will be enabled by technology that removes the self-interference and users will then enjoy full duplex services. This implies that 5G users should be able to transmit messages and receive data at the same time. This technology is explored comprehensively in Chapter 2. Furthermore, while LTE devices provide azimuth (horizontal) transmission and reception, 5G systems, using 3D beamforming, will provide users with services in horizontal and vertical dimensions. In practice, this implies that the LTE service would be limited to horizontal dimension while 5G service will cover users in multi-story building as well. This technology is known as full-dimension multiple-input multiple-output (FD-MIMO) technology and is investigated thoroughly in Chapter 2.

According to the United Nations (UN), as of 2016, more than 54.5% of the world's population live in urban settlements, and this is projected to increase to 60% by 2030. Furthermore, the UN habitat organization predicted that by 2017, cities will cover 'less than 2% of the earth surface and consumed 78% of world energy and produced 60% of CO_2 and greenhouse gas emissions'. Consequently, cities are facing a magnitude of challenges, not only because they generate most of the air pollution but also because they put a strain on available resources. Therefore, appropriate solutions are needed to resolve these challenges. Such solutions have to transform traditional cities into *smart and resilient cities* by integrating the communication technologies and the IoT to manage the city's assets.

EU smart city reference model is based on a city performing well in six characteristics: smart economy, smart mobility, smart environment, smart people, smart living, and smart governance. These characteristics are evaluated by 31 factors and 74 indicators. The reference comprised three versions: version 2.0 and 3.0 published in 2013 and 2014, respectively, and version 4.0 in 2015. Versions 2.0 and 3.0 EU smart cities model were defined for medium-sized cities and version 4.0 is defined for large EU cities. Since the publication of the reference model, many EU collaborative projects have been funded to investigate innovative schemes that brand EU cities smarter using the 5G information technologies (i.e. make traditional networks and services smart and efficient such as urban transport, water supply, waste disposal facilities, city administration, and city assets, to name few smart and efficient). A sample of these projects includes REViSITE (2010–2012), OPTIMS (2013–2016), BESOS (2013–2016), and CITYOPT (2014–2017).

In order for 5G to exploit the mmWave frequency bands, it is important to learn more about the channel characteristics and the radio propagation models. Contemporary propagation models were validated for frequencies up to about 2 GHz and are not adequate for small cells. In addition, such models are not suitable for hilly and heavily wooded terrains. Radio propagation in indoor environments is not influenced by terrain profiles as the outdoors propagation but can be affected by the layout inside a building and influenced by various building materials used. Furthermore, atmospheric gaseous, clouds, fog, snow, and hail have to be accounted for in the mmWave propagation.

Channel models are widely used to predict the performance of the propagated signal under various constraints. There is a need for new channel models to support mmWave 5G operation. These models have to support large bandwidths and be suitable to medium- and high-mobility speeds. In addition, 5G channel models have to support 1D and 2D massive MIMO with dual polarization and are appropriate for beamforming design. Intensive research activities were carried out by a number of groups for channels modelling at mmWave bands, such as Mobile and wireless communications Enablers for the Twenty-twenty Information Society (METIS), International Telecommunication Union-Radio (ITU-R), COST, and 5G mmWave Channel Alliance, to name a few. So far, research investigations show two possible channel models: independent and identically distributed (i.i.d) Rayleigh channel is used widely for flat faded channels, and the correlation inspired models. The latter comprised two types: the Kronecker channel model, where correlation occurs at one end of the link, and the Weichselberger channel model, exhibiting joint correlation of both link ends.

The electromagnetic (EM) action by the array elements on each other instigates array mutual coupling. Such action takes the form of energy absorption by one antenna element when a nearby element is radiating EM energy. Energy absorption by mutual coupling reduces antenna performance efficiency.

Array coupling problems have attracted interest not only from antenna engineers and researchers but also from researchers in other disciplines, such as biomedical imaging, including magnetic resonance imaging (MRI) that operates antenna arrays. Array mutual coupling can be reduced by spacing the antennas as far apart as possible. In practice, due to constraints on the physical dimensions of available spaces, the distances between the multiple antennas are usually small and antenna elements interact with each other (i.e. they are mutually coupled). Consequently, to improve the MIMO system efficiency, we need to find ways to decouple transmit and/or receive antenna arrays.

Reliable mmWave communication depends decisively on 3D beamforming to aim the beam in the correct direction and the ability to separate the data into independent streams aimed at end users using a process called multiuser precoding. Beamforming needs the knowledge of the spatial channel state information (CSI) at the transmitter and receiver. Spatial CSI is obtained by an adequately accurate estimate of the channel. In time-division duplex (TDD) transmission protocol, pilot sequences are transmitted on the uplink to obtain UL channel estimates at the BS. The downlink (DL) is determined using the reciprocity property.

That is, in ideal circumstances the DL CSI are given by the reciprocal of the UL CSI, and theoretically, there is no need to estimate the DL channel. There are basically two possibilities for the spatial channel estimation in cellular networks, namely: channel estimation in noncooperative TDD networks and estimation using a coordinated cells MIMO system. In massive MIMO multiuser cellular systems, it will be necessary to code each user data before transmission in a process called multiuser precoding and decode the received signals at the receiver in a second process called equalization to equalize the coding at the transmitter.

The purpose of the book is to present the most recent knowledge about 5G enabling technologies as applied to the PHY layer in a single book and in a unified framework. Various techniques are analysed and evaluated. Details are cited in a reader-friendly writing approach in all chapters. The analysis and explanation rendered are concise and informative. As the case with any advanced engineering book, derivations and analysis apply advanced mathematics. To alleviate these difficulties from students and professionals, or any related skills that require advanced mathematical expertise, mathematical difficulties are resolved either in the main text or, if the solution is lengthy, in appendices.

1.2 Overview of Contemporary Cellular Wireless Networks

First-generation (1G) commercial network was used in the 1980s and built to provide only analogue voice services without security. Users commonly experienced the frustration of dropped calls and poor battery life.

Second-generation (2G) was commercialized in the 1990s and was an enormous improvement to the subscriber experiences compared with 1G. The motivation for 2G was to increase system capacity; add additional services such SMS and caller ID, etc.; improve security; and reduce cost. 2G, also known as GSM, or global system for mobile. The key features of the GSM cellular system specification include using a digital time-division multiple access (TDMA) approach. Both UL and DL frequency bands are in the 900 MHz and 1.8 GHz bands in Europe.

There are five different cell sizes in GSM network, namely: macro, micro, pico, femto, and umbrella cells. Clearly, a lot of microcells contribute to a large number of handoffs, especially when the speed of the mobile is too high. An umbrella cell covers several micro cells to reduce the number of handoffs. By adopting this technique, more users can be accommodated within the available bandwidth. The signalling waveform used was Gaussian minimum-shift keying (GMSK). GSM cellular network is a circuit-switched system adopting 200 KHz radio frequency (RF) channels, and these channels are time-division multiplexed to enable up to eight users to access each carrier. So, in a practical scenario, GSM is a TDMA/frequency-division multiple access (FDMA) system, and each user is transmitting at about 34 kb/s. The total channel data rate is 270.8 kb/s.

GSM provided the following services: digital voice, SMS, international roaming, conferencing, call waiting, call hold, call forwarding, call barring, caller number identification, closed user groups (CUGs), unstructured supplementary service data (USSD) services, and authentication billing based on the services provided to their customers (e.g. charges based on local calls, long-distance calls, discounted calls, real-time billing). GSM uses USSD to send data between a mobile system and an application in the network.

GSM was an exceptionally successful mobile communication system, initially intended for use within EU, but within a short time it became an internationally accepted system. Further development of GSM gave rise to the general packet radio service (GPRS) using packet-switched data rather that circuit-switched data. The maximum number of time slots that can be used is four, giving a possible maximum data transfer rate of 57.6 kbps (or 38.4 kbps on a GSM 900 network) per time slot. Two major concerns for operators are capital expenditure (CAPEX) and operational expenditure (OPEX) costs.

Another extension of GSM is the Enhanced Data rates for GSM Evolution (EDGE) technology that uses 8PSK modulation, which require small changes in the BS and could often be achieved by software upgrades. EDGE offers radio data rate per time slot equal 69.2 kb/s and maximum user data rate per time slot 59.2, which gives maximum user data when using eight time slots to be 473.6 kb/s. 2G standard was implemented using different technologies: EU standard implemented the GSM technology but other technologies were used as well – United States (IS-95) and Japan (JDC). The data transfer speeds of GSM systems were only 35–50 kb/s.

Third-generation (3G) systems aimed at increasing network capacity and data throughput to promote growth and support new applications. Universal mobile telecommunications system (UMTS) was the main establisher of the 3G, and its new standards in cellular system performance and functionalities. In 1992, WARC allocated 230 MHz of spectrum between 1885–2025 and 2110–2200 MHz for the next generation (3G). A number of international organizations accepted the demand for international standards for the next-generation mobile communication. EU ETSI support UMTS, ARIB in Japan; and many other proposals were submitted, but the ones that gained support were UMTS/code division multiple access (CDMA); CDMA2000 under Interim Standard IS-95 first deployed using CDMA technology; and TDS-CDMA (time division synchronous code division multiple access) developed by China.

A number of companies and standards organizations set the 3GPP to produce technical specifications for 3G with radio access including both TDD and frequency division duplex (FDD) technologies. 3GPP published releases 99, 4, 5, 6, and 7 in specifying 3G. So, 3G enrolled in the EU in 2001 and UK operator 3 launched 3G in 2003. Wideband code division multiple access (WCDMA) can adopt Alamouti 2 × 2 MIMO scheme, which is the only orthogonal space-time code with rate 1. However, for transmit antennas higher than two, several orthogonal codes have been found but with lower rates, so spectral efficiency can be degraded. 3G standards based on WCDMA technology do not support MIMO technology.

UMTS/WCDMA with radio interface based on WCDMA technology offered DL data speed 384 kb/s–2 Mb/s. High-speed packet access (HSPA) enhanced WCDMA speeds to 14.4 Mb/s in DL and 5.76 Mb/s in UL. An upgrade to HSPA (HSPA+) provided theoretical peak speeds of 168 Mb/s in DL and 22 Mb/s in UL. A second organization, 3GPP2, was created by North America and Asian operators to adopt CDMA2000 for 3G with the following technologies: One times radio transmission technology (1 × RTT) recommended DL speeds up to 144 kb/s; Evolution-Data Optimized (EV-DO) increased DL data speeds up to 2.4 Mb/s; Evolution-Data Optimized Revision-A (EV-DO Rev. A) increased the DL speeds up to 3.1 Mb/s and reduced latency; Evolution-Data Optimized (EV-DO Rev. B) offered 2–15 channels, each with DL peak rate at 4.9 Mb/s; and finally, Ultra Mobile Broadband (UMB) DL peak rate reached 288 Mb/s.

3G was launched in many countries, but the success of 2G was difficult to replicate because the advancement from 2G to 3G innovated only a few services – not sufficiently inspiring for users to change their operators. 3GPP incorporated the multimedia broadcast and multicast service (MBMS) in combination with the internet protocol multimedia system (IMS) with no adjustment to the access technology [1].

Since evolved HSPA (also known as HSPA+) was launched in 2010, data traffic driven by smart phones and tablets has grown rapidly. Nokia traffic model published in 2012 using two traffic assumptions predicted traffic increase of 1000x relative to the traffic in 2010 will come into being around 2020 [2]. Operators should be prepared to install networks that can cope with this momentous traffic upsurge. The development of 3GPP LTE technologies is based on a series of releases. Most current 4G networks are based on 3GPP Releases 8 and 9, offering peak DL speed of 100 Mb/s and peak UL 50 Mb/s. Nonetheless, so to appropriately respond to increasing traffic and capacity growth, the LTE radio access technology has continuously evolved. 3GPP LTE-Advanced introduced enhancement on the existing 4G LTE networks and raised the peak DL transmission speed to 1000 Mb/s and the UL peak speed to 500 Mb/s, as recommended by 3GPP Release 10.

Macro cells bear almost the entire data traffic in contemporary mobile network, complemented by small cells in hot zones. A key performance enhancer to macro cells capacity is to partition each cell into smaller cells. Networks comprise an overlay of cells of different sizes. For example, outdoor users are likely to be served by a combination of macro, micro, and pico cells. Low-power remote radio head (RRH) and pico cells may provide both outdoor and indoor coverage. They can be employed in hotspots/hot zones such as shopping districts, train stations, or shopping malls, with typical cell radius of up to 200 m. In addition, pico and femto cells can be used indoors in cells of no more than 10–25 m. The tendency for multilayer placement (i.e. small-cell densification) is forced by the need to maintain better service quality both indoors and outdoors.

Conventionally, small cells in pre-LTE operate independently of the overlaid macro layer and transmit all the cell signals comprising cell-specific reference signals, synch signals, and the system full information. In addition, a user communicates with small cells and macro cells in both UL and DL directions. Coverage provided by the macro cell can be enhanced by operating the macro- and small-cell layers in a more integrated approach. Most importantly, *dual connectivity* requires the user terminal to be simultaneously connected to the two nodes [3]. Dual connectivity can provide multiple benefits in addition to throughput aggregation of the two nodes. The macro node is an *anchor node* to provide a mobile internet protocol (IP) link to communicate with the mobile user, as shown in Figure 1.1.

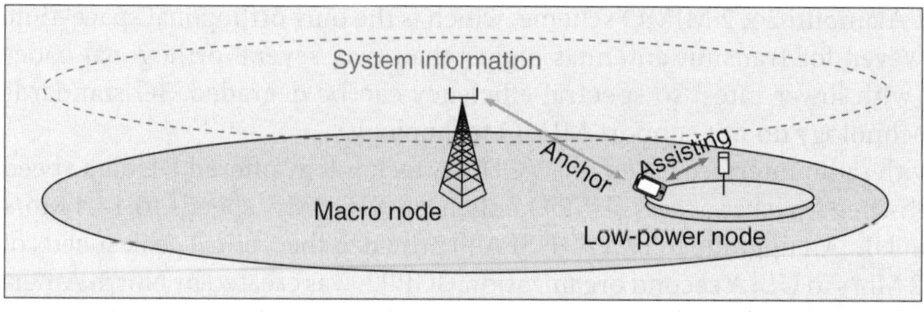

Figure 1.1 Soft cell: dual connectivity to the anchor node and assisting node [3].

Essentially, there exist two techniques to split a macro cell, in the horizontal (azimuth) domain and in the vertical domain. Combination of both sectorization approaches arrange for narrow beam targeting of single or more user(s).

1.3 Evolution of Wireless Communications in 3GPP Releases

3G technology is based on the novelty of the spread spectrum concept, a technology developed as an anti-jamming capability by the military establishments. The energy-spreading technique reduces the level of transmitted information below background noise level to provide data protection against any third-party intruder. The BW expansion is achieved using a coding process that is independent of the information being transmitted or the modulation being applied.

Commercial spread spectrum scheme generates procession gain to express the bandwidth expansion factor. For example, for 20 dB processing gain and bit rate of 1 Gb/s, the spreading clock has to generate over 100G chip per sec (Gc/s). 3G data rate can be increased by efficient spreading waveforms, adopting inter-cell interference cancellation and adaptive coding and modulation. These enhancements to HSPA eventually increased its data rates. These considerations conclude that there is an upper limit on the capacity of 3G networks enforced by the availability of suitable spreading clocks.

3GPP was set up in 1998 to work out specifications for future advanced mobile communications. The public and private collaborative project comprises seven standards organizations: market associations and many industrial companies. The aims of the collaboration were to develop a vision for an all IP wireless network covering radio access/core networks: service capabilities and interworking with Wi-Fi networks. The 3GPP led the campaign for LTE of wireless communications, and 3GPP standards are arranged in Releases. Release 14 was published in 2017; Release 15 was available at the end of 2018, and Release 16 is due at the end of 2019. Releases 14 and 15 concur with 5G trial service tests and 5G commercial service enrolled in 2020. The following section briefly describes some the key evolve technologies of each 3GPP release. 3GPP LTE continues developing enabling technologies for 5G, with releases shown in Figure 1.2.

1.3.1 3GPP Release 8

Release 8 was published in Dec. 2008. 3GPP introduced LTE technology for the very first time. Most of the subsequent releases added enhancements to the previously released technologies. Evolved Packet Core (EPC) is the IP-based core network defined by 3GPP in Release 8 and so circuit-switching data mode is rendered inoperative. EPC for LTE networks was reported by a group of venders in February 2009 to permit operators to adopt a range of access types using a common core network. LTE networks' deployments commenced in 2010 with a theoretically download speed of 100 Mb/s. However, initial speeds tended to be closer to 10 Mb/s. LTE

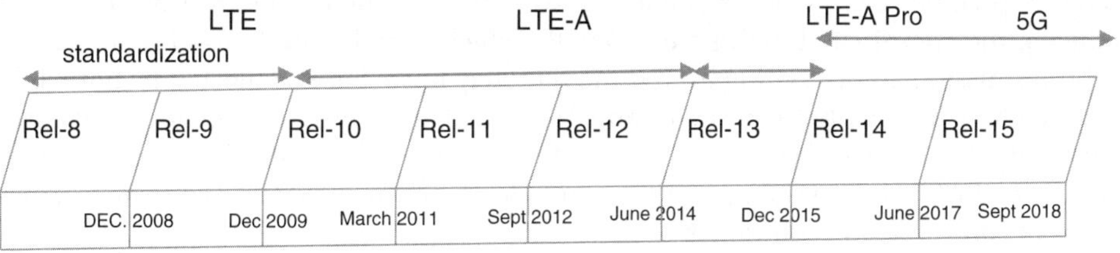

Figure 1.2 3GPP Long-term evolution development towards 5G.

networks adopt 4 × 4 MIMO system and flexible BWs 5–20 MHz in 5 MHz steps together with orthogonal frequency-division multiple access (OFDMA) in DL and single carrier frequency-division multiple access (SC-FDMA) in UL.

1.3.2 3GPP Release 9

Published in Dec 2009, Release 9 brought enhancements to LTE reported in Release 8 specifications. Three positioning schemes were specified namely: Assisted Global Positioning System (A-GPS); Observed Time Difference of Arrival (OTDOA): and Enhanced Cell ID (E-CID). These schemes can be employed to accurately determine a user's location during a natural disaster and emergency cases. Improvements were also added to the concept of self-organized network (SON), which was introduced in Release 8 with a focus on evolved Node B (eNB) self-configuration, while Release 9 added self-optimization as well. Evolved multimedia broadcast Multicast services (eMBMS) were outlined at physical layer in Release 8 but Release 9 completed the specifications for network layer and higher layer aspects. Support for single-layer (single-user) beamforming was described in Release 8 and Release 9 extended the beamforming to multi-layers (multi-users). In addition, Release 9 introduced Commercial Mobile Alert System (CMAS) in addition to the Earth and Tsunami Warning System (ETWS) introduced in Release 8.

1.3.3 3GPP Release 10

3GPP Release 10 described what is generally known as LTE-Advanced systems. Published in March 2011, it implements a combination of large (macro) cells as well as small cells, commonly known as heterogeneous networks (HetNet). In addition, it added enhancements in SON features for self-healing procedures. Release 10 increased the number of antennas in the MIMO system to up to 8 × 8 MIMO in DL and 4 × 4 in UL (user side). Furthermore, it provided worldwide roaming with high spectral efficiency. Release 10 introduced the clustered SC-FDMA in UL and permited selected carriers scheduling in UL, while Release 8 only allowed adjacent block of carriers for scheduling.

A key feature introduced in Release 10 is carrier aggregation (CA), which operators can use to employ their fragmented spectrum scattered across different or same bands to enhance user throughput by transporting data simultaneously over two or more carriers. LTE-Advanced handles BW of 100 MHz composed of five 20 MHz carriers. Another important feature described in Release 10 is the enhanced inter-cell interference coordination (eICIC) to manage the interference issues in HetNet. The eICIC alleviates interference on both data and control channels. Intra-frequency interference is mitigated using power, frequency, and time domain. The peak throughput in LTE-Advanced in DL is 1 Gb/s and 500 Mb/s in the UL. Release 10 recommended the relay nodes connected to a donor eNB to extend the coverage of the system eNB.

1.3.4 3GPP Release 11

Published in September 2012, Release 11 added further improvements to LTE-Advanced, such as introducing the coordinated multipoint (CoMP), which permits the transmitters to share data load transmission and reception even if the remotely hosted are connected by fibre link. An Enhanced Physical Downlink Control Channel (ePDCCH) was proposed to increase control channel capacity. The ePDCCH consumes some of the resources allocated to Physical Downlink Shared Channel (PDSCH) for sending control information, contrary to Release 8, which allows only the use of control region subframes.

Release 11 recommended new positioning schemes for UL added by employing sounding reference signals for time difference measured by a number of neighbouring eNBs. In addition, the release defines a method where user device (UE) informs the network whether it needs to operate in battery saving mode or normal mode. So, based on the user equipment (UE) request network, it can adjust its discontinuous reception (DRX) parameters. Further, Release 11 introduced radio access network (RAN) overload control, where in some scenarios of massive communications from a group of devices, the network can blockade these devices to send connection requests to the network. In addition, Release 11 recommended further enhancements to CA, including multiple time advances (TAs) for UL CA, nonadjacent intra band CA, and the CA support in TDD LTE physical layer.

1.3.5 3GPP Release 12

Published in June 2014, Release 12 recommended further enhancements to LTE-Advanced. Small cells were supported in Release 10 with structures that comprise eICIC. Release 12 recommended the optimization and enhancements of small cells, including their deployment in dense areas. In addition, the release introduced inter-site CA between macro and small cells, also known as dual connectivity. Release 12 focused on machine type communication (MTC) that created huge network signalling and capacity concerns, so a new UE category was defined for optimized MTC operations. One study item in Release 12 recommended operating in unlicensed spectrum, as it delivers many benefits to operators such as increased network capacity and improved performance.

1.3.6 3GPP Release 13

Release 13, also known as initial LTE-Advanced Pro release, was published in December 2015. Release 13 introduced three important technology categories. The first was FD-MIMO. The key idea of FD-MIMO was to employ a large number of antennas arranged in a 2D antenna array panel and additional frequency resources by adopting CA of up to 32 component carriers (CCs) compared to only up to 5 CCs in Release 10. Further, it supported up to 8 antenna MIMO as per previous releases but promised to examine high-order MIMO systems with up to 64 antenna MIMO. Release 13 proposed possible enhancements for DL multiuser transmission using superposition coding. The third were new enhancements recommended to MTC, including a low-complexity UE category, which was defined to support reduced BW, power, and a long battery life. Narrowband Internet of Things (NB-IoT) was standardized for providing wide-area connectivity for massive machine-type communications (mMTCs). Release 13 focuses on a combination of primary cell from licenced spectrum with a secondary cell from an unlicensed spectrum to accommodate the increasing data traffic demand.

1.3.7 3GPP Release 14

Published in June 2017, Release 14 is known as the initial start of 5G standardization. Some of the main technology areas include latency reduction to 1 ms or less for improved end-user experience compared to 5 ms in Release 8 for IP packets in ideal radio conditions. In addition, Release 14 provides full support for UL transmissions in unlicenced spectrum to build on the dual-connectivity framework recommended in Release 13.

Further, Release 14 discussed new use cases such as intelligent transport systems (ITSs) inclusive of vehicular-to-vehicular and vehicular-to-infrastructure communication. LTE technology greatly improves traffic safety on the roads and also makes it possible to include a wide range of road users, e.g. pedestrians and cyclists, in an overall traffic safety work. The existing device-to-device framework can serve as a basis for the work on vehicular-to-vehicular communication. Release 14 introduced more improvements to the MTC capacity and added new features for MTC devices, such as MBMS support for delivering software upgrades and device-to-device relaying for coverage extension. Release 14 focused on extending the MIMO framework to even larger number of antennas. Release 14 recommended a new radio access named as 'NR' to be standardized. The LTE evolution and the 'NX' technologies will build 5G radio access.

1.3.8 3GPP Release 15 (5G phase 1)

3GPP 5G New Radio (NR) specifications included in Release 15, on which work began in June 2016 and was set to complete in September 2018. 5G NR is a new air interface, i.e. the part of the circuit between the mobile device (UE) and the active BS. Active BS is used to imply the fact that BSs change as the UE moves and the BSs change accompanied by a handoff process. The non-standalone (NSA) mode of the NR was approved by the 3GPP in December 2017. The standalone (SA) mode is to be completed by September 2018 and entails user and control planes capability using the new 5G core network architecture.

Achieving the 5G expectation, it is essential that 5G NR must be able to deliver numerous and varied services across a different set of devices with different performance and latency needs. Optimized orthogonal frequency-division multiple access (OFDM) or OFDM variants are to be used for signalling waveforms and multiple access in 5G NR design and should enable lower latency. 5G NR should support a range of deployment models such as macro cells, small cells, and device to device (D2D) connections, with extremely low device cost and high levels of power and deployment efficiencies. The core 5G NR will comprise three key elements: enhanced mobile broadband (eMBB) including technologies Gb/s LTE, massive MIMO, mmWave technologies, and advanced channel coding; ultra-reliable low-latency communications (uRLLC) for services like autonomous deriving with technologies involved such cellular vehicle-to-everything (C-V2X) and cellular drone communications; and mMTC technologies involve such over-the-air broadcast firmware (software that makes hardware work) updates. Release 15 was released in late 2018.

1.3.9 3GPP Release 16 (5G phase 2)

The target completion date for Release 16 is 2020, when complete 5G standards will be published.

1.4 Multiuser Wireless Network Capacity Regions

Wireless channels often hold more than one user transmission. These channels are referred to as multiuser channels. At the network physical channel, there can be three categories of channels: multiple access, broadcast, and interference multiuser channels. The concept of a user is clear in a single user communication, so what constitutes a user in multiuser communication? A user in a multiuser channel can be defined as the amount of information component that can be reliably detected by one or more receivers in the multiuser channel. In other words, a user is channel and information. Furthermore, a single user can be decomposed into a number of subusers. Each subuser in the multiuser channel is described by a channel and stream of subuser data, so a single user rate is the sum of the rates of the streams.

The multiple access channel (MAC) draws signals from a number of individual users transmitted from physically separated transmitters, each with its own message vector. A single receiver processes the single channel output as a vector MAC. MAC in wireless communications is developing when several user transmitters share a common frequency band in transmitting UL to an access point or BS. Let us denote the ith user transmits a symbol vector x_i where $i = 1, 2, \ldots, K$. The receiver output a single vector $\mathbf{y}$. The MAC is shown in

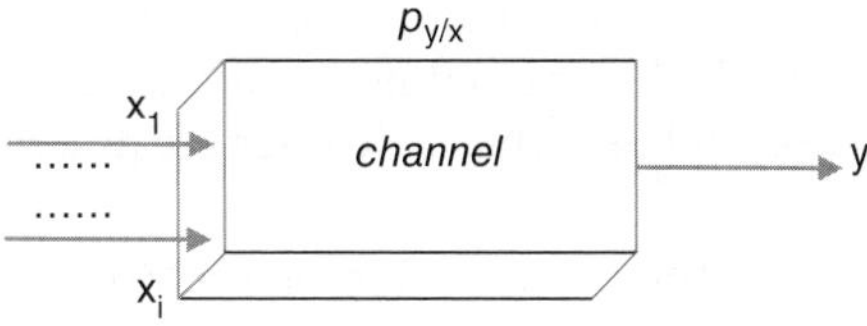

Figure 1.3 The multiple-access channel (MAC).

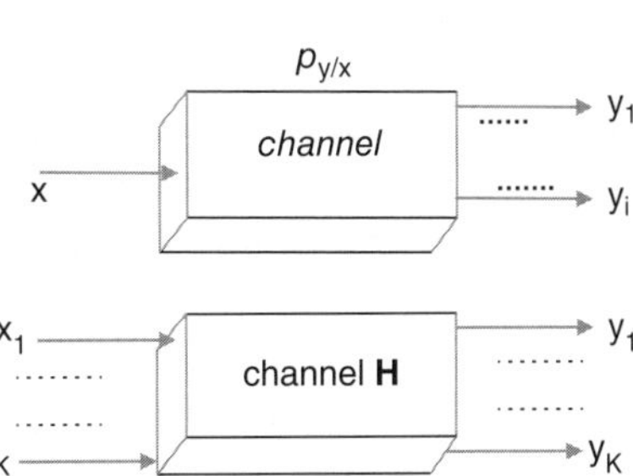

Figure 1.4 The broadcast channel (BC).

Figure 1.5 The interference channel (IC).

Figure 1.3 where $p_{y/x}$ is the conditional probability for **x** conditioned to observations **y** and the channel = $[h_K \ldots.. h_2\, h_1]$.

The dual of the MAC is the broadcast channel (BC) as shown in Figure 1.4. The input to the BC is generated by a single transmitter and the BC output has intended to physically separate users. The BCs are the opposite direction of the MACs. However, some transmission systems may have BC but no reverse MAC like channel such TV and radio broadcasting. When a certain user is decomposed into more than one user with cooperating receivers, then the channel is no longer a BC. If the user is decomposed into two or more isolated receivers then the channel remains BC with increased users. The BC is shown in Figure 1.4 as the channel = $[h_K \ldots.. h_2\, h_1]^T$.

The K-user interference channel (IC) is depicted in Figure 1.5 supporting K transmitters and K receivers. It is assumed that the receivers have independent additive complex white Gaussian noise with zero mean and given variance.

Each transmitter/receiver is equipped with a single antenna and it is assumed the transmitted signal is transported over a frequency flat channel. The flat channel gain received at the ith receiver coming from the jth transmitter is denoted as $h_{i,j}$ and the channel gain of the ith desired signal is $h_{i,i}$, where $i = j$. The interfering channel gain is $h_{i,j}$ where $i \neq j$. Accordingly, the interfering channel **H** can be mathematically modeled as K x K square matrix presented in (1.1a). The diagonal elements of the **H** matrix give the channel gain of the desired signals, and the non-diagonal elements are the interfering signal gain.

$$\mathbf{H} = \begin{bmatrix} h_{K,K} & ------- & h_{K,1} \\ h_{K-1,K} & ------- & h_{K-1,1} \\ & ------------ & \\ & ------------ & \\ & ------------ & \\ & ------------ & \\ h_{2,K} & --------- & h_{2,1} \\ h_{1,K} & --------- & h_{1,1} \end{bmatrix} \tag{1.1a}$$

When the interference channel operates in an additive Gaussian noise environment, the received K signal vector **y** can be expressed as,

$$\mathbf{y} = \mathbf{H}\,\mathbf{x} + \mathbf{n} \tag{1.1b}$$

where the K-elements vector **x** represents the transmitted signals and the K elements vector **n** represents the K-receivers complex white noise.

A key merit in multiuser channels [4] is the maximum available data rates for each user. Obviously, the latter is a function of the data rates used by the other users sharing the channel. The MAC and BC achievable rates are described in the so-called *capacity regions*.

1.4.1 The Capacity Region for Multiuser Channel

The multiuser capacity region can be determined by a search algorithm. However, the algorithm has to search over all $(K!)^K$. The search complexity can be managed for two or three users but for sizable number of users it demands a lengthy and complex computation. The MAC input is the collection of individual users' inputs made into a single input vector. The MAC and mutual information for the MAC channel from **x** to **y**, i.e. $\mathbf{I}(\mathbf{x};\mathbf{y})$ brings together the sum of users' rates. The sum-rate has no priority to the distribution of the rates among the users on the MAC. Using the chain rule of mutual information, we get Eq. (1.2):

$$\mathbf{I}(\mathbf{x};\mathbf{y}) = \mathrm{I}(\mathrm{x}_1;\mathbf{y}) + \mathrm{I}(\mathrm{x}_2;\mathbf{y}/\mathrm{x}_1) + \dots\dots\dots + \mathrm{I}(\mathrm{x}_K;\mathbf{y}/[\mathrm{x}_1,\mathrm{x}_2,\dots\dots\dots,\mathrm{x}_{K-1}])$$

$$\mathbf{I}(\mathbf{x};\mathbf{y}) = \sum_{k=1}^{K} \mathrm{I}(\mathrm{x}_k;\mathbf{y}/[\mathrm{x}_1,\mathrm{x}_2,\dots\dots\dots,\mathrm{x}_{k-1}]) \tag{1.2}$$

The BC is a multiuser system; its rates limit is given by a *capacity region*. The latter is the region that contains all user rates that are simultaneously achievable. The boundary of the capacity region corresponds to the maximum sum-rates that a transmitter can transmit messages to the receivers. In other words, the capacity region is the union of all the achievable rate vectors. Clearly, there should be a coding scheme for the receivers to be able to correctly decode their messages. An efficient decoding system is that the decoding errors go minimal as the coder block length increases.

In single-antenna BC, the sum-rate can be achieved by transmitting only to the 'strongest' receiver. Such a transmission scheme is called an 'opportunistic transmission'. A capacity region is for a general BC call for arranging the different users in the BC from the strongest to the weakest channel. Such a BC is referred to as a degraded channel, and users can be ordered according to their signal power to noise power ratio (SNRs, better known as signal-to-noise ratio). The commonly used coding scheme in the degraded Gaussian BC is the superposition coding.

In superposition, the BC transmits to all users intended receivers simultaneously and the receivers' messages are superimposed on top of one another. The receivers decode the signals starting from the weakest receiver to the strongest receiver. In other words, the weakest receiver decodes only its own message, while the strongest receiver decodes all the messages.

Let us consider a Gaussian noise BC system comprising a transmitter provided with M antennas and there are n users (receivers), each with N antennas. The channels to each user are assumed with coherence transmission interval T, i.e. the channels remain constant for T channel uses. After T uses, the channels change to an independent set of values and remain constant for another coherence interval. During the coherence interval, the signal received by the ith user is

$$\mathbf{y}_i(t) = \sqrt{P}\,\mathbf{H}_i\,\mathbf{s}(t) + \mathbf{n}_i \tag{1.3}$$

where $\mathbf{y}_i(t) \in \mathbb{C}^{N \times 1}$ is the vector of received signals at time $t = 1, \dots, T$, P is the total transmit power, $\mathbf{H}_i \in \mathbb{C}^{N \times M}$ is the channel matrix constant during coherence interval, $\mathbf{s}(t) \in \mathbb{C}^{M \times 1}$ is the transmit symbol, $\mathbb{E}\|\mathbf{s}(t)\|^2 = 1$, and $\mathbf{n}_i \in \mathbb{C}^{N \times 1}$ is additive Gaussian noise with random variable entries with zero mean and unit variance.

1.4.2 Analysis of Degraded BC with Superposition Coding

We assume that there is only a single transmit antenna (M = 1), during each coherence transmission interval T. In this scenario, then the users can be ordered according to their SNRs [5]. The channel to the ith user can be expressed as $\mathbf{h}_i \in \mathbb{C}^{N \times 1}$ where $\mathbf{h}_i = [h_{1,i}\ h_{2,i} \dots \dots \dots \dots \ h_{N,i}]^T$ and $|\mathbf{h}_i|^2$ is two-norm of the vector, i.e. $|\mathbf{h}_i|^2 = h_{i,1}^2 + h_{i,2}^2 + \cdots + \dots . h_{i,N}^2$. Assuming the receiver Gaussian noise variance is unit, the $SNR_i = |\mathbf{h}_i|^2$. Ordering the users according to their SNR can be expressed as

$$|\mathbf{h}_1|^2 \geq |\mathbf{h}_2|^2 \geq \dots \dots \dots \dots \geq |\mathbf{h}_n|^2 \tag{1.4}$$

Since the BC system is degraded, the capacity region can be achieved by *superposition coding and interference cancellation*. The key idea here is to divide the transmit signal into n independent components containing the messages of each user:

$$\sqrt{P}\, \mathbf{s}(t) = \sum_{i=1}^{n} \sqrt{P_i}\ s_i(t) \tag{1.5}$$

such that $\sum_{i=1}^{n} P_i = P$. Each $s_i(t)$ is constructed from an independent Gaussian codebook. Assuming the channels are not fading during the coherence interval and therefore $\mathbf{h}_i$s are all fixed, we arrange the powers and rates such that the weakest user can only decode its own message, whereas stronger users can decode their own and all weaker messages. This means that the set of rates that can be achieved are

$$R_1 < \log(1 + P_1|\mathbf{h}_1|^2) \tag{1.6a}$$

$$R_2 < \log\left(1 + \frac{P_2|\mathbf{h}_2|^2}{1 + P_1|\mathbf{h}_2|^2}\right) \tag{1.6b}$$

$$\dots \dots \dots \dots \dots \dots$$

$$\dots \dots \dots \dots \dots \dots$$

$$R_j < \log\left(1 + \frac{P_j|\mathbf{h}_j|^2}{1 + \sum_{i=1}^{j-1} P_i|\mathbf{h}_j|^2}\right) \tag{1.6j}$$

$$\dots \dots \dots \dots \dots \dots$$

$$\dots \dots \dots \dots \dots \dots$$

$$R_n < \log\left(1 + \frac{P_n|\mathbf{h}_n|^2}{1 + \sum_{i=1}^{n-1} P_i|\mathbf{h}_n|^2}\right) \tag{1.6n}$$

When the channels are fading, the users' ordering becomes a random process. In this case, coding over large block lengths that are adapted to the fading states on a block by block basis is required. We now investigate the ergodic capacity region when the channels are fading. The BC ergodic capacity region is defined in [6] as the set of all users' average rates that can be achieved with negligible probability of error when the average is taken with respect to all fading states. The ergodic (Shannon) sum-rate capacity is

$$\sum_{i=1}^{n} R_i \leq C_{sum} = \mathbb{E}\left\{ \max_{P_i \geq 0, \sum_{i=1}^{n} p_i = P} \log(1 + P_i|\mathbf{h}_i|^2) \right\}$$

$$\sum_{i=1}^{n} R_i \leq C_{sum} = \mathbb{E}\left\{ \log(1 + P \max_{1 \leq i \leq n} |\mathbf{h}_i|^2) \right\} \tag{1.7}$$

where the expectation is over the fading channels $\mathbf{h}_i$. It is thus clear that the scheme that maximizes the sum rate is the one that, during each coherence interval, transmits only to the user with the best SNR (i.e. only one of the P_i's is nonzero). This is referred to as *opportunistic transmission.*

1.4.3 The Capacity Region for Multiuser MIMO Channel

Unlike the single antenna system model we considered in deriving the channel matrix expressed in (1.1a), we now explore a cellular system with the BS equipped with M antennas and each of the K (mobile) users has N antennas. We examine the two-essential multiuser MIMO channels in the cellular system: the MIMO multiple access channel (MAC) and the MIMO broadcast channel (BC). The downlinks (DLs) of the system constitute the MIMO BC and the uplinks (ULs) constitute the MIMO MAC. The DL channel matrix from the BS to the ith user is denoted as $\mathbf{H}_i$. Assuming reciprocity hold for TDD transmission protocol, then the UL matrix of the ith user is $\mathbf{H}_i^H$.

Let $\mathbf{u}_k \in \mathbb{C}^{N\,x\,1}$ represents a typical k^{th} user transmitted signal on the MAC, i.e. $\mathbf{u}_k$ is a vector of N elements, each element symbolizes the signal component transmitted from each of the user N antennas. Denote the received signal $\mathbf{r} \in \mathbb{C}^{M\,x\,1}$ and the noise vector $\mathbf{w} \in \mathbb{C}^{M\,x\,1}$ where $\mathbf{w} \sim \tilde{\mathcal{N}}(0, \mathbf{I})$ is circularly symmetric complex Gaussian with identity covariance. The received signal at the BS is,

$$\mathbf{r} = \mathbf{H}_1^H\,\mathbf{u}_1 + \mathbf{H}_2^H\,\mathbf{u}_2 + \ldots\ldots\ldots\ldots + \mathbf{H}_K^H\,\mathbf{u}_K + \mathbf{w} \tag{1.8}$$

where the conjugate transpose of the channel is denoted as $\mathbf{H}^H$. Expression (1.8) can be re-written as,

$$\mathbf{r} = [\mathbf{H}_1^H\ \mathbf{H}_2^H\ \ldots\ldots\ldots\ldots\mathbf{H}_K^H\ ]\begin{bmatrix}\mathbf{u}_1\\ \mathbf{u}_2\\ -\\ -\\ \mathbf{u}_K\end{bmatrix} + \mathbf{w} \tag{1.9}$$

Each user transmit power is constraint to P_k. The transmit covariance matrix of user k is defined to be $\mathbf{Q}_k$:

$$\mathbf{Q}_k \triangleq \mathbb{E}[\mathbf{u}_k\,\mathbf{u}_k^H] \tag{1.10}$$

The power constraint can be expressed as

$$\mathrm{tr}(\mathbf{Q}_k) \leq \mathrm{P}_k \text{ for } k = 1, \ldots\ldots\ldots, \mathrm{K}.$$

On the other hand, let us denote the transmitted signal from the BS towards the user over the BC be $\mathbf{x} = \epsilon\mathbb{C}^{1\,x\,M}$. Let $\mathbf{y}_k\ \epsilon\mathbb{C}^{N\,x\,1}$ be the received signal at the kth user's receiver. The noise at receiver k is represented by $\mathbf{n}_k\epsilon\mathbb{C}^{N\,x\,1}$ and is assumed to be circularly symmetric complex Gaussian noise $\mathbf{n}_k \sim \mathcal{N}(0, \mathbf{I})$. The received signal of the kth user is given in Eq. (1.11):

$$\mathbf{y}_k = \mathbf{H}_k\,\mathbf{x} + \mathbf{n}_k \tag{1.11}$$

The transmit covariance matrix of the input signal is

$$\Sigma_{\mathbf{x}} \triangleq \mathbb{E}[\mathbf{x}\,\mathbf{x}^H] \tag{1.12}$$

The BS average power is constraint to P, which is expressed as $\mathrm{tr}(\Sigma_{\mathbf{x}}) \leq P$.

1.4.4 The MIMO MAC Capacity Region

We assume the MAC is nonvarying during the analysis. The capacity region of an MAC can be modelled as the convex closure of the union of rate regions corresponding to the product of every input distribution $p(u_1)\ldots\ \ldots\ \ldots\ldots p(u_K)$ satisfying the user-by-user power constraints.

However, considering Gaussian MIMO MAC, it is appropriate to consider the Gaussian elements so the entire convex operation is not needed. Accordingly, for any set of powers

$$\mathbf{P} = (P_1, \dots \dots \dots, P_K) \tag{1.13}$$

The capacity of the MIMO MAC is given as

$$C_{MAC}(\mathbf{P}; \mathbf{H}^H) \triangleq \cup_{\{\mathbf{Q}_i \geq 0, \mathrm{tr}(\mathbf{Q}_i) \leq P_i \; for\ all\ i\}} \left\{ \begin{array}{c} (R_1, \dots \dots, R_K) : \\ \sum_{i \varepsilon S} R_i \leq \frac{1}{2} \log |\mathbf{I} + \mathbf{H}_i^H \mathbf{Q}_i \mathbf{H}_i| \\ for\ all\ S \subseteq \{1, \dots, K\} \end{array} \right\} \tag{1.14}$$

The ith user transmits a zero-mean Gaussian signal with spatial covariance matrix $\mathbf{Q}_i$. Each set of covariance matrices $(\mathbf{Q}_1, \dots \dots \dots \dots, \mathbf{Q}_K)$ corresponds to a K-dimensional polyhedron. In geometry, a polyhedron is simply a 3D solid that consists of a collection of polygons, usually joined at their edges.

The MAC capacity region is equal to the union over all covariance matrices satisfying the power constraints of all such polyhedrons. The corner points of each pentagon can be achieved by successive decoding where users' signals are successively decoded and subtracted out of the received signal (i.e. successive interference cancellation). To explain the MAC capacity region in depth, let us consider a simple two-users scenario. Each set of covariance matrices corresponds to a pentagon, similar in form to the capacity region of the scalar Gaussian MAC.

The corner point B corresponds to decoding user 2 first (i.e. in the presence of interference from user 1) and decoding user 1 last (without interference from user 2). The performance of user 1 can achieve a single user rate as

$$R_1 = \log|\mathbf{I} + \mathbf{H}_1^H \mathbf{Q}_1 \mathbf{H}_1| \tag{1.15}$$

The slope A-B corresponds to the sum rate:

$$R_1 + R_2 \leq \log|\mathbf{I} + \mathbf{H}_1^H \mathbf{Q}_1 \mathbf{H}_1 + \mathbf{H}_2^H \mathbf{Q}_2 \mathbf{H}_2| \tag{1.16a}$$

At corner point B, the rate is

$$R_1 + R_2 = \log|\mathbf{I} + \mathbf{H}_1^H \mathbf{Q}_1 \mathbf{H}_1 + \mathbf{H}_2^H \mathbf{Q}_2 \mathbf{H}_2| \tag{1.16b}$$

$$R_2 = \log|\mathbf{I} + \mathbf{H}_1^H \mathbf{Q}_1 \mathbf{H}_1 + \mathbf{H}_2^H \mathbf{Q}_2 \mathbf{H}_2| - R_1 \tag{1.16c}$$

$$R_2 = \log\left|\frac{\mathbf{I} + \mathbf{H}_1^H \mathbf{Q}_1 \mathbf{H}_1 + \mathbf{H}_2^H \mathbf{Q}_2 \mathbf{H}_2}{\mathbf{I} + \mathbf{H}_1^H \mathbf{Q}_1 \mathbf{H}_1}\right| \tag{1.16d}$$

$$R_2 = \log\left|\mathbf{I} + \frac{\mathbf{H}_2^H \mathbf{Q}_2 \mathbf{H}_2}{\mathbf{I} + \mathbf{H}_1^H \mathbf{Q}_1 \mathbf{H}_1}\right| \tag{1.16e}$$

$$R_2 = \log|\mathbf{I} + (\mathbf{I} + \mathbf{H}_1^H \mathbf{Q}_1 \mathbf{H}_1)^{-1} \mathbf{H}_2^H \mathbf{Q}_2 \mathbf{H}_2| \tag{1.16f}$$

Corner point A concurs with decoding user 1 first (in the presence of interference from user 2) and decoding user 2 last in interference free signal. The performance of user 2 achieves a single user rate as

$$R_2 = \log|\mathbf{I} + \mathbf{H}_1^H \mathbf{Q}_1 \mathbf{H}_1| \tag{1.17}$$

Using similar method to above to derive user 1 rate as

$$R_1 = \log|\mathbf{I} + (\mathbf{I} + \mathbf{H}_2^H \mathbf{Q}_2 \mathbf{H}_2)^{-1} \mathbf{H}_1^H \mathbf{Q}_1 \mathbf{H}_1| \tag{1.18}$$

Accordingly, successive interference cancellation reduces a complex multiuser detection problem into a series of single-user detection processes. The capacity region of an MIMO MAC for users with single transmit antenna each when two users sharing the band is shown in Figure 1.6. In a scenario N = 1, the covariance matrix of each transmitter is a scalar quantity equal to the transmitted power, and each user transmits at full power. Thus, the capacity

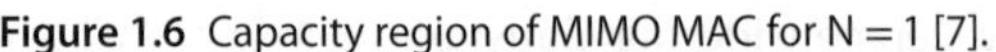

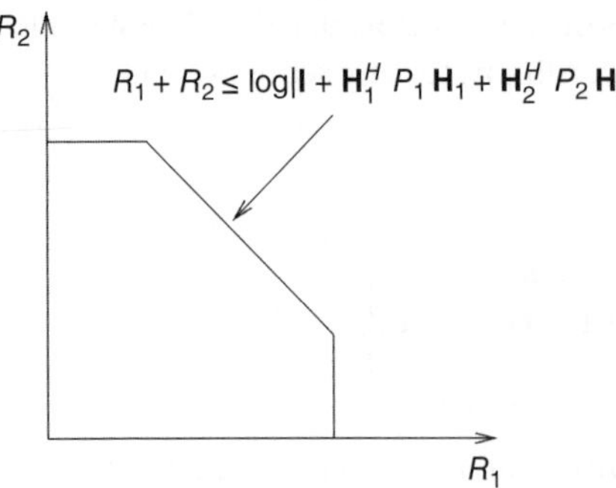

Figure 1.6 Capacity region of MIMO MAC for N = 1 [7].

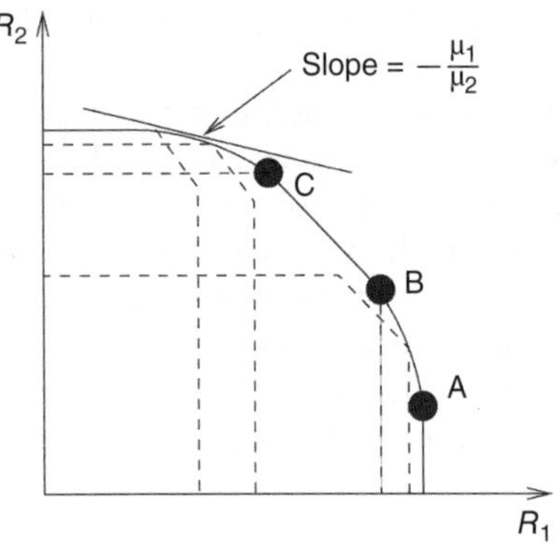

Figure 1.7 Capacity region of MIMO MAC for N > 1 [7].

region for a K-user MAC for N = 1 is the set of all rate vectors $(R_1, \dots \dots, R_K)$ satisfying equation (1.19):

$$\sum_{i \varepsilon S} R_i \leq \frac{1}{2} \log|\mathbf{I} + \mathbf{H}_i^H \boldsymbol{p}_i \mathbf{H}_i| \textit{ for all } S \subseteq \{1, \dots, K\} \tag{1.19}$$

This reduces the two-user capacity region to a simple pentagon as shown in Figure 1.7. While N > 1, a union must be taken overall covariance matrices. Evidently, different sets of covariance matrices maximize users' rates and are different from the set of covariance matrices that maximize the sum rate. So, let us examine a MAC capacity region when users are equipped with N > 1. As we already emphasised, the capacity region is equal to the union of pentagons. Each pentagon corresponding to a different set of transmit covariance matrices as depicted in Figure 1.7. A few are shown with dashed lines in the figure.

The generic boundary of the capacity region is curved, and only straight lines at the sum rate point. Each point on the curved part of the boundary is achieved by a different set of covariance matrices.

At point A, user 1 is decoded last and achieves single-user capacity, choosing $\mathbf{Q}_1$ as a water-fill of the channel $\mathbf{H}_1$ independent of $\mathbf{H}_2$ or $\mathbf{Q}_2$. User 2 is decoded first in the presence of interference from user 1, $\mathbf{Q}_2$ is chosen as a water-fill of the channel $\mathbf{H}_2$ and the decoding is degraded by the presence of interference from user 1.

The corner points B and C are the two corner points of the pentagon corresponding to the sum rate with optimal covariance matrices $\mathbf{Q}_1^{\text{sum}}$ and $\mathbf{Q}_2^{\text{sum}}$. At point B, user 1 is decoded last, hence achieving a single user rate while at point C user 2 is decoded last. Thus, points B and C are achieved using the same optimal covariance matrices but different decoding orders.

Next, we investigate the features of the optimal covariance matrices $(\mathbf{Q}_1, \dots \dots \dots \dots, \mathbf{Q}_K)$ that achieve different points on the capacity region boundary of the MIMO MAC. Since the

MAC capacity region is convex, it is well known that the boundary of the capacity region can be maximized by the function $\mu_1 R_1 + \dots \dots \dots \dots .+ \mu_K R_K$ overall capacity region rate vectors and for all nonnegative priorities $(\mu_1, \dots \dots \dots ., \mu_K)$ such that $\sum_{i=1}^{K} \mu_i = 1$ where μ is the water-fill level. Let us define a fixed set of priorities $(\mu_1, \dots \dots \dots ., \mu_K)$, then a point on the capacity region boundary is corresponds to the tangent to a line whose slope is defined by the priorities as shown in Figure 1.7.

The boundary of the MAC capacity region indicates that all points of the capacity region are corner points of polyhedrons corresponding to different sets of covariance matrices. Additionally, the corner point should correspond to successive decoding with interference cancellation in order of *increasing* priority, i.e. the user with the highest priority should be (interference free) decoded last. Accordingly, the problem of locating the capacity region boundary point associated with priorities assumed to be in descending order (arbitrarily renumbering the users to satisfy this condition) can be written as [7].

$$\max_{\mathbf{Q}_1,\dots\dots\dots\dots,\mathbf{Q}_K} \left(\mu_K \log \left| \mathbf{I} + \sum_{l=1}^{K} \mathbf{H}_{l=1}^{H} \mathbf{Q}_l \mathbf{H}_l \right| \right) + \sum_{i=1}^{K-1} (\mu_i - \mu_{i+1}) \log \left| \mathbf{I} + \sum_{l=1}^{i} \mathbf{H}_{l=1}^{H} \mathbf{Q}_l \mathbf{H}_l \right| \tag{1.20}$$

Maximizing the above function subject to satisfying the power constraints defines the optimal covariance matrices.

The most interesting aspect of the optimization problem is that the objective function is concave in the covariance matrices. An efficient numerical technique to find the sum rate maximizing covariance matrices, is called iterative water-filling. The technique is based on the Karush–Kuhn–Tucker (KKT) optimality conditions for the sum rate maximizing covariance matrices. The KKT optimization method is introduced in the tutorials on theory and techniques of optimization mathematics, Appendix 1.C at the end of the chapter. These conditions indicate that the sum rate maximizing covariance matrix of any user in the system should be the single-user water-filling covariance matrix of its own channel with noise equal to the actual noise plus the interference from the other $K - 1$ transmitters.

1.4.5 The MIMO BC Capacity Region [6, 8, 9]

When the transmitter (i.e. BS) has more than one antenna, the Gaussian BC is typically nondegraded. This is because users receive different strength signals from different transmit antennas, which make it difficult to order the users according to their SNRs. In this section, we assume the MIMO BC is a constant channel and BS supported with M antennas, and each receiver is equipped with N antennas. We consider the optimum coding scheme, i.e. dirty paper coding (DPC) [10, 11] applied at the MIMO transmitter that achieves the sum rate capacity region of BC. The basic concept of DPC is that if the transmitter has perfect knowledge of additive Gaussian interference in the channel, then the capacity of the channel is the same as if there was no additive interference. The DPC allows a known interference to be pre-subtracted at the transmitter through selecting codewords for different receivers, keeping the transmit power unchanged.

Let us consider the MIMO BC with DPC applied to two users: user 1 is referred to single antenna receiver 1 and user 2 to single antenna receiver 2. First, the transmitter arbitrary selects a codeword (i.e. x_1) for receiver 1. The transmitter then chooses a codeword for receiver 2 (i.e. x_2) with full knowledge of the codeword intended for receiver 1.

Therefore, the codeword of user 1 can be pre-subtracted and receiver 2 does not see the codeword intended for receiver 1 as interference. Similarly, when three users are considered, the codeword for receiver 3 is chosen such that receiver 3 does not see the signals intended for receivers 1 and 2 (i.e., presubtract $\mathbf{x}_1 + \mathbf{x}_2$) as interference. This presubtraction process continues for all K receivers.

If user $u(1)$ is encoded first, followed by user $u(2)$, etc., the rate vector can be achieved:

$$R_{u(i)} = \log \frac{\left| \mathbf{I} + \mathbf{H}_{u(i)} \left(\sum_{k \le u(i)} \mathbf{\Sigma}_k \right) \mathbf{H}_{u(i)}^H \right|}{\left| \mathbf{I} + \mathbf{H}_{u(i)} \left(\sum_{k < u(i)} \mathbf{\Sigma}_k \right) \mathbf{H}_{u(i)}^H \right|} \quad i = 1, \ldots \ldots \ldots, K \tag{1.21}$$

In deriving Eq. (1.21), we use the identical method that is used in deriving Eq. (1.16). The dirty paper region is defined as the convex hull, (i.e., convex boundary in which a given set of rate points is contained) of the union of all such rates vectors over covariance matrices $\mathbf{\Sigma}_1, \ldots \ldots \ldots \ldots \ldots, \mathbf{\Sigma}_K$ such that $\mathrm{tr}(\mathbf{\Sigma}_1 + \ldots \ldots + \mathbf{\Sigma}_K) \le P$ and overall permutations $(u(1), \ldots \ldots \ldots \ldots, u(K))$, the DPC capacity region of BC C_{DPC} $(P, \mathbf{H})$ is given by (1.22) and $\mathbf{R}_{u(k)}$ is given by equation (1.21),

$$C_{\mathrm{DPC}}(P, \mathbf{H}) \triangleq Co \left(\bigcup_{u, \mathbf{\Sigma}_k} \mathbf{R}_{u(k)} \right) \tag{1.22}$$

where Co implies a coordinated joint detection system. The DPC rate region for two-user channel $\mathbf{H}$ with power constraint $P = 10$ is depicted in Figure 1.8. The transmitter operates with two antennas and each receiver has a single antenna. The channel $\mathbf{H}$ matrix is

$$\mathbf{H} = \begin{bmatrix} \mathbf{H}_1 \\ \mathbf{H}_2 \end{bmatrix} = \begin{bmatrix} 1 & 0.5 \\ 0.5 & 1 \end{bmatrix}$$

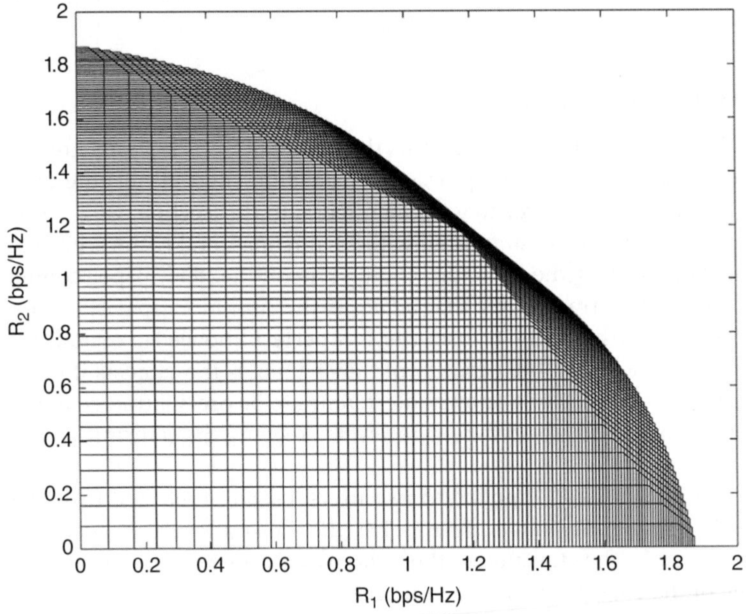

Figure 1.8 Dirty paper capacity region, $\mathbf{H}_1 = [1\ 0.5]$, $\mathbf{H}_2 = [0.5\ 1]$, $P = 10$ [7].

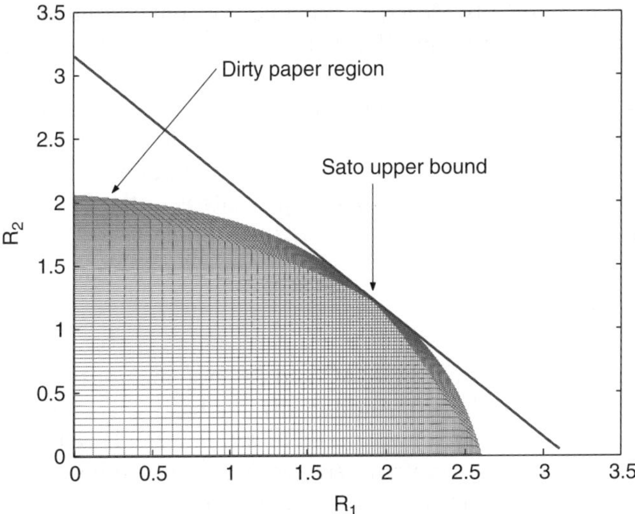

Figure 1.9 Dirty paper broadcast region – achievability and converse [13].

It is worth noting that Sato [12] presents an upper bound on the capacity region of general BC. This bound utilizes the capacity of the coordinated system as defined above. The capacity of the coordinated system is an upper bound on the BC sum rate $C_{DPC}(P, \mathbf{H})$.

The DPC BC capacity regions for a two single antenna receivers BC and two antenna transmitters as well as the Sato upper bound and the power constraint $P = 10$ are shown in Figure 1.9. The BC channel matrix H is given by

$$\mathbf{H} = \begin{bmatrix} \mathbf{H}_1 \\ \mathbf{H}_2 \end{bmatrix} = \begin{bmatrix} 1 & 0.5 \\ 0.2 & 0.8 \end{bmatrix}$$

1.5 Fading Wireless Channels

In wireless (mobile) communications, the transmitted signal strength is varying due to various variables such as time, location, signal obstruction, and so on. The signal variation causes the signal in most cases to randomly change, and this is called fading. There are different classifications of fading such as flat fading; fast fading, slow fading; and frequency selective fading. Generally, the signal energy propagation can be of two types: direct line of sight energy propagation between transmitter and receiver (LOS), but the most common type of propagation is due to scattering of the energy by objects in the way and the scattering mechanism generates multiple copies of the signal with different amplitude and phase (i.e. multipath signal) arriving at the receiver. The signal power level range over time is called *multipath intensity profile,* which is defined by multipath spread. The reciprocal of the multipath spread is called the channel coherence bandwidth. When the signal bandwidth is much less than the coherence bandwidth of the channel, the amplitudes of the signal frequency components are flat, and such channels are called frequency nonselective or flat fading. The flat fading is of two types: slow fading occurs if the transmitted symbol duration is much smaller than the coherence time of the channel (channel coherence time is the reciprocal of channel Doppler spread). On the contrary, a rapidly fading channel is one in which the symbol duration is much greater than the channel coherence time. When the transmitted signal bandwidth is greater than the channel coherence bandwidth, the frequency components exceed the channel coherence bandwidth, and are exposed to different gains and phase shifts, and the channel is said to be a frequency selective fading channel [14].

In most publications, researchers assume the transmission channel is subjected to flat fading when considering the analysis/simulations, and this is what we have assumed in this book. The capacity region analysis for flat-fading channels demands a number of assumptions regarding the perfect knowledge of CSI at transmitter / receiver or both. The analysis of the sum rate regions is dealt with in detail in the following chapters.

1.6 Multicell MIMO Channels

In Section 1.4, a single cell is assumed for analysing the MAC and BC system. Nonetheless, a cellular system, by definition, entails many cells [15]. Additionally, the basic characteristic of wireless energy propagation in a cellular system is not limited to within a single cell. This means that users as well as BSs in adjacent cells encounter interference from each other. However, BSs are not mobile, they can communicate through optical fibre links capable of high data rates. This gives a good opportunity for BSs to cooperate on processing different users' signals. However, capacity of the cellular network, with multiple users and multiple antennas, is a very difficult problem. The single cell multiuser capacity is important and can be used as a benchmark to assess the efficiency of a practical multicell system in similar approach to the capacity of a single user link serves to measure the performance merits of practical multiusers scheme. The assumption of perfect BSs cooperation is the key to the extension of a single cell with results to multiple cells systems. Theoretically, multiple BSs cooperation can be treated as composite BS with physically distributed antennas.

Explicitly, let use consider a group of B coordinated cells, each with M antennas and serving K mobiles, each with N antennas. Let us also define $\mathbf{H}_{i,j} \in \mathbb{C}^{\mathrm{N\,x\,M}}$ to be the *DL* channel of user i from base station j, then the *composite* DL channel of user i is $\mathbf{F}_i = [\mathbf{H}_{i,1} \cdots \ldots . \mathbf{H}_{i,B}]$ and the composite UL channel is given as $\mathbf{F}_i^H$. The received signal of user i can then be written as $\mathbf{Y}_i$,

$$\mathbf{Y}_i = \mathbf{F}_i \mathbf{W} + \mathbf{n}_i \tag{1.23}$$

where $\mathbf{W}$ is the composite transmitted signal defined as $\mathbf{W}^T = [\mathbf{W}_1^T \cdots \cdots \cdots \mathbf{W}_B^T]$. Here, we let $\mathbf{W}_j$ *to* represent the transmit signal from base j.

On the DL, since the base stations can cooperate perfectly, DPC can be used over the entire network transmit signals (i.e. across base stations) in a straightforward manner. This requires perfect data and *power cooperation* between the base stations. Denote $\mathbf{x}_{i,j}$ to represent the transmitted vector for user i from base station j, the composite transmit vector intended for user i is $\mathbf{X}_i^T = [\mathbf{x}_{i,1}^T \cdots \cdots \cdots . \mathbf{x}_{i,B}^T]$. Thus, the composite covariance of user i is defined as $\Sigma_i = \mathbb{E}[\mathbf{X}_i \mathbf{X}_i^T]$. The covariance matrix of the entire transmitted signal is $\Sigma_{\mathbf{x}} = \sum_{i=1}^{K} \Sigma_i$.

Although data cooperation is a justifiable assumption, in practice each BS has its own power constraint. The *per-BS power constraint* can be expressed as $\mathbb{E}[\mathbf{W}_j^H \mathbf{X}_j] \leq P_j$, where P_j is the power constraint at BS j. Consequently, power cooperation, or pooling the transmit power for all the BSs over the BC to have one overall transmit power constraint, is not realistic. On the UL the BSs are only receiving signals and, therefore no power cooperation is required.

1.7 Green Wireless Communications for the Twenty-First Century

Practically every human on this planet uses mobile communications. This growth in mobile phone subscribers comes with enormous energy consumption due to mobile networks.

The intensive concerns for the environment and the impacts of global warming on the climate necessitate an appropriate action on the energy efficiency of these systems. It is estimated that 80% of the energy needed for the operation of a wireless cellular network is consumed at the BS sites. Furthermore, to increase the energy efficiency in mobile communication systems there is a need to reduce the BS power consumption at instants of low network loads. Researchers around the world put in great efforts to find a solution to climate change challenge. One of these efforts is the EU funded research project Energy Aware Radio and network technologies (EARTH) [16].

When considering green energy for wireless communications, the first few questions that come to mind are how much energy is required to operate a wireless network and how to model the *energy efficiency evaluation framework* (E^3F). The EARTH E^3F suggested a power model for various BS types in addition to large-scale long-term traffic models. The BS power model relates the RF energy radiated at the antenna elements to the total supply power of a BS site. The proposed traffic model mirrors the spatial distribution of the traffic demands over large geographical regions in addition to urban and rural areas, as well as time-based variations between peak and off-peak hours. The performance of a wireless network at the system level can be evaluated for system throughput, quality of service (QoS) metrics, and cell edge user throughput.

The EARTH E^3F is depicted in Figure 1.10. The EU project considered for the performance modelling comprising various types of BSs, wireless channels, mobile devices and their power consumption.

The global E^3F is illustrated in three steps. The system evaluations over small-scale short-term scenarios models the statistical traffic as a File Transfer Protocol (FTP) file download or Voice over Internet Protocol (VoIP) calls. In addition, small-scale deployment includes

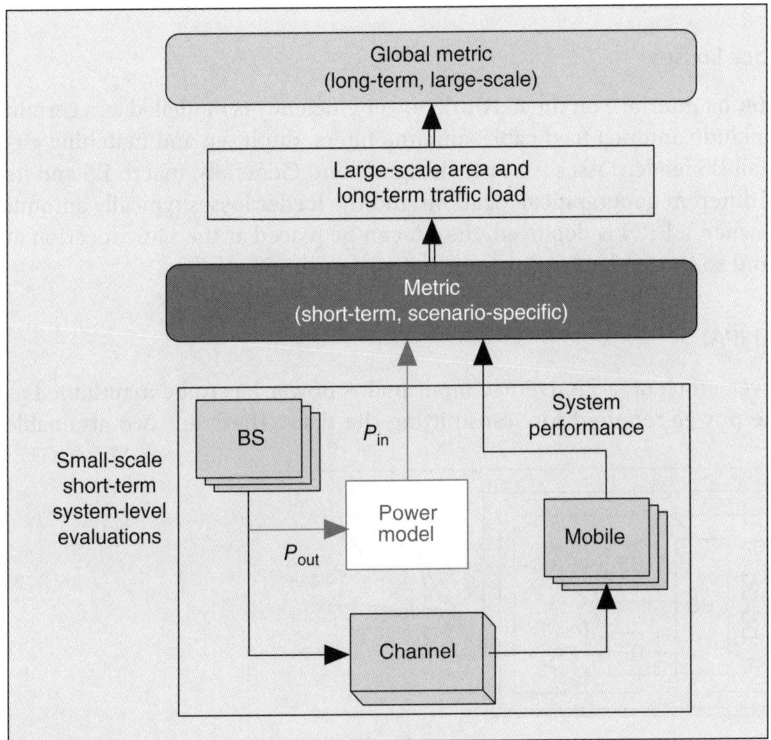

Figure 1.10 EARTH Energy Efficiency Evaluation Framework (E^3F) [16].

urban macro-cell network incorporating 57 cells with uniformly distributed users to enable short-term system-level evaluations. These evaluations are assisted by a BS power consumption model. The small-scale short-term evaluations are broadened to a global scale to cover countrywide geographical areas and ranging over a full day or week. The small-scale short-term evaluations cover cities, suburbs, and villages, and a set of network loads. The evaluations' results include energy consumption, throughput, QoS over a range of traffic loads. For a given daily/weekly traffic profile of each deployment, performance metrics over a day/week are determined from the short-term evaluation results and extended to large-scale deployment over large geographical areas.

1.7.1 Network Power Consumption Model

The power consumption model creates a connection between the system components and system levels that enable the calculation of the energy saving on specific components required to enhance the energy efficiency at the network level. System components are characterized by the BS type in terms of radiated power (BS output power), constraints on cost, and BS size. So, we need a power model designed for an individual BS type. However, we can account for the power consumption based on a generalized infrastructure of all BS types, including macro, micro, pico, and femto BSs. A BS consists of a number of transceivers (TRXs) equal to the number of transmit antennas. A block diagram of a BS transceiver is shown in Figure 1.11. A TRX is composed of a power amplifier (PA), RF TRX module, BS engine encompassing UL receiver and DL transmitter section, DC-DC power supply, an active cooling system, and an AC-DC unit for connection to the main supply. In the following section, we analyse the power consumption of each component of the BS TRX.

1.7.2 Antenna Interface Losses

The influence of the antenna interface on the network power efficiency is modelled as a certain amount of losses that include antenna feed cable, antenna filters, duplexer, and matching circuits. The loss from small BS feeder losses is mostly insignificant. Generally, macro BS and its antennas are located at different geographical locations and the feeder losses typically amount to −3 dB. Nonetheless, when a RRH is deployed, the PA can be placed at the same location at the transmit antenna, and so the feeder losses can be reduced or diminished.

1.7.3 Power Amplifier (PA)

With the aim to be power-efficient, a PA average input mains power has to be maintained as close as possible to the power required for transmitting the data. There are two attainable

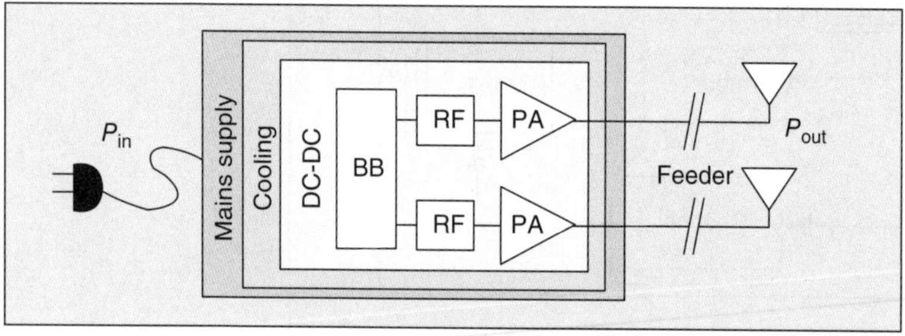

Figure 1.11 Block diagram of a base station transceiver [16].

solutions to this problem: either introduce digital predistortion (DPD) to the signal before transmission or dynamically adapt the amplitude of the input voltage of the amplifier to the transmission power requirement. A linearization scheme is commonly associated with the DPD to activate a compensation for the nonlinear part of the amplifier output. Actually, the compensation covers the adjacent channels interference (ACI) and all the precoding changes introduced with time and temperature signal characteristics. The linearization may use a Doherty PA technique, which applies two PAs driven in parallel, and the final power output is the combined output power of the two amplifiers. One amplifier is operated in the linear region in class A/B continuously until its nonlinear region. The second amplifier operates in class C to provide additional power when the amplifying device enters its the nonlinear region. Theoretically, Doherty amplifiers are capable of high efficiency. Nonetheless, when used in cellular networks working at high frequencies and obliged to output high power, Doherty amplifiers achieved an average efficiency of 25–30%. In addition, Doherty amplifiers are narrowband due to the matching required between the two internal amplifiers. DPD and Doherty techniques are applied by Huawei to construct low-power GSM BSs to operate with 500 W, which reduces CO_2 emissions considerably (claims to be equivalent to burning 1.7 tons of coal per year per BS) [12]. When considering there are 100 000 Huawei BSs operating worldwide, then the reduction of CO_2 emissions is equivalent to burning 170 000 tons of coal per year.

The second solution to enhance the PA efficiency is to adapt the amplifier voltage amplitude to the required transmit power, which was widely known as envelope-tracking technology proposed by Bell Labs in the late 1930s, but it did not take off until recently with the availability of low-power transistor amplifiers. The key concept of envelop tracking can be explained as follows: adapting the amplitude of the input voltage of the amplifier to the transmission requirement is accomplished by dynamically changing the supply voltage of the PA to match the signal to be amplified. Accordingly, envelope tracking guarantees the output of the PA is always in its most efficient operating region. The amplifier inefficiency when a constant supply of voltage applied is demonstrated in Figure 1.12. With no envelope tracking, the difference between the constant supply voltage and the RF waveform amplitude is consumed as heat by the power transistor. Figure 1.13 shows the situation when envelope tracking is used. The supply voltage (i.e. input power) is dynamically adjusted to be close to the RF output power, which makes an

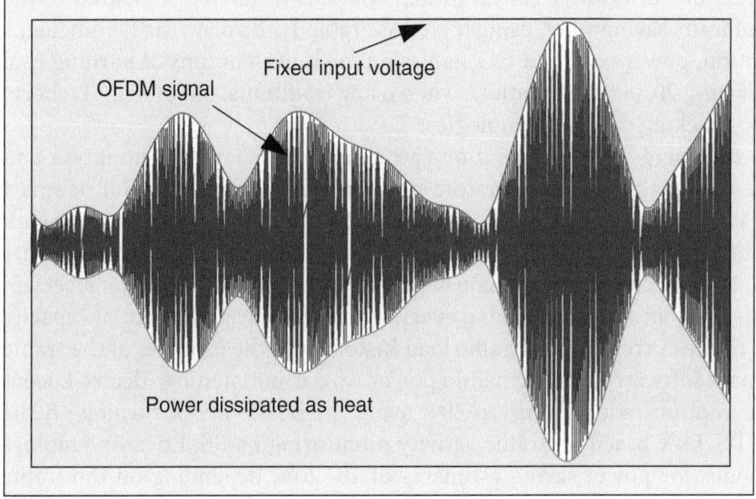

Figure 1.12 Power amplifier with constant input voltage supply [17].

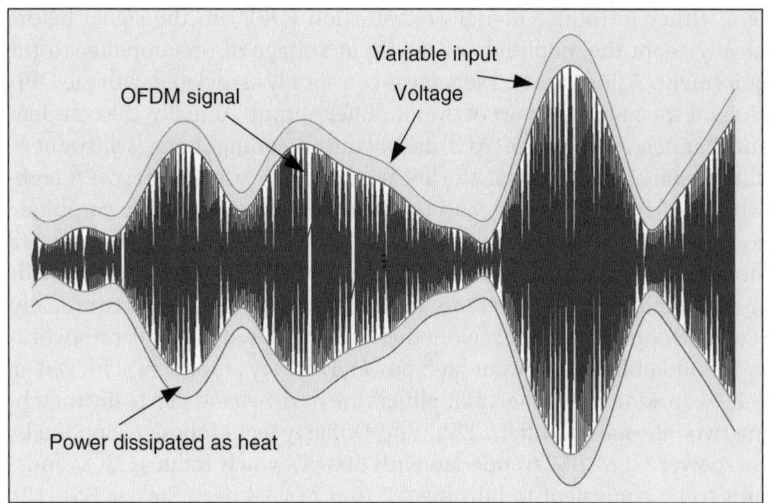

Figure 1.13 Power amplifier with input voltage supply using envelope tracking [17].

Table 1.1 Efficiency, costs, and environmental impact of a 20 000-BS network with different power amplifier technologies [17].

	Traditional technology	Doherty technology	Envelope tracking (HAT)
Power ampl. effic. (%)	15	25	45
Power consumption (MW)	51.7	27.2	16.1
Power cost (M)	\$54.3	\$28.6	\$17.0
CO_2 emission (tons)	194 600	102 400	60 800

impressive reduction in the energy converted to heat. Naturally, envelope tracking comes with CPU cost and a few watts of extra power consumption, which is negligible compared to the tens or even hundreds of watts saving when using high PAs. Table 1.1 displays the PA efficiency and its power consumption, power cost, and CO_2 emission in equivalent tons of burning coal for a large network consisting 20, 000 base stations when using traditional technology, Doherty technology, and envelop-tracking design technologies.

In practical systems, the energy-efficient solutions are related to energy consumption and *traffic load*. To achieve this goal, wireless operators have proposed the adaptation of smart software platforms to enable energy saving at BSs. For example, Nokia Siemens network employs a service quality manager (SQM) known as NetAct software traffic load capacity management at the BS. NetAct commands BS consumption automatically based on a certain setting. NetAct schedules system's capacity tuning every short interval and the actual capacity is chosen as a function of the current traffic, traffic load history, and the estimate of the traffic in the next interval. Smart software called dynamic power save is initiated by Alcatel-Lucent and claims power consumption reduction up to 30% using the possibility of turning off the amplifier power in the BS TRX based on traffic activity monitoring by BS. Ericsson employs standby operational modes for power saving estimates of 10–20%, depending on the traffic patterns. China mobile proposed a green action plan based on a low-carbon economy. The strategy is defined by the following six Rs: *Recovery & Recycle* implies recycling materials that need low-energy cost for their recycle life. *Right & Reduce* refers to small and lightweight carton design to reduce packaging and transport costs. *Returnable & Reuse* denotes the use of efficient recycling system to extend packaging material life cycle. Huawei announced a cut of over 6000 tons in CO_2 emissions when applying the six Rs strategy for a greener environment.

1.8 BS Power Model

PAs operating in the nonlinear region with input OFDM modulation (nonconstant envelope signals) generate a high level of ACI that causes considerable performance degradation at the receiver. Accordingly, the PA has to operate in the linear region (i.e. 6–12 dB below the saturation point). Nonetheless, such high operating back-off induces a reduction in the power efficiency, which converts to an increase in the PA power consumption. The objective of the analysis is to derive a power model that links the RF output power radiated at the array elements P_{out} to the total supply power at BS site P_{in}, as shown in Figure 1.10. Let us denote the power efficiency η_{PA}, PA power consumption P_{PA} and the BS feeder loss σ_{feed}. BS feeder loss can be as high as 3 dB for macro cells and can be considered negligible for smaller cells. The PA power consumption can be expressed as

$$P_{PA} = \frac{P_{out}}{\eta_{PA}(1 - \sigma_{feed})} \tag{1.24}$$

As we highlighted in Section 1.7.3, DPD combined with Doherty PAs can improve the power efficiency while keeping ACI closely controlled. However, extra feedback is necessary for the DPD, and additional signal processing is assumed to be essential in macro and micro BSs. In smaller BSs, the PA power consumption account for a smaller percentage of the total BS power consumption and advanced PA architectures are not required.

1.8.1 Small-Signal RF Transceiver

The transceiver consists of a transmitter and a receiver for the UL and DL communications. Linearity and blocking of the RF modules differ greatly depending on the BS type and therefore affect the RF architecture. For example, low intermediate frequency (IF) or even super-heterodyne design can be chosen for macro/micro BSs and zero IF architecture is acceptable for pico/femto BSs. The parameters that have the highest impact on the RF power consumption are: appropriate bandwidth, accepted signal to noise and distortion (SiNAD) and A/D resolution.

1.8.2 Baseband (BB) Unit

At the baseband (BB) unit, the following signal processing normally carried out: filtering, modulation / demodulation, predistortion processing, signal detection including synchronization, channel estimation, equalization, and channel coding /decoding. For large BSs, the digital BB unit encompasses the power consumed by the link to the backbone network and the MAC operation consumption. Denote the BB power consumption as P_{BB}.

1.8.3 Power Supply and Cooling

Losses encountered by the DC-DC supply, mains supply, and active cooling (i.e. using fans to reduce the heat of components) are only enforced to macro BSs. The losses may be approximated by loss factors σ_{DC}, σ_{Mains}, and σ_{cool}, respectively. Assume the BS power consumption

Table 1.2 Base station power consumption at maximum load for different LTE BS types.

		Macro	RRH	Micro	Pico	Femto
BS	Max Tx power, dBm	43.0	43.0	38.0	21.0	17.0
	(average) P_{max}, W	20.0	20.0	6.3	0.13	0.05
	Feeder loss σ_{feed}, dB	−3	0	0	0	0
PA	Back-off, dB	8.0	8.0	8.0	12.0	12.0
	Max PA out (peak), dBm	54.0	51.0	46.0	33.0	29.0
	PA eff. η_{PA}, %	31.1	31.1	22.8	6.7	4.4
Total PA, $\frac{P_{max}}{\eta_{PA}(1-\sigma_{feed})}$, W		128.2	64.4	27.7	1.9	1.1
RF	P_{TX}, W	6.8	6.8	3.4	0.4	0.2
	P_{RX}, W	6.1	6.1	3.1	0.4	0.3
Total RF, P_{RF}, W		12.9	12.9	6.5	1.0	0.6
BB	Radio (inner Rx/Tx), W	10.8	10.8	9.1	1.2	1.0
	Turbo code (outer Rx/Tx), W	8.8	8.8	8.1	1.4	1.2
	Processors, W	10.0	10.0	10.0	0.4	0.3
Total BB, P_{BB}, W		29.6	29.6	27.3	3.0	2.5
DC-DC, σ_{DC}, %		7.5	7.5	7.5	9.0	9.0
Cooling, σ_{cool}, %		10.0	0.0	0.0	0.0	0.0
Mains supply, σ_{ms}, %		9.0	9.0	9.0	11.0	11.0
Total per TRX chain, W		225.0	125.8	72.3	7.3	5.2
Sectors		3	3	1	1	1
Antennas		2	2	2	2	2
Carriers		1	1	1	1	1
Total N_{TRX} *chains*, P_{in}, W		1350	754.8	144.6	14.7	10.4

An LTE system with 10 MHz bandwidth and 2 × 2 MIMO configuration is assumed [16].

will grow proportionally with the number of transceivers N_{TRX}. Accordingly, the BS power consumption at maximum load (i.e. $P_{out} = P_{max}$) can be expressed as [16]

$$P_{in} = N_{TRX}\frac{\frac{P_{out}}{\eta_{PA}(1-\sigma_{feed})} + P_{RF} + P_{BB}}{(1-\sigma_{DC})(1-\sigma_{mains})(1-\sigma_{cool})} \tag{1.25}$$

The power efficiency is defined as

$$\eta \triangleq P_{out}/P_{in} \tag{1.26}$$

while the loss factor is defined:

$$\sigma \triangleq 1-\eta \tag{1.27}$$

The power consumption at maximum load for different LTE BS types as of 2010 are given in Table 1.2.

1.8.4 BS Power Consumption at Variable Load

In an LTE system DL, the BS load defined by $\frac{P_{out}}{P_{max}}$ is proportional to the amount of consumed resources, including both data and control signals. Furthermore, the BS load also depends on power control settings in terms of the spectral power density on an LTE UL transmission. Power consumption for various BS types as a function of the variable load is displayed in Figure 1.14.

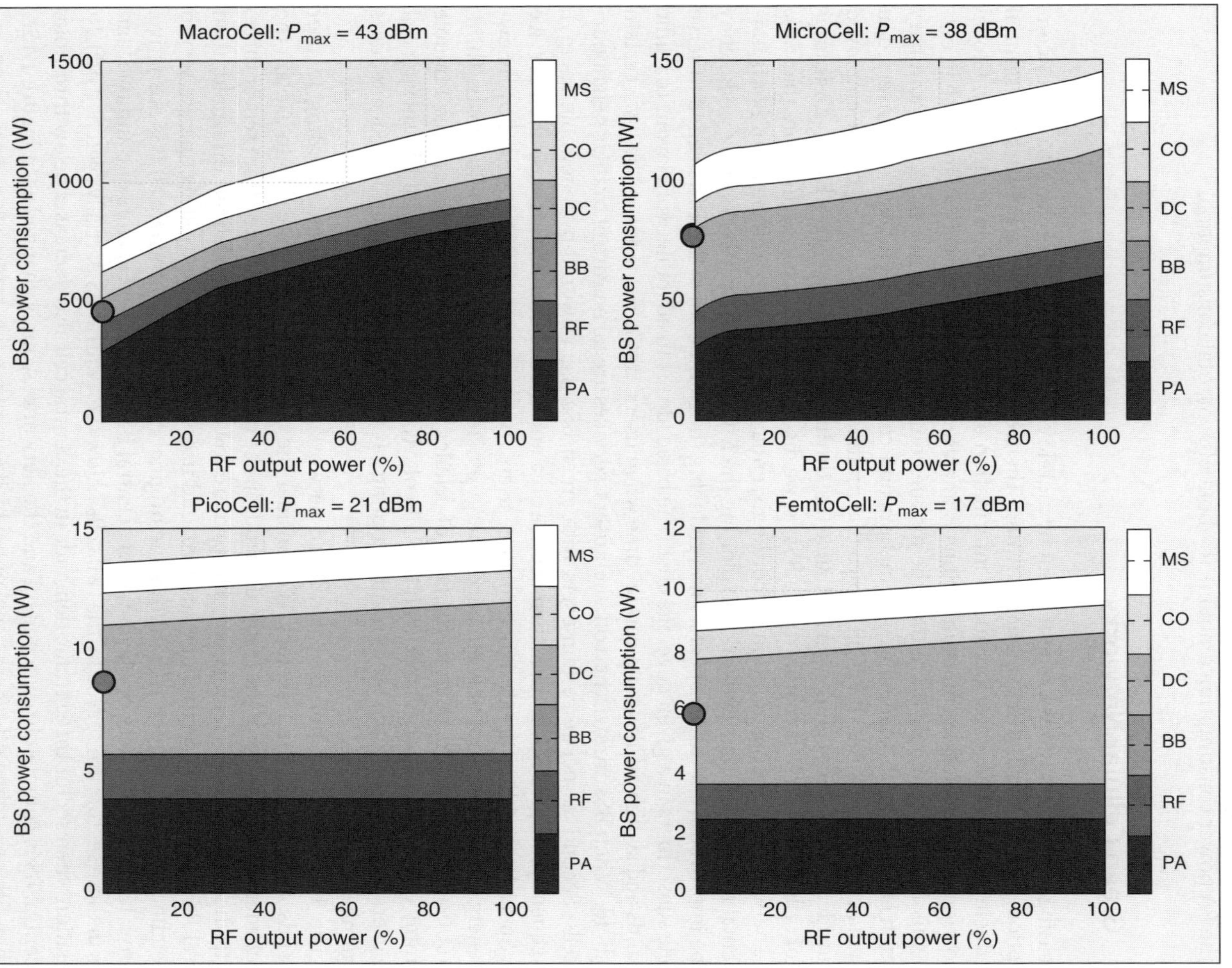

Figure 1.14 Power consumption for various BS types as a function of the load [16]. PA: power amplifier, RF: small signal RF transceiver, BB: baseband processor, DC: DC-DC converters, CO: macro BS cooling, MS: mains power supply.

Inspection of the power consumption versus BS load affirms that the PA power consumption mainly grows with BS load and the amount of growth largely depends on the BS type. Accordingly, in macro BSs the power consumption is load-dependent, but to a lesser extent for micro BSs, and load dependency is negligible for pico and femto BSs. Arguably, for low-power BSs the PA power consumption account is less than 30% of total power consumption, while for macro BSs the PA power consumption amounts to about 60% of total consumption.

1.9 Green Cellular Networks

Increasing carbon emissions brought about by mobile communication is related to energy consumption, which contributes to global warming. Thence, greening cellular networks is essential to reducing the carbon emission due to information and communication technology. Global carbon footprint of mobile communications is analysed in [18]. It predicts an increase in carbon emissions considerably more than the mobile communication traffic rates. Extraordinary research efforts in both academia, information, and communications industry are invested in reducing carbon emissions. It is well known that multicell cooperation (MCC) can be employed to considerably enhance cellular performance (i.e. throughput and coverage). Energy consumption of cellular networks is mainly due to BSs, as we discussed, and accounts for more than 50% of the network energy consumption. Accordingly, enhancing the energy efficiency of BSs is fundamental to green cellular networks. Exploiting the MCC, cellular networks energy efficiency can be improved from three aspects. First, reducing the number of active BSs needed to serve mobile subscribers in a given coverage area. The basic idea is to adapt network layout to traffic demands and switch off BSs for a certain time when users' traffic demands are below a certain threshold. When BSs are switched off, their users' services are taken care of by their neighbouring cells. Second, associate certain users with green BSs (i.e. powered by renewable energy). Subsequently, on-grid BSs reduce their service area and the off-grid BSs increase their service area. Special algorithms are designed to guide users to associate with BSs powered by renewable energy. This aspect not only enhances energy efficiency but also the operation costs. Third, exploit CoMP to raise the energy efficiency of cellular networks. This aspect creates a double whammy effect. On the one hand, with the help of MCC, energy efficiency of BSs serving cell edge users increased. On the other hand, adopting MCC expands coverage and hence reduces the number of active BSs.

Figure 1.15 depicts a number of cellular network layout adaptations according to user demands. Each BS has three sectors. We consider the central cell in Figure 1.15 to explain the layout adaptation. If most of the traffic demands in the central cell originated from sector three, and traffic demand from sectors one and two are lower than the predefined threshold, then central BS switched-off sectors one and two, and users in the switched-off sectors are served by neighbouring BSs. Figure 1.15a shows the original network layout where all BSs are active. Figure 1.15b shows the network layout where central BS switched-off sectors one and two, and users in these sectors are served by the two cells below (i.e. central BS partially switched off). In a situation when sector three traffic demands in the central cell also decreases below threshold, the central BS is completely switched off and the network layout is changed as in Figure 1.15c.

When BS is switched off, energy consumed by its elements (i.e. transceivers, processing circuits, cooling, etc.) are saved. Accordingly, network layout adaptation is capable of saving a considerable amount of energy in cellular networks. Nonetheless, there are certain constraints that have to be satisfied such as coverage requirement and QoS requirements of all users. These constraints can be guaranteed with MCC, or else the service suffers from high blocking probability and severe QoS degradation.

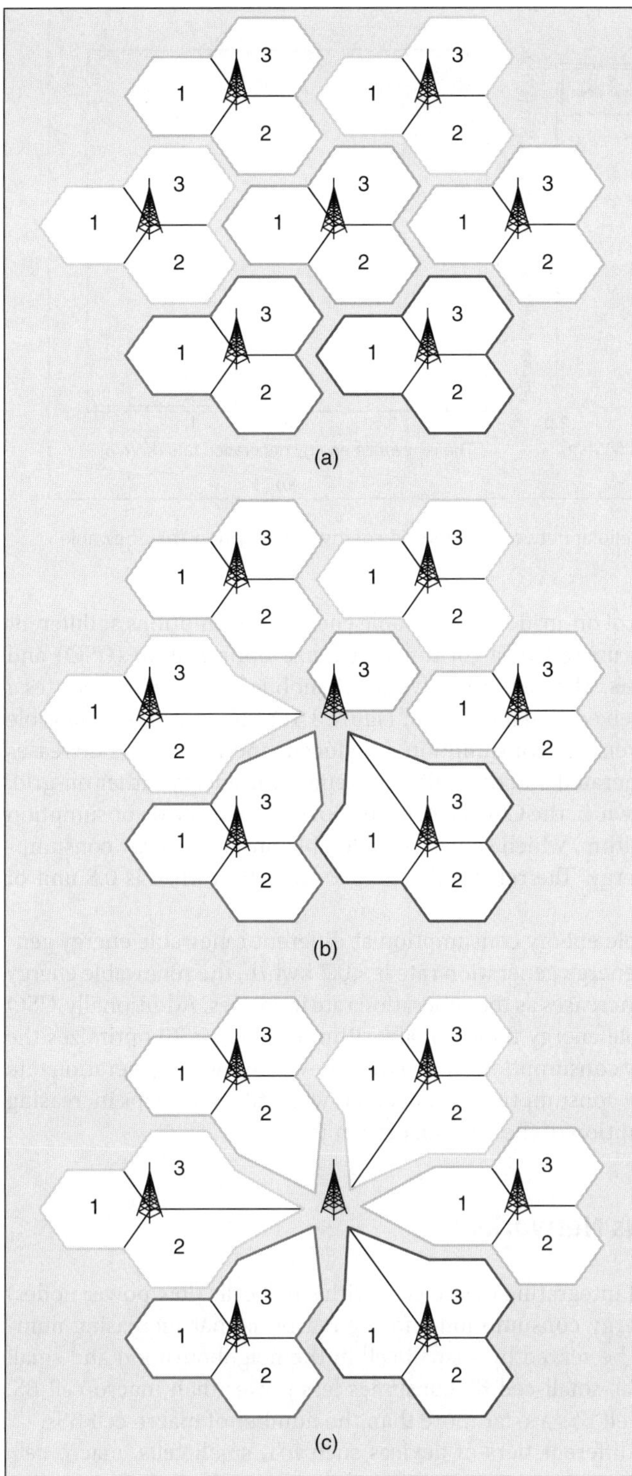

Figure 1.15 Block diagram showing network layout adaptations for macro cellular sites: (a) original network layout; (b) BS partially switched off; (c) BS entirely switched off [19].

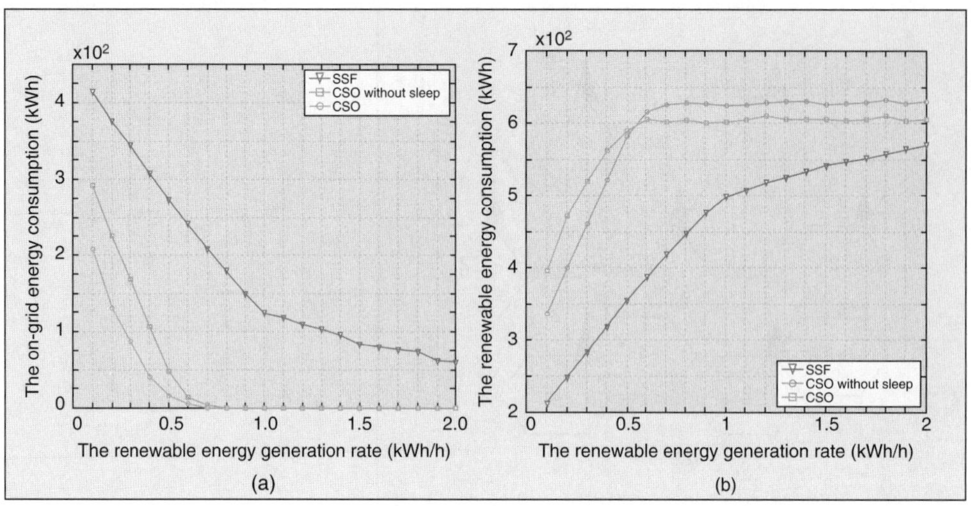

Figure 1.16 Energy consumption of the cellular network: (a) on-grid energy consumption; (b) renewable energy consumption [19].

Figure 1.16 depicted comparisons of on-grid and renewable energy consumptions at different renewable energy generation rates using two algorithms: cell size optimization (CSO) and strongest signal first (SSF) algorithms. The network serving 400 mobile users. SSF associates a user with the BS with strongest received signal strength. Figure 1.16a shows as the renewable energy rates increases, the on-grid energy consumption is reduced, and as the rate increases more, more renewable energy is generated and more BSs use renewable energy rather on-grid. When *renewable energy rate* ≥0.8 kwh/h, the CSO algorithm shows zero on-grid consumption due to CSO compared to SSF algorithm, which shows considerable on-grid energy consumption. Note that kwh is a unit of energy. The renewable energy rate is 0.8 kwh/h is 0.8 unit of energy per hour.

Figure 1.16b displays the renewable energy consumption at different renewable energy generation rates. When the renewable energy generation rate is <0.7 kwh/h, the renewable energy consumption, for both algorithms, increases as the generation rate increases. Additionally, CSO algorithm consumes more renewable energy than SSF algorithm because CSO optimizes the cell size to reduce the on-grid energy consumption. When the renewable energy generation rate is >0.7 kwh/h, the renewable energy consumption is almost flat while the SSF keeps increasing because the on-grid energy consumption of the CSO algorithm is zero.

1.10 Green Heterogeneous Networks

Heterogeneous networks (HetNets) integrating macro cells with small cells (low power nodes) can be used to reduce network energy consumption and the reason is that increasing number of users' equipment is likely to be served by a small cell in the neighbourhood and small energy radiation is needed as well as small-cell BS consumes less power than macro-cell BS. Nonetheless, the number of small-cell BSs are far more than the number of macro cell BSs.

Multi-tier HetNet composed of different tiers of devices such IoT, small cells, macro cell, etc., which makes the interference management and mitigation more complicated. A promising solution is based on cooperative transmission, which improves intercell interference but will not achieve energy efficiency. What is needed to perform cooperative transmission that enhances signal strength while it minimizes the energy consumption. Such technologies comprised transmit power control; BSs cooperation; BS sleep mode, etc. Since user-generated traffic varies with time and a great part of the energy consumption associated with BSs operations, it seems sensible to switch off underutilized BSs to minimize network consumption. BS sleep mode with dynamic clustering of CoMP provides a complete solution for performance improvements. In addition, hybrid energy resources (on-grid and renewable energy harvesting by BSs) can be used to enhance the green HetNets.

Green multicell cooperation (GMC) in HetNets with hybrid energy sources was investigated in [20]. The goal is to reduce the grid energy consumption in HetNets as much as possible. Accordingly, we need an efficient solution to deal with the spatial and temporal characteristics of the GMC problem simultaneously. A greedy decomposition is posed to solve the GMC problem by simplifying it by means of two independent subproblems: MCC and green energy planning subproblems. Further, the authors derived an upper bound for the greedy grid energy consumption achieved by the integrated solution and evaluated the time complexity of the proposed scheme.

1.11 Summary

In the introductory chapter, Section 1.1 we made a case for the 5G and motivation for writing the book and described the theme of the book. Most importantly, considering that 4G LTE development in coverage and speed is continuing, why do we need 5G now?

It is worth remembering the essential role of LTE technologies on enabling 5G services. 4G will not have the capabilities to support IoT and many other projected services for 2020. The key distinction for 5G services is not the expected high data delivery but latency. Current LTE latency is around 45 ms, compared to less than 1 ms in future 5G services.

5G is not an incremental advance of 4G. 5G will have a paradigm shift with massive BW, high BS and terminal devices densities, and massive MIMO systems. 5G will be formed as heterogeneous network (HetNet) comprising two layers of cells, macro cells, and small cells. A single macro cell comprises many small cells. In such a system, a dual-connectivity is normally employed: simultaneous connection of terminal to macro cell and small cell.

Contemporary channel and propagation models are not adequate for 5G small cells. In addition, 5G channel models have to support lD and 2D massive MIMO with dual polarization and should be appropriate for beamforming design.

Section 1.2 presents an overview of the contemporary cellular networks. Since the 1980s, we have witnessed four generations of mobile communications. 2G, also known as GSM, is a TDMA/FDMA system. During its long service duration, GSM was developed to GPRS to use packet-switched data and EDGE to increase data rates.

The motivation of 3G was the aspiration for higher data rate delivery for multimedia and improved robustness. Few factors help to advance W-CDMA technology in the 3G systems. A key attribute was how a wideband has the ability to resolve multipath components and with the help of rake receiver combines the signals of each component. Another key factor was the ability to spread data with orthogonal variable spreading factors (OVSF).

Nokia traffic model published in 2012 predicted that an increase of 1000X relative to the traffic in 2010 will come into being around 2020. So, to appropriately respond to increasing traffic and capacity growth, the LTE radio access technology has continuously evolved. Most current 4G LTE networks are based on 3GPP Releases 8 and 9, offering peak DL speed of 100 Mb/s and peak UL 50 Mb/s. 3GPP LTE-advanced introduced enhancement on the existing 4G LTE networks and raise the peak DL transmission speed to 1000 Mb/s and the UL peak speed to 500 Mb/s as recommended by 3GPP Release 10.

Section 1.3 summarized a few of the key developments in 3GPP LTE through Releases 8–15. The aims of the 3GPP were to develop a vision for an all IP wireless network covering radio access/core networks; service capabilities and interworking with Wi-Fi networks. 3GPP introduced the EPC to Release 8 specifications as IP-based core network for 4G LTE and also grants access to legacy networks such as 2G, 3G, and non-3GPP and fixed-access networks. So, circuit-switching data mode was operative no longer. EPC is adequate for fast processing for high data rates with low latency. The νEPC moves the core network components that traditionally run on a dedicated hardware to software run on low-cost servers standardized by ETSI. Release 9 defined three positioning schemes to accurately determine user location in emergency cases and further improvements were added to SON. In addition, Release 9 introduced CMAS added to ETWS introduced in Release 8. Release 10 introduced LTE-Advanced and defined the HetNet, where the coverage is achieved by a combination of large (macro) cells and small cells and improved the MIMO system to up to 8×8 in DL and 4×4 in UL. Releases 11 and 12 added in further improvement to LTE-Advanced. Release 13 introduce three important technologies: full-dimension MIMO; CA of up to 32 CCs compared to only 5 in release 10; and new enhancement to MTC including narrow BW low complexity user devices with long battery life for NB-IoT. Release 14 is known as the initial 5G standardization. Releases 15 and 16 are identified as 5G phase 1 and 2, respectively.

The capacity regions of multiuser cellular network are introduced in Section 1.4 by defining three basic channels in the multiuser network, namely: MAC, BC, and IC, followed by the analysis of ergodic sum-rate capacity of the degraded BC with superposition coding. The aim is presenting an overview of basic technologies in communication systems to the readers. We then consider the multiuser channels in an MIMO network to analyse the MIMO MAC capacity region and MIMO BC capacity region for $N = 1$ and $N > 1$ and we compared their regions with the achievable rate region of the optimum coding scheme, i.e. DPC.

In Section 1.5, we present a brief overview of the different types of fading channels, which are most commonly used in practical mobile communications, and in Section 1.6 we elucidated a possible extension of the single cell analysis carried out in Section 1.4 to multicell, multiuser analysis. In Section 1.7, we investigated in some detail an important issue associated with mobile communications, that is the impact of CO_2 emissions by active mobile networks. There are two key global initiatives: the EU energy consumption project and China green plan action. We consider in detail the EU project EARTH, which derived a mathematical model for computing the energy consumption and then briefly outlined China's mobile green plan action achievements. The main element in the network that consumed most of the energy is the BS. However, contemporary networks comprise various BS types in addition to different types of traffic models. The large-cell BS is the main consumer of energy in the system. In particular, key elements that may consume noticeable amounts of energy are PAs, transceivers (equal to the number of antennas), DC-DC power supply, active cooling system, AC-DC units, and the power loss in antenna feed cables. The network energy consumption was analysed in depth.

In Section 1.8, we derived a BS power consumption model that includes contribution from PA, small signal RF transceivers, BB units, power supply, and system active cooling. We then considered the power consumption at variable traffic load. We found that the consumption grows with load, but the growth is dependent of the BS type.

In Section 1.9, we considered the carbon emissions due to cellular networks. Using MCC, cellular networks can improve their energy efficiencies from three aspects: switching off some BSs when traffic demands have fallen below a certain threshold; using renewable energy to drive some BSs; and adopting MCC to expand coverage and hence reduce the number of active BSs.

In Section 1.10, we examined the possibilities of green heterogeneous network. A promising solution comprising BS sleep mode with dynamic clustering of CoMP is likely to provide a complete solution for performance improvements. In addition, hybrid energy resources (on-grid and renewable energy harvested by BSs) can be used to enhance the green HetNets.

1.A Tutorials on Theory and Techniques of Optimization Mathematics: Basics

Optimization of a mathematical function is a process of choosing the appropriate input to maximize/minimize the value of the function. In real wireless networks, the optimizations of resources are frequently required to satisfy certain conditions. In general, there are two types of optimizations: *unconstrained optimization* and *constrained optimization*. Before we proceed further into the mathematics of the optimization, let us define certain vocabularies that we need to use in what follows:

- Critical point: Any point in the function domain such that the function is not differentiable or its derivative is zero.
- Local minimum: At local minimum $x^*, f(x^*) < f(x)$ when $|x - x^*| < \in$ and $\in > 0$.
- Local maximum: At local maximum $x^*, f(x^*) > f(x)$ when $|x - x^*| < \in$ and $\in > 0$.
- Global minimum: At global minimum $x^*, f(x^*) \leq f(x)$ for all x.
- Global maximum: At global maximum $x^*, f(x^*) \geq f(x)$ for all x.
- Stationary point: The value of the function $f(x^*)$ calculated at the critical point x^*.
- Hessian matrix: A square matrix of second-order partial derivatives of a function of multiple variables (assuming the derivatives exist). Hessian matrix describes the curvature of the function.

The illustration in Figure 1.A.1 shows $f(x)$ on the y-axis and x on the x-axis and global max and min and the local max and min.

1.A.1 Optimization of Unconstrained Function with a Single Variable

Let get started with unconstrained optimization by considering a function $f(x)$ of a single independent variable such that

$$f(x) = \sin x \tag{1.A.1}$$

A plot of $f(x)$ vs x is shown in Figure 1.A.2. At the critical point x^*, we have

$$\frac{df}{dx} = 0 \tag{1.A.2}$$

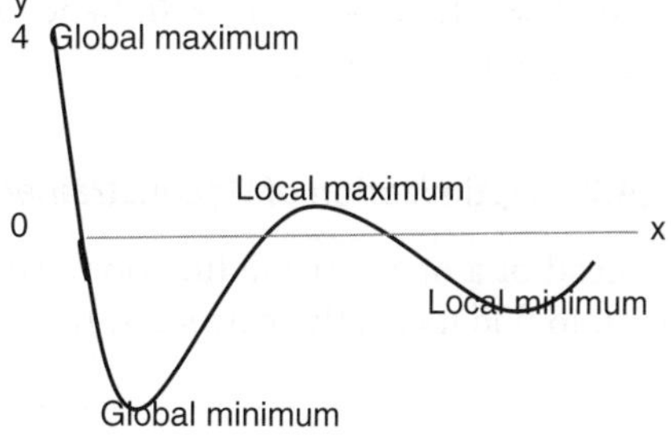

Figure 1.A.1 Illustration showing the global and local max and min.

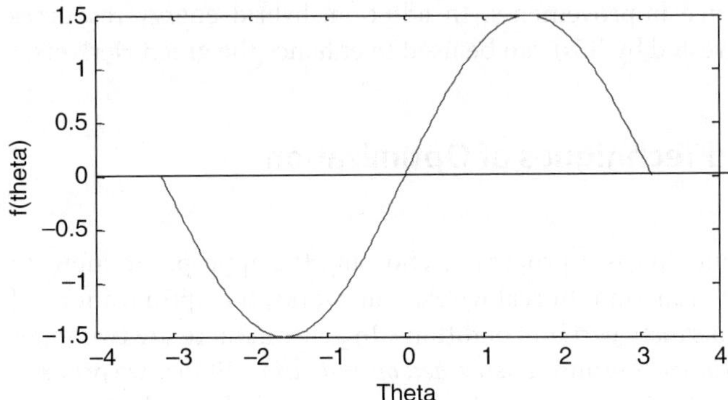

Figure 1.A.2 Plot of sine (theta) verses theta.

The gradient of $f(x)$ is changing from positive to negative before and after the maximum point. In other words, the gradient of $f(x)$ is decreasing. Consequently, the second-order derivative of the function at the critical point is negative. Similarly, it can be shown that at the minimum, the second-order derivative of the function $f(x)$ at the critical point is positive.

Worked Example 1 Investigate the maximum/minimum of the following single-variable function:

$$f(x) = 5x^3 + 6x^2 - 3x + 7$$

The first-order derivative of the function is

$$\frac{df}{dx} = 15x^2 + 12x - 3$$
$$= 3(5x^2 + 4x - 1) = 3(5x - 1)(x + 1)$$

Thus, at

$$\frac{df}{dx} = 0 \quad x_1^* = \frac{1}{5} \quad and \quad x_2^* = -1$$

Now

$$\frac{d^2f}{dx^2} = 30\,x + 12$$

At $x_1^* = \frac{1}{5}$, $\frac{d^2f}{dx^2} = 18 > 0$; hence $x_1^* = \frac{1}{5}$ is a relative minimum point and at the minimum point, $(x_1^*) = 6.68$.

At $x_2^* = -1$, $\frac{d^2f}{dx^2} = -18 < 0$; hence $x_2^* = -1$ is a relative maximum point and at the maximum point, $f(x_2^*) = 11$.

1.A.2 Optimization of Unconstrained Function with Multiple Variables

Instead of a one-variable function, we now consider a function of multiple variables and we wish to find its relative extremes. Let $\mathbf{x}$ be a vector of $n \times 1$ elements such that

$$\mathbf{x} = [x_1, x_2, x_3, \dots\dots\dots\dots\dots, x_n]^T \tag{1.A.3}$$

We assume that $f(\mathbf{x})$ is differentiable and that the derivatives are continuous. Let the Gradient of $f(\mathbf{x})$ denoted $\nabla f(\mathbf{x})$ such that

$$\nabla \mathrm{f}(\mathbf{x}) = \begin{bmatrix} \frac{\partial \mathrm{f}(\mathbf{x})}{\partial \mathrm{x}_1} \\ \frac{\partial \mathrm{f}(\mathbf{x})}{\partial \mathrm{x}_2} \\ \frac{\partial \mathrm{f}(\mathbf{x})}{\partial \mathrm{x}_n} \end{bmatrix} \tag{1.A.4}$$

1.A.3 The Hessian Matrix

The Hessian matrix (or simply Hessian) is a square matrix of second-order partial derivatives of a function (assuming it exists) used in optimization to determine whether the optimal point $\mathbf{x}^*$ is a maximum or minimum or *saddle point*. The Hessian of $f(\mathbf{x})$ is $\mathrm{n} \times \mathrm{n}$ matrix denoted by $\mathbf{H}_{f(\mathbf{x})}$ or simply $\mathbf{H}_f$ where

$$\mathbf{H}_f = \left[\frac{\partial^2 f(\mathbf{x})}{\partial \mathrm{x}_i \mathrm{x}_j} \right] \qquad i, j = 1, 2, \ldots \ldots \ldots \mathrm{n} \tag{1.A.5}$$

Using (1.A.4) and (1.A.5), the $\mathrm{n} \times \mathrm{n}$ Hessian of $f(\mathbf{x})$ is $\mathbf{H}_f$

$$\mathbf{H}_f = \begin{pmatrix} \frac{\partial^2 f(\mathbf{x})}{\partial \mathrm{x}_1{}^2} & \cdots & \frac{\partial^2 f(\mathbf{x})}{\partial \mathrm{x}_1 \partial \mathrm{x}_n} \\ \vdots & \ddots & \vdots \\ \frac{\partial^2 f(\mathbf{x})}{\partial \mathrm{x}_n \partial \mathrm{x}_1} & \cdots & \frac{\partial^2 f(\mathbf{x})}{\partial \mathrm{x}_n{}^2} \end{pmatrix} \tag{1.A.6}$$

Matrix $\mathbf{H}_f$ is symmetric such that $\frac{\partial^2 f(\mathbf{x})}{\partial \mathrm{x}_i \partial \mathrm{x}_j} = \frac{\partial^2 f(\mathbf{x})}{\partial \mathrm{x}_j \partial \mathrm{x}_i}$. We can now summarize the conditions for a maximum or a minimum critical point for differentiable multiple variable function $f(\mathbf{x})$ as follows:

- For maximum or minimum of $f(\mathbf{x})$ then $\nabla f(\mathbf{x}^*) = \mathbf{0}$.
- For maximum $\mathbf{x}^*$: leading principle minor of $\mathbf{H}_f$, order k has sign $(-1)^k$.
- For minimum $\mathbf{x}^*$: leading principle minor of $\mathbf{H}_f$, order k is >0.
- For saddle $\mathbf{x}^*$ when neither of the above conditions hold.

Note: To create leading principal minors of $\mathbf{H}_f$ ($\mathbf{h}_1, \mathbf{h}_2, \ldots \ldots \ldots .., \mathbf{h}_n$), start with upper left (1,1) element of $\mathbf{H}_f$ for $\mathbf{h}_1$, gradually add more rows and columns to the right of selected ones, and compute the determinant of the new matrix for ($\mathbf{h}_2, \ldots \ldots \ldots .., \mathbf{h}_n$). Consider a general 3×3 matrix $\mathbf{A}$:

$$\mathbf{A} = \begin{bmatrix} \mathrm{a}_{11} & \mathrm{a}_{12} & \mathrm{a}_{13} \\ \mathrm{a}_{21} & \mathrm{a}_{22} & \mathrm{a}_{23} \\ \mathrm{a}_{31} & \mathrm{a}_{32} & \mathrm{a}_{33} \end{bmatrix}$$

1st order principal minors: $|\mathrm{a}_{11}|$ formed by deleting the last two rows and cols. $|\mathrm{a}_{22}|$ formed by deleting the first and third rows and cols. $|\mathrm{a}_{33}|$ formed by deleting the first two rows and cols.

2nd order principal minors: $\begin{vmatrix} a_{11} & a_{12} \\ a_{21} & a_{22} \end{vmatrix}$ formed by deleting col 3 and row 3; $\begin{vmatrix} a_{11} & a_{13} \\ a_{31} & a_{33} \end{vmatrix}$ formed by deleting col 2 and row 2; $\begin{vmatrix} a_{22} & a_{23} \\ a_{32} & a_{33} \end{vmatrix}$ formed by deleting col 1 and row 1; 3rd order principal minors $|\mathbf{A}|$ formed by determinant of matrix.

Worked Example 2 Consider the optimization problem of function $f(\mathbf{x})$ and $\mathbf{x}$ is 2×1 vector:

$$f(\mathbf{x}) = 2x_2^3 - 3x_1^3 + 7\, x_1 x_2$$

$$\frac{\partial f(\mathbf{x})}{\partial x_1} = -9x_1^2 + 7x_2$$

$$\frac{\partial f(\mathbf{x})}{\partial x_2} = 6\, x_2^2 + 7\, x_1$$

At $\nabla f(\mathbf{x}) = 0$, the critical points $(x_{1,}^* \, x_2^*)$ are:

$$(0, 0), (-0.8903, 1.0191)$$

We now investigate whether the critical points are maximum or minimum, so we have to compute the Hessian matrix $\mathbf{H}_f$ of $f(\mathbf{x})$ as follows:

$$\nabla^2 f(\mathbf{x}) = \begin{bmatrix} \frac{\partial^2 f(\mathbf{x})}{\partial x_1^2} & \frac{\partial^2 f(\mathbf{x})}{\partial x_1 \partial x_2} \\ \frac{\partial^2 f(\mathbf{x})}{\partial x_2 \partial x_1} & \frac{\partial^2 f(\mathbf{x})}{\partial x_2^2} \end{bmatrix} = \begin{bmatrix} -18x_1 & 7 \\ 7 & 12x_2 \end{bmatrix}$$

We compute the leading principal minors (two of these minors): First, the leading principal minor $= -18x_1$ and the second leading principle minor $= -18 * 12\, x_1 x_2 - 49$.

At critical point (0,0), leading principal minors = 0 and −49. At critical point (0.8903, 1.0191), leading principal minors = 16.0254, 146.9778. Hence critical point (0, 0) a saddle of $f(\mathbf{x})$ and critical point (−0.8903, 1.0191) is a local minimum of $f(\mathbf{x})$.

Since at the critical point (−0.8903, 1.0191), both leading principal minors are positive definite, in fact (−0.8903, 1.0191) is a global minimum.

1.B Theory of Optimization Mathematics

Although the optimization theory and techniques cover a large area of mathematical analysis, in its simple definition in wireless networks, optimization proposes methods of finding the *best available* resource that is distributed fairly among users to make the network more efficient. Consequently, optimization is a process of choosing a level of resources for individual users, subject to certain constraints, for the total available resources. This involves finding maximum or minimum resource allocations that satisfies the given criteria.

Constraints have to be satisfied by the solution. There are two types of constraints namely: equality constraints and inequality constraints. In the following deliberations we present some aspects of the theory of optimization relevant to wireless networks, and then we proceed to use it to optimize wireless networks.

1.B.1 Constrained Optimization

Optimization problems may have constraints in diverse ways such as variables are integer or the signal may need to be positive. A generic constrained optimization may look like given a function of a variable **x** which has *n* values,

$$\min f(\mathbf{x}) \tag{1.B.1}$$

The optimization must satisfy the following conditions,

$$\text{subject to (s.t)} \quad \mathbf{h}(\mathbf{x}) = \mathbf{0} \tag{1.B.2a}$$

$$\mathbf{g}(\mathbf{x}) \leq \mathbf{0} \tag{1.B.2b}$$

where $\mathbf{h} = [\mathrm{h}_1\ \mathrm{h}_2\ - - - -\ \mathrm{h}_m]$ is *m*-element vector that is function of the *n*-element variable $\mathbf{x} = [\mathrm{x}_1\ \mathrm{x}_2 - - - -, \mathrm{x}_n]$, and $\mathbf{g} = [\mathrm{g}_1\ \mathrm{g}_2 - - - - - \mathrm{g}_p]$ is *p*-element vector that is function of the *n*-values variable **x**. The constraint $\mathbf{h}(\mathbf{x}) = \mathbf{0}$ is called equality constraint and $\mathbf{g}(\mathbf{x}) \leq \mathbf{0}$ is inequality constraint.

The optimization problem can be articulated as: find that values of the variable **x** that minimize $f(\mathbf{x})$. The equality constraints $\mathbf{h}(\mathbf{x}) = \mathbf{0}$ generates a set of m constraints, $\mathrm{h}_1(\mathbf{x}) = 0$; $\mathrm{h}_2(\mathbf{x}) = 0, \ldots\ldots \mathrm{h}_m(\mathbf{x}) = 0$.

Let $\mathbf{x}^*$ be a regular point of these constraints and then there are *m* real values denoted by an *m*-element vector $\boldsymbol{\lambda} \in \mathbb{R}^m$ such that,

$$\nabla \mathrm{f}(\mathbf{x}^*) + \nabla \mathbf{h}(\mathbf{x}^*)\boldsymbol{\lambda} = 0 \tag{1.B.3}$$

where the symbol ∇ characterizes the gradient expressed as a vector differential operator and $\lambda = [\lambda_1\ \lambda_2\ - - - - \lambda_m\ ]$ is called Lagrange multiplier provide an indication of how much the constraint cost when $\mathbf{x}^*$ is in the neighbourhood of the optimum solution.

1.B.2 Bordered Hessian Matrix H^B

A bordered Hessian matrix $\mathbf{H}^B$ comprises the Hessian of a multivariate function $\mathbf{H}_f$ bordered by the gradient of a multivariate function **h** and is used in certain constrained optimization problems. When the function $f(\mathbf{x})$ is continuous so the border Hessian matrix is square and symmetric, the Border Hessian form looks like

$$\mathbf{H}^B = \begin{bmatrix} 0 & \dfrac{\partial \mathbf{h}(\mathbf{x})}{\partial \mathrm{x}_1} & \dfrac{\partial \mathbf{h}(\mathbf{x})}{\partial \mathrm{x}_2} & \cdots\cdots & \dfrac{\partial \mathbf{h}(\mathbf{x})}{\partial \mathrm{x}_n} \\ \dfrac{\partial \mathbf{h}(\mathbf{x})}{\partial \mathrm{x}_1} & \dfrac{\partial^2 f}{\partial^2 \mathrm{x}_1} & \dfrac{\partial^2 f}{\partial \mathrm{x}_1 \partial \mathrm{x}_2} & \cdots\cdots & \dfrac{\partial^2 f}{\partial \mathrm{x}_1 \partial x_n} \\ \dfrac{\partial \mathbf{h}(\mathbf{x})}{\partial \mathrm{x}_2} & \dfrac{\partial^2 f}{\partial \mathrm{x}_2 \partial \mathrm{x}_1} & \cdots\cdots & & \dfrac{\partial^2 f}{\partial \mathrm{x}_2 \partial x_n} \\ \cdots & \cdots & \cdots & \cdots & \cdots \\ \cdots & \cdots & \cdots & \cdots & \cdots \\ \cdots & \cdots & \cdots & \cdots & \cdots \\ \cdots & \cdots & \cdots & \cdots & \cdots \\ \cdots & \cdots & \cdots & \cdots & \cdots \\ \dfrac{\partial \mathbf{h}(\mathbf{x})}{\partial \mathrm{x}_n} & \dfrac{\partial^2 f}{\partial \mathrm{x}_n \partial \mathrm{x}_1} & \dfrac{\partial^2 f}{\partial \mathrm{x}_n \partial \mathrm{x}_2} & \cdots\cdots & \dfrac{\partial^2 \mathrm{f}}{\partial^2 \mathrm{x}_n} \end{bmatrix} \tag{1.B.4}$$

Worked Example 3 Consider variable **x** as a vector of four elements: $\mathbf{x} = [x_1\ x_2\ x_3\ x_4]$ and function $f(\mathbf{x})$expressed as,

$$f(\mathbf{x}) = 2x_1^2 + 5{x_2}^2 + 4{x_3}^2 + 6{x_4}^2$$

We wish to find the optimal elements $\mathbf{x}^*$, λ^* that maximize $f(\mathbf{x})$ subject to the constraint

$$\mathbf{h}(\mathbf{x}) = x_1 + 3x_2 - 5x_3 - 8x_4 - 10$$

Solution

As there is just a single constraint, we use only one Lagrange multiplier. The Lagrange function $L(\mathbf{x}, \lambda)$ is

$$\mathbf{L}(\mathbf{x}, \lambda) = f(\mathbf{x}) - \mathbf{h}(\mathbf{x})\ \lambda$$

$$\mathbf{L}(\mathbf{x}, \lambda) = 2x_1^2 + 5{x_2}^2 + 4{x_3}^2 + 6{x_4}^2 + \lambda(-x_1 - 3x_2 + 5x_3 + 8x_4 + 10)$$

Using (1.B.4), we get

$$\frac{\partial L}{\partial x_1} = 4\,x_1 - \lambda = 0$$

$$\frac{\partial L}{\partial x_2} = 10\,x_2 - 3\,\lambda = 0$$

$$\frac{\partial L}{\partial x_3} = 8x_3 + \lambda(5) = 0$$

$$\frac{\partial L}{\partial x_4} = 12x_4 + \lambda(8)$$

$$\frac{\partial L}{\partial \lambda} = (x_1 + 3x_2 - 5x_3 - 8x_4 - 10) = 0$$

We now have to solve the five equations above to find the five unknowns. Solving these five equations, we get

$$\lambda^* = -1200/127$$

$$x_1^* = -300/127$$

$$x_2^* = -360/127$$

$$x_3^* = 750/127$$

$$x_4^* = -800/127$$

$$f(\mathbf{x}^*) = 425.3368$$

$$\mathrm{h}(\mathbf{x}^*) = 0$$

$$\mathrm{h}(\mathbf{x}) = x_1 + 3x_2 - 5x_3 - 8x_4 - 10$$

$$\frac{\partial h}{\partial x_1} = 1$$

$$\frac{\partial h}{\partial x_2} = 3$$

$$\frac{\partial h}{\partial x_3} = -5$$

$$\frac{\partial h}{\partial x_4} = -8$$

The bordered Hessian matrix for the example is

$$\mathbf{H}^{\mathrm{B}} = \begin{bmatrix} 0 & 1 & 3 & -5 & -8 \\ 1 & 4 & 0 & 0 & 0 \\ 3 & 0 & 10 & 0 & 0 \\ -5 & 0 & 0 & 8 & 0 \\ -8 & 0 & 0 & 0 & 12 \end{bmatrix}$$

Hence, the leading principle minors are:

$$\mathrm{H}_1^{\mathrm{B}} = -1$$
$$\mathrm{H}_2^{\mathrm{B}} = -46$$
$$\mathrm{H}_3^{\mathrm{B}} = -1368$$
$$\mathrm{H}_4^{\mathrm{B}} = -36896$$

Hence $\mathbf{x}^* = (0.2602, 0.3122, 0.6505, 0.6939)$ is local minimum.

1.C Karush–Kuhn–Tucker (KKT) Conditions

In mathematics, the KKT conditions (also known as Kuhn–Tucker) conditions are essential conditions for determining a solution to optimum problem provided certain constraints are satisfied. KKT scheme generalizes the Lagrange method, which is applicable to only equality constraints. Let $\mathbf{x}^*$be a local minimum and assume that has to satisfy both equality and inequality constraints, then there is a Lagrange multiplier vector $\lambda \in \mathbb{R}^m$ and a Lagrange multipliers vector $\boldsymbol{\mu} \in \mathbb{R}^P$ with

$$\boldsymbol{\mu} \geq \mathbf{0} \tag{1.C.1}$$

Consider the situation when there are inequality constraints. The optimization problem can be expressed as follows:

$$\begin{aligned} &\text{Optimize} && f(\mathbf{x}) \\ &\textit{subject to} && \mathbf{h}(\mathbf{x}) = \mathbf{0} \\ &\mathbf{g}(\mathbf{x}) \ \leq \ \mathbf{0} \end{aligned}$$

where

$$\mathbf{g}(\mathbf{x}) = [\mathrm{g}_1(\mathbf{x}) \ \ \mathrm{g}_2(\mathbf{x}) \ \ \mathrm{g}_3(\mathbf{x}) \ \ \ldots\ldots\ldots\ldots\ldots .\mathrm{g}_{\mathrm{p}}(\mathbf{x})]$$

As in a previous consideration of equality constraints, we assume $f(\mathbf{x})$, $\mathbf{h}(\mathbf{x})$ and $\mathbf{g}(\mathbf{x})$ are differentiable and that their derivatives are continuous. Additionally, $\mathbf{g}(\mathbf{x})$ provides p inequality constraints. If the equality and inequality constraints problem is transformed to the corresponding unconstrained problem using KKT conditions, then there exists a solution $\mathbf{x}^*$ if and only if there is $\boldsymbol{\mu}$ such that:

$$\boldsymbol{\mu} \ \geq \mathbf{0} \tag{1.C.2}$$

$$\mathbf{g}(\mathbf{x}^*)^{\mathrm{T}} \ \boldsymbol{\mu} = \mathbf{0} \tag{1.C.3}$$

$$\nabla f(\mathbf{x}^*) + \nabla \mathbf{h}(\mathbf{x}^*)^T \boldsymbol{\lambda} + \nabla \mathbf{g}(\mathbf{x}^*)^T \boldsymbol{\mu} = \mathbf{0} \tag{1.C.4}$$

where $\mathbf{x}^*$ is a regular point, λ and $\boldsymbol{\mu}$ are Lagrange multipliers for equality and inequality constraints.

Worked Example 4 In this example we have one equality constraint and one inequality constraint.

Consider the function

$$f(\mathbf{x}) = x_1^2 + x_2^2 + x_3^2 + x_4^2$$

We are asked to find the minimum of $f(\mathbf{x})$ subject to the following two constraints:

$$\min f(\mathbf{x}) \tag{1.C.5}$$

subject to

$$\mathbf{h}(\mathbf{x}) = x_1 + x_2 + x_3 + x_4 - 1 \tag{1.C.5a}$$

$$x_4 \leq g(\mathbf{x}) \tag{1.C.5b}$$

Solution

The Lagrange function is

$$\mathbf{L}(\mathbf{x}, \lambda, \mu) = x_1^2 + x_2^2 + x_3^2 + x_4^2 - \lambda(x_1 + x_2 + x_3 + x_4 - 1) - \mu(x_4 - g(\mathbf{x}))$$

We now apply the KKT conditions:

$$\begin{aligned}
&\frac{\partial \mathbf{L}}{\partial \mathbf{x}} = \mathbf{0}\\
&x_4 \leq g(\mathbf{x})\\
&\mu \geq 0\\
&\mu(x_4 - g(\mathbf{x})) \leq 0
\end{aligned} \tag{1.C.5c}$$

$$\frac{\partial \mathbf{L}}{\partial \mathbf{x}} = \mathbf{0}$$

$$\frac{\partial \mathbf{L}}{\partial \mathbf{x}} = \begin{pmatrix} 2x_1 - \lambda \\ 2x_2 - \lambda \\ 2x_3 - \lambda \\ 2x_4 - \lambda + \mu \end{pmatrix} = \begin{pmatrix} 0 \\ 0 \\ 0 \\ 0 \end{pmatrix}$$

Therefore,

$$x_1 = x_2 = x_3 = \frac{\lambda}{2}; x_4 = \frac{\lambda - \mu}{2}$$

$$x_1 + x_2 + x_3 + x_4 = 4\left(\frac{\lambda}{2}\right) - \frac{\mu}{2} = 1$$

Thus,

$$\lambda = \frac{\mu + 2}{4}$$

and

$$x_1 = x_2 = x_3 = \frac{\mu + 2}{8}$$

$$x_4 = \frac{\lambda - \mu}{2} = \frac{1}{4} - \frac{3\mu}{8} \tag{1.C.5d}$$

Accordingly,

$$\frac{1}{4}-\frac{3\mu}{8}\leq g(\mathbf{x}),$$

that is,

$$\frac{3\mu}{8}\geq\frac{1}{4}-g(\mathbf{x}) \tag{1.C.5e}$$

There are three possible cases for (1.C.5c)

Case 1 $g(\mathbf{x}) > \frac{1}{4}$ and since $\mu \geq 0$ so (1.C.5d) is satisfied.

Case 2 $g(\mathbf{x}) = \frac{1}{4}$ and (1.C.5c) is also satisfied.

Case 3 $g(\mathbf{x}) < \frac{1}{4}$ then (1.C.5e) is satisfied, but (1.C.5c) require $\mu = 0$ and (1.C.5d) gives $x_4 = \frac{1}{4}$ which violates $x_4 \leq g(\mathbf{x})$. Therefore $g(\mathbf{x}) = \frac{1}{4}$.

References

1 Frattasi, S., Fathi, F., Fitzek, F.H.P. et al. (2006). Defining 4G technology from the users prospective. *IEEE Network* 20 (1): 35–41.

2 Deployment Strategies for Heterogeneous Networks – White Paper- Nokia Solutions and Networks, 2014.

3 Astely, D., Dahlman, E., Fodor, G. et al. (2013). LTE release 12 and beyond. *IEEE Communications Magazine* 51 (7): 154–160.

4 Tse, D. and Viswanath, P. (2012). *Fundamentals Wireless Communication*. UK: Cambridge University Press.

5 Vanka, S., Srinivasa, S., Gong, Z. et al. (2012). Superposition coding strategies: design and experimental evaluation. *IEEE Transactions on Wireless Communications* 11 (7): 2628–2638.

6 Caire, G. and Shamai, S. (2003). On the achievable throughput of a multiantenna Gaussian broadcast channel. *IEEE Transactions on Information Theory* 49 (7): 1691–1706.

7 Goldsmith, A., Jafar, S.A., Jindal, N., and Vishwanath, S. (2003). Capacity limits of MIMO channels. *IEEE Journal on Selected Areas in Communications* 21 (5): 684–702.

8 Spencer, Q.H., Peel, C.B., Swindlehurs, A.L., and Haardt, M. (2004). An introduction to the multi-user MIMO downlink. *IEEE Communications Magazine* 42 (10): 60–67.

9 Hassibi, B. and Sharif, M. (2007). Fundamental limits in MIMO broadcast channels. *IEEE Journal on Selected Areas in Communications* 25 (7): 1333–1344.

10 Lin, S.-C. and Su, H.-J. (2007). Practical vector dirty paper coding for MIMO Gaussian broadcast channels. *IEEE Journal on Selected Areas in Communications* 25 (7): 1345–1357.

11 Costa, M.H.M. (1983). Writing on dirty paper. *IEEE Transactions on Information Theory* 29 (3): 439–441.

12 Sato, H. (1978). An outer bound on the capacity region of the broadcast channel. *IEEE Transactions on Information Theory* 24 (3): 374–377.

13 Vishwanath, S., Jindal, N., and Goldsmith, A. (2002). On the capacity of multiple input multiple output broadcast channels. *IEEE International conference on communications (ICC)* 3: 1444–1450.

14 Biglieri, E., Proakis, J., and Shamai, S. (1998). Fading channels: information: theoretic and communications aspects. *IEEE Transactions on Information Theory* 44 (6): 2619–2692.

15 Li, L. and Goldsmith, A. (2001). Capacity and optimal resource allocation for fading broadcast channels. I. Ergodic capacity. *IEEE Transactions on Information Theory* 47 (3): 1083–1102.

16 Auer, G., Giannini, V., Desset, C. et al. (2011). How much energy is needed to run a wireless network? *IEEE Wireless Communications* 18 (5): 40–49.
17 Mancuso, V. and Alouf, S. (2011). Reducing costs and pollution in cellular networks. *IEEE Communications Magazine* 49 (8): 63–71.
18 Fehske, A., Fettweis, G., Malmodin, J., and Biczok, G. (2011). The global footprint of mobile communications: the ecological and economic perspective. *IEEE Communications Magazine* 49 (8): 55–62.
19 Han, T. and Ansari, N. (2013). On greening cellular networks via multicell cooperation. *IEEE Wireless Communications* 20 (1): 82–89.
20 Chiang, Y.-H. and Liao, W. (2016). Green multicell cooperation in heterogeneous networks with hybrid energy sources. *IEEE Transactions on Wireless Communications* 15 (12): 7911–7925.

2

5G Enabling Technologies: Small Cells, Full-Duplex Communications, and Full-Dimension MIMO Technologies

2.1 Introduction

The amount of data to be transported by wireless networks is expected to increase more than 1000X by 2020, and by more than 10 000X by 2025. The current increase of data is mainly driven by the use of devices such as smart phones, tablets, and videostreaming. In addition, the number of connected devices and the need for higher data rates will continue to increase exponentially. Consequently, it was clear to researchers and industry that an incremental approach to system improvements as a response to these challenges would not achieve its targets by 2020. There were enormous demands on academia and industry to find smart solutions that would meet these challenges through innovative technologies that are practical and efficient.

The fifth generation (5G) wireless networks are developing rapidly through intensive research and field trials. 5G is expected to be deployed by 2020, when the standardization process is expected to be finalized. However, there is a wide acceptance that some of the currently researched technologies will likely play a crucial role in the design of 5G wireless networks. We do not claim that we are presenting a complete survey of these (enabling) technologies in this book; we instead focus on those proposed solutions that meet stringent requirements and are widely expected, by stakeholders and network's operators, to take a key role within the 5G paradigm.

Currently, while academia is engaged in collaborative research projects, the industry is deriving 5G standardization works. The initial interest and discussions on a possible 5G system with a vision of 'everything everywhere and always connected' captured the interest and imagination of all academic researchers and industrial engineers.

In order to understand the challenges facing the 5G systems, let us consider the key expectations from 5G systems: higher *data rate* and lower *latency*, *energy*, and *cost*. The expected aggregate data rate (i.e. network area capacity) in bits/s/unit area is expected to increase by 1000X current 4G optimal data rate, although the *worst data rate* that a user can expect (known as edge rate, or 5% rate) ranges from 100 Mb/s to as much as 1 Gb/s. That is an increase of 100X compared to the edge rate of 4G, which is roughly 1 Mb/s. However, the edge rate depends on many factors, such as system's load, cell size, and transmission environment. The best data rate (peak rate) that a user can expect is likely to be in the range of tens of Gb/s. To summarize, 5G is expected to increase links data rate between 10X and 100X the 4G data rate. The 4G round trip latency is in the order of about 45 ms compared with 5G, which is expected to support a round trip latency of fraction of 1 ms to provide for applications such as gaming, wearable devices, i.e. an order of magnitude faster than 4G [1].

5G is likely to reduce (or at least not increase) the energy consumption and for this reason it is expected to reduce the running cost of the system. Since the per-link data rate is expected to rise by 100X, so the energy (in joules) per bit together with the operation's cost per bit will

5G Physical Layer Technologies, First Edition. Mosa Ali Abu-Rgheff.

fall by not less than 100X. It is worth noting that the cost of 5G spectrum (possibly mmWave spectrum) is likely to be 10–100X cheaper per Hz than 4G operating in the spectrum below 3 GHz. In addition, the cost of small-cell base stations (BSs) should be 10–100X cheaper than macro cell BSs. However, due to BS densities and increased BW, the cost of the 5G backhaul from the edges of the network into the core will be more expensive compared to 4G. The low power consumption will likely extend the battery life by as much as 10X.

A diverse set of devices will be connected to 5G, so it must efficiently support large portions of these devices, such as machine-to-machine communication (M2M), macro cells, small cells, and high rate mobile devices. The design of the 5G network management and control plane responsible for signalling and routing) for such diverse devices is highly challenging.

So as to achieve such high provisions, 5G is supported by a combination of new and well-known technologies. The set of *enabling technologies* provides unlimited access to information and the ability to share data anywhere anytime by anyone for the comfort of people and to the advantage of business and society.

Self-organizing networks (SON) allow BSs to automatically configure themselves have to be developed for deployment within 5G systems. Advanced SON techniques will be applied to the physical elements as well as to enable operators to balance data load in a multi radio access environment and dynamic spectrum allocation. A set of enabling technologies for cellular 5G make it possible to generate advanced services such as *smart cities* and applications such as *Internet of Things* (IoT), both of which will be presented in this book.

BS densification is a key enabler for the 5G and is associated with a *densification gain* defined by the effective increase in data (edge/aggregate) rate relative to the increase of network density.

Consider a 5G system with BS density is λ_1 BSs/km^2 and achievable data rate R_1 and let the BS density increase to λ_2 to achieve data rate R_2. The *densification gain*, $\rho(\lambda_1, \lambda_2) > 0$, is the slope of the rate increase over the rate density increase:

$$\rho(\lambda_1, \lambda_2) = \frac{(R_2 - R_1)/R_1}{(\lambda_2 - \lambda_1)/\lambda_1} \tag{2.1}$$

If we double the BS density and assume that some of the added BSs were lightly loaded so that the edge rate increases by 50%, then the densification gain $\rho = 0.5$. It is possible that $\rho < 0$, which is a case generally called *the tragedy of the commons*, i.e. a situation within a shared-resource system where individual users act independently according to their own self-interest and contrary to the common good of all users by depleting the resource through their collective action. In general, cellular networks with centralized MAC protocol can operate with $\rho > 0$.

5G cellular networks may consist of various different kinds of BSs such as macro BSs, and a range of smaller cells such as pico BSs, and femto or home BSs. Reducing certain downlink subframes helps reduce interference towards macro users. The fraction of muted subframes is called *muting ratio*. The *muting strategy* is commonly known as enhanced inter-cell interference coordination (*eICIC*). The selection of *optimal muting* ratio is a compromise between the achievable data rate of macro cell users and the data rate achieved by home users.

The utilization of *eICIC* in 5G is necessary to address the interference problem and *eICIC* features comprise two parameters, which need to be optimized, namely: the *cell range extension* (CRE) of small cells; and the *almost blank subframe ratio* (ABSr).

The CRE is adjusted by load balancing algorithm. The ABSr is optimized by maximizing the proportional fair utility of user throughputs. These algorithms converge at the optimum solution [2]. At the 5G mmWave frequencies, the communication is mainly noise limited because the interference tends to reduce to zero, which may increase the SINR very impressively.

Recently it was shown in [3], considering urban multiple buildings, which block the mmWave signal at 72 GHz, how the cell edge rate changes with BSs density. Increasing the BS numbers from 36 to 72 and finally to 96 increase the edge rate (i.e. 5% of user rate) from 24.5 Mb/s to 479 Mb/s and finally to ~1.4 GHz, respectively.

The key enabling technologies for 5G to address the expected level of performance can be arranged as follows [4, 5]:

- Densification of current cellular networks with the addition of a large number of small cells and providing for peer-to-peer (P2P) communications such as device-to-device (D2D) or machine-to-machine (M2M). This enables multi-tier heterogeneous networks.
- Cloud-based radio access network C-RAN.
- Cache-enabled small-cell network (CE-SCN).
- Full-dimension MIMO in LTE-Advanced and 5G.
- Full duplex (FD) communications (simultaneous transmission and reception).
- Network sharing and virtualization of wireless resources.
- Millimetre Wave Energy Harvesting for greener communications.
- Massive MIMO and mmWave communication technology.

The enabling technologies are presented in four separate but related chapters; each focusing on a few technologies to dig deep into the concepts. The enabling technologies are presented in the chapters as follows:

Chapter 2: Small cells; full duplex; and full-dimension mmWave MIMO technologies
Chapter 3: Network virtualization and wireless energy harvesting
Chapter 4: Internet of Things (IoT) and smart cities
Chapter 5: mmWave massive MIMO technology

2.2 The Rationale for 5G Enabling Technologies

Let us consider the facts we already know about the growth of wireless system capacity. We recognize, from experience and practice, that such a growth is attributed to three main factors, namely: *improvement in link efficiency, increased use of radio spectrum,* and *increase in wireless infrastructure BSs.* To understand the rationale behind these enabling technologies, consider the capacity of a conventional system with an additive white Gaussian noise (AWGN) channel shared by a number of users. The throughput of the channel with a given SNR is governed by two integer parameters: the *load factor, n,* and the *spatial multiplexing factor, m.* Using Shannon's formula for digital communications, the user upper-bounded throughput R is

$$R < C = m\left(\frac{\mathrm{B}}{n}\right)\log\left(1+\frac{S}{I+N}\right) \tag{2.2}$$

where B is BS signal bandwidth, n is number of users sharing the BS, commonly known as *load factor, m* is the number of spatial streams between a BS and the users device(s), often known as *spatial multiplexing factor*. The symbols S, N, I are the signal, noise, and interference power at the receiver, respectively. It is clear from Eq. (2.2) that in order to increase the throughput of the user under certain user link SNR, the m and B have to increase, while n and I have to be reduced [6].

These observations imply that increasing B using additional spectrum contributes to a linear increase in user throughput. However, spectrum is a valuable resource and is limited, so every time a new application comes into usage, spectrum has to be allocated and be used exclusively for this application. In addition, the current wireless networks operate exclusively on spectrum at 6 GHz and below. Achieving such data rates with very high availability and reliability is only possible if the available spectrum is much higher than 6 GHz.

Different bands (including those below 6 GHz if available) can serve different applications within a harmonized global framework. In addition, higher operating frequencies mean the propagation path loss is likely to increase linearly, so we have to find a means to compensate such high path loss. Fortunately, 5G architecture offers solutions to these challenges, as we will find in the following chapters.

Reducing the load factor,n, suggests deploying a larger number of smaller BSs (instead of a single macro cell BS). Each cell is serving a limited number of users and ensuring that user traffic is distributed as evenly as possible amongst BSs. These BSs have a much smaller coverage area, usually known as *small cells*, a term that conventionally implies the deployment of micro and pico cells. Small cells generally have a reduced path loss between a user device and the closest small BS. Increasing the spatial multiplexing factor m, can be achieved by using a large number of antennas with suitable characteristics at user device and/or BSs, a technique commonly known as *massive MIMO system*.

The reduced path loss when using small cells increases the desired signal, S as well as interference I levels such that, for appropriate link SNR, $\frac{I}{N} \gg 1$. Consequently, interference mitigation is required to improve link efficiency. This requires a combination of adaptive resource coordination at the small cells BS transmitters with advanced signal processing at the receivers to ensure $\frac{I}{N} \ll 1$.

There are three key ways of increasing network capacity, namely: *buying more spectrum, making spectrum more efficient*, and/or employing a scheme known as *network densification*. The necessary features for wireless capacity enhancement are discussed in the next section and come under a common title of network densification.

2.3 Network Densification

Network densification implies adding more cell sites to increase the available network capacity. These cells are placed in areas where they are most needed, i.e. capacity-strained areas. Heterogeneous cellular networks, or simply HetNets, are networks with a variety of BSs, radio access, transmission solutions, and power levels of BSs. (HetNets) is 3GPP proposed solution as part of LTE-Advanced.

A conventional single-tier cellular network comprises a BS located near the centre of the cell, relay, and user terminals. On the other hand, the 5G cellular will be a multitier heterogeneous network constituting macrocells together with a number of small power nodes, namely small cells, remote radio heads (RRHs), relays, pico cells, femto cells, and C-Ran. The network will provide for point-to-point (P2P) such as machine-to-machine (M2M) and device-to-device (D2D) communications. The arrangements of heterogeneous small cells will be highly dense as compared to the conventional single-tier macrocell networks. Employing different classes of BSs such as marcocells and small cells provides flexible 5G coverage areas. Reducing the size of the cell increases the spectrum reuse and thus improves the area spectral efficiency. Small cells are widely employed in indoors, such as homes and office buildings, while D2D and M2M are used for communications between user terminals and autonomous sensors and actuators to significantly increase the universal spectrum and energy efficiency of the network. Finally, it may be concluded that since the intra and inter interference in the multi-tier 5G cellular networks can be managed, the 5G network performance is expected to be significantly better than the single-tier conventional network in terms of coverage, capacity, spectral efficiency, and power consumption [4].

The macro cell transmit power can be up to 40 W compared to transmit power of up to 1 W for small cells. The small-cell coverage's range is about 300 m compared to the macro cell range of thousands of metres. In addition, small cells serve tens of users in its coverage area while macro cell serves thousands of users. The small-cell BSs considered by 3GPP are pico BS (PBS); femto BS (FBS); relay BS; all under an umbrella of small cells but PBS are most commonly chosen because they can improve the capacity and usually have the same backhaul as macro BS (MBS).

Outdoor pico cells are frequently deployed by operators, since small cells can be commissioned at lower capital expenditure (CAPEX) and low operational expenditure (OPEX) compared to macro cells. At locations without wired backhaul access, relay nodes may be deployed instead of pico cells. That is, a relay node using wireless/cellular spectrum to provide access to mobile users and also to backhaul data to an anchor BS with wired backhaul. Such a technique facilitates the relay node to appear in functioning as a pico BS to the user it serves but behaves as a user device to its anchor BS. Whenever macro cells and small cells share the same operating frequency, new design challenges are to be expected and have to be resolved.

Meanwhile, handoff boundary between cells is based on the level of the received signal power at the user device. Many devices that are close to a small cell will be in the service area of macro cell as well as leading to severe uplink interference at the pico cell. Furthermore, high power received from the macro cell reduces the pico cell coverage and forces *underutilization* of small cells. Cell range expansion (*CRE*) [7], [8] can be used to extend the pico cell range *virtually* by adding a *bias value* to the pico cell received power instead of increasing its transmit power. The CRE technique is capable of extending the coverage. The cell-edge terminals throughput and the complete network throughput are greatly improved. Therefore, when *CRE* is used, inter-cell interference coordination (*ICIC*) is needed to reduce/eliminate the interference. However, inter-cell interference cancellation is a challenging problem because the macro BS strong transmitted power generates a high level of interference.

The conventional method is to use a common bias value amongst all users sharing the pico cell. This is because varying the bias levels would need measurement at the users distributed in the coverage area, which could be a very demanding problem. In addition, the *optimal bias values* depend on not only location of users but also the BSs used, which may differ from one user to the other. Due to the difficulty of setting appropriate bias value for each user, *ICIC* approach is usually used. We use ICIC to reduce/eliminate the interference from macro BS (MBS) to pico BS (PBS). ICIC is realized by dividing the radio resource between the macro BSs (MBSs) and the pico BSs (PBSs) and stopping the MBSs transmission on some radio resources to avoid the interference between them. Although each PBS can interfere with other PBSs signal, since all PBSs have nearly the same transmit power, the interference between them can be tolerated. Depending on the ratio of radio blocks (RBs) allocated to MBS and PBSs, the appropriate bias values may change, and it is difficult to set optimal bias values; consequently, the bias value is mostly determined by trial and error.

Recently, a technique was proposed to determine the user optimal bias value using *Q-learning algorithm* where individual users learn their individual bias values from their past experience independently using costs as return feedback instead of rewards. Such a technique minimizes the number of outage users. So, the aim of the technique is to minimize the costs. Notes on the Q-leaning, which may be useful to some readers, are presented in 2.A.

An intelligent agent collects scalar values (called cost c_t) when observing the environment state s_t and takes action a_t. At the start of the learning, Q returns a fixed value chosen by the designer. The value of Q is iteratively updated each time the agent selects an action and observes a cost for a new state. The Q-learning algorithm calculates the function $Q(s_t, a_t)$ of the state-action, based on cost feedback returns, using the formula in Eq. (2.3):

$$Q(s,a) = \mathbb{E}\{c(s,a)\} + \mathbb{E}\left\{\sum_{t=1}^{\infty} \gamma^t\, c(s_t, a_t) |\, s_0 = s, a_0 = a\right\} \tag{2.3}$$

where γ, $c(s_t, a_t)$, s_0, a_0 are discounted factor $0 \leq \gamma \leq 1$, cost of the set of state s_t and action a_t, the initial state, and initial action, respectively. The Q-table is updated every fixed time interval $\Delta t = t_1, t_2, \dots \dots, T$ using the formula

$$Q(s_t, a_t) \leftarrow (1-\alpha)Q(s_t, a_t) + \alpha[c_{t+1} + \gamma \min_a Q(s_{t+1}, a)] \tag{2.4}$$

where α is the learning rate ($0 \leq \alpha \leq 1$) and $\leftarrow$ indicates *update.*

The Q-learning scheme is a multi-agent system and all users can be agents in the system. Agents set up a Q-table to store the states, costs, actions and Q-values that represent the effectiveness of the sets. The aim of the agents is to minimize costs by choosing the appropriate actions. The costs here mean cumulative costs that are represented as Q-value. The $Q(s_t, a_t)$ is updated according to the following three steps:

Step (1) Agents observe their states from the environment and work out the sets that have the state in the Q-table. This also involves costs from the environment when the selected actions are evaluated.

Step (2) Using the state and cost that are known at step (2), the Q-value selected at the previous state and action is updated.

Step (3) Following a given action selection policy, i.e. ε- greedy policy, an action is selected making use of the Q-values of observed states at step (1).

Having outlined the update of the Q-table, we describe the process whereby the best bias value for each user device is selected. Each user receives pilot signals from each BS and chooses the strongest signal received from macro and pico ones to observe the user's current *state.* The received power is quantized to converge faster, and each user compares these pilot signal powers with *Q-table's states.* If there are no equal received powers on each user Q-table, users add new received powers to their own Q-tables. Amongst those sets whose received powers are equal to the pilot signal powers, users usually choose one set that has the *lowest Q-value* or (rarely) choose one set randomly to avoid local minima as ε-greedy policy. Each user uses chosen *bias value as an action.* Each user then compares macro received power with pico received power plus certain bias value. User devices usually connect to the cell with larger receive power. All actions described above are carried out by each user.

BSs allocate each user to each radio block (RB) randomly. However, in most cases each user can use only one RB. RBs are split between MBS and PBSs. BSs calculate the number of *outage* users and pass this information to each outage's user as a *cost.* Each user reevaluates the chosen set's Q-value and updates the Q-table. Repeating the steps makes Q-value of all sets of states and actions converge, and then agents can make the right actions.

It is worth noting that discovering new states and adding them to the Q-table, will increase the size of the table to the level that it cannot be accommodated by the storage capacity and also makes the learning time longer. To solve this problem, common bias values are used to facilitate faster convergence. The chosen bias values can be checked by trial and error before starting to learn, and although they are not the optimal bias values, they are close to the best values and can be used as initial values so the convergence becomes faster. For readers who wish to carry out Q-learning simulation, typical parameters given in [8], as shown in Table 2.1.

Table 2.1 Parameters for Q-learning simulation.

Macro cell radius	289 m	
Pico cell radius	40 m	
Carrier frequency	2.0 G Hz	
Bandwidth	1 0 M H z	
R B s	5 0	
Thermal noise density	−174 dBm/Hz	
Macro BSs	1	
Pico BSs	2	
Hot spots	2	
UEs inside macro cell	50	
UEs inside hot spot areas	25	
Macro BS transmit power	46 dBm	
Pico BS transmit power	30 dBm	
Macro path loss model	128.1 + 37.6 log10(R) dB (R [km]) where coverage range R in km	
Pico path loss model	140.1 + 36.7 log10(R) dB (R [km])	
Velocity of UEs	3 km/h	
Channel Rayleigh fading		
trials	500 000	
Learning rate (α)	0.5	
Discount factor	(γ)	0.5
Epsilon for ε-greedy policy	0.1	

2.4 Cloud-Based Radio Access Network (C-RAN)

Radio access network (RAN) is the key technology for operators to provide high data rate, high-quality services that are available to subscribed users at any time and every day. Traditional RAN architecture has the common feature that each BS only connects to a fixed number of antennas to transmit /receive signals in its coverage area. The system capacity is interference-limited and the BSs are built on a specific technology platform. This feature gave rise to many difficulties. For example, extending the coverage area requires deploying more BSs, which provokes an increase in CAPEX and OPEX. Under normal conditions, BSs average utilization rate is low and the BS' processing power can't be shared with other BSs in the network, which makes it difficult to improve spectrum efficiency, system capacity and network upgrading.

However, due to increasing demand on data services, operators need to upgrade their networks frequently and operate on multiple standard network (i.e. offering services from 2G, 3G, and 4G. etc.). In short with the advancement in wireless technology, traditional RAN becomes too expensive for operators to be competitive in the future wireless world. It lacks the flexibility to support services for innovative applications to generate new revenue from new services. It is important for future RAN to provide direct internet access to subscribers at low cost with high spectral and energy efficiencies.

Traditionally, a BS cabinet is placed in a dedicated room with its power support facility, backup battery, cooling system, environment control system, and backhaul transmission equipment. An evolved BS consists of a baseband unit (BBU), which is placed in the BS room. The BBU is comprised of a DSP to process baseband signals such as voice signals for transmission to a mobile terminal and to process signals received from the mobile terminal. The BBU has modular design, small size, low power consumption, and could be quickly deployed. The BBU system platform supports multi-mode by plugging in different processing boards, each board supports multistandard operation. The BBU platforms also have enabling modules, that can be shared between BBU boards, such as control module, timing module, remote radio head (RRH) I/O modules. However, different boards can't share processing resources and so an upgrade needs adding to new board hardware.

Distributed BSs consist of the remote radio head (RRH), which contains the BS RF circuitry plus the A/D and D/A converters and up/down frequency converters. The RRH is normally installed on the top of a BS tower close to the antenna(s). The RRH and the BBU are connected largely by fibre link. Centralized baseband (BB) pool processing greatly reduces the required BS space. Cooperative radio with distributed antennas, each equipped by remote radio head (RRH) can provide high spectral efficiency.

Cloud infrastructure RAN (C-RAN) provides low-cost, high-performance network architecture and confronts a lot of future challenges facing the operators. C-RAN is applicable to most traditional RAN like macro cell, microcell, pico cell, and indoor coverage. Most recently, BBU and RRH products also can be software-defined radio (SDR) enabled to support different standards [6].

C-RANs have been proposed as a cost-efficient method for deploying small cells where an intelligent configuration is used for the fronthaul network between centralized BBUs and RRHs deployed as small cells to deliver the performance and energy savings to the RRHs and the BBU pool [9]. The fronthaul deployed is logically reconfigurable to implement appropriate transmission schemes in different parts of the network and so responds effectively to both heterogeneous users as well as the dynamic traffic patterns and consequently satisfies the maximum traffic demand on the RAN while simultaneously optimizing the resource usage. The C-RAN concept is depicted in Figure 2.1. where the BBUs migrated to a data centre (i.e. BBU pool).

The fronthaul part of the network carries the cellular signals between the BBUs and the RRHs with bandwidth tens of GHz, which is significantly higher than that for the backhaul. The exact bandwidth requirement depends on the nature of the transported signals (i.e. digital/analogue, layer 1 and layer 2). Although BBUs and RRHs are decoupled geographically, there exists a one-to-one logical mapping between them, such that one BBU is assigned to transmit (receive) signal to (from) an RRH. This mapping scheme may change over time but the principle of one-to-one mapping allows generating a definite frame for each RRH (i.e. small cell).

However, a fixed one-to-one mapping is not optimal in a practical cellular system for two reasons: *first,* mobile users belonging to small cells have to support frequent handoffs that penalize the performance; and *second,* several BBUs have to be active, which consume energy in the BBU pool, while a single BBU may be adequate to serve the offered load.

Conventionally, a fronthaul network comprises an optical fibre with wavelength multiplexing to distribute signals from BBU pool to RRHs as either digitized radio signals over common public radio interface (CPRI) or as analogue radio signal via radio over fibre (RoF). In the dynamically reconfigurable fronthaul with intelligence, two schemes can be deployed, namely: *fractional frequency reuse* (FFR) is a radio resource management (RRM) scheme to address the inter-cell interference and *distributed antenna system* (DAS) to provide larger coverage. Since interfering cells operate possibly on different bands in FFR, different frames have to be generated for each cell, thus requiring a one-to-one logical configuration between a BBU and RRH.

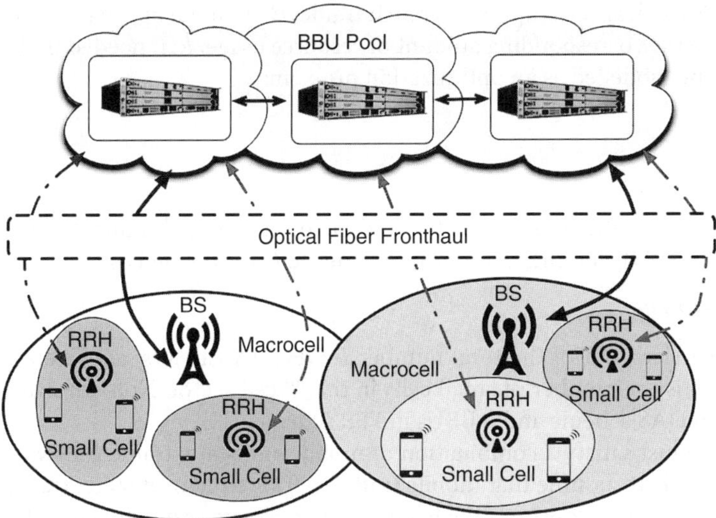

Figure 2.1 Sketch of C-RAN architecture [9].

Figure 2.2 Macro plus small-cell network deployment [9].

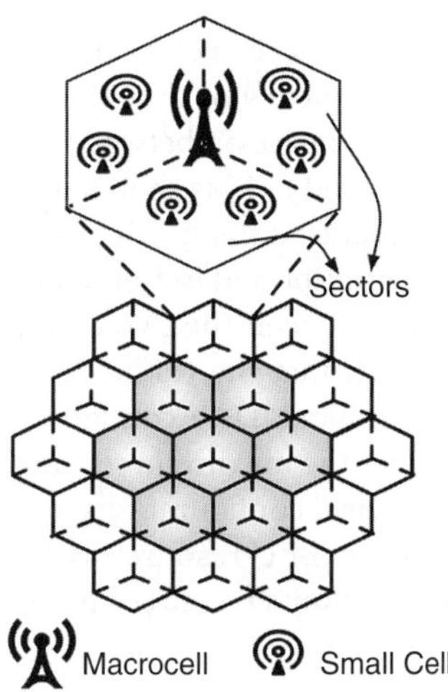

This is the conventional configuration currently used. However, with DAS a single frame is transmitted by multiple RRHs, which, in turn, can be achieved using a single BBU, thus requiring a one-to-many configuration. Clearly, a one-to-one signal configuration between BBUs and RRHs is greatly suboptimal. Consider a network model consisting of a large number of small cells deployed as an under-lay macro cell network as shown in Figure 2.2.

2.4.1 Resource Management Between Macrocells and Small Cells

Two types of radio resource management (RRM) can be used to reduce the macro-cell interference amongst small cells: macro cell and small cells can operate on *different carrier frequencies*; or both can use the *same frequency* but at different time slots. Both schemes are equivalent. Most common scenario is when macro cell and small cells operate on different carrier frequencies. Using FFR and DAS, in combination in different parts of the network, the goal is to find a fine balance between them. Specifically, to support as much traffic, D, with the optimal configuration, D_{opt} and to minimize the corresponding amount of resource usage, RU, needed in the BBU pool; this problem can be modelled as an optimization problem:

$$\min_{conf} RU_{conf} \tag{2.5a}$$

$$s.t \qquad D \geq \mathcal{F}.D_{opt} \tag{2.5b}$$

where $conf$ represents a possible configuration, $\mathcal{F}$ is the fraction of the optimal traffic demand that must be satisfied (i.e. $\mathcal{F} = 0.99$), and the resource usage metric RU is defined as

$$RU(b_i, n_i) = b_i + (B - b_i).n_i \tag{2.6}$$

where b_i is the number of carriers, out of the total number B OFDMA carriers, allocated to its DAS configuration, and n_i is the number of small cells in the ith cell sector. Note for every carrier, one BBU is needed in DAS scheme and n BBUs in FFR scheme.

The optimal configuration must satisfy a combination of mobile and static traffic flow with certain priorities. It is reasonable to assume that mobile traffic is prioritized over static traffic. However, minimization of the resource consumption is subject to satisfying as much as possible of the traffic demand, thus a fine balance should be found to satisfy such a constraint. Depending on the user's environment and the traffic load in a given macro cell sector, an appropriate transmission strategy employing a flexible combination of DAS and FFR (i.e. *hybrid configurations*) in each sector has to be configured. Hybrid configurations can be multiplexed either in the frequency or time domain but not both at the same time.

A comparison between time multiplexing and frequency multiplexing hybrid configurations (DAS and FFR) is shown in Figure 2.3. Hybrid configurations are appropriately multiplexed in a frequency domain when multicarrier spectrum resource is used. When hybrid configurations are multiplexed in frequency, operators can divide the spectrum into separate carriers that are split between the two configurations. The fraction of carriers allocated to the configurations is such that the traffic load is satisfied with a minimum number of BBUs resources. On the other hand, DAS configuration minimizes the use of BBU resources. This creates a situation where the maximum resource is allocated to the DAS configuration that can be supported only at hybrid configurations update within the order of minutes. If configurations are multiplexed in time, BBU resource assignment has to switch between DAS and FFR each 10 ms or so and hence limits resource and energy saving in a BBU pool.

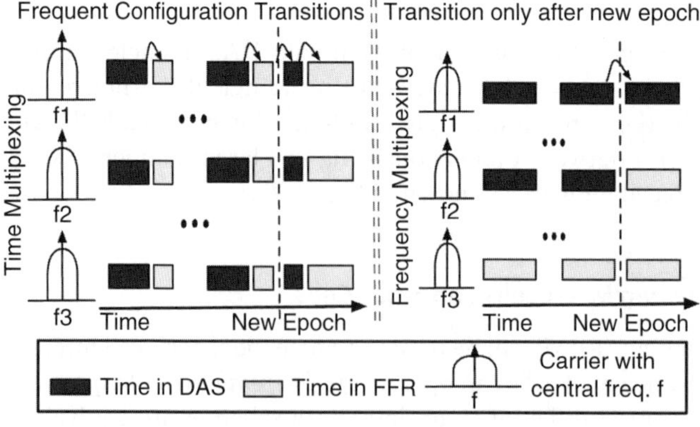

Figure 2.3 Comparison between time multiplexing and frequency multiplexing hybrid configurations (DAS and FFR) [9].

At locations where network traffic load is low, traffic from multiple sectors can be supported jointly with a single DAS configuration to reduce resource usage in the BBU pool. A clustering of two neighbouring sectors i and j can be supported at certain times until either their net offered traffic load cannot be supported or the resource usage metric RU cannot improve, i.e.

$$RU(b_{i\cup j}, n_i + n_j) > RU(b_i, n_i) + RU(b_j, n_j) \tag{2.7}$$

where $b_{i\cup j}$ represents the carriers split between sectors i and j. Sharing carriers and using DAS implies sharing carriers between users without reuse, which is straightforward. Dynamic FFR over a large number of cells is computationally intensive. So, for large clusters, FFR solution can run on each constituent sector. Offered traffic load in a sector/cluster can be scheduled on any carrier operating on DAS/FFR. It is worth noting that DAS configuration is important to reduce resource usage and also improves performance, especially for mobile users. Mobile users' traffic load in a sector/cluster puts a constraint on the minimum number of carriers needed to be allocated to its DAS configuration.

The sequence of actions in the proposed solution in at every epoch is as follows: every sector obtains aggregate traffic demand from each of its small cells and determines the number of carriers needed for the DAS and FFR configuration based on mobile and static traffic demand. This will enable the determination of the optimal frequency multiplexing, classification of the suitable traffic that needs to be scheduled, and hence the determination of the RU metric for the sector. So, arrange sectors in a cluster (2 sector /cluster) at a time based on their RU metric until either their offered traffic cannot be supported or the cluster metric cannot improve. Finally, we apply the configurations for each cell in the cluster.

2.4.2 BBU-RRH Switching Schemes

Mobile data traffic usually changes over time during day / night, and over geographical locations (urban, suburban, rural). In addition, traffic load, under normal conditions, peaks during the day in business areas and peaks at night in residential areas. Hence, traffic imbalance grows between times and locations.

BBU-RRH switching schemes for centralized RAN is an idea that is very close to cloud-RAN (C-RAN), proposed in [10], for reducing network costs due to significant reduction in BSs energy in addition to site costs. So, the C-RAN architecture is composed of a centralized BBU and locally deployed RRHs. The scheme assigns one BBU to one or more RRHs depending on the traffic loads that BBU can manage. This forces a reduction in the number of BBUs and BB' resources utilized. Moreover, the power consumption can be reduced considerably, which reduces CAPEX cost and OPEX.

Consequently, as few as possible BBUs are assigned to local RRHs. What is more, their RRHs neighbour can be serviced by the same BBU when possible. Such a scheme depends on introducing some collaboration controls between neighbouring RRHs (i.e. Coordinated Multi-Point (CoMP). C-RAN has a BBU pooling function to modify the combinations of the BBUs and RRHS so as not to exceed the total BBU resources upper limit.

The proposed concept comprises *semi-static* and *adaptive* switching schemes. The main difference between the two schemes is the switching time interval. The switching interval of the semi-static switching scheme can be as long as a day, while for the adaptive BBU-RRH switching scheme is as short as an hour. Further, the complexity of the semi-static switching is low because of the long interval, while the adaptive BBU-RRH switching scheme complexity is high since in a short interval the adaptive scheme has to assign a large number of BBUs-RRHs allocations. The concept behind the proposed schemes is to determine the best combinations between BBUs and RRHs that can deliver traffic load for certain interval time using a minimum number of BBUs.

The semi-static BBU-RRH allocation determines the appropriate combinations of the BBUs and RRHs to attempt to deliver peak-hour traffic load (average traffic that occurs during hour within peak traffic) for all RRHs within the interval. The goal is to assign BBUs from those who have near upper limit of BBU resource, to a RRH so that neighbouring RRHs can be accommodated on the same BBU (if possible). This process is repeated for all RRHs; hence, a semi-static scheme is suitable for all traffic assigning BBUs-RRHs according to restrict priorities.

The adaptive BBU-RRH switching scheme is launched when the BBU *resource usage exceeds* the upper limit (called case A) or *falls below* lower limit (case B). In case A, a target RRH is selected according to two priorities: It has the highest number of *neighbouring RRHs* and it has the lowest *traffic load*.

The BBU is assigned to the target RRH according to the following three priorities:

1. There are BBUs already assigned to *neighbouring RRHs* of the target RRH and the BBUs *resource usages* are lower than the upper limit.
2. Assignable BBUs must have their resources lower than the upper limit.
3. Assignable BBUs must not be already assigned.

Finally, if the BBU resource usage doesn't exceed the upper limit, the BBU-RRH procedure is completed; otherwise, repeat the procedure. The procedure continues until all BBUs and RRHs are selected. In case B, when BBU resource usage falls below the lower limit, all RRHs connected to the BBU are switched to other neighbouring BBUs whose resource usages are lower than the upper limit.

In conclusion, the number of switching events of an adaptive BBU-RRH switching scheme is significantly higher than that of the semi-static BBU-RRH switching scheme. However, the BBU reduction initiated by the adaptive scheme greatly increases compared with the semi-static scheme.

Figure 2.4 illustrates the % traffic load for 24 hours for two typical traffic profiles, one is an office (business) area shown in Figure 2.4a, and the other is a residence area shown in Figure 2.4b. The traffic load of each RRH is generated from a random series based on measured traffic load 24 hours earlier. The initial assignment of BBUs to RRHs is determined using semi-static scheme. The upper and the lower limits of BBU resource usage are assumed to be 0.9 and 0.5. The time interval for BBU-RRH switching in the adaptive scheme is assumed to be 1 hour. It is clear from Figure 2.4 that the office area peak traffic load occurs during the day between 11 a.m. – 5 p.m. and the traffic profile of the residence peaks at night at 11 p.m.–midnight.

Figure 2.5 shows the number of BBUs allocated by semi static and adaptive schemes for the office (Figure 2.5a) and residence (Figure 2.5b) traffic loads over 24 hours. The number of BBUs allocated for a semi-static scheme are 77 and 73, respectively. On the other hand, the number of BBUs assigned under an adaptive scheme changes according to the traffic load and so the maximum number of BBUs are assigned during the peak hour of the traffic load. Generally, the adaptive scheme seems to reduce the number of allocated BBUs for both office and residence areas. Compared to the semi-static, which is far from optimum, the adaptive scheme was very close to the optimum results in many instants.

2.4.3 Mobile Small Cells

The concept of *mobile small cells* was introduced in [5, 11], where a mobile device takes the role of an access point or remote radio unit (RRU). The new concept seems very interesting since it is capable of providing many advantages such as a systems capacity increase, higher data rates, lower power consumption due to less devices in use and shorter transmission links.

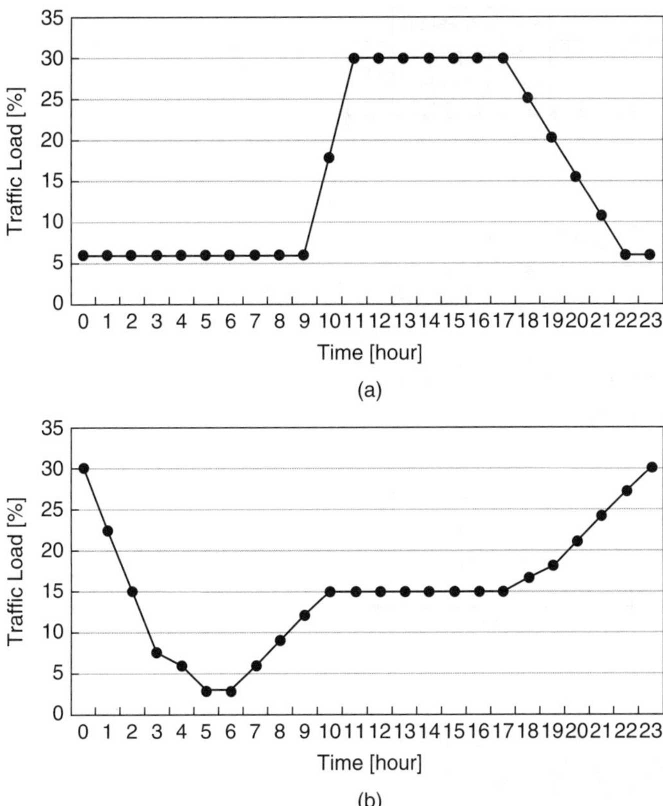

Figure 2.4 Traffic profile data for (a) office and (b) residence [10].

Another thing, there will be no network planning or hardware deployment. However, there are certain design issues that have to be addressed. The concept of on-demand mobile small cells (i.e. mobile user handsets used as small cells) raises a number of research and design difficulties. A selected number of these issues follow:

- Design of interference management between randomly deployed on-demand mobile small cell is not straightforward.
- There must be an efficient node discovery scheme.
- The choice of mobile handsets to act as small cells is not very clear.

The possible criteria for the mobile handset selection may depend on metrics like battery level, handset capability, or mobility patterns, for example. In addition, the proposed scheme requires the user data to be relayed through other user handsets, which raises privacy issues; hence, security becomes a major concern in such a scenario. Finally, mobile users use their own resources to serve others, so some incentives will be needed to encourage users to allow their handsets to be used as small cells, and some may incur costs.

In general, macro BS provides the signalling service for the whole coverage area and the mobile small cells deliver *high-rate data* transmission service with light overhead and possibly suitable *mmWave air interface.* One vision of 5G is based on the concept of the on-demand mobile small cells, which can be RRHs or user mobile headsets, to realize very dense cellular networks. In general, the RRHs (i.e. mobile small cells) transmit the RF signals to the users on the DL and transmit the information signals from users to the BBU pool on the UL. The edge link

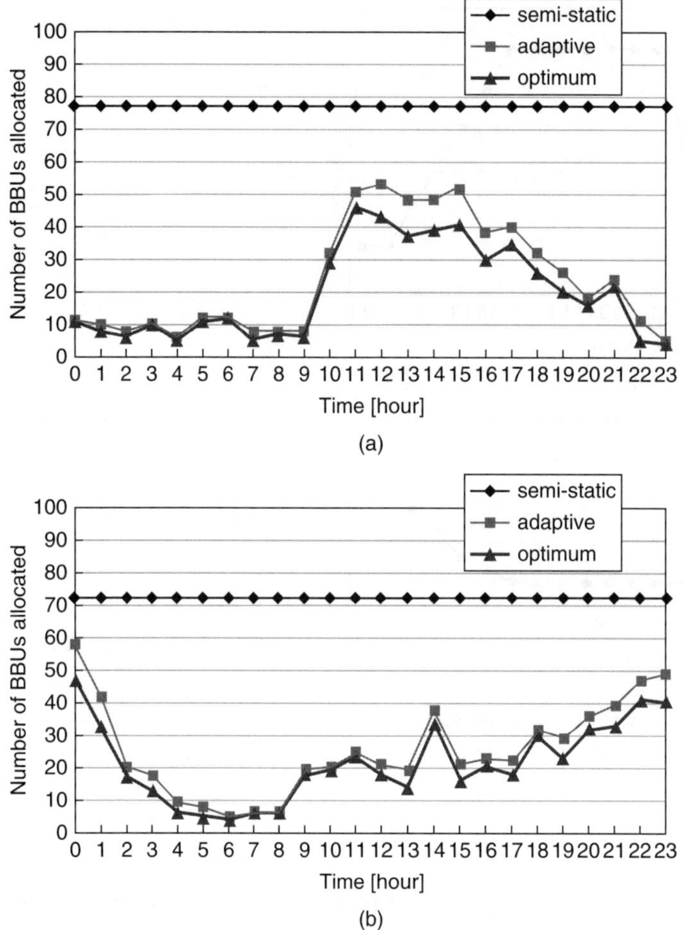

Figure 2.5 The number of BBUs assigned for (a) office and (b) residence [10].

that connects the small cells to the rest of the network can use fibre cable (we highlighted above) or employ wireless such as mmWave or microwave. The fronthaul connects the RRHs with the BBUs with fibre as transmission medium. However, wireless fronthaul solutions appeal for small-cell deployment since fibre is too expensive or just not available at many small-cell sites.

2.4.4 Automatic Self-Organising Network (SON)

The proposed mobile small cell is further enhanced using a (SON) scheme that dynamically selects a frequency band for optimal FFR and adapts appropriate transmit power for best user performance.

The average user throughput in (Mb/s) received in both deployments, with and without small cells, is shown in Figure 2.6. The average user throughput received when small cells are deployed is higher compared to average throughput received without small cells. It can be seen from Figure 2.6 that for the F(x) percentage of the users, 50% of the users receive an average throughput of only 2 Mb/s without using small cells compared to 12 Mb/s when users used the small cells. The set of results presented in the figure shows the throughput gain achieved when small cells are deployed within the macro cells coverage area.

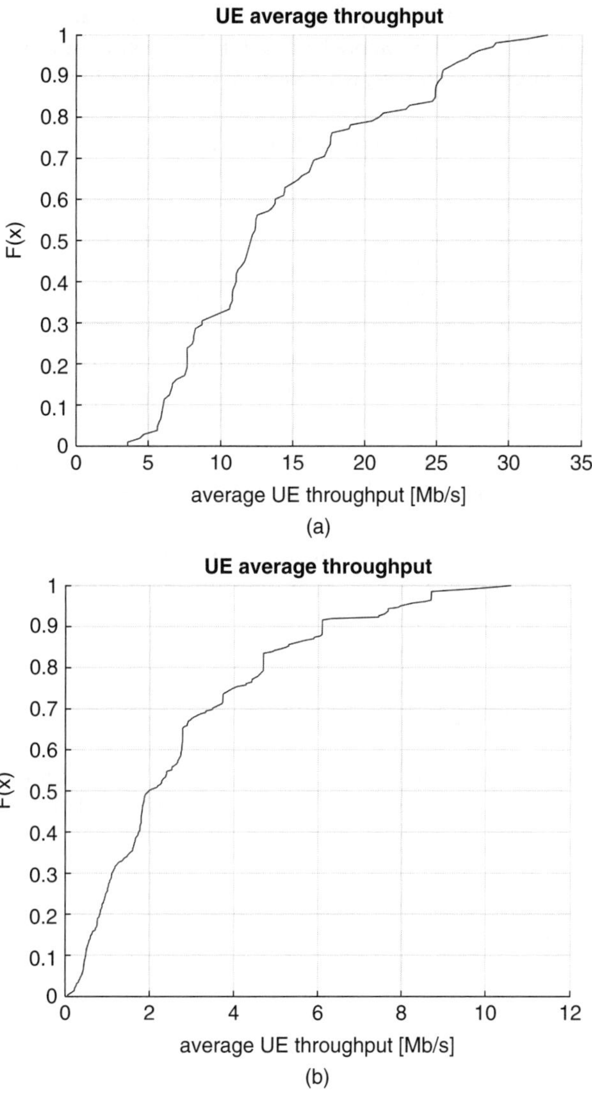

Figure 2.6 (a) Average throughput received by users with small-cells deployment. (b) Average throughput received by users without small-cell deployment [5].

2.5 Cache-Enabled Small-Cell Networks (CE-SCNs)

This section examines cache-enabled small-cell networks (CE-SCNs) [12, 13]. Caching users' content in nodes of the network, preferably near the intended users, permits the user to have instant access to the content. Furthermore, predicting human behaviour and proactively caching the contents at the edge of the network, BSs, and user terminals greatly improves user satisfaction and lowers infrastructure cost.

We consider the caching concept when applied in a scenario where small BSs (SBSs) are randomly distributed over the coverage area and have limited backhaul capacity. The aims are to derive a system model and define the performance metrics based on system parameters such as:

system SINR, BS location, bit rate used for target file delivery, cache storage size, and cached file popularity distribution. The performance of the cache-enabled SBSs (CE-SBSs) system model depends generally on the system physical layer parameters, which reveal key design insights of the CE-SBSs concept.

So as to deeply understand the theoretical issues in the concept, let us think about a cellular network consisting of SBSs distributed as Poisson Point Process (PPP), Φ, with cells' intensity λ and a wired backhaul link is used to connect the SBSs to a central scheduler (CS) to provide broadband connection to the end users. Assume the total backhaul link capacity is fixed at finite value. Denote the backhaul link capacity of each SBS by $C_{bh}(\lambda)$ and it is reasonable to assume that $C_{bh}(\lambda)$ is a finite decreasing function of λ for the following reason: As the number of SBSs increases (i.e. increasing the density λ), the backhaul link capacity $C_{bh}(\lambda)$ has to be split amongst SBSs, hence decreasing the backhaul capacity of each individual SBS. In addition, we assume the storage capacity unit with each SBS S is used mainly for caching the most popular (i.e. mostly used) files and all files have equal length L and are transported at equal bit rate R. Besides, the distribution of a user file its popularity is defined by a decreasing PDF $f_{pop}(f, \gamma)$ that is identical amongst all users, f is a point in the corresponding file and $\gamma > 1$ is the shape parameter of the file popularity *power law distribution.*

Therefore, for given popular files PDF, the SBSs are able to store as many files as their storage capacity takes, in a convenient time (possibly off-peak times or at night) before user requests arrive. So, to analyse the system performance during file delivery, we consider the network model shown in Figure 2.7 and we neglect the uplink request overhead.

2.5.1 File Delivery Performance Analysis of CE-SCN

The CE-SCN model sketch in Figure 2.7 shows a single SBS serving six mobile users (three males and three females) and on the top right side of the plot is a number SBSs randomly distributed.

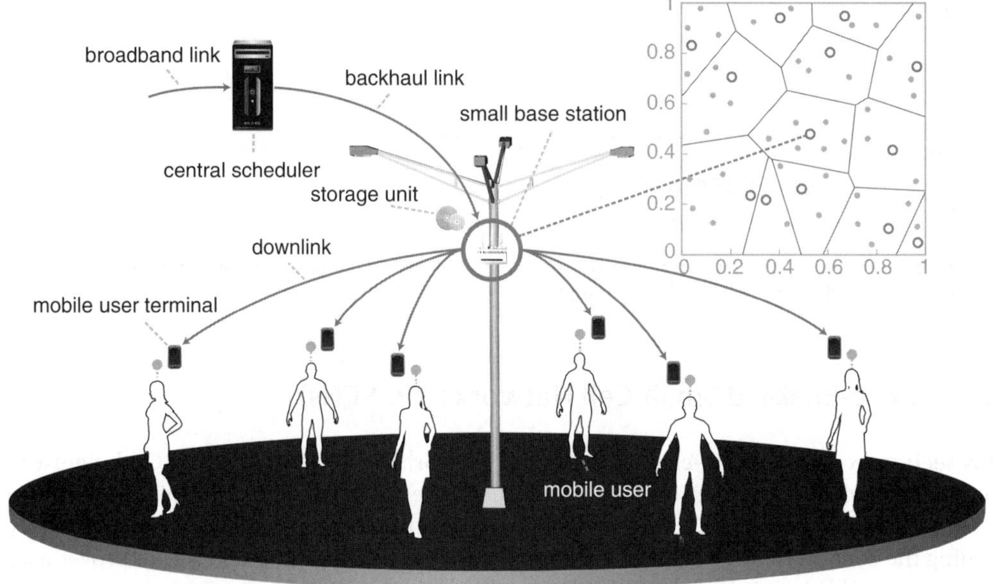

Figure 2.7 An illustration of the network model. The plot on the top right side is a snapshot of PPP per unit area with randomly deployed SBSs from Voronoi tessellation. Communication structure of a SBS equipped with a storage unit, is highlighted in the main figure [12].

Each user is associated with the nearest SBS. SBSs and their users, the top right-side plot, form what is known in mathematics as a Voronoi diagram, or Voronoi tessellation.

On the DL, the tagged SBS transmits with constant power $\frac{1}{\mu}$ and the path loss exponent $\alpha > 2$. The link between the tagged SBS and its associated user experiences Rayleigh fading with mean 1. The received power of the tagged user at distance d metre from its SBS is $h\,d^{-\alpha}$ where the random variable h follows an exponential distribution, $h \sim Exp(\mu)$ with mean $\frac{1}{\mu}$.

Following the user association, the request of a file is randomly selected according to the file popularity distribution $f_{pop}\,(f, \gamma)$. When this request arrives at the tagged SBS via the uplink, the requested file is retrieved from the internet or read from the local cache and the user is served instantly via the downlink. If the requested file is read from the local cache, a *cache hit* event is recorded. However, if the requested file is not found in the local cache, a *cache miss* event has occurred.

Clearly the performance of the SBS system during the file delivery depends greatly on the cache hit and miss events. For example, when a cache hit occurs, and assuming the requested file is read at an acceptable rate, the performance is then defined by the file transmission rate over the DL. While in the cache miss scenario, the file has to be fetched from the internet using the backhaul, and the performance is decided by the backhaul link rate and the additional cost of its usage. Two key metrics define the SBS system performance: *outage probability* and *average rate* of file delivery.

2.5.2 Outage Probability and Average File Delivery Rate in CE-SC System

Consider a DL cellular CE-SCN and assume the user is located at the origin *user o* (location of a typical user). The rate transmission in the DL is the factor limiting the small-cell system performance, and the DL rate depends on the SINR. Assume *user o* is associated with SBS B_o, at a random distance d so the SINR is defined as

$$SINR \triangleq \frac{h\,d^{-\alpha}}{\sum_{i\in\Phi/B_o} g_i r_i^{-\alpha} + \sigma_o^2} \tag{2.8}$$

where r_i is random distance of the ith interfering user from its associate SBS B_o and g_i Interference fading. Let us describe the *success probability* as the probability of DL with higher than target file bit rate R when a cache hit event occurred. Then we define the outage probability of a typical user from its tagged SBS as $\mathbb{P}_{out}(\lambda, R, \alpha, S)$, which is the complement of the success probability $\mathbb{P}_{success}(\lambda, R, \alpha, S)$ given by

$$\mathbb{P}_{success}(\lambda, R, \alpha, S) \triangleq \mathbb{P}[\log 1 + SINR \geq R, f_o \in \Delta_{B_o}] \tag{2.9}$$

where $f_o \in \Delta_{B_o}$ is the requested file by *user o* stored in the cache Δ_{B_o} of serving SBS B_o. Consequently, the outage probability is given as

$$\mathbb{P}_{out}(\lambda, R, \alpha, S) \triangleq 1 - \mathbb{P}[\log(1 + SINR) \geq R, f_o \in \Delta_{B_o}] \tag{2.10}$$

The outage probability expressed in (10) is simplified in [12, theorem 1] to

$$\mathbb{P}_{out}(\lambda, R, \alpha, S) = 1 - \pi\lambda \int_0^{\infty} \int_0^{\frac{S}{L}} e^{-\pi\lambda v\beta(R,\alpha) - \mu(e^R - 1)\sigma^2 v^{\frac{\alpha}{2}}} f_{pop}\,(f, \gamma) df dv \tag{2.11}$$

where $\beta(R, \alpha)$ is

$$\beta(R, \alpha) = \frac{2(\mu(e^R - 1))}{\alpha}\, \mathbb{E}_g\left[g^{\frac{2}{\alpha}}\left(\Gamma\left(-\frac{\alpha}{2}, \mu(e^R - 1)g\right) - \Gamma\left(-\frac{\alpha}{2}\right)\right)\right] \tag{2.12}$$

The derivations of Eqs. (2.11) and (2.12) are presented in Appendix 2B.

The exact values of the outage probability and *average rate* of file delivery can be obtained using numerical integration of (2.11) and averaging over the PPP and Rayleigh fading distributions and summation over the PDF $f_{pop}(f, \gamma)$ to estimate the cache hit ratio. Such an estimation process requires a high computation power.

An alternative approach is to consider special cases that can be used for the estimation without numerical integration. Suppose the following assumptions are satisfied in the cache-enabled small-cell model:

(a) backhaul capacity $C_{bh}(\lambda)$ can be defined as

$$C_{bh}(\lambda) \triangleq \frac{C_1}{\lambda} + C_2 \tag{2.13}$$

where $C_1 > 0$, $C_2 \geq 0$ are design parameters chosen to make $C_{bh}(\lambda) < R$.
(b) The noise power $\sigma^2 > 0$ and $\alpha = 4$.
(c) Interference is assumed to be Rayleigh fading, g_i, and $g_i \sim Exp(\mu)$.
(d) The file popularity distribution, in (2.11), described by power law as

$$f_{pop}(f, \gamma) \triangleq \begin{Bmatrix} (\gamma - 1)f^{-\gamma} & f \geq 1 \\ 0 & f < 1 \end{Bmatrix} \tag{2.14}$$

where $\gamma > 1$ the distribution shape parameter, and $C_{bh}(\lambda) < R$ where we assumed the backhaul link is high-speed fibre cable and so $C_{bh}(\lambda)$ in densely SBS is less than the bit rate of the requested files. Taking all the assumptions into consideration, the outage probability of *user o* from its tagged SBS [12] is

$$p_{out}(\lambda, R, 4, S, \gamma) = 1 - \frac{\pi^{\frac{3}{2}}}{\sqrt{\frac{e^R - 1}{SNR}}} \exp\left(\frac{(\lambda\pi(1 + \rho(R, 4)))^2}{\frac{4(e^R - 1)}{SNR}}\right) \times Q\left(\frac{\lambda\pi(1 + \rho(R, 4))}{\sqrt{\frac{2(e^R - 1)}{SNR}}}\right)\left(1 - \left(\frac{L}{L + S}\right)^{\gamma - 1}\right) \tag{2.15}$$

where $\rho(R, 4) = \sqrt{(e^R - 1)}\left(\frac{\pi}{2} - \arctan\left(\frac{1}{\sqrt{(e^R - 1)}}\right)\right)$, $Q(x)$ function is the tail probability of standard normal distribution and can be obtained from standard published tables, and L is file length. Using the same assumptions, the average delivery bit rate of *user o* from its SBS B_o is

$$\overline{R}(\lambda, R, 4, S, \gamma) = \frac{\pi^{\frac{3}{2}}\lambda}{\sqrt{\frac{e^R - 1}{SNR}}} \exp\left(\frac{(\lambda\pi(1 + \rho(R, 4)))^2}{\frac{4(e^R - 1)}{SNR}}\right) \times Q\left(\frac{\lambda\pi(1 + \rho(R, 4))}{\sqrt{\frac{2(e^R - 1)}{SNR}}}\right) \times \left(R + \left(\frac{C_1}{\lambda} + C_2 - R\right)\left(\frac{L}{L + S}\right)^{\gamma - 1}\right) \tag{2.16}$$

The impact of the SBS storage size on the outage probability and on average bit rate are displayed in Figure 2.8 where 1 *bit* = ln 2 = 0.693 *nats*. It is clear from Figure 2.8a that the probability of outage decreases rapidly with increasing SBS storage capacity, and for further storage capacity increase, the outage probability tends to a limit. This limit increases as target bit rate increases. The average delivery rate in nats/sec/Hz, is demonstrated in Figure 2.8b and increases rapidly with increasing storage capacity and with increasing delivery target bit rate. In addition, in both plots of Figure 2.8, the computed theoretical values are very close to the simulated results, implying that the system model is a good approximate to the real system.

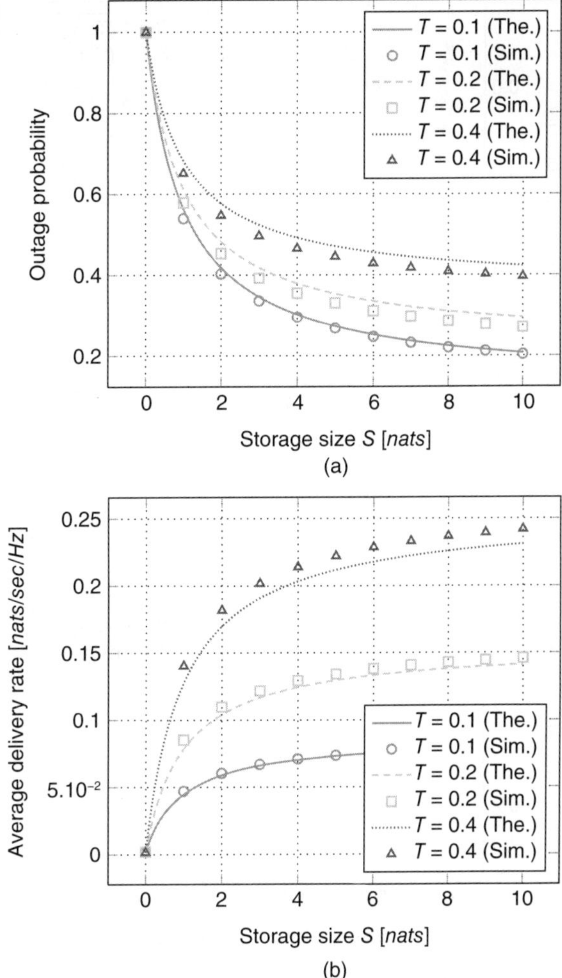

Figure 2.8 (a) Development of outage probability with respect to storage size, (b) Increase of average delivery rate with respect to storage size. SNR = 10 dB, $\lambda = 0.2$, $\gamma = 2$, $L = 1$, $\alpha = 4$, $C_1 = 0.0005$, $C_2 = 0$ [12].

Figure 2.9 shows the impact SBS intensity on the outage probability for various levels of storage capacity. The outage probability seems to decrease sharply with increasing SBS intensity and tends to a limit at an intensity greater than 0.2. Further, the higher the storage capacity, the lower the outage probability limit.

Figure 2.10 determined the impact of the target bit rate on the user outage probability for various levels of storage capacity. As the bit rate increases, the outage probability increases. For any bit rate, the outage probability can be reduced by increasing the storage capacity.

2.6 Full-Duplex (FD) Communications

In mobile communications, radio transceiver simultaneous transmission and reception generates a large local interference (e.g. multiuser, multicell, self-interference), so a long-held assumption in the wireless system design is to operate the system in half duplex (HD), i.e. transmit or receive but not both at the same time. Furthermore, half-duplex operation is used because of the self-interference in a multi-antenna system. So, to enable full duplex, we need

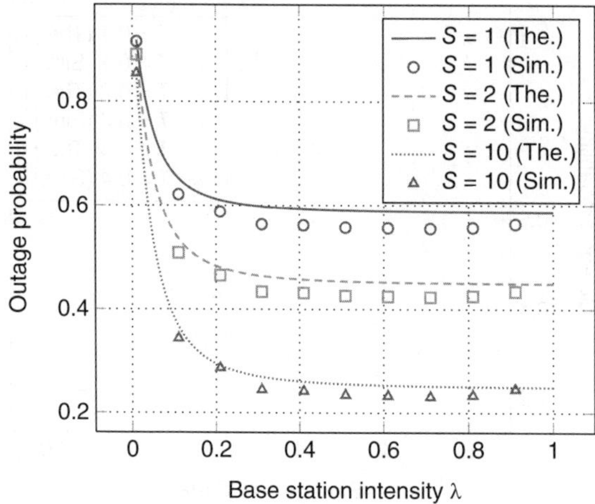

Figure 2.9 The decrease of outage probability with respect to BS intensity. SNR = 10 dB, T = 0.2, $\gamma = 2$, L = 1, $\alpha = 4$, $C_1 = 0.0005$, $C_2 = 0$ [12].

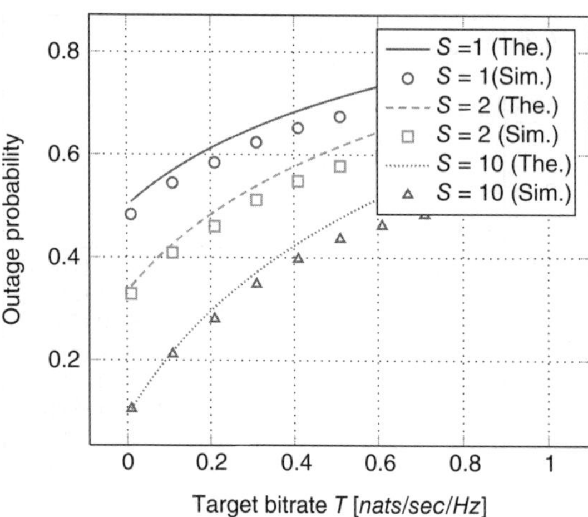

Figure 2.10 The increase of outage probability with respect to target bitrate. SNR = 10 dB, $\lambda = 0.2$, $\gamma = 2$, L = 1, $\alpha = 4$, $C_1 = 0.0005$, $C_2 = 0$ [12].

to reduce/eliminate the *self-interference*, the interference at the receiver, that is generated from the node's own transmission. The technique is commonly known as *self-interference cancellation* (SIC), which can be used to make full-duplex communications work as an enabling technology for 5G.

SIC can be developed in many ways to support the evolution of 5G enabling technologies. In addition to the FD transmission that doubles the link capacity with respect to HD, FD eliminates the distinction between TDD and FDD, makes TDD obsolete, and replaces it with *in-band FD*. On the other hand, FDD improves greatly by becoming adaptive and improves carrier aggregation. Furthermore, SIC enables the reuse of the same operating frequency for backhaul and access, which implies an instantaneous retransmission with high throughput operation for heterogeneous networks. Most importantly, simultaneous reception of feedback control signalling while transmitting data reduces interface delay substantially and tightens synchronization that is required in CoMP transmission.

At the first instant, one may think (possibly naively) that cancellation of self-interference looks straightforward since the transceiver knows the transmitted signal and hence the interference can be subtracted at the receiver by a simple operation at a minimum cost. In reality, elimination of the self-interference is much more complex. In practice, this subtraction is incorrect for the following reasons: the transmitted RF signal does not have the same properties as its

BB copy. The transmit circuits create higher-order distortion caused by the power amplifier and possibly the detuned oscillator frequency introduces different amounts of delay at different frequencies within the received signal BW for example. To sum up, the transmitted signal is a noisy nonlinear copy of the ideal transmitted signal. The receiver circuit elements contribute undesired effects known as receiver saturation, which must be dealt with before cancellation of the nonlinear self-interference. Receiver saturation is created by the ADC resolution due to the limited receiver dynamic range. Interference remaining after the cancellation may include harmonic components much higher than the background noise. To fully understand the receiver saturation, we need to understand the way signals are processed at the receiver. The received signal is first amplified by a low noise amplifier with AGC and down converted to either IF or BB, filtered, and sampled using ADC to create digital samples. The *accuracy of the digital bits depends on the resolution of the* ADC. The AGC adjusts the gain of the received signal to match the maximum level of the ADC to get maximum resolution. The received signal can be decoded correctly if the desired signal is stronger than the interfering signal plus noise. ADC dynamic range (DR) in dB is a function of the number of ADC bits as

$$DR = 20 \log_{10} 2^N$$

where N is the number of ADC bits. This equation can be expressed as

$$\mathrm{DR} = 6.0206 * N$$

Thus for 8-bit ADC, DR = 48 dB, and for 12-bit ADC, DR = 72 dB.

For 8-bit ADC, if the weaker signal is 40 dB lower in power than the stronger signal, it only gets 1-bit resolution. A stronger signal has ADC DR 48 dB and the weaker has DR 40 dB; thus the difference in DR is 8 dB, which represents N bits where

$$N = \frac{\text{difference in DR}}{6.0206} = 1.3288\ \text{bit} \sim 1\ \text{bit}$$

Theoretically, FD doubles the throughput; however, experiments on real nodes show that the FD scheme *achieves* 84% median physical layer throughput gain, when both schemes are compared to traditional single-hop network [14].

2.6.1 Analysis of FD Communication

Consider a half duplex (HD) operating on a single antenna link A-B of length l where both nodes A and B have packets to send to each other. Let P be the transmit power and P_n *be* the white noise power. Assume the link is symmetric at both ends so the signal-to-noise power ratio at node A or B is

$$SNR_A = SNR_B = \frac{P\, b\, l^{-\alpha}}{P_n} \tag{2.17}$$

where b and α are signal propagation parameters. The half-duplex rates, r_{HD}, are

$$r_{HD} = r_{AB} = r_{BA} \tag{2.18}$$

The HD rates are equal on the two links A-B and B-A. Using Shannon's formula

$$rate = \frac{1}{2} \log(1 + SNR)\ \text{b/s/Hz} \tag{2.19}$$

gives an achieved rate as

$$r_{HD} = \frac{1}{4} \log\left(1 + \frac{P\, b\, l^{-\alpha}}{P_n}\right) \approx \frac{1}{4} \log\left(\frac{P\, b\, l^{-\alpha}}{P_n}\right) \tag{2.20}$$

since typically $\frac{P\,b\,l^{\alpha}}{P_n} \gg 1$. The extra factor of 1/2 in (2.20) is a consequence of the two links, A-B and B-A having to operate on time-share using the medium access, since they cannot simultaneously transmit.

Now consider the full-duplex operation where the transmit and receive antennas are at distance d. The self-interference generated is $b\,d^{-\alpha}$. Denote the fraction of interference successfully cancelled as γ (e.g. $\gamma = 25\,dB$) with analogue RF signal interference cancellation, and the remaining RF interfering signal is considered as RF noise and is cancelled using RF noise cancellation circuits.

The residual self-interference is then $\frac{P\,b\,d^{-\alpha}}{\gamma}$. Again, the scenario is symmetric and we have

$$r_{FD} = \frac{1}{2}\log\left(1 + \frac{P\,b\,l^{-\alpha}}{\frac{P\,b\,d^{-\alpha}}{\gamma} + P_n}\right) \approx \frac{1}{2}\log\left(\frac{P\,b\,l^{-\alpha}}{\frac{P\,b\,d^{-\alpha}}{\gamma} + P_n}\right) \tag{2.21}$$

But $\frac{P\,b\,d^{-\alpha}}{\gamma} \gg P_n$, so

$$r_{FD} \approx \frac{1}{2}\log\left(\frac{l^{-\alpha}}{d^{-\alpha}}\gamma\right) \tag{2.22}$$

It is clear from Eq. (2.22) that the rate performance of the FD scheme depends on the propagation attenuation parameter α (and hence on carrier frequency) and it is independent of the transmit power P. Since α tends to increase with frequency, the rate achieved with FD typically decreases with increasing frequency.

2.6.2 FD Transmission Between Two Nodes

A great deal of work, in both industry and academia, has been made to find the best self-interference cancellation (SIC) method and several research groups have demonstrated their new SIC schemes in real world environments. Researchers have built a CMOS integrated circuit (IC) to mitigate the interference between Bluetooth and IEE 802.11 radios operating in close proximity [15]. The IC is tuneable and controlled by a closed loop for adaptive cancellation. It consists of a reference signal that is correlated with the interference signal to initiate the cancellation, followed by a filtering process. It is reported that the proposed method achieved 15–30 dB interference cancellation.

Microsoft Research groups joined with researchers from top US universities to propose an FD scheme for indoor WiFi networks using low transmit power, to improve on battery life time, and operating in the 2.5 and 5 GHz unlicensed spectrum [16]. The authors developed the Quellan QHx220 noise canceller to cancel most of the self RF analogue interference, i.e. to cancel 30 dB using the RF analogue interference cancellation. Researchers have also investigated the challenges facing FD MAC, such as end-to-end delay and network congestion [17]. These scenarios, in enabling the FD network, are mainly due to the finite resolution of the ADC devices.

To fully recognize the self-interference implications on the performance of FD communication, consider FD communication between two nodes shown in Figure 2.11. Assume the two-way link is symmetric but the transmit and receive antennas are physically different.

The signal transmitted from node-1 antenna T1 and received by node-1 antenna R1 will be 20 to 100 dB higher than the signal of interest received from node-2 transmit antenna T2 and received by node-1 antenna R1. The voltage generated at node-1 receive antenna R1 is, the sum of the two received signals, and is converted down to BB and scaled so the *sum occupied the full range* of the ADC used when converted to digits ($\pm$1 bits).

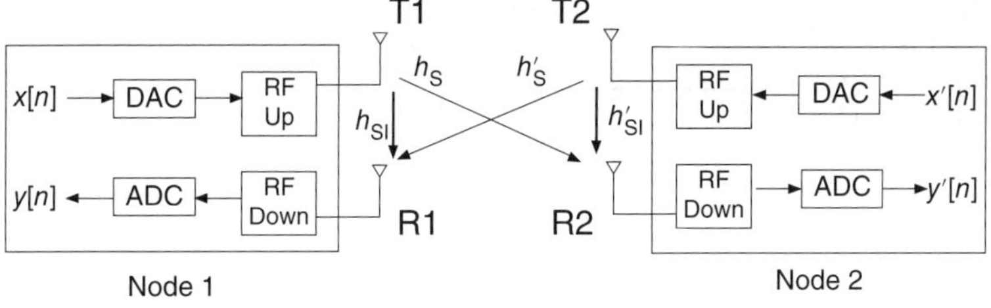

Figure 2.11 A full-duplex transmission between two nodes [17].

Since the received signal of interest from node-2 antenna T2 is much smaller than the self-interfering signal received from node-1 antenna T1, it gets fewer bits compared to the case when the signal of interest arrives at the ADC by itself. That is because when the desired signal arrives free of interference, the AGC scales the desired signal to the full dynamic range of the ADC. So even if the SNR of the received signals is individually high, the effective SNR of the received digital desired signal will be low. Consequently, the FD communication performance should be determined in terms of *SINR*.

The authors of [18] indicated that if the self-interference analogue signal is cancelled before the interfering signal is received by the receiver front-end, then the FD system can achieve throughput rates higher than the rates achieved by an HD system with identical resources. They proposed the use of a combination of three interference cancellation schemes, namely: *antennas separation, RF analogue interference cancellation,* and *digital interference cancellation.* In particular, they also experimented with the schemes: *antennas separation and digital cancellation*; a*ntennas separation and RF analogue cancellation*; and the combination of *antennas separation, RF analogue cancellation,* digital cancellation. Furthermore, the authors proposed the use of two antennas at each node – one antenna used for transmission and the other antenna used for reception. The two antennas are separated by appropriate distances to secure the required reduction in the interfering signal by antennas separation due to expected path propagation loss. In their experiments, the authors considered transmission powers from −5 to 15 dBm, distances of 20 and 40 cm between interfering antennas, and a distance of 6.5 m between nodes. More specifically, they implemented a combination of the following *three self-interference cancellation mechanisms*: passive cancellation using antennas separation; active RF analogue cancellation; followed by active BB digital cancellation to achieve more than 70 dB reduction in the interfering signal power.

2.6.3 Principles of Self-Interference

A sophisticated three stages infrastructure to enable FD transmission by reducing the SIC by 70 dB or more is proposed in [19], [20], [21] and [22]. Again, the infrastructure includes the flowing stages: *antennas cancellation, hardware cancellation,* and *digital cancellation.* The basis for high self-interference can be explained using Figure 2.12, which consists of two transceivers. Transmitter TX is transmitting desired data to the receiver RX in the other transceiver at a certain distance away. The desired signal is received as a weak signal. At the same time, another transmitter TX is transmitting a desired signal but received by its own receiver as self-interfering signal, which is about 70 dB stronger than the desired receive signal from the first transmitter TX.

Ideally, the antenna cancellation employs two identical transmitters sending similar signals to the receiver distance, d and $d + \frac{\lambda}{2}$, respectively, as shown in Figure 2.12b. The signals sent by the two transmitters arrive at the receive antenna with 180° phase difference, so theoretically they

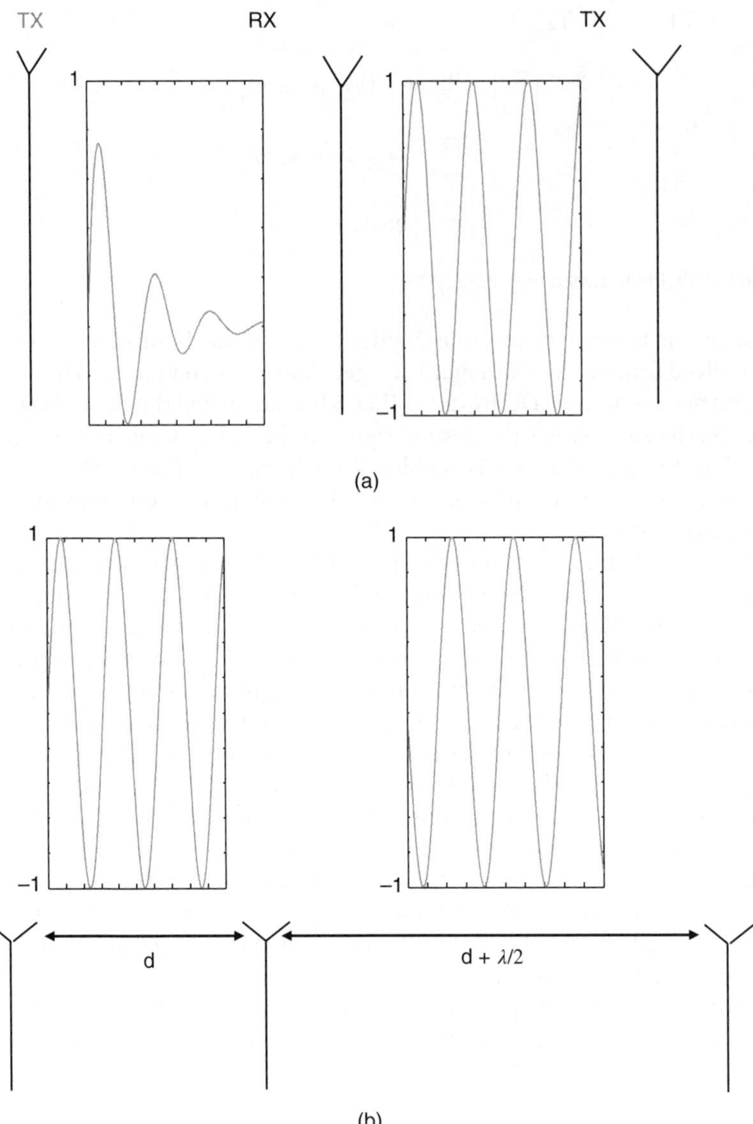

Figure 2.12 Block diagram showing (a) the self-interference principle; (b) ideal cancellation of self-interference.

should be cancelled out. However, complete interference cancellation requires the signal amplitudes from the two transmit antennas to arrive at the receive antenna to be exactly equal, which has proved to be complicated to achieve in practical systems. The self-interference cancellation (SIC) architecture is depicted in Figure 2.13. The architecture comprises two stages of self interference cancellations (SICs) namely the RF analogue cancellation before it hits the LNA to prevent receiver saturation and the digital cancellation that operates on the digital BB IQ samples between the transceiver and the BB modem to fully cancel the remaining interference. The RF analogue cancellation employs a special RF cancellation chip. Figure 2.13 exhibits the receive R antenna of a given transceiver (TRX) and receives the desired weak signal (grey line) and the same antenna receives a strong interfering signal from its own transmitter simultaneously.

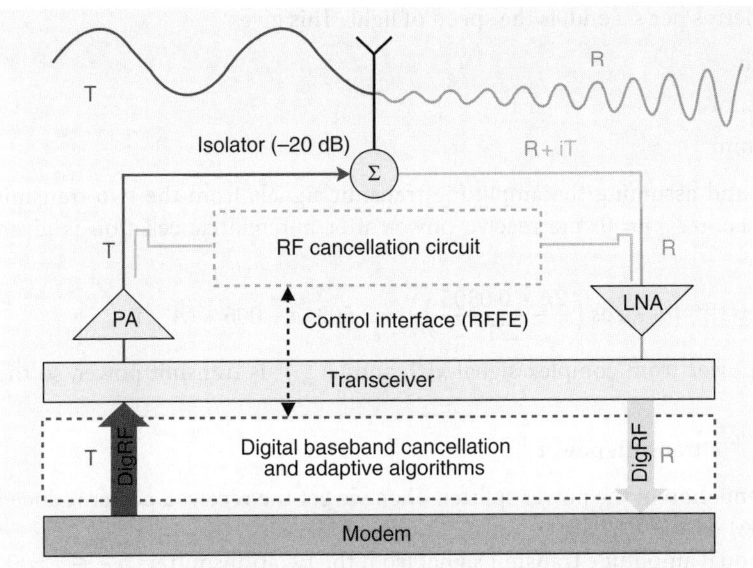

Figure 2.13 Block diagram of a wireless full-duplex transmission [20].

The complete received signal is subjected to RF cancellation before it hits the input of the LNA and followed by digital BB cancellation.

2.6.4 Theoretical Example Analysis of Antenna Cancellation

Consider the antenna cancellation in a ZigBee system 5 MHz signal bandwidth operating within frequency band 3 (2.4–2.4835 GHz). Let the system be equipped with two transmit antennas positioned at $-d$ and d from a receive antenna placed at the centre with one transmit signal identical to other but with π phase shift. Theoretically, if the transmitters are positioned at the receive antenna placement, then receive power is zero since transmit powers cancelled at the receive antenna. Let the receive antenna positioned at the centre and operated at 2.48 GHz. Thus, the signal has frequency components between 2.4775 and 2.4825 GHz, as shown in Figure 2.14.

In the following, we consider the effect of the received signal bandwidth on antenna self-interference cancellation. The difference in wavelengths can be mapped as an error in receiver position. For example, let δ be the difference in wavelengths. Then it corresponds to a $\delta/4$ error in receiver position. Let $\in_{ant}^{d}$ be the error in receiver antenna position compared to the ideal case where the signals from the two antennas arrive π out of phase of each other. Thus, $\in_{ant}^{d}$ for 2.4775 GHz in this case would be

$$\in_{ant}^{d} \sim \frac{1}{4}\left(\frac{c}{2.4775 * 10^{9}} - \frac{c}{2.48 * 10^{9}}\right) \tag{2.23}$$

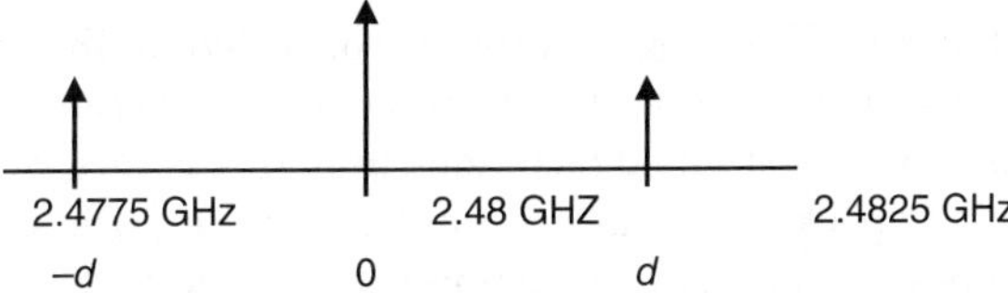

Figure 2.14 The receive antenna positioned at the centre and operated at 2.48 GHz, and signal with frequency components between 2.4775 GHz and 2.4825 GHz.

where $c = 299792458$ metres per second, is the speed of light. This gives

$$\in^{\mathrm{d}}_{\mathrm{ant}} = 0.0305\ \mathrm{mm}$$
$$\lambda_{\mathrm{f1}} = 121.0060\ \mathrm{mm}$$
$$\lambda_{\mathrm{f2}} = 120.8841\ \mathrm{mm}$$

Using Appendix 2C, and assuming the amplitude transmit signals from the two transmit antennas are matched i.e. $(\in^{A}_{ant}) = 0$, the receive power after antenna cancellation is given by (2.C.7) as

$$P_r = 2A_{ant}(A_{ant})|x[t]|^2\left(1 - \cos\left(\frac{2\pi * 0.0305}{121}\right)\right) = 2.5084e - 006 * (A_{ant})^2$$

where $|x[t]|^2$ is a unit power from complex signal $x[t]$, and $(A_{ant})^2$ is transmit power, so the received signal is

$$\mathrm{P_r} = 2.5084 * 10^{-6} * \text{transmit power}$$

Now we add 6 dB attenuation of the power splitter. Then we get the received power reduced by 10*log10(2.5084e-006) + 6 dB = 62 dB.

For $\in^{d}_{ant} = 1\ mm$, for equal amplitude transmit signal from the two transmitters, i.e. $\in^{A}_{ant} = 0$

$$2A_{ant}(A_{ant} + \in^{A}_{ant})|x[t]|^2\left(1 - \cos\left(\frac{2\pi \in^{d}_{ant}}{\lambda}\right)\right) + (\in^{A}_{ant})^2\,|x[t]|^2$$

Substitute $\in^{A}_{ant} = 0$ and $|x[t]|^2 = 1$

$\lambda = 12.1\ cm$ for centre frequency 2.48 GHz.

$$= 2A_{ant}{}^2\left(1 - \cos\left(\frac{2\pi * 1\ mm}{120.8841\ mm}\right)\right)$$

The received power is reduced by 31.5284 dB.

Thus, we can assume that the antenna cancellation would be ~30 dB.

2.6.5 Infrastructure for FD Transmission

The complete infrastructure for FD transmission was depicted in Figure 2.13. The antenna cancellation scheme reduces the interference from the transmitter's own node but the transmit power from the two antennas are unequal such that the scheme causes a maximum difference of transmit powers of 6 dB at any receiver location. When the transmit powers are unequal, the received powers are different and the resulting signal is not zero power. However, in a practical network, transmit antennas diversity gains would offset the 6 dB degradation.

The second stage in SIC is the *hardware cancellation*, also known as *RF interference cancellation* using the interference cancellation circuit based on QHx220 noise cancellation chip. The inputs to the chip are both the known self-interference signal and the received signal, the output is the received signal with self-interference taken off. The chip enables adjustment of the amplitude and phase of the reference interference signal and matches the reference interference to the interference in the received signal. It was reported that the hardware cancellation achieves ~ 20 dB reduction in self-interference in addition to the antenna cancellation discussed above. The chip RF analogue output is converted to BB, sampled using ADC and converted to digits.

The final stage in the SIC process is the *digital interference cancellation*. Digital cancellation is a long-established technique used by the receiver to extract a desired packet after it collides with a packet from an unwanted transmitter. The strategy used then was as follows: decodes the unwanted packet first, remodulates it, and then subtracts the analogue unwanted signal from the received collide signal. In FD case, the transmitted symbols causing the self-interference are known as constructing a clean signal is not required. Furthermore, typical digital cancellation requires compensation for clock drift between transmitter and receiver, but since the user transceiver shares the same clock, there should be no clock drift in FD systems. However, transmitter and receiver may use separate PLL system, thus some jitter is expected and digital cancellation achieves 10–20 dB reduction in addition to the other two stages, which gives a total cancellation of ~70 dB.

In addition to in-band FD technology in cellular networks, it was reported in [23] and [24] that the design and testing of laboratory prototype mobile devices supporting simultaneous transmission and reception (i.e. FD) at the same operating frequency. The authors explored the key challenges in implementing mobile in-band FD devices and offered solutions to these problems.

A Two-tier HetNet system model is illustrated in Figure 2.15. The model comprises a macro cell (MC) and a set of S small cells (CSs). The MC BS is equipped with a few transmit antennas M and massive receive antennas N (i.e. $N \gg M$). In addition, each MC transmit antenna transmit a single stream of data to the SC BS such that $M = S$. Each SC BS is equipped with a single antenna and N_{sc} receive antennas. Communications between the system tiers is achieved in two phases. Figure 2.15 (a and b) describes the first phase and the second phase communications respectively. Figure 2.15 (c) featured the coherence interval of T symbols. The T symbols consists of τ_p symbols are used for the channel training with the help of pilots broadcasting. The remaining coherence slots are divided equally into the first phase and the second phase data transmissions. Here we assume TDD transmission protocol can be applied to precode data for transmission.

During the first phase time, each SC BS transmits its data on UL to the MC BS. Concurrently, the kth antenna at the MC BS transmits independent stream of data to the kth SC BS as shown in Figure 2.15 (a). At the time of the first phase, the MC BS receives the desired data streams from the SC BSs on the UL as well as its own transmissions causing self-interference attributed to the FD mode of operation. Likewise, each SC BS receives the desired data from the MC BS as well as its own transmissions and the unwanted transmissions from nearby SC BSs causing SI and SC 2 SC interference respectively. The second phase is described by Figure 2.15 (b). The MC BS precodes its data and transmits independent stream of data to every SC BS. At the same time, each SC BS precodes its data and sends to a corresponding receive antenna on the MC BS. As we describe above, the kth antenna at the MC BS receives the desired data from the kth SC BS as well as its own transmissions causing SI interference. In a like manner, each SC BS is subjected to SI as well as SC 2 SC interferences. Furthermore, the DL in the second phase is determined by the conjugate transpose of the UL for the first phase and vice versa.

The two-tier system model using FD transmission method is clearly constrained by the SI and SC-2 – SC interferences which introduces great limitation of the UL and DL transmission rates [35]. Figure 2.16 (a) depicted a graph of the theoretical UL sum rate vs SI for FD and HD modes. The figure shows for FD the sum rate decreases expeditiously as the SI coefficient σ_m^2 increases from 0 to 1 but for the HD mode the sum-rate is fixed since HD mode does not generate SI interferences. Furthermore, for low SI interference, the FD outperforms the HD but for high SI interference the HD starts to outperform the FD system and there exists a crossover point between FD and HD modes.

Figure 2.16 (b) shows a similar graph plotted for DL sum rate vs. SI. For the DL sum rate, both the SI strength σ_s^2 and SC 2 SC interference σ_c^2 have to be taken into account. In Figure 2.19 (b), the strength of both interferences is adjusted concurrently and their joint effect on the achievable DL sum rate is depicted. Just like the UL, a crossover point at which FD and HD achieve equal DL sum-rate. The parameters used in the calculation of Figure 2.16 are: τ_p is the

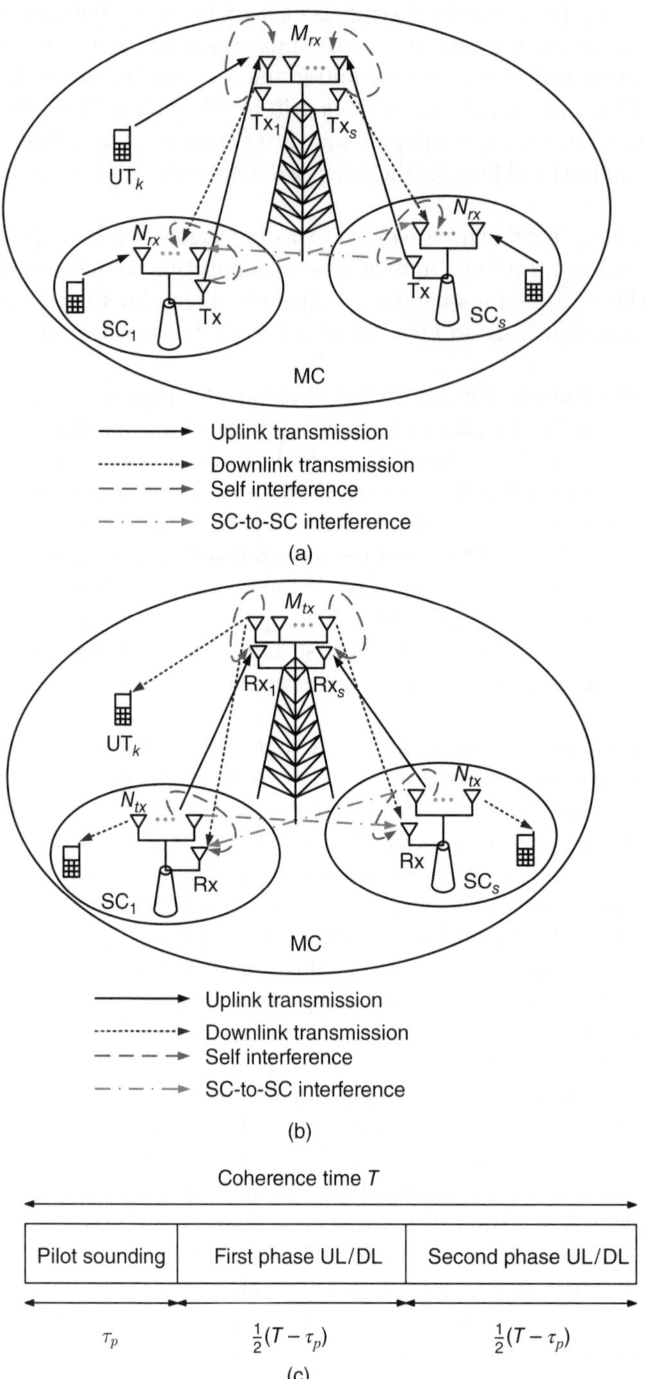

Figure 2.15 Two-tier HetNet system model (a) First phase, (b) Second phase, (c) Coherence interval of length T [35].

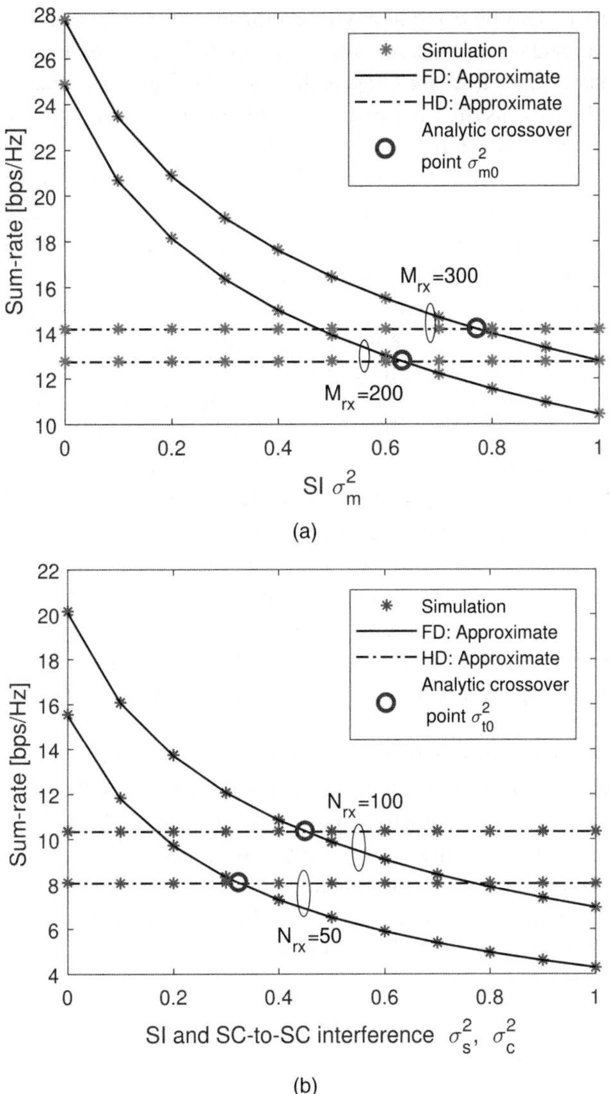

Figure 2.16 (a) UL sum rate versus SI; (b) DL sum rate versus SI and SC 2 SC [35].

pilot sequence length, p_τ is the pilot sequence power, β_1 and β_2 are the variance of the complex large scale fading for UL and DL respectively, and σ_m^2 the strength of the SI at the MC BS.

2.6.6 Full-Duplex MAC (FD-MAC) Protocol

The challenges in designing full-duplex MAC (FD-MAC) protocol and the modification required in the physical layer (PHY) to engage in a full duplex mode are discussed in [17]. Cellular wireless networks generally have their communications generated either from access point (AP) or a destination source. There are three challenges to FD-MAC protocol design. In any multi-node network, there will be multiple data flows at random instants and with random arrivals. So, the *first challenge* is the identification of the nodes that engage in FD mode. The *second challenge* is required by the PHY. The FD can be performed either synchronously between two nodes exchanging packets or can be done in asynchronously when the FD node *receives a packet while transmitting* a packet to another node. The MAC protocol has to implement these constraints in the design. The *third challenge* is that FD-MAC protocol has to provide equal opportunities to all nodes to access the medium whether using HD or FD modes. At any instant, there will be a maximum of two active flows amongst FD capable nodes, as depicted in Figure 2.17. Figure 2.17a shows the simplest scenario when an AP and a mobile user M_1 are exchanging packets and in Figure 2.17b both nodes are connected to an AP but out of range of each other. The AP is sending/receiving packets simultaneously for M_1 and M_2, which are not in the radio range of one another, Figure 2.17c shows the nodes in the radio range of each other so there is the possibility of four flows in the network at the same time as all nodes are in radio range of each other: $AP \rightleftarrows M_1$ and $AP \rightleftarrows M_2$ and Figure 2.17d shows node M_3 and node M_2 are hidden to node M_1 so the latter and a single hidden node may send packets to AP at the same time and end up colliding at AP since AP can only receive one packet at a time. So, it is important to have a mechanism to avoid such collisions. We consider case (a) and (b) first. In general, FD-MAC is a random-access protocol employing most key elements of 802.11 with CSMA/CA collision avoidance to select the FD opportunities for transmission. When AP and M_1 in Figure 2.17a have many packets for each other using FD mode, they should not be continuously capturing the medium to allow other nodes to send and receive packets from AP. Ensuingly, they should agree on a *shard random back-off* to allow other nodes to gain access to the medium. If the medium is not used at the end of the back-off period, then AP and M_1 can resume their FD transmission. In FD-MAC, nodes decode headers of all ongoing transmissions to allow them to initiate FD opportunistically. This scheme is known as *snooping to discover FD opportunities.*

Anode starts to send packets in random slots of time to decrease the probability of collisions. However, collisions may still occur if two or more nodes select the same back-off slot. When this happens, these nodes have to re-enter the competition with possibly an exponentially backoff, which incurs a high collision probability and degrades channel utilization, especially in congested scenarios. The *contention resolution algorithm* is to provide a fair channel access and a balance usage of the FD mode with access to all nodes in the network. The 802.11 infrastructure does not use the *request to send* and *clear to send* (RTS/CTS) flow control, but FD-MAC can be used with or without RTS/CTS flow control.

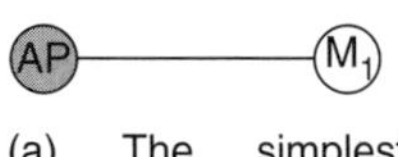

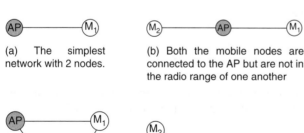

(a) The simplest network with 2 nodes.

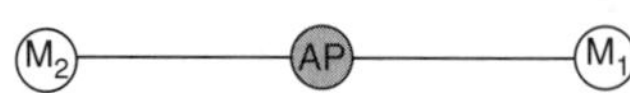

(b) Both the mobile nodes are connected to the AP but are not in the radio range of one another

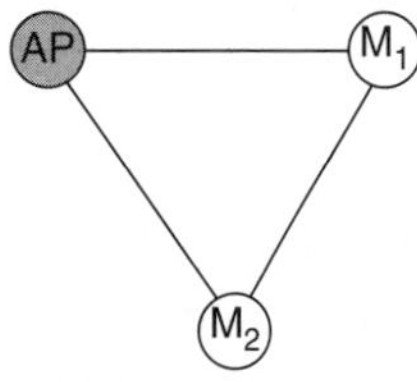

(c) All three nodes are in radio range of each other

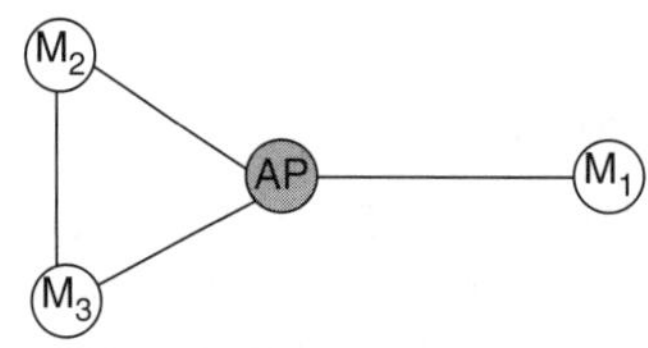

(d) M_2 and M_3 are hidden to M_1

Figure 2.17 A line connecting any two nodes indicates that they are in radio range of one another [17].

The hidden node scenario shown in Figure 2.17b is considered next. Nodes M_1 and M_2 are connected to the AP but are not in radio range of one another. The possible flows that take place are as follows: two FD flows $AP \rightleftarrows M_1$, $AP \rightleftarrows M_2$, in addition to $\{M_2 \rightarrow AP, AP \rightarrow M_1\}$ and $\{AP \rightarrow M_2, M_1 \rightarrow AP\}$. Each of these four FD modes, whether two-way exchange or otherwise, *starts with HD mode*. However, to ensure that transition to all four FD modes, FD-MAC has to ensure that the HD mode required to kick start is possible. As mentioned above the HD modes contend amongst themselves using 802.11 type contention scenarios and the two FD modes have a period of shared random back-off with opportunities for other HD modes to occur. The FD-MAC protocol therefore allows all modes to occur and to switch between various modes must have the required mechanisms to support and build in the protocol.

Consider Figure 2.17c where there are four flows in the network, $AP \rightarrow M_1$, $AP \rightarrow M_2$, $M_1 \rightarrow AP$, $M_2 \rightarrow AP$. The four flows bring the possibility of two full-duplex scenarios $AP \rightleftarrows M_1$ and $AP \rightleftarrows M_2$. Each of the FD transitions is enabled by features introduced by the FD-MAC protocol, which make the transition possible only through HD modes. This is because the *first packet in FD exchange is always HD packet*.

Consider the mode $AP \rightleftarrows M_1$. From this FD mode, the network can be changed to HD mode owing to the following different events: (i) when either AP or M_1 has no more packets in the buffer to send to the other. Both AP and M_1 are expected to give up FD mode (ii) when DATA or ACK packets are not decoded correctly, other nodes start 802.11 type contention (iii) the packets towards M_2 destination win the virtual contention resolution and the network will break away from the FD and enter HD mode $AP \rightarrow M_2$ (iv) M_2 wins the physical contention during silent shared back-off period initiating $M_2 \rightarrow AP$. Remember that all HD can switch amongst each other using 802.11 protocol. Consider the HD mode $M_1 \rightarrow AP$, the only possible transition is to FD mode $AP \rightleftarrows M_1$. The *FD-MAC protocol* allows all modes to occur by switching between various modes through scheme introduced by FD-MAC together with existing 802.11 capability.

Consider now the case where multiple nodes are snooping of ongoing transmission by AP to initiate FD opportunistically, as shown in Figure 2.17d, both M_2 and M_3 are hidden from M_1. Furthermore, M_2 and M_3 unaware of how many nodes there are, which may contend and hence when each detects FD opportunity, sends a packet to AP with probability p_i , where p_i is computed based on the current maximum back-off window as

$$p_i = \frac{\beta}{CW_{max}} \tag{2.24}$$

where β is a pre-chosen constant. The popular choice is to use $p_i \propto \frac{1}{CW_{max}}$ since each node can then use its current *maximum contention window*, CW_{max}, as an indicator for amount of expected competition in the system. In fact, one could fix $p_i = p$ where p is pre-chosen to allow equal opportunity for each node.

Packet structure to be used for FD-MAC has a new FD header, as shown in Figure 2.18. All these fields except FD header are identical to the corresponding fields in 802.11 packets. Each FD-MAC packet contains the following headers: *PHY header, MAC header (802.11MAC header and FD header), Payload (DATA) header*, and a cyclic redundancy check (*CRC*). We now briefly explain the fields in the proposed FD-MAC packet. The PHY header contains a preamble and the training symbols required for the functioning of the PHY layer. The existing elements of the MAC header proposed to be used in FD-MAC protocol are: duration of the FD-MAC packet (DUR); source address (SA); destination address (DA); and fragments of the current packet are still in line for destination (FRAG). The FD-MAC header differentiates between data packet and acknowledgement by referring to data packets as DATA and acknowledgement as ACK.

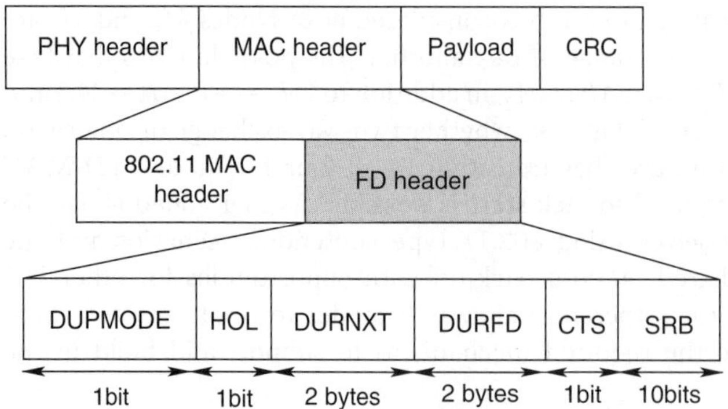

Figure 2.18 Structure of the packet being used for the FD-MAC protocol [17].

The FD header contains six fields and they are: a one-bit field (DUPMODE) to specify the packet mode type (HD or FD); one-bit field *head of line* (HOL) indicating the next packet in the buffer is for the destination of the current packet. HOL is used to avoid conflict with other using this field and corresponds to the one-bit field more data in 802.11 MAC header; the next is of 2 bytes field duration of the next packet (DURNXT); the next 2 bytes field discloses the duration of the FD opportunity (DURFD); the CTS is represented by a one-bit field to indicate that the destination of the current packet is clear and can send a packet to the source of the current packet; and finally a 10 bits field indicating the shared random back off (SRB). Field DURNXT and DURFD are required to counter the hidden node problem but they are *optional* and if not used, the FD header is only 13 bits.

The conclusions one could draw from these presentations are that current SIC solutions generally require costly hardware design and complex computations, and the use of FD technology as an efficient enabler in 5G, is requiring more work to provide practical FD devices that are cost efficient with low complexity algorithms [23], in addition to high performance low complexity practical FD protocols.

2.7 Review of Reference Signals, Antenna Ports, and Channels

It is important to comprehend the relationships between antenna ports, physical antennas, and transmission channels. 3GPP standards introduced what currently been known as antenna ports. These antenna ports are not akin to physical antennas. Antenna ports are originally related to a logical concept representing specific channel models created by transmission of reference signals assigned to the antenna ports. Each antenna port has its own reference signal(s). Each reference signal is assigned one or several antenna ports but no different reference signals are assigned the same port. In addition, multiple antenna ports can transmit on a single antenna.

A reference signal is a special sequence of bits that is used only in the physical layer. Unlike other information signals, reference signals carry no information and their use delivers to the UE reference point for the DL.

An important task of the reference signal is to enable the receiver to demodulate the received signal. Since the reference signal is known to both the transmitter and the receiver, we can compare the decoded received reference signal with the defined reference signal and use the results to estimate the communication channel and decode the received data. The implementation of the reference signal runs through two steps: First generation of the signal (i.e. sequence bits) and second resource allocation. To each bit of the sequence transmission, specific resource elements are allocated. Conventionally, the DL reference signals are generated using Gold sequences while the UL reference signals are generated using Zadoff-Chu sequences. Details on the antenna ports with the generation of their assigned reference signals are given in 3GPP Releases.

To date, 3GPP standardized the LTE systems in 15 Releases. The number of antenna ports increases with almost every new release. In addition, 3GPP defined the data sequences transmitted or received over a channel in a MIMO transmission as *stream*. Increasing antenna ports enables an increase in the number of different data streams to be simultaneously transmitted from each antenna port, which maximizes the DL system capacity.

2.7.1 DL and UL Physical Channels

There are six DL physical channels in Releases 8, 9, and 10, specifically:

1. Physical DL Shared Channel (PDSCH) is the main data bearing channel allocated to the user terminal on a dynamic and opportunistic basis.
2. Physical Broadcast Channel (PBCH) carries system parameters in a master information block of 14 information bits for user terminal requiring to access the network.
3. Physical Multicast Channel (PMCH) is used for multicasting a packet to a set of UEs, instead of a single UE.
4. Physical Control Format Indicator Channel (PCFICH) is used to inform the user terminal about the format of the signal being received, this is important because the user has no prior knowledge of the size of the control message.
5. Physical DL Control Channel (PDCCH) to carry DL resource, UL power control, UL resource grant, and paging scheduling.
6. Physical Hybrid ARQ Indicator Channel (PHICH) to report the ARQ status whether a transport block has been correctly received.

A key enhancement introduced by 3GPP in Release 11 is a new DL control channel. The Enhanced Physical DL Control Channel (EPDCCH) increases the control channel's capacity and supports frequency-domain inter-cell interference coordination (ICIC) and beamforming and / or diversity. Importantly, it coexists with legacy devices. Release 13 included a further important enhancement in the introduction of the machine-type communications (MTC) Physical DL Control Channel (MPDCCH), which replaced the EPDCCH. MPDCCH is used to transport common or user-specific signalling transmitted once or with repetition numbers that are configured by higher layer several different signalling messages, depending on the applications.

There are three physical UL channels, namely: Physical UL Shared Channel (PUSCH) used to transport radio resource control (RRC) signalling messages and application data; Physical UL Control Channel (PUCCH) to transport UL control information; and Physical Random-Access Channel (PRACH) to transport random access preamble a user terminal sends to access the network.

2.7.2 DL Reference Signals and Antenna Ports

Release 8 [25] contains a set of three *DL reference signals* (RS), specifically:

1. *Cell-specific reference signals* (C-RS) are transmitted in all DL subframes in a cell supporting physical PDSCH transmission and support a configuration of one, two, or four antenna ports

with antenna port number $p = 0, p \in \{0, 1\}$ and $p \in \{0, 1, 2, 3\}$. Each antenna port has a unique C-RS associated with it. C-RSs are used for cell search and initial acquisition, downlink channel estimation for coherent demodulation/detection at user's receiver, and downlink channel quality measurements.

2. *Multicast / Broadcast data over single frequency network reference signals (*MBSFN-RS) are only transmitted during MBSFN subframes, MBSFN reference signals on $p = 4$. Multiple cells that transmit the same content to multiple users form a MBSFN area. Each cell could be part of 8 MBSFN areas. There could be 256 different MBSFN areas.
3. *UE-specific reference signals (UE-RS)* are supported for single antenna port transmission of PDSCH and are transmitted on $p = 5$.

3GPP introduced an important enhancement in Release 9 [26] to the UE-RS (i.e. the spatial multiplexing). UE-RS are supported for a single antenna port of PDSCH (i.e. single layer beamforming) and are transmitted on $p = 5, 7, 8$. UE-RS also supported spatial multiplexing on $p = 7$, 8 to the UE-RS for transmission on the PDSCH (double-layer beamforming). positioning reference signal (P-RS) is another enhancement of the LTE Release 9 to determine the location of the UE based on radio access network information. P-RS are transmitted on $p = 6$. The P-RS is transmitted periodically in certain frames and occupies certain resource elements, as defined by the P-RS parameters.

3GPP launches a significant feature in Release 10 [27], the channel state Information reference signals (CSI-RS). CSI reference signals are transmitted on one, two, four, or eight antenna ports using $p = 15, p = 15, 16, p = 15, \ldots, 18$ and $p = 15, \ldots, 22$, respectively. CSI-RSs are known to BS and UE and used to measure the channel state, e.g. for multiple antenna cases. The measured CSI are feedback to BS, based on a separate set of reference signals. CSI reference signals are regularly transmitted from all antennas at the BS.

3GPP initiates an important feature in Release 11 [28] while keeping the same number of ports as in [27], introduces the demodulation reference signal associated with EPDCCH and transmitted on antenna port $p \in \{107,108,109,110\}$ as associated EPDCCH physical resource.

Release 13 [29] increases the CSI-RS, adding more antenna ports so $p = (15, \ldots \ldots, 26)$, $p = (15, \ldots \ldots, 30)$, and initiated the MTC demodulation reference signals for MPDCCH, demodulation reference signal associated with EPDCCH or MPDCCH and transmitted on the same antenna port $p \in \{107,108,109,110\}$ as associated with EPDCCH or MPDCCH physical resource. These reference signals are generated from a Gold sequence with length of 31, where each reference signal is in initialized differently.

2.7.3 UL Reference Signals [30]

UL reference signals supported are of two types: demodulation reference signals (D-RS) associated with transmission of PUSCH and PUCCH, and sounding reference signal. The *same set of base sequences* is used for demodulation and sounding reference signals. These sequences are generated based on roots of Zadoff-Chu sequence and the base sequences are divided into 30 groups, and the sequence lengths of each set of groups are 12 and 24 sequence elements.

2.7.3.1 UL Reference Signal Sequence Generation

UL reference signal sequence $r_{u,v}^{(\alpha,\delta)}(n)$ is defined by a cyclic shift α of a *base sequence* $\bar{r}_{u,v}(n)$ according to

$$r_{u,v}^{(\alpha,\delta)}(n) = e^{j\alpha\left(n+\delta\frac{\varpi \bmod 2}{2}\right)}\bar{r}_{u,v}(n), \quad 0 \le n < M_{sc}^{RS} \tag{2.25}$$

where $\delta = 1$ is set for demodulation reference signal associated with PUSCH transmission and $\delta = 0$ otherwise. Multiple reference signal sequences are defined from a single base sequence through different values of α. M_{sc}^{RS} is the length of the reference signal sequence, and $\varpi \in [0, 1]$ associated with the cyclic shift in the UL-related DL control information. Different values of α define multiple reference signal sequences. Base sequences $\bar{r}_{u,v}(n)$ are divided into groups, where $u \in \{0, 1, \ldots, 29\}$ is the group number and v is the base sequence number within the group. For reference signal sequence of length $M_{sc}^{RS} \geq 3N_{sc}^{RB}$, the base sequence $\bar{r}_{u,v}(0), \ldots\ldots, \bar{r}_{u,v}(M_{sc}^{RS} - 1)$ is given by

$$\bar{r}_{u,v}(n) = x_q(n \, mod \, N_{ZC}^{RB}), \quad 0 \leq n \leq M_{sc}^{RS} \tag{2.26}$$

where $x_q(m)$ is the qth root Zadoff-Chu's sequence. When the reference signal sequence of length $M_{sc}^{RS} = N_{sc}^{RB}$ and $M_{sc}^{RS} = 2N_{sc}^{RB}$, $M_{sc}^{RS} = \frac{N_{sc}^{RB}}{2}$, and $M_{sc}^{RS} = \frac{3N_{sc}^{RB}}{2}$ the base sequence is given by

$$\bar{r}_{u,v}(n) = e^{j\,\varphi(n)\frac{\pi}{4}}, \quad 0 \leq n \leq M_{sc}^{RS} - 1 \tag{2.27}$$

where N_{sc}^{RB} is the number of resource blocks allocated expressed in subcarriers.

The sequence-group number u in slot n_s is defined by a group hopping pattern $f_{gh}(n_s)$ and a sequence-shift pattern f_{ss} according to

$$u = (f_{gh}(n_s) + f_{ss}) \, mod \, 30 \tag{2.28}$$

The group-hopping pattern $f_{gh}(n_s)$ may be different for PUSCH, PUCCH, and SRS and is given by

$$f_{gh}(n_s) = \begin{cases} 0 & \textit{if group hopping is disabled} \\ \left(\sum_{i=0}^{7} c(8n_s + i).2^i\right) mod \, 30 & \textit{if group hopping is enabled} \end{cases} \tag{2.29}$$

where $c(i)$ is defined by pseudo-random sequences, which are defined by a length-31 Gold sequence.

2.7.3.2 Demodulation Reference Signal for PUSCH

The PUSCH demodulation reference signal sequence $r_{PUSCH}^{(\lambda)}(.)$ associated with transmission layer $\lambda \in \{0, 1, \ldots \ldots, v-1\}$ is given by

$$r_{PUSCH}^{(\lambda)}(m \,.\, M_{sc}^{RS} + n) = w^{(\lambda)}(m)\, r_{u,v}^{(\alpha_\lambda\, \delta)}(n) \tag{2.30}$$

where

$$\mathrm{m} = 0, 1$$

$$\delta = 0 \text{ or } 1 \text{ as before}$$

$$\mathrm{n} = 0, \ldots\ldots, \mathrm{M}_{\mathrm{sc}}^{\mathrm{RS}} - 1$$

and

$M_{sc}^{RS} = \frac{M_{sc}^{PUSCH}}{2}$ if higher -layer parameter is set and $M_{sc}^{RS} = M_{sc}^{PUSCH}$ otherwise. The sequences $r_{u,v}^{(\alpha_\lambda\, \delta)}(0), \ldots\ldots\ldots, r_{u,v}^{(\alpha_\lambda \delta)}(M_{sc}^{RS} - 1)$ and sequences $w^{(\lambda)}(m)$ are given in Release 13. The vector of reference signals is precoded according to

$$\begin{bmatrix} \tilde{\mathrm{r}}_{\mathrm{PUSCH}}^{(0)} \\ \tilde{r}_{\mathrm{PUSCH}}^{(p-1)} \end{bmatrix} = W \begin{bmatrix} r_{\mathrm{PUSCH}}^{(0)} \\ r_{\mathrm{PUSCH}}^{(v-1)} \end{bmatrix} \tag{2.31}$$

where p is the number of antenna ports used for PUSCH transmission using a single antenna port, $p = 1$, $W = 1$ and $v = 1$, for spatial multiplexing $p = 2$ or $p = 4$, and W is precoding matrix of size p x v.

2.7.3.3 Demodulation Reference Signal for PUCCH

The PUCCH demodulation reference signal sequence, $r_{PUCCH}^{(\tilde{p})}(.)$ is given by

$$r_{PUCCH}^{(\tilde{p})}(m' N_{RS}^{PUCCH} M_{SC}^{Rs} + mM_{SC}^{Rs} + n) = \frac{1}{\sqrt{P}} \overline{w}^{(\tilde{p})}(m)z(m)r_{u,v}^{(\alpha_{\tilde{p}},\delta)}(n) \tag{2.32}$$

where

$$\mathrm{m} = 0, \ldots\ldots\ldots, \mathrm{N}_{\mathrm{RS}}^{\mathrm{PUCCH}} - 1$$
$$\mathrm{n} = 0, \ldots\ldots\ldots, \mathrm{M}_{\mathrm{SC}}^{\mathrm{Rs}} - 1$$
$$\mathrm{m}' = 0, 1$$

and p is the number of antenna ports used for *PUCCH* transmission. The value of $z(m)$ depends on the PUCCH format used.

2.7.3.4 Sounding Reference Signal SRS

The sounding reference signal sequence $r_{SRS}^{(\tilde{p})}(n)$ is given by

$$r_{SRS}^{(\tilde{p})}(n) = r_{u,v}^{(\alpha_{\tilde{p}}\delta)}(n) \tag{2.33a}$$

where $r_{u,v}^{(\alpha_{\tilde{p}})}(n)$ is given by (2.25) and the cyclic shift $\alpha_{\tilde{p}}$ is given as

$$\alpha_{\tilde{p}} = 2\pi \frac{n_{SRS}^{cs,\tilde{p}}}{n_{SRS}^{cs,max}} \tag{2.33b}$$

$$n_{SRS}^{cs,\tilde{p}} = \left(n_{SRS}^{cs} + \frac{n_{SRS}^{cs,max}\,\tilde{p}}{N_{ap}} \right) \bmod n_{SRS}^{cs,max} \tag{2.33c}$$

$$\tilde{p} \in \{0, 1, \ldots\ldots.., N_{ap} - 1\} \tag{2.33d}$$

where $n_{SRS}^{cs} = \{0, 1, \ldots\ldots.., n_{SRS}^{cs,max} - 1\}$ is configured by higher-layers and N_{ap} is the number of antenna ports used for sounding reference signal transmission. It should be noted that $n_{SRS}^{cs,max} = 8$ or 12, configured separately for periodic sounding.

2.7.3.5 Random-Access Channel Preambles

The random-access preambles are generated from Zadoff-Chu sequences with zero correlation zone, generated from one or several root Zadoff-Chu sequences. The network configures the set of preamble' sequences the UE is allowed to use. There are up to two sets of 64 preambles available in each cell, Set 1 and set 2 correspond to higher layers PRACH configurations. The set of 64 preamble sequences is found by including all the available cyclic shifts of a root Zadoff-Chu sequence. The uth root Zadoff-Chu sequence is defined by

$$x(n) = e^{-j\frac{\pi un(n+1)}{N_{ZC}}}, \quad 0 \leq n \leq N_{ZC} - 1 \tag{2.34}$$

where N_{ZC} is the length Zadoff-Chu sequence is (839 or 139) for preamble format 0–3 and 4, respectively. The random-access preambles of length $N_{ZC} - 1$ are defined by cyclic shift according to

$$x_{u,v}(n) = x_u((n + C_v) \bmod N_{ZC}) \tag{2.35a}$$

where the cyclic shift is given by

$$C_v = \begin{cases} vN_{Cs} & v = 0, 1, \ldots, \left\lfloor \frac{N_{ZC}}{N_{Cs}} \right\rfloor - 1, N_{Cs} \neq 0 \\ 0 & N_{Cs} = 0 \\ d_{start} \left\lfloor \frac{v}{n_{shift}^{RA}} \right\rfloor + (v \bmod n_{shift}^{RA}) N_{Cs} & v = 0, 1, \ldots, n_{shift}^{RA} n_{group}^{RA} + \bar{n}_{shift}^{RA} - 1 \end{cases}$$

The parameters defined above are

$$n_{shift}^{RA} = \left\lfloor \frac{d_u}{N_{Cs}} \right\rfloor \tag{2.35b}$$

$$d_{start} = 2d_u + n_{shift}^{RA} N_{Cs} \tag{2.35c}$$

$$n_{group}^{RA} = \left\lfloor \frac{N_{ZC}}{d_{start}} \right\rfloor \tag{2.35d}$$

$$\bar{n}_{shift}^{RA} = \max(\lfloor (N_{ZC} - 2d_u - n_{group}^{RA} d_{start}) / N_{Cs} \rfloor, 0) \tag{2.35e}$$

where the variable d_u is the cyclic shift corresponding to Doppler shift and is given by

$$d_u = \begin{cases} P & 0 \leq P < \frac{N_{ZC}}{2} \\ N_{ZC} - P & otherwise \end{cases} \tag{2.35f}$$

and P is the smallest integer such that $(Pu) \bmod N_{ZC} = 1$ and N_{cS} is the cyclic shift value used for random-access preamble generation.

2.8 Full-Dimension MIMO Technology

Full-dimension MIMO (FD-MIMO) technology is one of the key enabling technologies for the future 5G [31–33]. Work by 3GPP on the technology was started by a study item in 2012 on '3D channel model for elevation beamforming'. More recently, another study item has been completed in June 2015 to initiate the formal standardization of the technology in Release 13. One of the key features of FD-MIMO systems, as distinct from the MIMO systems in current LTE standards, is the implementation of 2D active antenna array with a large number of antennas at BS [34]. We have shown in the following chapters that as the number BS antennas increases, the development of 3D channel model comprises both elevation and azimuth areas, in addition to the expansion of the multiuser MIMO signal dimensions. These advancements bring about the possibility that any two random channel realizations became orthogonal – i.e. the cross correlation between them tends to zero so that DL linear precoding is sufficient to control the inter-user interference and UL multiuser interference can be eliminated also using a simple receiver-combiner. However, such benefits can only be secured if perfect CSI is available at the BS.

A successful commercialization of the FD-MIMO system relies heavily on the development of suitable solutions to the issues of CSI acquisitions. The two main problems related to the CSI acquisition process are: degradation of CSI accuracy due to the limitation of feedback resources; and the increase of pilot overhead. The user device employs a reference preamble signal (RS), transmitted from the BS, to estimate the channel. These RSs are assigned in an orthogonal way and the RS overhead grows linearly with the number of antennas. A recent study shows that 48% of the UL resources that are originally assigned for data transmission are used by the RS transmission.

A key attribute to the FD-MIMO system is the launch of 2D active antenna systems. Active 2D antenna system usually includes active components such as power amplifier (PA) and low noise amplifier (LNA) that can be used to control the gain and phase of each antenna element. In addition, the fact that the active antenna system has 2D structure implies that the radio signal on both vertical (elevation) and horizontal (azimuth) can be controlled simultaneously and eventually a transmitted beam can be efficiently designed. This mechanism is known by the research community as 3D beamforming.

Another important advantage of 2D antenna array is that it can accommodate a large number of antennas on a rooftop tower or antenna masts compared to linear antenna array. A 64 2D antenna array can be arranged as a square (8 × 8 array) with a relatively smaller space requirement.

A conceptual diagram realizing high-order MU-MIMO in vertical and horizontal directions employing 2D active antenna array is depicted in Figure 2.19. Here, *high order* refers to the use of a large number of antennas at BS serving a large number of user devices by 3D user specific beamforming in both vertical and horizontal directions.

A practical FD-MIMO system is expected to maximize the network performance with regard to throughput; spectral efficiency; and peak data rate. However, there are various issues to consider in the design of the system including the characterization of the 3D channel model; upgrading the BS architecture to support 2D active antenna array; and efficient strategy for the pilot transmission and the feedback overhead. These issues are considered in the following chapters.

FD-MIMO BS achieves 3D beamforming towards terminals in both azimuth and elevation areas, as shown in Figure 2.20 (a), beamforming for 3D macro cell and micro cell are depicted in Figure 2.20 (b) and (c) respectively. It is generally equipped with 2D active antenna array panels. Terminal-specific 3D beamforming enables the FD-MIMO BS to efficaciously convey signals to the terminals being served, while at the same time minimizing the interference to unintended terminals and extensively increasing the spectral efficiency considerably.

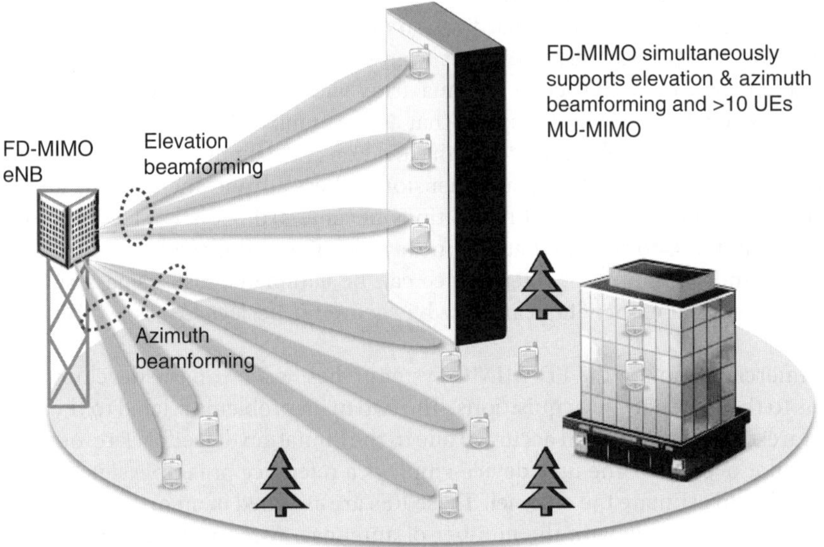

Figure 2.19 Full-dimension MIMO deployment scenario [34].

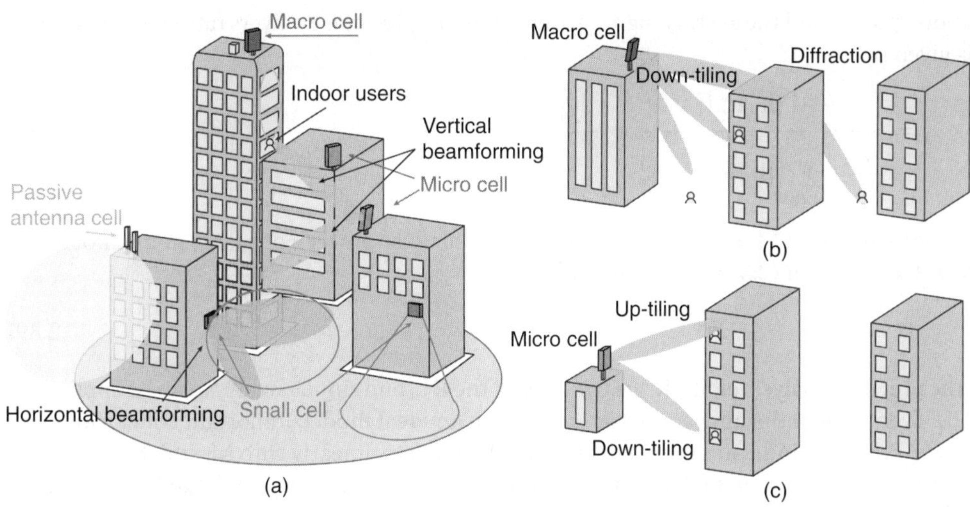

Figure 2.20 FD-MIMO deployment scenarios: (a) 3D macro cell site (placed over the rooftop) and 3D micro cell site (placed below the rooftop) with small cell; (b) beamforming for 3D macro cell; and (c) beamforming in 3D micro cell [37].

However, a number of design challenges have to be dealt with to comprehend the FD-MIMO gain in practical systems. Amongst the challenges are the need for low complexity and robust precoding methods that can maintain a balance of power efficiency and interference suppression.

In TDD systems, BS can obtain downlink CSI using uplink channel estimates if channel reciprocity holds. In practice, the phase and amplitude mismatch in different transmitter and receiver paths have to be calibrated and compensated. Several calibration schemes that can be used with large antenna array are considered in Chapter 10, and precoding schemes that can be employed with massive MIMO systems are considered in Chapter 11.

2.8.1 Full-Dimension MIMO (FD-MIMO) Analysis

Consider FD-MIMO system with M transmit antennas at the FD-MIMO BS transmits on the DL, with K single antenna co-scheduled terminals [35]. Let the total DL transmit power be P. Using the channel *conjugate precoding*, the received signal at the kth terminal is

$$y_k = \sqrt{\frac{P}{\mathrm{M}K}}\, \mathbf{h}_k \mathbf{h}_k^* \,\mathrm{x}_k + \left(\sqrt{\frac{P}{\mathrm{M}K}} \sum_{l=1, l\neq k}^{K} \mathbf{h}_k \mathbf{h}_l^* \, x_l + n_k \right) \tag{2.36}$$

where x_k is the transmitted signal for the kth terminal, $\mathbf{h}_k$ is the DL channel from BS for the kth user, and n_k is the kth terminal receiver additive Gaussian noise. As the number of BS antennas increases, the correlation between any two different random channel realizations tend to zero and the two random channels become orthogonal, i.e.

$$\lim_{\mathrm{M}\to\infty} \frac{\mathbf{h}_k \mathbf{h}_l^*}{\mathrm{M}} \Longrightarrow \delta_{kl} \tag{2.37}$$

where the Dirac delta function, δ_{kl}, is a real number line that is zero everywhere except at $k = l$, where it equals to 1. Expression (2.37) implies that with large M, the multiusers interference tends to null. Assuming a large M, the average interuser interference power is significantly larger than noise power, and the average signal-to-interference plus noise powers ratio SINR γ_k at each UE is given by

$$\gamma_k = \frac{\frac{P}{MK}\,|\mathbf{h}_k\,\mathbf{h}_k^*|^2}{\frac{P}{MK}\sum_{l=1,l\neq k}^{K}|\mathbf{h}_k\,\mathbf{h}_l^*|^2+\sigma_n^2} \approx \frac{\mathrm{M}}{K-1} \tag{2.38}$$

where we assumed all MU receivers contributed equal noise power. For a large number of users, i.e. $K \gg 1$, expression (2.38) can be simplified to

$$\gamma_k \approx \frac{\mathrm{M}}{K} \tag{2.39}$$

In the previous analysis of SINR, we considered the DL, and similar analysis can be applied for the UL. Even though the above analysis is based on an ideal model, important conclusions can be drawn. Eq. (2.38) indicates the SINR at each UE increases linearly with M, and if M increases at the rate as the scheduled UEs using the same resources, the same SINR can be maintained at each UE. For M to increase from 10 to 100 and the resource usage is kept the same, then the number of co-scheduled users can increase from 2 to 20, i.e. tenfold increase in system capacity with no sacrifice on SINR.

2.8.2 FD-MIMO System Design Issues

The conventional MIMO system design mostly takes into consideration the wave propagation in the horizontal direction only and the MIMO antennas construct a linear array. Although the FD-MIMO systems exploit 2D active antenna arrays the energy propagation in both vertical and horizontal directions have to be considered. In addition, the angular spread of transmitted energy in the vertical direction must be taken into account in the channel model. Likewise, the effect on the vertical energy propagation caused by the height difference between the transmitter and receiver should be considered in the channel model [36, 37].

Unlike passive antenna array in the conventional MIMO, an active antenna array system is capable of dynamically controlling the gain of each antenna element by applying a weighting scheme to the amplifiers attached to the array. Furthermore, a new precoding strategy is required to support the element level antenna structure.

A new transmitter architecture in the form of transceiver unit (TXRU) is required to be added to the BS transmitter. Active patch antenna array system and active devices are integrated on the same PCB signal path, which can easily be made between TXRUs and antenna elements. Since the design enables the control of gain and phase in both digital and analogue domains, one can have accurate control of the beamform direction.

2.8.3 3GPP Development of 3D Model for FD-MIMO System

2.8.3.1 Antenna Array Elements Radiation Patterns

The 3GPP activities as of 2015, have completed the 3D channel model and has published its technical report 3GPP TR 36.897 followed by formal standardization of Release 13. In characterizing the channel, it was assumed that the building height is distributed between four and eight floors of 3 m high per floor. The evaluation of the FD-MIMO system performance is carried out in a scenario where the antenna array and user devices are located at different heights and two scenarios are chosen: 3D urban macro scenario (3D Ma) and 3D urban micro scenario (3DMi), as shown in Figure 2.21. The transmit antennas for the case of UMa are placed over

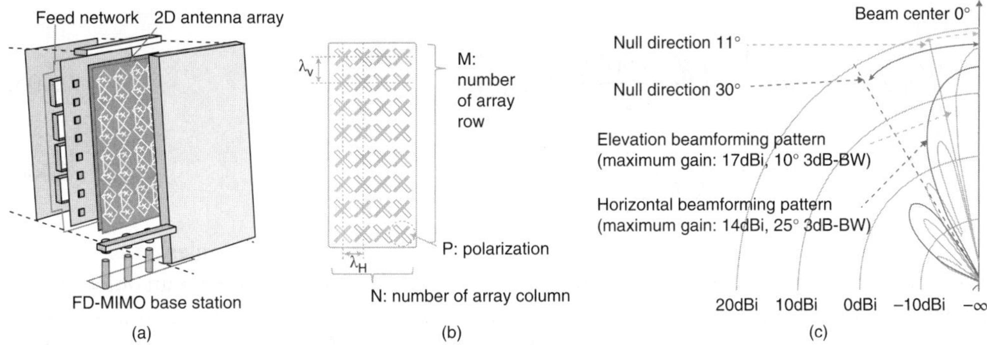

Figure 2.21 FD-MIMO systems: (a) concept of FD-MIMO systems; (b) 2D array antenna configuration; (c) vertical and horizontal beamforming patterns [37].

the rooftop and in the UMi are located below the rooftop. The height of the BS in UMa case was 25 m and that for the UMi case was 10 m. All outdoor users' height is assumed $h = 1.5m$ and the indoor user height is $h = 3(f_{floor} - 1) + 1.5m$, where f is the floor number uniformly distributed within $[1, N_{floor}]$ and N_{floor} is uniformly distributed along floors 4, 8. The FD-MIMO system concept, the 2D array antenna configuration, and the beamforming patterns are shown in Figure 2.21.

2.8.3.2 Antenna Configurations

The 2D antenna array is rectangular with cross polarization denoted as $(x - pol)$ and co-polarization denoted as $(co - pol)$ is shown in Figure 2.22 with M rows and N columns. Each antenna element in Figure 2.22 has a directional antenna radiation pattern, which includes both azimuth and elevation angles:

$$A''(\theta'', \phi'') = -\min[-A_{E,V}(\theta'') + A_{E,H}(\phi''), A_m] \tag{2.40a}$$

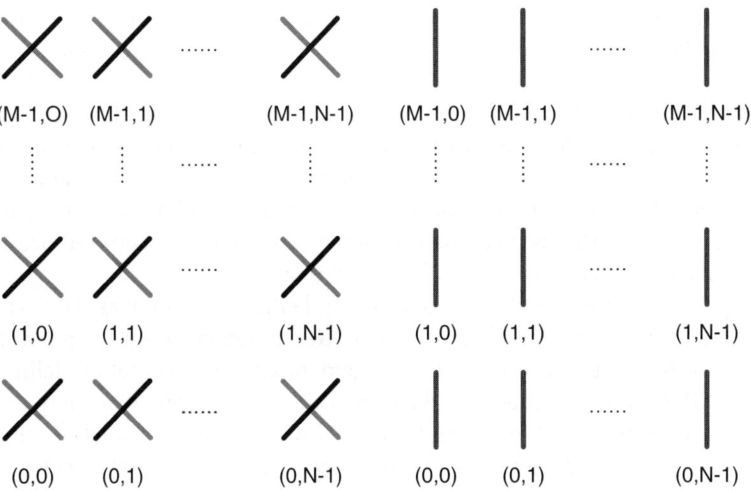

Figure 2.22 2D planar antenna array [33].

where

$$A_{E,V}(\theta'') = -min\left[12\left(\frac{\theta'' - 90^0}{\theta_{3dB}}\right)^2, SLA_V\right] \tag{2.40b}$$

$$A_{E,H}(\phi'') = -min\left[12\left(\frac{\phi''}{\phi_{3dB}}\right)^2, A_m\right] \tag{2.40c}$$

and $\theta_{3dB} = \phi_{3dB} = 65^0$, $SLA_V = A_m = 30\ dB$, θ'', ϕ'' are zenith and azimuth angles. This is related to the elevation angle (the angle from the line of sight to the horizontal plane for an object above the horizontal) while the zenith is the zenith angle = 90°, the elevation angle. Azimuth angle is the angle measured clockwise in the horizontal plane from a reference direction (i.e. direction due north) to the direction of the object of interest.

The antenna configuration in Figure 2.21b implies that there are three key parameters characterizing the antenna array structure (M, N, P): M is an element in the vertical direction; N is the number of elements in the horizontal direction; and the polarization degree P ($P = 1$ is for co-polarization and $P = 2$ is for dual-polarization). A standard 2D array setting is suggested using dual polarized antenna with $M = 8$ and $N = 4$. The setting indicates the null direction, an angle to make the magnitude of the beam pattern is zero, occurs at the elevation angle 11° and azimuth angle 30° reference to Figure 2.21c. This is the null in the vertical direction and is much smaller than in the horizontal, which can control the interuser interference by scheduling the users in the vertical direction.

2.8.3.3 FD-MIMO Development

The DL CSI acquisition in TDD systems is to some extent straightforward as a result of channel reciprocity. However, this is not the case for FDD systems. In contrast, the DL CSI in FDD systems is measured by the UE, quantized and sent to the BS. The TDD may seem to provide some benefits using channel reciprocity. It is clear from measurements carried out by various reach groups that reciprocity scheme for estimating the DL CSI does not reflect the interference from adjacent cells and co-scheduled UEs. So, considering the DL CSI feedback and the degradation of the DL pilot transmission, it seems that CSI feedback is important to both TDD and FDD scenarios. Clearly, the volume of CSI feedback has to be scaled with the number of antennas at the BS. So, for massive MIMO the amount of feedback would be of concern for FD-MIMO. The DL CSI feedback overhead can drain up to 48% of the UL resources allocated for data transmission.

A key feature of FD-MIMO is introduced when deploying active antenna systems. Unlike passive antenna that is used in a conventional MIMO system, active antenna systems transmitted signal's gain and phase are controlled by PA and LNA attached to each antenna element. In active 2D planar array, it is possible to control the transmitted wave on both elevation (vertical) and azimuth (horizontal) directions. That is it can be feasible to control the transmitted beam in 3D space. This control mechanism is known as 3D beamforming.

FD-MIMO system design maximizing the throughput, spectral efficiency and peak data rate faces a number of challenges, such as design of a new transmitter architecture for supporting 2D active antenna array, pilot transmission, feedback arrangement and 3D channel modelling according to these changes. The channel model should consider the angular spread of energy in the vertical direction. It is worth noting that in conventional MIMO systems, only the horizontal propagation is considered. So, wave propagation in both vertical and horizontal directions are required in FD-MIMO systems. That is, 3D wave propagation has to be specified. Furthermore, the radiation pattern depends on the antenna arrangement and the number of antenna elements and antenna spacing. Clearly FD-MIMO requires a new precoding scheme. In addition, new transmitter architectures are required. This can be achieved by adding a transceiver unit (TXRU) architecture in the design of the transmitter. The TXRU architecture provides hardware connection between BB signal path and the antenna array elements and hence facilitates the control of phase and gain of digital and analogue signals, ensuring accurate control of the beam direction.

A power-feeding network between TXRU and the antenna array elements is referred to as TXRU architecture. The architecture is composed of three constituents: TXRU array; antenna array; and radio distribution networks (RDNs). The RDN is to deliver the transmit signal from PA to antenna array elements and the received signals from antenna array to LNA. There are two main schemes for arranging the antenna elements with TXRUs: *array partitioning* and *array connected* architectures. In an array partition, array elements are divided into multiple groups and each TXRU is connected to a single group. Compare this to an array connected architecture, where RF signals from multiple TXRUs are delivered to the single antenna element. The difference between the array architectures is directly related to CSI-RS transmission. Each TXRU in array partition architecture transmits its own CSI-RS to the UE to measure the channel information of all TXRUs where the reference signal (RS) refers to pilot signal. So, N_T array elements partitioned into L groups of TXRUs and an orthogonal CSI-RS is assigned to each group. In the array connected architecture, each antenna element is connected to a number of TXRUs (say $L^{'}$) out of L TXRUs and orthogonal CSI-RSs are assigned to each TRXU with $N_T \frac{L'}{L}$ dimension weight vector transmitted to the UE to measure the CSI using the pilot signal (CSI-RS).

3GPP introduces two identities related to FD-MIMO architecture: antenna-port virtualization and TXRU virtualization. For antenna-port virtualization, a stream on an antenna port is precoded on a number of transceiver units. In TXRU virtualization, a signal on a TXRU is precoded on a number of antenna elements.

Precoding the data stream endures three phases: antenna-port virtualization followed by TXRU virtualization and finally, the antenna element pattern where the antenna element directional gain is constructed.

It is worth noting that in a traditional transceiver architecture model, a static one-to-one mapping is always assumed between antenna ports and TXRUs. In addition, each antenna port in LTE is associated with RS. In contrast, in 5G antenna port virtualization, a digital process of precoding is used in a high-selective environment.

Two TXRU virtualization schemes are possible: 1D TXRU virtualization and 2D TXRU virtualization. In 1D TXRU virtualization, M_{TXRU} TXRUs are associated with a column of M array elements that have the same polarization. With N columns in 2D antenna array and dual polarization $P = 2$, the total number of TXRUs $Q= M_{TXRU}. N. P$ associated with any of $M. N. P$ antenna elements. Various adjustments can be invoked between these different TRXU virtualizations in terms of hardware cost and complexity, power efficiency, and performance.

2.8.4 Beamformed CSI-RS Transmission

Two CSI-RS transmission schemes are studied in 3GPP: extension of conventional non-precoded CSI-RS; and beamformed CSI-RS. In the first scheme, the UE detects the non-precoded CSI-RS transmitted from the passive antennas. In the second, the BS *transmits* multiple beamed CSI-RS and the UE selects the preferred CSI-RS amongst them and feeds back its precoding weight index to enable the BS to determine the DL CSI. The CSI feedback information in FD-MIMO systems consists of three values, that is, rank indicator (RI), precoding matrix indicator (PMI), and channel quality indicator (CQI), where RI and PMI are used to assist the beamforming operation at the BS. So, UE needs to measure three values: CQI, RI, and PMI. These indicators are used to optimize resource allocation amongst the various UEs that are requesting services. The CQI informs the BS transmitter it can select 1 of 15 modulation and rate combinations for transmission. The RI informs the BS transmitter of transmission layers for the current MIMO channels that can be used. The PMI sends the BS the codebook index of the preferred precoding matrix.

Finding a jointly optimized solution for the three values is demanding, considering the limited signal processing of the UE hardware. Instead, reducing the computation complexity may be an achievable aim. A local independent optimal can be found by separating the optimization process into several steps [38]. All these three values are converted into bits and coded into 20 bits per subframe and are transmitted by UE on the PUCCH channel to the BS.

The beamformed CSI-RS scheme is better than non-precoded CSI-RS, especially when BS is equipped with a massive antenna array. The benefit of using the beamformed scheme is that it requires less overhead on UL and DL. The RS (pilot) overhead of the new beamformed CSI-RS is proportional to the number of RSs N and independent of the number of antennas at BS M [39]. In addition, if transmit power used is P, the power need for each non-precoded CSI-RS is $\frac{\mathrm{P}}{\mathrm{M}}$ while $\frac{\mathrm{P}}{\mathrm{N}}$ is used for the beamformed CSI-RS, which provides power gain over non-precoded CSI-RS equal $\frac{\mathrm{M}}{\mathrm{N}}$. For example, let M = 32 and N = 12, beamformed CSI-RS gain is $\frac{\frac{P}{12}}{\frac{P}{32}} = \frac{32}{12}$, which is equal 4.3 dB in signal power over non-precoded CSI-RS. The DL precoder for data transmission $\mathbf{W}_{\text{data}}$ is given by

$$\mathbf{W}_{\text{data}} = \mathbf{W}_{\mathrm{T}}\, \mathbf{W}_{\mathrm{P}}\, \mathbf{W}_{\mathrm{U}}$$

where

$\mathbf{W}_{\mathrm{T}} \in \mathbb{C}^{N_T \times L}$ is the precoder between TXRU to antenna element.

$\mathbf{W}_{\mathrm{P}} \in \mathbb{C}^{L \times N_P}$ is the precoder between antenna port to TXRU.

$\mathbf{W}_{\mathrm{U}} \in \mathbb{C}^{N_P \times r}$ is the precoder weight index fed back by UE to BS.

N_T is the number of antennas, N_P is the number of antenna ports, L is the number of groups contained in the antenna elements, and r columns of the feedback precoding.

The beamforming process is directly related to how accurately the direction of incoming signal and receiver direction are matched to each other. In adaptive beamforming, there is a possibility of slight beam direction error because of insufficient knowledge about the AoA or inaccurate phase shifting between transmit and receive antenna elements [40]. The power loss in dB is investigated for the elevation dimension mismatch for an array size 8 × 8. The azimuth angle of the receive beams is fixed at 0 ° and the elevations vary between 75° and 30° in steps of 15°. In each step the deviation on the azimuth is bounded to quite wide range ±20. It is clear that the power loss curves are not symmetric around the azimuth 0 °, and this is because the array factor widens towards azimuth 0 °. For the elevations considered in the investigation, the half power beamwidth (HPBW) are 13.4 °, 14.9 °, 18.3 °, and 25.8 °, respectively.

It should be noted that power loss at azimuth 0 ° is assumed to be zero dB (loss = 1) and that at ±20 deviations, the losses for 30 ° elevation are not as large as the losses encountered for 75 ° and again this is due to the beam directed to 30 ° elevation is wider and hence not attenuated as sharp as the 75 ° elevation beam.

2.8.5 CSI Feedback for FD-MIMO Systems

There are a number of possible feedback schemes that are appropriate for full-dimension MIMO systems. We summarize the key features of these schemes here:

- *Partial CSI-RS:* In the partial CSI-RS, only a subset of the antennas is used for the CSI measurement. This can be done by partitioning the 2D array into horizontal and vertical ports, for example, N_H ports in the row and N_V ports in the column. Then the total number of CSI-RS is reduced from $N_H \times N_V$ to $N_H + N_V$ [37], [41]. The overall CSI can be constructed based on a smaller number of ports. However, such schemes may not be able to deliver the SCI accuracy required.
- *Composite codebook:* Here the codebook is divided into two codebooks, vertical and horizontal codebooks, and the CSI is separately delivered to the eNB. By combining the two codebooks, the eNB reconstructs the channel CSI by using the Kronecker product of the two codebooks ($\mathbf{W}_U = W_{U,V} \otimes \mathbf{W}_{\mathrm{U,H}}$). The eNB reconstructs the whole CSI once it has received the two codebooks. To reduce the feedback overhead, and since the angular spread of the vertical direction is smaller than the horizontal direction, a new design for the vertical codebook is needed to achieve better tradeoffs between performance and feedback overhead.
- Hybrid CSI-RS transmission: This scheme benefits from the (coded) beamformed CSI-RS and the conventional non-precoded CSI-RS transmissions, simultaneously. First, the eNB transmits N_T non-precoded CSI-RS to the UE. After receiving sufficient long-term channel statistics from UE, the eNB transmits the beamformed CSI-RS for short-term feedback at UE. Short-term feedback overhead can be reduced significantly in addition to the long-term reduction in feedback overhead.
- *Beam index feedback:* When the eNB attempts to obtain the UE channel direction information (CDI) from the beamformed CSI-RSs, the eNB transmits multiple beamformed CSI-RS towards the UE. If the channel rank is one, the feedback of a beam index and corresponding channel quality indicator (CQI) is enough to generate the CSI at eNB. When channel rank is two, co-phase information is required for dual-polarization. One possibility to reduce feedback overhead is to use the same CSI-RS resource to transmit multiple spatially separated beams.
- *Flexible codebook transmission:* In this scheme, a master codebook is designed for multiple TXRUs. Consider 2D antenna array and 16 TXRUs. Then we can derive the following specific codebook: (2 × 8), (4 × 4), or (1 × 16). This information has to be conveyed to the UE by DL signalling channel so that the UE can construct the actual codebook. This scheme can be made flexible so the eNB configures suitable codebook layouts considering transmission terrains.

A joint RF beamforming and BB precoding (or hybrid beamforming) for the mmWave communication systems is proposed in [42], [43], and [34] to combine the benefits from both schemes. Digital beamforming demands a large number of costly ADCs/DACs in addition to increased complexity. The hybrid beamforming is the best balance between performance and complexity for moderate numbers of MIMO streams combined with large eNB array antenna elements that generate high gain. The RF beamforming controls phase and magnitude of input signal to each antenna so that the transmit/receive output signals from/to the antennas in such a way to form a directive beam(s) in a particular direction(s). Once the transmit RF beams are selected, the best BB precoder (or codebook) must be selected to achieve the highest channel capacity over the beams and MIMO channel.

Consider the mmWave system illustrated in Figure 2.23 consisting of an eNB with M antennas transmitting N_s data streams to a UE equipped with N_u antennas. We assume that the eNB transmitter is equipped with $\mathrm{M^c}$ chains such that $N_s \leq \mathrm{M^c} \leq \mathrm{M}$. The UE receiver then supports N_u^c chains such that $N_s \leq N_u^c \leq N_u$. Clearly, the number of chains that can be supported is $N_s \leq \min(\mathrm{M^c}, N_u^c)$. The eNB transmitter uses $\mathrm{M^c} \times N_s$ BB precoding matrix $\mathbf{P}$, followed by $(\mathrm{M} \times \mathrm{M^c})$ beamforming weight matrix. The BB precoding weights are implemented in the

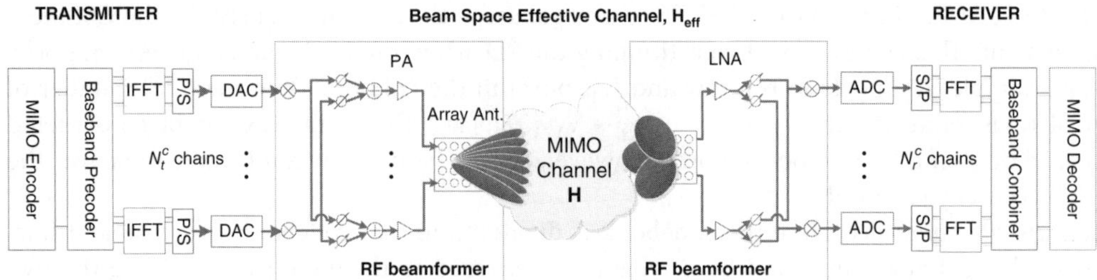

Figure 2.23 Block diagram of hybrid beamforming structure [42].

digital domain and the RF beamforming is implemented by variable gain amplifiers and phase shifters connected with the antennas.

2.9 Summary

The sets of enabling technologies aim to provide unlimited access to information and the ability to share data anywhere anytime by anyone for the comfort of people and to the advantage of business and society. In this chapter we presented a comprehensive coverage on some of the key enabling technologies that derive 5G networks.

We introduced the rationale for these technologies in Section 2.2, applying the principles of digital communication pioneered by Shannon's theory. 5G is expected to employ heterogeneous cellular networks (HetNet) where a multi-tier heterogeneous network consisting of macro cells together with a large number of low power small cells, relays, remote radio heads (RRHs) and the provision of D2D and M2M communications are commissioned to provide 5G services. The heterogeneous technology is considered in depth in Section 2.3.

The RAN technologies will be based on cloud-based (C-RAN) technology deploying cost efficient small cells. C-RANs detached BBU from RRH are permitting for centralized operation BBUs and scalable RRHs as small cells. Intelligent configurations are proposed for the front haul. The fronthaul is logically re-configurable to implement appropriate transmission schemes These technologies were explained and analysed in Section 2.4.

Cache-enabled small-cell networks where frequently used (popular) files are cached in nodes near the intended users so that instant access to the files by the intended users is possible. The probability of outage on the DL cache-enable SBSs was analysed with respect to the cache storage, and SBSs intensity were illustrated in Section 2.5.

Full-duplex (FD) transmission is expected to play a key role in 5G since it is capable of doubling the link data rate. However, the main challenge to FD is the self-interference generated at the node transceiver antennas. Schemes for reducing self-interference, together with the infrastructure for FD transmission on the DL, and the FD MAC protocol for the UL, are all analysed in Section 2.6.

In order to determine the characteristic channel for an antenna port, a UE must carry out a separate channel estimation for each antenna port. Separate RSs (pilot signals) that are suitable for estimating the respective CSI are defined in 3GPP standards for each antenna port. Each DL has specific RSs and corresponding antenna ports. UL has its own specific physical channels RSs. The reference signals, antenna ports, and physical channels are reviewed in section 2.7.

Another important enabler to 5G is the FD-MIMO scheme that is capable of supporting a large number of users and a large number of BS antennas. It allows the use of 2D active planar arrays (patch antennas) with 3D user-specific beamforming in both horizontal and vertical directions. The FD-MIMO technology was described and analysed in terms of individual user SINR to show the increase in system capacity. These issues relating to FD-MIMO technology were assessed in depth in Section 2.8.

2.A Notes on Machine Learning Algorithms

Machine learning technology can be categorized into three types: unsupervised learning, supervised learning, and reinforcement learning. Supervised learning is where you have input variables and an output variable of a process and the algorithm learns the mapping function from the input to the output. Supervised machine learning is the most effective but it is challenging to implement in practice due to the lack of availability of sufficient training data in the field. Unsupervised learning is where there is data but no corresponding outputs. The aim of the algorithm is to model the event to learn more about the input data. Unlike supervised learning, there is no correct output and no teacher, and the algorithms are left to their own devises to discover the model process. The reinforcement learning (RL) system applies experiences of intelligent agents that learn from the environment [8, 12, 13].

RL, also known as Q-learning, is a widely used algorithm for many applications such as artificial intelligence, automatic control, robotics, financial trading systems, navigations, cellular systems, and solutions to problems that require optimal policy. The reinforcement signal is specified by the environment as an evaluation to the action quality. Nonetheless, the environment gives a little bit of information. An intelligent agent makes use of the environmental information to revise its action to adapt to the environment. The environment changes to a new state after accepting the RL system action to obtain the rewards and punishment signal. The key concept of reinforcement learning is that if a specific RL system action generates a positive reward of the environment, this action will strengthen the trend, which is a positive feedback process; otherwise, the action will diminish the trend.

In a Markov environment, the interaction between the RL system and the environment may be regarded as Markov decision-making process defined by four factors: S, A, R, P. S denotes environment sate set, A denotes RL system action set, R is reward function, P is state transition probability, R^a_{ss} is the instantaneous reward value obtained by RL system when the environment state changes to a new state through action 'a', P^a_{ss} is a probability obtained by RL system when the environment state changes to a new state through action 'a'. Generally, P and R are unknown functions, so the RL system can only choose a strategy by trial and error each time, based on the uncertainty of the environment and long-term goals.

An *RL* system can be used to find the optimal action-selection policy for a finite Markov process. A policy is a set of rules an agent has to follow in selecting actions. The algorithm works by learning an action value function that gives a policy for the best action in a given state and follows the optimal policy thereafter. When the function is learned, the policy can be constructed by simply selecting the action with the highest value in each state. Q-learning algorithm can be used to solve problems with random transitions and rewards. Q-learning finds the optimal policy as the expected value of the average of the total rewards over all successful steps.

2.A.1 The Algorithm [44, 45]

The problem model consists of an intelligent agent, states S and a set of actions per state A. When the agent performs action $a \in A$, the environment moves from state to state. Executing an action in a specific state provides the agent with a reward (a numerical score) or (computing) costs. A reward is an immediate score for performing the action in a state. The agent's aim is to maximize its total reward (minimize total costs).

Consider the interaction between the intelligent agent and the environment as shown in Figure 2.A.1. The goal of the algorithm is to learn an action strategy Π: S →A, the strategy enables the action to obtain the largest cumulative reward value defined as

$$\sum_{t=0}^{\infty} \gamma^t \, r_{t+k+1} \tag{2.A.1}$$

where $0 \le \gamma < 1$ is a discount factor. It is worth noting that if the discount factor equals 1, the cumulative reward sum in (2.A.1) tends to infinite and would not converge. The fact that the RL system is a kind of policy optimization and infinite reward sum is not good optimization criteria.

The interaction occurs at discrete time steps: $t = 0, 1, 2,..$ etc. For example, the state at step t is $s_t \in S$, the action taken at step t is $a_t \in A$, the resulting reward is $r_{t+1} \in \mathbb{R}$, and the next state is s_{t+1}. These sequences of state-rewards can be represented systematically as in Figure 2.A.2. Assume the sequence of rewards after step t are

$$r_{t+1}, r_{t+2}, r_{t+3}, r_{t+4}, \quad \dots\dots\dots, r_T \tag{2.A.2}$$

We aim to maximize the expected return, $\{\mathcal{R}_t\}$, where

$$\mathcal{R}_t = r_{t+1}, r_{t+2}, r_{t+3}, r_{t+4}, \quad \dots\dots\dots, r_T \tag{2.A.3}$$

and T is the final step at which a final state is reached ending the search. If rewards received in the far future are worth less than received sooner, they will be discounted (reduced), i.e. multiplied by an exponential multiplier denoted as discounted rate γ. The *discounted return* $\mathcal{R}_t^{disc}$ is

$$\mathcal{R}_t^{disc} = r_{t+1} + \gamma\, r_{t+2} + \gamma\, r_{t+3} + \gamma\, r_{t+4}, \quad \dots\dots\dots. = \sum_{k=0}^{\infty} \gamma^k \, r_{t+k+1} \tag{2.A.4}$$

The Q-value function is given by

$$Q(s,a) = \mathbb{E}\left\{ \sum_{t=0}^{\infty} \gamma^t \; r(s_t, a_t) | s_0 = s, a_0 = a \right\} \tag{2.A.5a}$$

where s_0, a_0 are initial state and initial action and $r(s_t, a_t)$ is the reward of the set of state s_t and action a_t, respectively. We can use (2.A.5a,b) to compute the sum reward if we know the

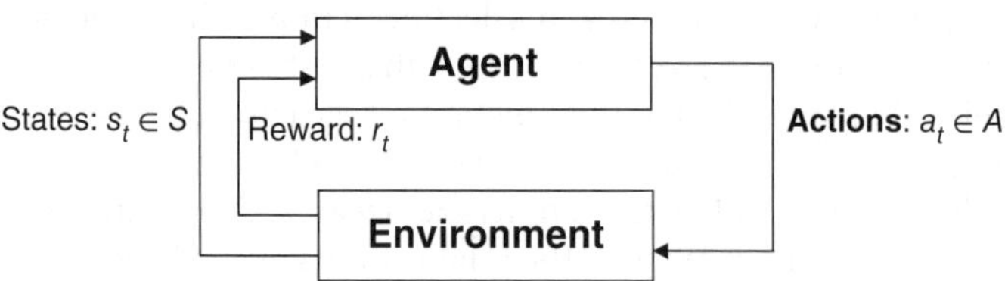

Figure 2.A.1 Interaction between the agent and the environment.

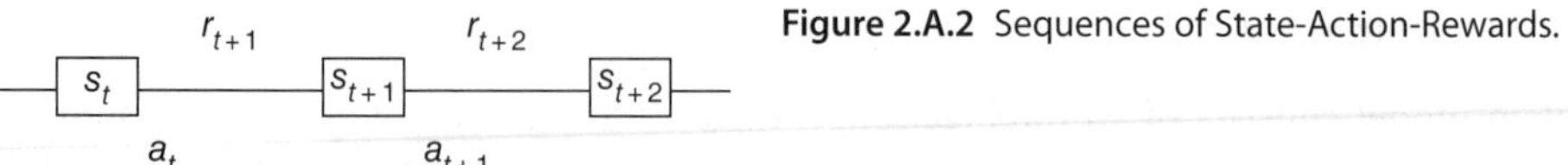

Figure 2.A.2 Sequences of State-Action-Rewards.

terminal state. In most cases, the final state is not defined but assumed as infinity and this one reason introduces the discount factor to limit the sum rewards. If $\gamma = 0$, agents don't care about future cost. Optimal policy is difficult to determine using (2.A.5a,b), so instead solving (2.A.5a,b) we apply Q-learning algorithm by rewriting (2.A.5a,b) as

$$Q(s,a) = \mathbb{E}\{r(s_t\} + \mathbb{E}\left\{\sum_{t=1}^{\infty} \gamma^t \; r(s_t, a_t)|s_0 = s, a_0 = a\right\} \tag{2.A.5b}$$

In the greedy policy, the agent selects the best Q value actions in a given state. In this case, the agent does not explore all actions. In typical Q-learning algorithms, "ε-greedy policy is used to determine the next action to be taken. In "ε-greedy policy, action with the best Q value is chosen with $1 - \varepsilon$ probability, and a random action is chosen with ε probability. The ε values can be increased or decreased to give preference to exploration vs. exploitation.

The algorithm uses a function to calculate the quantity Q of the state-action. At the start of the learning, Q returns a fixed value chosen by the designer. Then Q is updated each time the agent selects an action, and observes a reward and a new state, value iterative update. The function to calculate Q is

$$Q(s_t, a_t) \longleftarrow (1-\alpha)Q(s_t, a_t) + \alpha.\; (r_{t+1} + \gamma.\quad \min_a Q(s_{t+1}, a)) \tag{2.A.6}$$

where r_{t+1} is the reward observed after performing a_t, s_t and α is the learning rate with $(0 \le \alpha \le 1)$. The algorithm stops when state s_{t+1} is a final state. All final states s_f, $Q(s_f, a)$ are never updated but are taken to be equal to zero. The learning rate, a value of 0 makes the agent not learn anything, while a value of 1 makes the agent consider only the most recent information.

In deterministic environments, a learning rate of 1 is optimal. For random environments, the algorithm still converges on the learning rate, decreasing to zero. In practice, often a constant learning rate is used, such as 0.1 for all steps.

The discount factor γ determines the importance of future rewards. A value of 0 will make the agent only consider current rewards, while a value approaching 1 will make it strive for a long-term high reward. If the discount factor equal or exceeds 1, the action values may diverge.

2.B Outage Probability in CE-SC Networks

Using (2.10) in the text, the average outage probability of the o, connected to the nearest SBS (at distance d), conditioned to d is

$$\begin{aligned}\mathbb{P}_{out}(\lambda,\ R,\ \alpha,\ S) &\triangleq \mathbb{E}_d[1 - \mathbb{P}[\log(1+SINR) > R, f_o \in \Delta_{B_o} | d]] \\ &= 1 - \underbrace{\mathbb{E}_d[\mathbb{P}[\log(1+SINR) > R|d]]}_{term\ i}\ \underbrace{\mathbb{E}_d\mathbb{P}[f_o \in \Delta_{B_o} | d]}_{term\ ii}\end{aligned} \tag{2.B.1}$$

We proceed by analysing term by term starting with *term i*.

2.B.1.1 Analysis of *Term i*:

For a Poisson distribution, the probability of n SBSs within circular coverage area A with radius r is

$$\mathbb{P}(n) = \frac{(\lambda A)^n e^{-\lambda A}}{n!} \tag{2.B.2}$$

where $\lambda = mean = variance$ of Poisson distribution.

Consider the distance d separating *user o* from its associated SBS. Since each user communicates with the closest base station, no other base station can be closer than d. Consequently, all interfering SBSs are farther than d. For no SBS closer than r, then $n = 0$, the null probability is

$$\mathbb{P}(n) = \mathbb{P}\text{ (No SBS closer than } r\text{)} = \left.\frac{(\lambda A)^n e^{-\lambda A}}{n!}\right|_{n=0} = e^{-\lambda A} = e^{-\lambda\pi r^2} \tag{2.B.3}$$

The *cdf* is $\mathbb{P}(d \leq r) = 1 - e^{-\lambda\pi r^2} = F_d(r)$ and the PDF of d is

$$f_d(d) = \frac{\mathrm{d}F_d(d)}{\mathrm{d}(d)}$$

$$f_d(d) = \frac{\mathrm{d}(1 - e^{-\lambda\pi d^2})}{\mathrm{d}(d)} = 2\pi\lambda d\, e^{-\lambda\pi d^2} \tag{2.B.4}$$

We now examine (2.B.1) conditioned expectation and consider *term i*

$$\mathbb{E}_d[\mathbb{P}[\log(1 + SINR) > R|d]] = \int_{d>0} \mathbb{P}[\log(1 + SINR) > R|d] f_d(d)\, \mathrm{d}(d)$$

Substituting for $f_d(d)$:

$$= \int_{d>0} \mathbb{P}[\log(1 + SINR) > R|d]\, 2\pi\lambda d\, e^{-\lambda\pi d^2}\, \mathrm{d}(d) \tag{2.B.5}$$

The SINR is given in (2.8) in the text as

$$SINR \triangleq \frac{h\, d^{-\alpha}}{I_d + \sigma_o^2} \tag{2.B.6}$$

where the interference $I_d \triangleq \sum_{i\in\Phi/B_o} g_i r_i^{-\alpha}$ where r_i is random distance of the ith interfering user from its associate SBS B_o. Now have

$$\log(1 + SINR) > R \tag{2.B.7a}$$

$$1 + SINR > e^R \tag{2.B.7b}$$

$$\frac{h\, d^{-\alpha}}{I_d + \sigma_o^2} > e^R - 1 \tag{2.B.7c}$$

Substitute (2.B.7c) in (2.B.5) we get

$$\int_{d>0} \mathbb{P}\left[\frac{h\, d^{-\alpha}}{I_d + \sigma_o^2} > e^R - 1|d\right]\, 2\lambda\pi d\, e^{-\lambda\pi d^2}\, \mathrm{d}(d).$$

Then *term i* becomes:

$$\mathbb{E}_d[\mathbb{P}[\log(1 + SINR) > R|d]] = \int_{d>0} \mathbb{P}[h > d^\alpha(e^R - 1)(I_d + \sigma_o^2)|d]\, 2\lambda\pi d\, e^{-\lambda\pi d^2}\, \mathrm{d}(d) \tag{2.B.8}$$

We can now consider the probability of Rayleigh random variable h exceeding $d^\alpha(e^R - 1)(I_d + \sigma_o^2)$ conditioned on I_d, we get

$$\mathbb{P}[h > d^\alpha(e^R - 1)(I_d + \sigma_o^2)|d] = \mathbb{E}_{I_d}[\mathbb{P}[h > d^\alpha(e^R - 1)(I_d + \sigma_o^2)|\mathrm{d}, I_d]] \tag{2.B.9}$$

using the fact that $h \sim \mathrm{Exp}(\mu)$ is random variable with mean $\frac{1}{\mu}$. The probability of $h > d^\alpha(e^R - 1)(I_d + \sigma_o^2)$ can be written as

$$\mathbb{P}[h > d^\alpha(e^R - 1)(I_d + \sigma_o^2)|\mathrm{d}]$$

and exponential distribution in (2.B.9), we get

$$= \mathbb{E}_{I_d}[\text{Exp}(-\mu\, d^\alpha(e^R - 1)(I_d + \sigma_o^2))|d]$$

$$= e^{-\mu\, d^\alpha(e^R-1)\sigma_o^2}\ \mathbb{E}_{I_d}[\text{Exp}(-\mu\, d^\alpha(e^R - 1)I_d)|d] \tag{2.B.10}$$

Examining (2.B.10), we notice that the averaging over the random variable I_d can be expressed as

$$\mathbb{E}_{I_d}[\text{Exp}(-\mu\, d^\alpha(e^R - 1)I_d)|d] = \mathbb{E}_{I_d}[e^{-sI_d}] \tag{2.B.11}$$

Expression (2.B.11) is Laplace transform of I_d with Laplace number s:

$$s = \mu\, d^\alpha(e^R - 1) \tag{2.B.12}$$

Thus

$$\mathbb{E}_{I_d}[\text{Exp}(-\mu\, d^\alpha(e^R - 1)I_d)|d] = \mathbb{E}_{I_d}[e^{-sI_d}] = \mathcal{L}_{I_d}(\mu\, d^\alpha(e^R - 1))$$

where $\mathcal{L}_{I_d}(s)$ is Laplace of I_d.

Substituting in the simplified version of *term i* in 2.B.5 we get

$$\begin{aligned}\mathbb{E}_d[\mathbb{P}[\log(1 + SINR) > R|d]] &= \int_{d>0} e^{-\mu\, d^\alpha(e^R-1)\sigma_o^2}\ \mathcal{L}_{I_d}(\mu\, d^\alpha(e^R - 1))\, 2\lambda\pi d\, e^{-\lambda\pi d^2}\ \mathrm{d}(d)\\ &= \int_{d>0} e^{-\mu\, d^\alpha(e^R-1)\sigma_o^2}\ \mathcal{L}_{I_d}(s)\, 2\lambda\pi d\, e^{-\lambda\pi d^2}\ \mathrm{d}(d)\end{aligned} \tag{2.B.13}$$

Let use turn our attention to the Laplace transform of the interference in (2.B.13) and substituting for the interference we get

$$\mathcal{L}_{I_d}(s) = \mathbb{E}_{I_d}[e^{-sI_d}] = \mathbb{E}_{\Phi,\{g_i\}}\left[\text{Exp}\left(-s \sum_{i\in\Phi/B_o} g_i r_i^{-\alpha}\right)\right] \tag{2.B.14}$$

where SBSs are distributed as PPP Φ, g_i are interfering channels and r_i is the distance from the ith BS to the tagged receiver. Next, we expand (2.B.14) as follows:

$$\mathcal{L}_{I_d}(s) = \mathbb{E}_{\Phi,\{g_i\}}[\Pi_{i\varepsilon\Phi\backslash B_o}[\text{Exp}(-sg_i r_i^{-\alpha})]] \tag{2.B.15}$$

The multiplication of the Exp(.) is carried out for the SBSs except for B_o. Since the Exp(.) is function of g_i only, and g_i is independent of Φ, we arrange (2.B.15) as

$$\mathcal{L}_{I_d}(s) = \mathbb{E}_{\Phi}[\Pi_{i\varepsilon\Phi\backslash B_o}\ \mathbb{E}_{\{g_i\}}[\text{Exp}(-sg_i r_i^{-\alpha})]] \tag{2.B.16}$$

Since g_i are $i.\ i.\ d$ variables so $g_i \equiv g$.

$$\mathcal{L}_{I_d}(s) = \mathbb{E}_{\Phi}[\Pi_{i\varepsilon\Phi\backslash B_o}\ \mathbb{E}_g[\text{Exp}(-sgr_i^{-\alpha})]] \tag{2.B.17}$$

Using Campbell's theorem 3.3 for PPP, we have

$$\mathbb{E}(e^{tS}) = \text{Exp}\left(-\lambda \int_{\mathbb{R}^d} (1 - e^{tf(x)})\mathrm{d}x\right)$$

Substituting $t = 1$

$$\mathbb{E}[\Pi_{x\varepsilon\Phi}\ \mathrm{f(x)}] = \mathbb{E}(\mathrm{e^S}) = \text{Exp}\left(-\lambda \int_{\mathbb{R}^d} (1 - \mathrm{e^{f(x)}})\mathrm{dx}\right) \tag{2.B.18}$$

Denote the PDF of g as $f(g)$. Implement (2.B.18) in (2.B.17), we get

$$\mathcal{L}_{I_d}(\mu\, d^{\alpha}(e^{R}-1)) = \mathrm{Exp}\left(-2\pi\lambda\int_{v}^{\infty}(1-\mathbb{E}_{g}[\mathrm{Exp}(-sgv^{-\alpha})])v\, dv\right)$$

$$= \mathrm{Exp}\left[-2\pi\lambda\int_{0}^{\infty}\left(\int_{d}^{\infty}(1-e^{-\mu\, d^{\alpha}(e^{R}-1)gv^{-\alpha}})v\, dv\right)f(g)\, dg\right] \tag{2.B.19}$$

Let $v^{-\alpha} \longrightarrow y$, after some mathematical manipulations Laplace transform can be written as [13]:

$$\mathcal{L}_{I_d}(\mu\, d^{\alpha}(e^{R}-1)) = \mathrm{Exp}\left\{\lambda\pi d^{2} - \frac{2\pi\lambda(\mu(e^{R}-1))^{\frac{2}{\alpha}}\, d^{2}}{\alpha}\right.$$

$$\left.\times\int_{0}^{\infty} g^{\frac{2}{\alpha}}\left[\Gamma\left(-\frac{2}{\alpha},\mu((e^{R}-1)g)\right)-\Gamma\left(\frac{2}{\alpha}\right)\right] f(g)d(g)\right\} \tag{2.B.20}$$

Using the substitution $d^{2} \longrightarrow v$ with algebraic manipulation, *term i* reduced to:

$$\mathbb{E}_{d}[\mathbb{P}[\log(1+\mathrm{SINR}) > R|d]] = \pi\lambda\int_{0}^{\infty} e^{-\pi\lambda v\beta(R,\alpha)-\mu(e^{R}-1)\sigma\, v^{\frac{\alpha}{2}}}\, dv \tag{2.B.21}$$

where

$$\beta(R,\alpha) = \frac{2(\mu(e^{R}-1))}{\alpha}\mathbb{E}_{g}\left[g^{\frac{2}{\alpha}}\left(\Gamma\left(-\frac{2}{\alpha},\mu(e^{R}-1)g\right)-\Gamma\left(\frac{2}{\alpha}\right)\right)\right]$$

Now we consider the analysis of *term ii:*

Assume that every SBS caches the same popular files and SBSs have the same storage size, then cache hit probability becomes independent of distance d. Let $f_o \in \Delta_{B_o}$ is the requested file, by *user o*, stored in the cache Δ_{B_o} of user serving SBS B_o. Denote the file popularity distribution $f_{pop}\,(f,\gamma)$, then the average hit probability is

$$\mathbb{E}_{d}\mathbb{P}[f_o \in \Delta_{B_o}\,|d] = \int_{0}^{\frac{S}{L}} f_{pop}\,(f,\gamma)df \tag{2.B.22}$$

where S denotes storage unit capacity and L is length of every file.

2.C Signal Power at the Receive Antenna after Antenna Cancellation of Self-Interference [19]

In this appendix, we analyse the receive power at a receive antenna placed at distance d from one transmit antenna and the other transmit antenna is placed at $d+\frac{\lambda}{2}$ away from the receive antenna so that the signal from the transmit antennas destructively cause significant attenuation in the signal received at the receive antenna. Let the unit power BB signal be $x(t)$. At the transmit antennas the signal is scaled to two amplitudes A_1, A_2 and the transmitted signals are subject to path losses L_1, L_2 and their phase shifts when arrived at the receive antenna are denoted as ϕ_1, ϕ_2, respectively. So, the received signal is

$$r(t) = \frac{A_1}{L_1}x(t)\, e^{j(2\pi f_c t+\phi_1)} + \frac{A_2}{L_2}x(t)\, e^{j(2\pi f_c t+\phi_2)} \tag{2.C.1}$$

In the ideal case, $\frac{A_1}{L_1} = \frac{A_2}{L_2}$ but in practical systems, it would be impossible to get the two signals to match perfectly at the receive antenna, let us denote $\frac{A_1}{L_1} = A_{ant}$ and the amplitude mismatch

$\in^A_{ant}$ so that $\frac{A_2}{L_2} = A_{ant} + \in^A_{ant}$. Likewise, ideally the two signals are exactly π out of phase at the receive antenna and $\phi_2 = \phi_1 + \pi$ but in practical systems the transmitted signal contains a band of frequencies so we assume a phase shift error between the two signals to be $\in^\phi_{ant}$ so $\phi_2 = \phi_1 + \pi + \in^\phi_{ant}$. Therefore, the received signal in (2.C.1) becomes

$$r(t) = A_{ant}\, x(t)\, e^{j(2\pi f_c t+\phi_1)} + (A_{ant} + \in^A_{ant})\, x(t)\, e^{j(2\pi f_c t+\phi_1+\pi+\in^\phi_{ant})} \tag{2.C.2}$$

Eq. (2.C.2) can easily be simplified to

$$r(t) = A_{ant}\, x(t)\, e^{j(2\pi f_c t+\phi_1)}\,(1 - e^{j\in^\phi_{ant}}) - \in^A_{ant}\, x(t) e^{j(2\pi f_c t+\phi_1+\in^\phi_{ant})} \tag{2.C.3}$$

The instantaneous power of the complex receive signal is $r(t)\,(r(t))^*$ where $(r(t))^*$ is the complex conjugate of $r(t)$. It can be shown that (2.C.3) can be simplified to

$$\begin{aligned} r(t)(r(t))^* &= A^2_{ant}|x(t)|^2\,(2 - (e^{j\in^\phi_{ant}} + e^{-j\in^\phi_{ant}})) \\ &\quad + A_{ant} \in^A_{ant} |x(t)|^2(2 - (e^{j\in^\phi_{ant}} + e^{-j\in^\phi_{ant}})) + (\in^A_{ant})^2\, |x(t)|^2 \end{aligned} \tag{2.C.4}$$

$$r(t)(r(t))^* = A^2_{ant}|x(t)|^2\,(2 - 2\cos \in^\phi_{ant}) + A_{ant} \in^A_{ant} |x(t)|^2(2 - 2\cos \in^\phi_{ant}) + (\in^A_{ant})^2\, |x(t)|^2 \tag{2.C.5}$$

Therefore, the received power after antenna cancellation of self-interference is

$$2A_{ant}\,(A_{ant} + \in^A_{ant})|x(t)|^2(1 - \cos \in^\phi_{ant}) + (\in^A_{ant})^2\, |x(t)|^2 \tag{2.C.6}$$

Now the phase shift between transmit and receive antennas, ϕ, can be expressed in terms of the distance d between transmit and receive antennas as $\phi = \frac{2\pi d}{\lambda}$. Hence, the phase error, $\in^\phi_{ant}$, can be expressed as $\frac{2\pi \in^d_{ant}}{\lambda}$ where $\in^d_{ant}$ is the error in the receive antenna. The receive power in (2.C.6) becomes

$$2A_{ant}\,(A_{ant} + \in^A_{ant})|x(t)|^2 \left(1 - \cos \frac{2\pi \in^d_{ant}}{\lambda}\right) + (\in^A_{ant})^2\, |x(t)|^2 \tag{2.C.7}$$

For unit power transmitted signal $|x(t)|^2 = 1$. So, substitute for $|x(t)|^2$ in 2.C.7 we get
Received signal power $=2A_{ant}\,(A_{ant} + \in^A_{ant})\left(1 - \cos \frac{2\pi \in^d_{ant}}{\lambda}\right) + (\in^A_{ant})^2$

References

1 Andrews, J.G., Buzzi, S., Choi, W. et al. (2014). What Will 5G Be? *IEEE Journal on Selected Areas in Communications* 32 (6): 1065–1082.

2 Tall, A., Altman, Z., and Altman, E. (2014) 'Self organizing strategies for enhanced ICIC (eICIC)', International Symposium on Modelling and Optimization in Mobile, Ad Hoc, and Wireless Networks,' 318–325.

3 Larew, S.G., Thomas, T.A., Cudk, M., and Ghosh, A. (2013). Air interface design and ray tracing study for 5G Millimeter Wave Communications. *IEEE GlobeCom Workshops* 117–122.

4 Hossian, E. and Hasan, M. (2015). 5G cellular: key enabling technologies and research challenges. *IEEE Instrumentation and Measurement Magazine* 18 (3): 11–21.

5 Radwan, A., Saidul Haq, K.M., Mumtaz, S. et al. (2016). Low-cost on-demand C-RAN based mobile small-cells. *IEEE Access* 4: 2331–2339.

6 C-RAN, The Road Towards Green RAN (2011) White Paper Version 2.5, China Mobile Research Institute, available: http://labs.chinamobile.com/cran/wpcontent/uploads/CRAN_white_paper_v2… · PDF file.

7 Bhushan, N., Li, J., Malladi, D. et al. (2014). Network densification: the dominant theme for wireless evolution into 5G. *IEEE Communications Magazine* 52 (2): 82–89.

8 Kudo, T. and Ohtsuk, T. (2013). Cell range expansion using distributed Q-learning in heterogeneous networks. *EURASIP Journal on Wireless Communications and Networking* 2013 (61): 1–10.

9 Sundaresan, K., Arslan, M.Y., Singh, S. et al. (2016). FluidNet: a flexible cloud-based radio access network for small cells. *IEEE /ACM Transactions on Networking* 24 (2): 915–928.

10 Namba, S., Warabino, T., and Kaneko, S. (2012). BBU-RRH Switching Schemes for Centralized RAN. *IEEE International ICST Conference on Communications and Networking in China* 762–766.

11 Jungnickel, V., Manolakis, K., Zirwas, W. et al. (2014). The Role of Small Cells, Coordinated Multipoint, and Massive MIMO in 5G. *IEEE Communications Magazine* 52 (5): 44–51.

12 Baştuğ, E., Bennis, M., and, Debbah, M. (2014) 'Cache-enabled small-cell networks: Modelling and tradeoffs', International Symposium on Wireless Communications Systems (ISWCS), 649–653

13 Bastug, E. (2015) 'Distributed cashing Methods in Small Cell Networks', PhD Thesis, University of Paris-SAClay.

14 Radunovic, B., Gunawardena, D., Key, P. et al. (2010). Rethinking indoor wireless: low power, low frequency, full-duplex. *IEEE Workshop on Wireless Mesh Networks, Pages* 1–6.

15 Raghavan, A., Gebara, E., Emmanouil, M. et al. (2005). Analysis and design of an interference canceller for collocated radios. *IEEE Transactions on Microwave Theory and Techniques* 53 (11): 3498–3508.

16 Radunovic, B., Gunawardena, D., Key, P., et al. (2009) Microsoft Research Cambridge, Technical Report MSR-TR-2009-148.

17 Sahai, A., Patel, G. and Sabharwal, A. (2011) 'Pushing the Limits of Full-Duplex: Design and Real-Time Implementation', Department of Electrical and Computer Engineering, Rice University, Technical Report TREE1 .1. 04.

18 Duarte, M. and Sabharwal, A. (2010) 'Full-duplex wireless communications using off-the-shelf radios: Feasibility and first results', Asilomar Conference on Signals, Systems and Computers, Proceedings of the conference,1558–15620

19 Choi, J-II, Jain, M., Srinivasan, K., et al. (2010) 'Achieving Single Channel, Full Duplex Wireless Communication' ACM Annual International Conference on Mobile Computing and Networking: Proceedings of the sixteenth annual international conference on Mobile computing and networking, 1–12.

20 Hong, S., Brand, J., Choi, J.-I.I. et al. (2014). Applications of Self-Interference Cancellation in 5G and Beyond. *IEEE Communications Magazine* 52 (2): 114–121.

21 Sabharwal, A., Schniter, P., Guo, D. et al. (2014). In-Band Full-Duplex Wireless: Challenges and Opportunities. *IEEE Journal on Selected Areas in Communication* 32 (9): 1637–1652.

22 Choi, J-II., Jain, M., Srinivasan, K., et al. (2010) 'Achieving Single Channel, Full Duplex Wireless presentation slides, Stanford University, available at: http://sing.stanford.edu/pubs/mobicom10-duplex-slides.pdf.

23 Zhang, Z., Chai, X., Long, K. et al. (2015). Full duplex techniques for 5g networks: self-interference cancellation, protocol design, and relay selection. *IEEE Communications Magazine* 53 (5): 128–137.

24 Korpi, D., Tamminen, J., Turunen, M. et al. (2016). Full-duplex mobile device: pushing the limits. *IEEE Communications Magazine* 54 (9): 80–87.

25 3GPP TS 36.211 version 8.9.0 Release 8 (2009) 'Physical channels and modulation'.
26 3GPP TS 36.211 version 9.1.0 Release 9 (2010) 'Physical channels and modulation'.
27 3GPP TS 36.211 version 10.4.0 Release 10 (2011) 'Physical channels and modulation'.
28 3GPP TS 36.211 version 11.1.0 Release 11 (2013) 'Physical channels and modulation'.
29 3GPP TS 36.211 version 13.5.0 Release 13 (2016) 'Physical channels and modulation'.
30 3GPP TS 36.211 version 14.2.0 Release 14 (2017) 'Physical channels and modulation'.
31 Kim, Y., Ji, H., Lee, J. et al. (2014). Full Dimension MIMO (FD-MIMO): The Next Evolution of MIMO in LTE Systems. *IEEE Wireless Communications* 21 (2): 26–33.
32 Nam, Y.-H., Ng, B.L., Sayana, K. et al. (2013). Full-dimension MIMO (FD-MIMO) for Next Generation Cellular Technology. *IEEE Communications Magazine* 51 (6): 172–179.
33 Nam, Y-H., Rahman, M.S, Li, Y., Xu, G., et al.(2015) 'Full dimension MIMO for LTE-Advanced and 5G', Information Theory and Applications Workshop (ITA), 143–148
34 Xu, G., Li, Y., Yuan, J., Monroe, R. et al. (2017). Full Dimension MIMO (FD-MIMO): Demonstrating Commercial Feasibility. *IEEE Journal on Selected Areas in Communications* 35 (8): 1876–1886.
35 Anokye, P., Ahiadormey, R.K., Song, C., and Lee, K.-J. (2018). Achievable Sum-Rate Analysis of Massive MIMO Full-Duplex Wireless Backhaul Links in Heterogeneous Cellular Networks. *IEEE Access* 6: 23456–23469.
36 Xu, G., Li, Y., Yuan, J. et al. (2017). Full Dimension MIMO (FD-MIMO): demonstrating commercial feasibility. *IEEE Journal on Selected Areas in Communications* 35 (8): 1876–1886.
37 Ji, H., Kim, Y., and Lee, J. (Samsung Electronics, Korea), Onggosanusi, E., Nam, Y. and Zhang, J. (Samsung Research America), Lee, B. (Purdue University), Shim, B. (Seoul National University) (2017). Overview of Full-Dimension MIMO in LTE-Advanced Pro. *IEEE Communications Magazine* 55 (2): 176–184.
38 Schwarz, S., Mehlfuhrer, C., and Rupp, M. (2010) ' Calculation of the Spatial Preprocessing and Link Adaption Feedback for 3GPP UMTS/LTE', Wireless Advanced 2010, 1–6.
39 Zhang, W., Xiang, J., Li, Y.-N.R. et al. (2015). Field Trial and Future Enhancements for TDD Massive MIMO Networks. *IEEE 26th Annual International Symposium on Personal, Indoor, and Mobile Radio Communications (PIMRC)* 2339–2343.
40 Yong, S.K., Sahink, M.E., and Kim, Y.H. 'On the Effects of Misalignment and Angular Spread on the Beamforming Performance' Samsung Advanced Institute of Technology (SAIT), Communication Lab, Available at: http://wcsp.eng.usf.edu/papers.
41 Lee, B., Choi, J., Seol, J.-Y. et al. (2015). Antenna Grouping Based Feedback Compression for FDD-Based Massive MIMO Systems. *IEEE Transactions on Communications* 63 (9): 3261–3274.
42 Kim, T., Park, J., Seol, J-Y., et al. (2013) 'Tens of Gbps Support with mmWave Beamforming Systems for Next Generation Communications', IEEE Globecom - Wireless Communications Symposium, 3685–3690.
43 Yang, H., Herben, M.A.J., Akkermans, I.J.A.G., and Smulders, P.F. (2008). Impact Analysis of Directional Antennas and Multiantenna Beamformers on Radio Transmission. *IEEE Transactions on Vehicular Technology* 57 (3): 1695–1707.
44 Qiang, W. and Zhongli, Z. (2011). Reinforcement learning model, algorithms and its application. *IEEE International Conference on Mechatronic Science, Electrical Engineering and Computer* 1143–1146.
45 Watkins, C. (1989) 'Learning from Delayed Rewards', PhD Thesis, King's College, Cambridge, UK.

Further Reading

Tzanidis, I., Li, Y., Xu, G. et al. (2015). 2D Active Antenna Array Design for FD-MIMO System and Antenna Virtualization Techniques. *International Journal of Antennas and Propagation* Article ID 873530, Hindawi Publishing Corporation.

Nam, Y-H., Li, Y., (Charlie) Zhang, J. (2014) '3D Channel Models for Elevation Beamforming and FDMIMO in LTE-A and 5G', IEEE Asilomar Conference on Signals, Systems and Computers, 805–809.

Cisco (1992-2007) 'Antenna Patterns and Their Meaning', White paper.

Xu, G., Li, Y., Nam, Y-H., et al. (2014) 'Full-Dimension MIMO: Status and Challenges in Design and Implementation – SAMSUNG.

Liu, C., Wang, W., Li, Y., et al. (2015) 'Dual Layer Beamforming with Limited Feedback for Full-Dimension MIMO Systems', IEEE International Symposium on Personal, Indoor and Mobile Radio Communications - (PIMRC): Workshop on Advancements in Massive MIMO', 2329–2333.

Furuskog, J., Werner, K., and Riback, M. 'Field trials of LTE with 4×4 MIMO', Ericsson Review. 1 2010.

Ji, H., Kim, Y., Kwak, Y., and Lee, J. (2014) 'Effect of three-dimensional beamforming on full dimension MIMO in LTE-Advanced last', IEEE Globecom. Workshop, 821–826.

Jassal, A., Khanfir, H., and Lopez, S.M. (2014) 'Preliminary system-level simulation results for the 3GPP 3D MIMO channel model, IEEE Vehicular Technology Conference (VTC2014-Fall), 1–5.

Matsuno, H., Nakano, M. and Yamaguchi, A. (2013) 'Slim Omnidirection al orthogonal polarization MIMO antenna with halo and patch antennas on cylindrical ground plane', European conference on antennas and propagation,720–724.

Wu, H., Cai, J., Xiao, H., et al.(2016) High-Rank MIMO Precoding for Future LTE-Advanced Pro, *IEEE Vehicular Technology Conference (VTC Spring)*,1–6

3

5G Enabling Technologies: Network Virtualization and Wireless Energy Harvesting

3.1 Introduction

Virtualization, in its simplest form, is duplication of a hardware platform such as server, storage device, network resource, etc. using software with all functionalities separated from the hardware and reproduced as 'virtual instance' and operating similar to the traditional hardware elements. Clearly, there will be a need for some 'off the shelf' physical devices to support the virtual instances. In addition, a single hardware platform (network) can support multiple virtual devices that can increase/decrease as needed. Consequently, a virtualization solution for enhancing network capacity and coverage performance is more scalable and cost effective than a traditional physical based solution. In practice, network virtualization produces overlay networks on top of the physical hardware and creates virtual networks with the virtualized infrastructure.

Network virtualization recapitulates networking connectivity and services, hardware-based traditionally, to be delivered using a logical virtual network that runs on top of a physical network, so that virtualization is extended beyond the physical layer (PHY) to include layer 2 and layer 3 (switching and routing) and including transport layer as well. Network virtualization helps network on-demand provision and simplifies network scaling, workload and resource adjustments. In the early phase of network virtualization, it was modest but then evolved by 3rd Generation Partnership Project (3GPP) to a key enabling technology in future 5G networks during the last two decades. Examples of early-phase network sharing arrangements include single base-station (BS) resources sharing between multiple operators. However, contemporary virtualization is based on infrastructure sharing and spectrum sharing. Therefore, virtualization can be thought of as two separate network operational processes: infrastructure providers (InPs) to construct and manage BSs, spectrum access points (APs) mobility management entities, and service providers (SPs) providing various services to user equipments (UEs).

Energy harvesting (EH) is important to support mobile UEs with recharged batteries to keep their connectivity with 5G networks. Energy harvesting implies a process of deriving energy from external sources such as solar, wind, and thermal energy. Energy harvested provides recharging energy for mobile UEs batteries. On the other hand, wireless power transfer is the transfer of far-field electromagnetic (EM) energy over a relatively longer distance using beamforming techniques to a target receiver where the received signals are converted to DC and stored in power small devices such as implanted medical devices or mobile UEs.

5G Physical Layer Technologies, First Edition. Mosa Ali Abu-Rgheff.

3.2 Network Sharing and Virtualization of Wireless Resources

3.2.1 Earlier Network Sharing

Network operators used network sharing opportunities to share the costs for mobile networks, particularly in the early phase of deployment. Early network sharing involves sharing basic assets such as sites, masts, power, etc., which are defined as elements of passive sharing. Since network sharing as a concept brings saving in capital expenditure (CAPEX) and operational expenditure (OPEX) for operators, active sharing is appealing to operators in the sharing networks equipment and possibly radio resources. The terms related to network sharing are defined in [1]. *Traditionally a mobile network* is a public land mobile network (PLMN) composed of radio access network (RAN), core network (CN) and one serving operator to provide services to its subscribers. Subscribers of other operators can receive services from other operators who have international roaming agreements with their home provider. A *core network operator is an operator* that shares at least a RAN with multiple operators and provides services to subscribers and also subscribers of other operators who have a roaming agreement with their home provider.

Gateway core network (GWCN) is a network-sharing configuration in which parts of the CN (*MSCs/SGSNs/MMEs*) are also shared. *Multi-operator core network* (MOCN) is a network-sharing configuration in which only the RAN is shared. *Supporting/non-supporting UE* means a UE supports/does not support network sharing, meaning it is able to select a CN operator within a shared network/does not act in response to network sharing information.

Network sharing in 5G has evolved from sharing the actual physical resources into sharing virtual wireless networks elements of the physical networks. The concept is called *virtualization,* and it is a key enabler technology that is more in demand than ever before.

There are many types of network sharing possibilities, depending on operator strategies and on national legislations adopted in different countries. The popularity of network sharing stems from the fact that wireless area coverage can be achieved quickly and at best value for monetary solution. Accordingly, sharing networks and infrastructure become an integral part of modern 3GPP systems. It is worth noting that *earlier* 3GPP systems do not support full network sharing between different operators. However, limited support exists in Release 99 that permits operators to share a common UMTS Terrestrial Radio Access Network (UTRAN) and certain parts of the core networks, as illustrated in Figure 3.1.

The limited network sharing is used, albeit without Universal Mobile Telecommunications System (UMTS) license by, for example, 2G operators to provide its users with 3G services using another operator's allocated spectrum. Clearly, operators' cooperation is essential to the success

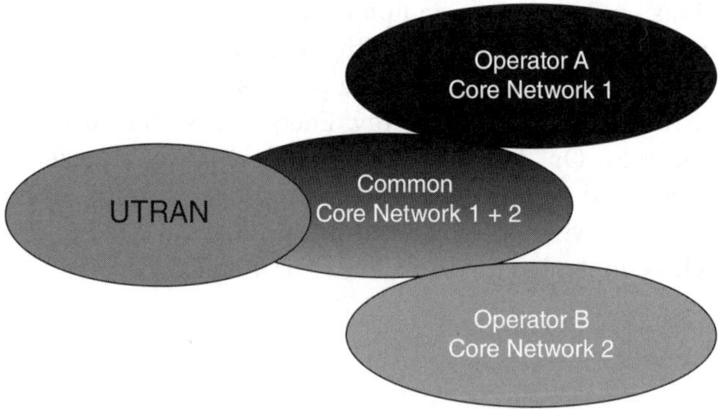

Figure 3.1 Two operators sharing the same UTRAN, and part of the core network is shared as well [2].

of such scenarios. It is important to understand that such network sharing schemes are certainly not optimized. The mobile industry has seriously considered infrastructures and scenarios for better solutions and formed a collaborative group of telecommunication associations called the 3GPP that developed a dedicated network for the future in an evolved system, generally named as long-term evolution, in short (LTE). Meanwhile, the following are some of the scenarios considered [2]:

- *Scenario 1*: Multiple core networks sharing common RAN (Release 99). Operators that have licensed multiple frequency allocations can use the 3GPP Release 99 to share the RAN elements but not the radio frequencies. Operators are able to connect to their dedicated carriers in the shared radio network controller (RNC) as illustrated in Figure 3.2.
- *Scenario 2*: Geographically split networks sharing. Two (or more) operators, with their own RANs and individual 3G licenses, can provide coverage of the entire country, which can have two possible cases: the first case when two or more operators take use of users national roaming so that only one core network will be associated with each RAN, but coverage regions may overlap. This shared-network scenario is shown in Figures 3.3 and 3.4. A possible second case is when operators can have their individual core networks connected to both RANs throughout the entire coverage area but operating at the different operator's allocated spectrum in different parts of the coverage area. The connection of the core networks to RANs can be achieved either by connecting the RNCs to both operators' core network elements or by sharing parts of the core network, e.g. Serving GPRS Support Nodes (SGSNs) and/or mobile switching centres (MSCs) or employing common core network parts.
- *Scenario 3*: Common network sharing. In this scenario, one operator provides coverage in a specific geographical area, and other operators are allowed to use this coverage for their subscribers. Outside this geographical area, individual coverage is provided by each of the individual operators. For example, consider the case of two operators, where a third-party could provide coverage to subscribers of operators A and B in areas with *high population density*. In *less-dense areas*, the subscribers should connect to the access network of their operator.
- *Scenario 4*: Common spectrum network sharing. Common spectrum network sharing is applicable when one operator has a 3G license and shares the allocated spectrum with other operators or a number of operators decide to pool their allocated spectrums and share the total spectrum (operators without allocated spectrum may also share this pooled spectrum).
- *Scenario 5*: Multiple RANs sharing common core network. Multiple RANs share a common network and can belong to different PLMNs and network operators. Common core network elements such as HSS/home location registers (HLR), SGSN etc. can be shared. The scenario is depicted in Figure 3.5. More details can be found in [2].

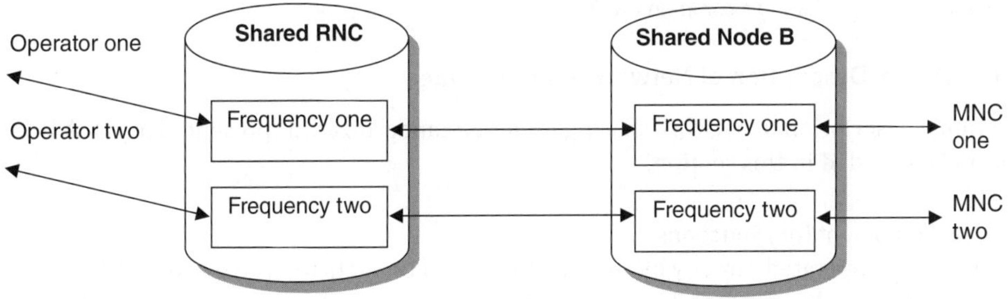

Figure 3.2 3GPP Release 99 framework has dedicated carrier layers in the RAN for multiple operators. The operators transmit their own mobile network code (MNC) on their dedicated carrier [2].

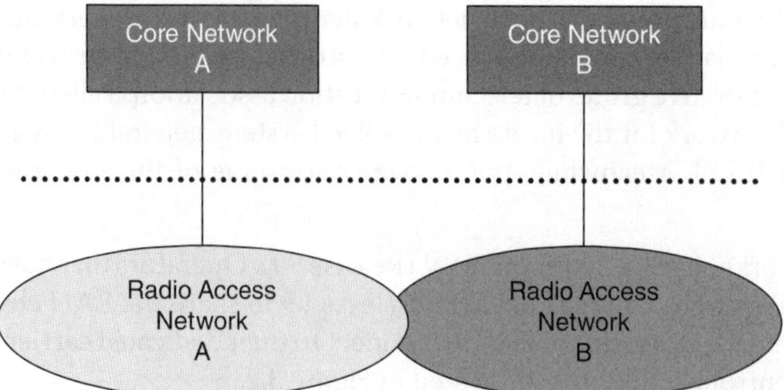

Figure 3.3 Geographically split network using national roaming between operators [2].

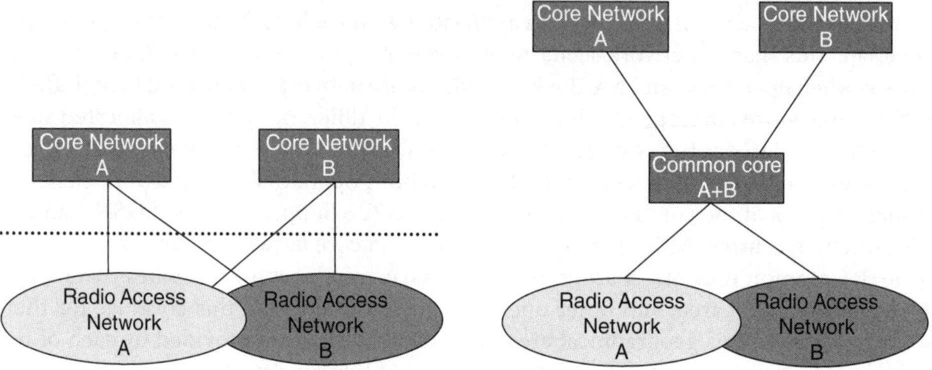

Figure 3.4 Geographically split shared radio networks scenarios with dedicated or common core networks [2].

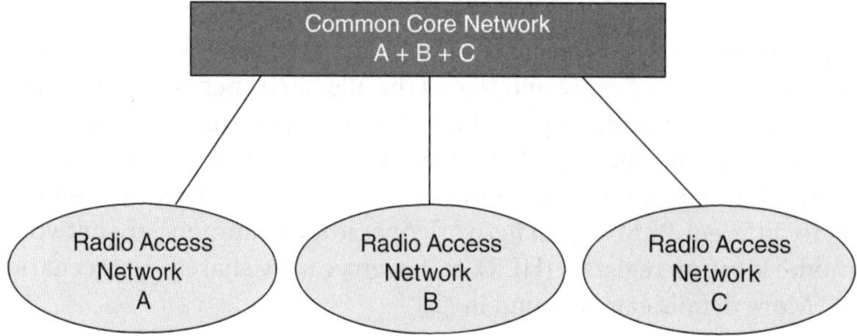

Figure 3.5 Multiple RANs sharing a common CN [2].

3.2.2 Functional Description of Network Sharing Nodes

The functional description of network sharing nodes is standardized, as presented in [1]. A brief description is provided in this section.

3.2.2.1 User Equipment (UE) Functions

A UE is any device operated directly by the end-user and connects to evolved Node B (eNB) to communicate. UE can be of a mobile equipment, a laptop computer, a desktop computer, etc. Two categories of UEs are defined relative to network sharing. Those UEs that operate with a shared network are called supporting UEs and the other category is called nonsupporting UEs.

A supporting UE selects the core network operator and provides the network identification PLMN-id of the serving operator for routing purposes and to indicate which of the core network operators is selected. PLMN-id is made up of two sets of numbers (MCC + MNC) where MCC is mobile country code and MNC is the mobile network code.

Nonaccess stratum (NAS) is a protocol for messages passed between the UE and core nodes [3]. NAS establishes and maintains the communication sessions as UE moves. Other typical NAS messages include authentication messages, service requests messages, and messages update. Once the UE establishes a radio connection, the UE uses the radio connection to communicate with the core nodes to coordinate the service

3.2.2.2 Radio Network Controller (RNC) Functions

RNC is an element in the RAN responsible for controlling the eNB that is connected to it. RNC responsibility includes radio resource management; some functions of the mobility management; and it is the element where data encryption is carried out before data transmission. RNC transmits network-sharing information and contains a list of the available core network operators in the shared network. Information relating to NAS is transmitted to a UE in dedicated signalling. The RNC sends the information for both supporting and nonsupporting UEs and the individual UE selects the appropriate information to use.

3.2.2.3 Evolved Node B (eNB) Functions

eNB is the hardware element in the UMTS radio access that is connected to the network and communicates directly wirelessly with mobile UEs, like a base transceiver station (BTS) in GSM networks. Traditionally, Node B is controlled by RNC. However, with an eNB, there is no separate controller element that simplifies the architecture and allows shorter response times. The infrastructure equipment eNB indicates the selected core network operator at the initial network layer signalling and uses the selected core network operator information as provided by the UE or by the mobility management entity (MME) to select target cells for future handovers and possibly radio resources in general, as appropriate.

3.2.2.4 Base Station Controller (BSC) Functions

If system information is transmitted to a supporting UE to support handover to other type of radio technology accessing the core network, the base station controller (BSC) indicates the PLMN-id of the core network operator towards which the UE already has signalling connection. If the UE is nonsupporting the BSC indicates the selected CN operator for nonsupporting the UEs.

3.2.2.5 Mobile Switching Centre (MSC) Functions

The MSC is the key element of a network switching subsystem. It is linked to communications switching functions such as call startup, release, and routing. BSs are connected to MSC and MSC is connected to the public switched telephone network (PSTN). MSC plays a key role in handovers. When a mobile user approaches the edge of the cell, a BSC requests a handover assistance from its MSC, and the MSC provides a list of adjacent cells and their corresponding BSC to facilitates the handover to the appropriate BSC.

When a UE accesses MSC for the first time, the MSC retrieves the international mobile subscriber's unique identity number (IMSI) from another MSC/visitor locator register (VLR) or from the UE. For GWCN, the MSC determines a serving CN operator based on the UE selected operator indicated by the RAN, or, for a nonsupporting UE, based on the RAN selected CN operator. In the case of a MOCN, an MSC may not be able to provide service to the UE. The UE may then have to be redirected to MSC of another core network operator. The MSC/VLR that finally serves the UE assigns a network resource indicator (NRI) to the UE. This will allow the RAN to route any subsequent UE accesses to the serving MSC/VLR.

3.2.2.6 Mobility Management Entity (MME) Functions

MME plays an important role in LTE evolved packet core (EPC). It is the main signalling node and is responsible for initiating paging and authentication of the mobile device. MME connects to the eNB through MME interface. Multiple MMEs can be grouped in a pool to meet increasing signalling load in the network. MME also plays a key role in handover signalling between LTE and 2G/3G networks.

When a UE accesses an MME for the first time, it is not yet known by the MME; the MME verifies whether the UE is permitted to access the selected PLMN. For that purpose, the MME retrieves the IMSI from another MME/SGSN or from the UE. The MME stores the identity of the selected core network operator. The MME indicates the selected core network operator (PLMN-id) to the UE in the Globally Unique Temporary Identity (GUTI).

Wireless network sharing has been developed by a number of standardization organizations including 3GPP, which defined two architectures for network sharing [2, 4], namely: MOCN, where only RAN with corresponding resource are shared, as depicted in Figure 3.6. This figure shows three different CN operators (A, B, C) connected to the same RAN (operator X). GWCN allows the sharing of the whole network, including the core network as shown in Figure 3.7 where RAN and its elements and CN elements are all shared as well.

3.2.3 Single BS Shared by a Set of Operators

Consider a single BS shared by a set of operators, $\mathcal{S}$, with cardinality (number of elements in $\mathcal{S}$) $|\mathcal{S}|$, and a set of active users served by the BS denoted by $\mathcal{K}$, with its cardinality $|\mathcal{K}|$. Let the number of users associated with operator m be $\mathcal{K}_m$ as illustrated in Figure 3.8. Assume a constant bandwidth and time slots allocation. Consider an arbitrary time slot. At every time slot, the scheduler at the BS decides how to split the radio blocks (RBs) among the operators. There is no fixed resource allocation scheme over time, as illustrated in Figure 3.9.

As an infrastructure-shared network, we assume that InP and operators have negotiated an acceptable sharing ratio priori, which is a service-level agreement that sets an individual sharing ratio $\widetilde{S}_m \in [0, 1]$ for each arbitrary operator m. Note that sharing ratio is a measure of your sharing of the available resources and can be constant, continuous, or dynamic. Commonly sharing comes as discrete blocks, depending on the allocated number of RBs. In the following analysis, for simplicity and without loss of generality, we set the sharing ratios to be continuous.

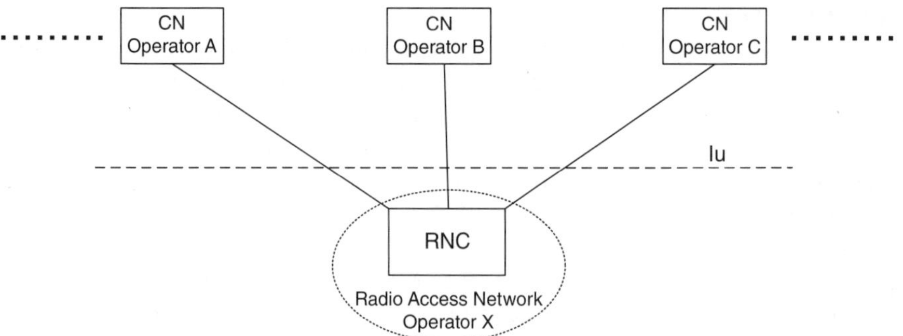

Figure 3.6 A multi-operator core network (MOCN) in which multiple CN nodes are connected to the same RNC and the CN nodes are operated by different operators [1].

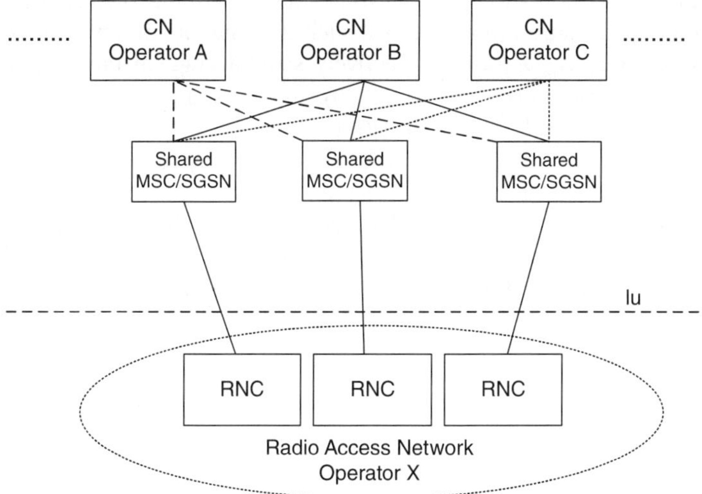

Figure 3.7 Gateway core network (GWCN) configuration for network sharing. Besides shared radio access network nodes, the core network operators also share core network nodes [1].

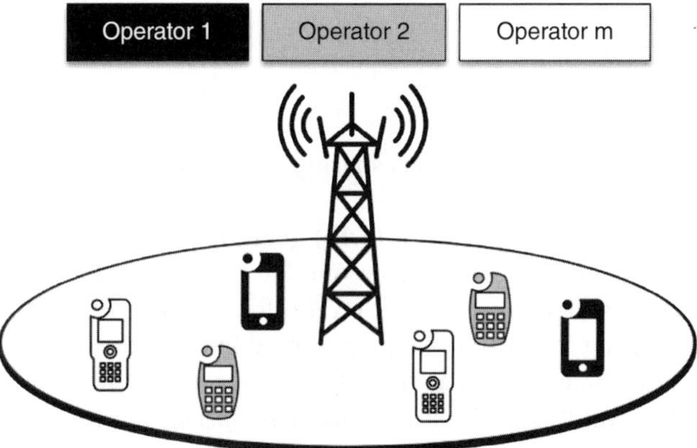

Figure 3.8 A single base station shared by multiple operators [7].

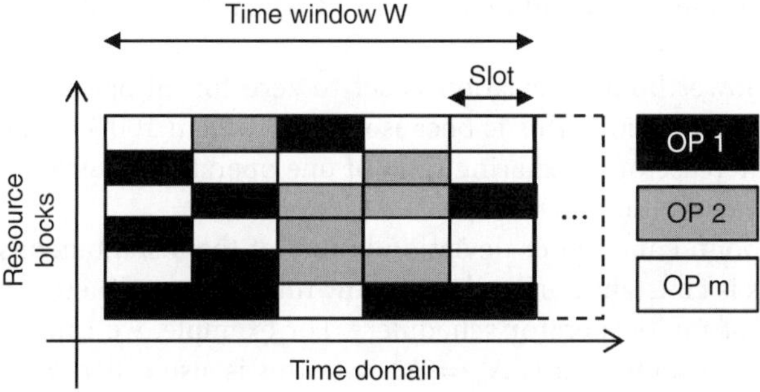

Figure 3.9 Resource allocation over time [7].

Furthermore, for dynamic sharing, we allow the scheduler to deviate from the agreed sharing ratios for certain slots. We limit the deviation to an interval such that the maximum positive and negative deviations are $\Delta_m^+ \in [0, 1]$, $\Delta_m^- \in [0, 1]$, respectively, so an average instantaneous sharing ratio $s_m[n] \in [0, 1]$ may vary between $[max\,(\widetilde{S}_m - \Delta_m^-, 0), \min(\widetilde{S}_m + \Delta_m^+, 1)]$. Define the instantaneous sharing ratio assigned to the kth user $s_k[n] \in [0, 1]$.

The scheduler aims to maximize the sum of the continuous concave utility function $f(r_k[n], s_k[n]\,)$ for scheduling policies (i.e. *maximum-rate*), which depends on the spectral efficiency $r_k[n] \in \boldsymbol{R}$ in bits/sec/Hz for user k at time slot n. Consider a general scheduling problem for multiple operators, maximizing the sum of utility function over all users using their fractional allocated resources $\boldsymbol{s}[n]$ $(i.e. s_1, \dots\dots . s_{|\mathcal{K}|})$ per time slot n can be expressed through Eqs. (3.1a) to (3.1f):

$$\max_{0 \le s \le 1} \sum_{k \in \mathcal{K}} f(r_k[n], s_k[n]\,) \tag{3.1a}$$

$$\text{s.t} \sum_{k \in \mathcal{K}} s_k[n] = 1 \tag{3.1b}$$

$$\sum_{k \in \mathcal{K}} s_k[n] = s_m[n] \qquad \textit{for all } m \in \mathcal{M} \tag{3.1c}$$

$$\varepsilon_m[n] = \left(\frac{1}{W} \sum_{i=n-W+1}^{n} s_m[i] \right) - \widetilde{S}_m \quad \textit{for all } m \in \mathcal{M} \tag{3.1d}$$

$$-\Delta_m^- \le \varepsilon_m[n] \le \Delta_m^+ \quad \textit{for all } m \in \mathcal{M} \tag{3.1e}$$

$$s_k[n] \ge 0 \quad \textit{for all } k \in \mathcal{K} \tag{3.1f}$$

where W is a time-window, $s_k[n] \in [0, 1]$ is the instantaneous sharing ratio allocated to the kth user at time slot n and $\mathcal{M}$ is the total number of operators.

We now interpret all the constraints on (3.1a). Constraint (3.1b) guarantees the sum of all resource components allocated to users of any operator to be one (i.e. complete use of the resources). Constraint (3.1c) ensures the instantaneous sharing ratio for operator m at time slot n, i.e. $s_m[n]$ is equal to sum of the sharing fraction of the users served by operator m. Constraint (3.1d) defines the resource ratio deviation $\varepsilon_m[n]$ as the difference between the average sharing ratio of operator m over time interval W and the *sharing ratio for operator*, i.e. $\widetilde{S}_m$. The implication of constraint (3.1d) is that operator m is using more resources than its share ratio when $\varepsilon_m[n] > 0$, and using less than its share ratio when $\varepsilon_m[n] < 0$ and using exactly its share when $\varepsilon_m[n] = 0$. Constraint (3.1e) sets the lower and upper limits of the deviation of the instantaneous sharing ratio to the moving average $\varepsilon_m[n]$, and constraint (3.1f) confirms that users sharing ratios to be non-negative.

It is interesting to note that when lower bound deviation is set to zero for all operators, this automatically prevents any positive deviation. This is because no more than 100% of the resources can be shared, so that any increase of the sharing ratio of one operator requires an equal ratio reduction of the sharing ratio of other operator(s).

So far, we have assumed the lower bound and upper deviation bound of the instantaneous sharing ratio of resources by operators is constant during the window time interval. There are other formations that allow a variety of multi-operator schedulers. For example, we can set the instantaneous sharing ratios to be constant, that is $\Delta_m^- = \Delta_m^+ = 0$ (this is also called *fixed sharing*), which means $\varepsilon_m[n]$ is forced to zero for every operator in every time slot and also the instantaneous sharing ratio, $s_m[n]$, always coincides with $\widetilde{S}_m$. On the other hand, free sharing strategy allows *opportunistic scheduling* to choose any sharing ratio for the operators as long as it maximizes the utility function in (3.1a). This strategy is carried out by choosing $\Delta_m^- = \Delta_m^+ = 1$

and applies no restriction of the scheduler, but it provides no assurance on the reliability at which a sharing ratio is kept. Note in this case $\varepsilon_m[n]$ is $[-1, 1]$ instead of $[0, 1]$ as defined in (3.1e) above.

3.3 Evolved Resource Sharing

Resource sharing is an old idea, originated by IBM during the 1960s, to logically divide the available resources to mainframe computers between different applications, but it didn't take off until the economic considerations became the dominant factor in system deployment. However, wired networks have taken the evolved resource idea further such that it is now a pivotal design practice in networked Information and communication technology (ICT) systems. Nonetheless, the new idea of evolved sharing in wireless networks is motivated by the great benefits gained previously in the wired networks and also can be considered as a natural start of wireless network virtualization.

Wireless networks virtualization is the process of building up a set of logical architectures using a given set of physical entities, that are transparent to the end users. As an example, a physical (hardware) server can be used to make a set of virtual servers comprising logical processors, network interface cards, memory, and storage that are similar in functions to the physical server. The key challenge in virtual wireless networks is to maximize the physical entities utilization and at the same time, deliver the required performance to end users. Wireless networks virtualization has to demonstrate three important properties, namely *end users' isolation, provision of customized services,* and *maximum system utilization.* Wireless networks virtualization can support the separation of traffic flows in terms of quality of service (QoS) and security.

Evolved sharing of the available spectrum provides better utilization and efficiency to operators in terms of infrastructure ownership (i.e. capital and operation expenditure) and also service provisions.

Wireless networks evolved sharing comprise *infrastructure sharing* and *spectrum sharing,* which is more complex than in wired network virtualization. This is due to the fact that there are many different wireless network topologies (i.e. infrastructure, ad hoc, single and multi-hop, etc.) and different spectrum bands (from several hundred MHz to several GHz, licensed and unlicensed, etc.), different coverage schemes (wide, metro, local, and personal), and finally different mobility managements.

Network sharing as a concept, whether used in wired or wireless networks, can be viewed as a process of splitting the entire network into InPs and SPs. InPs construct and manage the BSs, MMEs, spectrum, and APs. SPs provide various services to subscribers. Wireless network spectrum sharing and splitting implies breaking the resources into *slices.* In the advanced resource-sharing evolution, each virtual slice has its own virtual core network and virtual access network and hence appears to the user as an entire system by itself.

Recent research work investigated various analytical and experimental models for wireless evolved sharing and evaluated virtual architectures. Based on the concept of InPs and SPs, three paradigms are proposed in [5], namely:

1. *Universal sharing,* where the whole radio access path is considered as a cloud of constituents and the sharing is pervasive. The cloud comprises heterogeneous BSs with macro, pico, femto cells, relays, and other APs as well as wired backbones that are transparent to the user. The SP of a specific service chooses a package of network components such as links and spectrum and configures them in the desired layout to operate dynamically in an on-demand fashion.

2. *Cross-infrastructure sharing* is based on the assumption that sharing is possible across all InPs (inter-InP) and within individual InPs (intra-InP). This paradigm enables InPs in a given geographical area to allow their resources to be shared across SPs.
3. *Limited intra-infrastructure sharing* considers sharing with a single InP, which has spectrum used by different radio access technologies (RATs), and spectrum sharing occurs between SPs and across RATs. However, these paradigms are not yet well analysed or evaluated to check their efficiency and performance.

Conversely, wireless network virtualization, a key enabler for 5G, can be used to integrate heterogeneous wireless networks and to coordinate network resources appropriately. A model of wireless network virtualization proposed in [6] consists of three planes: the *data plane*, the *cognitive plane*, and the *control plane*, depicted in Figure 3.10. The functions of each plane are summarized as follows:

- The *data plane* manages the transmission of user traffic over the virtual networks. Heterogeneous RANs and corresponding wireless resources, are sliced into virtual networks, which carry user traffic.
- The *cognitive plane* functions to control the *state* of services, network resources, and user requirements. The information about the state is collected from UEs, RANs, and SPs,

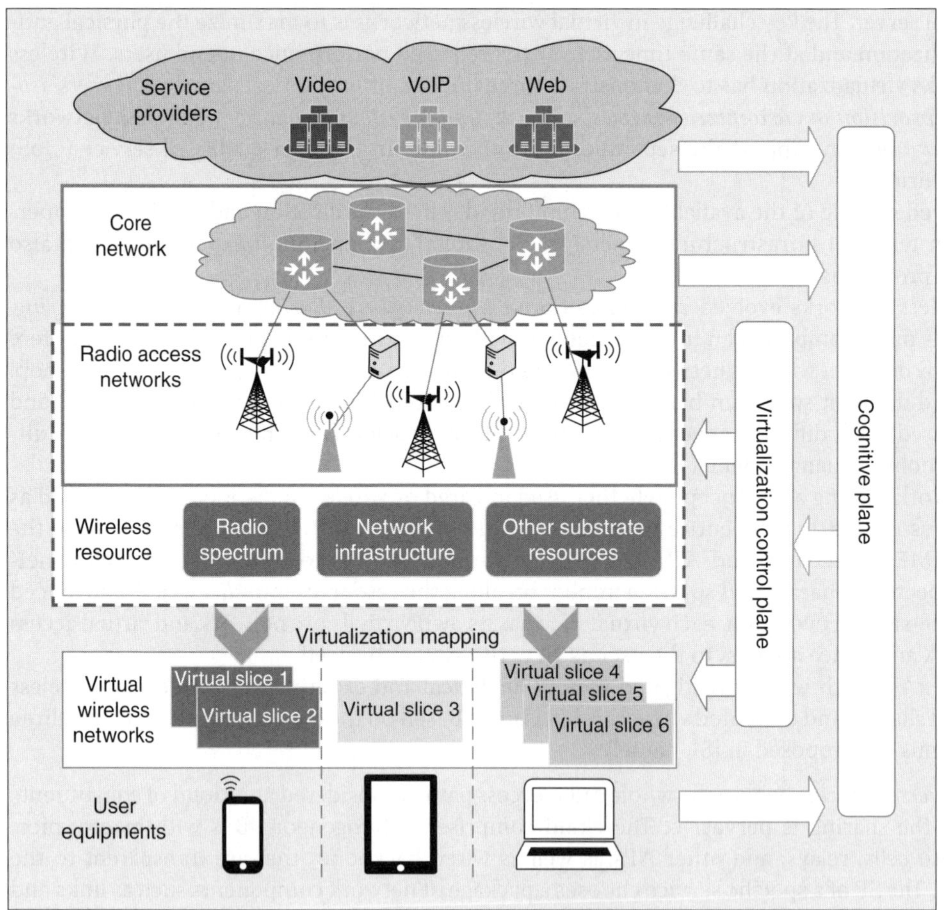

Figure 3.10 Model of wireless network virtualization [6].

and transmitted to the control plane, to properly slice and allocate the infrastructure and resources to the UEs.

- The *control plane* takes care of integrating all the available RANs and resources together and slicing them into virtual networks. Virtual networks are independent, and there is no interference and conflict between them. Each virtual network comprises network infrastructure and wireless resources to be allocated to UEs. Accordingly, network virtualization fundamentally is a process of *sharing a network system and its resources* with an optimized approach.

3.3.1 Principle of Cellular Network Evolved Resource Sharing

As discussed previously, the network infrastructure is separated from the services it provides; that is, we have InPs and the mobile network operators representing the SPs. BS resources, antennas, power, and spectrum are provided by an InP. The latter divides the physical resources into isolated slices. Each slice is allocated to an operator, who then allocates the resource to its users.

Basically, there are three players in such a scheme for resource allocation: InP who own the resource, the end users who benefits by employing these allocated resources, and middlemen acting between these two players. Most resource allocation used in pre-LTE is achieved by a single-level scheme. That is, the InP directly allocates physical resources to users of different mobile network operators according to certain predetermined resource-sharing ratios without involving middlemen. Sharing wireless resources among network operators saves costs by efficiently using the resources (hardware and spectrum).

3.3.2 Single-Level Resource Allocation Among Operators

There has been intensive work to optimize the *single-level resource allocation* among multiple operators. A widely used model for resources sharing is to let operators define a sharing ratio that the resource provider has to carry through. The ratio is a fraction of all resources per cell. The sharing ratio may be used in two ways. *Constant sharing ratio* implies that an operator is served precisely according to its sharing ratio. This approach guarantees that the operator always gets the resource share it pays for, but it is inefficient when the operator does not fully deploy its allocated share. *Dynamic sharing* prevents wasting unused resources by allowing the scheduler to deviate from the sharing ratio for a short time; thus unused resources can be made available to other users in need. The sharing ratio may still hold again over some longer span. Dynamic resource sharing ratio is considered in [7]. Another dynamic approach that maximizes resource usage is described in [8] where resources are divided into a number of frequency (F) and time domain (T) slices and the total number of resources available is $F \times T$.

In the *static version* of the approach, successful network request receives the same set of resources for every time slot for the duration specified. To explain this approach, consider a substrate of 5×5 resources available as illustrated in Figure 3.11. The static scheme is illustrated in Figure 3.11a. During time slot1, two requests, A and B, each request 2×3 resources arrived and are accepted. At the time slot2, the resources allocated to request A are released, and at the same time a new request C for 3×3 resources arrives. However, under the static approach, request C will be rejected even though there are enough resources available. This is because the static embedding scheme cannot change the assigned resource of the embedded request B. The solution to this problem is *dynamic re-assigning* the existing allocation at every time slot, as shown in Figure 3.11b. Thus, dynamic embedding enables resources to be much more efficient, and any additional requests that are within the available resource substrate can be accepted.

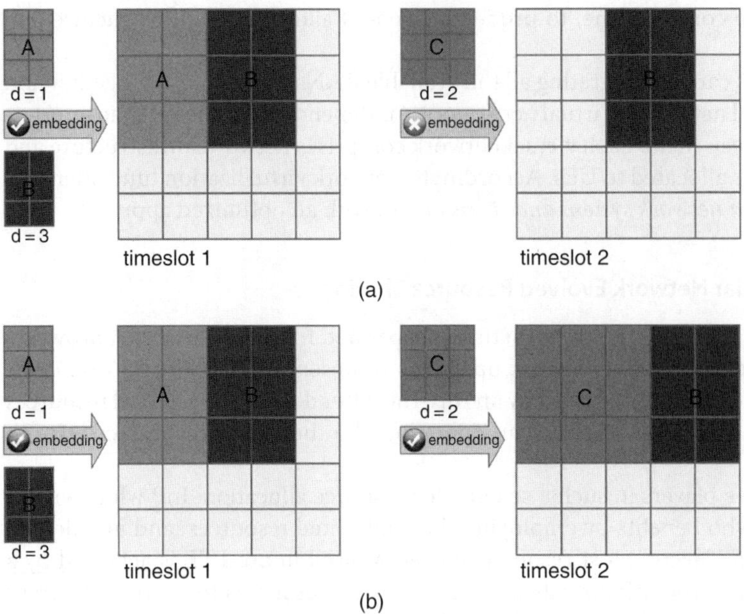

Figure 3.11 Using static embedding, requests A and B can be embedded in time slot 1; however, request C is rejected in time slot 2. When using dynamic embedding, request C can be accepted in time slot 2. (a) Static embedding (b) Dynamic embedding [8].

It is worth noting that dynamic resource allocation deploys more complex hardware than the static and involves larger overhead penalty. Dynamic allocation algorithms that can be used are Karnaugh-map (static and dynamic) algorithms and Greedy algorithms. It turns out that the greedy allocation provides an increase in revenue and lowers request rejection rate compared to Karnaugh-map (static and dynamic) embedding.

A dynamic slicing with flexible scheduling scheme, based on the SPs service contracts and fairness needs of users at cell centres and at the cell edge, is described in [9]. To analyse the slice scheduling, let us consider the downlink (DL) of a single cell in LTE system, as illustrated in Figure 3.12, and use it to provide detailed analysis of the dynamic slicing approach. We assume that different SPs are isolated by proper allocation of the physical resource blocks (PRBs) to

Figure 3.12 Resource slice scheduling for the downlink in a single cell in LTE system model [9].

the users that are served by the SPs. Each SP is assigned a certain amount of resource from the available PRBs based on the service contract with the InP.

Denote the minimum amount of PRBs allocated to SP m as ρ^m_{min} i.e. the *contracted resource share of SP m*. Let the system comprise $\mathcal{M}$ SPs, each SP m serving K_m users (i.e. SP1 serves K_1 users, SP 2 serves K_2 users, and so on). Further, assume there are C subchannels available, each with bandwidth B. Specifically, consider T subframes in the time domain so a total of available PRBs for scheduling within duration T will be TC. In addition, we assume the BS of the target cell has a total transmit power of P_{total} and the power allocated to user k served by SP m be p^{mk}_{tC}, denoted as the power allocated to user (m, k), in subchannel C at time t. Assume the BS receives perfect channel state information (CSIs) from all users served by all SPs and the channel gain of user (m, k) over subchannel C at time t as h^{mk}_{tC}. The rate r^{mk}_{tC} linked to each PRBs (t, C) for user (m, k) is then shown in Eq. (3.2):

$$r^{mk}_{tC} = \frac{B}{T}\log_2\left(1+\frac{p^{mk}_{tC}\mathrm{h}^{mk}_{tC}}{N_0B}\right) \tag{3.2}$$

where N_0 is the noise spectral density. Let $\mathcal{R}^m_k$ be the data rate achieved by user k served by SP m for $m = 1, 2, \ldots, \mathcal{M}$ and $=1, 2, \ldots \ldots K_m$. Let us introduce the assignment binary variable x^{mk}_{tC} which expresses user (m, k) assignment $PRB(t, C)$ during a scheduling round:

$$x^{mk}_{tC} = \begin{cases} 1 & \mathit{if\,PRB}(t, C) \text{ is assigned to user (m, k)} \\ 0 & \mathit{otherwise} \end{cases} \tag{3.3a}$$

In addition, we have

$$\sum_{m,k} x^{mk}_{tC} = 1 \tag{3.3b}$$

Each SP is assigned a *minimum number of PRBs* N^m_{PRB} to satisfy a minimum acceptable level of service for its users such that $N^m_{PRB} \geq \rho^m_{\min}\ TC$. We can now formulate the optimization of the sum rate $\mathcal{R}^m_k$ of the system as

$$\begin{aligned}\mathcal{R}^m_k &= \sum_{t=1}^{T}\sum_{C=1}^{C} x^{mk}_{tC}\, r^{mk}_{tC} \\ &= \sum_{t=1}^{T}\sum_{C=1}^{C} \frac{B}{T}x^{mk}_{tC}\ \log_2\left(1+\frac{p^{mk}_{tC}\mathrm{h}^{mk}_{tC}}{N_0B}\right) \quad \mathit{for\ all\ m, k}\end{aligned} \tag{3.4}$$

The proposed method for scheduling the PRBs slices is based on a proportional fairness to optimize the sum rate under certain constraints. The optimization is accompanying the maximizing variables x^{mk}_{tC} and p^{mk}_{tC}. The resource allocation that maximizes of the sum rate can be modelled as

$$\max_{x^{mk}_{tC}, p^{mk}_{tC}} \sum_{m=1}^{\mathcal{M}}\sum_{k=1}^{K_m} \mathcal{R}^m_k = \max_{x^{mk}_{tC}, p^{mk}_{tC}} \sum_{m=1}^{\mathcal{M}}\sum_{k=1}^{K_m}\sum_{t=1}^{T}\sum_{C=1}^{C} \frac{B}{T}x^{mk}_{tC}\ \log_2\left(1+\frac{p^{mk}_{tC}\mathrm{h}^{mk}_{tC}}{N_0B}\right) \tag{3.5a}$$

subject to the following constraints:

C1 Total power constraint

$$\sum_{m=1}^{\mathcal{M}}\sum_{k=1}^{K_m}\sum_{C=1}^{C} p^{mk}_{tC} \leq P_{total} \quad \mathit{for\ all\ t} \tag{3.5b}$$

$$p^{mk}_{tC} \geq 0 \quad \mathit{for\ all\ m, k, t, C} \tag{3.5c}$$

C2 Orthogonality constraint

$$\sum_{m=1}^{\mathcal{M}} \sum_{k=1}^{K_m} x_{tC}^{mk} = 1 \quad \textit{for all } t, C \tag{3.5d}$$

$$x_{tC}^{mk} \in \{0, 1\} \quad \textit{for all } m, k, t, C \tag{3.5e}$$

C3 Service contract constraint

$$N_{PRB}^{m} = \sum_{k,t,C} x_{tC}^{mk} \geq \rho_{\min}^{m} \; TC \quad \textit{for all } m \tag{3.5f}$$

where $\rho_{\min}^{m} \in [0, 1]$ *for all m* and $\sum_{m=1}^{\mathcal{M}} \rho_{\min}^{m} \leq 1$

C4 Fairness constraint

Define ς is a small number to ease the fairness constraint and ζ_k^m is a set of predetermined values to ensure the proportional fairness among users served by each SP so the fairness constraint can be expressed as

$$(1 - \varsigma)\frac{\mathcal{R}_1^m}{\zeta_1^m} \leq \frac{\mathcal{R}_k^m}{\zeta_k^m} \leq (1 + \varsigma)\frac{\mathcal{R}_1^m}{\zeta_1^m} \quad \textit{for all } m, k \tag{3.5g}$$

The proportional fairness is necessary to control the capacity ratios among users for a given available total transmit power.

The optimization problem expressed in Eqs. (3.5a–3.5g) is a nonlinear programming problem, with decisions that are a mixture of binary and continuous variables and is generally hard to solve. The complexity of finding a solution is extremely high for a network with a realistic size, taking into account fast-changing channel conditions and a very short time to decide.

One approach is to break down the optimization problem into two distinct problems namely: the PRB allocation problem and power distribution problem. Then two different methods can be used to find the optimal solutions. The solution to these problems is found separately and alternates between them until convergence. This is called *iterative coordinate search*. In each iteration for a given power distribution, an optimal PRB allocation is found and then for the found PRS allocation the optimal power distribution is determined, and so on.

3.3.3 Opportunistic Sharing-Based Resource Allocation

Although dynamic slicing schemes can achieve high resource utilization, the mobile virtual network operators (MVNOs) (the middlemen) are not involved in the allocation and intra-slice customization cannot be achieved. In optimal resource allocation, the InP has to provide directly for all users, which includes high cost of computation complexity. On the other hand, new schemes are proposed for resource allocation to MVNOs, such as opportunistic sharing-based resource allocation scheme in [10], which allows various users to opportunistically access to shared resources, a fundamental concept in cognitive radio (CR).

Network's traffic fluctuates all the time, such that each network traffic may contain a baseline part that always exists and a fluctuant part that occurs randomly with certain probability. Multiple networks may opportunistically share the same physical spectrum resource for their probabilistic fluctuant traffic, while channels of the same network are not allowed to share a single physical channel. The networks sharing the same physical channel (albeit opportunistically) create the possibility of traffic collision (i.e. the fluctuant traffic of two or more networks occurs simultaneously and collides), leading to worse performance and experience for users.

Let us introduce a collision probability threshold p_{th} such that only when the collision probability among multiple networks less than p_{th} channels can share the same physical channel. The economic model for a concave pricing strategy for the opportunistic spectrum sharing is based on the expected resource requirement of the ith network, e_i, and is given as

$$e_i = \mathcal{u}_i + p_i \, \mathcal{v}_i \tag{3.6}$$

where $\mathcal{u}_i, \mathcal{v}_i, p_i$ are baseline resources (fixed), fluctuant resources, and occurrence probability, respectively. Since most common concave function is logarithmic function, the linear expression (3.6) can be easily computed as logarithmic function,

$$\phi_i(e_i) = \varphi_i(\mathcal{u}_i, \mathcal{v}_i, p_i) = a \, \log(\mathcal{u}_i + p_i \, \mathcal{v}_i) \tag{3.7}$$

where a is a constant and, without loss of generality, we can set $a = 1$. Eq. (3.7) is a concave pricing strategy, such that when the networks requirements for resources increases, the price per unit resource decreases. Nonetheless, the physical resources available are limited and hence the price should rise as requirements increase. Finally, no requirement means no charge. For simplicity, denote the resource allocation unit to correspond to one wireless channel, which is either assigned exclusively to a single network or shared by multiple networks with a collision probability threshold p_{th}. In addition, assume there are a total of n_v networks. Each network resource requirement Υ_i is modelled by three parameters as

$$\Upsilon_i = (\mathcal{u}_i, \mathcal{v}_i, p_i) \tag{3.8}$$

and the resource requirements set ξ could be denoted as

$$\xi = \{\Upsilon_i, i = 1, 2, \ldots., n_v\} \tag{3.9}$$

As before, we define the resource assignment binary variable, $\mathcal{X}_i = [0, 1]$, where $\mathcal{X}_i = 1$ if the bid for resources is successful; otherwise, the bid has failed. The objective of this optimization problem is to maximize the revenue of the physical wireless networks. Mathematically the problem can be formulated as

$$\max \sum_{i=1}^{n_v} \mathcal{X}_i \cdot \; \varphi_i(\mathcal{u}_i, \mathcal{v}_i, p_i) \tag{3.10}$$

$$\text{s.t} \sum_{i=1}^{n_v} \mathcal{u}_i \cdot \; \mathcal{X}_i + \mathcal{L}(\mathcal{v}, p, \mathcal{X}) \leq C \tag{3.11}$$

where C is total physical wireless channels available and $\mathcal{L}$ is the number of physical channels opportunistically shared by fluctuant requirements of all successful networks.

$$\mathcal{X}_i = \begin{cases} 1 \; \textit{successful} \\ 0 \; \textit{failed} \end{cases} \quad \text{where} \;\; i = 1, 2, \ldots \ldots, n_v \tag{3.12}$$

The optimization expressed in (3.10) is a nondeterministic polynomial-time-hard (NP-hard) problem and there is no optimal solution in polynomial time, so often such problems are solved using heuristic (approximate) or greedy algorithms. The dynamic programming algorithms show that opportunistic spectrum sharing improves the system performance compared to non-sharing, and achieves better revenue, utility rate, and successful bids.

3.4 Network Functions Virtualization (NFV)

The network functions virtualization (NFV) idea was presented at the SDN World Congress in October 2012 by a group of SPs who were concerned about the cost and time it took to add new network functions or applications [11]. Networks broadly encompass a range of hardware devices. In addition, providing a new service most often requires yet another device. Accordingly, the space and energy to house and to operates these devices become increasingly difficult, costly, and complex to integrate and deploy in a network. Furthermore, hardware devices lifecycles are becoming shorter as technology is rapidly changing due to enhanced innovations, which give rise to reduction in return on investment of initiating new services.

NFV aspires to find solutions to these problems by employing virtualization technology to build up many network equipment types onto industry-standard high-volume software-based devices. That is, many network functions are implemented in software that can run on industry-standard server hardware. Operators may start a new service without installing a single piece of new equipment. NFV changes the conventional network model from hardware-centred to software-based functions installed into standard servers; storage; and switches. In essence, NFV eliminates the need for specific hardware and simplified the process by which network operators add, remove, change, or upgrade network serve provisions. Essentially, NFV eliminates the need for new specific hardware as they are able to add, remove, or change network functions at the server level in a simplified administering process.

Fourteen major international mobile network operators have met to address their concerns, i.e. how best to establish effective collaboration virtualization solution. The important objective for this meeting is to outline the benefits, and challenges for NFV and the basis for promoting an international collaboration to speed up the development of these enabling technologies. The issues discussed in the meeting are published as a nonproprietary white paper presented at a conference in Darmstadt, Germany in 2012. A follow-up white paper was published in 2013. In order to expedite the development of their solution, they proposed a network NFV that virtualises a network node built from independent, software-based blocks (or modules) and a proposal that can be chained (or combined) to compose specific network services based on the concept of virtualized network function (VNF) forwarding graphs. VNF forwarding graphs add the logical connectivity between the function blocks that need to be chained and allocated onto the actual NFV infrastructure. Actually, the VNF forwarding graphs are akin of connecting existing physical devices via cables. A forwarding graph expresses the sequence of network functions that process different end-to-end data flows in the network.

Operator-led industrial specification group (ISG) is being set up under the supervision of European Telecommunications Standards Institute (ETSI) to study the technical challenges for NFV as described in the white papers. ISG in ETSI was founded by end of 2012 by seven of the world's network operators. After four years and a huge amount of publications, the NFV evolved from prestandardization to detailed specification and early proof of concepts. Up to date, ETSI ISG published three releases to update the NFV over the period December 2014 to February 2018. Each release targets many aspects of the NFV challenges. A basic illustration of NFV architecture is depicted in Figure 3.13.

The NFV foresees the execution of noise figures (NFs) as software-only entities that run over the NFV infrastructure shown in Figure 3.13. The NFV framework is composed of three main domains identified as follows:

1. Orchestration and NFV management encompass the orchestration and lifecycle management of physical and/or software resources that provide for the infrastructure virtualization and the lifecycle management of VNFs.
2. VNF is the software application of a network function capable of running over the NFV infrastructure.
3. Network functions virtualization infrastructure (NFVI) encloses several hardware resources such compute, storage, and network and how these resources can be virtualized.

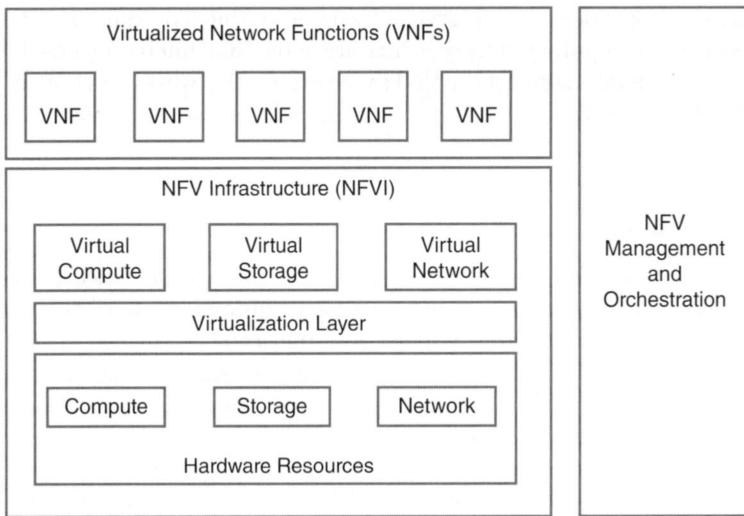

Figure 3.13 A simplified look at NFV architecture [12].

The benefits developed out of NFV to network operators (and subscribers) are enormous and summarized as follows:

- Operator CAPEX and OPEX are reduced drastically due to reduction in physical equipment and energy costs.
- Time-to-market to deploy a new service is hugely reduced.
- Return on investment on the new service has increased.
- Flexibility to scale service up, scale down, or introduce evolved service is considerably enhanced.
- Trial before deploying a service can be carried out at lower risk.
- There is greater openness to virtualization market and software applications.

NFV virtualizes a network node into independent [12] software-based blocks chained together to compose specific network services. ETSI proposed the concept of VNF forwarding graphs. VNF forwarding graphs add the logical connectivity between the function blocks and are allocated onto the actual NFV infrastructure. A forwarding graph expresses the sequence of network functions that process different end-to-end flows in the network. In contrast with contemporary cellular networks, where a particular feature is initiated network-wide, forwarding graphs enable features per service demand. Service design is taken from service implementation while the network functions are virtualized using a separate virtualization layer. Network functions that can be virtualized include:

- Evolved packet core (νEPC) functions, which cover mobility management; serving gateway (SGW), and packet data gateway.
- Baseband (BB) processing units' functions that cover multiple access control (MAC), radio link control (RLC), and radio resource control (RRC) actions.
- Switching function (νswitch)
- Traffic load balancing
- Service centres

The NFV architecture supports a wide range of services by orchestrating (coordinating) the VNF and operation across diverse computing, storage, which are pooled and interconnected by network resources. Other resources interconnect the VNFs with external networks and nonvirtualized functions enabling the integration of contemporary technologies with virtualized 5G network function.

3.4.1 Virtualized Network Functions

Virtual network function suggests a different way to design, deploy, and manage network services. NFV dislodge network functions such as caching, firewall, intrusion detection, domain name service to name few, from registered hardware devices so they can be provided in software. Such an approach supports a fully virtualized infrastructure including virtual servers, switches, and storage. It can be employed by data plane processing as well as control plane functions in both wired and wireless infrastructures.

NFV and software-defined networking (SDN) are two technologies overlapping though separate independent concepts. SDN separates the combination of the software-based control plane from the hardware-based data plane (e.g. data packets forwarding) while in NFV, networks are realized in software elements (virtual network functions) and employed servers in place of hardware to reduce CAPEX [13]. Therefore, in SDN the data plane communicates with the control plane through a specific communication protocol. OpenFlow is a communication protocol standard that is used by the controller to exploit data plane operations.

Although both NFV and SDN are complementary, they are seeking different ends. NFV adds flexibility and agility to implement network functions. SDN comprises a chain of automated network objects, like firewalls, switches, and routers. Combining the evolved NF with SDN will take some time yet. This is an evolved process since contemporary and virtualized functions have to be integrated so that systems are able to handle both functions simultaneously.

A functional block (FB) within a network infrastructure is provided with a definite external interface and distinct functional action. Relevant to the virtualization process is the virtual machine (VM), which is a virtualized environment and acts very alike a physical computer server. The virtualized interface defined in a logical and abstract way between two virtualized FBs.

3.4.2 Principles of the Network Functions Virtualization Infrastructure (NFVI)

A lot of network systems are represented in FBs including those specified by 3GPP where components of the overall system are defined as FBs and the interactions between the FBs act as a defined interface. An important principle of the system engineering is based on the concept that a complete system specification is provided by the specification of the component blocks and their interconnections. In addition, all system components are consistently FBs. A conventional function blocks diagram is illustrated in Figure 3.14.

A FB's operation is autonomous and its features are determined by the input it received in its interfaces, its dynamic state, and its static transfer function. The aim of NFV is to separate

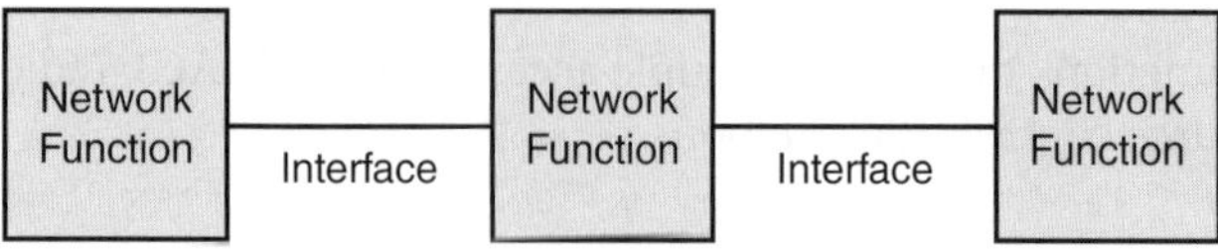

Figure 3.14 Functional block architecture [12].

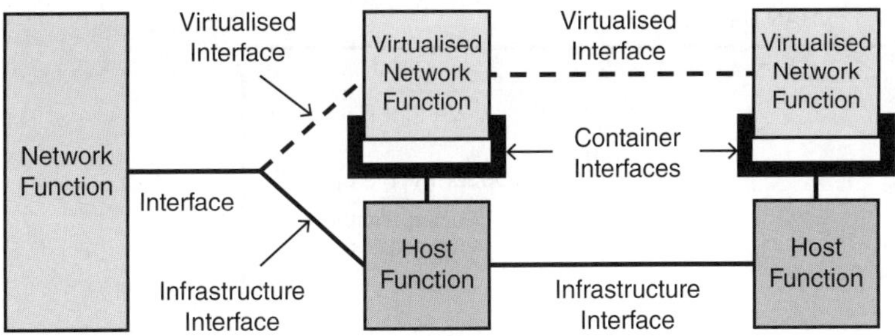

Figure 3.15 Virtualization of network function [12].

software that defines the VNF from hardware and the generic hosting NFVI sets the VNF in operation. Accordingly, it is important that VNFs and the NFNI be specified separately.

Let us consider the case when two of the three FBs in Figure 3.14 have been virtualized as shown in Figure 3.15.

In Figure 3.15 there are elements that need to be defined. The host functional block (HFB) is a functional block that can be configured or programmed or both. HFB can host one or more virtual functions blocks (VFBs). Within each HFB is an environment that can be configured or realized VFB called container interface. It must be noted that the container interface is not an interface between FBs. Between each two HFBs there is an infrastructure interface. The VNF needs the host function for its existence, i.e. VNF can't exist independently as a FB since VNF is a configuration of the host function.

3.5 *v*RAN Supporting Fronthaul

A collaborative project called Telecom Infra Project (TIP) was launched in February 2016. Tip community is made up of operators, technology providers, suppliers, developers, integrators, and start-ups, etc. The key aim of TIP is to collaborate on developing new technologies, to explore new business approaches, and encourage new investment in telecom sectors [14].

The *v*RAN fronthaul project launched in 2017 is aimed at developing *v*RAN ecosystem that can be employed over less-than-ideal transport links. The *v*RAN solution provides a variety of benefits for both end users and operators. Critical decisions have to be made on where to split the RAN architecture. For example, a split that makes BB processing centralized allows better performance (i.e. improved inter-site coordination and pooling for centralized baseband unit [BBU] processing) while the requirements on fronthaul (i.e. connection between remote radio unit [RRU] and the virtualized BB unit) become challenging to meet. The central point of the TIP *v*RAN is to make the transmission between *v*BBU and RRU operate efficiently over nonideal fronthaul. Nonideal fronthaul is defined by TIP project group in term of latency generally between 2 and 60 ms and throughput up to 10 Gb/s compared to the ideal backhaul defined by 3GPP as one-way latency under 2.5 μs and throughput up to 10 Gb/s. This definition of nonideal fronthaul enables an operator to employ existing infrastructure under a wide range of cases with bandwidth (BW) and higher latency compared to common public radio interface.

The *v*RAN BS architecture is shown in Figure 3.16. The radio elements in the BS are split at the PHY. The lower PHY is placed in the RRU but the upper PHY along with layer2 and layer3 are placed in the *v*BBU. The open interface infers the RRU and *v*BBU can be provided by different vendors.

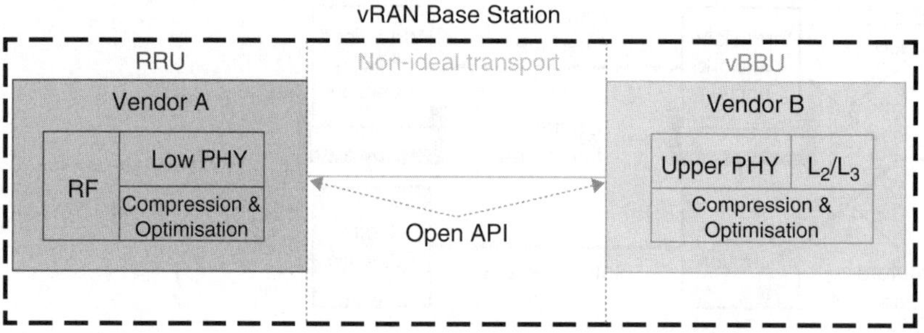

Figure 3.16 TIP vRAN base station architecture [14].

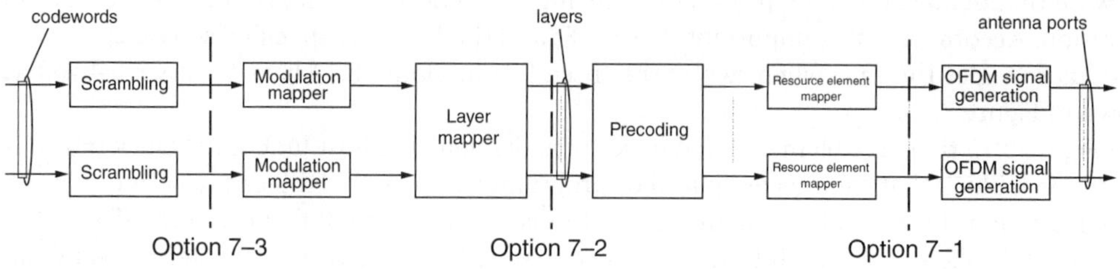

Figure 3.17 Downlink split injection points [14].

3.5.1 Splitting the Architecture

3.5.1.1 Downlink (DL)

Figure 3.17 displays a general DL with several potential functional split options that can be used as signals injection points, i.e. option 7-1, option 7-2, and option 7-3. In all these options, the encoding is performed at the BBU. Compression and optimization processing are needed in both the RRU and *v*BBU to enable functionality over nonideal fronthaul.

Split 7–3 when supported by *v*RAN fronthaul interface has generally the utmost benefit from fronthaul application aspects. Split 7–3 allows modulation mapping; layer mapping; precoding; resource element mapping; and OFDM signal generation to be performed at the RRU side. Accordingly, DL dynamic parameters are all available at the RRU to carry out such operations as modulation orders, spatial schemes needed to define the associated beamforming weights.

Split 7–1 includes OFDM signal generation while split 7–2 includes precoding and OFDM generation. Split 7–2 or 7–1 can be used for injecting DL signals that don't have fixed constellations and therefore can be injected as complex samples. Such an arrangement makes a lower split better conformed to distribute DL multiple-input multiple-output (MIMO) transmission. An example is coordinated multipoint (CoMP) where the same data is transmitted across multiple distributed RRU with appropriate beamforming weights so multiple spatial layers are multiplexed on same time-frequency tones.

3.5.1.2 Uplink (UL)

Figure 3.18 depicts a conventional Uplink (UL) chain with two extraction points that can be used as functional splits for UL. Split 7–2 is supported by the interface with most of the processing performed at the RRU side such as spatial processing or compression where signals from receive antennas are processed to acquire per-spatial layer streams transferring signals per antenna to signals per layer. This is a beneficial feature from a fronthaul aspect applied every time the number of spatial layers is much smaller than the number of receive antennas as in massive MIMO

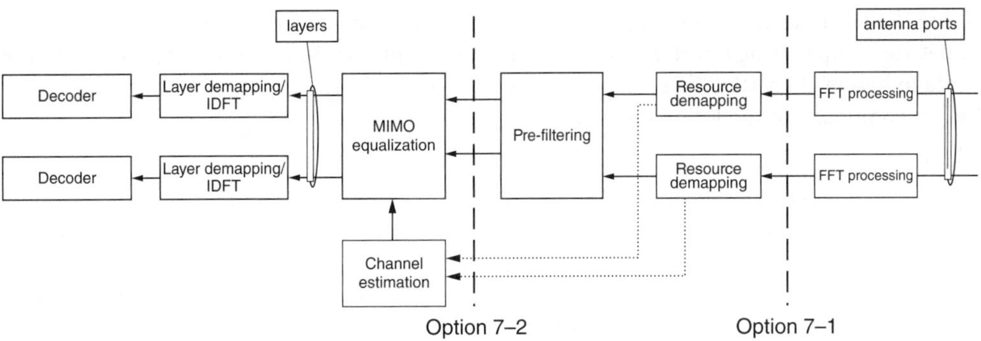

Figure 3.18 Uplink split extraction point [14].

cases. Spatial processing needs to be dynamically known to the RRU and cover antennas selection, linear equalization where the equalizer weights being selected by the *v*RAN and arranged to the RRU. For example, there could be two physical uplink shared channels (PUSCH) with notably different coding and modulation formats, spatial features, or resource allocation and both received by the BB in the same subframe. The RRU may receive a request to carry out different processing/compression for the two signals.

The fronthaul interface can be considered as application programming interface (API) to efficiently transfer structured signals from/to RRU. So, it is simply a set of standard interfaces designed to allow one software programme to interact with another. Key to the design is an open interface that is essential to commercial success. Accordingly, it provides for features such as scheduling algorithms, multiuser MIMO, CoMP schemes, and selectivity of spatial beams within co-scheduled signals in the same subframe.

An important benefit of the examined splits is total cost reduction to enable operators to pick and choose elements from different vendors. For this to succeed, the interoperability between different vendors is vital to make available multivendor solutions in the same system. Furthermore, support of a range of ideal and nonideal transport links as fronthaul requires the fronthaul protocol to be noncommittal to the physical and link layer of the transport, in contrast to other interfaces that operate on layer 1 or layer 2 of the transport link. Therefore, the layer 3 packet-switched interface was picked to ensure seamless operation over a variety of physical and link layer protocols.

To sum up, the low-PHY functional split between *v*BBU and RRU leads to both benefits and challenges. A key benefit is the centralization of all MAC component, as well as a good part of PHY. *v*BBU could exploit from the centralized functions to enable important features for both DL and UL such as CoMP, distributed MIMO transmission, and reception over physically separated RRUs. Such techniques provide important performance benefits particularly in dense heterogeneous degraded by intercell interference.

3.6 Virtual Evolved Packet Core (*v*EPC)

The EPC is standardized by 3GPP Release 8. It is an all-internet protocol (IP) mobile core network designed by 3GPP for LTE and also grants access to legacy networks such as 2G, 3G, and non-3GPP and fixed-access networks [15]. EPC is capable of fast processing of packets to allow high data rates with low latency. Contrast to 2G and 3G, which have mobile core functions being realized in two subdomains namely: circuit-switched (CS) for voice and packet-switched (PS) for data, EPC unified these two subdomains as a single IP domain. The main features of the LTE

EPC are the separation of control plane from user (data) plane, removal of the CS domain in core network, supporting multi-access networks 3GPP and non-3GPP, and raises the maximum DL rate to higher than 100 Mb/s.

EPC is composed of the following:

- MME, which is the main control plane entity
- SGW, which is responsible for routing and forwarding user data packets from and to the BS
- Packet data network (PDN) Gateway (PGW), which ensures connectivity between the user (data) plane and external networks such internet
- A home subscriber server (HSS), which is the central user subscription information, and as the policy and charging rules function

A block diagram depicted the EPC architecture is shown in Figure 3.19.

A recent study sponsored by Affirmed Networks shows that *v*EPC can save mobile operators a lot of money. The study considered a large SP scaling up to 6 million subscribers to compare CAPEX and OPEX costs of virtualized solution versus traditional EPC solution. Another study carried out by International Data Corporation (IDC) for Cisco systems in 2016 reveals that *v*EPC enables the mobile network infrastructure to operate at a higher utilization rate of up to 87%, resulting in OPEX cost efficiency of up to 25% [16]. In addition, the virtualization and the control/user plane separation – driven architecture can result in possibly OPEX cost saving to the extent of 20–40% over a 5-year period. Most importantly, the IDC study found that 67% reduction in time to market and launch a new service such as MVNO, mobile network, internet of things (IoT) service, etc.

The concept of 'network slices' sets up a service-based end-to-end appropriated virtual network employing two techniques: slice using the NFV and SDN. Slice concept is likely be a key feature of 5G networks. Nonetheless, the resources allocated to each slice cannot be moved and hence blocked off. Accordingly, a *slice-per-service* design with slice resources for every service in order to guarantee performance constraints would initiate a loss of multiplexing gain. Furthermore, operators have to set up and operate a massive number of slices for diverse services resulting in an increase in OPEX.

An adequate method to reduce such high OPEX is to construct slices so that each slice accommodates a group of similar network requirement so as to reduce the total number of slices. This technique is known as *per-group slicing* based on mapping a service group to an appropriate slice and reduces OPEX and multiplexing gain loss and applied to various 5G services.

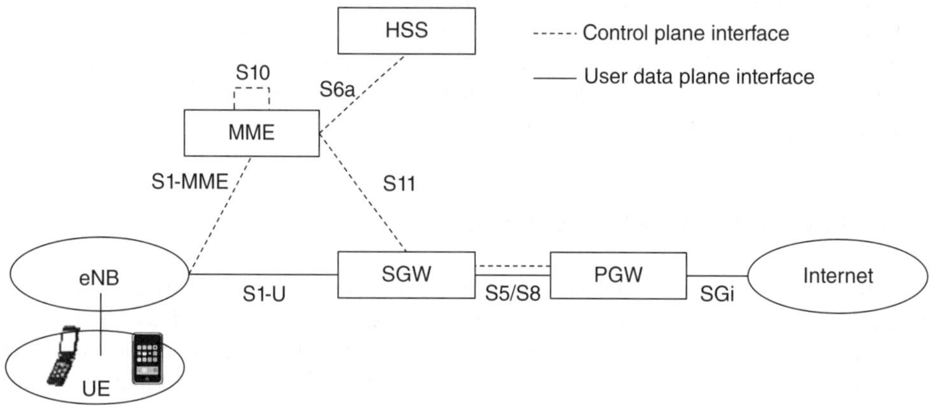

Figure 3.19 EPC architecture [15].

3.7 Virtualized Switches

The data path inhabits the switch itself and a separate controller makes routing decisions. Both switch and controller communicate using OpenFlow networking protocol proposed by Stanford University in 2009 [17]. OpenFlow improves network performance by separating the control plane and the user data plane. In addition to the networking protocol, another two elements are necessary to enable a whole OpenFlow system: the virtual (or physical) switch that supports OpenFlow and the controller, which delivers flow-setting packets to control the switch flow table. The OpenFlow switch-monitoring system is proposed in [18] to manage all hosts and traffic pass through switches under the controller.

OpenFlow is widely employed in applications such as IP-based mobile networks, and virtual switches are included in all virtual solutions to deal with networking. A virtual switch (vSwitch) is a software programme that permits one virtual machine (VM) to communicate with another. In addition, it can intelligently direct communication on the network by inspecting packets before allowing them to pass on. Currently, the top vendors providing network switching solutions are Cisco and Nokia. Nokia virtualized service router provides a basic set of an IP with IP4 and IP6 support for common routing protocols [19, 20].

3.8 Auction in Resource Provision

An auction is a process to buy/sell goods/services, and such methods are interdisciplinary processes bridging engineering requirements and business process management. Auction methods are extensively used in applications such as CR and cellular networks for radio resource allocation (subchannels, time slots, and transmit power levels) etc.

Auction methods involve four principal elements [21–24]:

1. *Bidder*. This is who wants to buy radio resources.
2. *Seller*. The seller owns and wants to sell licenses, subchannels, time slots. Bidders and sellers are the main players in an auction.
3. *Auctioneer*. As the intermediate *trade agent*, the *auctioneer* hosts the auction process. In wireless systems, BS or AP sometimes conducts resource auctions by its auction controller.
4. *Valuation*. Each good and service has a value at which the buyer/seller wants to buy/sell. Valuation is monetary evaluation of the assets, and buyers/sellers have reserved valuation on their own. Different buyers and sellers have different valuations, which might be higher or lower than the basic valuation depending on their priorities.

The price that a seller can submit is called an *ask price*. A buyer can submit a *bidding price* and a *hammer price* is settled by the auctioneer. There are many kinds of auction designs but examples of simple types are forward auction is a seller-side auction where buyers bid for goods/services made available by the sellers as illustrated in Figure 3.20a; in reverse auction is a buyer-side auction where sellers compete to sell goods to buyers, as shown in Figure 3.20b; in single-sided auction: only the buyer or seller make bids/ask prices as in Figure 3.20a,b; and double-sided auction: both buyer and seller make bids and ask prices as shown in Figure 3.20c.

The most common auctions are *Vickrey-Clarke-Groves* (VCG) auction, *Combinatorial* auction, and *Cooperative* auction. In VCG auction, the seller offers a set of M_s items denoted as $\boldsymbol{T} = t_1, \ldots, t_m, \ldots, t_{M_s}$ to sale to a set of N_b buyers, denoted as $\boldsymbol{B} = \{b_1, \ldots, b_n, \ldots, b_{N_b}\}$.

Each buyer $b_n \in \boldsymbol{B}$ submits a bid $v_n(\boldsymbol{T}^{(b_n)})$ where $\boldsymbol{T}^{(b_n)} = \{t_1^{(b_n)}, \ldots, t_m^{(b_n)}, \ldots, t_{M_s}^{(b_n)}\}$ is a set of items buyer b_n wants to buy from the auctioneer. After all bids are collected, the auctioneer

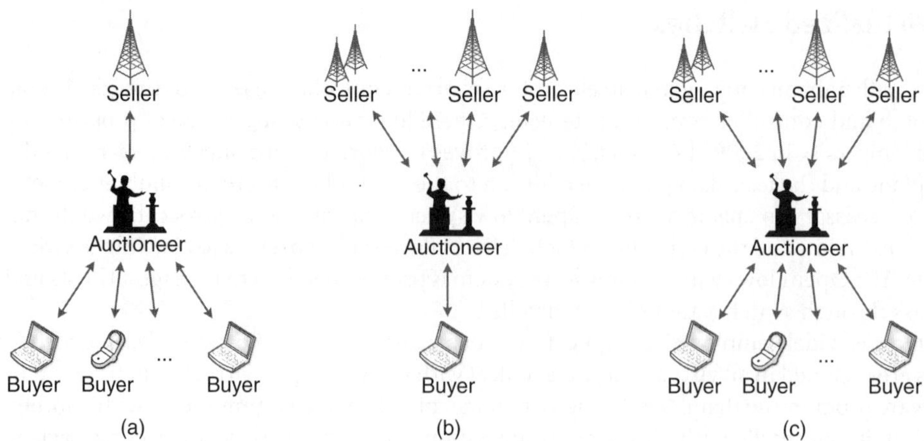

Figure 3.20 Different types of auctions: (a) forward auction with a single seller; (b) reverse auction with a single buyer, (c) double auction. The arrows indicate transactions of commodities and money among auction players [21].

computes the *optimal items allocation*, $A = \{\widehat{T}^{(b_1)}, \ldots.., \widehat{T}^{(b_n)}, \ldots., \widehat{T}^{(b_{N_b})}\}$ that maximize total revenue. The set allocated to buyer b_n is $\widehat{T}^{(b_n)} = \{\widehat{t}_1^{(b_n)}, \ldots., \widehat{t}_m^{(b_n)}, \ldots.., \widehat{t}_{M_s}^{(b_n)}\}$. The costs are taken from buyer b_n is $D_n = U_{op}^{\check{B}\backslash\{b_n\}} - \sum_{j=1,j\neq n}^{N_b} v_j(\widehat{T}^{(b_j)})$ where $U_{op}^{\check{B}\backslash\{b_n\}}$ is the *optimal sell revenue* when buyer b_n is not making a bid, and $\sum_{j=1,j\neq n}^{N_b} v_j(\widehat{T}^{(b_j)})$ is the *sum of the buyers' valuations* for the allocated items excluding buyer b_n.

In most cases, buyers request a set of multiple items but each buyer will be satisfied if only some of the items are being received. However, in certain cases, the buyer needs the complete set of items otherwise he will not be satisfied. Such bidding deals come under the *combinatorial auctions* where each bid for the complete set is either accepted or rejected.

A cooperative auction is a group auction where sellers offer buyers price discount, depending on the number of requested items (i.e. the more are requested the lower the price). This scheme results in a win-win strategy since sellers want to sell more items and buyers want to buy at low prices.

A *buyer utility* is defined as the difference between the valuation and the hammer (sell) price, while a *seller utility* is defined as the difference between hammer (sell) price and the seller's valuation. The sum of all auction participants' utilities is defined as *social welfare* showing the profit generated to the market. A utility that is higher than zero shows gained profit, which encourages more participants to join the auction; otherwise, too few participants interested in the auction may cause the auction to collapse.

3.9 Hierarchical Combinatorial Auction Models

A two-level resource allocation using combinatorial auction model where the InP is responsible for allocating the resources to each network operator, and each operator manages the resource allocation to the users, which makes the allocation problem a hierarchical (i.e. *two-level*) problem. In fact, we have two hierarchical auction models: a *single-seller multiple-buyers* model, as shown in Figure 3.21, is the top level of the auction, and a *multiple-seller multiple-buyer* model, as shown in Figure 3.22, is the lower level of the auction [25].

In the lower-level auction, the users are the bidders and each operator acts as a seller, while in the upper level auction the operators act as the bidders and the InP acts as the seller. The optimal amount of resources allocated to each bidder is determined using a *winner determination problem* (WDP) algorithm.

In the analysis of the proposed hierarchical combinatorial auction, we consider the following system model. Specifically, we focus on the *DL transmission* of a cellular system with an InP providing BSs and resources to a set of $\mathcal{M} = \{1, 2, \ldots., m, \ldots\ldots, M_{\mathcal{M}}\}$ operators. Each operator (say m) then provides services to K_m subscribed users in the considered cell. InP owns a set of $C = \{1, 2, \ldots., C\}$ subchannels, each with bandwidth B as per our consideration of single-level resource allocation considered in (Section 3.3.2). In addition, frequency reuse is considered for all cells. BS in each cell is equipped with M antennas, and each user's device has a single antenna (i.e. *multiuser MIMO system*). We assume M to be large (e.g. several hundred antennas) to achieve massive MIMO effect by scaling up the traditional MIMO by orders of magnitude. We assume that BSs and UEs in the system are perfectly synchronized in time and the system model is operated in time division duplex (TDD). For achieving all the benefits of *massive MIMO,* the BSs require knowledge of the CSI for precoding processing. Also, we employ channel reciprocity to determine the DL channel using the conjugate transpose of the UL channel. The latter can be estimated by the BS from the UL pilots transmitted by each user.

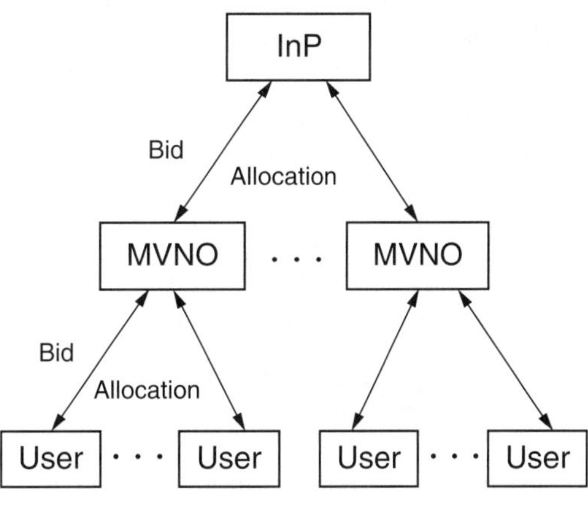

Figure 3.21 A single-seller multiple-buyer hierarchical auction model [25].

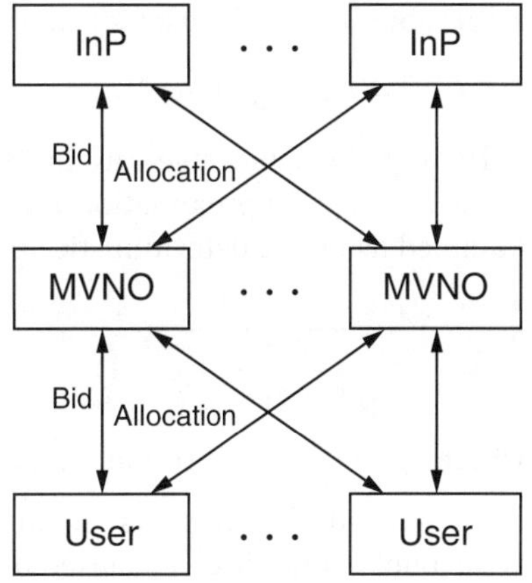

Figure 3.22 A multiple-seller multiple-buyer hierarchical auction model [25].

Consider the *DL* received signal at user k of operator m on subchannel n in a given cell. Let p_k^n be the transmit power for the link of bandwidth B from the BS to user k. Denote L_k and g_k to represent the path loss and the shadowing gain, respectively. The fading UL between the BS and k^{th} user is $\mathbf{h}_k(n) \in \mathbb{C}^{M_m \times 1}$ where M_m is the number of antennas allocated to operator m. Consider the ideal scenario where the DL is given by $\mathbf{h}_k^H(n)$. Let the multiuser precoding used by the BS be denoted as $\boldsymbol{f}_k(n) \in \mathbb{C}^{M_m \times 1}$. The additive white Gaussian noise (AWGN) is $\mathrm{n}_k^n \sim \mathcal{CN}(0, N_0)$ where N_0 is noise power spectral density and the transmitted signal is s_k^n. The DL received signal, $\widehat{s}_k^n$, is given by

$$\widehat{s}_k^n = \sqrt{p_k^n L_k g_k}\, \mathbf{h}_k^H(n) \boldsymbol{f}_k(n)\, s_k^n + \sum_{j \neq k} \sqrt{p_j^n L_k g_k}\, \mathbf{h}_k^H(n) \boldsymbol{f}_j(n)\, s_j^n + \sum_{l \neq 0} \sqrt{p_l^n L_{kl} g_{kl}}\, \mathbf{h}_{kl}^H(n) \boldsymbol{f}_{kl}(n)\, s_l^n + \mathrm{n}_k^n \tag{3.13}$$

where the first term in the right-hand side of (3.13) is the *desired signal* for user k, the second term is the reuse *interference within the cell,* and the third term is the interference from *other cells,* respectively. The received signal-to-interference-plus-noise ratio (SINR) γ_k^n for the kth user using subchannel n is given as

$$\gamma_k^n = \frac{p_k^n L_k g_k \boldsymbol{f}_k^H(n)\ \mathbf{h}_k(n)\, \mathbf{h}_k^H(n) \boldsymbol{f}_k(n)}{B N_0 + I_{reuse} + I_{inter-cell}} \tag{3.14}$$

where I_{reuse} and $I_{inter-cell}$ are the channel reuse interference within the cell and inter-cell interference power corresponding to the second and third term in the right-hand side of (3.13), respectively. When M_m, K_m grow infinitely large while keeping a finite ratio $\frac{K_m}{M_m}$, the user channels tend to become orthogonal and the reuse channel interference, I_{reuse}, tends to be minimal, i.e. ($I_{reuse} \ll I_{inter-cell}$), so it can be ignored and (3.14) is simplified as

$$\gamma_k^n = \frac{p_k^n L_k g_k \mathbf{h}_k^H(n) \boldsymbol{f}_k(n) \mathbf{h}_k(n) \boldsymbol{f}_k^H(n)}{B N_0 + I_{inter-cell}} \tag{3.15}$$

Eq. (3.15) also can be rewritten as

$$\gamma_k^n = \frac{p_k^n L_k g_k \mathbf{h}_k^H(n) \boldsymbol{f}_k(n) \mathbf{h}_k(n) \boldsymbol{f}_k^H(n)}{B N_0 + \alpha I_{inter-cell}} \tag{3.16}$$

Clearly, $I_{inter-cell}$, is a random variable and α is the inter-cell interference factor a random variable representing the intensity of the inter-cell interference where $0 \leq \alpha \leq 1$. The ergodic achievable DL rate for user k using sub-channel n is given by Shannon formula

$$r_k^n = B\, \mathbb{E}[\log_2(1 + \gamma_k^n)] \tag{3.17}$$

The ergodic achievable rate in (3.17) is difficult to calculate for a massive MIMO system. Instead, a tight approximation for the achievable rate of the massive MIMO system can be obtained using the deterministic approximation of the SINR given in [25]:

$$\widehat{\gamma}_k^n = \frac{1}{\frac{\overline{L}}{\rho_k(n)\, A_m} + \alpha(\overline{L} - 1)} \tag{3.18}$$

where $\rho_k(n) = \frac{p_k^n}{B N_0}$ represents the transmit signal power to noise power ratio (SNR), $\overline{L} = 1 + \alpha(N_c - 1)$ and N_c represents the number of cells, α is the inter-cell interference factor, $\alpha(\overline{L} - 1)$ is the pilot contamination caused by the reuse of pilot sequences in other cells. The derivation of (3.18) can be found in [26, Eq. (3.29)]. The approximate achievable DL rate, r_k for user k can be expressed as

$$r_k = \sum_{n \in C_m} y_k(n)\, B \log(1 + \hat{\gamma}_k^n) \tag{3.19}$$

where $y_k(n) = 1$ if subchannel n is assigned to user k and $y_k(n) = 0$ otherwise, and C_m is the set of subchannels allocated to operator m.

3.10 Energy-Harvesting Techniques

Traditionally, fixed energy sources (batteries) are used to power devices in wireless networks and sensor networks. However, batteries have a limited operation time and have to be either replaced or recharged. These options can sometimes prove quite inconvenient and costly or unfeasible (e.g. batteries in medical devices implanted under the patient's skin for recording body data). Recently, energy-harvesting techniques have been proposed to recharge the battery from an external energy source such as solar, wind, motion, vibration, temperature, or thermoelectric sources. The efficiency of energy-harvesting systems using these resources depends on the available energy levels, which may vary over time due to weather conditions or device location. However, it is of utmost importance that mobile subscribers have mobility and connectivity anywhere, anytime, as dictated by the vision of 5G networks. So, it is essential to ensure the availability of sufficient energy at the user device whenever required. This reality encourages researchers and industry to develop wireless-powered cellular networks (WPCNs) where BSs take care of both energy harvesting and users' information transmission to/from the subscribers.

The concept of wireless power transfer/wireless energy (WE) harvesting is explained with the help of the system depicted in Figure 3.23. The system consists of a wireless generator and transmit antenna on the BS side. The harvesting node captures the wireless energy that propagates through the channel between the BS and the harvesting node and enters the wireless energy to DC conversion circuit consisting of receive antenna(s), matching network/bandpass filter (BPF), rectifying circuit, and lowpass filter (LPF). The matching network/BPF ensures the impedance matching of the receive antenna to the rectifying circuit and eliminates any harmonics generated by the rectification process. The rectifying circuit can have a variety of forms but essentially has to use a number of diodes and capacitors. The LPF removes any harmonic frequencies from the DC output.

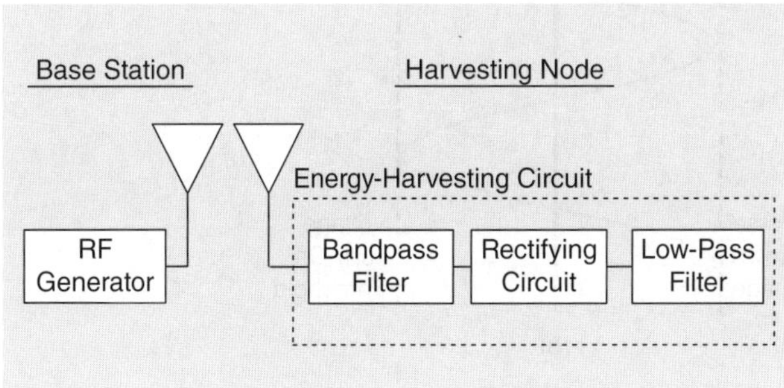

Figure 3.23 Harvesting wireless energy model [36].

3.10.1 Fundamentals of Wireless Energy Harvesting

Semiconductor-based rectifying diodes are of low cost with small form factor and can handle a relatively small amount of power, so that they are suitable for wireless applications. The nonlinear I-V relationship of diodes is shown in Figure 3.24, which shows three regions: low voltages below the reverse breakdown voltage (V_{br}) where the diode conducts in the reverse direction, between V_{br} and V_T (turn on voltage) the diode is off, and above V_T where the diode is forward biased and conducts current depending on the connected load R_L. When the diode is applied as a rectifier, the maximum DC voltage across the diode, $V_{o,DC}$ is

$$V_{o,DC} = \frac{V_{br}}{2} \tag{3.20}$$

The maximum DC power, P_{DCmax} is

$$P_{DCmax} = \frac{V_{br}^2}{4R_L} \tag{3.21}$$

The energy-harvesting circuit is characterized by two different metrics: efficiency and sensitivity. The output DC power P_{outDC} is

$$P_{outDC} = \frac{V_{outDC}^2}{R_L} \tag{3.22}$$

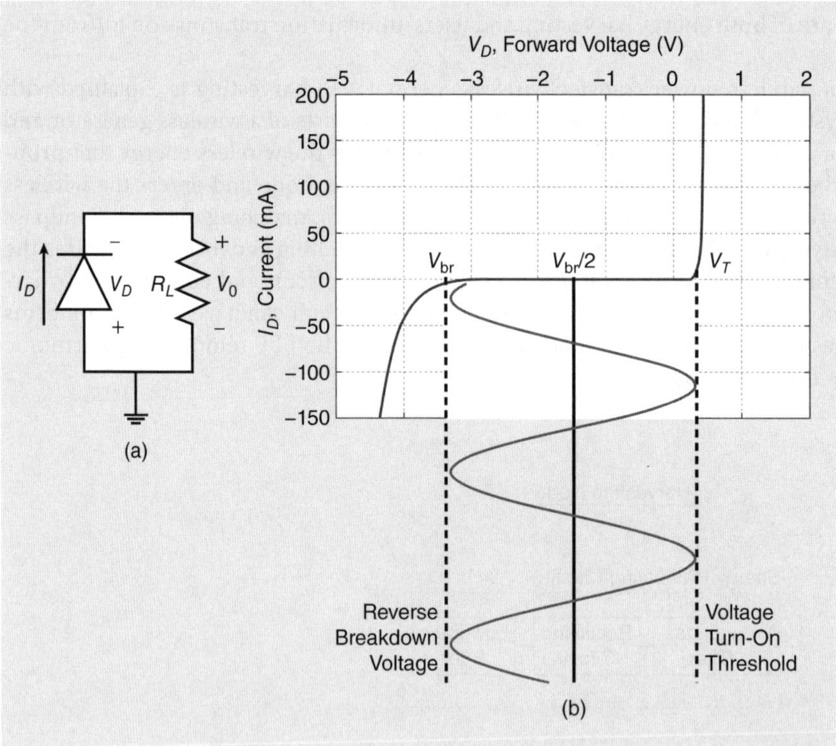

Figure 3.24 A typical diode I-V curve with annotated breakdown and turn-on voltages [36].

where V_{outDC} is the voltage across the load R_L. If the received wireless power at the input of the energy-harvesting circuit is P_{inEH}, then the power conversion efficiency, η_{PC}, is

$$\eta_{PC} = \frac{P_{outDC}}{P_{inEH}} = \frac{\frac{V_{outDC}^2}{R_L}}{P_{inEH} - P_{reflected}} \tag{3.23}$$

where $P_{reflected}$ is reflected power due to mismatch. When power reflected is neglected, we get Eq. (3.24):

$$\eta_{PC} = \frac{\frac{V_{outDc}^2}{R_L}}{P_{inEH}} = \text{Maximum power conversion efficiency.} \tag{3.24}$$

In energy-harvesting systems, Schottky diodes are most frequently chosen because of their low-voltage threshold and lower junction capacitance as compared to PN diodes. The low threshold allows for more efficient operation at low powers, and the low junction capacitance increases the maximum frequency at which the diode can operate. A traditional Schottky diode model is shown in Figure 3.25. It consists of a resistance R_s in series with a parallel connection of variable junction resistance R_j and a variable junction capacitance C_j that changes as a function of input power. The nonlinearity of Schottky diodes in energy-harvesting circuits makes the design of an impedance matching a very complex undertaking.

Figure 3.26 shows several interesting properties for the harvesting system using the Schottky diode. Figure 3.26a illustrates the effect of varying the forward bias V_f of the Schottky diode on the energy-harvesting efficiency for various input power. As V_f decreases, the energy conversion efficiency, at given power, increases. For −20 dBm input power when $V_f = 0.4$ the energy conversion efficiency is almost zero, but when $V_f = 0.04$ efficiency is increased to about 40% and $V_f = 0.004$ the efficiency jumps to about 80%. As the input power increases, the efficiency increases up to the maximum level with increasing input power. However, after the maximum efficiency level, the efficiency rapidly decreases for all bias voltages for increased input power from 15 dBm and beyond. Figure 3.26b demonstrates the effects of varying the reverse breakdown voltage on the efficiency for various input power levels. As the reverse breakdown voltage increases, the maximum efficiency level increases with increments up to certain maximum

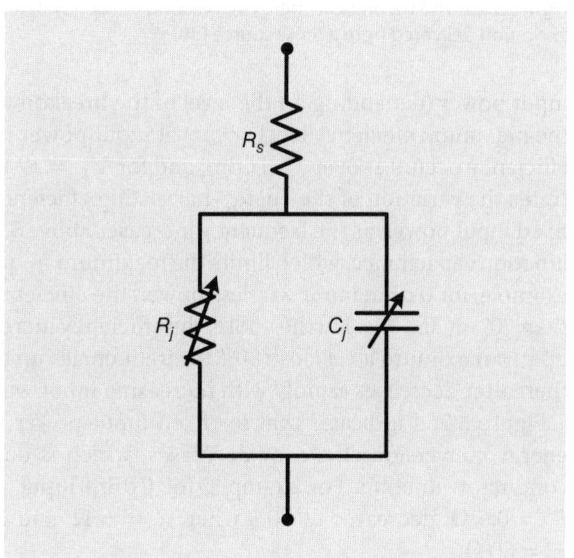

Figure 3.25 A standard Schottky diode model [36].

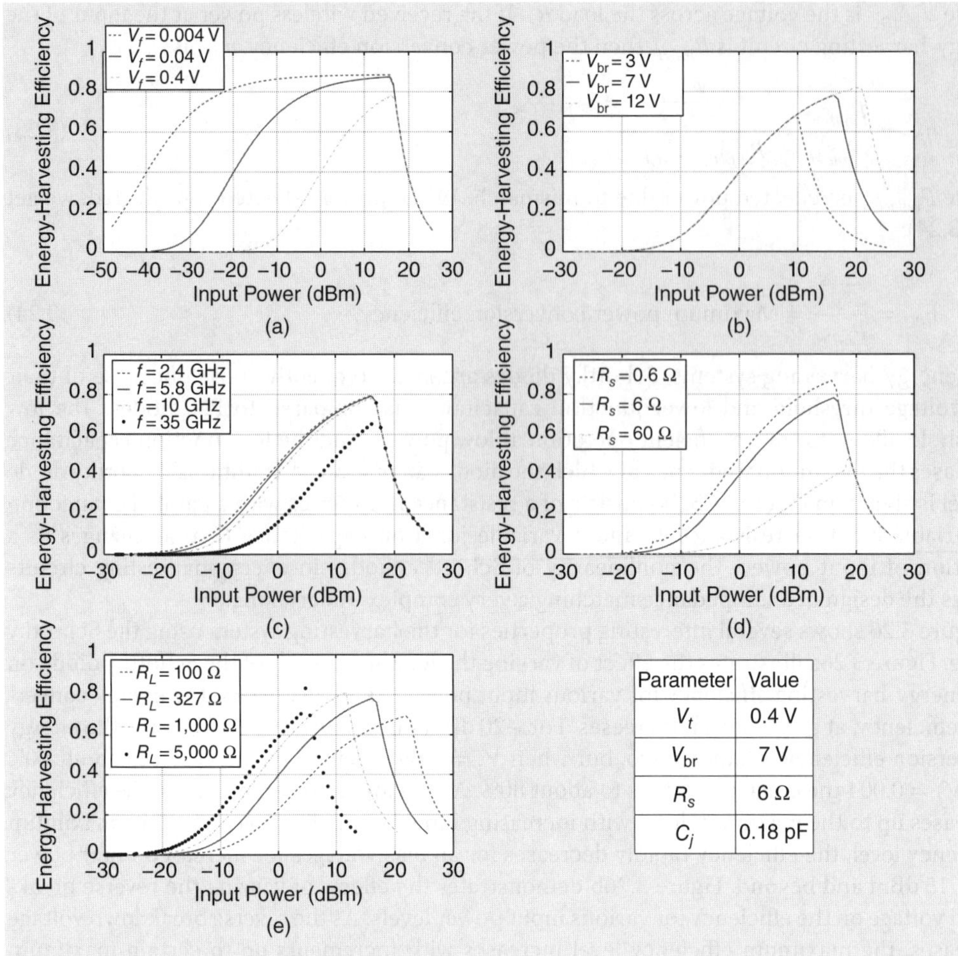

Parameter	Value
V_t	0.4 V
V_{br}	7 V
R_s	6 Ω
C_j	0.18 pF

Figure 3.26 The theoretical energy-harvesting efficiency of a single, silicon-nickel Schottky diode under CW excitation delivered from a 50-Ω source [36].

input power (depending on the level of the breakdown voltage); for example, for $V_{br} = 3$ volts, the maximum efficiency level occurs at input power is 10 dBm, for $V_{br} = 7$ volts, the maximum efficiency occurs at over 15.5 dBm, and for $V_{br} = 12$ volts, it is over 20 dBm. Figure 3.26c illustrates the variation of the energy-harvesting efficiency with system operating frequency. For a fixed input power, as the frequency increases above 5.8 GHz, the efficiency decreases due to the junction capacitance, which limits the maximum frequency for which a diode can operate. As an example, for 0 dBm input wireless power, the efficiency is over 40% at 5.8 GHz but drops to less than 20% at 35 GHz. Furthermore, the efficiency increases with increasing wireless input power up to a maximum level close 80% for frequencies up to 10 GHz and about 65% for 35 GHz, and thereafter decreases rapidly with increasing input wireless power.

Figure 3.26d indicates that for fixed input power when the series resistance increases, the energy conversion efficiency decreases, which is due to usual resistance losses in the semiconductor junction. For example, for 0 dBm input power, the efficiency is about 50% when $R_s = 0.6\ \Omega$, decreases to 40% when $R_s = 6\ \Omega$, and further decreases to just over 10% when $R_s = 60\ \Omega$.

Figure 3.26e displays the energy conversion efficiency dependence on the load resistance R_L. For fixed input power, as this resistance increases the efficiency increases, but as the input power increases the maximum efficiency level decreases due to the larger resistance producing the same voltage for a smaller current, this limits the input power for maximum efficiency level.

3.10.2 Wireless Powered Communications

The EM wave radiation can be used to harvest energy over the air by distance-wireless receivers to power wireless devices (WDs) with stable energy so they can operate without interruption. The important application of wireless energy-harvesting/transfer (WET) is to use harvested energy to transmit/decode information to/from other devices. Examples of these applications can be found in IoT systems, large-scale wireless sensor networks (WSNs) for environment monitoring, smart power grid, and wireless powered mobile devices.

In the previous section, we introduced the wireless to DC conversion circuit, which enables the WET, and in this section, we investigate how to implement the WET in large wireless cellular networks not only for energy harvesting but also for information transport [27, 28]. We start with a simple wireless network model that illustrates the basic concepts of the wireless powered communication (WPC) with a number of energy APs, information APs, co-located energy/information APs and a set of distributed WDs, as shown in Figure 3.27, where the flow of energy, information, and interference are shown.

Consider the DL transmission; the energy APs (e.g. AP2) transmits energy to a set of WDs that can be used to transmit/receive information to/from information APs (e.g. AP3) in the UL and DL, respectively. On the other hand, the co-located energy/information APs (e.g. AP1) both transmit energy and provide data access to the WDs. There are three modes of operation that can be considered: energy transfer in the DL only (e.g. AP1 to WD1 and AP2 to WD5); energy and information transfer in the DL (e.g. AP1 to WD4); and energy transfer in the DL and information transfer in UL (e.g. AP1 to WD3).

A wireless energy broadcast network with receiver structure, depicted in Figure 3.28, consists of an energy transmitter sending wireless energy to multiple energy receivers (ERs), each transmitter/receiver is equipped with a number of antennas. The transmitted energy signals could be pseudo random signals or unmodulated sinusoidal tones. Modulated energy signals can be designed to avoid spikes in the spectral power density (SPD) and to satisfy interference requirements. However, modulated energy signals occupy certain BW, as do information signals. A simplified energy-harvesting (EH) circuit model is represented in Figure 3.28. Denote

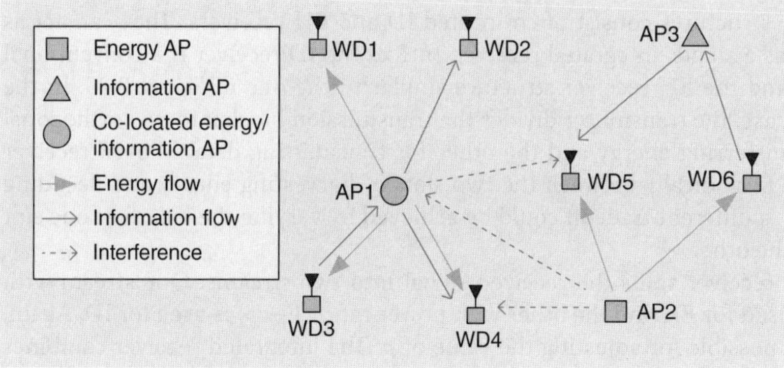

Figure 3.27 A network model for wireless powered communication [27].

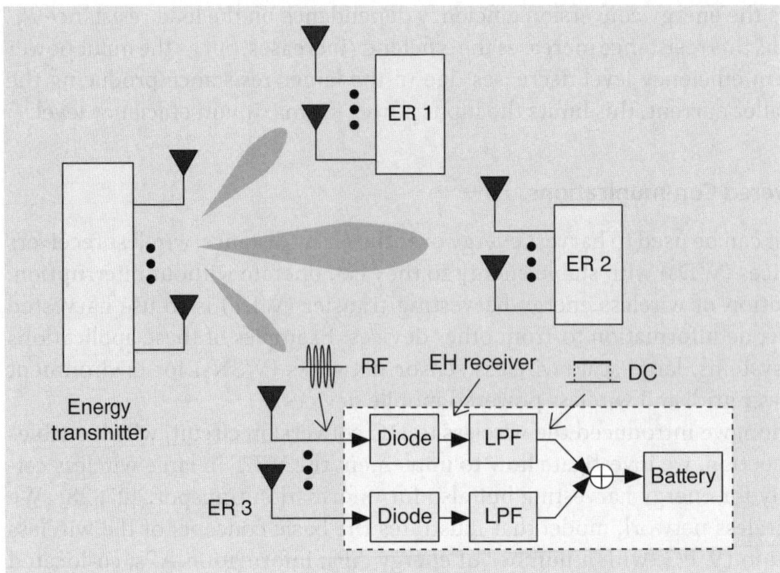

Figure 3.28 A wireless energy broadcast network and the energy receiver structure [27].

the harvested energy per unit symbol time as Q; then

$$Q = \eta P_r = \eta P_t D^{-\alpha} G_A \tag{3.25}$$

where η is the receiver energy conversion efficiency, $0 < \eta < 1$, P_t is transmit power, D is the distance (normalized with respect to certain reference distance) from the ER to the transmitter, $\alpha \geq 2$ is the path loss factor, and G_A is the combined gain of transmitter and receiver antennas.

Multiple antennas used in the energy transmitter system not only increase the antenna power gain but also enable energy beamforming to focus the transmit energy in a small region to significantly improve the energy transfer efficiency. The energy harvested and users' information can either be transmitted together or separately in orthogonal time/frequency domains. However, a joint (simultaneous) wireless information and power transfer (SWIPT) network is more efficient in spectrum usage. A SWIPT network model with the information detection/energy harvesting (ID)/(EH) receiver structures is illustrated in Figure 3.29. The practical ID/EH receiver structure elements are: time switching (TS); power splitting (ps); integrated ID/EH receiver (IntRx); and antenna switching (AS).

All of these receiver structures consist of co-located ID and EH receivers. They can act as independent units or as a single, integrated receiver, IntRx. The ID receiver is a conventional information decoder and the EH receiver structure similar to the one in Figure 3.29. In the TS receiver structure case, the transmitter divides the transmission block into two orthogonal time slots, one for transferring energy and the other for transmitting data. The TS receiver switches its operation periodically between the two slots of harvesting energy and decoding information. However, a different tradeoff could be achieved by varying the length of one slot at equal reduction to the other.

The power splitting receiver splits the received signal into two streams. One stream with power ratio $0 \leq \rho \leq 1$ used for EH and the other with power ratio $(1 - \rho)$ is used for ID. Again, different tradeoffs are possible for adjusting the value of ρ. The integrated receiver combines the RF front ends of ID and EH receivers and splits the signal after conversion to DC current. The DC current is divided into two streams for battery charging and information decoding.

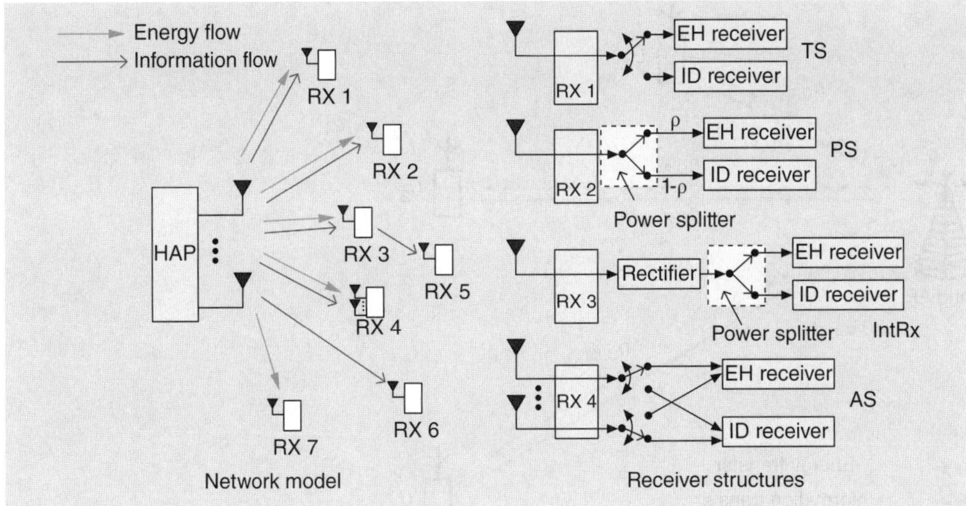

Figure 3.29 A simultaneous wireless information and power transfer (SWIPT) network model [27].

The use of MIMO technology in the SWIPT network model significantly mitigates channel fading for both energy transmission (energy beamforming) and information transmission (spatial diversity/multiplexing). A multi-antenna hybrid access point (HAP) could exploit the spatial Dof to focus the antenna radiation on specific location to enhance the EH and mitigate the interference. The antennas switching (AS) receiver applies a sub set of antennas for energy harvesting. The rest of the antennas are used for information detection.

3.10.3 Full-Duplex Wireless-Powered Communication Network

Full-duplex (FD) technology application for improving the throughput of wireless-powered communication network is investigated in [29]. A FD wireless-powered communication network is illustrated in Figure 3.30, where HAP running in FD mode broadcasts wireless energy (WE) to a set of users in DL and at the same time receives autonomous information from users using TDMA in the UL and both links are transmitting on the same channel. This network protocol is called HAP and operates in FD mode but users operate in time division half duplex mode in the UL. Harvested energy is transmitted on the DL when users are not transmitting.

The system model consists of one AP serving K users denoted by U_i, $i = 1, \ldots, K$ operating over the same frequency band. We assume that users are within adequate distance separation to reduce interference. Each UE is supported with a single antenna. The HAP is equipped with two antennas, one antenna is used for transmitting energy to users on in DL and the other antenna for receiving information for users in UL. Consequently, users recharge their batteries using the received WE signal in the DL and then used their batteries to transmit information in the UL. We assume the HAP is provided with a stable energy supply and each user depends totally on the transferred energy on DL for operating the UE. A new protocol for the FD transmission that enables efficient simultaneous energy transfer in DL and information transfer in UL over the same BW is shown in Figure 3.31. Each transmit block of duration T is divided into $K+1$ slots, each with a duration of $\tau_i T$. The 0th slot $\tau_0 T$ is allocated to energy transfer only, to ensure energy transfer is available even if only one user is active. The rest of the slots can be used for both DL energy transfer and UL for information transfer. During the ith slot, HAP broadcasts

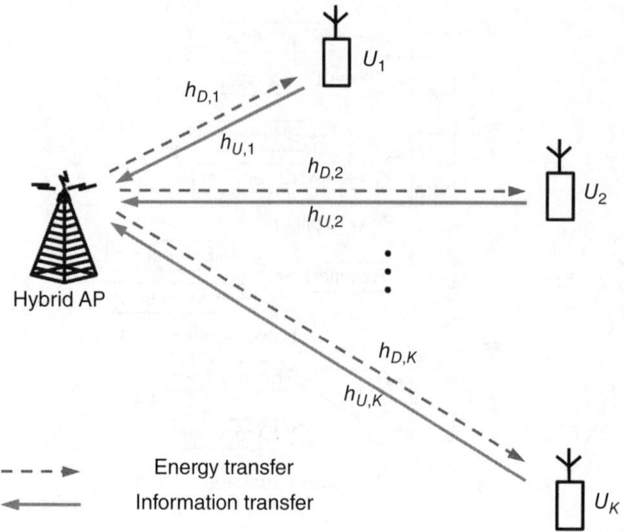

Figure 3.30 Wireless-powered communication network (WPCN) with DL WET and UL WIT [29].

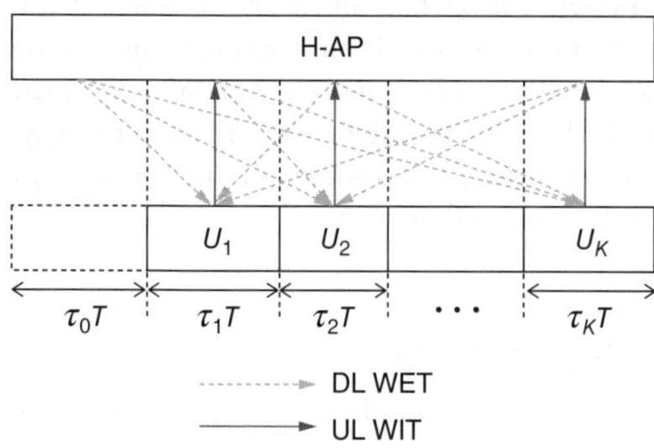

Figure 3.31 Simultaneous DL energy and UL information transmission in FD – Wireless powered communication network (WPCN) [29].

wireless energy to all users in the network with transmit power P_i so we have

$$\sum_{i=0}^{K} \tau_i = 1 \tag{3.26}$$

The average transmitted power of the HAP over $K + 1$ slots in each block P_{avg} and the peak transmit power at each slot, P_{peak} are given by

$$\sum_{i=0}^{K} \tau_i P_i \leq P_{avg} \tag{3.27}$$

$$P_i \leq P_{peak}, \quad i = 0, 1, \dots \dots K \tag{3.28}$$

When the *half-duplex scheme* is used in wireless-powered communications, the *harvest-then-transmit* protocol, proposed by the same authors, can be employed to achieve orthogonal energy transfer in DL and information transfer of active users in UL as shown in Figure 3.32 and in this scenario energy is broadcasted in DL during the 0th slot only while the ith slot, $i = 1, \dots \dots K$ is used for information transfer of users in UL only.

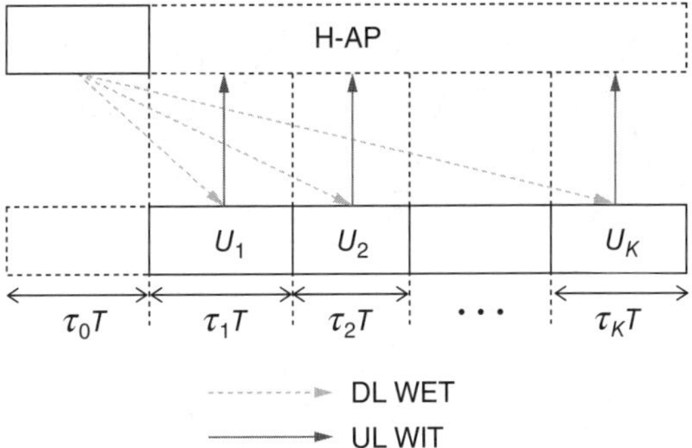

Figure 3.32 Orthogonal DL energy and UL information transmissions in HD-WPCN [29].

3.10.4 Wireless Power Transfer in Cellular Networks [30]

So far, we have presented energy harvesting for users distributed within a single cell network. Next, we investigate WET in cellular networks. A hybrid network architecture that overlays a UL cellular network with randomly deployed stations, called power beacons (PBs) that delivers energy wirelessly to mobile devices, is shown in Figure 3.33. The hybrid network under an outage constraint on data links is explored in [30] based on a stochastic geometry model of a single antenna BSs modelled as an independent Poisson point processes (PPPs) with density λ_B and the mobile users are associated with their nearest BSs. We assumed that the mobile users' density is much larger than the BS density so that there is at least one mobile user in each cell with probability = 1 (i.e. almost surely). The PBs forming an independent PPPs with density λ_P. Single-antenna mobile devices are uniformly distributed in Voronoi tessellation cells. The hybrid network model is shown in Figure 3.34 where the BSs, PBs, and mobiles are marked with triangles, circles, and stars, respectively. The Voronoi cells with respect to BSs are plotted with solid lines.

Each mobile device consists of a transmitter and a wireless-powered receiver. The receiver contains two energy storage units to empower wireless power transfer delivery. When a mobile device is active one unit is used for energy harvesting and the other for powering the transmitter as shown in Figure 3.35. Their roles are switched when the latter unit is fully discharged. When

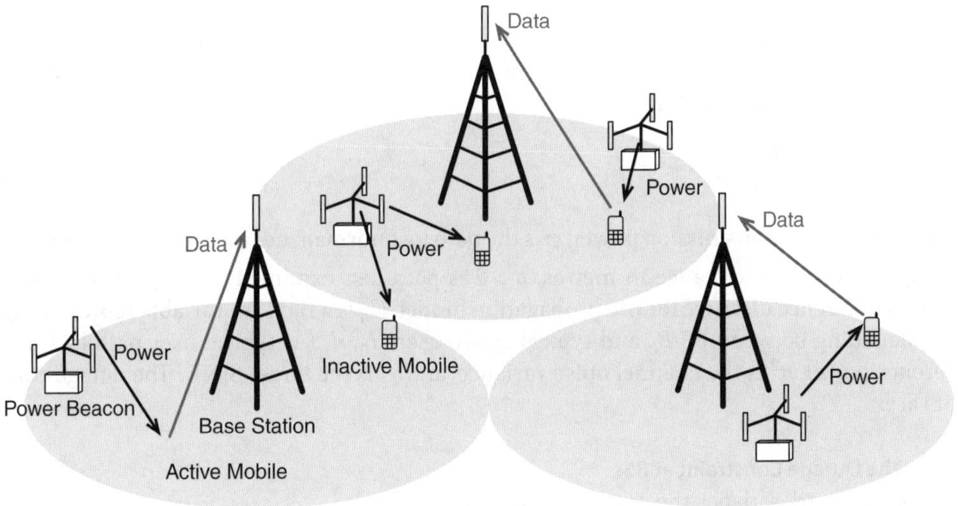

Figure 3.33 Enabling wireless power transfer in cellular networks with randomly deployed beacon [30].

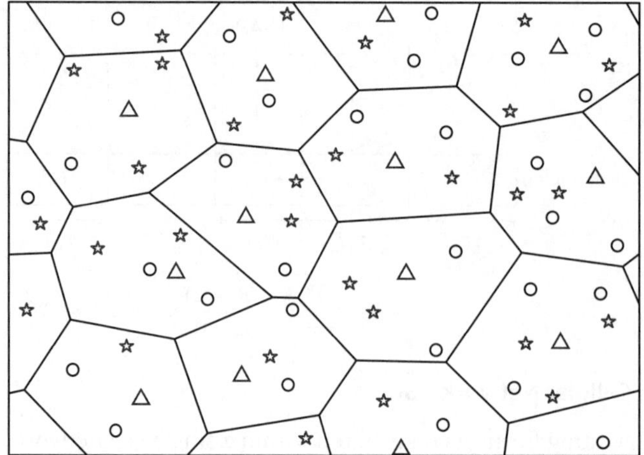

Figure 3.34 Hybrid-network model where BSs, PBs, and mobiles are marked with triangles, circles, and stars, respectively. The Voronoi cells with respect to BSs are plotted with solid lines [30].

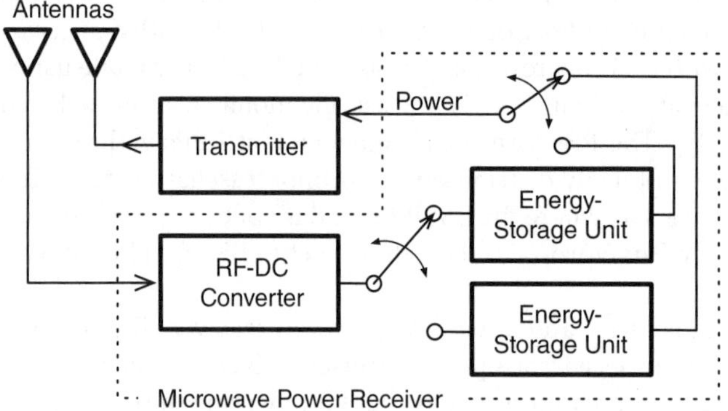

Figure 3.35 Power – beacon network model [30].

the transmitter is inactive, the wireless-power receiver remains active until all units are fully recharged. The cellular network anticipated performance is assessed by the outage probability $\Pr_{out}$ defined for a typical BS B_0 as

$$\Pr_{out} = \Pr\left(\frac{p\,\varphi\,G_0\,d_U^{-\alpha}}{\mathrm{I} + \sigma^2} < \gamma_t\right) \tag{3.29}$$

where p is the mobile transmission power, φ is the path loss normalized factor defined as $\varphi = \frac{L_0}{d_0^{-\alpha}}$ where d_0 is a reference distance in metres, $\alpha > 2$ is path loss exponent, L_0 is path loss measured at the reference distance for the propagation model, G_0 is a random variable representing small scale fading between BS B_0 and typical active user U_0, $d_U^{-\alpha}$ is typical user path loss, I is interference power, σ^2 is the channel noise variance, and γ_t is the target SINR. The outage constraints are:

3.10.4.1 The Outage Constraint at BSs

The cellular network satisfies the outage constraint at BSs as

$$\Pr_{out} \le \varepsilon \text{ for } 0 < \varepsilon < 1 \tag{3.30}$$

The mobile UEs store energy in either large units (e.g. rechargeable batteries) or in small ones (e.g. super capacitors). The latter can store 10 to 100 times more energy per unit volume than electrolytic capacitors and deliver energy much faster than rechargeable batteries and tolerate many more charge/discharge cycles than rechargeable batteries.

3.10.4.2 The Power Outage Constraint at PBs

At times, the instantaneous received power at each UE falls below the mobile UE transmission power p. Such an event is called *power outage* and it occurs with a probability called the power outage probability. The PBs should satisfy the following restrictions:

- For the case of large energy storage, the power-outage probability should be zero.
- For the case of small energy storage, the power-outage probability should be no larger than a constant δ with $0 < \delta < 1$.

3.10.4.3 Hybrid Network Mobiles with Large Energy Storage

In this scenario, we assume that the energy storage capacity at mobile UEs is infinite to provide reliable transmission. Infinite energy storage powered by energy harvested transfer implies that UEs can have transmit power p with probability [30]:

$$\Pr(P \geq p) = \left\{ \begin{matrix} 1 & if\ \mathbb{E}[P] \geq \omega p \\ \frac{\mathbb{E}[P]}{\omega p} & otherwise \end{matrix} \right\} \tag{3.31}$$

where $0 \leq \omega \leq 1$ is the duty cycle for mobile transmission and P is the instantaneous mobile received power. Consequently, a mobile can transmit continuously with power up to $\frac{\mathbb{E}[P]}{\omega}$. The average received power at a UE expected from PBs is from two types of transmission: *Isotropic energy transfer*, where energy radiates from a PB equally in all directions so a mobile may receive similar powers from multiple nearby PBs, and *directed energy transfer*, where the nearest PB to UE is the dominant energy source due to beamforming.

3.10.4.4 Hybrid Network Mobiles with Small Energy Storage

In this scenario, the hybrid network has to satisfy both constraints underlined above. The distribution function of the instantaneous received power at UE has to be summed up over a Poisson point that might have no closed-form solution. Instead, we consider an upper bound where only the nearest PB for a UE is considered. For isotropic energy transfer:

$$q\varphi' \nu^{-\beta} \geq p \tag{3.32}$$

where φ' is the normalized path loss factor like φ, q is PB transmission power, $\nu \geq 1$ is a constant, and $\beta > 2$ is path loss exponent. The power outage probability is upper bounded as

$$\Pr(P < p) \leq e^{-\pi \lambda_p \left(\frac{q\varphi'}{p}\right)^{\frac{2}{\beta}}} \tag{3.33}$$

In addition, for directional energy transfer

$$z_m q\varphi' \nu^{-\beta} \geq p \tag{3.34}$$

where z_m denotes the array gain. Here we have a tradeoff between $(q, \lambda_P, \lambda_B)$ such that the beamforming gain increases, the PB q by a factor of z_m. The upper bounded power outage probability is

$$\Pr(P < p) \leq e^{-\pi \lambda_p \left(\frac{z_m q\varphi'}{p}\right)^{\frac{2}{\beta}}} \tag{3.35}$$

The derivation of (3.33) and (3.35) are provided in [30].

3.10.5 Harvested Energy Calculation

The spectral efficiency of users using harvested energy on DL and UL vary significantly depending on user locations. Wireless powered cellular networks (WPCNs) can have three configurations as illustrated in Figure 3.36. In configuration 1, the harvested energy comes from a FD BS. The FD BS operation can be identified in two scenarios:

1. In-band full-duplex (IBFD) where a FD BS broadcasts energy to users and receives information from the users using the same channel at the same time. The critical issue here is self-interference.
2. Out-of-band full duplex (OBFD), which allows users to harvest energy and transfer information in different channels at the same time.

Half-duplex (HD) OBFD has no issues with self-interference. Configuration 2 is identified as harvesting energy from PBs. Users harvest energy from PBs symmetrically located in a circular pattern at a certain distance from a BS located at the centre of the cell. All users harvest energy from the PBs until their time slot starts. Power beacons (PBs) can deliver energy to users in both isotropic and directed transfer mode. Configuration 3 is identified as co-located PB and distributed antenna setup. In this configuration, we employ the distributed antenna elements of the BS, such that a user can harvest energy from the nearest PB in directed mode and transmit information to the nearest BS antenna element.

We consider a system mode comprised macro cell of radius R with a BS located in the cell centre and powered from fixed energy source. The macro cell overlaid with M_{p} PBs and serving K users uniformly distributed within the macro cell coverage area. Each user is equipped with a single antenna that can either transmit information or receive harvested energy at a given time. Users transmit to the BS for a predefined duration τ in a successive manner. Therefore, the total time frame $T_{frame} = T + \tau K$ where T is the duration needed to ensure energy transfer for the user who is scheduled to transmit in the next time slot.

3.10.5.1 Energy Harvested from a FD BS (configuration 1)

The time frame T_{frame} is shown in Figure 3.36. The slots allocated for UL transmission are identified as $k_1, \ldots \ldots \ldots \ldots \ldots \ldots, K$. It is reasonable to assume that the energy harvested by a user from the information transmission of other users is negligible. The energy harvested by a UE

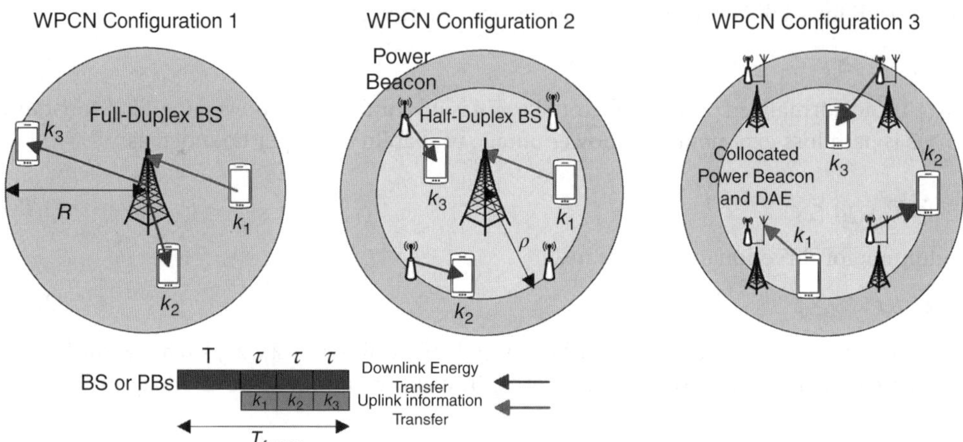

Figure 3.36 Graphical illustration of the considered WPCN configurations and the time frame structure for uplink information and downlink energy transfer [37].

is defined as the product of its received power in the DL and the time duration in which the power is received. Accordingly, the energy harvested by an arbitrary user, k, scheduled in time slot k is given as

$$E_k = \eta\, P_{total}\,(T + \tau(k-1))\, r^{-\alpha}\, \mathrm{h}_{DL} \tag{3.36}$$

where P_{total} is the maximum transmit power of the IBFD/OBFD FD BS with which is transmitted throughout the duration T_{frame}, r is the distance of the UE from the BS, α is the path-loss exponent, η is the receiver harvesting efficiency (assumed to be fixed), and h_{DL} denotes the DL composite (multipath and shadowing) fading channel. Note in practical systems, η may vary, depending on the harvested energy levels. Each UE can harvest energy while waiting for its UL transmission time slot to begin. The total time for the energy transfer of a UE k scheduled in time slot k is given as $T + \tau(k-1)$. The energy harvested by a UE k who is scheduled to transmit in slot k gaining a sufficient transmit power P_k during transmission time τ as

$$P_k = \frac{\xi E_k}{\tau} \tag{3.37}$$

where ξ is a factor represents the fraction of harvested energy levels that can be applied for transmission. It is worth noting that not all harvested energy can be used for transmission because there will be circuit power consumption from the user device itself in addition to the energy storage losses. We assume the users are uniformly distributed within the cell, so the distribution of the distance of a given arbitrarily located user from the BS is derived as follows: The uniform distribution $f(r)$ is proportional to the distance r, i.e.

$$f(r) = C.r \tag{3.38}$$

where C is a constant. Now integrate $f(r)$ over the interval $(0, R)$ and equate to 1, we get $C = \frac{2}{R^2}$; hence, $f_r(r) = \frac{2r}{R^2}$ where R is the radius of a circular macro cell with BS at the cell centre. Denote the distance of a user who is located at the maximum distance from the BS as r_k so

$$r_{(k)} = max\{r_1, \ldots\ldots\ldots\ldots.., r_k\} \tag{3.39}$$

The energy harvested by the farthest user scheduled in time slot k can thus be defined by modifying (3.36) to,

$$E_k = \eta\, P_{total}\,(T + \tau(k-1))\, {r_{(k)}}^{-\alpha}\, \mathrm{h}_{DL} \tag{3.40}$$

Users are selected arbitrarily to transmit in each time slot so a specific user (whether arbitrarily located or located at the cell edge) has equal probability of transmission in each time slot.

3.10.5.2 Energy Harvested from PBs (configuration 2)

In this scenario, the UEs harvest energy from PBs that are symmetrically distributed in a circular manner at a distance d from the BS. UEs continuously harvest energy until the beginning of their time slot for transmission. The PBs deliver energy to UEs in both the isotropic as well as directed energy transfer modes.

Isotropic Mode In the isotropic mode, PB radiates energy in an omnidirectional method so that all UEs are provided with energy at the same time. Consequently, the harvested energy of a user is scheduled to transmit in time slot k as shown in Eq. (3.41):

$$\overline{E}_k = \eta\, P_{PB}\,(T + \tau(k-1)) \sum_{i=1}^{M_p} d_i^{-\alpha} \mathrm{h}_i \tag{3.41}$$

where η is the receiver harvesting efficiency as defined previously, $P_{PB} = \frac{P_{total}}{M_p}$ is the transmit power of a PB with M_p is the number of PBs distributed over the macro cell, P_{total} and T, α and τ are defined previously, d_i is the distance between a given UE and the ith PB, the transmit power of a PB, P_{PB} ensures same total power consumption as is the case in full duplex configuration, h_i denotes the composite fading channel between a given UE and the ith PB, and α denotes the path-loss exponent of the energy transfer links from PBs. The path-loss exponent α may change with the frequencies that are used.

Denote d the distance between symmetrically deployed PBs and the BS. Now the distance d_i is given as

$$d_i = \sqrt{d + r^2 - 2rd\cos(\theta - \theta_i)} \tag{3.42}$$

where θ is the angle of the user from $x - axis$ and θ_i is the angle between the $x - axis$ and the line connecting the BS to the ith PB. The energy harvested by the farthest UE scheduled to transmit in time slot k can be defined by (3.41), replacing r in the expression of d_i with $r_{(k)}$.

Directed Mode In this energy transfer mode, a UE harvests energy only from the *nearest* PB. Since using efficient beamforming ensures the power from the side lobes of other users' beams is negligible, we assume the harvested energy by a given user can be modelled for using the beamforming response of the main lobe ω of the beamforming, which is assumed to be fixed. The energy harvested by arbitrary UE k in the directed mode $\dddot{E}_k$ is

$$\dddot{E}_k = \frac{\eta P_{total}}{K}(T + \tau(k-1))\,\omega d_{(1)}^{-\alpha}\,\mathrm{h}_{DL} \tag{3.43}$$

where the transmit power of a $P_{PB} = \frac{P_{total}}{M_p}$ is further divided by the average number of users $\frac{K}{M_p}$ associated with PB. This ensures a finite power for each UE associated to a PB, to carry on energy transfer until time slot for UL transmission starts. $d_{(1)}$ is the distance between an arbitrary UE and its nearest PB calculated using (3.42) with θ_i replaced with $\theta_{(1)}$ the angle between the $x - axis$ and the line connecting the BS and the nearest PB.

3.11 Integrated Energy and Spectrum Harvesting for 5G Communications

Contemporary wireless network architectures and their technologies have treated energy and spectrum harvesting as two separate cases. CR technology can be used to harvest the underutilized spectrum to address the spectrum scarcity. The energy-harvesting CR architecture is illustrated in Figure 3.37. Research activities have been attempting to address these cases at the same time, through developing architectures and technologies for 5G multi-tier networks, where users in different tiers may have different needs for channel access. An integrated spectrum and energy harvesting for 5G wireless networks is proposed in [31]. The proposed architecture efficiently collects spectrum and energy in an intelligently managed method. The main elements of energy and spectrum harvesting are CR devices; picocell networks; femtocell networks; and macro cell network. The *energy harvesting* CR *devices* communicate with other devices to periodically sense and access the available licensed channels. These CR devices are equipped with energy harvesters to convert ambient energy to recharge mobile users' batteries. The device to device (D2D) network set up by CR technology devices can either harvest energy from nearby primary transmission or transmit information if such transmission does not cause interference

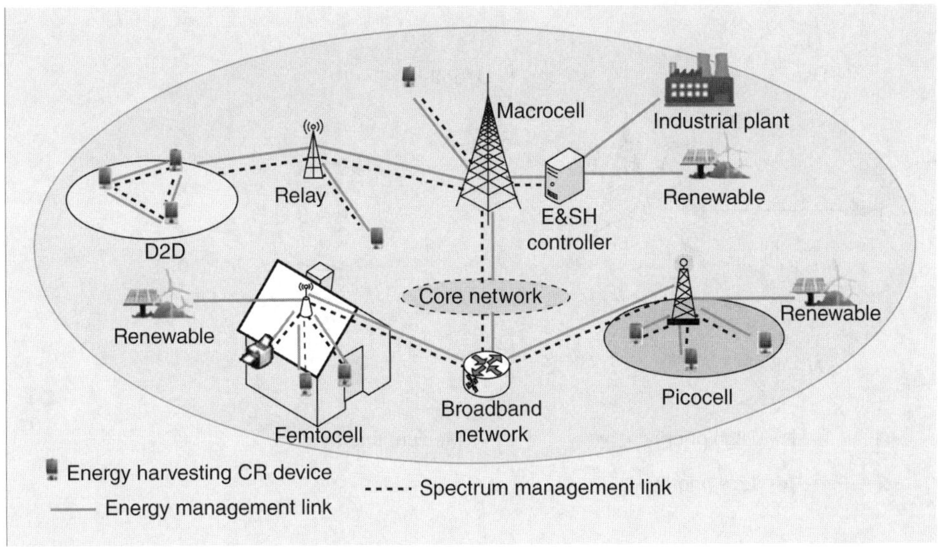

Figure 3.37 Energy-harvesting cognitive radio architecture [31].

to the primary network. The D2D networks allow direct communications between these devices without involving the BS. Furthermore, they can store harvested energy in rechargeable batteries to be used for subsequent transmissions. The D2D have sensing ability to discover unused licensed spectrum which can then be used for transmission. Small cell networks like picocell network comprising a BS and energy harvesting CR devices, and femtocell network including AP and energy harvesting CR based end users devices, which are commonly employed in 5G networks. It is reasonable to assume there may not be enough energy harvested from ambient wireless signals. So, energy from environmental resources (e.g. solar, wind, etc.) may have to be harvested as well and stored in batteries to be consumed by small cell networks. Macrocell network incorporates a BS with energy and spectrum harvesting controller to jointly manage the harvesting and usage of both energy and spectrum in the macrocell network.

The strategy for energy and spectrum management is shown in Figure 3.38. Prioritized spectrum access is likely to be used in 5G multi-tier networks. Traffic-based priority occurs as a result of different requirements of users (e.g. reliability, latency requirement, energy constraints). Tier-based priority can be employed for users belonging to different network tiers. Users can switch between different modes, fit in different tiers to satisfy different service requirements.

In a network with a large number of small-cell networks simultaneous access to a great number of devices is a challenge for spectrum and energy harvesting. The BS/AP in a small-cell network is responsible for controlling the massive access and energy harvesting of each device. There are several schemes that can be utilized for allowing mobile users to share communication system resources for the UL transmissions (i.e. multiple access). These schemes can be categorized into: contention-based, contention-free, or hybrid access schemes. A contention-based multiple-access scheme allows communication collisions to occur when multiple users transmit information at the same time (i.e. devices compete for the channel), while a contention-free access scheme does not allow collisions to occur (e.g. reserved-based scheme). A hybrid access scheme makes use of both contention-based and contention-free access during network operation to perform different functions.

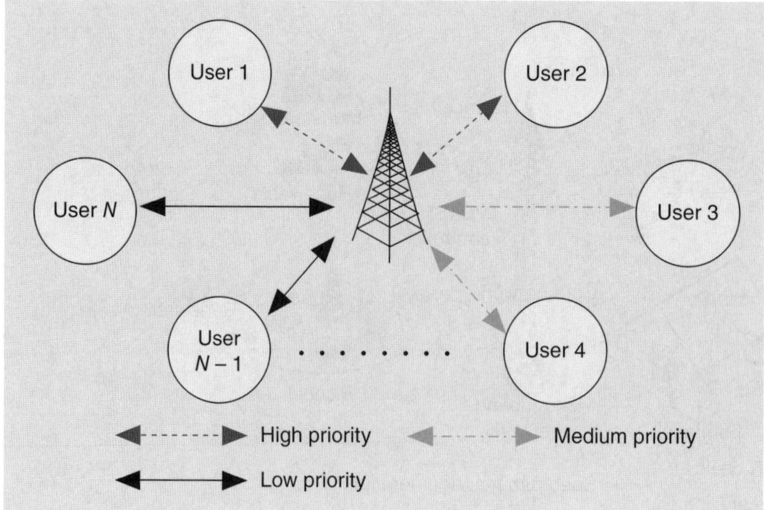

Figure 3.38 An energy and spectrum management strategy for a small-cell/macro-cell network [31].

In wireless networks, contention-based UL random access scheme is most popular due to its efficiency to allow an uncertain number of mobile users to send information. It is also simple to implement with a flexible operating model and contributes low overhead. A user device that wins the contention can access the UL for transmission. However, transmission collisions are imminent when a large number of devices request to communicate over the UL all at once.

3.12 Energy and Spectrum Harvesting Cooperative Sensing Multiple Access Control (MAC) Protocol

CR cooperative spectrum sensing arises when a group of CRs shares the sense information it acquired. Clearly, this provides a better way of spectrum usage over the area where the CRs are sited. Cooperative sensing is not only advantageous compared to noncooperative sensing, but it becomes necessary if interference is to be avoided between CR users and primary users (PUs). There are two main approaches to cooperative sensing: *centralized* and *distributed.*

In the centralized approach, there exists a master node within the CR network that collects the sensing information from all cooperating sense nodes and analyses the sensing information to determine the frequencies that can/cannot be used. The master node (called the controller) also organizes the cooperated nodes to undertake different measurements at different times, i.e. detect channel signal levels, measure the signal levels on adjacent channels, etc. In the distributed approach, no one node takes control. Communication takes place between different nodes and they share sense information.

With the CR cooperative spectrum sensing, a hidden node problem is considerably reduced, enabling the sensing to be more accurate and giving better options for the channel. Among such options is an accurate channel signal detection that reduces the number of false alarms. CR cooperative spectrum sensing necessitates certain requirements to be provided: control channel is to be used for communication between cooperating nodes, and this takes up a portion of the system BW. The nodes have to be synchronized during the spectrum sensing process, which can be very challenging. Furthermore, cooperating nodes should have suitable geographical spread to tackle the hidden node problem.

When CR cooperative spectrum sensing is used in energy harvesting for 5G nodes, sufficient energy is required for CR nodes' operations. Therefore, energy-efficient cooperative sensing strategies are required for battery-equipped secondary user (SU) nodes configuring the CR networks. Two main issues arise in cooperative spectrum sensing and energy-harvesting CR network, namely: how many SU nodes should participate in the cooperative sensing and energy harvesting, and how much energy should be consumed by the cooperative sensing nodes. A cooperative sensing MAC protocol for energy-harvesting CR network is proposed in [31]. The proposed protocol selects SU nodes for cooperative sensing. |Nodes with an energy-harvesting capability are deployed to continually sense the PU licensed channels until an available channel that is not used by PU is discovered or the SU nodes energy exceeds predetermined threshold. Furthermore, energy-harvesting abilities can be extended through receiving energy from the environment to operate different SU nodes.

The proposed MAC protocol time frame is divided into three phases: in the contention phase, when master SU node has data to transmit, it listens to the control channel. If no packet is transmitted, master SU node sends its message to the intended receiver to reserve the control channel. In the sensing phase, the master SU broadcasts SENSE-REQUEST messages to require cooperative sensing volunteers and waits for other SUs to respond with SENSER-JOIN messages. The reply messages must contain their energy level information that can be used for sensing. When the timer expired, the master SU determines the number of cooperative SUs and the amount of energy each SU is able to consume in sensing. Then the master SU starts cooperative sensing. Once all the PU licensed channels are sensed, the master SU exchange messages with cooperative SU nodes for the list of available channels. In the transmission mode master node transmits the data on the discovered channels.

3.13 Millimetre Wave (mmWave) Energy Harvesting

The mmWave bands are widely expected to be used for 5G cellular networks, they have two major attractions for wireless energy harvesting, namely: massive number of antennas, and highly dense BSs deployments. Energy-harvesting techniques are empowering low power devices/BSs to extract energy and/or information from mmWave signals. A key challenge to the energy harvesting in urban areas is due to the mmWave propagation sensitivity to building blockage. The harvesting techniques performance is measured in terms of metrics, namely: *energy coverage probability*; *average harvested energy*; and the *overall energy and information coverage probability* at a typical wireless-powered device. Energy harvesting depends on system parameters such as *BS density*; *antenna geometry parameters* such as beamwidth to maximize energy coverage for a given user population, and *wireless CSI*.

3.13.1 mmWave Network Model

In order to analyse the energy harvesting, we need an appropriate analysis platform representing the mmWave network model. In this section, we follow the analysis method presented in [32] to address various parameters effecting the energy harvesting. In the analysis, we consider the transmission on DL unless stated otherwise. Let us consider such a network consisting of mmWave BSs serving a finite number of wireless-powered devices/UEs. The UEs operate by extracting energy from the received mmWave signal. Assume the mmWave BSs are located according to a homogenous PPP denoted as $\Phi_B(\lambda_B)$ and λ_B represents BSs intensity. The users' population is drawn from another homogeneous PPP, denoted as $\Phi_u(\lambda_u)$ where λ_u users intensity. Furthermore, we assume Φ_B and Φ_u are two independent processes. In theory, mmWave

BSs and users can be located indoors or outdoors. Research studies have shown that mmWave signals suffer high penetration loss from most building materials. Therefore, we assume that the building blockage is impenetrable and focus on the outdoors BSs only. The BS-user link is of two types: either line-of-sight (LOS) or non-line-of-sight (NLOS), depending on whether it is intersected by a building blockage. Building blockage analysis in [33] is based on Cowan's theory on 'objects arranged randomly in space' published in 1989 [34]. The building blockage theory symbolises the buildings as objects drawn from a Boolean random point process. Let us define the probability function $p(r)$ of LOS link of length r defined by a power law function as

$$p(r) = e^{-\beta r} \tag{3.44}$$

where β is a constant depends on the geometry and density of the *building blockage*. We assume that a BS-receiver link of length r declared LOS with a probability $p(r)$ independent of other links. Based on this representation, we split the BS PPP into two independent but nonhomogeneous PPPs: *LOS and NLOS BSs.* On the other hand, we consider the user population to consist of two types of users, *connected* and *nonconnected*. A connected user is tagged with a BS, either LOS or NLOS, with a perfect beam alignment between a BS and its UE to maximize the directivity gain. In addition, to simplify the analysis without loss of generality, we assume that a BS serves only one user at a time. A nonconnected user is not tagged with any BS and has no beam alignment with a BS. Owing to the limited available resources, the mmWave network may serve a fraction of the population as connected users leaving the rest in the nonconnected mode.

Let ε be the probability of randomly chosen node to be a connected user independent of other nodes. Based on this, we can split the user homogeneous PPP, Φ_u, into two independent PPPs: $\Phi_{u,\,con}$ and $\Phi_{u,\,ncon}$ with intensities $\varepsilon\lambda_u$ and $(1-\varepsilon)\lambda_u$, respectively. A UE' receiver, connected or nonconnected, encounters *energy outage* if the received power falls below a set threshold. Clearly, the threshold depends on the receiver sensitivity and power consumption requirements. Define the minimum received energy required to activate the harvesting circuit to be ψ_{min}. The receiver energy outage threshold ψ has to be greater than ψ_{min} to operate the energy-harvesting circuit. Denote the harvesting rectifier efficiency as ξ.

Consider a connected user with an outage threshold ψ_{con}, and define the energy coverage probability as $P_{con}(\lambda_B, \psi_{con})$. Similarly, for the nonconnected user, the energy coverage probability is $P_{ncon}(\lambda_B, \psi_{ncon})$. Note the energy coverage probability is defined as the probability that the energy harvested (which is a random variable) at a typical receiver (in unit time) is greater than the energy outage threshold. Therefore, for all users (connected and nonconnected), the overall energy coverage probability, $\Lambda(\varepsilon, \lambda_B, \psi_{con}, \psi_{ncon})$, is given by

$$\Lambda(\varepsilon, \lambda_B, \psi_{con}, \psi_{ncon}) = \varepsilon P_{con}(\lambda_B, \psi_{con}) + (1-\varepsilon)P_{ncon}(\lambda_B, \psi_{ncon}) \tag{3.45}$$

We can see that the overall energy coverage probability expressed in (3.45) is a function of various parameters: BS density λ, probability of randomly selecting a connected user ε, energy coverage probability of connected user and nonconnected user, and outage energy threshold for connected user and nonconnected user.

3.13.2 mmWave Channel Model

The LOS mmWave signals propagate in the same way as it would in free space and the NLOS mmWave signals show a higher path loss exponent plus shadowing. Denote the path loss exponents, α_L, α_N for the LOS and NLOS links, respectively. Define the path loss in LOS link, for a user positioned at distance r_l from the lth BS as

$$g_l(r_l) = C_L r_l^{-\alpha_L} \tag{3.46}$$

where C_L is path loss constant for LOS link. Similarly, for the NLOS link,

$$\overline{g}_l(r_l) = C_N r_l^{-\alpha_N} \tag{3.47}$$

Note that the path loss formulas (3.46) and (3.47) also include the distance dependent path loss due to building blockage but does not include additional shadowing loss. Denote the fading channel as h_l of the link tagged to l^{th} BS, $\in\Phi_B$. Assuming independent fading for each link, the fading power is given as

$$P_l^f = |h_l|^2 \tag{3.48}$$

The fading power expressed in (3.48) is random variable and for LOS link, it can be modelled as normalized gamma variable

$$P_l^f \sim \Gamma\left(N_L, \frac{1}{N_L}\right) \tag{3.49a}$$

Similarly, for NLOS case,

$$\overline{P}_l^f \sim \Gamma\left(N_N, \frac{1}{N_N}\right) \tag{3.49b}$$

where N_L and N_N are fading parameters assumed as integer.

3.13.3 Antenna Model

Let us assume that the BSs and user's devices are equipped with M, N antenna element arrays, respectively. To simplify our analysis, we consider a sectored antenna model depicted in Figure 3.39 where the directivity gains for the main lobe (G) with half-power beamwidth (θ), while the side lobe directivity gains (g) with half-power beamwidth ($\overline{\theta}$). The beam pattern of the antenna under consideration is represented by the notation $A_{G,\, g,\theta,\overline{\theta}}(\varphi)$ where φ is the angle from the axis of the antenna maximum gain (i.e. the boresight direction).

Consider a randomly chosen BS in Φ_B and an energy-harvesting user in Φ_u with antenna beam patterns as $A_{G_t g_t \theta_t \overline{\theta}_t}(.)$ at the BS and $A_{G_r g_r \theta_r \overline{\theta}_r}(.)$ at the user receiver, respectively. The total directivity gain, δ_l, a random variable for the link between the lth BS and a typical user is

$$\delta_l = A_{G_t g_t \theta_t \overline{\theta}_t}(\varphi_t^l)\; A_{G_r g_r \theta_r \overline{\theta}_r}(\varphi_r^l) \tag{3.50}$$

where φ_t^l (φ_r^l) is the angle of departure (angle of arrival), both are uniformly distributed in $[0,\ 2\pi]$. There are five possible combinations for the alignment between a user and its tagged BS. To investigate these combinations, let us denote the random variable $\delta_l = D_i$ with a probability p_i where $i \in \{0, 1, 2, 3, 4\}$ so $D_i \in \{0, G_tG_r, G_tg_r, g_tG_r, g_tg_r\}$. Note $D_0 = 0$ represents the case where there is no beam alignment between BS and user beams. Each element of D_i occurs with specific probability, p_i depending on its beamwidth with corresponding probabilities; $p_i \in \{q_0, q_t\, q_r, q_t\, \overline{q}_r, \overline{q}_t\, q_r, \overline{q}_t\, \overline{q}_r\}$. The latter elements are defined as follows:

$$q_t = \frac{\theta_t}{2\pi}, \overline{q}_t = \frac{\overline{\theta}_t}{2\pi}, q_r = \frac{\theta_r}{2\pi}, \overline{q}_r = \frac{\overline{\theta}_r}{2\pi},\quad q_0 = 2 - q_t - \overline{q}_t - q_r - \overline{q}_r \tag{3.51}$$

Figure 3.39 Sectored antenna model for mmWave energy harvesting [32].

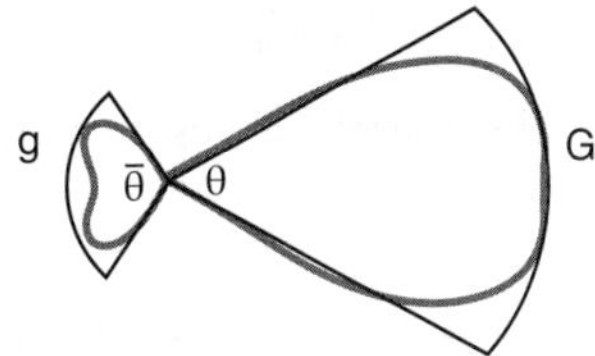

It is worth noting that for perfect alignment between typical user and its BS, the directivity gain is $G_t G_r$.

3.14 Analysis of mmWave Energy-Harvesting Technique

Consider mmWave network model that includes an energy-harvesting circuit within each UE's receiver. Since the user detects at least one LOS BS, denote the probability density function (PDF) $\tau_L(x)$ of the distance between an energy-harvesting user and its nearest LOS BS given in [33, theorem 8], [32, lemma 1] as

$$\tau_L(x) = 2\pi\ \lambda_B\ B_L^{-1}\ x\ p(x) e^{-y_L} \quad x > 0 \tag{3.52}$$

where $y_L = 2\pi\ \lambda_B \int_0^x v\ p(v)dv$ and B_L is the probability that the user receiver observes at least one LOS BS:

$$B_L = 1 - e^{-\overline{y}_L} \text{ where } \overline{y}_L = 2\pi\ \lambda_B \int_0^{\infty} v\ p(v)dv$$

Similarly, the PDF $\tau_N(x)$ of the distance between the user and its nearest NLOS BS is

$$\tau_N(x) = 2\pi\ \lambda_B\ B_N^{-1}\ x\ (1 - p(x)) e^{-y_N} \quad x > 0 \tag{3.53}$$

where $y_N = 2\pi\ \lambda_B \int_0^x v(1 - p(v))dv$ and B_N is the probability that the user receiver observes at least one NLOS BS:

$$B_N = 1 - e^{-\overline{y}_N} \text{ where } \overline{y}_N = 2\pi\ \lambda_B \int_0^{\infty} v(1 - p(v))dv$$

Let ϱ_L, ϱ_N denote the probability that the energy-harvesting UE is connected to a LOS mmWave BS and NLOS BSs, respectively. Then ϱ_L is given in [32, lemma 2] as

$$\varrho_L = B_L \int_0^{\infty} e^{-2\pi \lambda_B \int_0^{\rho_L(x)} (1-p(v))v dv}\ \tau_L(x)dx \tag{3.54}$$

where

$$\rho_L(x) = \left(\frac{C_N}{C_L}\right)^{\frac{1}{\alpha_N}} x^{\frac{\alpha_L}{\alpha_N}} \tag{3.55}$$

and

$$\varrho_N = 1 - \varrho_L \tag{3.56}$$

Consider a typical energy-harvesting UE positioned at the origin $\boldsymbol{O}$ of $\mathbb{R}^2$ of a rectangular blockage and denote the BS transmitted power P_t. The received power P_r at the typical user is

$$P_r = \sum_{l \in \Phi(\lambda_B)} P_t\ \delta_l\ P_l^f g_l\ (r_l) \tag{3.57}$$

The energy harvested at a typical user receiver per unit time γ is given as

$$\gamma = \xi\ P_r\ \mathbb{1}_{\{P_r > \Psi_{\min}\}} \tag{3.58}$$

where ξ is the rectification circuit efficiency and $[0 < \xi \leq 1]$.

3.14.1 Connected User Case

According to [32, theorem 1], the energy coverage probability, $P_{con}(\lambda_B, \psi) = \Pr(\gamma > \psi)$, at a connected user where ψ is the energy threshold and the mm Wave network with BS density λ_B is given as

$$P_{con}(\lambda_B, \psi) = P_{con,L}(\lambda_B, \overline{\psi})\varrho_L + P_{con,N}(\lambda_B, \overline{\psi})\varrho_N \tag{3.59}$$

where $\overline{\psi} = \max\left(\frac{\psi}{\xi}, \psi_{\min}\right)$ and $P_{con,L}(.)$, $P_{con,N}(.)$ are conditional energy coverage probabilities for serving BS LOS/NLOS, respectively.

3.15 Summary

Network sharing as a concept, whether used in wired or wireless networks, can be viewed as a process of splitting the entire network into InPs and SPs. InPs construct and manage the BSs, MMEs, spectrum and APs. SPs provide various services to subscribers. An earlier version of the network sharing deals with sharing basic assets such as sites, masts, power, etc. This type of network sharing is commonly called passive sharing, which can bring some saving in CAPEX and OPEX for network operators. This has since been developed by 3GPP into active sharing, where equipment and possibly some radio resources sharing is very appealing to operators. The challenges related to sharing nodes and BS were established in Section 3.2.

Spectrum sharing is carried out by InP. The latter divides the physical resources into isolated slices. Each slice is allocated to an operator, who then allocates the resource to its users.

There are three players in such a scheme for resource allocation: InP who grants the resource, the end users who benefits employing these allocated resources, and middlemen acting between these two players. Most resource allocation used in pre-LTE is achieved by a single-level scheme, that is the InP directly allocates physical resources to users of different mobile network operators according to certain predetermined resource sharing ratios without involving middlemen. In such resource allocation, the InP has to provide directly for all users, which includes high cost of computation complexity. On the other hand, new schemes are proposed for resource allocation to operators, such as opportunistic sharing-based resource allocation scheme, which allows various users to opportunistically access to shared resources, a fundamental concept in CR. Network resource-sharing options were thoroughly dealt with in Section 3.3.

NFV changes the conventional network model from hardware-centred to software-based functions installed into standard servers; storage; and switches. In essence, NFV eliminates the need for specific hardware when it introduces a new service or upgrade running services. In fact, NFV simplified the process by which network operators add, remove, or change network serve provisions. The benefits of NFV to network operators (and subscribers) are enormous. Using NFV technology, operator CAPEX and OPEX are reduced drastically due to a reduction in hardware and energy-consumption costs. The challenges and benefits of NFV network design were examined in depth in Section 3.4.

Virtualized RAN provides a variety of advantages for both end users and operators. The important decisions related to νRAN include the split between νBBU and RRU in a virtualized RAN to operate efficiently over nonideal fronthaul. These issues were investigated in Section 3.5.

The main features of the LTE EPC are the separation of control plane from user (data) plane, removal of the CS domain in core network. A recent study shows that νEPC can save mobile operators a lot of money in CAPEX and OPEX costs compared to the traditional EPC solution.

νEPC enables the mobile network infrastructure to operate at a higher utilization rate of up to 87% resulting in OPEX cost efficiency of up to 25%. In addition, the virtualization and the control/user data plane separation – driven architecture can result in OPEX cost saving in the extent of 20–40% over 5-year. νEPC was very carefully examined in Section 3.6.

A virtual switch (νSwitch) is a software programme that permits a VM (server) to communicate with another. It can direct communication on the network by inspecting packets before passing them on. νSwitch was briefly presented in Section 3.7.

Resource provisions employed various auction methods. The most common auctions are VCG auction, *combinatorial* auction, and *cooperative* auction were presented in Sections 3.8 and 3.9.

It is important to ensure the availability of sufficient energy at the user device to experience the 5G vision to provide user connectivity *anywhere, anytime,* and *to anyone.* Batteries have finite life and must be recharged or replaced. Wireless energy-harvesting nodes convert the received wireless energy to DC to recharge the wireless node's battery. The principle of wireless energy harvesting consists of a wireless energy conversion unit (e.g. filtering and rectification) to convert wireless energy to DC stored energy. Wireless-powered cellular communication is expected to play an important role in 5G communications and was investigated thoroughly in Section 3.10.

Integrating energy harvesting and spectrum harvesting is driven by two important technologies namely: wireless energy-harvesting technology and CR technology and was addressed in details in Section 3.11. While the wireless energy harvesting provides DC energy to recharge mobile batteries, the CR technology harvests the unused spectrum from a nearby PU that can be used for energy harvesting or information communication. Energy-harvesting-based cooperative sensing provides distributed sensors to enhance the UL MAC protocol was investigated in Section 3.12. In addition, energy harvesting from received mmWave signals seems very appropriate for devices communicating within the same spectrum band. The energy harvesting using mmWave transmitted signals were considered in Sections 3.13 and the analysis of the technique was presented in Section 3.14.

References

1 ETSI TS 123 251 v13.1.0 (2016-03), Universal Mobile Telecommunications System (UMTS); LTE Network Sharing: Architecture and Functional Description (3GPP TS 23.251 version 13.1.0 Release 13).

2 ETSI TR 122 951 v12.0.0 (2014-10), Universal Mobile Telecommunications System (UMTS); LTE Service aspects and Requirements for Network Sharing (3GPP TR 22.951 version 12.0.0 Release 12).

3 3GPP TS 23.122 v4.1.0(2001-06) Technical Specification Group Core Network; NAS Functions related to Mobile Station (MS) in idle mode (Release 4)

4 3GPP TS 23.251 v10.2.0 (2011-06) '*Technical Specification Group Services and System Aspects*': Network sharing - Architecture and Functions Description (Release 10)

5 Wang, X., Krishnamurthy, P., and Tipper, D. (2013). Wireless network virtualization. *Journal of Communications (Open Access)* 8 (5): 337–344.

6 Feng, Z., Qiu, C., Feng, Z. et al. (2015). An effective approach to 5G: wireless network virtualization. *IEEE Communications Magazine* 53 (12): 53–59.

7 Malanchini, I., Valentin, S., and Aydın, O. (2014). 'Generalized resource sharing for multiple operators in cellular wireless networks', IEEE International Wireless Communications and Mobile Computing Conference (IWCMC), 803–805.

8 Belt, J. van de, Ahmadi, H., and Doyle, L.E. (2014). 'A dynamic embedding algorithm for wireless network virtualization', IEEE 80th Vehicular Technology Conference (VTC2014-Fall), 1–6.

9 Kamel, M.I, Le, L.B., and Girard, A. (2014). 'LTE wireless network virtualization: dynamic slicing via flexible scheduling', IEEE 80th Vehicular Technology Conference (VTC2014-Fall), 1–5.

10 Yang, M., Li, Y., Jin, D., et al. (2013). 'Opportunistic spectrum sharing based resource allocation for wireless virtualization', Seventh International Conference on Innovative Mobile and Internet Services in Ubiquitous Computing, 51–58.

11 Chiosi, M., Clarke, D., Willis, P. et al. (2012). *Network Functions Virtualization-Introductory White Paper*, Issue 1. Darmstadt, Germany: SDN and OpenFlow World, Congress http://portal.etsi.org/NFV/NFV_White_Paper.pdf.

12 ETSI GS NFV-INF 001 V1.1.1 (2015-01) - Network Functions Virtualization (NFV); Infrastructure Overview.

13 Zaidi, Z., Friderikos, V., Yousaf, Z., et al. (2018) 'Will SDN be part of 5G?'. (Early Access).

14 Creating an ecosystem for νRAN supporting non-ideal fronthaul, White paper Telecom Infra Project (TIP).

15 Shimojo, T., Sama, M.R., Khan, A., and Iwashina, S. (2017). 'Cost-efficient method for managing network slices in a multi-service 5G core network', IFIP/IEEE IM 2017 Workshop: 2nd International Workshop on Management of 5G Networks, 1121–1126.

16 Atreyam, S, (2016) 'Economic Benefits of Virtualized Evolved Packet Core', International Data Corporation (IDC) - Sponsored by: Cisco Systems, 1–10.

17 Tsai, P-W., Lai, Y-T., Cheng, P-W., et al. (2013) 'Design and develop an OpenFlow Testbed within virtualized architecture', Asia-Pacific Network Operations and Management Symposium, 1–3.

18 Yang, C-T., Chen, W-S., Su, Y-W., et al. (2013). 'Implementation of a virtual switch monitor system using OpenFlow on cloud', IEEE International Conference on Innovative Mobile and Internet Services in Ubiquitous Computing, 283–290.

19 Tseng, H-M., Lee, H-L., Hu, J-W., et al. (2011) 'Network virtualization with cloud virtual switch', *IEEE International Conference on Parallel and Distributed Systems,* 998–1003.

20 Network functions virtualization and optimization for cloud-era networks, Nokia Strategic white Paper, Introducing virtualized service routing, 1–15, 2016.

21 Zhang, Y., Lee, C., Niyato, D., and Wang, P. (2013). Auction approaches for resource allocation in wireless systems: a survey. *IEEE Communications Surveys & Tutorials* 15 (3): 1020–1041.

22 Tang, W. and Jain, R. (2012). Hierarchical auction mechanism for network resource allocation. *IEEE Journal on Selected Areas in Communications* 30 (11): 2117–2125.

23 Wang, Q., Ye, B., Xu, T. and Lu, S. (2011). 'An approximate truthfulness motivated spectrum auction for dynamic access', IEEE Wireless Communications and Networking Conference, 257–262.

24 Fu, F. and Kozat, U.C. (2013). Stochastic game for wireless network virtualization. *IEEE/ACM Transactions on Networking* 21 (1): 84–97.

25 Zhu, K. and Hossain, E. (2016). Virtualization of 5G cellular networks as a hierarchical combinatorial auction. *IEEE Transactions on Mobile Computing* 15 (10): 2640–2654.

26 Hoydis, J., Brink, S.T., and Debbah, M. (2013). Massive MIMO in the UL/Dl of cellular networks: how many antennas do we need? *IEEE Journal on Selected Areas in Communications* 31 (2): 160–171.

27 Bi, S., Ho, C.K., and Zhang, R. (2015). Wireless powered communication: opportunities and challenges. *IEEE Communications magazine* 53 (4): 117–125.

28 Ulukus, S., Yener, A., Erkip, E. et al. (2015). Energy harvesting wireless communications: review of recent advances. *IEEE Journal on Selected Areas in Communications* 33 (3): 360–381.
29 Ju, H. and Zhang, R. (2014). Optimal resource allocation in full-duplex wireless-powered communication network. *IEEE Transactions on Communications* 62 (10): 3528–3540.
30 Huang, K. and Lau, V.K.N. (2014). Enabling wireless power transfer in cellular networks: architecture, modelling and deployment. *IEEE Transactions on Wireless Communications* 13 (2): 902–912.
31 Liu, Y., Zhang, Y., Yu, R., and Xie, S. (2015). Integrated energy and spectrum harvesting for 5G wireless communications. *IEEE Network* 29 (3): 75–81.
32 Khan, T.A., Alkhateeb, A., and Heath, R.W. (2016). Millimeter wave energy harvesting. *IEEE Transactions on Wireless Communications* 15 (9): 6048–6062.
33 Bai, T., Vaze, R., and Heath, R.W. (2014). Analysis of blockage effects on urban cellular networks. *IEEE Transactions on Wireless Communications* 13 (9): 5070–5085.
34 Cowan, R. (1989). Objects arranged randomly in space: an accessible theory. *Advances in Applied Probability* 21 (3): 543–569.
35 ETSI GS NFV-REL 006 V3.1.1 (2018-02) - Network Functions Virtualization (NFV) Release 3
36 Valenta, C.R. and Durgin, G.D. (2014). Harvesting wireless power. *IEEE Microwave Magazine* 15 (4): 108–120.
37 Tabassum, H. and Hossain, E. (2015). On the deployment of energy sources in wireless-powered cellular networks. *IEEE Transactions on Communications* 63 (9): 3391–3404.

Further Reading

ETSI GS NFV-PER 002 V1.1.2 (2014-12) - Network Functions Virtualization (NFV); Proof of Concepts; Framework.

ETSI GS NFV-EVE 003 V1.1.1 (2016-01) - Network Functions Virtualization (NFV); Ecosystem; Report on NFVI Node Physical Architecture Guidelines for Multi-Vendor Environment.

ETSI GS NFV-EVE 004 V1.1.1 (2016-03) - Network Functions Virtualization (NFV); Virtualization Technologies; Report on the application of Different Virtualization Technologies in the NFV Framework.

ETSI GS NFV-REL 003 V1.1.2 (2016-07) - Network Functions Virtualization (NFV); Reliability; Report on Models and Features for End-to-End Reliability.

ETSI GS NFV-IFA 003 V2.4.1 (2018-02) - Network Functions Virtualization (NFV) Release 2; Acceleration Technologies; *v*Switch Benchmarking and Acceleration Specification

ETSI GS NFV-SOL 002 V2.4.1 (2018-02) - Network Functions Virtualization (NFV) Release 2; Protocols and Data Models; RESTful protocols specification for the Ve-Vnfm Reference Point.

ETSI GS NFV-IFA 004 V2.4.1 (2018-02) - Network Functions Virtualization (NFV) Release 2; Acceleration Technologies; Management Aspects Specification

ETSI GS NFV-IFA 002 V2.4.1 (2018-02) - Network Functions Virtualization (NFV) Release 2; Acceleration Technologies; VNF Interfaces Specification

ETSI GS NFV-IFA 005 V2.4.1 (2018-02) - Network Functions Virtualization (NFV) Release 2; Management and Orchestration; Or-Vi reference point - Interface and Information Mod Specification

ETSI GS NFV-SEC 013 V3.1.1 (2017-02) - Network Functions Virtualization (NFV) Release 3; Security; Security Management and Monitoring specification

ETSI GS NFV-IFA 018 V3.1.1 (2017-07) - Network Functions Virtualization (NFV); Acceleration Technologies; Network Acceleration Interface Specification; Release 3

ETSI GS NFV-EVE 001 V3.1.1 (2017-07) - Network Functions Virtualization (NFV); Virtualization Technologies; Hypervisor Domain Requirements specification; Release 3

ETSI GR NFV-EVE 012 V3.1.1 (2017-12) - Network Functions Virtualization (NFV) Release 3; Evolution and Ecosystem; Report on Network Slicing Support with ETSI NFV Architecture

4

5G Enabling Technologies: Narrowband Internet of Things and Smart Cities

4.1 Introduction to the Internet of Things (IoT)

The number of physical objects connected to the internet is increasing at great speed. Example of such objects include devices intended for use in monitoring and control systems to regulate the operation of heating, ventilation, and air conditioning (HVAC), transportation; health care; industrial automation; and emergency response to natural/man-made disasters, to name a few. Internet of Things (IoT) enables the physical objects to see, hear, think, and perform jobs by communicating to each other and sharing information to coordinate decisions [1, 2]. IoT enables technologies in such cases: sensor networks, embedded devices, communication technologies, ubiquitous and pervasive computing, and internet protocols (IPs).

IoT transforms the traditional physical objects into smart objects with the capabilities to support the economy to grow and also to improve our lives. On the other hand, physical objects (things) need to be developed to fit the individual requirements and to be available anywhere anytime. IoT needs new communication protocols [3] compatible between the heterogeneous things.

IoT architecture standardization is important to enable manufacturers to deliver IoT high-quality products. Traditional internet architecture needs to be developed to match IoT challenges, including the large number of objects attempting to connect to the internet using various possible protocols. Catering for a large number of smart objects requires a large addressing space i.e. IPv6 has to be used. Other important requirements for IoT are security and privacy to be able to monitor and control the heterogeneous objects connected to the internet.

IoT provides huge opportunities for manufacturers, internet service providers (ISPs), and application software developers, given that smart object numbers are expected to be more than 200 billion deployed globally by 2020 in addition to huge traffic volume. Specifically, health-care markets like IoT-based services (e.g. m-health) and telecare (e.g. medical wellness, prevention, diagnosis, treatment, and monitoring) services are expected to create growth of trillions of US dollars annually. This significant growth potential comes with a remarkable opportunity for manufacturers to transform their traditional products into *smart things*. In order to understand the functionality of IoT, we consider the IoT building blocks illustrated in Figure 4.1. The building blocks include object identification, sensing, communication, computation, and service protocols.

Identification is a process used by IoT system to identify and match services with requesters demand. Methods used with the identification process include electronic product code (EPC), which is a universal identifier that provides a unique identity for every physical object anywhere in the world. A ubiquitous code (uCode), is used to identify things uniquely and is made of 128 bits, meaning there are (2^128 = 3.4 × 10^38) different codes. Beside object ID, it is important to find the object address within the communication network. The address of an object is done

5G Physical Layer Technologies, First Edition. Mosa Ali Abu-Rgheff.

Figure 4.1 The IoT elements [1].

using 6LoWPAN, which is an IP-based technology to compress IPv6 header to make it suitable for low-power wireless network.

Sensing is the process of collecting data from objects and sending them to a data warehouse (DW). DW is a system that reports and analyses the received data. DW is a key component of business intelligence. DW stores current and previous data in the same place and produces an analytical report to be distributed throughout the enterprise. IoT sensors can be *smart sensor, actuators,* or *wearable sensing devices.*

Communication protocols have to be developed explicitly for IoT nodes to operate with low power and in most cases over noisy narrowband links. IoT operates with various communication schemes such as 4G and 5G. The computation ability of IoT devices is limited due to finite available resources and various hardware platforms are developed to run IoT applications. Cloud platforms can be key computational part of IoT. These platforms provide facilities for smart objects to send their data to be processed in real time and extract knowledge from the collected data. IoT services can be grouped into four categories, namely:

1. *Identity-related services* are basic services available to most popular applications.
2. *Information aggregation services* collect and summarize briefly raw sensor measurements that have to be processed and reported to the IoT application.
3. *Collaborative-aware services* operate on the data obtained from the information aggregation so as to make decisions accordingly.
4. *Ubiquitous services* aim to provide collaborative-aware services *anytime, anywhere, to anyone* who needs them.

IoT semantics are related to the ability to extract knowledge smartly to provide the required services. Semantic is defined as the technique to extract knowledge that implies discovering/using resources and modelling information. In other words, analysing the data to make the correct decision to provide the exact service. In this sense the *semantic* can be thought as the *brain* of the IoT.

4.2 IoT Architecture

There is increasing interest in designing IoT architecture, mainly due to its capability to deliver services across a large number of applications. However, due to the variety of applications, this has resulted in a range of IoT architectures with varied components and functionalities and is often limited interoperability [4].

The key technologies needed to enable the interconnection among IoT objects (i.e. things) are fundamental to the IoT ecosystem and realization of the IoT applications, and have to guarantee interconnections among heterogeneous devices (things) that in most cases have very low resources (e.g. computation and energy capacity).

The two main components that enable the interoperability in the IoT domain include IoT gateways and device management techniques [5]. IoT gateways act as a bridge to connect the sensor networks with conventional communication networks. The key role to the IoT gateway is to enable technological interoperability. IoT gateway is usually positioned at the edge of the system so it can process the incoming data instantly, thus reducing the IoT response delay but it requires bandwidth (BW) for transmitting the data for processing. Furthermore, IoT gateway processors manage communications in addition to signal processing. Examples of current IoT gateway initiatives are: home gateway initiative (HGI) and open service gateway initiative (OSGi).

The HGI was founded in 2005 by leading broadband service providers (Deutsche Telekom, France Telecom [Orange], KPN Netherlands, Nippon Telegraph and Telephone Corporation (NTT) Japan, Swiss Telecom, Telenor Norway, Telecom Italia, and TeliaSonera Sweden) in association with leading vendors of digital home equipment, silicon, software, and services. HGI defined specifications are for use in home gateways, home networks, and smart homes. The specifications address typical concerns such as connectivity, security, and interworking with external networks, regulation for exchanged content, and remote device access.

The OSGi is an industrial initiative for standard procedure to connect devices (home appliances and security systems) to the Internet. The modular assembly of applications built with Java platform technologies. The service gateway is an application server (computer) between the internet and home or small business network of devices.

An OSGI application module (called bundles) is dynamically assembled for interconnecting through appropriate interfaces together with appropriate bundles.

The number of devices (things) connected to IoT requires suitable device management (device bug fixes, software update, device replacement or repair, etc.). IoT system is required to address a key set of device management policies that include provisioning and authentication; configuration and control; monitoring and diagnostics; and software updates and maintenance. Next, we briefly present these policies.

4.2.1 Provisioning and Authentication

Provisioning is a process of enrolling a device into the system and authentication is the act of securely establishing the identity of a device to ensure that it can be trusted. Authentication is part of the provisioning so that only devices with proper credentials can be registered with the system. In other words, a device should be supplied with a certificate/key stored in a secure memory that recognizes it as authentic, and the device knows the server URL to connect in order to enrol itself.

4.2.2 Configuration and Control

Once the device is connected to a home network and presents its credentials such as model and device serial number, it receives further configuration for specific settings such as name, location, and application. A system with control capability would be able to remotely reset the device to achieve a certain state or recover from errors or even implement configuration changes.

4.2.3 Monitoring and Diagnostics

IoT systems with a large number of remote devices demand a smooth and secure operation of each device to reduce the operational expenditure (OPEX). In addition, even small operational issues can have damaging consequences on the business outcomes. Monitoring system operation includes monitoring devices computing, storage, networking, and system I/O, and compares their data with the nominal values to detect system faults and to enable low cost OPEX.

4.2.4 Software Updates and Maintenance

It is important to securely update and maintain remote device software, and such practises should be a long-term running process. In particular, when a device communicates via wireless, the link may not be reliable when the device is moving, so maintenance should take a short time to save costs. Software update has to be implemented with minimum business impact.

4.3 Layered IoT Architecture

An IoT layered architecture, proposed in [1], is capable of connecting very large numbers of heterogeneous objects via the internet. The proposed architecture has three or five layers, as illustrated in Figure 4.2. The basic three-layer architecture in Figure 4.2a consists of the perception, network, and application layers. More recently, five-layer architecture has been proposed, as shown in Figure 4.2b. Next, we present an outline for each of the five-layer architectures. Objects layer, also known as devices layer or perception layer, constitutes the physical sensors that collect and process information to perform various functionalities, such as querying location, weight, motion, vibration, and humidity. The information is digitized and the data is transformed to object abstraction layer through secure channels to extract useful information. The object abstraction layer sent received data from the objects layer to the service management layer over secured channels. This layer is also responsible for cloud computing and data management processing.

The service management layer, also known as the pairing layer, pairs a service with its requester using addresses and names. Furthermore, it processes received data to make decisions and deliver services over network wire protocols such as simple object access protocol (SOAP) and Java wire protocol. In addition, this layer enables IoT applications to work with all heterogeneous devices using the same hardware platform.

The application layer provides services requested by users. It enables high quality 'smart' services that meet users' needs. Applications include smart health care, smart buildings, smart homes, transportation, and industrial automation. Typical home applications are temperature and humidity measurements when user requests the data. The business layer manages all IoT system activities and services, including establishing a business model with graphs, flowcharts, etc. based on the data received from the application layer. In addition, it designs, analyses,

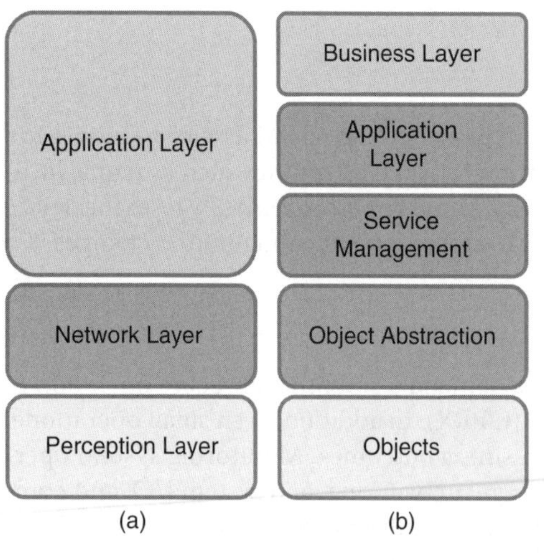

Figure 4.2 The IoT architecture. (a) three-layer (b) five-layer [1].

evaluates, monitors, and develops IoT system elements, thus supports the decision-making process. Moreover, it compares the output of each layer to improve users' privacy.

4.4 IoT Security Issues

IoT introduces the vision of a future internet that an object possessing computing and sensorial capabilities is able to communicate with other devices using IP-based communication protocols. Appropriate mechanisms are required to provide secured communications with such devices in areas as diverse as healthcare (e.g. remote patient/elderly people monitoring), smart electricity grid, home automation (e.g. heating/lightning control) and smart cities (e.g. pollution monitoring, smart lightning systems), and many others. Standardization bodies such as the IEEE and the Internet Engineering Task Force (IETF) are conducting extensive efforts towards the design of security technologies for the IoT. Such technologies, supported by industry, form the communication protocol stack for the IoT that meets the criteria of reliability and power-efficiency and internet connectivity.

The key elements [3] of the security technique are designed to protect IoT communications and to provide a guaranteed level of protection assessed in terms of *confidentiality*, *integrity*, *authentication*, and *nonrepudiation* (i.e. a transferred message has been sent and received by the parties claiming to have sent and received the message) of information flow. In addition, security schemes are required for IoT communications with sensing devices. For example, wireless sensor network (WSN) can be exposed to attacks such as denial of service (DoS), which damages the *availability* and *resilience* of the WSN. Furthermore, protection must be provided against *threats* to normal functioning of IoT communications (e.g. *fragmentation attacks*), *privacy*, *anonymity* (i.e. the need of being anonymous), *liability*, and *trust* are necessary to meet users' acceptance.

4.5 Narrowband IoT

4.5.1 NB-IoT Modes of Operation

Machine type communications (MTCs) in combination with connectivity provide attractive solutions with sensors, and meters used for water, gas, electric mains, and car parking, and other appliances. IoT is composed of numeral networks designed for specific objectives such as local area network (e.g. small community or a single home) in contrast with wide area coverage network. 3GPP introduced key features for IoT based on existing cellular systems GSM and long-term evolution (LTE). Another category of IoT is the narrowband internet of things (NB-IoT) that offers deployment flexibility such that operators launch NB-IoT using a small portion of their available spectrum (180 kHz for downlink [DL] and uplink [UL]), which are much smaller than 1.4–20 MHz LTE BW.

NB-IoT is a narrowband system to provide cellular connectivity to the IoT. This system was launched in 3GPP Release 13 and further enhanced in Release 14. The system is based on LTE technology with significant simplifications to reduce devices complexity. However, NB-IoT is not fully compatible with existing 3GPP. NB-IoT devices' design is optimized to increase coverage, reduce overhead, reduce power consumption to increase battery life up to 10 years, and increase the number of connected devices capacity. In addition to wide area coverage, devices have to be of low cost and latency is eased to10 seconds. NB-IoT is designed to have premium coexistence performance with systems operating with legacy technologies (GSM, LTE).

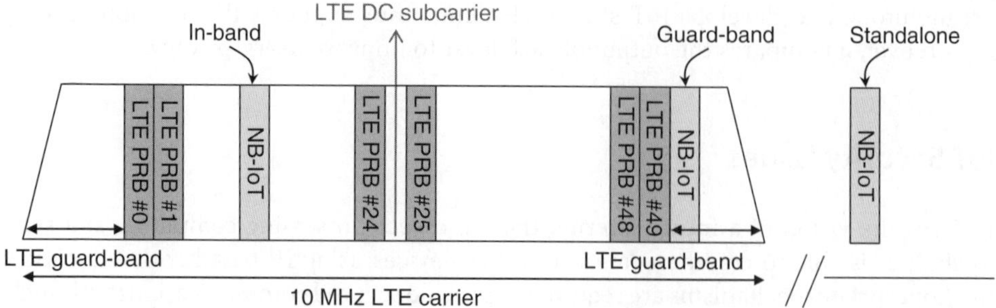

Figure 4.3 Examples of NB-IoT standalone deployment and LTE in-band and guard-band deployments [6].

NB-IoT can operate in any of three modes: in-guard band, in-band standalone in LTE, as shown in Figure 4.3. NB-IoT can operate as a standalone carrier or within LTE system spectrum as LTE carrier (in-band) or in the LTE guard band. In the in-band operation mode, LTE physical resource blocks (PRBs) are reserved for NB-IoT. In addition, the evolved node B (eNB) base station (BS) transmits power that is shared between LTE and NB-IoT with the prospect to use power spectral density (PSD) boosting for NB-IoT. For in-guard band operation, NB-IoT is placed within the guard band of an LTE spectrum.

4.5.2 NB-IoT Transmission Options

4.5.2.1 DL Transmission Method

NB-IoT DL transmission is based on OFDMA with 15 kHz subcarrier spacing as LTE. Durations of slot, subframe, and frame are 0.5, 1, and 10 ms, respectively, which are similar to those in LTE. NB-IoT carrier uses one LTE PRB corresponding to 12 subcarriers with each 15 kHz spacing for a total of 180 kHz maximum NB-IOT BW. When NB-IoT operates inside LTE BW, orthogonality between NB-IoT PRB and another LTE PRBs is maintained in the DL.

4.5.2.2 UL Transmission Method

The UL of NB-IoT supports both multicarrier and single carrier transmission. UL multicarrier transmissions (i.e. 3, 6, 12 subcarriers) based on single-carrier frequency-division multiple access (SC-FDMA) using 15 kHz carrier spacing as LTE and 12 subcarriers make up the 180 kHz channel. NB-IoT durations are similar to those for the DL for slot, subframe, frames: 0.5, 1, and 10 ms, respectively, matching those in LTE.

Single subcarrier transmission, UL NB-IoT supports two modes: 15 and 3.75 kHz. The 15 kHz subcarrier spacing coexists perfectly with LTE in the UL. The 3.75 kHz uses 2 ms slot duration.

The narrowband IoT Physical Uplink Shared Channel (NPUSCH) has two formats of subcarriers spacing (SCS). Format 1 has two SCS options: 15 kHz and 3.75 KHz. The 3.75 kHz SCS comprises 1 subcarrier and 16 slots. The 15 kHz SCS option has 4 choices of subcarrier numbers, 12 with 2 slots, 6 with 4 slots, 3 with 8 slots, and 1 with 16 slots. Both formats are obligatory for user equipment (UE).

4.6 DL Narrowband Physical Channels and Reference Signals

4.6.1 DL Physical Broadcast Channel (DPBCH)

Master information block (MIB) contains vital information that is broadcasted by the narrowband IoT BS using a Downlink Physical Broadcast Channel (DPBCH). Unlike LTE, NB-IoT physical channels and reference signals are essentially multiplexed in time. The narrowband master information block (MIB-NB) contains 34 bits and is transmitted in subframe #0 of every frame carried by DPBCH and remain unchanged over a time period of 640 ms, i.e. 64 radio frames.

The following information is provided by MIB-NB:

- 4 bits indicating the most significant bits (MSBs) of the system frame number (SFN)
- 2 bits indicating the two least-significant bits (LSBs) of the frame number
- 4 bits for system information block SIB1-NB scheduling and size, transmitted over Narrowband Physical Downlink Shared Channel (NPDSCH)
- 5 bits indicating the system information value tag
- 1 bit indicating whether access class barring is applied
- 7 bits indicating the operation mode with the mode specific values
- 11 spare bits reserved for future extensions

In addition to the essential system information defined in MIB-NB, a set of seven system information blocks are defined as follows [7]:

SIB1-NB as defined above.
SIB2-NB defines the radio resource configuration (RRC) information.
SIB3-NB defines the cell reselection information for intra-frequency and inter-frequency.
SIB4-NB defines neighbouring cell-related information relevant to intra-frequency cell reselection.
SIB5-NB defines neighbouring cell-related information relevant for inter-frequency cell reselection.
SIB14-NB defines access barring parameters.
SIB16-NB defines information related to global positioning system (GPS) time and coordinated universal time (UTC)-same content as in LTE describing time information.

UEs fully use these SIBs but ignore those from LTE. It should be noted that it is mandatory for a UE to have a valid MIB-NB, SUIB1-NB through SIB5-NB. The other SIBs have to be valid if they are required for the operation, e.g. if access barred is indicated in MIB-NB, then UE needs to have a valid SIB14-NB.

A 16-bit cyclic redundancy check (CRC) is appended to the MIB-NB, which consists of 34 bits giving 50 bits in total, which are then encoded using a 1/3 tail-biting convolutional code (TBCC) for channel encoding. The encoded MIB-NB has three streams of 50 bits each. An interleaving followed by rate matching and scrambling are applied to the streams. The rate matching is a repetition code where the three streams of size 150 bits (50×3) is repeated 16 times to get 2400 bits. The repetition rate is very high in view of the fact that the MIB contains very vital information the UE cannot afford to lose.

A single scrambling code rate one is used. Scrambled data size of 2400 bits is scrambled again using the same scrambling code that is reinitializing at the start of every update cycle, a cyclic shift of one-bit shift to the scramble code and applied separately onto each of the three output streams. A null shift (no shift) is applied at the beginning of MIB-NB update cycle, followed by shift 1, shift 2, shift 3, and shift 4 [8]. The encoded and scrambled MIB-NB are quadrature phase shift keying (QPSK) modulated over these 2400 bits to obtain 1200 complex QPSK symbols. The QPSK MIB symbols are stored in a number of sub-buffers and the contents of each sub-buffer is transmitted over the PBCH every frame (every 10 ms) in subframe 0 and all frames and repeated four times and then updated with new data. When the last sub-buffer is transmitted on a narrowband PBCH (NPBCH), a new MIB arrived from higher layers and the MIB basic process above is continued.

A UE receiver then performs blind decode to obtain the decoded bits in addition to de-scrambling and de-interleaving and de-rate matching, which match the number of symbols received to the number of bits in a transport block. The CRC 16 bits are scrambled with an antenna-specific mask. The CRC mask specifies the number of narrowband reference signal (NRS) ports [9].

4.6.2 Repetition Code SNR Gain Analysis

We now show how the repetition code improves coverage for UEs [10]. We will choose a simple system model together with some assumptions of a peer to peer link where the same symbol, x, is repeatedly transmitted R times over additive white Gaussian noise (AWGN) channel. The received signal $\boldsymbol{y}$ is

$$\mathbf{y} = \mathbf{h}x + \mathbf{n} \tag{4.1}$$

where $\mathbf{y}$, $\mathbf{h}$, $\mathbf{n}$ are $R \times 1$ vectors of received signal, channel coefficients and additive Gaussian noise, $n_i \sim C(0, 1)$ respectively. Assume $\{||x||^2\} = 1$, $\mathbb{E}\{||\mathbf{h}(i)||^2\} = \gamma, i = 1, \ldots\ldots\ldots, R-1$. Further, we choose an ideal channel estimate, $\widehat{\mathbf{h}}$, such that $\widehat{\mathbf{h}} = \mathbf{h}$. Assuming the received signals can be combined coherently. it can easily be shown that Eq. (4.1) can be written as

$$\frac{\mathbf{h}^*}{||\mathbf{h}||}y = ||\mathbf{h}||\, x + \frac{\mathbf{h}^*}{||\mathbf{h}||}\mathbf{n} \tag{4.2}$$

The received signal power to noise power ratio (SNR) symbol be γ so the effective SNR after combining is $\text{SNR}_e = R\ \gamma$. A maximum number of repetitions of NB-IoT is equal to 2048 for DL and 128 for UL channels are specified in Release 13 [11]. This implies that over 30 dB gain can be achieved using the repetition code when transmission is performed over a good channel. Channel estimation can be affected by events such as clock drift and carrier offset causing the channel estimate to be in error. Let us now consider the case of nonideal channel estimate $\widehat{\mathbf{h}}$ and denote the channel estimate error as $\mathbf{h}_e$ with a variance γ_e, $\{||\mathbf{h}_e(i)||^2\} = \gamma_e$, we get

$$\widehat{\mathbf{h}} = \mathbf{h} \pm \mathbf{h}_e \tag{4.3}$$

The nonideal channel estimate changes expression (4.1) to

$$y = \widehat{\mathbf{h}}\, x + \mathbf{n} \tag{4.4}$$

Assume that $\mathbf{h}$ and $\mathbf{h}_e$ are independent. Multiply with $\widehat{\mathbf{h}}^*$ and divide by $||\widehat{\mathbf{h}}||$ both sides of (4.4) we get

$$\frac{\widehat{\mathbf{h}}^*}{||\widehat{\mathbf{h}}||}y = \frac{\widehat{\mathbf{h}}^*\,\widehat{\mathbf{h}}}{||\mathbf{h} \pm \mathbf{h}_e||}x + \frac{\widehat{\mathbf{h}}^*}{||\widehat{\mathbf{h}}||}\mathbf{n} \tag{4.5}$$

In addition, we assume the estimate error relatively small, i.e. $\mathbf{h}_e \ll \mathbf{h}$ such that

$$\frac{\widehat{\mathbf{h}}^*}{||\widehat{\mathbf{h}}||}y \approx \frac{\mathbf{h}\,(\mathbf{h}^* \pm \mathbf{h}_e^*)}{||\mathbf{h} \pm \mathbf{h}_e||}x + \frac{\widehat{\mathbf{h}}^*}{||\widehat{\mathbf{h}}||}\mathbf{n}$$

$$\frac{\widehat{\mathbf{h}}^*}{||\widehat{\mathbf{h}}||}y \approx \frac{(||\mathbf{h}||^2 \pm \mathbf{h}_e^*\,\mathbf{h})}{||\mathbf{h} \pm \mathbf{h}_e||}x + \widetilde{\mathbf{n}} \tag{4.6}$$

where $\widetilde{\mathbf{n}} = \frac{\hat{\mathbf{h}}^*}{\|\hat{\mathbf{h}}\|}\mathbf{n}$. The average signal power can be approximated as

$$\mathbb{E}\left\{\frac{\|\mathbf{h}\|^4}{(\|\mathbf{h} \pm \mathbf{h}_e\|^2)}\right\} \approx \frac{1}{\mathbb{E}\left\{\frac{1}{\|\mathbf{h}\|^2}\right\} + \mathbb{E}\left\{\frac{\|\mathbf{h}_e\|^2}{\|\mathbf{h}\|^4}\right\}}$$

$$\approx \frac{1}{\frac{1}{R\gamma} + \frac{R\gamma_e}{2(R\gamma)^2}} = \frac{1}{\frac{2R\gamma + R\gamma_e}{2(R\gamma)^2}} = \frac{2(R\gamma)^2}{2R\gamma + R\gamma_e}$$

$$= \frac{R\gamma}{1 + \frac{\gamma_e}{2\gamma}} \tag{4.7}$$

$$\mathbb{E}\left\{\frac{\|\mathbf{h}\mathbf{h}_e^*\|^2}{(\|\mathbf{h} + \mathbf{h}_e\|^2)}\right\} \approx \frac{\mathbb{E}\{\|\mathbf{h}\mathbf{h}_e^*\|^2\}}{\mathbb{E}\{\|\mathbf{h} + \mathbf{h}_e\|^2\}} = \frac{\gamma\gamma_e}{\gamma + \gamma_e} \tag{4.8}$$

The approximate average SNR is

$$SNR = \frac{R\gamma}{1 + \frac{\gamma_e}{2\gamma}} + \frac{\gamma\gamma_e}{\gamma + \gamma_e} \tag{4.9}$$

The SNR gain due to repetition coding is show in Table 4.1

4.6.3 Narrowband Physical DL Shared Channel (NPDSCH) and Control Channel (NPDCCH)

In addition to Narrowband Physical Broadcast Channel (NPBCH), there are two more DL narrowband physical channels, namely: NPDSCH and narrowband physical downlink control channel (NPDCCH). The NPDCCH is similar to the LTE Physical Downlink Control Channel (PDCCH). It consists of two sets of control channel elements and in a single PRB (12 subcarriers), one set corresponding to the upper 6 and the other set corresponds to the lower 6 subcarriers. The maximum number of repetitions of the NPDCCH transmission for the UE depends on its coverage level and could be chosen from $\{1, 2, 4, \dots \dots \dots, 2048\}$ [9]. The NPDCCH bears scheduling information for both DL and UL data channels. It also holds the hybrid automatic repeat request (HARQ) acknowledgment information for UL as well as paging and random-access response (RAR) scheduling information. The size of the control information is fixed at 23 bits and is encoded over one subframe [12] and 15 bits are used for paging.

The narrowband physical DL shared channel (NPDSCH) bears the data destined to a specific UE as well as paging message and RAR message. There are a variety of block sizes supported in NPDSCH ranging from 256 bits up to a maximum of 680 bits. Repetition coding is used, and up to 2048 transmission repetitions can be used to carry identical copies of the data to enhance coverage. The transport blocks are mapped over a number of subframes. The number of both subframes and repetitions are registered in the downlink control indicator (DCI), which is held

Table 4.1 NB-IoT SNR gain in dB using repetition code.

$\gamma = 10$ dB			
$\gamma_e = 0$ dB		$\gamma_e = 10$ dB	
R	*SNR* dB	*R*	*SNR* dB
4	15.9	4	15
1024	39.9	1024	38.3

in NPDCCH. The information data is interleaved and encoded using TBCC for an optimized reception at the UE.

The NPDSCH has no automatic acknowledgment to a transmission. However, the UE transmits the acknowledgment on NPUSCH using one subcarrier binary phase shift keying (BPSK) modulation with a length of four slots (NRUSCH format 2). The instantaneous peak data rate for NPDSCH transmission is 170 kbps for in-band operation. In contrast the corresponding sustained peak data rate is about 26.2 kbps.

4.6.4 Narrowband Reference Signal (NRS)

In order to find a suitable cell, or if a cell change is needed, the UE activates the cell re-selection procedure to measure the received power and the quality of the received reference signal and compares these values to cell-specific thresholds provided by the NB-SIBs that are broadcasted by Narrowband Broadcast Channel (NBCH). NB-SIBs provide relevant information to the UE such as UE access to cells, perform cell selection, information about intra-frequency, inter-frequency cell selection.

The important NB-SIBs are SIB1, SIB2, and SIB3. If both values of the received NRS signal power and the signal quality are above these thresholds, the UE considers itself to be in coverage of that cell and it is attached on it.

NRS is transmitted in subframe #0 and subframe #4, and in subframe #9 only if it is not containing narrowband secondary synchronization signal (NSSS) of every frame. NRS is employing one or two antenna ports ∈{2000, 2001}. The NRS sequence $r_{l,n_s}(m)$ is defines by [13] as

$$r_{l,n_s}(m) = \frac{1}{\sqrt{2}}\{(1 - c(2m)) + j(1 - 2c(2m + 1))\} \tag{4.10}$$

where $m = 0, 1, , \ldots \ldots \ldots, 2N_{RB}^{max,DL} - 1$ where $N_{RB}^{max,DL}$ is the largest DL bandwidth configuration expressed in multiples of resource block size N_{sc}^{RB} expressed in a number of subcarriers, n_s is a slot number within a frame and l is OFDM symbol number within the slot, $c(i)$ is a length-31 Gold sequence. The NRS sequence $r_{l,n_s}(m)$ is mapped to complex-valued QPSK symbols $a_{k,l}^{(p)}$ used for antenna port p in slot n_s where (k, l) denotes resources used for transmission of NRS. The resource element R_p used for NRS transmission on antenna port p is illustrated in Figure 4.4.

4.6.5 NB-IoT Primary Synchronization Signal (NPSS)

The DL synchronization signals in NB-IoT consist of two periodically DL broadcast signals, namely: narrowband primary and secondary synchronization signals (NPSS) and NSSS. NPSS provides the initial timing and frequency information while NSSS provides refinement to the information obtained by NPSS. Upon power-up (or wake up) in response to a data transfer request such as meter reading reporting, UE searches for the DL synchronization signals in NB-IoT system to detect the presence of a cell at this frequency and acquire the accurate timing and frequency of the system e.g. symbol and frame timing, carrier frequency, sampling clock, etc. Using these parameters, UE can decode the information block transmitted on the NPBCH to obtain the information needed to set up a communication link within the system. A failed detection of the synchronization signal will trigger another detection procedure. NPSS is transmitted in subframe #5 every 10 ms NB-IoT frame. The NPSS time domain waveform

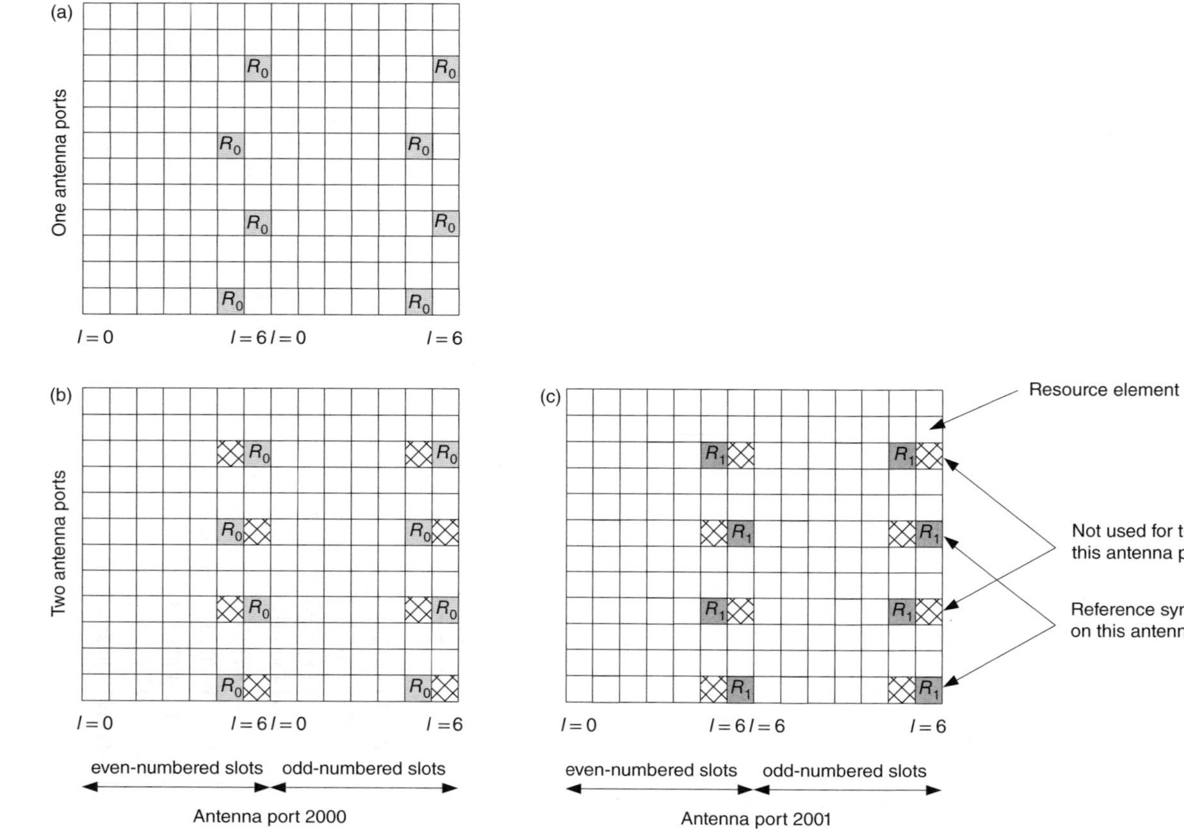

Figure 4.4 NRS resource element for one antenna port and two antenna ports. [13]. (a) NRS resource elements for one antenna port. (b) NRS resource elements for two antenna ports, port 2000. (c) NRS resource elements for two antenna ports, port 2001.

is generated from the frequency domain sequence with $K = 11$ subcarriers (out of 12) using the last $M = 11$ OFDM symbols in the subframe. It is then transformed into a time domain by means of inverse Fourier transform (IFT) and transmitted in each NB-IoT radio frame. The time domain NPSS signal $\boldsymbol{X}$ can be expressed as [14]

$$\boldsymbol{X} = [\boldsymbol{X}_0\, \boldsymbol{X}_1 \ldots \ldots \ldots \boldsymbol{X}_m \ldots \ldots \ldots \ldots \boldsymbol{X}_{M-1}] \tag{4.11}$$

where $0 \leq m \leq M-1$, and

$$\boldsymbol{X}_m = c_m\ \boldsymbol{s}$$

$$\boldsymbol{X}_m = \{\, x_{m,i} = c_m s_i\ ,\quad 0 \leq i \leq K-1\} \tag{4.12}$$

where $\boldsymbol{s}$ is the NPSS base waveform in time domain, generated from Zadoff-Chu (ZC) sequence in frequency domain using the IFT of K = 11 subcarrier of an OFDM symbol, and subcarrier 12 is unused.

$$s_i = \frac{1}{\sqrt{K}} \sum_{k=0}^{K-1} u_k\, e^{j2\pi \frac{k}{K} i} \tag{4.13}$$

where

$$u_k = e^{-j\frac{\pi\mu k(k+1)}{K}}\ ,0 \leq k \leq K-1, \mu = 5; K = 11 \tag{4.14}$$

Values of expression (4.14) are given in Table 4.2.
Hence,

$$X = \begin{bmatrix} c_0\ s_0 & c_1\ s_0 & c_m\ s_0 & c_{10}\ s_0 \\ c_0\ s_1 & c_1\ s_1 & c_m\ s_1 & c_{10}\ s_1 \\ & & & \\ c_0\ s_{10} & c_1\ s_{10} & c_m\ s_{10} & c_{10}\ s_{10} \end{bmatrix} \tag{4.15}$$

Table 4.2 u_k values generated for NPSS.

k	u_k
0	1
1	−0.9595 − j0.2817
2	−0.6549 − j0.7557
3	−0.1423 + j0.9898
4	−0.9595 + j0.2817
5	0.1454 + j0.9096
6	−0.9595 + j0.2817
7	−0.1423 + j0.9898
8	−0.6549 − j7557
9	−0.9595 − j0.2817
10	1

where, $0 \le i \le K-1$ is a time domain sample index of an OFDM sample, $0 \le m \le M-1$ is the OFDM symbol index. The NPSS base waveform in time domain $\boldsymbol{s}$ generated from Zadoff-Chu sequence in frequency domain with IFT given in (4.13):

$$\boldsymbol{s} = \begin{Bmatrix} s_0 \\ s_1 \\ s_2 \\ s_3 \\ s_4 \\ s_5 \\ s_6 \\ s_7 \\ s_8 \\ s_9 \\ s_{10} \end{Bmatrix} = \begin{Bmatrix} 0.3015 \\ 0.2536 - j0.1630 \\ 0.1253 - j0.2743 \\ -0.0429 - j2984 \\ -0.1974 - j0.2279 \\ -0.2893 - j0.0849 \\ -0.2893 + j0.0849 \\ -0.1974 + j0.2279 \\ -0.0429 + j0.2984 \\ 0.1253 + j0.2743 \\ 0.253 + j0.1630 \end{Bmatrix} \tag{4.16}$$

The base waveform $\boldsymbol{s}$ has a constant amplitude of 1, occupies the duration of one OFDM symbol, is repeated M (=11) times, and is located in the last 11 OFDM symbols of subframe #5. An 11-bit sequence $\boldsymbol{c}$ is called cover code and is used to avoid timing ambiguity arose from the repetitions

$$\boldsymbol{c} = \begin{bmatrix} 1 & 1 & 1 & 1 & -1 & -1 & 1 & 1 & 1 & -1 & 1 \end{bmatrix} \tag{4.17}$$

The UE applies matched-filter (MF) detector to perform correlation of the received NPSS signal with a local copy of the NPSS at every sample point within a duration equal to the period of the NPSS (i.e. 10 ms). In the ideal scenario (no frequency offset), the maximum correlation occurs when the received NPSS aligned itself with the local copy. The presence of additive Gaussian noise corrupts the received NPSS signal, as it does in all communication systems. The received NPSS signal power attenuation with a frequency offset Δf is $\left| sinc\left(\frac{\Delta f}{\frac{B}{MK}}\right)\right|^2$ where $B = 165$ kHz is the NPSS signal BW. For example, a frequency offset of 5 kHz causes 22.5 dB attenuation to the received NPSS signal.

4.6.6 NB-IoT Secondary Synchronization Signal (NSSS)

NSSS employs the last 11 OFDM symbols and transmits in subframe #9 in every even frame. NSSS periodicity is 20 ms and is generated from a frequency-domain ZC sequence of a length of 132 elements. Each NSSS element is mapped to a resource element. NB-IoT is designed for frequent and few bytes data transmission between the UE and the network; therefore, in most cases, a handover procedure is not likely to be required.

The NSSS sequence is used by UE to detect a target physical-layer cell identity (PCID) for the 504 PCID candidates and a super frame of eight frames (80 ms timing). NSSS is generated by element-wise multiplication between a ZC sequence and a binary scrambling sequence. There are 504 unique physical-layer cell identities. The 504 PCIDs are grouped into 168 unique PCID groups, each group containing three unique identities. The grouping is such that each PCID is part of one and only one PCID group. A PCID denoted as N_{ID}^{cell} can be expressed as

$$N_{ID}^{cell} = 3\,N_{ID}^{(1)} + N_{ID}^{(2)} \tag{4.18}$$

Thus, PCID is uniquely defined by a number $N_{ID}^{(1)}$ in the range of 0–167, representing the PCID group, and a number $N_{ID}^{(2)}$ is 0, 1, or 2, representing the PCID within the group. NSSS sequence is given in [11] as

$$d(n) = b_q(m).e^{-j2\pi\theta_f n}\, e^{-j\frac{\pi u\hat{n}(\hat{n}+1)}{131}} \tag{4.19}$$

where $n = 0, 1, \ldots, 131$ and $\hat{n} = n\ mod\ 131$. The term $e^{-j\frac{\pi u\hat{n}(\hat{n}+1)}{131}}$ represents a Zadoff-Chu sequence and parameter u indicates 126 root indices, which are given as

$$u = N_{ID}^{Ncell} mod 126 + 3 \tag{4.20}$$

In (4.20), N_{ID}^{Ncell} denotes a *PCID* and $b_q(m)$ represents the four Walsh Hadamard sequences with the sequence length of 128. Parameter m is given as $m = n\ mod\ 128$ and $q = \left\lfloor \frac{N_{ID}^{Ncell}}{126} \right\rfloor$, i.e. q is the largest integer less or equal $\frac{N_{ID}^{Ncell}}{126}$. Note $\lfloor x \rfloor$ means the floor of x, i.e. the largest integer less than or equal to x.

Moreover, $e^{-j2\pi\theta_f n}$ represents a cyclic shift value of the NSSS sequence according to frame number n_f (0, 2, 4, 6, 8), for establishing synchronization of the 80-ms super frame. Cyclic shift value θ_f is represented as

$$\theta_f = \frac{33}{132}\left(\frac{n_f}{2}\right) mod 4 \tag{4.21}$$

The ZC sequence in the frequency domain is mapped into the subcarrier and is converted to a time domain sequence by the Inverse Fast Fourier Transform (IFFT). After multiplying the binary scrambling sequence to the OFDM symbols including a cyclic pre-fix (CP), the NSSS sequence is generated. The binary sequence $b_q(m)$ is given by Table 4.3.

Table 4.3 $b_q(m)$ values used for generating NPSSS [11].

q	$b_q(0), \ldots \ldots \ldots \ldots \ldots, b_q(127)$
0	[1 1]
1	[1 −1 −1 1 −1 1 1 −1 −1 1 1 −1 1 −1 −1 1 −1 1 1 −1 1 −1 −1 1 1 −1 −1 1 −1 1 1 −1 1 −1 −1 1 −1 1 1 −1 −1 1 1 −1 1 −1 −1 1 −1 1 1 −1 1 −1 −1 1 1 −1 −1 1 −1 1 1 −1 1 −1 −1 1 −1 1 1 −1 −1 1 1 −1 1 −1 −1 1 1 −1 1 −1 −1 1 1 −1 1 −1 −1 1 1 −1 −1 1 −1 1 1 −1 1 −1 −1 1 −1 1 1 −1 −1 1 1 −1 1 −1 −1 1 −1 1 1 −1 1 −1 −1 1 1 1 −1 −1 1 −1 1 1 −1]
2	[1 −1 −1 1 −1 1 1 −1 −1 1 1 −1 1 −1 −1 1 −1 1 1 −1 1 −1 −1 1 1 −1 −1 1 −1 1 1 −1 −1 1 1 −1 1 −1 −1 1 1 −1 −1 1 −1 1 1 −1 1 −1 −1 1 −1 1 1 −1 −1 1 1 −1 1 −1 −1 1 1 −1 −1 1 −1 1 1 −1 −1 1 1 −1 1 −1 −1 1 −1 1 1 −1 1 −1 −1 1 1 −1 −1 1 1 −1 1 −1 −1 1 −1 1 1 −1 −1 1 1 −1 1 −1 −1 1 −1 1 1 −1 1 −1 −1 1 −1 1 1 −1 −1 1 1 −1 1 −1 −1 1 1]
3	[1 −1 −1 1 −1 1 1 −1 −1 1 1 −1 1 −1 −1 1 −1 1 1 −1 1 −1 −1 1 1 −1 −1 1 −1 1 1 −1 −1 1 1 −1 1 −1 −1 1 1 −1 −1 1 −1 1 1 −1 1 −1 −1 1 −1 1 1 −1 −1 1 1 −1 1 −1 −1 1 −1 1 1 −1 1 −1 −1 1 1 −1 −1 1 −1 1 1 −1 1 −1 −1 1 −1 1 1 −1 −1 1 1 −1 1 −1 −1 1 1 −1 −1 1 −1 1 1 −1 −1 1 1 −1 1 −1 −1 1 −1 1 1 −1 1 −1 −1 1 1 1 −1 −1 1 −1 1 1 −1]

4.6.7 Narrowband Positioning Reference Signal (NPRS)

Positioning is supported in 3GPP Releases 13 and 14 for NB-IoT. Some uses of NPRS are wearable smart things, safety monitoring, parking sensors, livestock tracking, environment monitoring, gas/water/electricity metering, etc. Positioning employs a technique known as 'observed time difference of arrival' (OTDOA). In this technique, the NB device measures the arrival time of several transmitted reference signals and by using the difference between the observed arrival times, the position of the NB device can be estimated.

The basic concept of OTDOA-based positioning is presented in [15–17] and depicted in Figure 4.5. Several time synchronized eNBs (BSs) simultaneously transmit positioning reference signals that will arrive with different time delays at the NB user equipment (NB-UE) corresponding to the distance to each BS. The NB-UE measures the time of arrival (TOA) of positioning reference signals received from multiple BSs. The NB-UE subtracts neighbour BSs TOA to formulate the reference signal time difference measurements, which is the OTDOA. The NB-UE sent these measurements to the location sever to calculate the NB-UE position.

Consider the case where a NB-UE located at position (x, y) receives simultaneously transmitted NPRS signals from BS $i = 0, 1, \ldots \ldots \ldots \ldots, B-1$ located at positions (x_i, y_i). Let us assume the signals are transmitted at time $t = 0$. The number of the engaged BSs B is known to the NB-UE. However, the NB-UE is not synchronized with the network so the NPRS transmit time is not known to the NB-UE. Denote the transmit power of BS i as p_i and path loss L_i. The NPRS transmitted signal from the ith BS passes through multipath fading channel where paths are denoted as $k = 0, 1, \ldots \ldots, L_B - 1$. The signal on the kth path coming from the ith BS arrives at the NB-UE after a time delay $\tau_{i,k}$, its TOA $n_{i,k}$ (in number of samples) is

$$n_{i,k} = \lfloor \tau_{i,k} f_s \rfloor \tag{4.22}$$

where f_s is the sampling frequency used by the NB-UE. Denote the line-of-sight (LOS) signal path arrives at NB-UE as $\tau_{i,0}$ such that time delay is

$$\tau_{i,0} = \frac{d_i}{c} \tag{4.23}$$

where c is speed of light and d_i is the distance between i^{th} BS and the NB-UE given by

$$d_i^2 = (x - x_i)^2 + (y - y_i)^2 \tag{4.24}$$

NB-UE is positioned at (x, y).

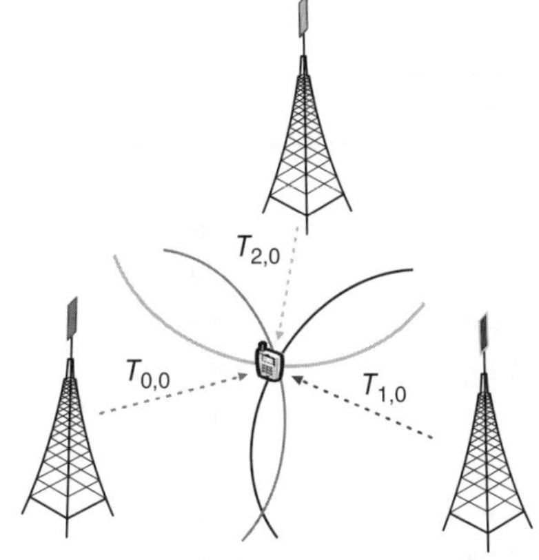

Figure 4.5 Basic principles of OTDOA-based positioning [17].

The received signal at the NB-UE depends on the complex channel coefficients, $h_{i,k}[n]$, which are unknown, and also depends on the transmission physical environment. Each path component is created by a transmitted signal being scattered by blockages such as buildings, trees, etc. The aim is to consider the LOS path component, which in most cases is not present in urban area coverage, and even if present, is weaker than the scattered components.

We consider 3GPP channel model for typical urban environment as given in Table 4.4 and we will take LOS path delay as a delay reference i.e. LOS path has zero delay. Table 4.4 indicates that power received by LOS urban channel is relatively weaker and so the NE-UE finds it difficult to detect the LOS signal. In addition, the NB-UE movements create a Doppler shift effect, and since the NB-UE is not synchronized with BSs, there will be a carrier frequency shift, causing the signal phase shift. However, since we are dealing with the amplitude of the received signal, the phase rotation can be ignored. The NB-UE estimates the delays $n_{i,0}$ of the received signal on the first path (reference path) from each BS. These received signals are OFDM symbols carrying NPRS, the NB-UE knows a priori, the NB-UE conducts a symbol by symbol correlation between received and known NPRS symbols. The average TOA delay estimates $\hat{n}_{i,0}$ samples given by the delay at the peak of the correlation functions. The TOA estimates are given by

$$\hat{\tau}_{i,0} = \frac{\hat{n}_{i,0}}{f_s} \quad i = 1, \ldots\ldots\ldots\ldots, B-1 \tag{4.25}$$

The time difference of arrival (TDOA) $\Delta\hat{\tau}_i$ can be calculated by subtracting the TOA from all other received signal TOA of BSs from a reference BS ($i = 0$)

$$\Delta\hat{\tau}_i = \hat{\tau}_{i,0} - \hat{\tau}_{0,0} \tag{4.26}$$

The TDOAs are sent to the network to calculate the position of the NB-UE by solving the OTDOA ($B - 1$) linear equations in (4.27) to determine the position of the NB-UE (x, y):

$$\left\{\begin{array}{ccc} c\Delta\hat{\tau}_1 & & \sqrt{(x-x_1)^2+(y-y_1)^2} - \sqrt{(x-x_0)^2+(y-y_0)^2} \\ c\Delta\hat{\tau}_2 & & \sqrt{(x-x_2)^2+(y-y_2)^2} - \sqrt{(x-x_0)^2+(y-y_0)^2} \\ & = & \\ & & \\ c\Delta\hat{\tau}_{B-1} & & \sqrt{(x-x_{B-1})^2+(y-y_{B-1})^2} - \sqrt{(x-x_0)^2+(y-y_0)^2} \end{array}\right\} \tag{4.27}$$

Table 4.4 3GPP urban tap delay channel model [18].

Excess tap delay [ns]	Relative power [dB]
0	−1.0
50	−1.0
120	−1.0
200	0.0
230	0.0
500	0.0
1600	−3.0
2300	−5.0
5000	−7.0

The accuracy of TDOA-based method just described is determined by a number of factors. The key two factors are the errors in the BS antenna coordinates (x_i, y_i) and the location of the NB-UE relative to location of the BSs. For three BSs (i.e. minimum number of BSs required), the best measurements can be obtained when the BSs form an equilateral triangle with the NB-UE at the centre. If the NB-UE is positioned near to the edges of the triangle, the error is increased, which increases the position estimate variance. Other factors that have influence on the measurement are the radio propagation environment (i.e. LOS path) and the interference between BSs that needs to be removed using some interference mitigation technique.

Interference cancellation requires the knowledge of channel state information (CSI). Finally, the case for weak LOS signal reception in urban environment has to be addressed. The OTDOA estimation with the *first arriving path detection* can be enhanced using a number of methods. An *iterative method* for estimating the TOA is described in [16]. In each iteration the strongest channel tap in the cross correlation $R_{xy}(\tau)$ between the received signal and the replica of the transmitted NPRS is identified. The time delay associated to the detected channel tap is stored. The channel coefficients (*or* $R_{xy}[\tau]$) are adjusted by excluding the contribution of the detected channel tap. The exclusion of the detected tap is done in the following method: Let $\overline{R}_{xy}^{l}[\tau]$ be the updated cross correlation (channel taps) in the lth iteration and denote the time delay associated with the strongest correlation in l^{th} iteration as $\overline{\tau}^{l}$ then the updated cross correlation is given by,

$$\overline{R}_{xy}^{l+1}[\tau]:=\overline{R}_{xy}^{l}[\tau] - \overline{R}_{xy}^{l}[\overline{\tau}^{l}]\, R_{xx}[\tau - \overline{\tau}^{l}], l \in \{0, 1, \dots, N\} \tag{4.28}$$

where := means equal by definition, N is the number of iteration and $R_{xx}[\tau]$ is the normalized autocorrelation function of the NPRS. The iteration is halted once the updated channel coefficients (or cross correlation) is too noisy to extract any useful information. Halting criteria is determined by the peak to an average ratio (PAR) of $|R_{xy}[\tau]|$ defined as

$$\text{PAR}_l := \frac{\max(|\overline{R}_{xy}^{l}[\tau]|)}{mean(|\overline{R}_{xy}^{l}[\tau]|)} \tag{4.29}$$

PAR_l is a consistently decreasing sequence and the iteration should halt when PAR_l is below a certain threshold PAR_{th}. Denote the set of N + 1 time delays estimated in N iterations as $:=\{\overline{\tau}^{0}, \overline{\tau}^{1}, \dots\dots\dots\dots\dots, \overline{\tau}^{N}\}$. Then the time delay of the first TOA is given as $\hat{\tau}^0 = \min\{\mathcal{T}\}$.

Figure 4.6 shows iterative 0, 1, 5 for one cell and the dashed line represents the exact time corresponding to the true distance. In addition, PAR_l decreases after each iteration, i.e. {*iteration* 0 (86.7), *iteration* 1 (50.9), *iteration* 5 (17.1)}.

Another technique to improve the accuracy of an NB-UE positioning is based on defining a threshold. The threshold can be based on the noise floor or the SNR of the received NPRS signal [15]. In the threshold technique, the channel was assumed to be invariant over a number of OFDM symbols and the expected absolute square of the cross correlation $|R_{xy}(\tau)|^2$ between the received signal and the replica of the transmitted NPRS was calculated and taken as TOA detection metric. It was found that outside the received signal arrival time region, only noise and interference remain in the correlation so this region is called the *noise floor*. The noise floor can be determined by averaging the correlation terms outside the signal region. The signal region can be identified by a moving time window. Example for the criteria is the SNR threshold, a fixed SNR_{th} for detection, which set to −13 dB for OTDOA measurement in 3GPP release 9. Other criteria known as the *adaptive threshold* denoted Λ_{th} is given by

$$\mathbb{E}\{|R_{xy}(\tau)|^2\} \geq \Lambda_{th} \tag{4.30}$$

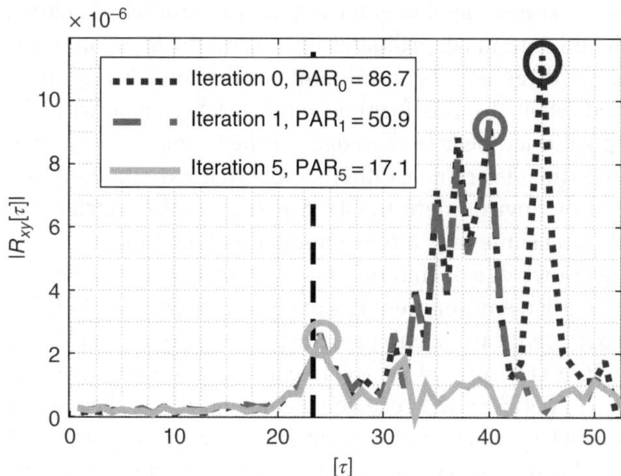

Figure 4.6 Iterations 0, 1, and 5 [16].

Figure 4.7 clearly shows that when the threshold is too low, side lobes before and after the main lobe start to appear. The side lobe before the main lobe may be detected as the first arriving path. However, with the help of smoothing average over three taps, the detected tap and its neighbouring taps, one can find the final estimated first arriving path as close to the true one.

NPRSs are transmitted in DL subframes configured for NPRS transmission only, meaning that no data is transmitted in the subframes allocated to NPRSs transmission and thus NPRSs received signals degraded by very low level of interference. NPRSs are defined for subcarrier spacing 15 kHz with normal CP only and transmitted on antenna port 2006. The complex NPRS sequence r_{l,n_s} is defined in [11, 13] as

$$r_{l,n_s}[m] = \frac{1}{\sqrt{2}}\{(1 - 2c(2m)) + j\,(1 - 2c(2m + 1))\} \tag{4.31}$$

where l is the OFDM symbol number in the time-domain in the slot number n_s within the frame, $c(i)$ is the pseudo random sequence generated by length-31 Gold sequence, $m = 0, 1, \ldots., .2\,N_{RB}^{max,DL}$, and $N_{RB}^{max,DL}$ is the maximum DL bandwidth, defined in a number of consecutive subcarriers, allocated for NPRS transmission. The pseudo random sequence is initiated at the start of each OFDM symbol.

For an NB-IoT carrier set for in-band, the NPRS is mapped to QPSK symbols, and a cell-specific frequency shift υ_{shift} given in terms of the narrowband positioning reference signal (NPRS) N_{ID}^{NPRS} as

$$\upsilon_{shift} = N_{ID}^{NPRS}\ mode\,6 \tag{4.32}$$

is applied to the NPRS pattern to avoid time-frequency PRS collisions for up to six neighbouring cells. The number of PBCH antenna ports used for transmission of NPRS is defined by high layers. A carrier configured for NPRS transmission in DL subframes is defined by the parameters T_{NPRS}, α_{NPRS}, N_{NPRS} denoting the carrier-specific subframe configuration period, carrier-specific starting subframe offset, and the number of consecutive DL subframes. The configuration is carried out by higher layers and transmitted to NB-UE. Mapping of NPRS to resource elements are given in Figure 4.8.

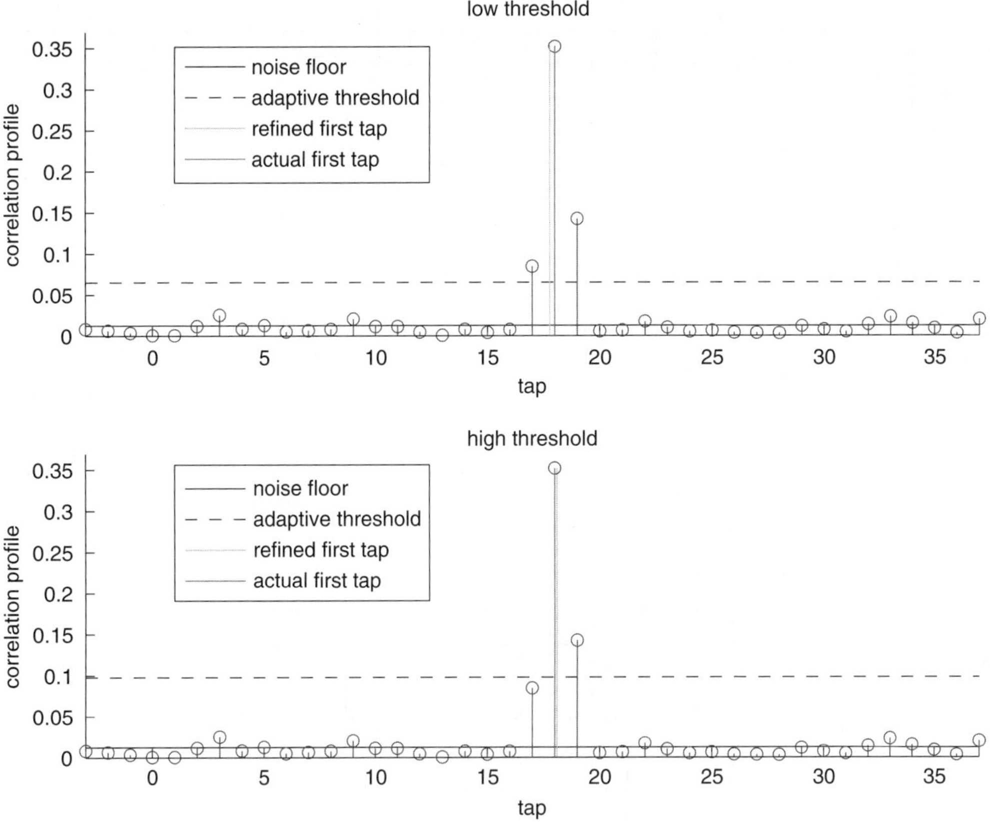

Figure 4.7 First tap detection with lower and higher threshold [15].

4.7 UL Narrowband Physical Channels and Reference Signals

This section discusses Narrowband Physical UL Shared Channels (NPUSH), Narrowband Physical Random Access Channels (NPRACH), and Demodulation Reference Signals (DMARS) [14].

4.7.1 Narrowband Physical UL Shared Channel (NPUSCH)

The NPUSCH carries both the data as well as control information using two defined formats. Format1 is used for carrying UL transport data encoded using turbo code for error correction. The maximum transport block size of format1 is 1000 bits. NPUSCH format1 supports both 15 kHz and 3.75 kHz numerology for SC-FDMA transmission. The UE can be supported with 12, 6, or 3 subcarriers. The 12 subcarrier's format is employed by the legacy LTE UEs, the 6 and 3 subcarrier formats are employed by the NB-IoT UEs. Furthermore, NPUSCH format1 reduces the peak to average power ratio (PAPR) caused by the single carrier transmission uses modulation format such as $\pi/2 - BPSK$ or $\pi/4 - QPSK$ with phase continuity between symbols can be used, and has seven OFDM symbols per slot, one of which is the demodulation reference symbol (DMRS).

Format2 is used for signalling HARQ ACK for DL data and uses repetition code of up to 128 and scrambling for error corrections with both formats. NPUSCH format2 uses SC-FDMA only, also has seven OFDM symbols per slot but uses three symbols DMRS for channel

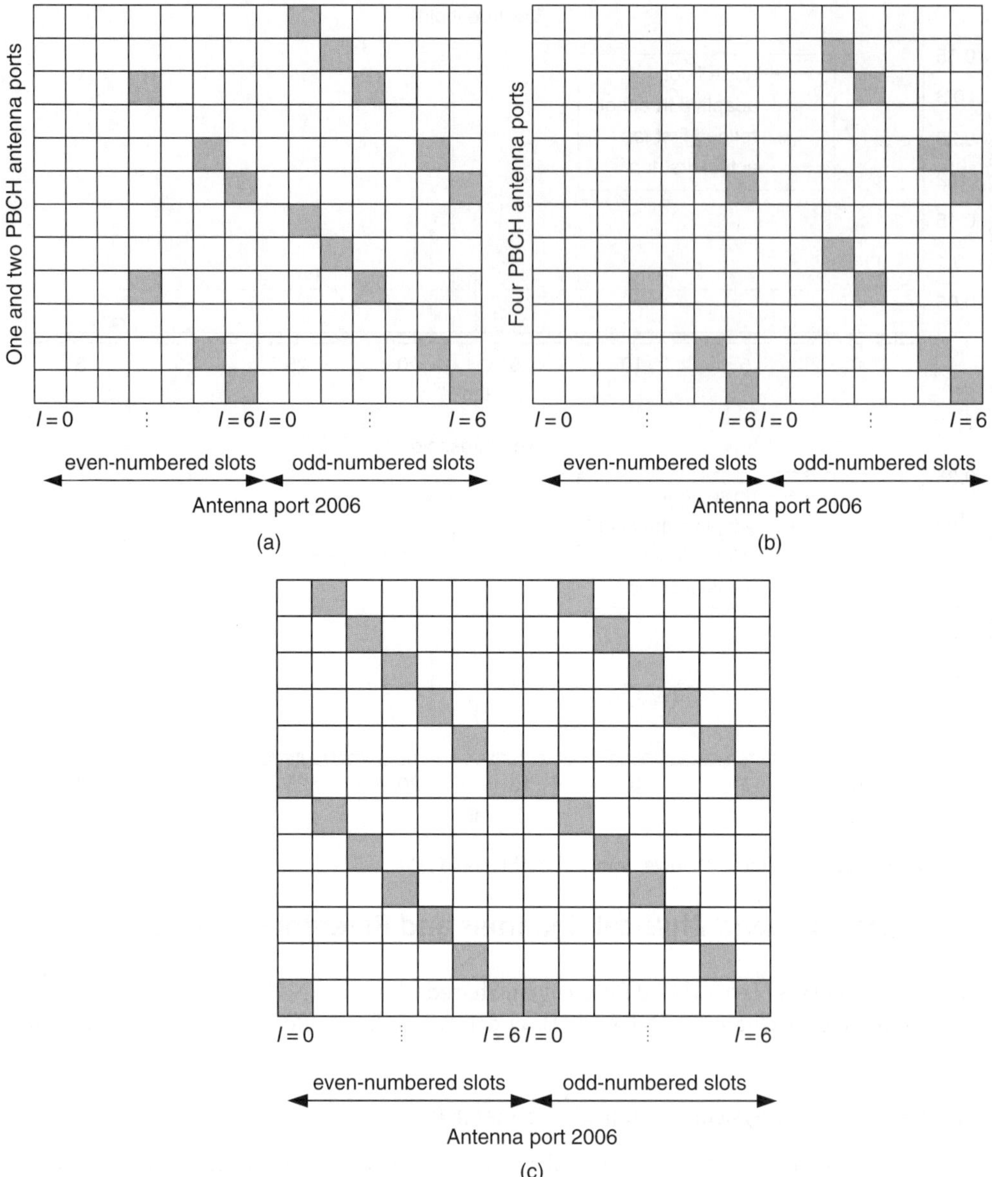

Figure 4.8 Mapping of NPRS to resource elements to standalone or guard-band configured [13]. (a) One and two PBCH antenna ports. Mapping of NPRS is set for in-band configuration. (b) Four PBCH antenna ports. Mapping of NPRS is set for in-band configuration. (c) Mapping of NPRS is set for standalone or guard-band configuration [13].

estimation. The instantaneous peak UL data rate is 250 kbps. In contrast, the sustained peak data rate is only 62.5 kbps.

4.7.2 Narrowband Physical Random-Access Channel (NPRACH)

The Narrowband Random-Access Channel (NPRACH) enables the UE to initial access and achieve synchronization with a BS. On the other hand, the BS uses the random-access preamble transmitted by a UE to estimate the UL timing delay. The random-access preamble is transmitted on SC-FDMA with frequency hopping within a configured NPRACH band constrained within 12 sub-carriers. Recall that a conventional OFDM symbol structure consists of a CP followed by OFDM data symbol. For the sake of maintaining orthogonality in the random-access transmissions among different UEs using different subcarriers, the CP has to be long enough to contain the initial UL timing uncertainty in the cell that can be as long as the maximum round trip delay (plus channel delay spread and possible DL synchronization timing error). For example, a large coverage NB-IoT may require a CP at least 233.3 μs [19]. The CP overhead can be reduced when the N-sample OFDM symbol is repeated and then the single CP is added. The larger number of OFDM symbol repetitions the smaller the CP overhead. However, the number of repetitions should be kept small enough so that the channel variations during transmission are kept negligible. The composite symbol of CP samples together with the OFDM symbol samples repetitions are called *symbol group*, denoted here as *symbolG* and shown in Figure 4.9. A random-access preamble consists of 4 symbol groups transmitted once or more on a single subcarrier at every *symbol group* transmission. According to [11, 13], the standardized design parameters are as follows: NPRAC subcarrier spacing is 3.75 kHz (i.e. symbol length 266.7 μs). Two CP lengths are provided to support different cell sizes: 66.7 μs for cell size 10 km and 266.7 μs for cell size 35 km. The number of symbols in a symbolG is specified as five. A basic unit standard of four symbolGs are defined, which can be repeated up to 128 times to enhance coverage.

The hopping pattern of NPRACH, depicted in Figure 4.10, consists of inner layer fixed size hopping and outer layer pseudo-random hopping. The outer hopping is applied between a group of four symbolGs. Inner hopping is applied within every four symbolGs. Specifically, the inner hopping is carried out as follows: single carrier hopping between first and second symbolGs, and between third and fourth symbolGs, while a six subcarriers hopping is used between the second and the third symbolGs.

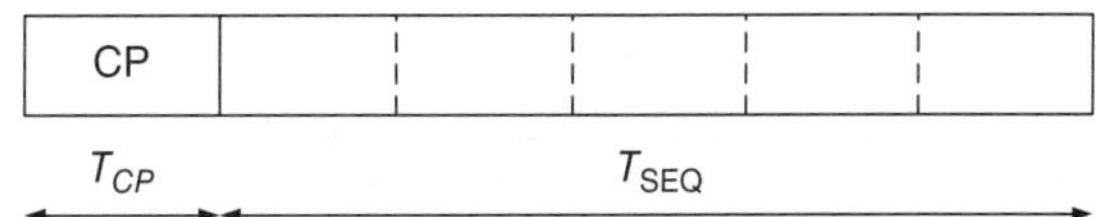

Figure 4.9 Narrowband Physical Random-Access Channel (NPRACH) symbol SymbolG consists of 1 CP + 5 OFDM symbols [13].

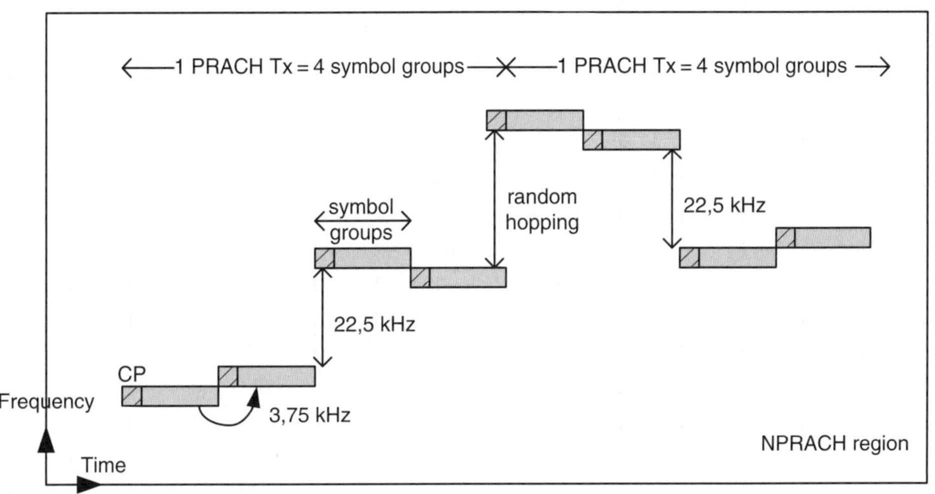

Figure 4.10 NPRACH hopping pattern [9].

A NB-IoT cell can configure an NPRACH band contains 12, 24, 36, or 48 subcarriers for each coverage. Consequently, there are 12, 24, 36, or 48 orthogonal preambles, each individually identified by its hopping pattern. In general, UE randomly selects one preamble to transmit. It is likely that the same preamble may have been selected by multiple UEs as well, resulting in a collision that has to be resolved in the random-access procedure.

4.7.3 Demodulation Reference Signals

DMRS is defined for UL in NB-IoT. It is multiplexed with data so that it is only transmitted in resource elements (Res) containing data transmission. There is no multiple-input multiple-output (MIMO) system defined for UL NB-IoT so DMRS transmissions use a single antenna port.

DMRS sequence for number of consecutive subcarriers in an UL resource unit (RU) for NB-IoT (N_{sc}^{RU}) = 1, denoted as $\bar{r}_u(n)$ is

$$\bar{r}_u(n) = \frac{1}{\sqrt{2}}(1+j)(1-2c(n))\,w(nmod16) \tag{4.33}$$

and $0 \le n < M_{rep}^{NPUSCH} N_{slots}^{UL} N_{RU}$ where M_{rep}^{NPUSCH} is scheduled # of repetitions of a NPUSCH transmission, N_{slots}^{UL} is a number of consecutive slots in an UL resource unit for NB-IoT, N_{RU} is number of scheduled UL resource units for NB-IoT, and $c(i)$ is a length-31 Gold sequence initialized with $c_{init} = 35$. The expression $w(n)$ is a random sequence of 16 elements and $u = n\ mod\ 16$ is given in Table 4.5

4.7.3.1 DMRS Sequence for NPUSCH Format1

The DMRS sequence for NPUSCH format1 $r_u(n)$ is given by

$$r_u(n) = \bar{r}_u(n) \tag{4.34}$$

Table 4.5 *w*(*n*) values for various values of *u* [13].

u	*w*(0),………………………………………………, *w*(15)															
0	1	1	1	1	1	1	1	1	1	1	1	1	1	1	1	1
1	1	−1	1	−1	1	−1	1	−1	1	−1	1	−1	1	−1	1	−1
2	1	1	−1	−1	1	1	−1	−1	1	1	−1	−1	1	1	−1	−1
3	1	−1	−1	1	1	−1	−1	1	1	−1	−1	1	1	−1	−1	1
4	1	1	1	1	−1	−1	−1	−1	1	1	1	1	−1	−1	−1	−1
5	1	−1	1	−1	−1	1	−1	1	1	−1	1	−1	−1	1	−1	1
6	1	1	−1	−1	−1	−1	1	1	1	1	−1	−1	−1	−1	1	1
7	1	−1	−1	1	−1	1	1	−1	1	−1	−1	1	−1	1	1	−1
8	1	1	1	1	1	1	1	1	−1	−1	−1	−1	−1	−1	−1	−1
9	1	−1	1	−1	1	−1	1	−1	−1	1	−1	1	−1	1	−1	1
10	1	1	−1	−1	1	1	−1	−1	−1	−1	1	1	−1	−1	1	1
11	1	−1	−1	1	1	−1	−1	1	−1	1	1	−1	−1	1	1	−1
12	1	1	1	1	−1	−1	−1	−1	−1	−1	−1	−1	1	1	1	1
13	1	−1	1	−1	−1	1	−1	1	−1	1	−1	1	1	−1	1	−1
14	1	1	−1	−1	−1	−1	1	1	−1	−1	1	1	1	1	−1	−1
15	1	−1	−1	1	−1	1	1	−1	−1	1	1	−1	1	−1	−1	1

4.7.3.2 DMRS Sequence for NPUSCH Format2

The DMRS sequence for Narrowband Physical Uplink Shared Channel (NPUSH) format 2 is given by

$$r_u(3n+m) = \overline{w}(m)\,\overline{r}_u(n) \tag{4.35}$$

where $m = 0, 1, 2$ and $\overline{w}(m)$ is given by

$$\overline{w}(0) = [1\,1\,1]; \overline{w}(1) = \left[1\,e^{\frac{j2\pi}{3}}\,e^{\frac{j4\pi}{3}}\right]; \overline{w}(2) = \left[1\,e^{\frac{j4\pi}{3}}\,e^{\frac{j2\pi}{3}}\right]$$

Denote N_{sc}^{RU} as number of consecutive subcarriers in an UL resource unit (RU). The DMRS sequence $r_u(n)$ for $N_{sc}^{RU} > 1$ and a cyclic shift α of the base sequence,

$$r_u = e^{j\alpha n}\, e^{j\phi(n)\,\pi/4} \tag{4.36}$$

where $0 \le n < N_{sc}^{RU}$.

For $N_{sc}^{RU} = 3$, the values for $\phi(n)$ for a given u are give in Table 4.6

A DMRS sequence-group hopping is enabled or disabled using the cell-specific parameter provided by the higher layers. Sequence-group hopping is essential for reducing the interference that corrupts the UL reference signal and the accompanying data.

Reference signal for NPUSCH format1 the sequence-group hopping can be enabled where the signal group number u in slot n_s is defined in [13] by a group hopping pattern $f_{gh}(n_s)$ and a sequence-shift pattern f_{ss} as

$$u = (f_{gh}(n_s) + f_{ss})\, mod N_{seq}^{RU} \tag{4.37}$$

where N_{seq}^{RU} is a number of reference sequences for the UL RU. The number of sequences N_{seq}^{RU} depends on the number of available subcarriers N_{sc}^{RU} as in Table 4.7.

The sequence-shift pattern f_{ss} is given by

$$f_{ss} = (N_{ID}^{Ncell} + \Delta_{ss})\, mod N_{seq}^{RU} \tag{4.38}$$

where N_{ID}^{Ncell} is the NB-IoT cell ID and $\Delta_{ss} \in \{0, 1, \dots \dots \dots, 29\}$ is given as a higher-layer parameter in the range 0–29 and if no value is signalled from BS, the UE assumed $\Delta_{ss} = 0$.

Table 4.6 $\phi(n)$ values for three resource blocks for various values of u [13].

u	$\phi(0)$	$\phi(1)$	$\phi(2)$
0	1	−3	−3
1	1	−3	−1
2	1	−3	3
3	1	−1	−1
4	1	−1	1
5	1	−1	3
6	1	1	−3
7	1	1	−1
8	1	1	3
9	1	3	−1
10	1	3	1
11	1	3	3

Table 4.7 Number of reference sequences for the UL resources [13].

N_{sc}^{RU}	N_{seq}^{RU}
1	16
3	12
6	14
12	30

Table 4.8 Frequency and time resource elements mapping for NPUSCH [13].

NPUSCH format	Values of index l	
	Subcarrier spacing $\Delta f = 3.75$ kHz	Subcarrier spacing $\Delta f = 15$ kHz
1	4	3
2	0, 1, 2	2, 3, 4

Sequence-group hopping can be enabled or disabled by means of the cell-specific parameter is provided by higher layers. The group-hopping pattern $f_{gh}(n_s)$ is generated using length-31 Gold sequence $c(n)$ given by

$$f_{gh}(n_s) = \left(\sum_{i=0}^{7} c(8\, n_s' + i) 2^i \right) mod N_{seq}^{RU} \tag{4.39}$$

where n_s' is the slot number for the first slot of the RU for $N_{sc}^{RU} = 1$ and $n_s' = n_s$ for $N_{sc}^{RU} > 1$. The mapping to resource elements is specified in units of (k, l) where index k indicates frequency domain, and index l time domain. The values of symbol index l are given in Table 4.8.

4.8 NB-IoT System Design

4.8.1 LTE System Specifications

NB-IoT network provides long range low rate communications to a massive number of battery operated NB-UEs using low transmission power. NB-IoT technology is supported by a licenced spectrum. Considering NB-IoT as narrow bandwidth, it can adequately get through small parts of networks of licenced spectrum. For example, within GSM band or within LTE spectrum. NB-IoT modes have to be visible to NB-UEs when first turn on and search for NB-IoT carrier. NB-IoT supports three different modes of operation, as shown in Figure 4.11.

- *In-band mode.* The IoT narrowband consists of one LTE resource block of 12 OFDM subcarrier, each with 15 kHz spacing, i.e. NB-IoT spectrum is 180 kHz for UL and DL channels. LTE and NB-IoT share the transmit power at the BS (eNB) as shown in Figure 4.11b.
- *Guard band mode.* In this mode the NB-IoT is operating within LTE guard band channel as shown in Figure 4.11c.

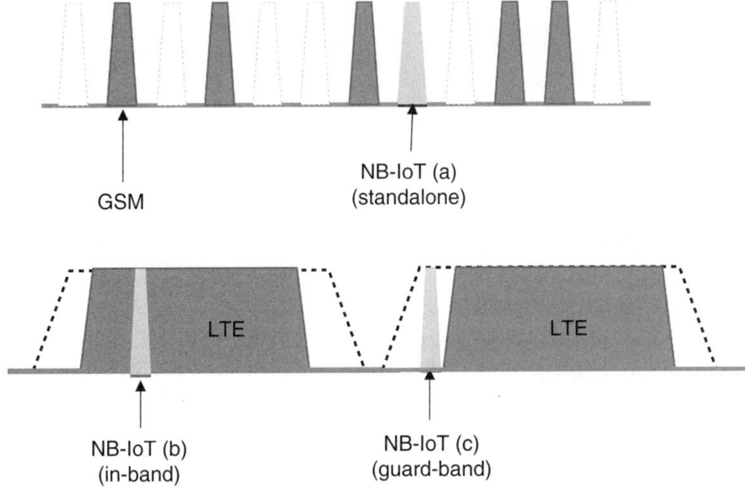

Figure 4.11 NB-IoT deployment options in GSM and LTE [10].

- *Standalone mode.* This mode is appropriately deployed in GSM occupying one GSM carrier of 200 kHz and all GSM BS transmit power is utilized for NB-IoT applications, which substantially improve the coverage as shown in Figure 4.11a.

The key challenges in the design of NB-IoT networks include NB-UE battery life, system capacity to accommodate a massive number of NB-UEs, and coverage to provide connectivity to not only urban but also to rural users. In addition, the NB-UE devices have to be of a low enough cost to be widely deployed. The cost constraint on a NB-UE device compromises the link performance, reduces the network coverage, and reduces the system capacity in terms of the number of connected UEs as well as reducing the data rate. Practically, NB-IoT applications require low power and lower data rate compared to LTE devices, with low time delay sensitivity. A low data rate transmission extends the transmission time interval, which increases the battery energy consumption – i.e. shortens the NB-UE battery life. Extending the battery life requires battery-efficient operation, meaning a high spectral efficiency i.e. higher b/s /Hz. Equivalently, higher b/s/Hz can also be achieved by employing larger numbers of NB-UE served per given bandwidth. It is important to note that NB-IoT networks are largely UL dominated [21], meaning data originated by the NB-UEs can be communicated to the network. Typical NB-IoT system specifications are summarized in Table 4.9.

4.8.2 Bandwidth Perspective-Effective BW

4.8.2.1 Capacity Extension Consideration

The obvious benefit of reducing the system BW by (say by Δ dB) is reduction in noise BW by the same amount, i.e. introducing a reduction in noise power by Δ dB. Accordingly, for the same receiver sensitivity, the BS received signal SNR will increase by Δ dB and so is the spectral efficiency. Assuming low power network operation, reducing system BW reduces the system achievable data rate but the increased spectral efficiency holds up the device data rate unchanged and hence reduction in system BW allowed more NB devices to be employed, i.e. some tradeoffs between data rate and SNR and receiver sensitivity have to be considered.

Table 4.9 LTE NB-IoT system specifications [20].

LTE NB-IoT	
Transmit power	23 dBm
Maximum coupling loss	160 dB
Receiver sensitivity	−137 dBm
Frequency band	Licenced
Minimum transmission BW	3.75 kHz
Modulation	$\frac{\pi}{2}$ – BPSK, $\frac{\pi}{4}$ – QPSK
Medium access control (MAC)	Single carrier-FDMA (SC-FDMA)
Data rate	Up to 100 kb/s
Roaming	yes
Standard	3GPP LTE Release 13 and Release 14

4.8.2.2 Coverage Extension Consideration

Let us convert Δ dB increase in spectral efficiency into Δ dB additional transmission range (due to reduction in receiver sensitivity by Δ dB). Therefore, spectral efficiency is kept unchanged but the data rate is reduced because of reduced BW. The tradeoffs between user data rate and SNR on the one hand and the receiver sensitivity on the other is shown in Figure 4.12. Conventionally, the user data rate is given by Shannon-Hartley formula for a system with BW B (dB Hz) and a given channel SNR dB and receiver sensitivity ρ as denoted by the length of the arrow in Figure 4.12 (*case nominal*). Now if the BW is reduced by Δ dB and channel SNR increased by Δ dB but data rate and receiver sensitivity are kept unchanged, we can serve more users due to the increased capacity, as shown in Figure 4.12 (*case capacity extension*). Next, considering the system with reduced BW and keeping SNR unchanged but reducing the receiver sensitivity by Δ dB we can serve the same number of users as in the capacity extension case but with

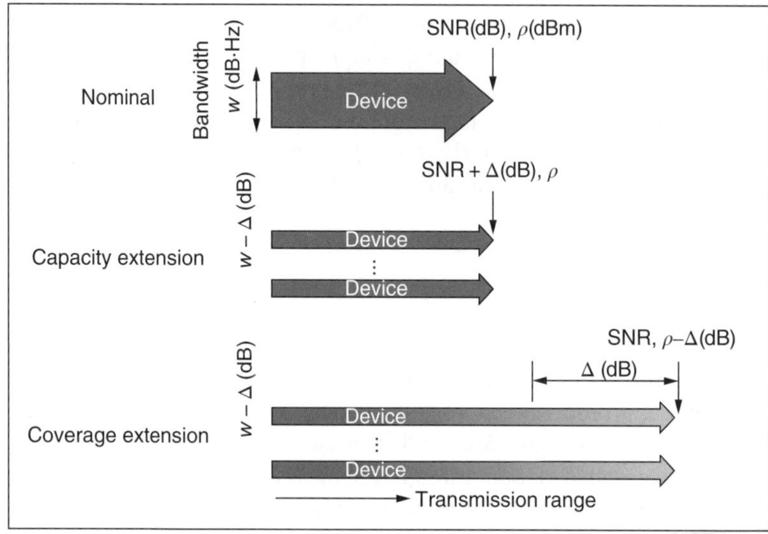

Figure 4.12 Graphical depiction of capacity and coverage extensions using narrow-band transmissions [20].

range extended by Δ dB, as shown in Figure 4.12 by the extended length of the arrow (*case coverage extension).* It is evident that for a given transmit power, the coverage (or transmission range) is fundamentally reciprocal to receiver sensitivity, so under certain transmit power, a given receiver sensitivity and coverage are interchangeable. The two scenarios are illustrated in Figure 4.12.

In the above discussion, we gave the impression that reducing a device transmission BW to infinitesimal would increase the system capacity without reducing the device's data rate and implied that the extra capacity comes at no cost. However, we assumed the NB-IoT is operating in a linear capacity region. Unfortunately, linearity assumption does not hold as system BW becomes *too small*. But *how small is too small?* In addition, extending the coverage reduces the device data rate, so how would that affect the battery life?

We need to choose the smallest BW for NB-IoT devices that maximizes the capacity and coverage while maintaining a long battery life. The new BW concept that brings together the device transmission BW together system capacity, coverage, and battery life is called *effective device BW* [20]. An important consideration should be given to the energy consumption of NB-IoT device during transmitting and receiving.

Next, we analyse the transmission time of data employing NB-IoT system. The transmission time causes two related effects namely: *time delay* and *battery energy consumption*. While time delay is relaxed in most NB-IoT applications, transmission time in NB-IoT when made relatively longer by using repetitive transmissions of the same signal to improve the range, impacts battery energy consumption, i.e. shortens battery life. So, transmission times are key to the NB-IoT devices connectivity and have to be optimized alongside coverage and system capacity.

Theoretically, the larger the BW, the higher the transmission data rate and the shorter the time required to transmit data. Accordingly, the BW allocated to NB-UE appears to be against the battery life. To investigate this conclusion further, let us consider the time $\mathcal{T}(\mathrm{B}, \rho, |\mathcal{M}|)$ required to transmit a message $\mathcal{M}$ of $|\mathcal{M}|$ bits using transmission bandwidth B, receiver sensitivity $\rho = \frac{\mathrm{P}}{\alpha}$ where P, α, β, η, N_0 are transmit power, coupling loss, degradation in realistic performance compared to theoretical limit, receiver noise figure, and thermal noise PSD (i.e. −174 dBm/Hz) respectively. Applying the Shannon-Hartley formula, the transmission time is given as

$$\mathcal{T}(\mathrm{B}, \rho, |\mathcal{M}|) = \frac{|\mathcal{M}|}{\mathrm{B}\log_2\left(1 + \frac{\beta\rho}{\eta\, B\, N_0}\right)} \tag{4.40}$$

It is shown in Appendix 4.A that the minimum time needed to transmit message $\mathcal{M}$ when B becomes very large, i.e. $\mathrm{B} \to \infty$ is

$$\mathcal{T}(\infty, \rho, |\mathcal{M}|) = |\mathcal{M}|\ \log 2\, \frac{1}{\frac{\beta\rho}{\eta\ N_0}} \tag{4.41}$$

Consequently, the use of a finite bandwidth B acquires an extra transmission time, or penalty, $\Delta\mathcal{T}(\mathrm{B}, \rho)$, which is given by

$$\Delta\mathcal{T}(\mathrm{B}, \rho) = \frac{\mathcal{T}(\mathrm{B}, \rho, |\mathcal{M}|)}{\mathcal{T}(\infty, \rho, |\mathcal{M}|)} - 1 \tag{4.42}$$

4.8.3 Battery Usage Efficiency

The transmission time penalty can be converted to a corresponding battery penalty. Let us define the ratio of the battery energy consumed on the transmission/reception of a message of an ideal case and a realistic case, as the battery usage efficiency $\gamma(B, \rho)$, i.e.

$$\gamma(\mathrm{B}, \rho) = \frac{\text{battery consuption ideal case (infinite B)}}{\text{battery consuption realistic case (finite B)}} \tag{4.43}$$

Denote I for the electric current drawn from the battery during usage, then (4.43) becomes,

$$\gamma(\mathrm{B}, \rho) = \frac{\mathrm{I}.\mathcal{T}(\infty, \rho, |\mathcal{M}|)}{\mathrm{I}.\mathcal{T}(\mathrm{B}, \rho, |\mathcal{M}|)} = \frac{1}{\Delta\mathcal{T}(\mathrm{B}, \rho) + 1} \tag{4.44}$$

The battery usage efficiency given in (4.44) defined $\gamma(\mathrm{B}, \rho) = 100\%$ for the most efficient and $\gamma(\mathrm{B}, \rho) = 0\%$ for the least efficient. In addition, the transmission time penalty in (4.42) means that a 10% penalty equal a 10% increase in transmission time relative to minimum transmission time (with an infinite BW), while 100% penalty corresponds to doubled transmission relative to minimum transmission time.

Time penalty expressed in (4.42) as a function of the transmission BW is plotted in Figure 4.13 at various receiver sensitivity. Examining Fig. 4.13, it can be noted that there exists a unique bandwidth for every coverage level such that a larger BW will have no effect on transmission time. This BW is named by [20] as 'effective BW' B_{eff}. In fact, for $B \gg B_{eff}$, the NB-UE is inclined to be more power-limited rather than BW-limited, and the NB-UE is working in the linear capacity region so allocating more BW to the NB-UE will be unsuccessful in reducing the transmission time and hence battery consumption.

Basically, the *effective BW* divides the capacity region into linear and nonlinear regions. The linear region is a power-limited capacity region while the nonlinear is bandwidth-limited region.

Figure 4.13 shows the penalty as a percentage for systems of different levels of receiver sensitivity in dBm.

$\Delta\mathcal{T}(\mathrm{B}, \rho) = 0$ meaning the NB-UE transmission BW is enough to minimizes the transmission time, when $\Delta\mathcal{T}(\mathrm{B}, \rho) = 1$ corresponds to transmission BW that makes NB-UE transmission time twice the required minimum transmission time required for the coverage level, i.e.100% time penalty.

Assuming we define the B_{eff} such that $\Delta\mathcal{T}(\mathrm{B}_{eff}, \rho)$ equals (for example) 10%, then it corresponds to ~90% battery usage efficiency.

Furthermore, B_{eff} increases as receiver signal level in dbm for sensing increases, that is the sensitivity decreases, which is due to the fact that a receiver needs more BW for successful detection assuming the noise is manageable.

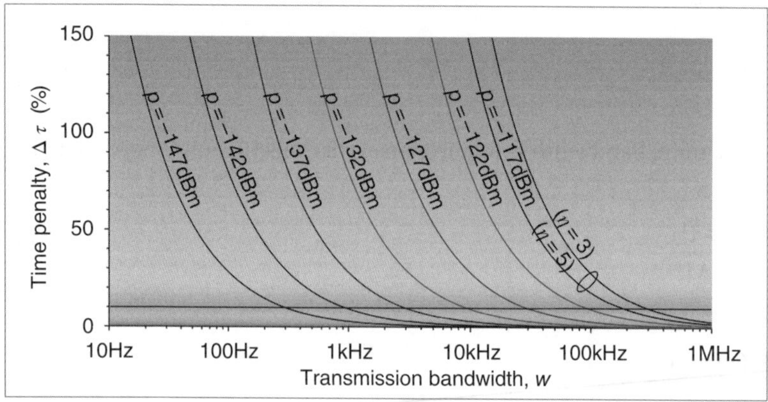

Figure 4.13 Time/battery penalty as a function of the transmission bandwidth at various sensitivities [20].

Table 4.10 Effective bandwidth for various coverage (sensitivity) levels in dBm [20].

Coverage/BS receiver sensitivity, ρ	−137 dBm	−142 dBm	−147 dBm
Effective BW, B_{ef}	3 kHz	1 kHz	300 Hz
Battery efficiency	90%	90%	90%

The values of B_{ef} for various coverage levels for time penalty 10% (i.e. 90% battery efficiency) are given in Table 4.10.

4.9 Smart Cities

According to the United Nations, as of 2016, an estimated 54.5% of the world's population lives in urban settlements. By 2030, urban areas are projected to house 60% of people globally and one in every three people will live in cities with at least half a million inhabitants. Furthermore, UN habitat organization predicts, by 2017, cities will cover less than 2% of the Earth surface but consume 78% of the world energy and produce 60% of carbon dioxide and greenhouse gas emissions. Consequently, cities are facing significant challenges, not only due to generated pollution but also the strain they impose on available resources. Therefore, appropriate solutions are needed to improve the cities' quality of life. Such solutions transform traditional cities into *smart and resilient cities*. These solutions can be developed by integrating the communication technologies and the IoT to manage the city's assets. Examples of these assets include local departments information systems, schools, libraries, hospitals. power plants, transport networks, water supply networks, waste management, etc. A *smart city* can be described as a *city that engages with and increases transparency to its citizens*, improves public resources use, and increases the services quality offered to the citizens while reducing the operational cost of public administration [22].

Before embarking on a smart city project, it is important to develop a roadmap procedure and carry out a background research about the community it is serving. For example, it is useful to gather information about the city geography, links between the city and countryside and flow of people between them, etc. The principles for a successful smart city initiative can be defined in three words: *people, processes,* and *technology*.

A city should have a good knowledge about its citizens and communities; know the processes, create policies and objectives to meet the citizens' needs, and technology can be implemented to improve the quality of life and create business opportunities.

A smart city roadmap consists of three key components:

1. Before deciding to build a smart city, you need to know why there is a need for smart city. So, evaluate the benefits of such a project, and study the community to know the city citizens' needs and their unique features: ages, education, hobbies, and city attractions.
2. Develop a smart city policy. This policy will lead the project and define key players and their responsibilities, objectives, and goals.
3. Engage the citizens through the use of e-government and create an effective governance that enhances efficiency and improves service delivery.

The most ambitious smart city's programme is planned by European Union (EU). It was set up in 2007 to run projects financed by private and public stakeholders to develop the EU smart city model. Most urban contemporary research tends to focus on the global metropolis (large cities) because they are usually the significant economic and culture centres of the country and an important hub for international connections (e.g. commerce and communication, etc.). However, the challenges facing a medium-sized city are rather different from those in a large city. Medium cities have to deal effectively with competition from large cities but appear to be less prepared in terms of critical mass, available resources and infrastructure capacity.

EU research on smart city (europeansmart version 1,2, 3) between 2007 and 2014 basically divides the EU cities into two categories:

1. Medium-sized cities with urban population between 100 000 and 500 000. The population in the city under consideration has to have a reasonable level of training by university institutions.
2. Large-sized cities with populations between 300 000 and 1 million. The EU research on larger cities started 2015 (europeansmart version 4) and cover almost 1, 600 cities.

4.10 EU Smart City Model

The EU smart city model is based on the city performance in *six key fields of urban development*: *smart economy*; *smart mobility*; *smart environment*; *smart people*; *smart living*; and *smart governance* [23]. It is important to note that currently there are no global standards to qualify a city to be smart. EU smart city's model is depicted in Figure 4.14, and a city has to excel in each of the key fields mentioned previously.

4.10.1 Smart Economy

Smart economy should be based on innovative ideas that increase productivity and reduce cost and make wide use of information and communication technology (i.e. digital economy); it has to be open and competitive, inspired by knowledge and innovation to generate high growth rate; with strong entrepreneurship to provide unique products or services that are not offered by other businesses in the same market; smart economy should be socially responsible to promote welfare for the city's citizens and green, based on sustainable energy resources. Furthermore, smart economy should be flexible in the labour market to give it the ability to adjust to fluctuations according to the increase or decrease in consumers' demands for their services and products.

4.10.2 Smart Mobility

Smart mobility elements cover disabled accessibility and people with special needs; information and communication technologies (ICT) based transport systems. Accessibility also deals with

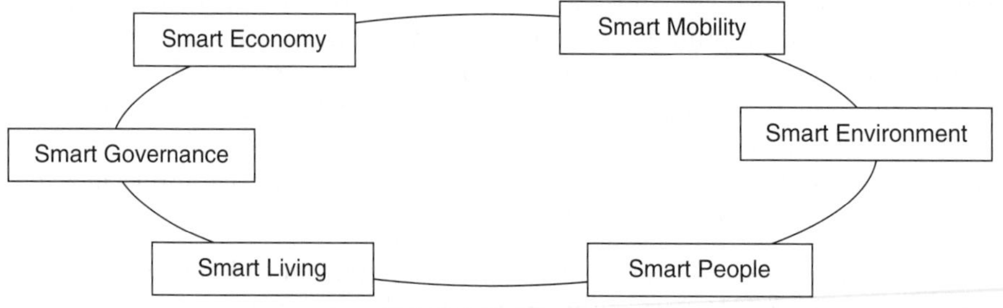

Figure 4.14 EU smart city model with high performance in six key fields of urban development [23].

the design of products, devices, services for people experiencing disabilities. In addition, accessibility includes disabled people access to workplaces such that employers have legal requirements to end discrimination against disable persons. Mobility access deals with issues concern accessible transportation for disabled persons such as low-floor vehicles to permit wheelchair access. Accessibility also includes citizens' access to ICT infrastructure and Wi-Fi services.

4.10.3 Smart Environment

Smart environment deals with ecological awareness; pollution and air quality; and sustainable resource management. We will discuss each of these issues in some detail.

Ecological awareness defines our knowledge of the measure of human demand on the Earth's resources (ecological footprint). In other words, the amount of natural resources needed to sustain our consumption and manage our waste. In order to minimize the effect of waste products on the environment, we have to *reuse what we use*, reusable products and recycling waste. Environmental awareness identifies environmentally friendly work practices for waste, energy, transport, and water issues. Renewable energy is energy, that is described as environmentally friendly, collected from natural processes such as solar, wind, tides, hydropower, and biomass (plants, trees, etc.), unlike nonrenewable energy such as fossil fuels (coal, gas, and oil) that harms the environment.

Air quality can be increased through reducing air pollution, such as smog hanging over cities. Generally, pollution is due to any substance that people introduce into the atmosphere that has a damaging effect on living things. Most sources of air pollution are due to combustion of fossil fuel that takes place in vehicle engines, long-haul aeroplanes, and power stations.

The common air pollutants are smoke deposits that leave soot on buildings and trees, causing them damage and making it difficult for living things to breathe; carbon monoxide, which is a poisonous gas; carbon dioxide (a greenhouse gas) causes Earth warming. Another pollutant related to climate change is sulphur dioxide, which causes acid rain.

Sustainable management of natural resources is key to sustainable livelihood of people on Earth. It is a technical scheme, which aims to make any production or consumption of natural resources sustainable as long as possible. In order to achieve this aim, sustainable management assessed two different quantities namely: *rate of consumption and rate of restoration.* In most cases, the goal is to keep these two quantities in equilibrium. In some cases, consumption/restoration can be controlled by sustainable management policies.

4.10.4 Smart People

Smart people are characterized by *education, lifelong learning*, open-*mindedness*, and *ethnic plurality.* Education is the process of acquiring general knowledge, skills, and a profession through schools and university training while lifelong learning, according to the European Commission is defined as '*all learning activities undertaken throughout life*, aim to enhance knowledge, skills, and competence' [24]. In spite of various definitions, there are three different forms of lifelong learning: *formal*, informal, and *nonformal* [25]. *Formal* learning entails structured periods of learning that leads to recognized awards and qualifications, generally delivered by further/higher education institutions. *Informal* learning is a friendly, or unofficial learning that does not lead to the award of academic credits. *Nonformal* learning covers structured periods of learning that may include formative assessment but does not lead to the award of academic credits.

Open-mindedness is the readiness and willingness of individuals to take on new ideas and opinions of others. It expresses people's approach to the views and knowledge of others. It is also an important attribute for effective team working.

Ethnic plurality or multiethnic society *is* a society that has, many different ethnic groups, each from a different race, history, and culture, within its nation. The majority of EU cities are multiethnic societies. The success of a multiethnic society depends on the sharing of a *single language.* For example, the EU uses English where all citizens of the European Union are brought together using this common language. The success of a multiethnic society depends on the understanding and the tolerance of all kinds of people.

4.10.5 Smart Living

Smart living covers issues such as the access to *cultural facilities: education facilities, tourism attraction, social cohesion, housing quality, health conditions,* and *individual safety.*

Cultural facilities are graded according to how likely they offer public access for recreation, heritage, and education. There are various indoor cultural places such as theatres museums concert halls arts centres, and galleries.

Education facilities should provide quality education to the community they serve. Quality education concludes that each learner should be healthy, well-nourished, and ready to participate and learn. In addition, the learning environments have to be safe, protective, and provide adequate resources and facilities. Furthermore, the learning outcomes should encompass knowledge, life skills, and are linked to national goals for education and positive participation in society. For a knowledge-based economy, the quality of higher education is increasingly regarded as strategically important for national economic development and competitiveness especially in the context of globalization.

Tourism attractions, whether natural beauty such as beaches, tropical resorts, coral reefs, mountains, forests, or man-made such as national parks, hiking, camping, botanical gardens, castles, traditional markets, and so on attract tourists that contributes to local economy and improve local community quality of life.

Social cohesion is built on three pillars: *social inclusion* that empowers marginalized people to take advantage of rising global opportunities, *social capital* where resources from people cooperating together for common ends have equal shares, and *social mobility* where individuals are able to move upwards in status based on wealth, occupation, education, etc.

Therefore, *social cohesion* works towards the well-being of all its members, fights exclusion and marginalization, promotes trust, and offers its members the opportunity for upward mobility.

Housing quality can be assessed using a number of indicators, such as location, site lay and landscaping, size, available services, and neighbourhood noise.

Smart living should ensure health and safety issues to its residents. In addition, it should provide assistive smart homes for the disabled and elderly population in the community. Smart cities with smart buildings/homes interconnected with smart technology will inevitably lead to smart living. A smart home is a dwelling where heterogeneous devices and appliances are networked to provide ubiquitous services to their residences.

Recently, ambient assisted living (AAL) is attracting a lot of interest from researchers in academia and industry to develop novel technologies to improve the quality of life for impaired people. An accessible street navigation system has been developed in [26]. The system is based on modular sensor box interfaced with an intelligent navigator. The sensing provides ultrasound, Light Detection and Ranging (Lidar), which is a remote sensing method that uses light in the form of a pulsed laser to measure ranges, and a 77-GHz mid-range radar. The system is capable to perform fast path replanning using real-time link to a remote server.

A novel approach to implementing assistive smart homes is proposed in [27, 28]. The new proposal is based on an intention recognition mechanism combined with an intelligent agent-based architecture. One of the most common intelligent agent implementation methods is that of *beliefs, desires,* and *intentions* (BDIs). Sensors monitor inhabitant activities and data from these sensors are collected and processed to identify activities of daily life. Analysing the daily-life activities, difficulties in task completion and so that a measured assistance can be offered through smart home services. The proposal addresses issues like privacy, reusability, scalability, and applicability.

In order to provide care at reduced costs for an elderly population, assistive robots in the smart home environment are suggested as a possible solution. However, the main challenge is the personalization of the robot to meet the elderly persons needs and then teach the robot to carry out behaviour in response to these activities. Authors of [29] proposed a design for the technical approach, using a commercially autonomous robot in a fully sensorized house, to the teaching-learning training of the robot, and smart home systems as an integrated unit.

A new smart parking system proposed in [30] is based on intelligent resources allocation, reservation, and pricing. The system offers guaranteed parking reservation with the lowest possible cost and searching time for drivers and the highest revenue utilization for parking business. The system is designed according to mathematical model based on mixed-integer linear programming. As homes become smarter and more technology dependent, the need for an adequate security mechanism, with minimum human intervention grows. Otherwise, internet-enabled smart homes can turn into vulnerable places to various ill-intentioned groups that infringes the residences privacy. For example, an eavesdropper can activate a particular device/appliance using public channels to interfere in the lives and privacy of the residents.

An anonymous secure framework (ASF) in connected smart homes is proposed in [31] to provide efficient authentication and key agreement between the smart home residents not to reveal to a third party (might be an eavesdropper) to enable devices anonymity. The key is for a one-time session to be renewed at the end of every device session. The proposed ASF has low computation complexity, while security can significantly be improved.

4.10.6 Smart Governance

Smart governance covers three important features namely: *political awareness, public and social services,* and *efficient and transparent administration*:

1. *Political awareness* can be defined as the ability of individuals to acquire and possess political knowledge through perception, reasoning, or intuition. In general, there are five levels of awareness: *illiterate,* who has no knowledge of the political issues concerning his community; *misinformed,* who knows little in every perspective but is likely to be loud in views and dogmatic; *general awareness,* who is generally educated and tends to be politically motivated; *activist, who* is well educated and aware of certain political issues; *expert,* who knows all political issues of importance to him/her and to the community. Smart city communities should be aware of political issues that are affecting their city with an appropriate level of awareness.
2. *Public service* is a service that is provided by the government to the people living within the community, either directly through the public sector or through a franchised service. Every service we use, including water, electricity, to transport, telecommunication, and education falls under this umbrella. Similarly, *social service* includes residential care and social care for disabled people.
3. *Transparent government administration* is the management and direction of a government that may include local government or the government of a city or town. It is paramount that the administration of local or national government be efficient, transparent, and accountable to regulatory bodies in terms of standards of practices.

4.11 Summary

Two key LTE applications that have a great impact on our society and our quality of life are: IoT and Smart City. The IoT converts dull devices into smart ones that can communicate with each other using the internet to support a secure and comfortable life for mankind. IoT provides connectivity among smart devices, systems, and services so as to make our new world a connected world.

These smart devices can be used, for example, in transportation, health care, industrial automation, emergency response to natural/man-made disasters, and in remote control home appliances such as HVAC, to name but a few. IoT technology was explained in detail in Sections 4.1–4.8.

Section 4.1 introduces IoT and examples of IoT applications. Massive numbers of devices will be connected using IP, and a strong case was made for narrowband IoT. Every smart device connected complies to a set of device management policies. IoT first has to securely establish the identity of the device to ensure that it can be trusted before enrolling into the system. Once the device is connected, it presents its credentials (i.e. model, serial number, etc.), and the system provides it with specific setting (name, location, application, etc.). IoT system with massive remote devices needs smooth and secure operation, using monitoring and diagnostics subsystems (i.e. monitoring devices computing, storage, networking, and system I/O, and comparing their data with the nominal values to detect system faults). In addition, software is securely updated and maintained regularly to minimize operation running cost. These issues were discussed in Section 4.2.

Section 4.3 presented the five-layer IoT architecture and described the function of each layer functions and protocols used (SOAP, Java). IoT is required to provide secured communications in applications such health care and smart grids, so security is a key element in IoT. IoT security issues were considered in Section 4.4.

Narrowband IoT was standardized recently by 3GPP in Releases 13 and 14, which can be implemented in three modes (in-band, in-guard band, or standalone) in LTE systems, as shown in Figure 4.3. Transmission options (DL and UL) resources were introduced in Section 4.5. Physical channels and reference signals generation and encoding for DL were dealt with in depth in Section 4.6, while physical channels and reference signal generation for UL were treated in Section 4.7. Finally, NB-IoT physical-layer system design was examined in Section 4.8.

Smart City concept transfers traditional cities into smart and resilient by integrating the communication technologies and the IoT to manage the city's assets i.e. local departments information systems, schools, libraries, hospitals, power plants, transport networks, water supply network, and waste management. Smart cities increase transparency to its citizens, improve public resource usages, and increase services quality to the citizens while reducing operational cost of public administration. The principle for a successful smart city initiative can be defined in three words: *people*, *processes*, and *technology*. The principles of *smart cities* were introduced in Section 4.9.

EU Smart City project is an advanced developed model it has six key elements these include: *smart economy*; *smart mobility*; *smart environment*; *smart people*; *smart living*; and *smart governance*. Each element in the EU smart cities project is considered individually with in-depth details in Section 4.10.

4.A [20] Minimum Time Required to Transmit Message M When B → ∞

The time $\mathcal{T}(B, \rho, |\mathcal{M}|)$ required to transmit a message $\mathcal{M}$ of $|\mathcal{M}|$ bits using transmission bandwidth B, transmit power P, and receiver sensitivity ρ and using is given by Shannon-Hartley theorem as,

$$\mathcal{T}(\mathrm{B}, \rho, |\mathcal{M}|) = \frac{|\mathcal{M}|}{\mathrm{B}\log_2\left(1 + \frac{\beta\rho}{\eta\,\mathrm{B}\,N_0}\right)} \tag{4.A.1}$$

where $\rho = \mathrm{P} - \alpha$ in dBm, α is the attenuation including path loss, antenna gains, and any cable losses. The attenuation also includes coupling loss, η is receiver noise figure, and β is a constant to adjust theoretical limit to a realistic system performance. Let logarithm function denote as, $\log_2(x)$ where,

$$x = 1 + \frac{\beta\rho}{\eta\,\mathrm{B}\,N_0} \tag{4.A.2}$$

Using the change of base for log

$$\log_2 x = \frac{\log x}{\log 2}$$

where $\log x$ is natural logarithm. Expression (4.A.1) can be written as

$$\mathcal{T}(\mathrm{B}, \rho, |\mathcal{M}|) = \log 2\frac{|\mathcal{M}|\,\mathrm{B}^{-1}}{\log x} \tag{4.A.3}$$

Take the derivative of both sides of (4.A.3) with respect to B we get

$$\frac{\mathrm{d}}{\mathrm{dB}}(B, \rho, |\mathcal{M}|) = |\mathcal{M}|\,\log 2\,\frac{\mathrm{d}}{\mathrm{dB}}\left(\frac{B^{-1}}{logx}\right) \tag{4.A.4}$$

Now we have

$$\frac{\mathrm{d}}{\mathrm{dB}}\left(\frac{B^{-1}}{\log x}\right) = \frac{\log x\frac{\mathrm{d(B^{-1})}}{\mathrm{dB}} - \mathrm{B}^{-1}\,\frac{\mathrm{d}(\log x)}{\mathrm{dB}}}{(\log x)^2} \tag{4.A.5}$$

Consider the derivative of the right-hand side of (4.A.5) term by term

$$\frac{\mathrm{d(B^{-1})}}{\mathrm{dB}} = -\mathrm{B}^{-2} \tag{4.A.6}$$

$$\frac{\mathrm{d}(\log x)}{\mathrm{dB}} = \frac{\mathrm{d}(\log x)}{\mathrm{d}x}\,\frac{\mathrm{d}x}{\mathrm{dB}}$$

$$\frac{\mathrm{d}(\log x)}{\mathrm{d}x} = \frac{1}{x}$$

Differentiate both sides of (4.A.2) with respect we get

$$\frac{\mathrm{d}x}{\mathrm{dB}} = -\frac{\text{fi}\rho\mathrm{B}^{-2}}{\eta\;N_0}$$

Therefore,

$$\frac{d(\log x)}{dB} = -\frac{1}{x}\frac{fi\rho B^{-2}}{\eta\ N_0} \tag{4.A.7}$$

Substitute (4.A.6) and (4.A.7) in (4.A.5) we get

$$\frac{d\left(\frac{B^{-1}}{\log x}\right)}{dB} = \frac{-\log x B^{-2} + B^{-1}\frac{1}{x}\frac{fi\rho B^{-2}}{\eta\ N_0}}{(\log x)^2} \tag{4.A.8}$$

Substitute (4.A.8) in (4.A.4),

$$\frac{d\mathcal{T}(B,\rho,|\mathcal{M}|)}{dB} = |\mathcal{M}|\ \log 2\left[\frac{-\log x B^{-2} + B^{-1}\frac{1}{x}\frac{fi\rho B^{-2}}{\eta\ N_0}}{(\log x)^2}\right] \tag{4.A.9}$$

To find the minimum of B that makes $\mathcal{T}(B,\rho,|\mathcal{M}|)$, we set (4.A.9) equal zero:

$$-\log x\, B^{-2} + B^{-1}\frac{1}{x}\frac{fi\rho B^{-2}}{\eta\ N_0} = 0$$

$$B^{-1} = \frac{\log x}{\frac{1}{x}\frac{\beta\rho}{\eta\ N_0}} \tag{4.A.10}$$

$$\mathcal{T}(B,\rho,|\mathcal{M}|) = \log 2|\mathcal{M}|\ \frac{1}{\frac{1}{x}\frac{\beta\rho}{\eta\ N_0}} \tag{4.A.11}$$

As $B\to\infty, x\to 1$, and (4.A.11) becomes

$$\mathcal{T}(\infty,\rho,|\mathcal{M}|) = |\mathcal{M}|\ \log 2\ \frac{1}{\frac{\beta\rho}{\eta\ N_0}} \tag{4.A.12}$$

References

1 Al-Fuqaha, A., Guizani, M., Mohammadi, M. et al. (2015). Internet of things: a survey on enabling technologies, protocols, and applications. *IEEE Communication Surveys and Tutorials* 17 (4): 2347–2376.

2 Amadeo, M., Campolo, C., Quevedo, J. et al. (2016). Information-centric networking for the internet of things: challenges and opportunities. *IEEE Network* 30 (2): 92–100.

3 Granjal, J., Monteiro, E., and Sá Silva, J. (2015). Security for the internet of things: a survey of existing protocols and open research issues. *IEEE Communication Surveys and Tutorials* 17 (3): 1294–1312.

4 Krco, S., Boris Pokriü, B., and Carrez, F. (2014). Designing IoT architecture(s): a European perspective. In: *IEEE World Forum on Internet of Things (WF-IoT)*, 79–84.

5 Gazis, V., Gortz, M., Huber, M. et al. (2015). A survey of technologies for the internet of things. In: *International Wireless Communications and Mobile Computing Conference (IWCMC)*, 1090–1095.

6 Wang, Y.-P.E., Lin, X., Adhikary, A. et al. (2017). A Primer on 3GPP Narrowband Internet of Things. *IEEE Communications Magazine* 55 (3): 117–123.

7 Rohde & Schwarz White Paper 1MA266_0e (2016) Narrowband Internet of Things, 1-24 Available: http://www.rohde-schwarx.com/appnote/1MA266.

8 Patent application publication, US 2017/0187488 A1, (2017) 'Physical Broadcast Channel (PBCH) and Master Information Block (MIB) Design', 1–16.
9 Ratasuk, R., Mangalvedhe, N., Zhang, Y. et al. (2016). Overview of narrowband IoT in LTE release 13. In: *IEEE Conference on Standards for Communications and Networking*, 1–7.
10 Beyene, Y.D., Jantti, R., Rittik, K., and Iraji, S. (2017). On the performance of narrow-band Internet things (NB-IoT). In: *IEEE Wireless Communications and Networking Conference (WCNC)*, 1–6.
11 3GPP TS 36.211 version 13.5.0 Release 13, LTE; Evolved Universal Terrestrial Radio Access (E-UTRA); Physical channels and modulation.
12 Adhikary, A., Lin, X., and Wang, Y.-P.E. (2016). Performance evaluation of NB-IoT coverage. In: *IEEE Vehicular Technology Conference* (ed. A.R. Alvarino, X. Wang, P. Gaal, et al.), 1–5.
13 3GPP TS 36.211 version 14.2.0 Release 14, LTE; Evolved Universal Terrestrial Radio Access (E-UTRA); Physical channels and modulation.
14 Wenjie, Y., Hua, M., Zhang, J. et al. (2017). Enhanced system acquisition for NB-IoT. *IEEE Access* 5: 13179–13191.
15 Huang, M. and Xu, W. (2013). Enhanced LTE TOA/OTDOA estimation with first arriving path detection. In: *IEEE Wireless Communications and Networking Conference (WCNC)*, 3992–3997.
16 Ryden, H., Zaidi, A., Razavi, S. et al. (2016). Enhanced time of arrival estimation and quantization for positioning in LTE networks. In: *IEEE International Symposium on Personal, Indoor, and Mobile Radio Communications (PIMRC)*, 1–6.
17 Hu, S., Berg, A., Li, X., and Rusk, F. (2017). Improving the performance of OTDOA based positioning in NB-IoT systems. In: *IEEE Global Communications Conference*, 1–7.
18 3GPP TS 36.104 version 12.5.0 Release 12, LTE; Evolved Universal Terrestrial Radio Access (E-UTRA); Base Station (BS) radio transmission and reception, *Table B.2-3 Extended typical Urban channel model*, p. 125.
19 Lin, X., Adhikary, A., and Wang, Y.-P.E. (2016). Random access preamble design and detection for 3GPP narrowband IoT systems. *IEEE Wireless Communications Letters* 5 (6): 640–643.
20 Yang, W., Wang, M., Zhang, J. et al. (2017). Narrowband wireless access for low-power massive internet of things: a bandwidth perspective. *IEEE Wireless Communications* 24 (3): 138–145.
21 Yu, C., Yu, L., Wu, Y. et al. (2017). Uplink scheduling and link adaptation for narrowband internet of things systems. *IEEE Access* 5: 1724–1734.
22 Zanella, A., Bui, N., Castellani, A. et al. (2014). Internet of things for smart cities. *IEEE Internet of Things Journal* 1 (1): 22–32.
23 EU Smart Cities The Smart City Model at www.smart-cities.eu.html.
24 Making a European Area of Lifelong Learning a Reality, European Commission, Brussels 2001.
25 Hyde, M. and Phillipson, C. (2014). *How Can Lifelong Learning, Including Continuous Training within the Labour Market, Be Enabled and Who Will Pay for this? Looking Forward to 2025 an 2040 how Might this Evolve?* Report University of Manchester UK.
26 Mancini, A., Frontoni, E., and Zingaretti, P. (2015). Embedded multi sensor system for safe point-to-point navigation of impaired users. *IEEE Transactions on Intelligent Transportation Systems* 16 (6): 3543–3555.
27 Rafferty, J., Nugent, C.D., Liu, J., and Chen, L. (2017). From activity recognition to intention recognition for assisted living within smart homes. *IEEE Transactions on Human-Machine Systems* 47 (3): 368–379.

28 Magherini, T., Fantechi, A., Nugent, C.D., and Vicario, E. (2013). Using temporal logic and model checking in automated recognition of human activities for ambient-assisted living. *IEEE Transactions on Human-machine systems* 43 (6): 509–521.
29 Saunders, J., Syrdal, D.S., Koay, K.L. et al. (2016). "Teach me–show me"—end-user personalization of a smart home and companion robot. *IEEE Transactions on Human-Machine Systems* 46 (1): 27–40.
30 Kotb, A.O., Shen, Y.-C., Zhu, X., and Huang, Y. (2016). iParker—A new smart car-parking system based on dynamic resource allocation and pricing. *IEEE Transactions on Intelligent Transportation Systems* 17 (9): 2637–2647.
31 Kumar, P., Braeken, A., Gurtov, A. et al. (2017). Anonymous secure framework in connected smart home environments. *IEEE Transactions on Information Forensics and Security* 12 (4): 968–979.

Further Reading

Ali, A. and Hamouda, W. (2017). On the cell search and initial synchronization for NB-IoT LTE systems. *IEEE Communications Letters* 21 (8): 1843–1846.
del peral-Rosado, J.A., Lopez-Salcedo, J.A., and Seco-Granados, G. (2017). Impact of frequency-hopping NB-IoT positioning in 4G and future 5G networks. In: *IEEE International Conference on Communications Workshops (ICC Workshops)*, 815–820.
Mangalvedhe, N., Ratasul, R., and Ghosh, A. (2016). NB-IoT deployment study for low power wide area cellular IoT. In: *International Symposium on Personal, Indoor, and Mobile Radio Communications (PIMRC)*, 1–6.
Miao, Y., Li, W., Tian, D. et al. (2017). Narrow band internet of things: simulation and Modelling. *IEEE Internet of Things Journal* 5 (4): 2304–2314. Early Access article.
Mostafa, A.E., Zhou, Y., and Wong, V.S. (2017). Connectivity maximization for narrowband IoT systems with NOMA. In: *IEEE International Conference on Communications (ICC)*, 1–6.
Ratasuk, R., Vejlgaard, B., Mangalvedhe, N., and Ghosh, A. (2016). NB-IoT system for M2M communication. In: *IEEE Wireless Communications and Networking Conference Workshops (WCNC)*, 428–432.
Ratasul, R., Mangalvedhe, N., Kaikkonen, J., and Robert, M. (2016). Data channel design and performance for LTE narrowband IoT. In: *Vehicular Technology Conference (VTC-Fall)*, 1–5.
Song, Z., Liu, X., Zhao, X. et al. (2017). A low-power NB-IoT transceiver with digital-polar transmitter in 180-nm CMOS. *IEEE Transactions on Circuits and Systems I: Regular Papers* 64 (9): 2569–2581.
Tavares, M., Samardzija, D., Viswanathan, H. et al. (2017). A 5G lightweight connectionless protocol for massive cellular internet of things. In: *IEEE Wireless Communications and Networking Conference Workshops (WCNCW)*, 1–6.
Wang, Y. and Wu, Z. (2016). A coexistence analysis method to apply ACLR and ACS between NB-IoTand LTE for stand-alone case. In: *International Conference on Instrumentation & Measurement, Computer, Communication and Control (IMCCC)*, 375–379.
Zayas, A.D. and Merino, P. (2017). The 3GPP NB-IoT system architecture for the internet of things. In: *IEEE International Conference on Communications Workshops (ICC Workshops)*, 277–282.
Zhang, L., Ijaz, A., Xiao, P., and Tafazolli, R. (2017). Channel equalization and interference analysis for uplink narrowband internet of things (NB-IoT). *IEEE Communications Letters* 99. Early Access article.

5

Millimetre Wave Massive MIMO Technology

5.1 Introduction

Multiple-input, multiple-output (MIMO) technology becomes increasingly matured and capable of providing reliable and high speed data and sum-rate capacity provisions. It has now been deployed in many wireless network standards, such as WiFi (802-11), WiMAX (802-16m) and long term evolution (LTE) 4G systems. MIMO schemes were initially studied for point-to-point link where two devices communicate with each other. A point-to-point layered space-time architecture, employing an equal number of antenna elements at both transmitter (TX) and receiver (RX) is described in [1, 2], which shows the link capacity increases linearly with a number of antenna elements (n) for both fixed bandwidth (BW) and fixed total transmitted power. A point-to-point link with $n = 2$ is studied in [3] and shows that such a system is providing highly reliable wireless communications.

When the channel is subjected to Rayleigh fading and its characteristics are not available at the transmitter but can be tracked by the receiver, with fixed transmit power, the layered space-time architecture can provide a capacity with equal antennas at both ends, for a large but practical, n that scales with increasing signal power to noise power ratio (SNR). For example, for $n = 1$ Shannon's formula scales as one more bit per cycle for every 3 dB increase in SNR. However, with a multi-antenna system, the scaling is about n more bits per cycle for each 3 dB increase in SNR. If the number of antennas is not equal on both sides, the increase in bits scales with the minimum number of antennas used as SNR increases.

The promising capacity increase observed by research carried out at Lucent technologies in the late 1990s on layered space-time architectures encouraged more investigations to advance the scope of the MIMO system to multi-user multiple-input multiple-output (MU-MIMO). MIMO system originally explored with an equal number of antennas at both base stations (BSs) and receiving terminals for mainly frequency- division duplex transmission. For these implementations, the BS typically employs fewer than 10 antennas producing an important improvement in spectral efficiency but still relatively modest considering the need for future many giga-bits-per-second (Gbps) data rates in future mobile environments. Furthermore, the multi-user diversity gain makes the system performance less sensitive to the fading environment compared to point-to-point MIMO communication system.

In a quest to achieve larger multiplexing gains and to simplify the needed signal processing, the massive MIMO (also known as large-scale antenna system) has been proposed in [4], making a clean break with the conventional MU-MIMO practice through future use of a BS with a large number of antenna (typically of the order of few hundred) simultaneously serving a group of single-antenna terminals. Although massive MIMO communication systems were first proposed in 2010, they have attracted considerable interest from both academia and industry in a very short time. The proposed antennas arrangement can only be used at the BS for a

5G Physical Layer Technologies, First Edition. Mosa Ali Abu-Rgheff.

cost-effective and practical solution with large gains. Furthermore, the average energy per bit is reduced with an increasing number of antennas, together with a system performance, which is less sensitive to the uncorrelated noise and the small-scale fading than in the point-to-point link MIMO system.

5.2 Capacity of Point-to-Point MIMO Systems

MIMO systems pre 5G employ between two to a maximum of eight antennas for wireless communications such as WiFi (802-11), WiMAX (802-16m) and LTE networks. We focus our study on the analysis and interpolation of sum-rate capacity. We start with MIMO-system employed in a point-to-point link and then extend the study to MU-MIMO systems.

5.2.1 Capacity of SIMO/MISO Links

Consider a point-to-point link with a single antenna at transmitter and N antennas at receiver (SIMO). The received signal at the ith receive antenna y_i is

$$\mathrm{y}_i = \mathrm{h}_i\mathrm{x} + \mathrm{n}_i \tag{5.1}$$

where x, h_i and n_i are input signal, complex channel gain between transmitter antenna and the ith receive antenna, and additive complex Gaussian noise $\mathcal{N}(0,\sigma^2)$, respectively. The channel vector $\mathbf{h} = [\mathrm{h}_1, \mathrm{h}_2, \ldots \ldots, \mathrm{h}_\mathrm{N}]^T$ and $\mathbf{n} = [\mathrm{n}_1, \mathrm{n}_2, \ldots \ldots, \mathrm{n}_\mathrm{N}]^T$. The capacity of the link channel is

$$C_{simo} = \mathbb{E}\left[\log_2\left(1 + \frac{P\|\mathbf{h}\|^2}{\sigma^2}\right)\right] \text{ b/s/Hz} \tag{5.2}$$

where $\|\mathbf{h}\|^2 = \mathrm{h}_1{}^2 + \mathrm{h}_2{}^2 + \ldots. + \mathrm{h}_\mathrm{N}{}^2$ and P is total transmit power. The channel received power is increased by the total channel gain $\|\mathbf{h}\|^2$ and hence SNR increased by same power gain maximizing the channel SNR.

The dual multiple input single output (MISO) channel can be derived in a similar method, albeit the number of antennas at the transmitter M and a single antenna at the receiver. The MISO capacity is given by (5.2).

The increase in the received SNR is due to the received signals added up coherently. MISO, would generate more transmit power to the link with better gain. The transmit strategy of aligning transmitted signals is known as transmit beamforming. In both strategies the net benefits are in power gain, which is translated into performance reliability.

5.2.2 Capacity of MIMO Links

Consider a point-to-point MIMO link consisting of M transmit antennas and N receive antennas connected by a channel such that each receive antenna receives signals from all transmit antennas. The channel received signal $\boldsymbol{y}$ is

$$\mathbf{y} = \sqrt{\rho}\mathbf{H}\mathbf{x} + \mathbf{n} \tag{5.3}$$

Where ρ denotes SNR at the transmitter i.e. $\rho = \frac{transmit\ power}{noise\ variance}$ and when noise variance $\sigma^2 = 1$, the $\rho = transmit\ power$, $\mathbf{H}$ is (N x M) complex channel matrix, additive noise $\mathbf{n}$ is a vector (N x 1) represents noise components, that are independent and identically distributed with zero mean and unit variance $\mathcal{N}(0,1)$, $\mathbf{y}$ is the received signal vector (N x 1) and $\mathbf{x}$ is the transmitted vector (M x 1). We assume there is symmetry between transmit antennas such that equal powers are allocated to each transmit antenna. Let us denote a common SNR at each receive antenna as SNR. The transmitted power is normalized so that

$$\mathbb{E}\|\mathbf{x}_i\|^2 = 1 \tag{5.4}$$

Assuming perfect knowledge channel state information (CSI) is available at the transmitter, the transmitter can use its knowledge of **H** to align its transmission with eigenmodes of the channel (i.e. multiple orthogonal transmission links). When the instantaneous channel realization is not known at the transmitter, i.e. only has knowledge of the channel fading distribution, it would be impossible to align the transmission to every channel realization. Furthermore, the transmitter will not be able to identify the channel eigenmodes and hence the directions in which the channel is stronger in terms of power. In such a scenario, the optimal transmission strategy is to allocate equal power to each direction. The average mutual information [5, 6] is

$$C_{mimo} = \mathbb{E}_{\mathbf{H}}\left[\log_2 \det\left(\mathbf{I}_{\mathrm{N}} + \frac{\rho}{\mathrm{M}}\mathbf{H}\mathbf{H}^H\right)\right] \text{ bits/channel use} \tag{5.5}$$

where $\mathbf{I}_{\mathrm{N}}$ denotes(N x N) identity matrix, det(.) denotes the determinant operation, and $\mathbf{H}^H$ denotes the Hermitian channel matrix **H**.

We now examine the limiting bounds of (5.5). A key restriction to achievable rates faced by users at the edge of the cell is commonly owing to low SNRs, so only beamforming gains are influential and (5.5) can be modified [5] to

$$C_{mimo}^{\rho\to 0} \approx \frac{\rho\, \mathrm{tr}(\mathbf{H}\,\mathbf{H}^H)}{\mathrm{M}\log 2} \tag{5.6a}$$

If we normalize the magnitude of the channel coefficients to be almost one, then $\mathrm{tr}(\mathbf{H}\,\mathbf{H}^H) \approx \mathrm{M}\det\left(\mathbf{I}_{\mathrm{N}} + \frac{\rho}{\mathrm{M}}\mathbf{H}\mathbf{H}^H\right)$ and (5.6a) can be simplified to

$$C_{mimo}^{\rho\to 0} \approx \frac{\rho\,\mathrm{N}}{\log 2} \tag{5.6b}$$

The achievable rate assigned to users at the cell edge is given by (5.6b), which is independent of M. In this case the multiplexing gains are lost and hence the MIMO system contributes no benefits.

Another limiting case arises when M grows large. Let us consider the case where N is fixed and assume the row-vectors of the channel matrix are asymptotically orthogonal as M increased towards infinity. So, we get

$$\left[\frac{\mathbf{H}\mathbf{H}^H}{\mathrm{M}}\right]_{\mathrm{M}\gg\mathrm{N}} \overset{a.s.}{\to} \mathbf{I}_{\mathrm{N}} \tag{5.7}$$

Therefore, substituting (5.7) in (5.5), the capacity in (5.5) tends to the following:

$$C_{mimo}^{\mathrm{M}\gg\mathrm{N}} \approx \log_2 \det(\mathbf{I}_{\mathrm{N}})(1+\rho) \tag{5.8a}$$

$$C_{mimo}^{\mathrm{M}\gg\mathrm{N}} = \mathrm{N}\log_2(1+\rho) \tag{5.8b}$$

The achievable rate in (5.8b) serves as upper bound for the MIMO system, and for $\rho \gg 1$ we have

$$C_{mimo}^{\mathrm{M}\gg\mathrm{N}} \approx \mathrm{N}\log_2(\rho) \tag{5.8c}$$

Finally, consider the restriction case where N grows large while keeping M constant and assume the column-vector of the channel matrix to be asymptotically orthogonal such that

$$\left[\frac{\mathbf{H}^H\mathbf{H}}{\mathrm{N}}\right]_{\mathrm{N}\gg\mathrm{M}} \overset{a.s.}{\to} \mathbf{I}_{\mathrm{M}} \tag{5.9}$$

Using the identity

$$\det(\mathbf{I} + \mathbf{A}\mathbf{A}^H) = \det(\mathbf{I} + \mathbf{A}^H\mathbf{A})$$

We can rewrite (5.5) as

$$C_{mimo}^{\mathrm{N}\gg\mathrm{M}} = \log_2 \det\left(\mathbf{I}_{\mathrm{M}} + \frac{\rho}{\mathrm{M}}\mathbf{H}^H\mathbf{H}\right) \tag{5.10}$$

Using (5.9) to substitute for $\frac{\rho}{\mathrm{M}}\mathbf{H}^H\mathbf{H}$ in (5.10) we get

$$C_{mimo}^{\mathrm{N}\gg\mathrm{M}} = \log_2 \det\left(\mathbf{I}_{\mathrm{M}} + \frac{\rho\mathrm{N}}{\mathrm{M}}\mathbf{I}_{\mathrm{M}}\right) \tag{5.11a}$$

$$C_{mimo}^{\mathrm{N}\gg\mathrm{M}} \approx \mathrm{M}\log_2\left(1 + \frac{\rho\mathrm{N}}{\mathrm{M}}\right) \tag{5.11b}$$

Examining the limitation on the number of antennas at the transmitter and receiver, we may conclude that a point-to-point link with large M antennas at the transmitter and a single antenna at the receiver does not improve the system capacity as clearly shown in (5.8c). However, in Section 5.5, we will examine the MIMO system multiuser single antenna users and show that indeed the MIMO system capacity does improve considerably in multiuser single antenna users. Let us leave the extreme cases and consider a conventional MIMO where the random non-negative definitive matrix $\mathbf{W}$ is

$$\mathbf{W} = \begin{Bmatrix} \mathbf{H}\mathbf{H}^H & \mathrm{N} < M \\ \mathbf{H}^H\mathbf{H} & \mathrm{N} \geq \mathrm{M} \end{Bmatrix} \tag{5.12}$$

The distribution of $\mathbf{W}$ in (5.12) is commonly known as the *Wishart distribution* with

$$\mathbf{H}\mathbf{H}^H = \mathbf{H}^H\mathbf{H} \tag{5.13}$$

The rank of $\mathbf{H}$ is at most min{N, M} and is nonzero. Let us represent the channel matrix $\mathbf{H}$ using singular values λ_ν where $\nu = 1, 2, \ldots, \min\{\mathrm{N}, \mathrm{M}\}$.

$$\mathbf{H} = \boldsymbol{\Phi}\lambda_\nu\boldsymbol{\psi}^H \tag{5.14}$$

where $\boldsymbol{\Phi}$ and $\boldsymbol{\psi}$ are (N, N) and (M, M) unitary matrices, respectively, and λ_ν is (N, M) diagonal matrix whose diagonal elements are the singular values. $[\lambda, \ldots\ldots\ldots, \lambda_{\min(\mathrm{N,M})}]$. Using matrix analysis, it can be shown that (5.5) can be expressed as

$$C_{mimo} = \sum_{\nu=1}^{\min(\mathrm{N,M})} \log_2\left(1 + \frac{\rho\,\lambda_\nu^2}{\mathrm{M}}\right) \tag{5.15}$$

The achievable rate given by (5.15) is the sum rate of min(N, M) parallel links where the νth link has a SNR of $\frac{\rho\,\lambda_\nu^2}{\mathrm{M}}$. To provide, a thorough understanding of the matrices analysis involved in the derivation of (5.15), we present a simple example with ($M = 3, N = 2$) in Appendix 5.A.

Using Jensen's inequality [7], we can simplify (5.15) as

$$\sum_{\nu=1}^{\min(\mathrm{N,M})} \log_2\left(1 + \frac{\rho\,\lambda_\nu^2}{\mathrm{M}}\right) \leq \min(\mathrm{M,N})\log_2\left(1 + \frac{\rho}{N_t}\left[\frac{1}{\min(\mathrm{M,N})}\sum_{\nu=1}^{\min(\mathrm{M,N})}\lambda_\nu^2\right]\right) \tag{5.16}$$

When all singular values λ_ν^2 are equal, the left-side and right-side of (5.16) give equal (maximum) capacity; otherwise the capacity given by the left-side is less than the capacity given by the right-side term under appropriate channel conditions. Let us express the matrix $\mathbf{H}$ in terms of singular values given in (5.14) and noting that $\psi^H\,\psi = 1$:

$$\mathbf{H}\mathbf{H}^H = \boldsymbol{\Phi}\lambda_\nu\psi^H\,\psi\,\lambda_\nu\boldsymbol{\Phi}^H = \boldsymbol{\Phi}\lambda_\nu^2\boldsymbol{\Phi}^H \tag{5.17}$$

The trace of the correlation matrix $\mathbf{HH}^H$ is given by

$$\mathrm{tr}(\mathbf{HH}^H) = \mathrm{tr}(\lambda_v^2) = \sum_{v=1}^{min(\mathrm{N,M})} \lambda_v^2 \tag{5.18}$$

The range of values for $\mathrm{tr}(\lambda_v{}^2)$ in (5.18) extends the *lowest value* under line-of-sight (LOS) propagation conditions when all singular values are equal to zero except one (LOS) term and the best value is when all min(N, M) singular values are equal when the elements of the channel matrix are independent and isolated random variables. These two scenarios bound the achievable rate limits as given in Eq. (5.19):

$$\log_2\left(1 + \frac{\rho\mathrm{tr}(\mathbf{HH}^H)}{\mathrm{M}}\right) \leq C_{mimo} \leq \min(\mathrm{M,N})\log_2\left(1 + \frac{\rho}{M}\left[\frac{\mathrm{tr}(\mathbf{HH}^H)}{\min(\mathrm{M,N})}\right]\right) \tag{5.19}$$

If we assume normalization of the channel coefficients to almost one as above we get

$$\mathrm{tr}(\mathbf{HH}^H) \approx \max(\mathrm{M,N})\min(\mathrm{M,N}) = \mathrm{MN} \tag{5.20}$$

Substituting (5.20) in (5.19), we get the upper and the lower bounds for the achievable rate as

$$\log_2(1 + N\rho) \leq C_{mimo} \leq \min(\mathrm{M,N})\log_2\left(1 + \frac{\rho}{\mathrm{M}}\max(\mathrm{M,N})\right) \tag{5.21}$$

5.3 Outage of Point-to-Point MIMO Links

An important concept that is commonly used with MIMO system is the outage probability in fading channels. Consider a wireless link with M transmit and N receive antennas. The channel outage is usually discussed for nonergodic fading channels i.e. the randomness of the channel gain cannot be averaged out (removed) over time. So, long-term constant bit rate cannot be supported. An outage is defined as an event that mutual information of the channel does not support the targeted data rate. Suppose we wish to communicate data at a target rate R b/s/Hz and assume the transmitter knows full CSI during transmission. Accordingly, a channel outage occurs when

$$\log_2\det\left(\mathbf{I}_{\mathrm{N}} + \frac{\rho}{\mathrm{M}}\mathbf{HH}^H\right) < R \tag{5.22}$$

The outage probability pr_{out}^{mimo} has to satisfy the following constraint [8]:

$$\mathrm{pr}_{out}^{mimo} = \mathrm{pr}\left[\log_2\det\left(\mathbf{I}_{\mathrm{N}} + \frac{\rho}{\mathrm{M}}\mathbf{HH}^H\right) < R\right] \tag{5.23}$$

Assuming that we want to minimize the probability of outage occurrence for given transmit power ρ, then we can formulate it as Eq. (5.24):

$$\mathrm{pr}_{out}^{mimo}(R) = \min_{\rho}\ \mathrm{pr}\left\{\log_2\det\left(\mathbf{I}_{\mathrm{N}} + \frac{\rho}{\mathrm{M}}\mathbf{HH}^H\right) < R\right\} \tag{5.24a}$$

$$= \min_{\rho}\ \mathrm{pr}\left\{\log_2\left(\mathbf{I}_{\mathrm{N}} + \frac{\rho}{\mathrm{M}}\lambda_v^2\right) < R\right\} \tag{5.24b}$$

The solution of (5.24) depends on the statistics of channel **H**. For deterministic **H**, the optimal solution can be found using a singular value decomposition (SVD) and an efficient water-filling algorithm to allocate power over the Eigen modes. In a fading environment, **H** is random and the optimal solution can be reached through an appropriate covariance matrix that satisfies the channel fades.

5.4 Diversity-Multiplexing Tradeoffs

MIMO concept provides the channels with spatial diversity gains to improve the reliability of link performances. The idea behind the diversity gain is offering the receiver with multiple independent, albeit, faded replicas of the same information so the probability that all received components fade simultaneously is reduced. The idea is also related to the number of transmit antennas used, which influences the probability of errors. For example, let us consider a simple un-coded PSK signal transmitted over a single receive antenna fading channel **H** with the SNR averaged over the fading channel gain. The probability of error $pr_e(\text{SNR})$ under high SNR is

$$\Pr_{\text{e}}(\text{SNR}) \approx \frac{1}{4}(\text{SNR})^{-1} \tag{5.25}$$

Assuming that the same signal is transmitted to a receive equipped with two antennas then the probability of error becomes equal to

$$pr_e(\text{SNR}) \approx \frac{3}{16}(\text{SNR})^{-2} \tag{5.26}$$

So, by using an extra receive antenna, the error probability decreases with SNR at an increased speed of $(\text{SNR})^{-2}$. The improvement in performance at high SNR is determined by the exponent of SNR of the probability of error. This exponent is called the *diversity gain,* which is equal to the number of independently faded paths reaching the receiver. This result can be generalized for a system with M transmit antennas and N receive antennas. There will be a total of $M \times N$ fading paths and the full (maximum) diversity gain provided by the channel is equal to M N. The ergodic capacity as a function of the channel SNR is averaged to the available rate given in (5.5). In case that the receiver or transmitter knows the CSI, the channel capacity can be approximated as [9]:

$$C_{mimo} = C(\text{SNR}) \approx \min(\text{N}, \text{M}) \log_2 \text{SNR} + O(1) \tag{5.27}$$

where $O(1)$ has a value not more than 1. Since the channel capacity increases linearly with log (SNR), we choose a data rate that also increases with SNR. Let us consider a family of codes $\{C(\text{SNR})\}$ of block length l where each code corresponds to a SNR level. For example, let $\mathcal{R}(\text{SNR})$ b/symbol be the rate of the code $C(\text{SNR})$. In other words, we have a system that can achieve a spatial multiplexing gain r if the supported date rate is

$$\mathcal{R}(\text{SNR}) \approx r \,\log \text{SNR}\; b/s/H \tag{5.28}$$

Under large SNR, the multiplexing gain, r can be defined as [8]

$$r = \lim_{\text{SNR}\to\infty} \frac{\mathcal{R}(\text{SNR})}{\log \text{SNR}} \tag{5.29}$$

The diversity order, d, is defined as

$$d = -\lim_{\text{SNR}\to\infty} \frac{\log \Pr_e(\text{SNR})}{\log \text{SNR}} \tag{5.30}$$

$\Pr_e(\text{SNR})$ can be expressed as an exponential function as SNR^{-d}. Furthermore, a general formulation of the diversity order can be defined in terms of probability of channel outage as [10, 11]

$$d = -\lim_{SNR\to\infty} \frac{\log \Pr_{out}(\text{SNR},\ \mathcal{R}(\text{SNR}))}{\log \text{SNR}} \tag{5.31}$$

When the channel is in use during outage, powerful codes can be exploited to make the block error probability very small. Therefore, the outage probability can be treated as an accurate approximation of the actual block error probability. Contemporary wireless systems operate at a target error probability ε. So, the objective is to maximize the data rate at a given SNR without exceeding the target ε. Denote the rate at the target error probability to be R_ε such that

$$R_\varepsilon(\text{SNR}) = \max_\zeta\{\zeta \,:\, \text{pr}_{out}(\text{SNR}, \zeta) \le \varepsilon\} \tag{5.32}$$

where ζ is the data rate. The previous equation contemplates a tradeoff between the outage and the rate at the given SNR.

In general, the outage – rate relationship displayed above offers a possible tradeoff that is complex to compute. Therefore, what is needed is a metric that concisely characterizes the tradeoff. Let us consider the diversity (as a proxy for the outage probability) and the multiplexing gain (as a proxy for the rate). The diversity order (d) is defined as the asymptotic slope of the log (outage) – log (SNR) curve as defined in (5.29) and (5.31). In this way, the diversity-multiplexing tradeoff is now expressed by concise terms that can be easily and accurately computed, i.e. (r, d) that can be achieved as SNR $\rightarrow \infty$.

5.5 Multi-User-MIMO (MU-MIMO) Single-Cell Systems

We have shown in the previous sections that MIMO technology offers point-to-point link a promising multiplexing gain when the link SNR is high combined with a suitable transmission environment. However, the performance of MIMO point-to-point link is demeaned when the terminal is at the edge of the cell. Adding more receive antennas improves the performance but this may be impractical and certainly increases the cost of the terminal. Consequently, such a model should not be replicated for MU-MIMO systems. A better MIMO system to simultaneously serve K terminals is to deploy a very large number of antennas M at the BS rather than deploying a multiple version of the MIMO point-to-point link model. In this system, $\text{M} \gg K$. Conveniently, each terminal will be equipped with a single antenna. In the analysis of the new approach, we dispense with the notation (N, M) that is used in the above derivation and use our new system model of MU-MIMO that includes one antenna per terminal and M antennas at the BS. In addition, the transmission protocol we use is time-division duplex (TDD) and we assume the reciprocity property is valid so the downlink (DL) channel is simply the transpose of the uplink (UL) channel. It is clear that the time consumed for transmitting on the DL is proportional to the number of BS antennas while transmission on the UL is proportional to the number of pilot symbols since each user terminal employs a single antenna. We consider a time-invariant and deterministic (i.e. constant) channel matrix **H** between the BS and each terminal so **H** is $K \times \text{M}$ matrix made of two components, namely: the small-scale fading $\text{M} \times \text{K}$ matrix $\boldsymbol{G}$ accounts for changes over lengths of a wavelength or less, and $K \times K$ diagonal matrix $\boldsymbol{\beta}$ representing the large-scale fading coefficients

$$\mathbf{H} = \sqrt{\boldsymbol{\beta}}\boldsymbol{G} \qquad k = 1, 2 \ldots \ldots, K \tag{5.33}$$

where the kth column vector of $\boldsymbol{G}$ describes the small-scale fading $\boldsymbol{g}_k$ between the BS and the kth terminal where $\boldsymbol{g}_k \sim \mathbb{C}\mathcal{N}(\mathbf{0}, \mathbf{I}_\text{M})$ is complex Gaussian variables with zero mean and unit variance. We assume $\boldsymbol{g}_k$ to be statistically independent across the users. The kth element of $\sqrt{\boldsymbol{\beta}}$ is the large scale fading β_k accounts for both attenuation and shadow fading expressed as

$$\beta_k = \frac{\alpha_k}{r_k^x} \tag{5.34}$$

where α_k is log-normal random variable represents shadow fading with standard deviation σ_{shadow}, r_k is the distance between k^{th} user and the BS antennas, and x is the path loss exponent. The BS array is compact such that terminals are subject to the same large-scale fading.

Under the most favourable transmission conditions, column vectors of the channel matrix are asymptotically orthogonal as the number of BS antennas increased towards infinity:

$$\left[\frac{\mathbf{H}^H\,\mathbf{H}}{\mathrm{M}}\right]_{\mathrm{M}\gg K} = \sqrt{\boldsymbol{\beta}}\frac{G^H G}{\mathrm{M}}\sqrt{\boldsymbol{\beta}} \approx \boldsymbol{\beta} \tag{5.35}$$

5.5.1 UL Channel Capacity

Consider K single antenna terminals collectively transmit $K \times 1$ vector of symbols $\mathbf{x}$, per each UL channel's use and the BS array receive a $\mathrm{M} \times 1$ vector $\mathbf{y}_U$:

$$\mathbf{y}_U = \sqrt{\rho_U}\mathbf{H}\mathbf{x}_U + \mathbf{n}_U \tag{5.36}$$

where $\mathbf{n}_U$ is the $\mathrm{M} \times 1$ vector of receivers' noise at the BS whose components are independent and distributed as $\mathbb{CN}(0, 1)$, $\mathbf{H} \in \mathbb{C}^{\mathrm{M} \times \mathrm{K}}$ is the UL channel matrix, ρ_U is the transmit SNR of the UL and since variances are each equal 1, then ρ_U is UL transmit power assumed constant for all UL users. Transmitted symbol from k^{th} user is the kth element of the vector $\mathbf{x}_U$. Assume the total power of the transmit signal is normalized:

$$\mathbb{E}\{|\mathbf{x}_k|^2\} = 1, \quad k = 1, \ldots\ldots, K \tag{5.37}$$

Furthermore, we assume the BS perceived the UL transmission SCI and as $\mathrm{M} \to \infty$ individual users channel vectors tend to be asymptotically orthogonal. Thus, applying (5.33) we have

$$\mathbf{H}^H\,\mathbf{H} = \boldsymbol{\beta}\,G^H\,G \approx \mathrm{M}\,\boldsymbol{\beta}\,\mathbf{I}_K$$

The global achievable sum rate of all UL users is [12, 13]

$$C_{sum-UL}^{\mathrm{M}\gg K} = \log_2 \det(\mathbf{I} + \rho_U\mathbf{H}^H\mathbf{H})\ \text{bits/channel use} \tag{5.38a}$$

$$= \log_2 \det(\mathbf{I}_K + \rho_U\mathrm{M}\boldsymbol{\beta}\mathbf{I}_K) \tag{5.38b}$$

$$= \sum_{k=1}^{K} \log_2(1 + \rho_U\mathrm{M}\beta_k) \tag{5.38c}$$

A simple matched filtering (MF) processing at the BS achieves the UL capacity in (5.38c). The BS processes the received signal vector by multiplying the conjugate-transpose of the channel as

$$\mathbf{H}^H\mathbf{y}_U = \mathbf{H}^H(\sqrt{\rho_U}\mathbf{H}\mathbf{x}_U + \mathbf{n}_U) \approx \mathrm{M}\sqrt{\rho_U}\boldsymbol{\beta}\,\mathbf{x}_U + \mathbf{H}^H\mathbf{n}_U \tag{5.39}$$

Since the channel vectors are asymptotically orthogonal when $\mathrm{M} \to \infty$, $\mathbf{H}^H$ does not change the Gaussian distribution of the noise $\mathbf{n}_U$. Considering $\boldsymbol{\beta}$ is a diagonal matrix, the signals from different users are separated into different streams and there will be no inter-user interference.

5.5.2 DL Channel Capacity

For each channel use, the BS transmits signal vector $\mathbf{x}_D \in \mathbb{C}^{\mathrm{M} \times 1}$ through its M antennas. The K single antenna terminals collectively receive $\mathrm{K} \times 1$ vector $\mathbf{y}_D$:

$$\mathbf{y}_D = \sqrt{\rho_D}\mathbf{H}^H\mathbf{x}_D + \mathbf{n}_D \tag{5.40a}$$

where $\mathbf{n}_D$ is the $\mathrm{K} \times 1$ vector of user receivers noise, assumed to be independent and distributed as $\mathbb{CN}(0, 1)$, ρ_D is the SNR at the BS transmitter. As for the UL case, the total transmit power for the DL is assumed to be one and independent of the number of antennas:

$$\mathbb{E}\{\|\mathbf{x}_D\|^2\} = 1 \tag{5.40b}$$

The BS learns the CSI corresponding to all users based on UL channel **H**. Therefore, the BS can perform power allocation to maximize the sum rate. We assume the MIMO system operates on TDD mode. The sum capacity for the system using power allocations is [8, 12]

$$[C_{DL}] = \max_{P} \log_2 \det(\mathbf{I}_{\mathrm{M}} + \rho_D \mathbf{H}\mathbf{P}\,\mathbf{H}^H) \tag{5.41a}$$

$$\text{s.t.} \sum_{k=1}^{K} p_k = 1, \qquad p_k \geq 0\, \textit{for all } k \tag{5.41b}$$

where **P** is a diagonal matrix with the power allocations $\{p_1, \ldots \ldots \ldots \ldots \ldots., p_K\}$ being the diagonal elements. Using the favourable transmission conditions such column vectors of the channel matrix are asymptotically orthogonal as the number of BS antennas increased towards infinity, the asymptotic sum rate of DL channel becomes

$$[C_{sum-DL}]_{\mathrm{M} \gg K} = \max_{\mathbf{P}} \log_2 \det(\mathbf{I}_K + \rho_D \sqrt{\mathbf{P}}\mathbf{H}^H\mathbf{H}\sqrt{\mathbf{P}}) \tag{5.42a}$$

$$\approx \max_{\mathbf{P}} \log_2 \det(\mathbf{I}_K + \mathrm{M}\rho_D \mathbf{P}\boldsymbol{\beta}) \tag{5.42b}$$

Assuming that a MF precoding is used at the transmitter and denote the source information vector as $\mathbf{s}_D \in \mathbb{C}^{K \times 1}$, then transmitted signal vector is

$$\mathbf{x}_D = \mathbf{H}\boldsymbol{\beta}^{-\frac{1}{2}}\mathbf{P}^{1/2}\mathbf{s}_D \tag{5.43}$$

And the received signal vector at all K users is

$$\begin{aligned} \mathbf{y}_{\mathrm{D}} &= \sqrt{\rho_D}\mathbf{H}^H\mathbf{H}\boldsymbol{\beta}^{-\frac{1}{2}}\mathbf{P}^{1/2}\mathbf{s}_D + \mathbf{n}_D \\ &\approx \sqrt{\rho_d}\mathrm{M}\boldsymbol{\beta}^{-\frac{1}{2}}\mathbf{P}^{1/2}\mathbf{s}_D + \mathbf{n}_D, \quad \textit{when}\ \mathrm{M} \to \infty \end{aligned} \tag{5.44}$$

Since **P** and $\boldsymbol{\beta}$ are both diagonal matrices, the users links can be treated as single-input single output (SISO) and inter-user interference is suppressed, so the achievable sum rate is given in (5.42b).

5.6 Multi-User MIMO Multi-Cell System Representation

Consider a cellular MU-MIMO system with N_c cells, each cell BS is equipped with M antennas serving K single antenna terminals. Denote the channel between the kth user in the nth cell to the mth antenna of the ith BS as $\mathrm{h}_{i,k,n.m}$, which is made of two components, namely: a complex small-scale (Rayleigh) fading coefficient $g_{i,k,n,m}$ and a large-scale (path loss and shadowing) fading coefficient $\sqrt{\beta_{i,k,n,m}}$ [13] such that

$$\mathrm{h}_{i,k,n.m} = g_{i,k,n,m}\sqrt{\beta_{i,k,n,m}} \tag{5.45}$$

We assume the large-scale fading coefficients are the same for different antennas at the same BS but are user dependent, while the small-scale fading coefficients are different for different users and for different antennas at the same BS. The channel matrix for all K terminals in the nth cell to the ith BS is

$$\mathbf{H}_{i,n} = \begin{pmatrix} \mathrm{h}_{i,1,n,1} & \mathrm{h}_{i,K,n,1} \\ \mathrm{h}_{i,1,n,\mathrm{M}} & h_{i,K,n,M} \end{pmatrix} = \boldsymbol{G}_{i,n}\sqrt{\boldsymbol{\beta}_{i,n}} \tag{5.46a}$$

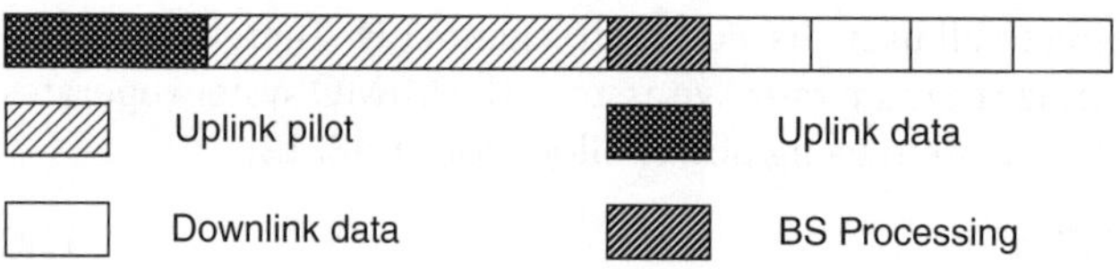

Figure 5.1 MU-MIMO TDD transmission protocol [13].

where

$$G_{i,n} = \begin{pmatrix} g_{i,1,n,1} & \cdots & g_{i,K,n,1} \\ \cdots\cdots & \cdots\cdots & \cdots\cdots \\ g_{i,1,n,\mathrm{M}} & \cdots & g_{i,K,l,\mathrm{M}} \end{pmatrix} \tag{5.46b}$$

and

$$\boldsymbol{\beta}_{i,n} = \begin{pmatrix} \delta_{i,1,n} & 0 & 0 \\ \cdots & \cdots\cdots & \cdots\cdots \\ 0 & \cdots & \delta_{i,K,n} \end{pmatrix} \tag{5.47}$$

for a single-cell, $N_c = 1$ MU-MIMO serving K single antenna users and a BS with M antennas. We apply the MU-MIMO TDD protocol shown in block diagram in Figure 5.1.

According to the TDD protocol, all users in a given cell first synchronously send uplink data signals, followed by sending uplink pilot sequences that are used by the BS to estimate uplink CSI to the users in their cells. The BS uses these estimates to decode the uplink data and to generate downlink beamforming vectors for downlink transmission. However, due to limited channel coherence time, pilot sequences used by users in neighbouring cells may no longer be orthogonal with those within the desired cell, causing interference, a phenomenon known as *pilot contamination problem.*

5.7 Sum Capacity of Broadcast Channels

5.7.1 Degraded BC

A broadcast channel (BC) is traditionally used in radio and TV where a transmitter sends the same information to multiple remote receivers. Present-day BC is more generalized in the sense that it does not necessarily send the same information to multiple receivers. BC in wireless communication was first introduced by Cover [14], who also recommended an achievable coding strategy based on superposition. In addition, MIMO antenna system is employed at the BS to provide spatial diversity gains. The fundamental rationale of the degraded BC is introduced in Chapter 1 Section 4.2. BC comprises the simultaneous transmissions of information from a single transmitter to multiple receivers as illustrated in Figure 5.2. The input signal x contains two messages, m_1, m_2, intended to receiver 1 and receiver 2, respectively.

The two-receiver BC is denoted by an input vector $\mathbf{x}$ and the output y_1 and y_2 and probability distributions $\Pr(\mathrm{y}_1 \mid \mathbf{x})$, $\Pr(\mathrm{y}_2 \mid \mathbf{x})$. The superposition coding strategy has been proved to be optimal for degraded BC. However, such a strategy is suboptimal when applied to nondegraded BC. Vector BC channel, where the input and output of the channel are vectors, is not degraded completely. A degraded BC can be categorized in two types: *physically degraded* BC and *stochastically degraded* BC. A BC is a physically degraded when

$$\Pr(\mathrm{y}_1, \mathrm{y}_2|\mathbf{x}) = \Pr(\mathrm{y}_1|\mathbf{x})\ \Pr(\mathrm{y}_2|\mathrm{y}_1) \tag{5.48}$$

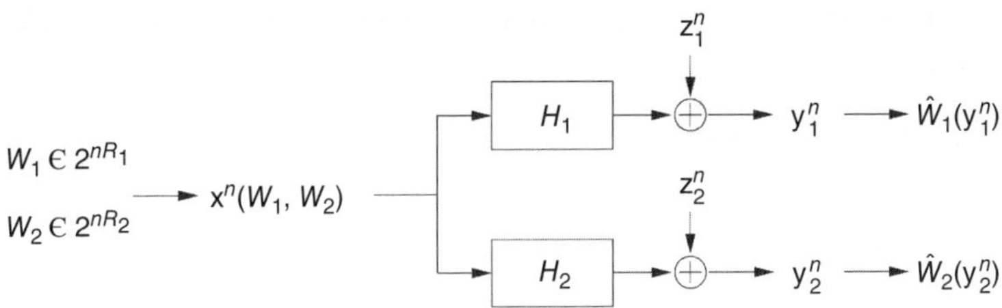

Figure 5.2 Gaussian vector broadcast channel [15].

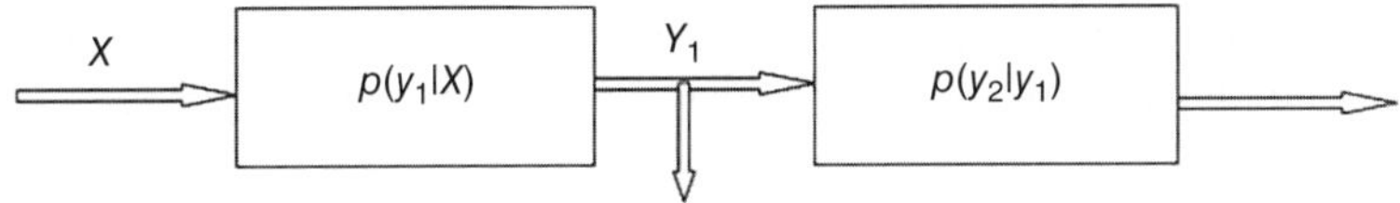

Figure 5.3 A physically degraded BC.

The physically degraded BC forms a Markov chain, i.e. the channel from the transmitter to the destination receiver is the cascade of the channel from the transmitter to the first receiver and the channel from the second receiver and the channel the third receiver and so on [16]. A physically degraded channel is shown in Figure 5.3 for two receivers. A stochastically degraded channel is one that is statistically equivalent to a physically degraded channel that means it has the same mean, variance, and same marginal distributions.

The receiver that operates on a stronger channel first decodes the message on a weaker channel and then decodes its own message. That is, one receiver's signal is a noisier version of the other receiver's signal. Consider the scalar physically degraded Gaussian BC in Figure 5.3. The input-output signals are given as

$$y_1 = X + n_1 \tag{5.49}$$

$$y_2 = X + n_2 \tag{5.50}$$

where X is the scalar transmitted signal and we assume it is subjected to power constraint P, y_1 and y_2 are received signals, $n_1 \sim \mathcal{N}(0, \sigma_1^2)$ and $n_2 \sim \mathcal{N}(0, \sigma_2^2)$ where the additive Gaussian noise variances $\sigma_1^2$and σ_2^2 are for receivers 1 and 2, respectively.

Note that we are transmitting a vector signal carrying independently two messages for the two receivers. We use this model to explain the basic idea underpinning the physically degraded BCs. For the sake of our analysis, we assume $\sigma_1 < \sigma_2$, i.e. receiver 1 operates over a stronger channel than receiver 2. If we combine (5.49) and (5.50) we get

$$y_2 = y_1 - n_1 + n_2 \tag{5.51a}$$

We can write (5.51a) as

$$y_2 = y_1 + n_2' \tag{5.51b}$$

where $n_2' = n_2 - n_1$, $n_2' \sim \mathcal{N}(0, \sigma_2^2 - \sigma_1^2)$, $n_2 = n_2' + n_1$ and n_1 is independent of n_2. Thus, y_2 can be considered as a degraded version of y_1.

The capacity region for a degraded BC is achieved using superposition coding and interference subtraction. Let us divide the total power P into $P_1 = \alpha P$ and $P_2 = (1 - \alpha)P$ where $0 \le \alpha \le 1$ and construct two independent Gaussian codebooks for the two messages with transmit power P_1 and P_2, respectively. The two individual messages are encoded using individual codes chosen from each of the codebooks and the coded messages are summed up and transmitted. Since receiver 1 operates over a stronger channel, receiver 1 can decode the message intended to receiver 2 and subtracts it from the received signal to get a cleaner receive signal y_1. The achievable rate-pair are given as

$$R_1 \le \frac{1}{2}\log_2\left(1+\frac{P_1}{\sigma_1^2}\right)=\frac{1}{2}\log_2\left(1+\frac{\alpha P}{\sigma_1^2}\right) \tag{5.52}$$

$$R_2 = \frac{1}{2}\log_2\left(1+\frac{P_2}{P_1+\sigma_2^2}\right)=\frac{1}{2}\log_2\left(1+\frac{(1-\alpha)P}{\alpha P+\sigma_2^2}\right) \tag{5.53}$$

Cover's scheme [15] using superposition coding and interference subtraction is optimal for the degraded Gaussian BC. The scalar degraded channels can be ordered completely according to their channel strength, while the vector BC has a vector input and a vector output and is not necessarily degraded, hence employing Cover's scheme does not achieve the BC capacity.

5.7.2 Nondegraded Gaussian Vector BC

In nondegraded Gaussian vector BC system with a single transmitter equipped with multiple transmit antennas, independent information is sent to multiple receivers. Although coordination is allowed among transmit units, it is unfeasible among receivers.

Consider a Gaussian channel shown in Figure 5.4 where a message is encoded by the transmitter X and the receiver y and the transmitter codebook $(n,\ 2^{nR})$. Let S be a Gaussian interfering signal whose realization is known to the transmitter only, and n is a Gaussian noise independent of $\mathcal{S}$ and X, y as input and output signals are described as

$$y = X + S + n \tag{5.54}$$

Costa [17] in his paper on dirty paper coding (DPC) demonstrated that when S and n are independent random variables, under fixed power constraint, the capacity of the channel with and without the interference is the same, i.e. the effects of the interference on the channel capacity are completely eliminated. Furthermore, the optimal transmit signal X is statistically independent of S.

Even though DPC is proven to give the achievable rate for a single antenna BC as presented in (5.51a) and (5.51b), however, for multiple-antenna BC the dirty paper strategy has to be optimized over different antennas to determine the optimal transmission policy and hence the corresponding sum rate capacity. The solution cannot be reduced to a point-to-point MIMO problem. Computation needed for the optimization of the MIMO BC sum rate is very complex. It is a nonlinear programming problem, and its optimization takes a lot of time to identify whether the problem has no solution or if the solution is global.

It turns out that the BC nonconvex problem can be transformed into a convex sum-power of the corresponding multiple access channel (MAC) using the BC-MAC duality that is a linear programming problem, and its optimization has a single solution, if one exists, from which the BC covariance matrices can be found. Convex optimization is much easier to deal with and is very time-efficient.

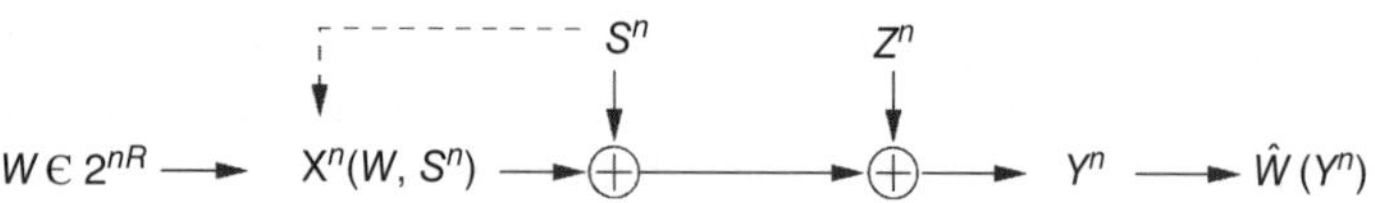

Figure 5.4 Gaussian channel with transmitter side information [15].

5.7.3 MIMO BC Sum Capacity Using DPC

Consider a system model for the MIMO BC consists of a single cell with a transmitter equipped with M serving K users, each with N antennas. Further, we examine the BC dual, i.e. the MIMO MAC serving the K users, each with N antenna. Each of the dual UL is a conjugate transpose of the corresponding MIMO BC. The received signal in the BC case is

$$\mathbf{y}_k = \mathbf{H}_k\mathbf{x} + \mathbf{n}_k \quad \text{where} \quad k = 1, \dots\dots\dots\dots, K \tag{5.55}$$

The received signal in the dual MIMO MAC case is

$$\mathbf{y}_{\text{MAC}} = \sum_{k=1}^{K} \mathbf{H}_k^H \mathbf{x}_k + \mathbf{n} \tag{5.56}$$

where $\mathbf{H}_k \in \mathbb{C}^{\text{N}X\text{M}}$ for k^{th} user in the MIMO BC, $\mathbf{x} \in \mathbb{C}^{\text{M}X1}$ is the BC transmitted signal, and $\mathbf{x}_k \in \mathbb{C}^{\text{N}X1}$ is MAC transmitted signal . We assume the channel matrices are fixed and known to the transmitter and to each of the receivers. The vectors $\mathbf{n}_k$ and $\mathbf{n}$ are independent additive Gaussian noise with unit variance on each vector. The sum power constraint of P in the MIMO BC is $\mathbb{E}[||\mathbf{x}||^2] \leq P$ and in the MIMO MAC is $\sum_{k=1}^{K} \mathbb{E}[||\mathbf{x}_k||^2] \leq P$.

The duality technique changes the non-convex BC (DL channels) optimization problem into a convex MAC (sum of UL channels), which is easier to solve and to explore the DL covariance matrices. The duality scheme is illustrated in Figure 5.5.

The sum rate capacity of the MIMO BC, $C_{\text{BC}}(\mathbf{H}_1, \dots\dots\dots, \mathbf{H}_K, P)$ was found to be achievable [18, 19], using DPC. The sum rate capacity of the MIMO BC is given by the following maximization problem:

$$C_{\text{BC}}(\mathbf{H}_1, \dots\dots\dots, \mathbf{H}_K, P) = \max_{\{\Xi_k\}_{k=1}^{K} : \Xi_k \geq 0, \sum_{k=1}^{K} \text{tr}(\Xi_k) \leq P} \log|+\mathbf{H}_1\Xi_1\mathbf{H}_1^{\text{H}}|$$

$$+ \log\frac{|\mathbf{I} + \mathbf{H}_2(\Xi_1 + \Xi_2)\mathbf{H}_2^H|}{|\mathbf{I} + \mathbf{H}_2\Xi_1\mathbf{H}_2^H|} + \dots\dots + \log\frac{|\mathbf{I} + \mathbf{H}_K(\Xi_1 + \dots\dots\dots + \Xi_K)\mathbf{H}_K^H|}{|\mathbf{I} + \mathbf{H}_K(\Xi_1 + \dots\dots\dots + \Xi_{K-1})\mathbf{H}_K^H|} \tag{5.57}$$

The maximization in (5.57) is carried out over DL covariance matrices $\Xi_1 + \dots + \Xi_K$. Each covariance is $\text{M} \times \text{M}$ matrix, which has to be determined to achieve the maximum sum rate in (5.57). The duality principle [20] states that with DPC, the capacity regions of MIMO BC and MIMO MAC are identical.

The sum capacity of the MIMO MAC $C_{\text{MAX}}(\mathbf{H}_1^K, \dots\dots\dots, \mathbf{H}_K^H, P)$ is given by the convex maximization problem:

$$C_{\text{BC}}(\mathbf{H}_1, \dots\dots\dots, \mathbf{H}_K, P) = C_{\text{MAX}}(\mathbf{H}_1^K, \dots\dots\dots, \mathbf{H}_K^H, P) \tag{5.58}$$

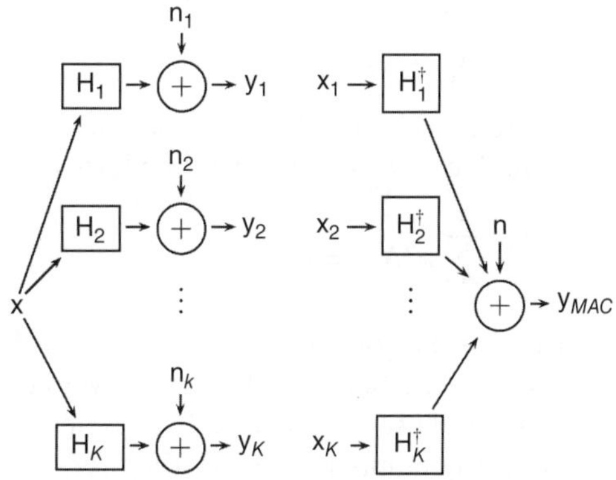

Figure 5.5 MIMO BC (left) and MIMO MAC (right) duality scheme [18].

Accordingly,

$$C_{\mathrm{MAX}}(\mathbf{H}_1^K,\dots\dots\dots.,\mathbf{H}_K^H,P)=\max_{\{\mathbf{Q}_k\}_{k=1}^K:\mathbf{Q}_k\geq 0,\sum_{k=1}^K \mathrm{tr}(\mathbf{Q}_k)\leq P}\log\left|\mathbf{I}+\sum_{k=1}^K \mathbf{H}_k^H\mathbf{Q}_k\mathbf{H}_k\right| \tag{5.59}$$

It is worth noting that the users in (5.59) have a joint power constraint instead of individual constraints as in the conventional MAC. The maximization is over $Q_1,\dots\dots\dots\dots,Q_K$, which are the covariance matrices of the MAC, each is an N × N matrix, which in this case is a convex optimization problem subject to power constraint P. The maximization in (5.59) is performed over the uplink covariance matrices. In an individual power constraint channel, each user's water-filling algorithm is determined by his-own power constraint and the covariance of each user could be updated individually at a time. In MIMO BC, there exists a sum power constraint on the BC covariances so it is necessary to update all the covariances simultaneously to maintain a constant water level for all users.

An iterative water-filling algorithm is proposed in [18] whereby all K covariances are simultaneously updated during each iterative step in (5.59). The nth iteration of the algorithm is detailed as follows:

1. Generate each user effective channel $\mathbf{G}_k^{(n)}$ as

$$\mathbf{G}_k^{(n)}=\mathbf{H}_k\left(\mathbf{I}+\sum_{j\neq k}\mathbf{H}_j^H\mathbf{Q}_j^{(n-1)}\mathbf{H}_j\right)^{-\frac{1}{2}}\quad \text{for } k=1,\dots\dots\dots\dots,K \tag{5.60}$$

2. Assume these individual channels as parallel (i.e. non-interfering channels), the new covariance matrices, $Q_1^{(n)},\dots\dots\dots\dots\dots\dots\dots,Q_K^{(n)}$, i.e. $\{Q_k^{(n)}\}_{k=1}^K$ can be obtained by water-filling with total power constraint algorithm as

$$\{Q_k^{(n)}\}_{k=1}^K=\arg\max_{\{\mathbf{Q}_k\}_{k=1}^K:\mathbf{Q}_k\geq 0,\sum_{k=1}^K \mathrm{tr}(\mathbf{Q}_k)\leq P}\sum_{k=1}^K\log|\mathbf{I}+(\mathbf{G}_k^{(n)})^H\mathbf{Q}_k\mathbf{G}_k^{(n)}| \tag{5.61}$$

The maximization (5.61) is equivalent to water-filling block diagonal channel with diagonals equal

$$\mathbf{G}_1^{(n)},\dots\dots\dots\dots\dots\dots,\mathbf{G}_K^{(n)} \tag{5.62}$$

Let us express the SVD of $\mathbf{G}_k^{(n)}(\mathbf{G}_k^{(n)})^H$as

$$\mathbf{G}_k^{(n)}(\mathbf{G}_k^{(n)})^H=\mathbf{U}_k\mathbf{D}_k\mathbf{U}_k^H \tag{5.63}$$

where $\mathbf{U}_k$ unitary matrix and $\mathbf{D}$ square and diagonal matrix. Then the updated covariance matrices are

$$\mathbf{Q}_k^{(n)}=\boldsymbol{U}_k\boldsymbol{\Lambda}_k\mathbf{U}_k^H \tag{5.64}$$

where $\boldsymbol{\Lambda}_k=[\mu\ \mathbf{I}-(\mathbf{D}_k)^{-1}]^+$ and the notation $[\mathbf{A}]^+$ denotes a component-wise maximum including zero. The water-filling level μ is chosen such that

$$\sum_{k=1}^K \mathrm{tr}(\boldsymbol{\Lambda}_k)=P \tag{5.65}$$

The above algorithm does converge to the sum rate capacity when applied to K users i.e. $K=2$. Nonetheless, it does not always converge to the optimum when employed for large number of users, i.e. $K>2$. What is needed is an algorithm that converges to the sum capacity to any number K users. The modified algorithm is illustrated in Figure 5.6. When the

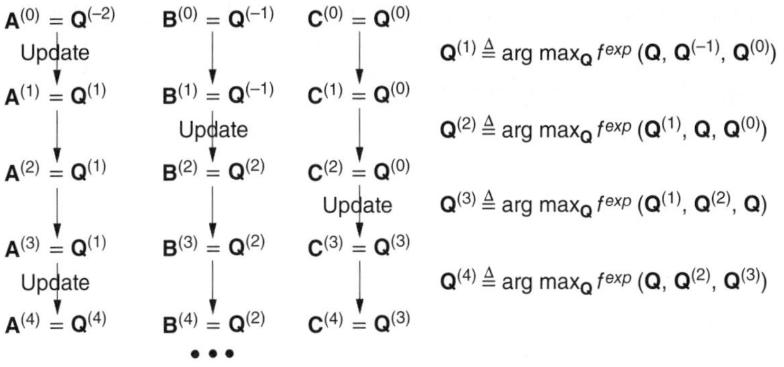

A	B	C	
$\mathbf{A}^{(0)} = \mathbf{Q}^{(-2)}$	$\mathbf{B}^{(0)} = \mathbf{Q}^{(-1)}$	$\mathbf{C}^{(0)} = \mathbf{Q}^{(0)}$	
Update ↓	↓	↓	$\mathbf{Q}^{(1)} \triangleq \arg\max_{\mathbf{Q}} f^{exp}(\mathbf{Q}, \mathbf{Q}^{(-1)}, \mathbf{Q}^{(0)})$
$\mathbf{A}^{(1)} = \mathbf{Q}^{(1)}$	$\mathbf{B}^{(1)} = \mathbf{Q}^{(-1)}$	$\mathbf{C}^{(1)} = \mathbf{Q}^{(0)}$	
↓	Update ↓	↓	$\mathbf{Q}^{(2)} \triangleq \arg\max_{\mathbf{Q}} f^{exp}(\mathbf{Q}^{(1)}, \mathbf{Q}, \mathbf{Q}^{(0)})$
$\mathbf{A}^{(2)} = \mathbf{Q}^{(1)}$	$\mathbf{B}^{(2)} = \mathbf{Q}^{(2)}$	$\mathbf{C}^{(2)} = \mathbf{Q}^{(0)}$	
↓	↓	Update ↓	$\mathbf{Q}^{(3)} \triangleq \arg\max_{\mathbf{Q}} f^{exp}(\mathbf{Q}^{(1)}, \mathbf{Q}^{(2)}, \mathbf{Q})$
$\mathbf{A}^{(3)} = \mathbf{Q}^{(1)}$	$\mathbf{B}^{(3)} = \mathbf{Q}^{(2)}$	$\mathbf{C}^{(3)} = \mathbf{Q}^{(3)}$	
Update ↓	↓	↓	$\mathbf{Q}^{(4)} \triangleq \arg\max_{\mathbf{Q}} f^{exp}(\mathbf{Q}, \mathbf{Q}^{(2)}, \mathbf{Q}^{(3)})$
$\mathbf{A}^{(4)} = \mathbf{Q}^{(4)}$	$\mathbf{B}^{(4)} = \mathbf{Q}^{(2)}$	$\mathbf{C}^{(4)} = \mathbf{Q}^{(3)}$	
	•••		

Figure 5.6 Structure of the modified iterative water-filling algorithm that converges [18].

original algorithm is applied for $K = 2$, a cyclic coordinate ascent algorithm is used to verify the convergence. It was found that this algorithm is identical to the sum rate iterative algorithm. For, $K > 2$ we can use the cyclic coordinate ascent algorithm to verify the convergence but in this case it is not identical to the original iterative water-filling algorithm but can be interpreted as the sum rate iterative with a memory of the covariance matrices generated in the previous $K - 1$ iteration.

As an example, we consider the convergence to the sum rate for $K = 3$ and the maximization given as

$$\begin{aligned}
&\max \frac{1}{3} \log|\mathbf{I} + \mathbf{H}_1^H \mathbf{A}_1 \mathbf{H}_1 + \mathbf{H}_2^{\mathrm{H}} \mathbf{B}_2 \mathbf{H}_2 + \mathbf{H}_3^{\mathrm{H}} \mathbf{C}_3 \mathbf{H}_3| \\
&+ \max \frac{1}{3} \log|\mathbf{I} + \mathbf{H}_1^H \mathbf{C}_1 \mathbf{H}_1 + \mathbf{H}_2^{\mathrm{H}} \mathbf{A}_2 \mathbf{H}_2 + \mathbf{H}_3^{\mathrm{H}} \mathbf{B}_3 \mathbf{H}_3| \\
&+ \max \frac{1}{3} \log|\mathbf{I} + \mathbf{H}_1^H \mathbf{B}_1 \mathbf{H}_1 + \mathbf{H}_2^{\mathrm{H}} \mathbf{C}_2 \mathbf{H}_2 + \mathbf{H}_3^{\mathrm{H}} \mathbf{A}_3 \mathbf{H}_3|
\end{aligned} \tag{5.66}$$

Subject to these constraints:

$$\begin{aligned}
&\mathbf{A}_i \geq 0 \\
&\mathbf{B}_i \geq 0 \\
&\mathbf{C}_i \geq 0
\end{aligned} \tag{5.67}$$

For $i = 1, 2, 3$ and

$$\begin{aligned}
&\mathrm{tr}(\mathbf{A}_1 + \mathbf{A}_2 + \mathbf{A}_3\ ) \leq P \\
&\mathrm{tr}(\mathbf{B}_1 + \mathbf{B}_2 + \mathbf{B}_3\ ) \leq P \\
&\mathrm{tr}(\mathbf{C}_1 + \mathbf{C}_2 + \mathbf{C}_3\ ) \leq P
\end{aligned} \tag{5.68}$$

Any solution to the maximization problem in (5.66) for $K > 2$ would also be a solution for the case when = 2. So as to maximize (5.66), we apply the cyclic coordinate ascent algorithm first with respect to $\triangleq(\mathbf{A}_1, \mathbf{A}_2, \mathbf{A}_3)$, then with respect to $\mathbf{B} \triangleq (\mathbf{B}_1, \mathbf{B}_2, \mathbf{B}_3)$, and again with respect to $\mathbf{C} \triangleq (\mathbf{C}_1, \mathbf{C}_2, \mathbf{C}_3)$, etc. In each iteration step, the convergence is guaranteed because of the uniqueness of the maximization method. However, the effective channel of each user depends on the last $K - 1$ covariance matrices. Figure 5.6 illustrates the modified algorithm for users after the nth iteration. Denote the objective function in (5.66) as f^{exp} $(\mathbf{A}, \mathbf{B}, \mathbf{C})$ and we start with initializing the three variables to $\mathbf{A}^{(0)}, \mathbf{B}^{(0)}, \mathbf{C}^{(0)}$ and express them in a general form that can be applied to arbitrary K and also call them as $\mathbf{Q}^{(-2)}, \mathbf{Q}^{(-1)}, \mathbf{Q}^{(0)}$.

In step 1, we update **A** and at the same time hold **B** and **C** fixed and define $\mathbf{Q}^{(1)}$ to be the updated $\mathbf{A}^{(1)}$:

$$\mathbf{Q}^{(1)} \triangleq \mathbf{A}^{(1)} = \arg \max_{\mathbf{Q}:\mathbf{Q}_i \geq 0, \sum_{i=1}^{3} \text{tr}(\mathbf{Q}_i) \leq P} f^{exp}(\mathbf{Q}, \mathbf{B}^{(0)}, \mathbf{C}^{(0)})$$

$$= \arg \max_{\mathbf{Q}:\mathbf{Q}_i \geq 0, \sum_{i=1}^{3} \text{tr}(\mathbf{Q}_i) \leq P} f^{exp}(\mathbf{Q}, \mathbf{Q}^{(-1)}, \mathbf{Q}^{(0)}) \tag{5.69}$$

In step 2, matrices **B** are updated with $\mathbf{Q}^{(2)} \triangleq \mathbf{B}^{(2)}$.

In step 3, matrices **C** are updated with $\mathbf{Q}^{(3)} \triangleq \mathbf{C}^{(3)}$, and in step 4, **A** is updated again to $\mathbf{A}^{(2)}$ and so on. It is worth noting that $\mathbf{Q}^{(n)}$ is defined as the set of matrices updated in the nth iteration. The general equation for $\mathbf{Q}^{(n)}$ is derived in [18, Appendix III] as

$$\mathbf{Q}^{(n)} \triangleq \arg \max_{\mathbf{Q}:\mathbf{Q}_k \geq 0, \sum_{k=1}^{K} \text{tr}(\mathbf{Q}_k) \leq P} f^{exp}(\mathbf{Q}, \mathbf{Q}^{(n-K+1)}, \ldots\ldots, \mathbf{Q}^{(n-1)})$$

$$\mathbf{Q}^{(n)} \triangleq \arg \max_{\mathbf{Q}:\mathbf{Q}_k \geq 0, \sum_{k=1}^{\text{K}} \text{tr}(\mathbf{Q}_k) \leq P} \sum_{k=1}^{K} \log|\mathbf{I} + (\mathbf{G}_k^{(n)})^H \mathbf{Q}_k \ \mathbf{G}_k^{(n)}| \tag{5.70}$$

And the effective channel of the kth in the nth step is

$$\mathbf{G}_k^{(n)} = \mathbf{H}_k \left(\mathbf{I} + \sum_{j=1}^{K-1} \mathbf{H}_{[k+j]_K}^H \mathbf{Q}_{[k+j]_K}^{(n-K+j)} \mathbf{H}_{[k+j]_K} \right)^{-\frac{1}{2}} \tag{5.71}$$

where $[\text{x}]_K = (\text{x} - 1)\,\text{mod}(K) + 1$. Evidently, the previous $K - 1$ iterations of the covariance matrix values i.e. $\mathbf{Q}^{(n-K+1)}, \ldots \ldots, \mathbf{Q}^{(n-1)}$ need to be stored to generate the effective channels $\mathbf{G}_k$.

The modified algorithm is almost identical to the original sum power iterative algorithm with the exception of generating the effective channels, as it now depends on the previous $K - 1$ steps, not just the previous, as in the original algorithm.

1) Generate effective channels

$$\mathbf{G}_k^{(n)} = \mathbf{H}_k \left(\mathbf{I} + \sum_{j=1}^{K-1} \mathbf{H}_{[k+j]_K}^H \mathbf{Q}_{[k+j]_K}^{(n-K+j)} \mathbf{H}_{[k+j]_K} \right)^{-\frac{1}{2}} \tag{5.72}$$

   for $k = 1, \ldots\ \ldots\ \ldots\ldots, K$.

2) Assume the effective channels as parallel, i.e. noninterfering channels determine the covariance matrices $\{\mathbf{Q}_k^{(n)}\}_{k=1}^K$ by water-filling with equal power P

$$\{\mathbf{Q}_k^{(n)}\}_{k=1}^K \triangleq \arg\max_{\{\mathbf{Q}_k\}_{k=1}^K : \mathbf{Q}_k \geq \mathbf{0}, \sum_{k=1}^{\text{K}} \text{tr}(\mathbf{Q}_k) \leq P} \sum_{k=1}^{K} \log|\mathbf{I} + (\mathbf{G}_k^{(n)})^H \mathbf{Q}_k \ \mathbf{G}_k^{(n)}| \tag{5.73}$$

Computing the sum rate capacity of the MIMO BC over sum power maximization in (5.68) requires mapping the given MAC covariance matrices, $\mathbf{Q}_1, \ldots\ \ldots\ \ldots\ \ldots, \mathbf{Q}_K$, to corresponding BC matrices, $\Xi_1 + \ldots\ \ldots\ \ldots + \Xi_{\text{K}}$ that attain the same sum rates on a user-by-user premise and hence also in terms of sum rate using the same sum power, i.e.

$$\sum_{k=1}^{K} \text{tr}(\mathbf{Q}_k) = \sum_{k=1}^{K} \text{tr}(\Xi_k) \tag{5.74}$$

Let us define the following two matrices:

$$\mathbf{A}_k \triangleq \mathbf{I} + \mathbf{H}_k \left(\sum_{l=1}^{k-1} \Xi_l \right) \mathbf{H}_k^H, \mathbf{B}_k \triangleq \mathbf{I} + \sum_{l=1+1}^{K} \mathbf{H}_l^H \mathbf{Q}_l \mathbf{H}_l \tag{5.75}$$

where $k = 1, \ldots \ldots \ldots \ldots \ldots \ldots, K$. Define matrix $\boldsymbol{F}_k$ in the following SVD:

$$\mathbf{B}_k^{-\frac{1}{2}} \mathbf{H}_k^H \mathbf{A}_k^{-\frac{1}{2}} = \boldsymbol{F}_k \boldsymbol{D}_k \boldsymbol{G}_k^H \tag{5.76}$$

where $\boldsymbol{D}_k$ is a square and diagonal matrix. Then the corresponding BC covariance matrices, Ξ_k, can be computed from the following transformation:

$$\Xi_k = \mathbf{B}_k^{-\frac{1}{2}} \mathbf{F}_k \mathbf{G}_k^H \mathbf{A}_k^{\frac{1}{2}} \mathbf{Q}_k \mathbf{A}_k^{\frac{1}{2}} \mathbf{G}_k \mathbf{F}_k^H \mathbf{B}_k^{-\frac{1}{2}} \tag{5.77}$$

Readers can find the derivation of the expressions used in the above equations in [18, 20].

5.7.4 DPC Scheme Research Development for Application in the MIMO BC

The achievable throughput for the optimal strategy of (nonlinear) dirty-paper coding and the suboptimal but with lower complexity linear precoding (such as zero-forcing and block diagonalization) are compared in [21] when both strategies have the same multiplexing gain. The sum rate difference between the two strategies is computed at asymptotically high SNR. The authors finding shows that when the total number of receive antennas is equal or slightly less than the number of transmit antennas, linear precoding suffers a high penalty relative to DPC. However, this penalty is reduced significantly when the number of transmit antennas is large relative to the number of receive antennas. Furthermore, the finding shows that allocating power is asymptotically optimal for DPC at high SNR. Channel modelling for the two strategies is complicated and at high SNR, no *closed-form solution seems to appear for either DPC or linear precoding when* $\mathrm{M} < K\mathrm{N}$, *where* M, N are a number of transmit antennas and user's receive antennas, respectively and K users in Gaussian MIMO BC.

The channel considered in most cases is with fading coefficients MIMO channel. The coefficients are either fixed or known to the transmitter and/or the receiver. In practical scenarios, the channel coefficients vary over time, and are usually estimated at the receiver and fed back to the transmitter. Consequently, the coefficients arriving at the transmitter are outdated and may include some errors. A generalization of the dirty paper approach, known as fading-paper approach is presented in [22] and the authors assumed an interference, which is known causally. The fading-paper approach is not guaranteed to be optimal.

The DPC considered by Costa in [17] is defined in (5.63) above as

$$\mathrm{y} = \mathrm{x} + \mathcal{S} + \mathrm{n}$$

where $\mathcal{S}$ is the interference, known to the transmitter but not to the receiver. The Costa channel was further developed to the MIMO Gaussian vector BCs, where $\mathbf{y}, \mathbf{x}, \boldsymbol{\mathcal{S}}$ and $\mathbf{n}$ are vectors replacing the above scalar equivalents. The dirty-paper transmission over nonfading MIMO BCs can be expressed as

$$\mathbf{y} = \mathbf{H}(\mathbf{x} + \mathcal{S}) + \mathbf{n} \tag{5.78}$$

where $\mathbf{x}$ and $\boldsymbol{\mathcal{S}}$ are M – dimentional vectors, $\boldsymbol{y}$ and $\boldsymbol{n}$ are N – dimentional vectors, and $\mathbf{H}$ is an $\mathrm{N} \times \mathrm{M}$ fixed channel matrix. The fading-paper channel can also be represented by (5.78). When $\mathbf{H}$ is random, only the statistics of $\mathbf{H}$ are assumed to be known to the receiver but not to the transmitter. The fading-paper receiver considers the channel fades fluctuate from one time instant to another, while the dirty-paper receiver assumes that $\mathbf{H}$ is fixed. In other words, the dirty-paper decoder searches for a codeword that is jointly typical with **y**, while the fading-paper searches for a codeword that is jointly typical with both **y** and **H**. For understanding the concepts of typicality and joint typicality are explained clearly in [23]. However, we can conclude that despite the above shortcomings, the dirty-paper transmission can still be applied to the fading-paper channel by simply considering that **H** is fixed at its average and treat the fluctuations as noise.

The scenario when the state (i.e. the interference) is multiplied by a fast fading process that is not known at the transmitter is considered in [24] to determine the inner and outer bound of the capacity. When the fading is not known to the receiver, the gap between the bounds is small for fading distributions such as Gaussian, uniform, and Rayleigh but does not include the log-normal fading distribution.

5.7.5 Review of the DPC Scheme for Massive MIMO Systems

As pointed out in the previous section, a MIMO system with M transmit antennas at the BS and N receive antennas at the user terminal in a point-to-point link has a capacity about min{M, N} times the SISO link. In modern cellular MIMO systems, the BS form factor provides enough space to install multiple antennas – unlike the mobile terminal, which has a tiny form factor that allows a much small number of antennas to be placed. Consequently, in most cases min{M, N} = N implies that the multiple transmit antennas have no influence on the capacity gain expected from MIMO systems. In the extreme case with N = 1, there will be no capacity gains at all whatever the number of transmit antennas. A solution to this problem is to serve not a single user at any instant but multiple users simultaneously. Implementing the solution requires an efficient precoding scheme to reduce the interference between user streams. An optimal coding scheme that can be used to eliminate the interference is the DPC just discussed. As a result, when number of $K > \mathrm{M}$ then $\min\{\mathrm{M}, K\mathrm{N}\} = \mathrm{M}$ for N = 1 and a linear increase in capacity in M can be achieved by DPC.

However, the high computation costs of successive precoding and decoding, particularly for large numbers of users, make the DPC scheme difficult to implement in practical systems. An alternative strategy is the beamforming (BF), albeit suboptimal, can also serve multiple users at the same time. The BF strategy reduces the computation complexity and has the capability to achieve the best part of DPC capacity even with single antenna users. In fact, if BF weight vectors are chosen optimally, BF scheme can achieve the sum rate of DPC as $K \to \infty$ [25, 26].

5.8 mmWave Massive MIMO Systems

5.8.1 Introduction

Strong interest is shown in investigating the spectrum 3–300 GHz for commercial mobile applications. This spectrum is commonly known as millimetre Wave (mmWave) bands with wavelengths of a few mm. The European Union has funded several projects to research various aspects of the challenges facing the mmWave systems. These projects include 5Gnow (www.5gnow.eu), METIS2020 (www.metis2020.com), MiWaves (http://www.miwaves.eu), and MiWEBA (http://www.miweba.eu) among others.

Massive MIMO system is a key enabler in 5G, since it not only provides large directivity gains but also reduces interference through beam shaping. mmWave signals are more sensitive to blockage, such as buildings, than microwaves. This sensitivity issue causes a sizable difference between LOS and non-line-of-sight (NLOS) transmission. The mmWave wireless communications is based on the following principles and facts. An increase in system carrier frequency from 3 to 30 GHz (10-fold) implies an increase of order of magnitude in free space path loss. However, an increase in carrier frequency means a decrease in wavelength, which enables the use of a larger number of antennas providing the BF gain needed to balance the increase in path loss and establish a link with reasonable SNR for acceptable performance.

mmWave massive MIMO is widely predicted to include at least 100 antennas at BS compared with the number of antennas used by networks pre 5G of no more than 10. For example, the number of antennas employed by LTE at both BS and terminals are 8×8.The optimum number of antennas in 5G depends on many design metrics including: details of propagation, the complexity of signal processing, and obviously on the cost of the antennas. On the other hand, multiple antennas at the terminal would increase terminal throughput proportionally. However, pilot resource (per terminal) would also increase at the same rate and therefore the number of terminals served would be reduced. Consequently, increasing the number of antennas at terminals would not increase cell throughput. One can justifiably expect the manufacturing of a large number of low-power 5G transceivers to be more cost effective than pre 5G fewer high power units.

Massive MIMO at BSs can deploy several different antenna configurations such as linear, rectangular, square, circular, or cylindrical shapes [27], as shown in Figure 5.7, and they can be deployed locally or in distributed modes. A huge improvement in both radiated energy efficiency and simultaneously an increase in capacity with a large number of antennas are attributed to the fact that the energy can be focused with extreme sharpness into small regions where the intended terminals are located. So as to achieve this benefit, the BS shapes the transmitted signals appropriately (i.e. beamforming) so that all waveforms add up constructively at the intended terminals but destructively anywhere else. This technique reduces the interference to other terminals considerably and further precoding at the BS can eliminate the interference completely.

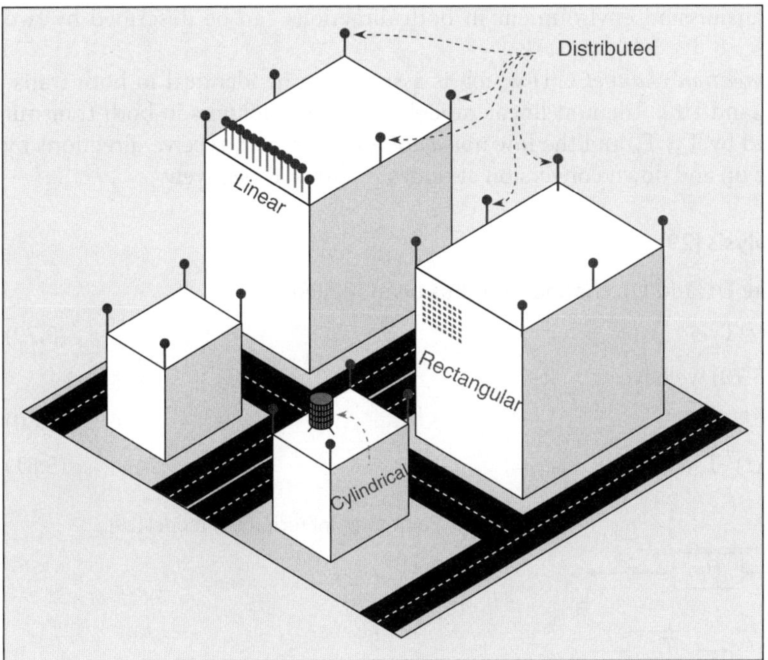

Figure 5.7 Some possible antenna configurations for a massive MIMO BS [27].

In normal circumstances, the transmitted waveforms at the BS antennas scattered by the BS surrounding obstructions, travel over multiple routes to reach the remote terminals and arrive at different delays (i.e. different phases). The multipath waveforms interfere destructively causing signal fading and the received signal dips. The terminal has to move to an alternative nearby location or to wait until the received signal is strong enough for the data to be decoded. The terminal wait is called *latency,* which is caused by channel fading. Massive MIMO together with beamforming reduces fading dips and significantly reduces the latency on the air interface.

5.8.2 Reciprocity Model for Point-to-Point Links

There are a number of efficient calibration solutions proposed in the literature.

In [30], an over-the-air reciprocity calibration is proposed that needs no extra hardware but the precision of the calibration is highly dependent on the link quality between the BS and the calibration terminal and therefore selecting an appropriate terminal is a very important issue. The authors proposed calibration schemes for both single-cell and CoMP transmission scenarios. In [31], the massive MIMO system calibration is carried out at the BS by sounding the M antennas one by one while receiving from the other (M − 1) antennas. This process generates M(M − 1) signals that are used for calibration. A number of LS estimators using a different number of received signals are evaluated to conclude that accurate calibration of BS antennas is possible using the mutual coupling between the antennas.

In practical TDD systems reciprocity is dependent on the propagation channel which is widely accepted to be reciprocal unless the propagation is subjected to an unexpected physical phenomenon, and the transmitter/receiver hardware which may not be reciprocal between uplink and downlink. These hardware chains need to be calibrated over the RF band of interest. A reciprocity model for a point-to-point link is shown in Figure 5.8, which shows TDD system with two devices A and B. The channel model encompasses a cascade of three linear filters. The cascade of filters featuring the transmit circuit, EM propagation path, and the receive circuit. Assume the transmission environment in both directions can be described by two fading AWGN channels.

Denote the *EM propagation channel* C(t) which is assumed to be identical in both transmission directions (DL and UL). The non-linear model of power amplifiers in both transmit directions is represented by T_A, T_B and the low noise amplifiers in both receive directions by R_A, R_B. Let f_A and f_B be up and down conversion at nodes A and B respectively.

5.8.3 Reciprocity Analysis [29]

At a given frequency, the UL and DL channel can thus be modelled as

$$G(t) = R_B e^{j\omega_B t}\, C(t) T_A e^{-j\omega_A t} \tag{5.79}$$

where $\omega = 2\pi f$. From (5.79) we have

$$C(t) = R_B^{-1} G(t) e^{(\omega_A - \omega_B)t} T_A^{-1} \tag{5.80}$$

$$H(t) = R_A\, e^{\omega_A t}\, C(t)^T T_B e^{-\omega_B t} \tag{5.81}$$

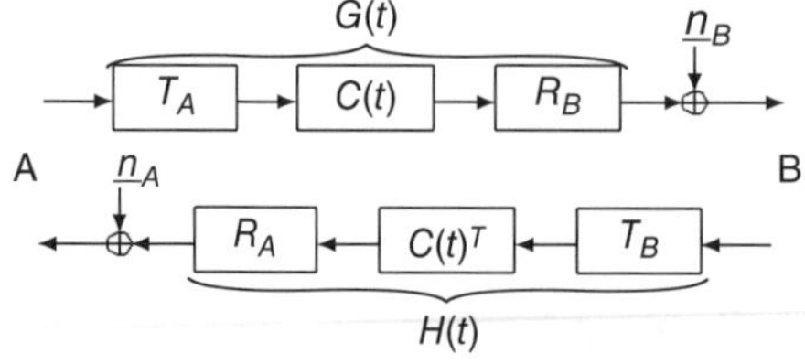

Figure 5.8 Point-to-point reciprocity model [29].

From (5.81),

$$C(t)^T = R_A^{-1}\, H(t)e^{(\omega_B-\omega_A)t}T_B^{-1} \tag{5.82}$$

hence,

$$C(t) = T_B^{-T}\, H(t)^T R_A^{-T} e^{(\omega_B-\omega_A)t} \tag{5.83}$$

Therefore, equating (5.80) with (5.83) we get

$$R_B^{-1}\, G(t)e^{(\omega_A-\omega_B)t}T_A^{-1} = T_B^{-T}\, H(t)^T R_A^{-T} e^{(\omega_B-\omega)t} \tag{5.84}$$

where $(f_A - f_B) = -((f_B - f_A)$ are the residual frequencies offset in both directions, which can be estimated and compensated for, and therefore we assume that it can be eliminated and (5.84) can be simplified to

$$R_B^{-1}\, G(t)T_A^{-1} = T_B^{-T}\, H(t)^T R_A^{-T} \tag{5.85}$$

Denote $P_A = R_A^{-T}T_A$ and $P_B = T_B^T R_B^{-1}$, the reciprocity relationship [29] for the model is

$$P_B\, G(t) - H(t)^T P_A = 0 \tag{5.86}$$

We can measure (simultaneously) the bidirectional channels $G(t)$ and $H(t)$ at times t_i where $i = 1, \ldots \ldots \ldots \ldots, T$.

Let us consider an optimal estimator for P_A and P_B based on the noisy channel $\widehat{G}(t_i)$, $\widehat{H}(t_i)$ so at any single frequency we need to minimize the reciprocity relationship (5.86). Denote the estimation error due noise as $\widetilde{G}(t)$, $\widetilde{H}(t)$ to be added to the measured values using the total least-squares (TLS) method [32, 33]:

$$(\widehat{P}_A, \widehat{P}_B) = \underset{(P_A, P_B, \widetilde{G}_i, \widetilde{H}_i)}{\operatorname{argmin}} \sum_{i=1}^{T} \|P_B(\widehat{G}(t_i) + \widetilde{G}_i) - (\widehat{H}(t_i) + \widetilde{H}_i)^T P_A\|_2^2 + \|\widetilde{G}_i\|_2^2 + \|\widetilde{H}_i\|_2^2 \tag{5.87a}$$

$$s.t. \qquad \|P_A\|_2^2 = 1 \tag{5.87b}$$

where the minimization with respect to $P_A, P_B, \widetilde{G}_i, \widetilde{H}_i$ such that$\|P_A\|_2^2 = 1$ to ensure the trivial solution $(P_A, P_B) = (0, 0)$ is avoided.

5.8.4 Reciprocity Analysis Extension to Multiple Users

Let node A be a BS engaged in bidirectional TDD communications with K users. Denote a user as u_k where $k = 1, \ldots \ldots, K$. Denote the gains of transmit and receive RF frontend of the user k as T_{u_k}, R_{u_k}, respectively. The measured channels between A and u_k are given by (5.79) and (5.81) as

$$G_k(t) = R_{u_k} C_k(t) T_A \tag{5.88}$$

$$H_k(t) = R_A C_k(t)^T T_{uk} \tag{5.89}$$

Denote $P_{u_k} = T_{u_k}^T R_{u_k}^{-1}$. We have, as proved above,

$$P_{u_k} G_k(t) - H_k(t)^T P_A = 0 \tag{5.90}$$

The optimal estimator for the multiuser case is given by

$$(\widehat{P}_A, \widehat{P}_{u_k}, \ldots \ldots \ldots \ldots, \widehat{P}_{u_k}) =$$

$$\underset{(P_A, P_{u_k}, \widetilde{G}_{k,i}, \widetilde{H}_{k,i})}{\operatorname{argmin}} \sum_{\substack{k=1 \\ i=1}}^{K,T} P_{u_k}(\widehat{G}_k(t_i) + \widetilde{G}_{k,i}) - (\widehat{H}_k(t_i) + \widetilde{H}_{k,i})^T P_{A2}^2 + \widetilde{G}_{k,i2}^2 + \widetilde{H}_{k,i2}^2 \tag{5.91a}$$

$$s.t. \qquad \|P_A\|_2^2 = 1 \tag{5.91b}$$

5.8.5 Reciprocity and Pilot Contamination

Pilot contamination, as a key phenomenon, is not limited to massive MIMO, but it has a profound effect on the massive MIMO compared to the traditional MIMO system. Orthogonal pilot sequences are usually used in multiuser system for channel estimation. However, when the available supply of the orthogonal sequences is exhausted, pilot sequences are reused from one cell to another. On the uplink, when the received pilot signal is correlated with the pilot sequence for the intended terminal, the channel estimate is corrupted by interference originating from other terminals using the same pilot sequences. The pilot contamination is caused by the reuse of the same (or linearly dependent) pilot sequences in different cells. On the DL, beamforming based on the corrupted channel estimates generate interference at those terminals that are reusing the same pilot sequence.

In order to get perfect CSI, the pilot contamination must be completely removed or at least minimized. Recently, several efficient techniques have been presented [34–37] to overcome pilot contamination. In [34], a coordinated approach was proposed between cells to remove the pilot contamination effect under certain conditions of the channel covariance matrix. In [35], an approach was suggested to mitigate the pilot contamination problem without the need for coordination among cells. The proposed method is called a blind scheme in the sense that it does not require pilot sequence to estimate the channel. The basic idea is based on SVD of the received signal and noise. It seems to work if specific constraints associated with the number of antennas to the coherence time and signal power to interference power plus noise power ratio (SINR) are satisfied. Additionally, the algorithm is very complex when used for massive antennas and the dominant complexity arises from the SVD of the received signal. In [36, 37], the analysis of pilot contamination is presented for MU-MIMO cellular system and a multi-cell minimum mean square error (MMSE) based precoding scheme was developed that moderates the pilot contamination.

5.9 MIMO Beamforming Schemes

5.9.1 Introduction to Beamforming

Beamforming (BF) technology enables an access point to effectively focus its signal at the intended target. This operation brings about a better SNR and achieves greater throughput by reducing the interference level and significantly improves system performance. In this scenario, the transmitter is the beamformer and the receiver is the beamformee. The principle behind BF transmission is to concentrate the energy towards a target receiver so that the overall interference in nearby receivers in the system is reduced. Transmitted/received signals are manipulated in magnitude and/or phase to form a beam in the desired direction. In practice, the BF dynamically alters the transmission pattern and can be changed on a per-frame basis. A transmit antenna generates stronger electromagnetic (EM) waves in the given direction to improve system performance and area coverage, as shown in Figure 5.9.

5.9.2 Analysis of Beamforming

Consider an array comprising N omni-directional receive antennas and M antennas transmit point sources. Let the information signal from the mth transmit source be $s_m(t)$ modulated on a carrier f_c and the complex modulated signal is

$$s_m(t)e^{j2\pi f_c t} \tag{5.92a}$$

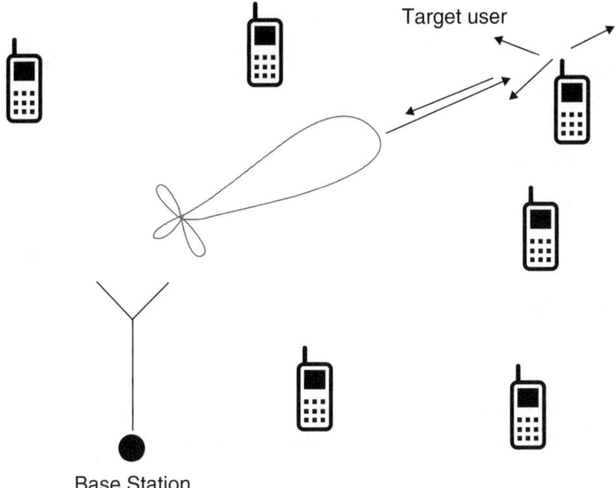

Figure 5.9 Beamforming in multiuser system.

Assume that the wavefront of the signal in (5.92a) transmitted from the mth source induced signal on the nth receive antenna element arrives after a delay $\tau_n(\theta_m)$, the receive signal can be expressed as

$$s_m(t)e^{j2\pi f_c(t+\tau_n(\theta_m))} \quad , \quad m = 1, \ \ldots\ldots \ , \mathrm{M} \tag{5.92b}$$

In deriving (5.92b) we assumed the system BW is narrow enough to keep the amplitude of the signal in (5.92a) almost constant during τ_n seconds. Denote $\mathrm{x_n(t)}$ for the total induced signal from all directions of the M sources so that

$$\mathrm{x}_n(\mathrm{t}) = \sum_{m=1}^{\mathrm{M}} s_m(t)e^{j2\pi f_c(t+\tau_n(\theta_m))} + \mathrm{n}_n(\mathrm{t}), \quad n = 1, \ \ldots\ldots\ldots \ , \mathrm{N} \tag{5.92c}$$

where $\mathrm{n}_n(\mathrm{t})$ is an additive noise on the nth receive antenna element, $\mathrm{n}_n \sim \mathcal{N}(0, \sigma_\mathrm{n}^2)$. Considering the narrowband beamformer structure in Figure 5.10, and denoting the beamformer weights by vector $\boldsymbol{\omega}$:

$$\boldsymbol{\omega} = [\omega_1, \ldots\ldots\ldots, \omega_\mathrm{M}]^T \tag{5.93}$$

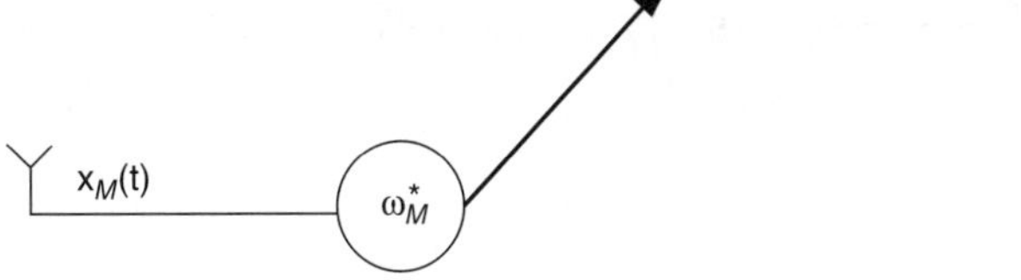

Figure 5.10 Narrowband beamformer structure [38].

The array output is given by $\mathbf{y}(t)$ as

$$\mathbf{y}(t) = \boldsymbol{\omega}^H \, \mathbf{x}(t) \tag{5.94}$$

where

$$\mathbf{x}(t) = [x_1(t), \ldots\ldots\ldots, x_n(t), \ldots\ldots\ldots, x_N(t)]^T \tag{5.95}$$

Without loss of generality, we ignore the noise. The received mean power $P(\boldsymbol{\omega})$ is given by:

$$P(\boldsymbol{\omega}) = \mathbb{E}[\mathbf{y}(t)\mathbf{y}^H(t)] = \boldsymbol{\omega}^H \, \mathbf{R} \, \boldsymbol{\omega} \tag{5.96}$$

where $\mathbf{R}$ is $N \times N$ correlation matrix of the receive signals:

$$\mathbf{R} = \mathbb{E}[\, \mathbf{x}\,\mathbf{x}^H] \tag{5.97}$$

The elements of matrix $\boldsymbol{R}$ denote the correlation between various received signals within $\mathbf{x}$. For example, R_{ij} denotes the correlation between the signals received by the ith and the jth antennas.

5.10 BF Schemes

5.10.1 The Delay and Sum BF

Consider a transmitted signal pinging two-element antenna at certain angle θ arriving at different times. Let the arrival time of one signal be taken as reference according to certain coordinate system and the other signal arrival delayed by t_d, as shown in Figure 5.11. Each received signal is multiplied by a factor 0.5 so that the total gain in the signal direction θ is equal unity.

The two elements are separated by a distance d. Assume the transmitted signal arrived from direction θ and the received signal at the output of the first antenna is $s(t)$. Thus, the delay of the other received signal is given by:

$$t_d = \frac{d}{c}\, cos\theta \tag{5.98}$$

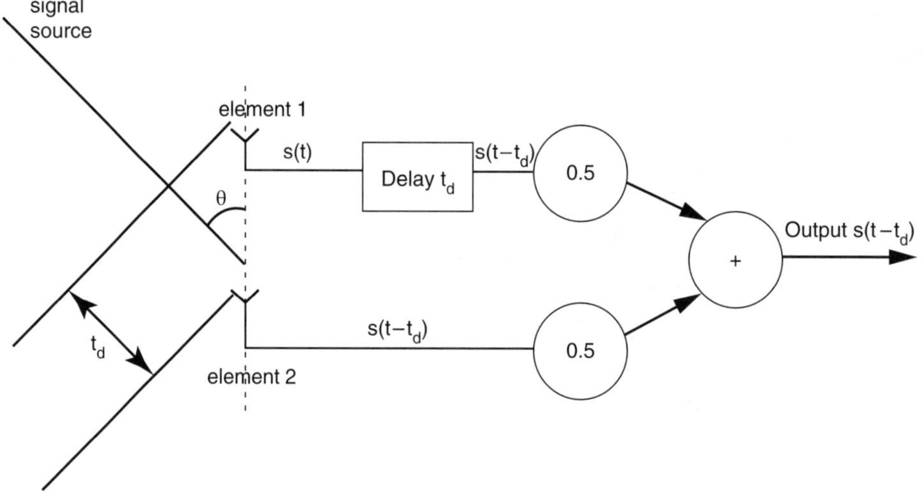

Figure 5.11 Delay-and-sum beamformer structure [38].

where c is the speed of light. Therefore, the signal at the output of the second antenna is $s(t - t_d)$. In order to combine the signals coherently, we delay the signal from the first element to get $s(t - t_d)$, multiply both signals with a scaling factor 0.5, and add them up to produce an output signal $s(t - t_d)$. This beamformer reduces the power of the uncorrelated noise at the array output and produces a gain in the SNR.

5.10.2 Null Steering Beamformers

This beamformer produces null in the radiation pattern in the direction of arrival of the interfering plane waveform signal, thus cancelling the interference in that direction. An early version of this scheme subtracts the estimated interference using a delay and sum beamformer, which is very effective for cancelling strong interference from a small number of interferers, but it becomes complicated as the number of interferers grows in applications like mobile cellular communications. A beam in the desired direction and simultaneous null in the interference directions can be designed by proper choice of BF weights. The steering vector in the ith direction θ_i is a complex vector $\boldsymbol{s}_i$ expressed as

$$\boldsymbol{s}_i = [\, e^{j\omega_0 \tau_1(\theta_i)}, e^{j\omega_0 \tau_2(\theta_i)}, \ldots\ldots\ldots, e^{j\omega_0 \tau_K(\theta_i)}] \tag{5.99}$$

where ω_0 is the radian frequency of the desired signal, and τ_i is the time delay measured from a reference receive antenna at the origin of the system. Denote the steering vector $\boldsymbol{s}_0$ in the direction of the desired signal where unity response is required and $\boldsymbol{s}_1, \boldsymbol{s}_2, \boldsymbol{s}_3 \ldots\ldots\ldots, \boldsymbol{s}_K$ are K steering vectors related to K directions. The principle behind the null BF is to choose the correct weights that make the receiver responses in the desired direction (i.e. steering vector $\mathbf{s}_0$) with unity and the response in the other K directions that are required to be annulled.

We can articulate these constraints by the following simultaneous equations:

$$\boldsymbol{\omega}^H \boldsymbol{s}_0 = 1 \tag{5.100a}$$

$$\boldsymbol{\omega}^H \boldsymbol{s}_i = 0 \qquad i = 1, 2, \ldots\ldots\ldots, K \tag{5.100b}$$

We can express (5.100) in matrix notation as

$$\boldsymbol{\omega}^H \, \mathbf{A} = \mathbf{c}_1^T \tag{5.101}$$

where matrix $\mathbf{A}$ contains columns that are being the steering vectors associated with these directions:

$$\mathbf{A} = [\mathbf{s}_0, \ldots\ldots\ldots\ldots, \mathbf{s}_K] \tag{5.102a}$$

$$\mathbf{c}_1 = [1, 0, \ldots\ldots\ldots\ldots, 0]^T \tag{5.102b}$$

Using (5.101), we get

$$\boldsymbol{\omega}^H = \mathbf{c}_1^T \mathbf{A}^{-1} \tag{5.103}$$

Assuming $\mathbf{A}$ is invertible. If it not invertible we can use its equivalent Moore-Penrose pseudo inverse and (5.103) can be expressed as

$$\boldsymbol{\omega}^H = \mathbf{c}_1^T \mathbf{A}^H \, (\mathbf{A} \, \mathbf{A}^H)^{-1} \tag{5.104}$$

5.10.3 Beamformer Using a Reference Signal

A schematic diagram showing a beamformer using a reference signal is shown in Figure 5.12. It estimates the weights of the beam and the array output is deducted from an available reference signal to create an error signal $\varepsilon(t)$ given as

$$\varepsilon(t) = \mathrm{r(t)} - \boldsymbol{\omega}^H \, \mathbf{x}(t) \tag{5.105}$$

where $\mathbf{x}(t)$ is given in (5.92c) and (5.95). The error $\varepsilon(t)$ is next used to control the weights. In most cases, the mean square error (MSE) between array output and the reference signal is used to minimize the error. The MSE is given by

$$\mathrm{MSE} = \mathbb{E}[|\varepsilon(t)|^2] = \mathbb{E}[\varepsilon(t)\, \varepsilon^H(t)] \tag{5.106}$$

Thence,

$$\mathrm{MSE} = \mathbb{E}[(\mathrm{r(t)} - \boldsymbol{\omega}^H \, \mathbf{x}(t))\, (\mathrm{r(t)} - \boldsymbol{\omega}^H \, \mathbf{x}(t))^H] \tag{5.107}$$

Accordingly, we have

$$\mathrm{MSE} = \mathbb{E}[\mathrm{r(t)r^*(t)}] + \mathbb{E}[\boldsymbol{\omega}^H \, \mathbf{x}(t)\, \mathbf{x}^H(t)\boldsymbol{\omega}] - \mathbb{E}[\mathrm{r(t)}\mathbf{x}^H(t)\boldsymbol{\omega} - \boldsymbol{\omega}^H \, \mathbf{x}(t)\mathrm{r^*(t)}] \tag{5.108}$$

Since $\boldsymbol{\omega}^H \, \mathbf{x}(t)\mathrm{r^*(t)}$ is symmetric, we have

$$\mathrm{r(t)}\mathbf{x}^H(t)\boldsymbol{\omega} = \boldsymbol{\omega}^H \, \mathbf{x}(t)\mathrm{r^*(t)} \tag{5.109}$$

Substituting (5.97) in the second term of (5.108) and making use of (5.109), we get

$$\mathrm{MSE} = \mathbb{E}[|\mathrm{r(t)}|^2] + \boldsymbol{\omega}^H \, \mathbf{R}\, \boldsymbol{\omega} - 2\, \boldsymbol{\omega}^H \mathbf{R}_{\mathrm{Xr}} \tag{5.110}$$

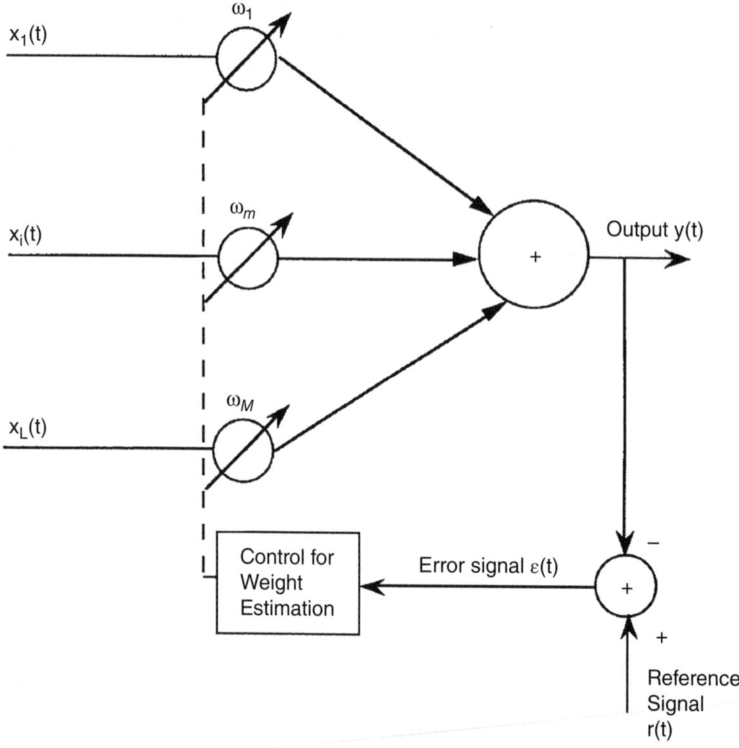

Figure 5.12 Beamforming using a reference signal [38].

where $\mathbf{R}_{xr} = \mathbb{E}[\mathbf{x}(t)\mathrm{r}^*(\mathrm{t})]$ is the correlation between the reference signal and the array signals.

The MMSE can be obtained as optimal solution to (5.110). The optimum Wiener solution $\boldsymbol{\omega}_{mmse}$ can be found by setting the partial derivative with respect to $\boldsymbol{\omega}$ to zero [39].

$$\frac{\partial}{\partial \boldsymbol{\omega}}[|\mathrm{r(t)}|^2 - 2\,\boldsymbol{\omega}^H \mathbf{R}_{sx} + \boldsymbol{\omega}^H \mathbf{R}_{xx}\boldsymbol{\omega}] = [-2\mathbf{R}_{xr} + 2\,\mathbf{R}_{xx}\boldsymbol{\omega}] = 0$$

Thus,

$$\boldsymbol{\omega}_{mmse} = \mathbf{R}_{xx}^{-1}\mathbf{R}_{xr} \tag{5.111}$$

Let us consider the case where some of the arriving signals include interference and noise, i.e.

$$\overline{\mathbf{x}} = \mathbf{x} + \mathbf{i} + \mathbf{n}$$

where $\mathbf{x}$, $\mathbf{i}$, $\mathbf{n}$ are desired, interference and noise signals, respectively, so that

$$\mathbf{R}_{\overline{xx}} = \mathbf{R}_{xx} + \mathbf{R}_{ii} + \boldsymbol{R}_{nn} \tag{5.112}$$

And $\mathbf{R}_{xx}$,$\mathbf{R}_{ii}$,$\mathbf{R}_{nn}$ are the correlation of the desired signal, the interference and the noise correlation, respectively. Assume the same reference signal is applied again. Furthermore, we assume the error is zero:

$$\varepsilon(t) = \mathrm{r}(t) - \boldsymbol{\omega}^H \mathbf{x}(\mathrm{t}) = 0$$

$$\mathrm{r}(t) = \boldsymbol{\omega}^H \mathbf{x}(\mathrm{t})$$

$$\mathbf{R}_{xr} = \mathbb{E}[\mathbf{x}\mathrm{r}^*(t)] = \mathbb{E}[\mathbf{x}\mathbf{x}^*\,\boldsymbol{\omega}^T]$$

$$\mathbf{R}_{xr} = \mathbb{E}[\mathbf{s}.\mathbf{s}^*]\boldsymbol{\omega}^T = \mathbb{E}[|\mathbf{x}|^2].\boldsymbol{a}_0 = \boldsymbol{S}.\boldsymbol{a}_0$$

where $\mathrm{p} = E[|\mathbf{x}|^2]$ and $\boldsymbol{a}_0$ is weight vector for the desired signal.

Thus

$$\boldsymbol{\omega}_{mmse} = \mathrm{p}\mathbf{R}_{xx}^{-1}\boldsymbol{a}_0 \tag{5.113}$$

An important parameter that contributes to the design of the beamformer is the estimate of the direction of arrival (AOA) of the desired signal. There are many algorithms that have been proposed and studied in the area of angle of arrival (AOA) such as multiple signal classification (MUSIC) and estimation of a signal parameter via rotational invariant technique (ESPRIT), Barlett and ML. Furthermore, there is large volume of papers in the literature describing improvement techniques on the proposed algorithms above. The simplest with highest accuracy is the MUSIC scheme [40], which is outlined in 5.B.

5.11 mmWave BF Systems

5.11.1 Introduction

BF technology is one of the key enablers for 5G cellular wireless communications. A beamformer concentrates the transmit/receive signal power in the desired direction to overcome the high path loss at mmWave frequency bands. Small wavelengths of mmWave frequencies enable the installation of a large number of antenna elements giving large array gains and producing highly directional beams. BF could be applied in the analogue or digital domain; nonetheless, digital BF offers better performance and provides a high degree of freedom. However, it is complex and costly particularly when used with OFDM systems since separate FFT/IFFT,

DACs/analogue-to-digital conversion (ADCs), are required per each RF chain. In contrast, analogue BF is simple and effective in generating high array gains but less flexible than digital BF. Therefore, a balance has to be struck between simplicity and flexibility/performance. A mix of analogue and digital BF i.e. a hybrid BF seems a reasonable choice especially for mmWave massive MIMO antennas. In a hybrid structure, baseband (BB) digital signals precoding/decoding are carried out together with others BB processing.

5.11.2 Hybrid Digital and Analogue BF for mmWave Antenna Arrays

Studies on hybrid beamforming schemes for massive antenna array have taken new research directions. Researchers at Nokia and Samsung independently investigated two different but related aspects of efficient solutions for BB beamforming to reduce the prohibitive complexity when implementing massive array precoding/decoding.

The hybrid precoding involves a joint optimization over a possible BB precoding/decoding at different subarray configurations. The work at Samsung considers a codebook-based precoding where different precoders are picked from a priori predetermined codebook with certain subarray architecture. However, the complexity of such approach scales exponentially with the number of subarrays used at the TX and the RX are due to the need to assign directions to each of the subarrays. Consequently, such an approach is often costly and prohibitive. To overcome this problem, researchers look for reduced complexity algorithms using some known facts inherited in mmWave channels, namely: mmWave channels are characterized by a sparse multipath structure that corresponds to a small dominant number of AOD at TX and AOA at RX. This fact implies that most of the signal power is captured in a small number of directions at both the TX and RX. This knowledge hints to the possibility of reduced complexity, in principle by considering a small but important set of directions.

BF is necessary in mmWave systems to overcome high pathloss and to achieve quality systems' performance. Multiple user(s) streams require precoding to reduce inter streams interference and further improve data rates and performance quality. Traditionally, MIMO processing is generally performed at BB to control both the signal phase and amplitude but this needs a dedicated RF chain per antenna and for massive antennas system this means a high hardware capital cost. The precoding/ BF is likely to be divided between the digital and analogue domain. BF in the RF domain after frequency upconversion is expected to be using *analogue phase shifters* and digital is employed for BB precoding. Such a model for a multiple streams single user mmWave system is considered in [41, 42] using layered BB/RF precoding. The developed precoding algorithm approximates the optimal precoding that can be implemented in RF hardware. The layered system model is shown in Figure 5.13 and consists of a BS equipped with M antennas communicates N_s data streams. The BS transmitter comprises N_{RF}^t RF chains such that $N_s \leq N_{RF}^t \leq$ M. The number of data streams, N_s, is limited by the transmitter and receiver number of chains, N_t^c, N_r^c, respectively, such that $N_s \leq \min(N_{RF}^t, N_{RF}^r)$.

The transmitter applies $N_{RF}^t \times N_s$ BB precoder matrix $\mathbf{F}_{BB}$ followed by an $\mathrm{M} \times N_{RF}^t$ RF beamformer weight matrix $\mathbf{F}_{RF}$ connected with the antennas. The transmitted signal is given as

$$\mathbf{x} = \mathbf{F}_{RF}\,\mathbf{F}_{BB}\,\mathbf{s} \tag{5.114}$$

where $\mathbf{s}$ is the stream symbols vector $N_s \times 1$ and $\mathbb{E}[\mathbf{s}\,\mathbf{s}^H] = \frac{1}{N_s}\mathbf{I}_{N_s}$. Assuming perfect timing and carrier recovery, the received signal is

$$\mathbf{y} = \sqrt{\rho}\,\mathbf{H}\,\mathbf{F}_{RF}\,\mathbf{F}_{BB}\,\mathbf{s} + \mathbf{n} \tag{5.115}$$

where $\mathbf{y}$ is $\mathrm{N} \times 1$ received vector, $\mathbf{H}$ is $\mathrm{N} \times \mathrm{M}$ channel matrix where $\mathbb{E}[\|\mathbf{H}\|_F^2] = \mathrm{MN}$, ρ is average received power, and $\mathbf{n} \sim \mathcal{CN}(0, \sigma^2)$. As we highlighted before, analogue phase shifters are used

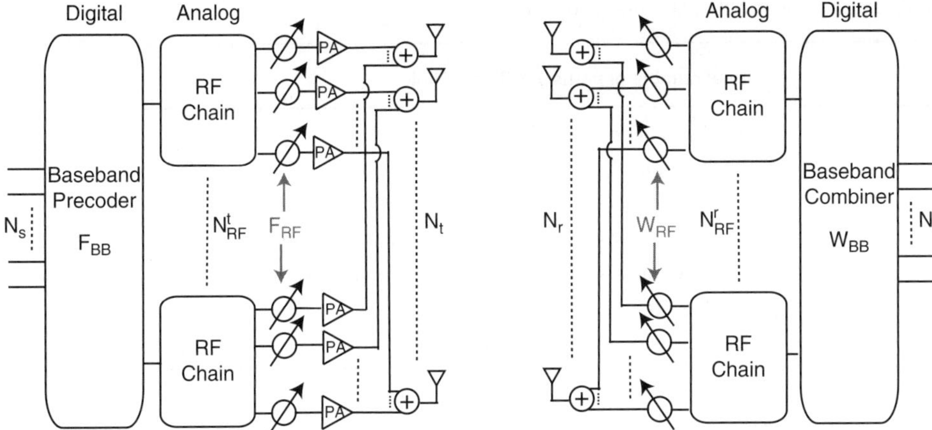

Figure 5.13 Block diagram of mmWave single user with layered digital baseband precoding followed by constrained radio frequency precoding [41].

to realize the RF precoder with elements constraint to satisfy

$$(\mathbf{F}_{RF}^{(i)}\ \mathbf{F}_{RF}^{(i)H})_{l,l} = \mathrm{M}^{-1} \tag{5.116}$$

where $\mathbf{F}_{RF}^{(i)}$ is the ith column of the $\mathrm{M} \times N_{RF}^t$ RF beamformer matrix $\mathbf{F}_{RF}$ and $(.)_{l,\,l}$ refer to the lth diagonal element so all element of $\mathbf{F}_{RF}$ have equal norm. The total transmit power constraint is imposed by applying $\|\ \mathbf{F}_{RF}\ \mathbf{F}_{BB}\|_F^2 = N_s$. So, to decode the transmitted data streams we need $N_{RF}^r \geq N_s$ and we assume the **H** CSI realizations are known at the transmitter and receiver. The processed received signal is given as

$$\tilde{\mathbf{y}} = \sqrt{\rho}\,\mathbf{W}_{BB}^H\ \mathbf{W}_{RF}^H\ \mathbf{H}\ \mathbf{F}_{RF}\ \mathbf{F}_{BB}\ \mathbf{s} + \mathbf{W}_{BB}^H\ \mathbf{W}_{RF}^H\mathbf{n} \tag{5.117}$$

where $\mathbf{W}_{BB}$ is the BB combining $N_{RF}^r \times N_s$ matrix and $\mathbf{W}_{RF}$ is the receive RF beamformer weight $\mathrm{N} \times N_{RF}^r$ matrix with norm entries.

Considering the channel **H** and noting that for large number of BS antennas M, the available space between antennas becomes small, which causes high antenna correlation so the popular spatially uncorrelated fading channel model is ill-suited. The channel model that is used by IMT-Advanced and Samsung researchers is similar to the 3D channel spatial channel model (SCM) presented in Chapter 9. For a given MIMO link between the transmitter and the receiver, the channel matrix is given by

$$\mathbf{H}(t) = \sum_{i=1}^{N_p} \sqrt{\frac{P_i}{\mathrm{MN}}} \sum_{j=1}^{N_{sp}} \mathbf{\Lambda}_{i,j} \otimes \mathbf{a}_r(\theta_{i,j}^{AoA}).\ \ \mathbf{a}_t^T(\theta_{i,j}^{AoD}) e^{j2\pi f_{i,j}^d t} \delta(t - t_i) \tag{5.118}$$

where the parameters used in (5.118) are defined in Table 5.1.

The array response vector depends on the array structure. For the simplest structure of uniform linear array, the array response vector is

$$\mathbf{a}(\theta) = \frac{1}{\sqrt{N}}[1\ \ e^{j\kappa d \sin\theta}\ \ldots\ldots\ldots\ldots\ e^{j\kappa d(\mathrm{N}-1)\sin\theta}\ ]^T \tag{5.119}$$

where $\kappa = \frac{2\pi}{\lambda}$ and λ is the wavelength at the operating frequency, N is the number of antennas, and $\frac{1}{\sqrt{N}}$ is the normalization factor such that $\|\mathbf{a}(\theta)\|_2^2 = 1$, and θ is path angle from antenna

Table 5.1 Parameters for an MIMO link, given Eq. (5.118).

N_p	**Total number of paths per channel link**
N_{sp}	Total number of subpaths per path
P_i	Power per the ith path
$\mathbf{\Lambda}_{i,j}$	Initial phase matrix for the jth subpath in the ith path N × M matrix
$\mathbf{a}_r(\theta_{i,j}^{AoA})$	Array response vector for the given AoA N × 1 vector
$\mathbf{a}_t^T(\theta_{i,j}^{AoD})$	Array response vector for the given AoD M × 1 vector
$f_{i,j}^d$	Doppler frequency
$\otimes$	Element-wise product operation

boresight angle. The achievable rate R assuming Gaussian signalling is given in [41] as

$$R = \log_2\left(\left|\mathbf{I}_{N_s} + \frac{\rho}{N_s}\mathbf{R}_\mathbf{n}^{-1}\,\mathbf{W}_{BB}^H\,\mathbf{W}_{RF}^H\,\mathbf{H}\,\mathbf{F}_{RF}\,\mathbf{F}_{BB}\,\mathbf{F}_{BB}^H\,\mathbf{F}_{RF}^H\,\mathbf{H}^H\,\mathbf{W}_{RF}\,\mathbf{W}_{BB}\right|\right) \tag{5.120}$$

where $\mathbf{R}_\mathbf{n} = \sigma_\mathbf{n}^2\,\mathbf{W}_{BB}^H\,\mathbf{W}_{RF}^H\,\mathbf{W}_{RF}\,\mathbf{W}_{BB}$ is the noise covariance after combining. Next, we aim to determine the layer joint RF beamformer weights/BB precoder that optimize the achievable rate in (5.120). The optimization problem can be expressed as

$$(\mathbf{F}_{RF}^{opt}\,\mathbf{F}_{BB}^{opt}) = \arg\max_{\mathbf{F}_{RF},\,\mathbf{F}_{BB}} \log_2\left(\left|\mathbf{I}_{N_s} + \frac{\rho}{\sigma_\mathbf{n}^2 N_s}\,\mathbf{H}\,\mathbf{F}_{RF}\,\mathbf{F}_{BB}\,\mathbf{F}_{BB}^H\,\mathbf{F}_{RF}^H\mathbf{H}^H\right|\right) \tag{5.121a}$$

$$\text{s.t. } \mathbf{F}_{RF} \in \mathcal{W},\ \|\,\mathbf{F}_{RF}\,\mathbf{F}_{BB}\|_F^2 = N_s \tag{5.121b}$$

where $\mathcal{W}$ is a set of phase shifters with equal gain elements. There is no general solution to the optimization problem in Eq. (5.121). The solution to jointly optimize $\mathbf{F}_{RF}$ and $\mathbf{F}_{BB}$ can be found using exhaustive iterative search using appropriate algorithm.

Maximum ratio combining (MRC) [43] is a conventional combining technique where signals from multiple paths are combined in phase to maximize the SNR, as shown in Figure 5.14. MRC is investigated for application in wireless system where signals from received antennas are weighted so that the sum of their SNR is maximized. Nonetheless, MRC techniques are traditionally used solely for applications at the receiver end. However, as wireless systems evolved, there may be a need for diversity at the transmitter or at both the transmitter and the receiver to cope with extreme fading effects. The Alamouti encoding scheme is an example of effective transmit diversity. A maximum ratio transmission scheme is explored for a single user link with

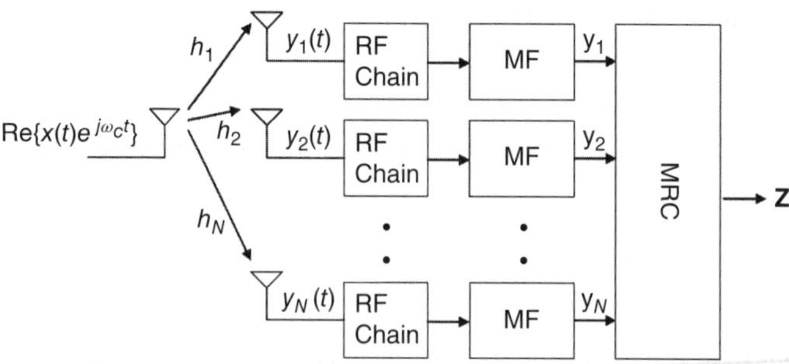

Figure 5.14 Conventional post detection MRC receiver [44].

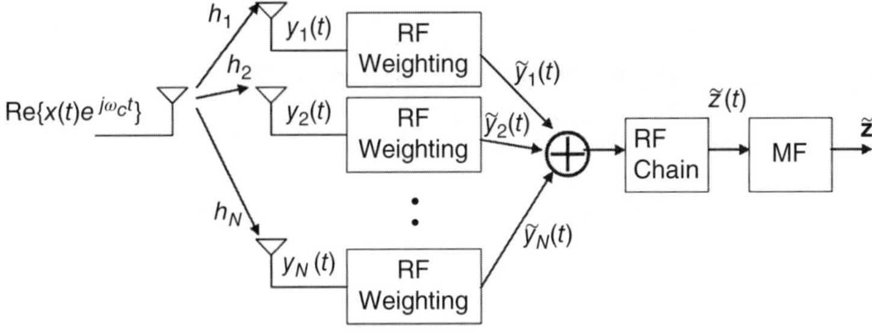

Figure 5.15 MRC technique RF combining at the receiver [44].

M at the transmit end and N antennas at the receive end. So order of diversity is expected to be MN and the probability of error is expected to decrease inversely with $(SNR)^{MN}$. Simulation results verify the predicted decrease in probability of error with SNR to the power of MN.

The maximum ratio transmission technique may be deployed in wireless MIMO systems for RF chain combining for analogue beamforming applications. The operation of MRC at RF level can be invested in MIMO system design. A conventional post-detection MRC system operated as a diversity system with N branches is depicted in Figure 5.15. Such a system comprises N RF chains, each chain consisting of an LNA, frequency converter to down-convert the RF signal to an appropriate detector, and A/D converter. An appropriate detector could be, for example, MF [44]. Accordingly, the system contains N RF chains working with N MFs. For large N, common concerns in employing such a system are the hardware complexity and its cost and the size associated with the multiple RF chains and MFs.

In comparison with UHF or microwave, the propagation attenuation of the mmWave is much higher and the radiated power is much lower. So, it is imperative to employ high-directive antennas to salvage sufficient power to achieve appropriate performance. So, for mobile users and users at disparate locations, mmWave has to deploy steerable directive antennas and configurable arrays. mmWave short wavelength makes it feasible to place a large number of antennas in a confined space. Although full digital array (RF frontend and digital BB) achieves capacity and flexibility, such an array is extremely costly and impractical because the available space is tight.

A hybrid array that comprises multiple analogue subarrays, each using its own digital chain, appears to be a highly attainable solution. An important characteristic of the mmWave propagation is multipath sparsity induced by the fact that energy of reflect mmWave signal diminishes rapidly so only limited number (1 or 2) reflections bear significant power. So, only a few paths arrive in concentrated directions. The mmWave diffraction is less important due to shorter radius of Fresnel zone. As a concept, multipath sparsity portrays a massive hybrid array as an attractive solution for mmWave communications. In the makeup of a hybrid array, the entire array is divided into several subarrays, and each subarray consists of antennas connected with adjustable phase shifters within the RF chain, as depicted in Figure 5.16. In addition, each subarray is connected to a BB processor by means of a DAC in the transmitter or ADC in the receiver. The signal at each antenna of the subarray is weighted by a discrete value of phase shifting from a quantized value set. These quantized sets are generated according to specified quantization bits. For example, for 4-bit quantization there will be 16 discrete values uniformly distributed within the range $[-\pi, \pi]$. The signals from the entire subarrays are interconnected and can be processed centrally in the BB processor, where the precoding/decoding and other BB processing are applied. Digital beamforming is initiated at the DACs and ADCs by weighing the signals to DACs or from the ADCs using complex values.

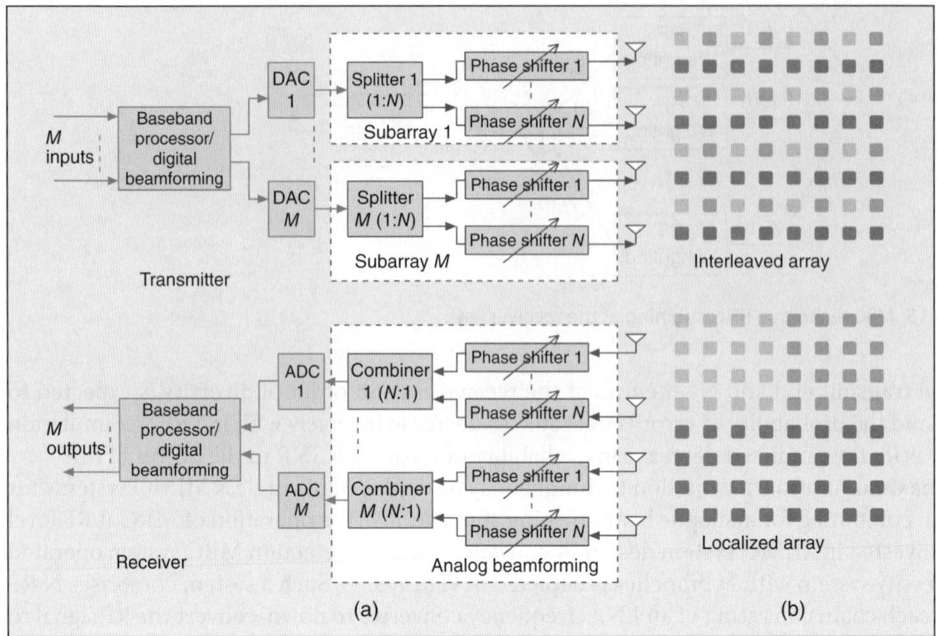

Figure 5.16 Hybrid array architecture (a) Hybrid array architecture for a transmitter and receiver; (b) two types of array configurations in hybrid uniform square arrays: interleaved (upper) and localized (bottom) configurations [45].

In contrast with MIMO system previously considered where the number of antennas used are M and N at transmitter and receiver, respectively, the hybrid antenna array comprises MN at both receiver and transmitter, which makes it complex and highly costly. Nonetheless, the distance between corresponding elements in adjacent subarrays (i.e. subarray spacing) is relatively longer compared with conventional system, which reduces the coupling between these array elements.

The hybrid array can be arranged into two uniform configurations: interleaved and localized modes [45], as illustrated in Figure 5.16 for a 16×4 uniform square array. In the localized configuration, antenna elements in each subarray are adjacent to each other, while in an interleaved configuration, antenna elements in each subarray distribute uniformly over the whole array. The two configurations have different properties and are tailored for contrasting applications. For example, interleaved configuration hardware is challenging to implement but concocts a narrow beamwidth for beamforming applications. In contrast, localized configuration is practical to implement and establishes a wide beamwidth suitable for MIMO applications.

Traditionally, analogue phase array applies integer multiples fixed value for its phase shifters regarding the signal direction. This approach is effective when signals are concentrated in one direction. However, the hybrid array phase shifting values selected are arbitrary from the quantized set to optimize the performance and allow each array to form multiple simultaneous beams.

The receive RF chain possible configurations associated with the variable phase shifter and magnitude attenuation denoted as blocks ϕ and α, respectively, are presented in Figure 5.17. The RF configuration in Figure 5.17a could be exploited when an LNA is shared between a number of antenna elements (i.e. corporate combining /splitting). In Figure 5.17b, the phase shifter is placed after the LNA to improve the receiver sensitivity. This configuration can be

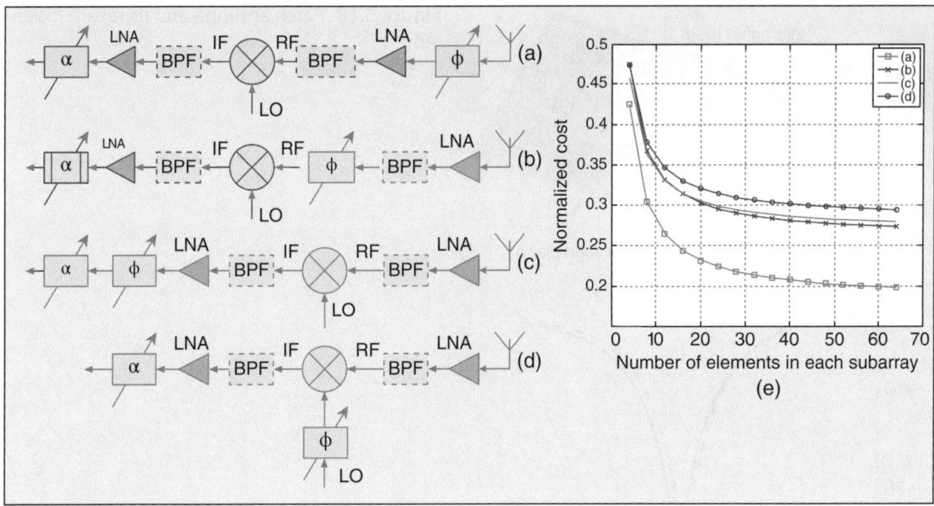

Figure 5.17 Block diagrams of the receive RF chain associated with each antenna element where the blocks ϕ and α denote variable phase shifter and magnitude attenuator, respectively [45].

implemented using either shared frequency converter or individual frequency conversion and combining at the intermediate frequency (IF). Figure 5.17c-d illustrates practical configurations to be implemented with commercially available phase shifters.

5.12 Massive MIMO Hardware

The mmWave transceivers are faced with practical hardware challenges. Components like ADCs have a tendency to be of higher power consumption and more expensive relative to microwave or lower frequencies. Consequently, conventional schemes that dedicate separate RF chain for each antenna can be expensive and exceedingly difficult to implement to mmWave transceivers. The analogue phase shifters, typically used in BF to steer the transmitted beam to the intended direction, have a number of limitations. For example, they can only change the transmitted signal phase within a finite set of quantized values. These constraints have an immense impact on the transceivers design. The design of antennas for the handheld device is one of the most critical points. The handset antennas have to be suitable for specific requirements, including operating frequency, BW, polarization, antenna size, and manufacturing costs, among others.

A patch array element with dimensions $L = 1.23$ mm and $w = 1.7$mm fabricated for use in mmWave applications is depicted in Figure 5.18. The patch antenna is etched on a dielectric substrate with $h_1 = 0.101$ mm and $\varepsilon_{r1} = 3.66$. On the opposite side of the substrate, a microstrip feed of width 0.21 mm is etched over the liquid crystal (LC) substrate of thickness $h_2 = 0.101$ mm. LC is matter that has properties of liquid (i.e. flow like liquid) and crystal, (.i.e. molecules oriented as in crystal). The LC dielectric constant changes its values with the applied bias voltage. At 0 V bias voltage, the dielectric constant is denoted as $\varepsilon_{ini} = 2.78$ and saturation bias voltage is denoted as $\varepsilon_{sat} = 3.37$.

LCs are favourable dielectric material for developing a variety of reconfigurable mmWave devices such as microstrip rectangular patch antennas, switches, and beamformers due to their large birefringence and moderate low loss. A four element patch antenna array with

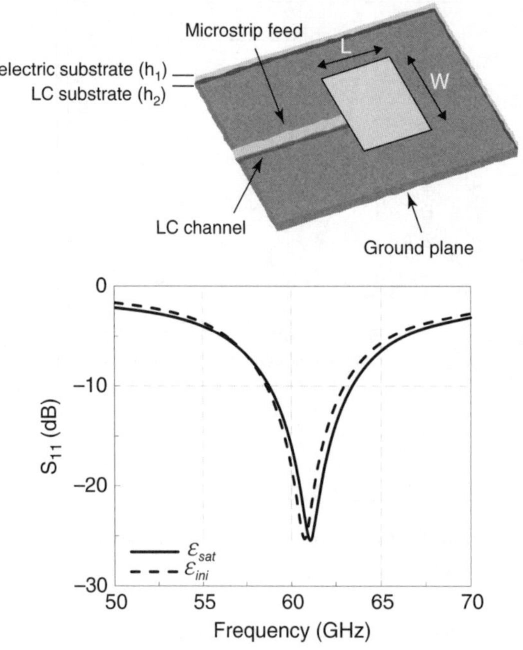

Figure 5.18 Patch antenna and its return losses [46].

beam-steering capability for 60 GHz and the phase shifters as LC-based meander lines is proposed in [46], each providing a maximum differential phase shift of 38°. The patch antenna return loss is depicted in Figure 5.18, which is about −25 dB at 60 GHz, and a four elements patch antenna array is shown in Figure 5.19.The antenna resonates around 61 GHz and exhibits a −10 dB BW of 4.3 GHz over 58.7–63 GHz. LC materials [47] can be found in a wide spectrum of applications: digital watches, calculators, cell phones, and laptop displays. Nowadays, LC attractive properties make it suitable for making devices [48] for mmWave and optics applications. Meander line mmWave LC phase shifter was proposed in [49].

Micro-electromechanical (MEM) techniques can be used to design mmWave phase shifters that have low insertion loss and good reliability, as suggested in [50]. The design technique is based on coplanar waveguide (CPW) transmission lines described earlier in [51]. The CPW transmission line is loaded periodically with the MEMS bridges, which act as shunt capacitors/varactors, as shown in Figure 5.20. A single voltage applied to the centre conductor of the CPW line yields the required phase shift. A proto-type MEMS phase shifter with width of centre conductor without/with bridge loaded is 200 and 30 μm, respectively, and 32 loaded metal bridges were built and tested at frequencies up to 60 GHz. The phase shift with bias voltages from 5 to 25 V at frequencies up to 60 GHz is shown in Figure 5.21 for centre conductor width without bridge loaded was 200 μm, and with bridge loaded 30 μm, and the maximum bridge height $h = 1.5$ μm.

Graphene is a single, tightly packed layer of carbon atoms that are bonded together in a hexagonal honeycomb lattice. It has high electrical conductivity and excellent transmittance at mmWave frequency bands. The properties of LC phase shifter with graphene films as electrodes are investigated in [48]. The proposed phase shifter can produce a maximum phase shift of 10.8° at a saturation voltage of 5 V with a 50 μm LC single cell. A schematic diagram of the grapheme/LS phase shifter is shown in Figure 5.22. Measurements of phase shifts versus bias voltage for 0.25, 0.50, 0.75, 1 THz (1 THz =1000 GHz) are shown in Figure 5.23.

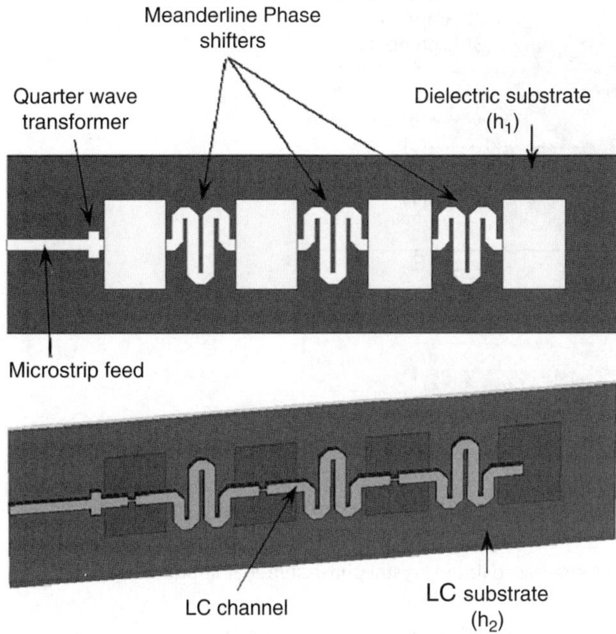

Figure 5.19 Four-element slotted patch antenna array including phase shifters [46].

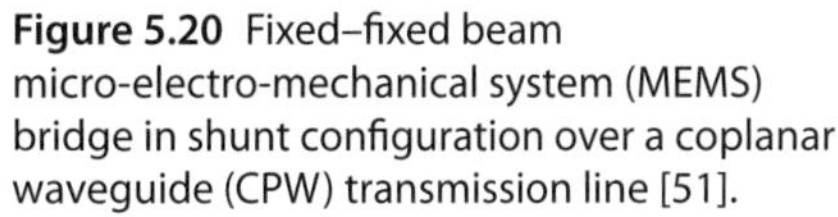

Figure 5.20 Fixed–fixed beam micro-electro-mechanical system (MEMS) bridge in shunt configuration over a coplanar waveguide (CPW) transmission line [51].

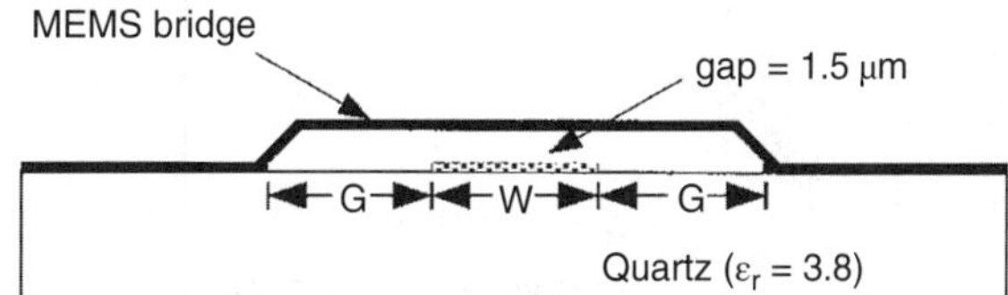

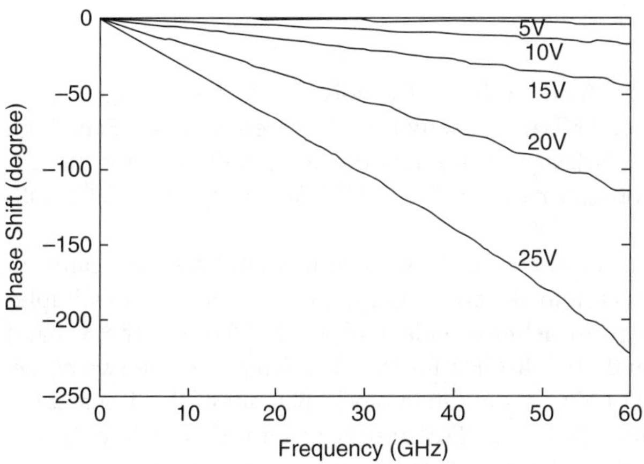

Figure 5.21 Phase shift with different control voltages for MEMS mmWave phase shifter [50].

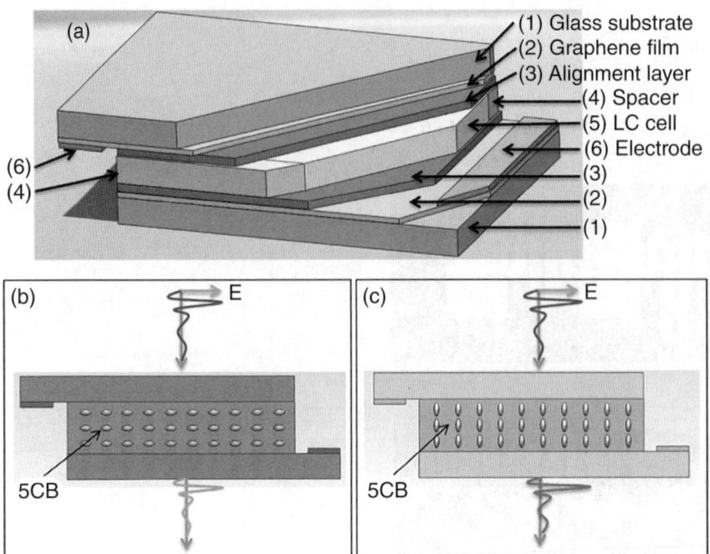

Figure 5.22 A schematic diagram of the graphene-based liquid crystal phase shifter at terahertz [48].

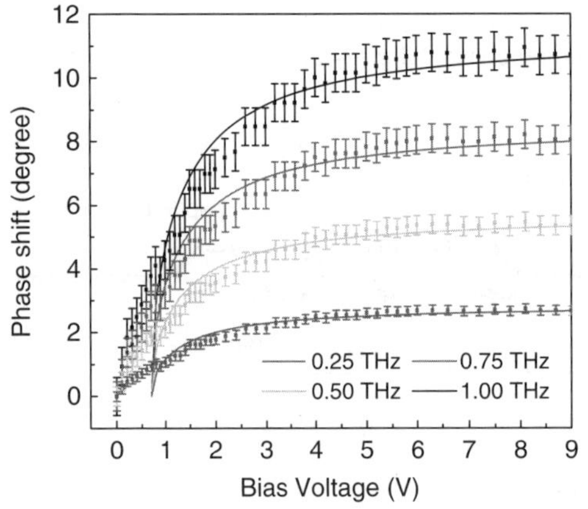

Figure 5.23 Voltage-controlled phase shifts at different frequencies [48].

LNAs are used in almost all wireless frontend receivers. The noise figure (NF) and gain of the LNA are important and determine the total NF and sensitivity of the receiver. Low NF and high gain (above 20 dB) are required in all LNA design to suppress the noise in the systems. Broadband mmWave LNAs realized in Silicon-Germanium (SiGe) [52] Bipolar CMOS (BiCMOS) technologies.

The design and performance evaluation of two broadband mmWave LNA fabricated in 0.25 µm and 0.13 µm SiGe BiCMOS technologies are presented in [53]. Both LNAs implemented in a T-type matching topology to achieve wide BW 47–77 GHz for the V-band (*V-band spectrum* 57–66 GHz) LNA and 70–140 GHz for the *W/F band (W-band spectrum* 75–110, *E-band spectrum* 90–140 GHz) LNA. A simplified block diagram of the T-matching network includes three transmission lines *TL1, TL2, TL3*, and a capacitor *C1* are described in Figure 5.24. The measured maximum gain is about 23 dB for both LNAs and the measured NF is between 7 and 7.2 dB. The measured and simulated gain versus frequency for the LNAs is shown in Figure 5.25.

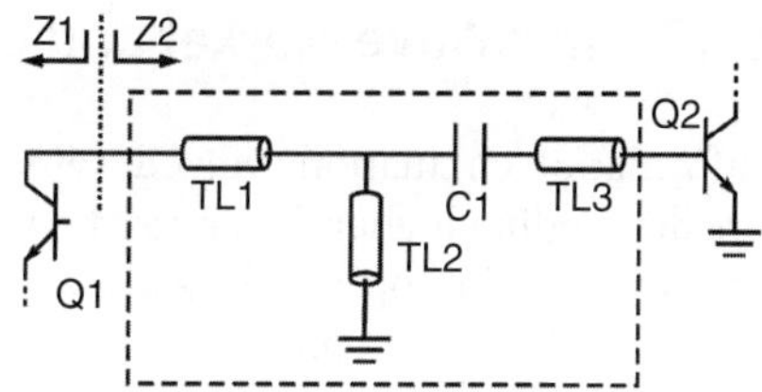

Figure 5.24 Simplified schematic of the T-type matching network [53].

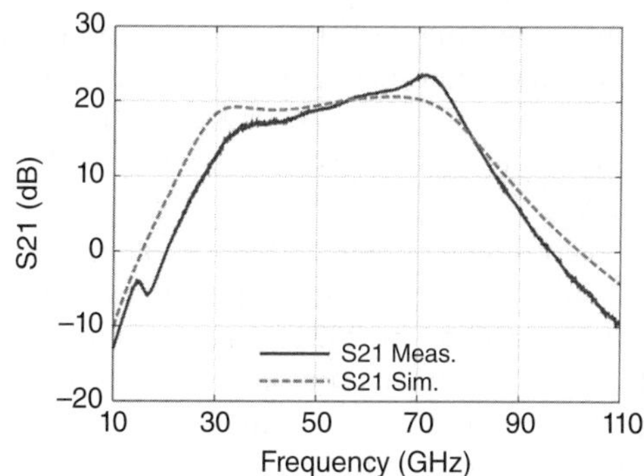

Figure 5.25 Measured and simulated S21 of the L-band gain of low noise amplifier (LNA) [53].

Power amplifiers (PAs) are used in cellular systems to boost RF power before radiation into space. However, PAs are nonlinear devices and impose other constraints on transceivers beside the tradeoff of bandwidth and transmission speed. Broadband PA at mmWave band is challenging because of the efficiency trades with the gain-bandwidth (GBW) product. The major limit is the capacitive parasitics between the driver and the power devices. A design for the interstage and output matching for larger BW and high efficiency are of main concern. A two-stage differential PA is presented in [54] and realized in 28 nm CMOS using low-power devices. The PA delivers 13 dBm saturated output over 40–67 GHz BW with peak power-added efficiency (PAE) approaching 16%. The measured output power (dBm), gain (dB), and PAE % is shown in Figure 5.26. The PAE (%) is defined as

$$PAE\ (\%) = \frac{RF\ output\ power - RF\ input\ power}{DC\ power\ used} \tag{5.122}$$

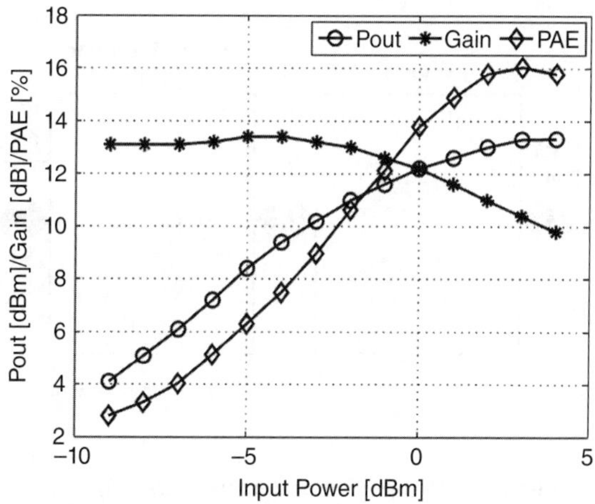

Figure 5.26 Measured output power, gain, and power-added efficiency (PAE) at 50 GHz [54].

5.13 mmWave Market and Choice of Technologies

Information circulation evolved from corporate data centres to mobile phones where end users are demanding higher data rates; hence, the demand for innovative technologies to meet these challenges. Fibre optic cables are capable of delivering scores of Tb/s (1Tb = 10^{12}b/s) but at extremely high cost compared with other available options. mmWave technology has the capability to offer higher rates at lower cost than fibre optics. The mmWave equipment market includes mmWave scanners, radars, backhauls; components into imaging, frequency and radio; applications into health care, telecommunications, and automotive applications, to name few. Key market players include NEC Corp, Alcatel-Lucent, LOEA Corp, Ericsson, Nokia Siemens Networks LLC, Huawei technology. China is expected to contribute to the mmWave equipment and devices market. North America and Europe may significantly contribute to future market growth.

According to a new market research report, published recently by Global Forecast to 2022, the global mmWave technology market is predicted to grow from \$346.8 million in 2015 to \$4632.8 million by 2022 at a compound annual rate of 42.99% between 2016 and 2022. The high growth rate in mmWave market is due to forecast increase in demand for products used in telecommunications (i.e. small cells), health care (such scanners), and defence (i.e. automotive radars). According to European Telecommunications Standards Institute (ETSI) [55], a number of semiconductor technologies that can operate at frequencies up to 90 GHz have been available since 2016. Each of the technologies has specific strengths/weakness with respect to various mmWave transmission applications. The key challenge to industry is to develop the semiconductor technology to devices and components that enables the 5G networks to achieve the expected performance, which can be translated to devices (processors, chipsets, etc.) in terms of transmit power, linearity, phase noise, noise figure, RF and BB BW, packaging, and support beam steering.

The semiconductor technology must address the demands in backhaul and fronthaul in both macrocell and small-cell scenarios. The essential components required from semiconductor technology are illustrated in Figure 5.27 and can be categorized into RF analogue, including the amplifiers (Power amplifiers, LNAs), RF filters, frequency generation and conversion, BB frontend (ADC, DAC, and PLL), Digital BB (FPGA, DSP), and packaging and assembly technologies.

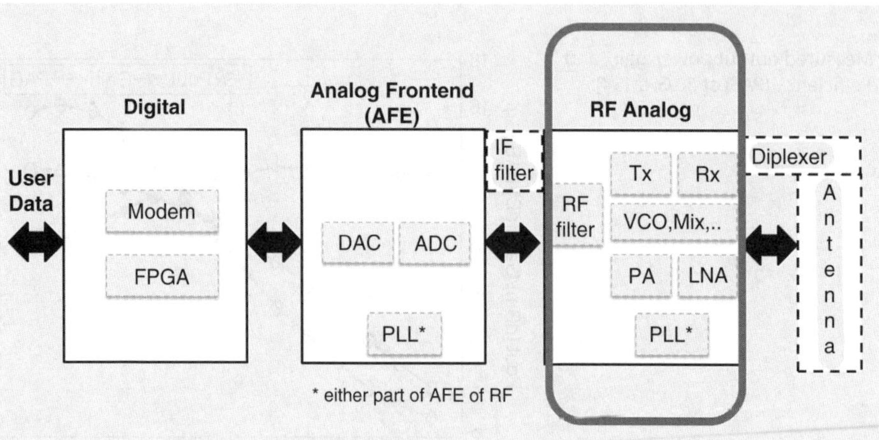

Figure 5.27 Semiconductor system overview [55].

Advances in device fabrication and continued scaling of the minimum feature size have extended the maximum operating frequency of a single CMOS transistor in excess of several hundred GHs. The commercial mmWave CMOS applications are operating within the 60 GHz standards (IEEE 802.11ad) for short-range communication with data rate of several Gb/s. Researchers have now demonstrated fully integrated mmWave CMOS LNAs and mixers, synthesizers, and transceivers. The design of mmWave receiver for use in low-power phased array massive MIMO systems is described in [56]. The receiver occupies 1.2 mm^2 and achieves an overall gain response of 20 dB (almost flat gain) across a 20 GHz BW with DSB NF of 7.8 dB.

Scaling of CMOS technologies can be used to accommodate analogue circuits in very small chip areas and to integrate analogue RF frontend, BB, and self-calibrated circuitry on a single die. This provides low-cost integrated solutions for high data rate applications for mass-market production. However, mass deployment of mmWave transceivers faces a great many challenges, such as high power consumption, large silicon area, and bandwidth limitations. A 60 GHz LNA in a 28 nm low-power bulk CMOS process and the mmWave circuit design aspects have been validated with the fabrication and characterization of the two stage amplifier shown in [57]. The LNA provides 13.8 dB flat power gain over the frequency range 55–70 GHz band with 3-dB BW of 18 GHz. The measured NF is 4 dB at 60 GHz.

Doherty power amplifiers support high data rate but silicon mmWave Doherty amplifiers exhibit limited power back-off efficiency. A single multiple mmWave bands Doherty PA around 28, 37, 39 GHz for massive MIMO applications is described in [58] with significantly enhanced power combiner. A prototype of the Doherty PA is implemented in 0.13 μm SiGe BiCMOS. Silicon-Germanium (SiGe) technology optimizes power consumption while the Bipolar CMOS (BiCMOS) offers high speed and gain, which are critical for high-frequency analogue sections. The multiple-band Doherty power amplifier achieves more than 16.8/17.1/17 dBm peak output power, respectively, over Class-B and class-A sections.

The key applications that derive mmWave technology development are the mmWave sensing, wireless communication, and fibre optics systems. A diversity of commercial silicon mmWave circuits are used in all of these applications. These circuits perform in the 30–100 GHz frequency band and are mmWave circuits expected to operate at frequencies that approach 300 GHz in the future. 5G networks are most likely to operate around 30 GHz, the connected wireless or fibre-optics backhaul will require greater than 100 Gb/s mmWave ICs. The newly developed single-chip SiGe transceiver chipsets for mmWave backhaul by Infineon enables data rates below 10 Gbps even with simple modulation schemes.

Future backhaul may need to move to 140 and 220 GHz for increased BWs to accommodate higher data rates. Two transmitter architectures that can be employed in high data rate mmWave transmitters are illustrated in Figure 5.28. A conventional linear architecture with linear up converter and linear power amplifier is shown in Figure 5.28a and fully digital RF power DAC transmitter architecture in Figure 5.28b. The critical building blocks of these transmitters include linear power amplifiers, RF power DAC, voltage-controlled LO, up-converter mixer, and linear power amplifiers. These building-block performances have to be certified in future technology nodes.

5.14 Summary

5G wireless networks will be a leap forward in mobile communications. It promises the ability to accommodate more users communicating at data speed 1000x contemporary network data rates with better reliability and less power consumption. Such a high data speed requires a lot of spectrum. Almost all present wireless communication systems operate in bands below

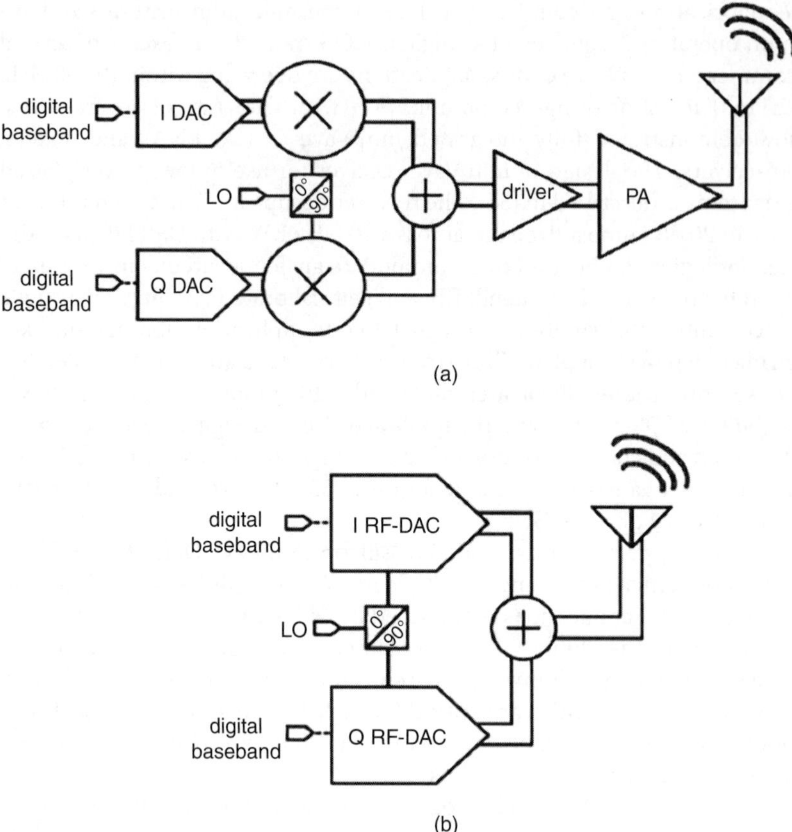

Figure 5.28 Traditional linear radio transmitter architecture with linear up-converter and linear power amplifier and fully digital RF power DAC radio transmitter architecture [59].

3 GHz and the available bands within this spectrum are becoming rapidly scarce. There are two possibilities for gaining access to more bandwidth. The first is repurposing spectrum for band below 3G, which is likely to be difficult since users are unwilling to relinquish complete control of bands already used. The other option is to share underutilized spectrum with cognitive radio technology; however, in most cases incumbents are not fully willing to cooperate. Intense interest is shown in exploring the spectrum 3–300 GHz, which is not yet exploited for commercial mobile wireless applications. This spectrum is commonly called mmWave bands with mm wavelengths. The European Union has funded several projects worldwide and particularly in the EU to carry out research on various aspects of the challenges facing the use of the new bands.

An overview of the capacity analysis for a point-to-point MIMO link system was presented in Section 5.2, followed by assessment of MIMO link outage in Section 5.3. The diversity gain is to provide the receiver with multiple independent, albeit, faded replicas of the information so the probability that all received components fade simultaneously is reduced. This is also related to the number of transmit antennas used, which influences the probability of errors. All these ideas together with the diversity-multiplexing tradeoffs were outlined in Section 5.4. The MIMO technology offers the point-to-point link promising multiplexing gains. Next we provided the analysis of multiuser MIMO single cell system in Section 5.5 and we derived the sum capacity of the UL and optimization of the sum capacity of the DL channels. We then extended the single cell representation into multiuser MIMO multi-cell systems in Section 5.6.

In-depth consideration of the BCs, including the distinctions between degraded and non-degraded BCs, the analysis of the MIMO BCs sum capacity using DPC, and the DPC scheme development and review of the DPC for massive MIMO systems was presented in Section 5.7. MAC and BC channel reciprocity is a central issue in mmWave massive MIMO system. Section 5.8 presented the reciprocity analysis for MIMO point-to-point link and multiuser, and the reciprocity with the possibility of pilot contamination. Then we considered another important issue in the massive MIMO systems, which are the MIMO BF schemes. We started with an introduction on the BF and analysed the beamforming in Section 5.9.

We examined possible types of BF schemes like delay and sum, null steering BF and the BF with a reference signal, were discussed and analysed in detail in Section 5.10.

Section 5.11 discussed mmWave BF systems, including massive hybrid antenna array. The hardware required for massive MIMO like materials for patch antenna, phase shifters, DACs, ADCs, power and LNAs and transceivers, are investigated in Section 5.12. Finally, the mmWave market and choice of technologies including key market players, predicted market size, and future growth, and industry semiconductor developments were assessed in Section 5.13.

5.A Derivation of Eq. (5.14) for M = 3, N = 2

Starting with $\mathbf{H} = \mathbf{\Phi}\lambda_v\boldsymbol{\psi}^H$ then

$$\mathbf{HH}^H = \mathbf{\Phi}\lambda_v\boldsymbol{\psi}^H\boldsymbol{\psi}\lambda_v{}^H\mathbf{\Phi}^H = \mathbf{\Phi}\lambda_v{}^2\mathbf{\Phi}^H \tag{5.A.1}$$

Assume $\mathbf{\Phi} = \begin{bmatrix} a_1 & a_2 \\ b_1 & b_2 \end{bmatrix}$ and λ_v where $v = 1, 2, \dots, \min(N, M)$. So, only two eigenvalues λ_v and $v = 1, 2$ exist.

$$\lambda_v = \begin{bmatrix} \lambda_1 & 0 \\ 0 & \lambda_2 \end{bmatrix} \text{ and } \lambda_v^2 = \begin{bmatrix} \lambda_1^2 & 0 \\ 0 & \lambda_2^2 \end{bmatrix}$$

Since $\mathbf{\Phi}$ is a unitary matrix, then

$$\mathbf{\Phi}^{-1} = \mathbf{\Phi}^H = \frac{1}{D}\begin{pmatrix} b_2 & -a_2 \\ -b_1 & a_1 \end{pmatrix} \tag{5.A.2}$$

where $D = a_1b_2 - a_2b_1$

Therefore, inserting $\mathbf{\Phi}, \mathbf{\Phi}^H, \lambda_v^2$ expressions in (5.A.1) we get

$$\mathbf{HH}^H = \frac{1}{D}\begin{bmatrix} a_1b_2\lambda_1{}^2 - a_2b_1\lambda_2{}^2 & -a_1a_2\lambda_1{}^2 + a_1a_2\lambda_2{}^2 \\ b_1b_2\lambda_1{}^2 - b_1b_2\lambda_2{}^2 & -b_1a_2\lambda_1{}^2 + a_1b_2\lambda_2{}^2 \end{bmatrix} \tag{5.A.3}$$

$$\mathbf{I}_2 + \frac{\rho}{\mathrm{M}}\mathbf{HH}^H = \begin{bmatrix} 1 + \frac{\rho}{\mathrm{M}}\frac{1}{D}(a_1b_2\lambda_1{}^2 - a_2b_1\lambda_2{}^2) & \frac{\rho}{\mathrm{M}}\frac{1}{D}(-a_1a_2\lambda_1{}^2 + a_1a_2\lambda_2{}^2) \\ \frac{\rho}{\mathrm{M}}\frac{1}{D}(b_1b_2\lambda_1{}^2 - b_1b_2\lambda_2{}^2) & 1 + \frac{\rho}{\mathrm{M}}\frac{1}{D}(-b_1a_2\lambda_1{}^2 + a_1b_2\lambda_2{}^2) \end{bmatrix} \tag{5.A.4}$$

It can easily be shown that the determinant of (5.A.4) can be simplified to

$$\det\left(\mathbf{I}_2 + \frac{\rho}{\mathrm{M}}\mathbf{HH}^H\right) = 1 + \frac{\rho}{\mathrm{M}}(\lambda_1{}^2) + \frac{\rho}{\mathrm{M}}(\lambda_2{}^2) + \left(\frac{\rho}{\mathrm{M}}\right)^2[\lambda_1{}^2\lambda_2{}^2] \tag{5.A.5}$$

We rewrite (5.A.5) to a format that is mathematically convenient to express in logarithm

$$\det\left(\mathbf{I}_2 + \frac{\rho}{\mathrm{M}}\mathbf{HH}^H\right) = \left[1 + \frac{\rho}{\mathrm{M}}(\lambda_1{}^2)\right]\left[1 + \frac{\rho}{\mathrm{M}}(\lambda_2{}^2)\right] \tag{5.A.6}$$

Taking the log of both sides of (5.A.6) we get

$$\log_2 \det\left(\mathbf{I}_2 + \frac{\rho}{\mathrm{M}}\mathbf{H}\mathbf{H}^H\right) = \log_2\left[1 + \frac{\rho}{\mathrm{M}}(\lambda_1{}^2)\right] + \log_2\left[1 + \frac{\rho}{\mathrm{M}}(\lambda_2{}^2)\right] \tag{5.A.7}$$

We express (5.A.7) as the sum of the terms as follows:

$$\log_2 \det\left(\mathbf{I}_2 + \frac{\rho}{\mathrm{M}}\mathbf{H}\mathbf{H}^H\right) = \sum_{\nu=1}^{\min(\mathrm{M},\mathrm{N})} \log_2\left[1 + \frac{\rho}{\mathrm{M}}\lambda_\nu{}^2\right] \tag{5.A.8}$$

5.B MUSIC Algorithm Used in Estimating the Direction of Signal Arrival

5.B.1 Introduction

Consider the scenario where multiple signal wavefronts impact on antennas in arbitrary locations and with arbitrary gain/phase/polarization in a noise/interference environment. It is required to find important information about these signals. Such a scenario could be found in sensor networks, commercial mobile wireless networks, military communications, and general signal direction finding. The problem is to determine characteristics of the wavefronts impinging on the array with such number signals; direction of arrival; level of noise and interference; polarization, etc. In this appendix, we introduce algorithms that can be used to determine most of these parameters, but we will focus on estimating the direction of arrival. Once the AOAs are estimated, the transceiver knows at which angles it should beamform. The AOAs estimates are required for the receive beamforming as well as for the transmit beamforming, assuming the reciprocity can be implemented.

There are three conventional categories of techniques to estimate directional information; specifically, spectral-based technique; subspace-based estimation method; and deterministic parametric estimation. The spectral-base category is further developed into Bartlett beamforming, and the subspace-based methods comprise the MUSIC, an algorithm proposed by Ralph Schmidt [40] in 1986. MUSIC estimates the AOA by searching for angles corresponding to multiple local maxima and provides statistically compatible estimates in contrast to the Bartlett method. The ESPRIT, exploits the invariant property of two displacement arrays and computes the AoA directly. So ESPRIT provides the estimate with less complexity compared to MUSIC. However, the accuracy of ESPRIT estimates are worse than that of MUSIC. In this appendix, we explore AOA estimation in hybrid antenna arrays using MUSIC algorithm. Nonetheless, MUSIC algorithm is designed for digital arrays and cannot be used in hybrid arrays directly.

5.B.2 MUSIC Algorithm for Estimating 1D Array AOAs

The main goal of MUSIC algorithm is to determine a *function* that when plotted versus angle gives an estimate of AOA at its maxima. This *function* is known as *pseudo-spectrum* $P_{\mathrm{MUSIC}}(\theta)$ [60, 61]. Consider a 1D linear array for receive beamforming in Figure 5.B.1 where each array element is fitted with an ADC. By tweaking the weight factors after the ADC we can steer the receive direction for the received signals. For analogue (RF) beamforming, each antenna is provided with an analogue phase-shifter and the beam direction can also be guided by fine-tuning the phase-shifter of each antenna.

Let the 1D array comprising N antennas and L signals arrive at the array from different L angles corresponding to the L paths of a transmitted signal $\mathbf{s}(k)$ where θ_l is the AoA of the lth path signal.

Figure 5.B.1 Uniform linear (1D) array antenna for receive beamforming [60].

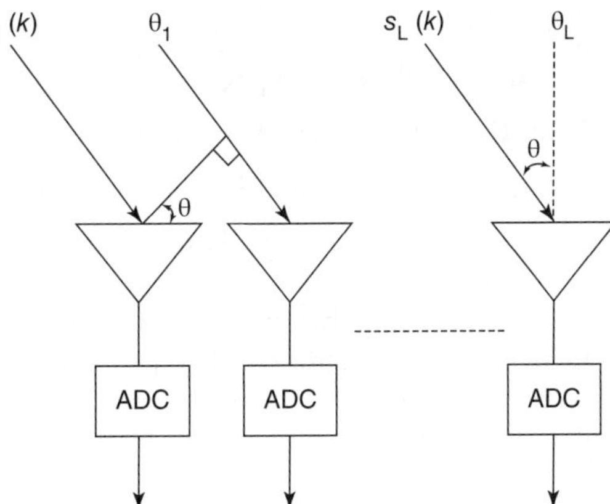

We assume L < N. Furthermore, at each instant of time (i.e. k) we have L incident complex signals (amplitude and phase) represented by a vector $\mathbf{s}(k)$

$$\mathbf{s}(k) = [\mathrm{s}_1(k), \dots\dots\dots\dots, \mathrm{s}_L(k)]^T \tag{5.B.1}$$

Directions of each path signal are defined by a steering vector $\mathbf{a}(\theta)$. So, the steering vector for the lth path signal $\mathbf{a}(\theta_l)$ is given as

$$\mathbf{a}(\theta_l) = [1 \quad \exp(\mathrm{j}\beta \mathrm{d}\sin\theta_l) \quad \exp(\mathrm{j}2\beta \mathrm{d}\sin\theta_l) \dots\dots \exp(j\beta d(\mathrm{N}-1)\sin\theta_l)]^H \tag{5.B.2}$$

where $l = 1, 2, \dots \dots \dots \dots, \mathrm{L}$, $\beta = \frac{2\pi}{\lambda}$ is wave number measured in radian per unit distance, and d is space between adjacent array elements. Define matrix $\mathbf{A} \in \mathbb{C}^{\mathrm{N\,X\,L}}$ as

$$\mathbf{A}(\theta) = [\mathbf{a}(\theta_1) \dots\dots\dots \mathbf{a}(\theta_l) \dots\dots\dots \mathbf{a}(\theta_L)] \tag{5.B.3}$$

Denote the signal at n^{th} receive antenna as $\mathrm{x}_n(k)$ where $n = 1, 2, \dots \dots, \mathrm{N}$, and $\mathbf{x}(k)$ as the signal vector received at the time instant k:

$$\mathbf{x}(k) = [\mathrm{x}_1(k), \dots\dots\dots\dots\dots, \mathrm{x}_N(k)]^T \tag{5.B.4}$$

Denote the channel noise $\mathbf{n}(k)$ expressed as

$$\mathbf{n}(k) = [\mathrm{n}_1(\mathrm{k}), \dots\dots\dots, \mathrm{n}_n(\mathrm{k}), \dots\dots\dots, \mathrm{n}_N(k)]^T \tag{5.B.5}$$

where $\mathrm{n}_n(\mathrm{k})$ is the noise observed in the nth receive antenna.

$$\mathbf{x}(k) = \mathbf{A}\,\mathbf{s}(k) + \mathbf{n}(k) \tag{5.B.6}$$

The correlation matrix of the received vector $\mathbf{x}(k)$ is

$$\mathbf{R_{xx}} = \mathbb{E}[\mathbf{x}.\mathbf{x}^H] = \mathbb{E}[(\mathbf{A}\,s(k) + \mathbf{n}(k))((s(k))^H \ \mathbf{A}^H + \mathbf{n}(k)^H)] \tag{5.B.7}$$

$$\mathbf{R_{xx}} = \mathbf{A}\,\mathbb{E}[s(k)s(k)^H]\mathbf{A}^H + \mathbb{E}[\mathbf{n}\mathbf{n}^H] \tag{5.B.8}$$

where we assumed that noise and arrived signals are uncorrelated. We can simplify (5.B.8) to

$$\mathbf{R_{xx}} = \mathbf{A}\,\mathbf{R_{ss}}\,\mathbf{A}^H + \mathbf{R_{nn}} \tag{5.B.9}$$

where

$$\mathbf{R_{ss}} = \mathbb{E}[s(k)s(k)^H] = \mathrm{diag}[\sigma_1^2, \ \dots\dots\dots, \sigma_L^2] \tag{5.B.10}$$

We comprehended $\mathbf{R_{xx}}$ is N × N matrix, $\mathbf{R}_{ss}$ is L × L arriving signals correlation matrix and $\mathbf{R_{nn}}$ is N × N noise matrix. Since we assumed AWG noise with zero mean and common variance $\sigma_{\mathbf{n}}{}^2$ so that $\mathbf{R_{nn}} = \sigma_{\mathbf{n}}{}^2\mathbf{I}_N$ and $\mathbf{I}_N$ is N × N identity matrix.

Accordingly,

$$\mathbf{R_{xx}} = \mathbf{A\,R_{ss}\,A}^H + \sigma_{\mathbf{n}}{}^2\mathbf{I}_N \tag{5.B.11}$$

Since $L < N$, the matrix $\mathbf{A\,R_{ss}\,A}^H$ is singular, i.e. its determinant is zero:

$$\det[\mathbf{A\,R_{ss}\,A}^H] = \det[\mathbf{R_{xx}} - \sigma_{\mathbf{n}}^2\ \mathbf{I}_N] = 0 \tag{5.B.12}$$

Clearly, if the arriving signals are uncorrelated, then $\mathbf{R_{ss}}$ has to be diagonal matrix, but if they are correlated in part, then $\mathbf{R_{ss}}$ is regular matrix i.e. its determinant is nonzero.

It is worth noting that over a given time window, we have taken K samples from each signal path over the time window. If the signals and noise are not defined, but the process is known (or assumed) to be ergodic, then we can approximate the statistical correlation by time-averaged correlation so that

$$\mathbf{R_{xx}} \approx \frac{1}{K}\sum_{k=1}^{K} \mathbf{x}(k)\,\mathbf{x}(k)^H, \tag{5.B.13}$$

$$\mathbf{R_{nn}} \approx \frac{1}{K}\sum_{k=1}^{K} \mathbf{n}(k)\,\mathbf{n}(k)^H, \tag{5.B.14}$$

$$\mathbf{R_{ss}} \approx \frac{1}{K}\sum_{k=1}^{K} \mathbf{s}(k)\,\mathbf{s}(k)^H \tag{5.B.15}$$

The implication of (5.B.12) is that $\sigma_{\mathbf{n}}^2$ is an eigenvalue of $\mathbf{R_{xx}}$. The N – dimensional space of $\mathbf{R_{xx}}$ is partitioned into L – dimensional signal subspace and (N – L) – dimensional noise subspace (null subspace). Let us carry out SVD of $\mathbf{R_{xx}}$ we get

$$\mathbf{R_{xx}} = [\mathbf{U_s}\ \ \mathbf{U_n}]\begin{bmatrix}\mathbf{\Sigma_s} & 0\\ 0 & \mathbf{\Sigma_n}\end{bmatrix}\begin{bmatrix}\mathbf{U_s^H}\\ \mathbf{U_n^H}\end{bmatrix} \tag{5.B.16}$$

where $\left[\mathbf{U_s}\ \ \mathbf{U_n}\right]$ is the $N\times N$ matrix containing the signal and noise singular vectors, the columns of the sub-matrix $\mathbf{U_s} \in \mathbb{C}^{N\times L}$ span the signal subspace, and $\mathbf{U_n} \in \mathbb{C}^{N\times(N-L)}$ correspond to its orthogonal (noise) subspace, $\mathbf{\Sigma_s} \in \mathbb{R}^{L\times L}$ and $\mathbf{\Sigma_n} \in \mathbb{R}^{(N-L)\times(N-L)}$ are diagonal matrices containing the eigenvalues. Since both $\mathbf{R_{xx}}$ and $\mathbf{A\,R_{ss}\,A}^H$ are non-negative definite, there are additional L eigenvalues $\lambda_n^2 > \sigma_{\mathbf{n}}^2 > 0$.

Denote $\mathbf{u}_n$ be the nth eigenvector of $\mathbf{R_{xx}}$ corresponding to λ_i^2 so

$$\mathbf{R_{xx}}\,\mathbf{u}_n = [\mathbf{A\,R_{ss}\,A}^H + \sigma_{\mathbf{n}}^2\ \mathbf{I}_N]\,\mathbf{u}_n = \lambda_n^2\ \mathbf{u}_n \quad \text{for } n = 1, \ldots\ldots, N \tag{5.B.17}$$

So, we get

$$\mathbf{A\,R_{ss}\,A}^H\,\mathbf{u}_n = (\lambda_n^2 - \sigma_{\mathbf{n}}^2)\mathbf{u}_n \quad \text{for } n = 1, \ldots\ldots, N \tag{5.B.18}$$

$$\mathbf{A\,R_{ss}\,A}^H\,\mathbf{u}_n = \begin{Bmatrix}(\lambda_n^2 - \sigma_{\mathbf{n}}^2)\mathbf{u}_n & \text{for } n = 1, \ldots\ldots, L\\ 0 & \text{for } n = L+1, \ldots\ldots, N\end{Bmatrix} \tag{5.B.19}$$

Accordingly, the N – dimensional space vector space of $\mathbf{R_{xx}}$ can be partitioned into the signal subspace $\mathbf{U_s}$ and the noise subspace $\mathbf{U_n}$

$$\left[\mathbf{U_s}\ \ \mathbf{U_n}\right] = \Bigg[\underbrace{\mathbf{u}_1 \ldots\ldots \mathbf{u}_L}_{\mathbf{U_s}\text{ eigenvalues}}\quad \underbrace{\mathbf{u}_{L+1} \ldots, \mathbf{u}_N}_{\mathbf{U_n}\text{ eigenvalues}}\Bigg]$$

The array steering vector $\mathbf{A}(\theta)$ is in the signal subspace and orthogonal the noise subspace eigenvector $\mathbf{U_n}$, which implies the following:

$$\mathbf{a}^H(\theta)\,\mathbf{U_n} = 0 \tag{5.B.20}$$

The Euclidean distance squared, d^2_{EUC} is given as

$$d^2_{EUC} = \mathbf{a}^H(\theta)\,\mathbf{U_n}\mathbf{U_n}^H\,\mathbf{a}(\theta) \approx 0 \tag{5.B.21}$$

If we express MUSIC algorithm, we find the AOAs as

$$\mathrm{Peak}^{\mathrm{L}}_{(\theta)}\ \frac{1}{\mathbf{a}^H(\theta)\,\mathbf{U_n}\mathbf{U_n}^H\,\mathbf{a}(\theta)} \tag{5.B.22}$$

where $\mathrm{Peak}^{\mathrm{L}}_{(\theta)}(.)$ produces L largest peaks of the function corresponding to L (θs) and the maximum number of detectable AOAs is N – 1 since $\mathrm{U_n}$ must contain at least one column.

5.B.3 MUSIC Algorithm for Estimating 1D Linear Hybrid Array AOAs [60]

Figure 5.B.2 illustrates the formation of a 1D linear hybrid antenna array for receive beamforming. We assume that the array elements have perfect omnidirectional radiation patterns. The array comprises multiple subarrays and each subarray shares the same ADC. Both the phase shifters and the digital weights can be used to drive the beam direction of the array. The digital weight can adjust both the amplitude and phase of the signal while the phase shifter can only adjust the signal phase.

The operation of the hybrid array is initiated when a signal is received at an antenna, phase shifted and summed with other phase-shifted signals in a given subarray. The beamed signal is sampled by an ADC and multiplied by a digital weight. Lastly, the multiplied signals are summed to give the output. To summarize, we first carry out analogue beamforming (phase shifting), and the outcome signal is then sampled to obtain digital beamforming.

Denote the number of subarrays (i.e. ADC) as $\mathrm{N_{ADC}}$ and the number of antennas in each subarray as N_s, then the total number of antennas used N is $\mathrm{N_{ADC}}\,\mathrm{N}_s$. Collecting all the received vectors from the subarrays into one vector to get an overall received vector and applying the procedure used in the conventional 1D array, the AOAs of hybrid array using MUSIC algorithm are given as

$$\mathrm{Peak}^{\mathrm{L}}_{(\theta)}\ \frac{1}{\bar{\mathbf{a}}^H(\theta)\,\mathbf{U_n}\mathbf{U_n}^H\,\bar{\mathbf{a}}(\theta)} \tag{5.B.23}$$

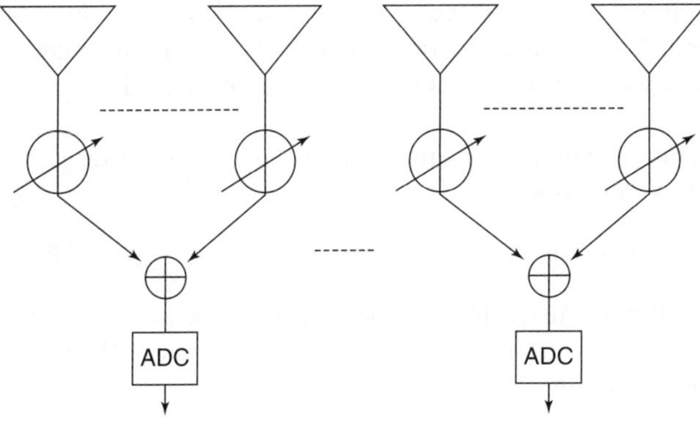

Figure 5.B.2 1D hybrid antenna array [60].

where

$$\bar{\mathbf{a}}(\theta) = \begin{bmatrix} \mathbf{B}_1^H\,\mathbf{a}(\theta) \\ \mathbf{B}_2^H\,\mathbf{a}(\theta)\exp(-j\beta \mathrm{N_s d}\sin\theta) \\ \ldots. \\ \ldots. \\ \mathbf{B}_{\mathrm{N_{ADC}}}^H\,\mathbf{a}(\theta)\exp(-j\beta \mathrm{N_s d(N-1)}\sin\theta) \end{bmatrix} \tag{5.B.24}$$

where $\mathbf{B}_{n_{\mathrm{ADC}}} = \left[\mathbf{b}_{n_{\mathrm{ADC}},1}\ \ \mathbf{b}_{n_{\mathrm{ADC}},2}\ \ \ \ \mathbf{b}_{n_{\mathrm{ADC}},K}\right]$, K is the number of successive observations to gather K beamed signal samples and the analogue beamforming matrix $\mathbf{B} \in \mathbb{C}^{\mathrm{N}\times\mathrm{N_{ADC}}}$ is given as

$$\mathbf{B}^H = \begin{bmatrix} \mathbf{b}_1^H & \mathbf{0} & & & & & & \mathbf{0} \\ \mathbf{0} & \mathbf{b}_2^H & & & & & & \mathbf{0} \\ & & \cdots & & & & & \\ & & & \cdots & & & & \\ & & & & \mathbf{b}_{n_{\mathrm{ADC}}}^H & & & \\ & & & & & \cdots & & \\ & & & & & & \cdots & \\ & & & & & & & \cdots \\ & \mathbf{0} & & & & & & \mathbf{b}_{\mathrm{N_{ADC}}}^H \end{bmatrix} \tag{5.B.25}$$

where $\mathbf{b}_{n_{\mathrm{ADC}}} = \left[\mathrm{b}_{n_{\mathrm{ADC}},1}\ \ \mathrm{b}_{n_{\mathrm{ADC}},2}\ \ \ldots\ \ldots\ \ldots\ \ldots\ \mathbf{b}_{n_{\mathrm{ADC}},\mathrm{N_s}}\right]^T$ is a beamformer vector formed by phase-shifter n_{ADC} and $\mathbf{a}(\theta)$ is given in (5.B.2). It can easily be seen that the estimation of the AOAs in the hybrid array is equivalent to the conventional AOAs estimation in the digital array (with the same number of antennas and the same beamforming matrix). More details can be found in [60].

5.B.4 MUSIC Algorithm for Estimating 2D Array AOAs.

Consider a narrowband wireless communication scenario where the signal at a receive array contributed by the lth multipath is expressed as follows [62, 63]:

$$\mathbf{x}_l(t) = \mathbf{c}(u_l, v_l)\,\mathbf{s}_l(t) + \mathbf{n}(t) \tag{5.B.26}$$

where $\mathbf{c}(u_l, v_l)$, $\mathbf{s}_l(t)$, $\mathbf{n}(t)$ are the steering vector, transmitted signal, and complex white Gaussian noise, respectively. Let a horizontal uniform rectangular array comprise N elements arranged in the $\mathrm{x-y}$ plane as $\mathrm{N = N_1 \times N_2}$. The antenna index can be composed as (n_1, n_2) with $1 \le n_1 \le \mathrm{N_1}$ and $1 \le n_2 \le \mathrm{N_2}$.

The response of the $(n_1, n_2)^{th}$ antenna element to an impinging wave on any of the array elements from a certain direction can be expressed as

$$\mathbf{c}_{n_1,n_2}(u_l, v_l) = \exp\{-\mathrm{j}(n_1 - 1)u_l + (n_2 - 1)v_l\} \tag{5.B.27}$$

The phase shifts caused by antenna displacements in 3D spherical coordinates are denoted by $(radius,\ \phi_l, \theta_l)$ where ϕ_l and θ_l are the elevation and the azimuth angles, respectively. We find the point at x and y on a horizontal plane as

$$\mathrm{x} = d_{\mathrm{x}}\ \cos\theta_l \sin\phi_l \tag{5.B.28a}$$

$$\mathrm{y} = d_{\mathrm{y}}\ \sin\theta_l \sin\phi_l \tag{5.B.28b}$$

The wave number is defined as the number of radians per unit distance i.e. $2\pi\lambda^{-1}$. The phase shifts induced by antenna at (x , y) directions u_l and v_l are given by

$$u_l = 2\pi\lambda^{-1}\, d_x\, \cos\theta_l \sin\phi_l \tag{5.B.29a}$$

$$v_l = 2\pi\lambda^{-1}\, d_y\, \sin\theta_l \sin\phi_l \tag{5.B.29b}$$

where, λ, θ_l, and ϕ_l denote the wavelength, azimuth AOA, and elevation AOA, respectively. The equidistance between two adjacent array elements are denoted as d_x and d_y in x and y directions, respectively. The equidistance is mostly assumed less/equal $\frac{\lambda}{2}$. Furthermore, we can articulate, $\mathbf{c}(u_l, v_l))$ as

$$\mathbf{c}(u_l, v_l) = \mathbf{c}(u_l) \otimes \mathbf{c}(v_l), 1 \le l \le L \tag{5.B.30}$$

where

$$\mathbf{c}(u_l) = [1, e^{-ju_l} \;, \dots\dots\dots\dots\dots\dots, e^{-j(N_1-1)u_l}\;]^T \tag{5.B.31}$$

$$\mathbf{c}(v_l) = [1, e^{-jv_l} \;, \dots\dots\dots\dots\dots\dots, e^{-j(N_{2-1})v_l}\;]^T \tag{5.B.32}$$

with $\otimes$ denoting the Kronecker product. The Kronecker product $\otimes$ causes the steering vectors to merge into a single long $(N_1 \times N_2) \times 1$ column vector. Additionally, each signal at the output of the receive array is sampled into K-points:

$$\mathbf{x}_l(k) = \mathbf{c}(u_l, v_l)\, s_l(k) + \mathbf{n}_l(k), 1 \le k \le K \tag{5.B.33}$$

Adding up the output signals, the total signal at the receiver is given by summing up the L paths to get

$$\mathbf{Y}(k) = \sum_{l=1}^{L} \mathbf{x}_l(k) = \mathbf{C}\,\mathbf{S}(k) + \mathbf{N}(k) \tag{5.B.34}$$

where $\mathbf{C} = [\mathbf{c}(u_1, v_1), \dots \dots \dots \dots, \mathbf{c}(u_L, v_L)]$, $\mathbf{S}(k) = [s_1(k), \dots \dots \dots, s_L(k)]^T$, and $\mathbf{N}(k) = [\mathbf{n}_l(k), \dots \dots \dots \dots, \mathbf{n}_L(k)]^T$.

The MUSIC algorithm is based on the spatial covariance matrix to estimate the AOAs of multipoint covariance given by

$$\mathbf{R} = \frac{1}{K} \sum_{k=1}^{K} \mathbf{Y}(k)\, \mathbf{Y}^H(k) \tag{5.B.35}$$

$$\mathbf{R} = \mathbb{E}[(\mathbf{C}\,\mathbf{S}(k) + \mathbf{N}(k))(\mathbf{C}\,\mathbf{S}(k) + \mathbf{N}(k))^H] \tag{5.B.36}$$

$$\mathbf{R} = \mathbb{E}[(\mathbf{C}\,\mathbf{S}(k) + \mathbf{N}(k))(\mathbf{S}^H(k)\mathbf{C}^H + \mathbf{N}^H(k))]$$

$$\mathbf{R} = \mathbf{C}\,\mathbb{E}[\mathbf{S}(k)\mathbf{S}^H(k)]\mathbf{C}^H + \mathbb{E}[\mathbf{N}(k)\mathbf{N}^H(k)] \tag{5.B.37}$$

$$\mathbf{R} = \mathbf{C}\mathbf{P}\,\mathbf{C}^H + \mathbb{E}[\mathbf{N}(k)\mathbf{N}^H(k)]$$

$$\mathbf{R} = \mathbf{C}\mathbf{P}\,\mathbf{C}^H + \sigma^2\, \mathbf{I}_{K \times L} \tag{5.B.38}$$

Assume the receiver noise is white with mean zero and variance σ^2, signals and noise are assumed to be uncorrelated and **P** is the covariance of the impinging signals and **P** is the signal variance given by

$$\mathbf{P} = \mathbb{E}[\mathbf{S}(k)\mathbf{S}^H(k)] \tag{5.B.39}$$

where $\mathbf{P} = \mathrm{diag}(p_1, \dots \dots \dots \dots, p_L)$ and rank of **P** defines the dimension of the signal subspace i.e. L. The elements of $\mathbf{S}(k)$ may be uncorrelated and in such case **P** is $L \times L$ matrix (square matrix) with full rank(**P**) = L and the determinant of **P** is nonzero. It is also likely in most applications that **S**(k) contains some correlated (or linearly dependent) elements. In general, **P** is a positive definite. When $L < N$ then **P** and hence $\mathbf{CP\,C}^H$ are singular. So, determinant of $\mathbf{CP\,C}^H$ is zero. i.e.

$$\det(\mathbf{CP\,C}^H) = \det(\mathbf{R} - \sigma^2\,\mathbf{I}_{K\times L}) = 0 \tag{5.B.40}$$

Equation (5.40) entails that σ^2 is an eigenvalue of **R** and there are $N - L$ eigenvalues like σ^2. We partition the N – dimensional space into the signal subspace $\mathbf{U_S}$ and the noise subspace $\mathbf{U_N}$

$$[\mathbf{U_S}\;\;\mathbf{U_N}\,] = \underbrace{\mathbf{u}_1 \ldots\ldots \mathbf{u}_L}_{\mathrm{U_S\ eigen\ values}}\quad \underbrace{\mathbf{u}_{L+1} \ldots\ldots.. \mathbf{u}_N}_{\mathrm{U_N\ eigen\ values}}$$

The eigenvalues decomposition is separated into two parts:

$$\mathbf{R} = \mathbf{U_S}\,\mathbf{\Lambda_S}\,\mathbf{U_S^H} + \mathbf{U_N}\,\mathbf{\Lambda_N}\,\mathbf{U_N^H} \tag{5.B.41}$$

where $\mathbf{U_S}$ and $\mathbf{U_N}$ eigenvectors correspond to the L large eigenvalues forming the signal subspace given by the diagonal matrix $\mathbf{\Lambda}_s$ and $N - L$ smaller eigenvalues from the noise subspace given by the diagonal matrix $\mathbf{\Lambda_N}$. So, $\mathbf{\Lambda}_s$ and $\mathbf{\Lambda}_n$ are given by

$$\mathbf{\Lambda_S} = \mathrm{diag}[\lambda_1^s, \ldots\ldots\ldots., \lambda_L^s\,] \tag{5.B.42a}$$

$$\mathbf{\Lambda_N} = \mathrm{diag}[\lambda_{L+1}^n, \ldots\ldots\ldots., \lambda_N^n\,] \tag{5.B.42b}$$

Since the noise subspace is theoretically orthogonal to the signal subspace, as we explained above, we must have

$$\mathbf{U_N^H}\,\mathbf{C} = 0 \tag{5.B.43}$$

The Euclidean distance square between $\mathbf{U_N^H}$ and **C** d_{Euc}^2 is given by

$$d_{\mathrm{Euc}}^2 = \mathbf{C}^H\,\mathbf{U_N}\mathbf{U_N^H}\,\mathbf{C} \tag{5.B.44}$$

The MUSIC spatial spectrum $\boldsymbol{P}_{\mathrm{MUSIC}}$ is the inverse of the Euclidean distance given by

$$\mathrm{Peak}_{(\theta,\phi)}^{L}\,\frac{\mathbf{C}^H\,\mathbf{C}}{\mathbf{C}^H\,\mathbf{E_N}\,\mathbf{E_N}\,\mathbf{C}} \tag{5.B.45}$$

Azimuth AOA (θ) and elevation AOA (ϕ) can be found by searching the angles that correspond to the spatial peaks in $\mathrm{Peak}_{(\theta,\phi)}^{L}$ (.).

References

1 Foschini, G.J. (1996). Layered space-time architecture for wireless communication in fading environment when using multi-element antennas. *Bell Labs Technical Journal* 1 (2): 41–59.
2 Foschini, G.J. and Gans, M.J. (1998). On limits of wireless communications in a fading environment when using multiple antennas. *Wireless Personal Communications* 6: 311–335.
3 Alamouti, S.M. (1998). A simple transmit diversity technique for wireless communications. *IEEE Journal on Selected Areas in Communications* 16 (8): 1451–1458.
4 Marzetta, T. (2010). Noncooperative cellular wireless with unlimited numbers of base station antennas. *IEEE Transactions on Wireless Communications* 9 (11): 3590–3600.
5 Rusk, F., Persson, D., Lau, B.K. et al. (2013). Scaling up MIMO: opportunities and challenges with very large arrays. *IEEE Signal Processing Magazine* 30 (1): 40–60.
6 Telatar, E. (1999). Capacity of multi-antenna Gaussian channels. *European Transactions on Telecommunications* 10 (6): 585–595.

7 Loka, S. and Kouki, A. (2001) The impact of correlation on multi-antenna system performance: correlation matrix approach, IEEE Vehicular Technology Conference Proceedings, 2, 533–537.

8 Zheng, L. and Tse, D.N.C. (2003). Diversity and multiplexing: a fundamental tradeoff in multiple-antenna channels. *IEEE Transactions on Information Theory* 49 (5): 1073–1096.

9 Hassibi, B. and Sharif, M. (2007). Fundamental limits in MIMO broadcast channels. *IEEE Journal on Selected Areas in Communications* 25 (7): 1333–1344.

10 Lozano, A. and Jindal, N. (2010). Transmit diversity vs. spatial multiplexing in modern MIMO systems. *IEEE Transactions on Wireless Communications* 9 (1): 186–197.

11 Prasad, N. and Varanasi, M.K. (2006). Outage theorems for MIMO block-fading channels. *IEEE Transactions on Information Theory* 52 (12): 5284–5296.

12 Viswanath, P. and Tse, D.N. (2003). Sum capacity of the vector Gaussian broadcast channel and downlink-uplink duality. *IEEE Transactions on Information Theory* 49 (8): 1912–1921.

13 Lu, L., Ye Li, G., Swindlehurst, A.L. et al. (2014). An overview of massive MIMO-benefits and challenges. *IEEE Journal of Selected Topics in Signal Processing* 8 (5): 742–758.

14 Cover, T.G. (1972). Broadcast channels. *IEEE Transactions on Information Theory* 18 (1): 2–14.

15 Yu, W. and Cioffi, J.M. (2004). Sum capacity of Gaussian vector broadcast channels. *IEEE Transactions on Information Theory* 50 (9): 1875–1892.

16 Bergmans, P. (1974). A simple converse for broadcast channels with additive white Gaussian noise. *IEEE Transactions on Information Theory* 20 (2): 279–280.

17 Costa, M.H.M. (1983). Writing on dirty paper. *IEEE Transactions on Information Theory* 29 (3): 439–441.

18 Jindal, N., Rhee, W., Vishwanath, S. et al. (2005). Sum power iterative water-filling for multi-antenna Gaussian broadcast channels. *IEEE Transactions on Information Theory* 51 (4): 1370–1580.

19 Caire, G. and Shamai, S. (2003). On the achievable throughput of a multiantenna Gaussian broadcast channel. *IEEE Transactions on Information Theory* 49 (7): 1691–1706.

20 Vishwanath, S., Jindal, N., and Goldsmith, A. (2002). Duality, achievable rates and sum rate capacity of Gaussian MIMO broadcast channel. *IEEE Transactions on Information Theory* 49 (10): 2658–2668.

21 Lee, J. and Jindal, N. (2007). High SNR analysis for MIMO broadcast channels: dirty-paper coding versus linear coding. *IEEE Transactions on Information Theory* 53 (12): 4787–4792.

22 Bennatan, A. and Burshtein, D. (2008). On the fading-paper achievable region of the fading MIMO broadcast channels. *IEEE Transactions on Information Theory* 54 (1): 100–115.

23 El Gamal, A. and Cover, T.M. (1980). Multiple user information theory. *Proceedings of the IEEE* 68 (12): 1466–1483.

24 Rini, S. and Shamai, S. (2015) On the dirty-paper channel with fast fading dirt, IEEE International Symposium on Information Theory, 2286–2290.

25 Yoo, T. and Goldsmith, A. (2006). On the optimality of multiantenna broadcast scheduling using zero-forcing beamforming. *IEEE Journal on Selected Areas in Communications* 24 (3): 528–541.

26 Sharif, M. and Hassibi, B. (2007). A comparison of time-sharing, DPC, and beamforming for MIMO broadcast channels with many users. *IEEE Transactions on Communications* 55 (1): 11–15.

27 Larsson, E.G., Edfors, O., Tufvesson, F., and Marzetta, T.L. (2014). Massive MIMO for next generation wireless systems. *IEEE Communications Magazine* 52 (2): 186–195.

28 Guillaud, M. and Kaltenberger, F. (2013). Towards practical channel reciprocity exploitation: Relative calibration in the presence of frequency offset. *IEEE Wireless Communications and Networking Conference* 2525–2530.

29 Kaltenberger, F., Jiang, H., Guillaud, M., and Knopp, R. (2010) Relative channel reciprocity calibration in MIMO/TDD systems, *Future Network and Mobile Summit*, Conference Proceedings 1–10.

30 Shi, J. , Luo, Q. and You, M. (2011) An efficient method for enhancing TDD over the air reciprocity calibration IEEE Wireless Communications and Networking Conference (WCNC), 339–344.

31 Vieira, J., Rusek, F., and Tufvesson, F. (2014) Reciprocity calibration methods for massive MIMO based on antenna coupling, IEEE Global Communications Conference (GLOBECOM), 3708–3712.

32 de Groen, P., 1998 An Introduction to Total Least Squares, https://arxiv.org/pdf/math/9805076.

33 Markovsky, I. and Huffel, S.V. (2007). Overview of total least-squares methods. *Signal Processing* 87: 2283–2302.

34 Yin, H., Gesbert, D., Filippou, M., and Liu, Y. (2013). A coordinated approach to channel estimation in large-scale multiple-antenna systems. *IEEE Journal on Selected Areas in Communications* 31 (2): 2664–2273.

35 Müller, R., Cottatellucci, L., and Vehkaperä, M. (2014). Blind pilot decontamination. *IEEE Journal of Selected Topics in Signal Processing* 8 (5): 773–786.

36 Ashikhmin, A., Marzetta, T. Pilot contamination precoding in multi-cell large scale antenna systems, IEEE International Symposium on Information Theory Proceedings (ISIT), pp. 1137–1141, 2012.

37 Ngo, H.Q and Erik G. Larsson, (2012) EVD-Based channel estimation in multicell multiuser MIMO systems with very large antenna arrays, IEEE International Conference on Acoustics, Speech and Signal Processing (ICASSP), 3249–3252.

38 Godara, L.A.L.C. (1997). Application of antenna arrays to mobile communications, part II: beam-forming and direction-of-arrival considerations. *Proceedings of the IEEE* 85 (8): 1195–1245.

39 Van Veen, B.D. and Bukley, K.M. (1988). Beamforming: a versatile approach to spatial filtering. *IEEE Acoustics, Speech, and Signal Processing Magazine* 5 (2): 4–24.

40 Schmidt, R.O. (1986). Multiple emitter location and signal parameter estimation. *IEEE Transactions on Antennas and Propagation* 34 (3): 276–280.

41 El Ayach, O., Abu-Surra, S., Ragagopal, S., and Pi, A. (2012) Low complexity precoding for large millimeter wave MIMO systems, IEEE ICC 2012-Signal Processing for Communications Symposium, 3724–3729.

42 Kim, T., Perk, J., Seol, J-Y. et al. (2013) Tens of Gbps support with mmWave beamforming systems for next generation communications, IEEE Globecom 2013- Wireless Communications Symposium, 3685–3690.

43 Lo, T.K.Y. (1999) Maximum ratio transmission, IEEE International Conference on Communications, 2, 1310–1314.

44 Kim, S.W. and Wang, Z. (2007) Maximum ratio diversity combining receiver using single radio frequency chain and single matched filter, IEEE Global Telecommunication Conference, 4081–4085.

45 Zhang, J.A., Huang, X., Dyadyuk, V., and Guo, Y.J. (2015). Massive hybrid antenna array for Millimeter-wave cellular communications. *IEEE Wireless Communications* 22 (1): 79–87.

46 Deo, P., Mirshekar-Syahkal, D., Seddon, L., et al. (2013) Liquid crystal based patch antenna array for 60 GHz applications, IEEE Radio and Wireless Symposium, 127–129.

47 Yu, C., Urszula, C., Parrott, E.P.J., Parka, J., Herma, J, Chigrinov, V.G., and Pickwell-MacPherson, E. (2013) Large birefringence liquid crystal in terahertz range with temperature tuning, IEEE International Conference on Infrared, Millimeter, and Terahertz Waves, 1–2.

48 Wu, Y., Ruan, X., Chen, C.-H. et al. (2013). Graphene/liquid crystal based terahertz phase shifters. *Optics Express* 21 (18): 1–8.

49 Bulja, S. and Mirshekar-Syahkal, D. (2010). Meander line mmWave liquid crystal (LC) based phase shifter. *Electronics Letters* 46 (11): 769–771.

50 Shouzheng, Z., Ying, W., Guanglong, W., and Zongsheng, L. (2000) Design of MEMS millimeter-wave phase shifters, IEEE International Conference on Microwave and Millimeter Wave Technology, 498–501.

51 Baker, S. and Rebeiz, G.M. (1998). Distributed MEMS true-time delay phase shifters and wide-band switches. *IEEE Transactions on Microwave Theory and Techniques* 46 (11): 1881–1890.

52 Infineon Technologies Single-Chip SiGe Transceiver Chipset for V-band Backhaul Applications from 57 to 64 GHz: Application note, Germany, 2014

53 Liu, G. and Schumacher, H. (2013). Broadband millimeter-wave LANs (47-77 and 70-140 GHz) using a T-type matching topology. *IEEE Journal of Solid-State Circuits* 48 (9): 2022–2029.

54 Bassi, M., Zhao, J., Bevilacqua, A. et al. (2015). A 40-67 GHz power amplifier with 13 dBm P(sat) and 16% PAE in 28 nm CMOS LP. *IEEE Journal of Solid-State Circuits* 50 (7): 1618–1628.

55 ETSI White Paper (July 2016) mmWave Semiconductor Industry Technologies-Status and Evolution.

56 Bhagavatula, V., Zhang, T., Suvarna, A.R., and Rudell, J.C.R. (2016). An ultra-wideband IF millimeter-wave receiver with a 20 GHz channel bandwidth using gain-equalized transformers. *IEEE Journal of Solid-State Circuits* 51 (2): 323–331.

57 Fritsche, D., Tretter, G., Carta, C., and Ellinger, F. (2015). Millimeter-wave low-noise amplifier design in 28-nm low-power digital CMOS. *IEEE Transactions on Microwave Theory and Techniques* 63 (6): 1910–1922.

58 Hu, S., Wang, F., and Wang, H. (2017) A 28GHz/37GHz/39GHz multiband linear Doherty power amplifier for 5G massive MIMO applications, *IEEE International Solid-State* Circuits *Conference,*32–33.

59 Voinigescu, S.P., Shopove, S., Bateman, J. et al. (2017). Slicon millimeter-wave, terahertz, and high-speed fiber-optic device and benchmark circuit scaling through the 2030 ITRS horizon. *Proceedings of the IEEE* 107 (6): 1087–1104.

60 Chuang, S.-F., Wu, W.-R., and Liu, Y.-T. (2015). High-resolution AoA estimation for hybrid antenna arrays. *IEEE Transactions on Antennas and Propagation* 63 (7): 2955–2968.

61 Mohanna, M., Rabeh, M.L., Zieur, E.M., and Hekala, S. (2013). Optimization of MUSIC algorithm for angle of arrival estimation in wireless communications. *NRIAG Journal of Astronomy and Geophysics* 2: 116–124.

62 Feng, R., Liu, Y., Huang, J., Sun, J., and Wang, C-X. (2017) Comparison of MUSIC, unitary ESPRIT, and SAGE algorithms for estimating 3D angles in wireless channels, IEEE/CIC International Conference on Communications in China (ICCC), 1–6.

63 Liu, W., Li, Y., Yang, F., Ding, L., and Zhi, C. (2017) Millimeter-wave channel estimation with interference cancellation and DOA estimation in hybrid massive MIMO systems, IEEE International Conference on Wireless Communications and Signal Processing (WCSP), 1–6.

6

mmWave Propagation Modelling: Atmospheric Gaseous and Rain Losses

6.1 Introduction

The rapid increase in demand for cellular data implies that future cellular networks may need to deliver as much as 1000 times the current capacity, which requires an additional huge spectrum. Additional spectrum that would be licenced after 2020 will likely be used for 5G. Furthermore, growth in wireless connectivity in the next decade is expected to be billions more people, devices and machines that use 5G networks to access online services and connect with each other. At the same time, mobile phones will become increasingly ubiquitous and each new user will send and receive far more data than they did with their previous handset. Consequently, more networks are expected to deliver data at multi-gigabits/sec peak throughputs and tens of Mb/sec cell edge rates. Such envision has a number of important challenges [1] that must be addressed such as the current limited spectrum. There has been growing interest in the mmWave bands (30–300 GHz), since spectrum at such high frequencies can easily deliver the expected cellular data traffic increase. In addition, the very small signal wavelength combined with advances in low-power CMOS technology enable multiple antennas to be fabricated with the phone chip, and such antenna arrays will generate very high array gain.

The chapter's main focus is mmWave propagation. An important feature of mmWave propagation is the multipath sparsity. In both time (temporal) and space (spatial) domains, i.e. the multipath channel is characterized by few dominant propagation paths, usually separated by a long time. Sparsity is observed in a number of communication applications such as cm wave frequency bands cellular communications and terrestrial high-definition television broadcasting channels. However, sparsity in mmWave systems is mainly caused by two propagation phenomena: the reflected mmWave signal energy decreases rapidly due to the high path loss and only few paths with significant energy survive, and the diffraction will be less important due to a negligible Fresnel radius. Exhaustive propagation measurements campaign in [2] concluded that mmWave signals are made of a few multipath components concentrated in certain directions made of mainly line-of-sight (LOS) components with lower power non-line-of-sight (NLOS) components.

As a consequence of the multipath sparsity, a massive hybrid array beamforming becomes an attractive solution for the mmWave communications. Multipath sparsity can also be exploited for tracking the mmWave channels [3, 4]. Furthermore multipath sparsity can be used to facilitate efficient low cost massive mmWave array beamforming for a limited number of TX and RX beams [5]. In large mmWave MIMO systems, low complexity precoding can be attained by exploiting the structure of channels using the propagation sparsity [6]. The multipath sparsity can also be used to dramatically increase the MIMO capacity by adapting the array configuration to the level of sparsity [7, 8] especially at very low signal power to noise power ratio (SNR) regime.

5G Physical Layer Technologies, First Edition. Mosa Ali Abu-Rgheff.

One of the key challenges to realizing the vision of mmWave cellular network is the free space path loss. According to Friis law, free space omni-directional path loss grows with the square of the frequency, implying that a large propagation path loss appears at mmWave frequency bands. However, it is also true that a small wavelength enables a proportional increase in the number of antenna systems bring about more gain, which partially neutralizes the free space path loss increase when suitable transmission is adopted. A significant concern is that mmWave signals are extremely vulnerable to shadow fades. One more important issue to consider is the outdoor BS mmWave transmission to an indoor mobile terminal inside a building since materials such as bricks attenuate the mmWave signal by as much as 40 dB and the human body itself attenuates the signal by more than 20 dB. Such materials are also an important scatterer of mmWave propagation. On the other hand, heavy rain causes fading, which is a common problem in long-range mmWave backhaul links, but it may not be an issue for small cells of 100–200 m radii.

Rapid channel fluctuations, in conjunction with shadowing, constitute a great problem to mmWave mobile communication. Since the relationship between Doppler spread and coherence time is reciprocal, i.e. the larger the Doppler spread the smaller the coherence time. For example, a mobile user travelling at 60 km/h within a 30 GHz cellular coverage will be subjected to 3.3354 kHz Doppler spread and a coherence time of nearly 300 μs. That is, the mmWave channel changes every 300 μs. When this change occurs during a high level of shadowing, it may cause dramatic swings in signal loss. The conclusion from the above presentation is that the mmWave connectivity will be highly infrequent (sporadic) and the communication will need to be rapidly adaptable.

An essential challenge to the mmWave transmission is attributed to the power consumed by the signal processing associated with the A/D and D/A conversion. This power consumption scales linearly in the sampling rate and exponentially in the number of bits/sample. For example, A/D conversion at a sampling rate of 100 Ms/sec with 12 bits/sample with 16 antennas will consume 250 mw of power, which drains the mobile device battery. Therefore, an efficient RF power and beamforming are needed.

6.2 Contemporary Radio Wave Propagation Models

In order to exploit the mmWave frequency bands for high-speed transmission systems, it is important to consider the channel characteristics in addition to radio propagation models. Propagation models vary with respect to terrains, operating frequencies, TX/RX antennas gains and heights, RF bandwidth, and effective isotropic radiated power (EIRP). Radio propagation in indoor environments are not influenced by terrain profiles as the outdoors propagation but can be affected by the layout in the building and the various building material used.

A radio propagation model is an empirical mathematical formulation for the characterization of radio wave propagation as a function of frequency, distance, and other conditions based on data collection. The purpose of the propagation model is to predict the path loss (PL) along a link or effective radio coverage area of a transmitter. The radiated power is computed from the transmit power multiplied by the antenna gains at both ends of the link. PL is commonly specified in dB ($PL\ (dB) = 10\ \log_{10}(PL)$). For the PL prediction and simulations in contemporary cellular environment, Hata-Okumura model is widely employed. This model is valid for frequency range 500–1500 MHz, mobile receiver positioned at distances greater than 1 km from BS and antenna heights greater than 30 m. Unfortunately, the Hata-Okumura model and other models may not be adequate to address the new feature such as small cells, shorter BS antenna heights, and higher frequencies. Furthermore, such models are not suitable for hilly and heavily wooded terrain.

6.2.1 AT&T Propagation Model

An extensive measurement campaign was carried out by AT&T to find ways to overcome the Hata-Okumura propagation model limitations, and a large volume of data was collected from 95 BS across USA [9]. A mobile receiver van equipped with the required equipment to measure and record local mean receive powers within the coverage area. A wide range of terrain categories was covered that can be classified into three: category A *hilly terrain with moderate to heavy tree densities (maximum path loss category)*; category B *mostly flat terrain with moderate to heavy tree densities or hilly terrain with light tree densities*; and category C *flat terrain with light tree densities (minimum path loss category).*

Using the measured data at 1.9 GHz over distance d between 0.1 and 3 km, a log-log plot of PL versus distance is shown in Figure 6.1. It was clear that PL is increasing with some power γ of the distance. A PL of A dB is measured at a reference distance $d_0 = 100$ m. At least squares regression curve is constructed through the scattered points in Figure 6.1 to minimize the deviation of the scattered points. This curve intercepts the PL scale at d_0 at a point (A) and for $d > d_0$ PL is assumed to be expressed in the flowing formula:

$$PL(d) = A + 10\gamma\log_{10}\left(\frac{d}{d_0}\right) + s, \textit{ for } d > d_0 \quad \text{dB} \tag{6.1}$$

where A is the path loss in dB at distance d_0 and s is a shadow fading term, random variable with standard deviation over the terrain of σ dB. It was proved that A is actually equal free space (FS) loss in dB at d_0.

The statistical path loss (PL) model can be represented by the following mathematical expression:

$$PL_{AT\&T}(d) = PL(d_0) + 10\gamma\log_{10}\left(\frac{d}{d_0}\right) + s \text{ dB} \tag{6.2a}$$

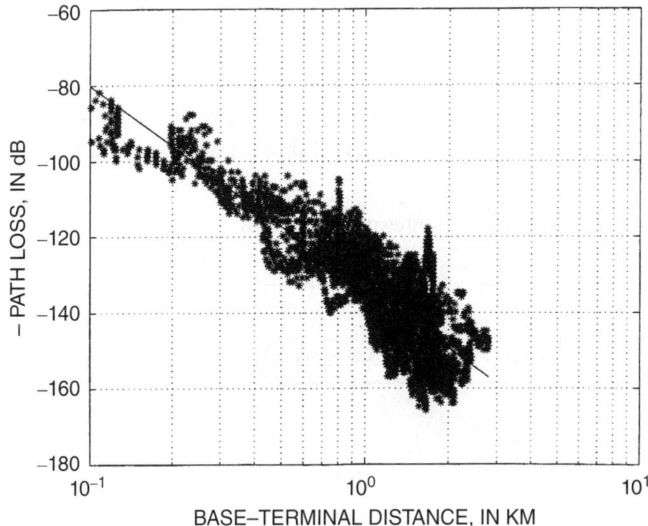

Figure 6.1 Scatterplot of path loss and distance for a macro cell (base antenna height was 25 m). The straight line represents the least-squares linear regression fit [9].

For reference distance d_0 (set to 0.1 km) where the free space PL is

$$A = PL(d_0) = 20\log_{10}\left(\frac{4\pi d_0}{\lambda}\right) \text{ dB} \tag{6.2b}$$

where λ is the wavelength, d_0 and d are in metres. The free space PL in (6.2b) can be simplified for use in mmWave frequency bands as

$$PL(d_0) = 32.4 + 20\log_{10}d_0 + 20\log_{10}f \tag{6.2c}$$

where d_0 in m and f now is defined in GHz. The PL exponent, γ is modelled as

$$\gamma = a_1 - a_2 h_{BS} + \frac{a_3}{h_{BS}} + \mathrm{x}\sigma_\gamma,\ 10\text{m} \geq h_{BS} \geq 80\text{m} \tag{6.3}$$

where h_{BS} is BS antenna height between 10 and 80 m; the mean value of γ is equal $a_1 - a_2 h_{TX} + \frac{a_3}{h_{TX}}$; σ_γ is standard deviation of γ and x is zero mean Gaussian variable with unit standard deviation $N(0, 1)$. The shadow fading component s varies randomly from one terminal location to another within any given cell. It is a zero mean Gaussian variable and can thus be expressed as

$$s = y\sigma \tag{6.4}$$

where y is a zero mean Gaussian variable with unit standard deviation $N(0, 1)$ and σ is standard deviation of s, which is itself a Gaussian variable such that

$$\sigma = \mu_\sigma + z\,\sigma_\sigma \tag{6.5}$$

where μ_σ is mean of σ and σ_σ is the standard deviation of σ, and z is a zero mean Gaussian variable of unit standard deviation $N(0, 1)$. The numerical values $PL_{\text{AT\&T}}(d)$ model parameters including a_1, a_2 in m^{-1}, a_3 in m, and the values of $PL_{\text{AT\&T}}(d)$ parameters σ_γ, μ_σ, and σ_σ are given in Table 6.1 for three terrain categories, Terrain A, Terrain B, and Terrain C.

6.2.2 Stanford University Interim (SUI) Propagation Model

The 802.16 IEEE access working group and researchers from University of Stanford teamed up to develop the AT&T PL model described in Section 6.2.1. Their aim is to account for the frequency of operation and antenna height and they came up with a model for the frequency band (2–11) GHz known as Stanford University interim (SUI) PL model [10, 11]. The SUI model is based on SU team measurements and published research literature. The SUI empirical model is verified on three terrain categories similar to those in Table 6.1 and two SUI models

Table 6.1 Numerical values for path loss model parameters for three different terrain categories [9].

Model parameter	Terrain category		
	Category (A)	Category (B)	Category (C)
a_1	4.6	4.0	3.6
a_2 *in* m^{-1}	0.0075	0.0065	0.0050
a_3 *in* m	12.6	17.1	20.0
σ_γ	0.57	0.75	0.59
μ_σ	10.6	9.6	8.2
σ_σ	2.3	3.0	1.6

are used for each terrain. The generic structure for the SUI MIMO Channel model has the following parameters: Cell size 6.4 km; BS antenna height 15 m; *customer premises equipment (CPE)* antenna height 3 m; base-station transceiver (BST) antenna beamwidth 120^0;CPE antenna beamwidth 50^0; vertical polarization only and for frequencies close to 2 GHz.

The SUI model, $PL_{SUI}(d)$ introduced two correction terms to the $PL_{\mathrm{AT\,\&\,T}}(d)$ so that

$$PL_{SUI}(d) = PL_{AT\&T}(d) + \Delta PL_{f_c} + \Delta PL_{h_{RX}} \text{ dB} \tag{6.6a}$$

where $PL_{\mathrm{AT\,\&\,T}}(d)$ is articulated by Eq. (6.2a). The correction for frequency of operation is

$$\Delta PL_{f_c} = 6\log_{10}\left(\frac{f_c}{2000}\right) \text{ dB} \tag{6.6b}$$

where f_c is the frequency in MHz. So, for $f_c = 30$ GHz, $\Delta PL_{f_c} \approx 7$ dB

The correction for receiver height is

$$\Delta PL_{h_{RX}} = -10.8\log_{10}\left(\frac{h_{RX}}{2}\right) \text{ dB for Terrain types A and B} \tag{6.6c}$$

where the CPE h_{RX} between 2 and 8 m. For a CPE height of 4m, $\Delta PL_{h_{RX}} = -3.25$ dB. Substituting (6.2a) and (6.6b and 6.6c) in (6.6a) we get

$$PL_{SUI}(d) = PL(d_0) + 10\gamma\log_{10}\left(\frac{d}{d_0}\right) + \Delta PL_{f_c} + \Delta PL_{h_{RX}} + s \text{ dB} \tag{6.7}$$

The use of mmWave bands reduces service coverage and impairs communication performance due to large path-loss, due not only to LOS but also for NLOS environments between transmitter and receiver. Researchers at Samsung Electronics show [12] that LOS PL at distance 400 m could increase by 20 dB, and NLOS increases by nearly 30 dB as the operating frequency increased from 1.8 to 28 GHz, as shown in Figure 6.2.

6.2.3 Modified SUI Model for mmWave Propagation

Extensive propagation measurement campaigns have been *carried out* at 28 and 38 GHz as described in [2, 13, 14] to develop the SUI PL model for predicting path loss at mmWave frequency bands and to work out the necessary correction needed. A broadband channel sounding

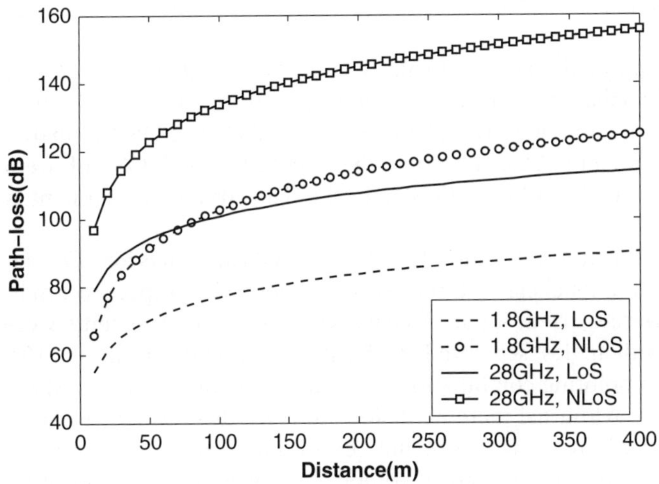

Figure 6.2 Path-loss of 1.8 and 28 GHz frequency bands [12].

is used for the measurement of the mmWave bands PL, angle of arrival (AOA), and Doppler characteristics measurements. The AOA is adopted to investigate individual signal paths. If the AOA pattern shows distinctive peaks, it is clear that the path loss can be reduced by steering receive antenna towards the reception peak.

Channel sounding is the process of transmitting RF signal over the channel and measuring the received signal to determine the propagation channel. The sounders type can be: the periodic pulse sounders that operate in t-domain and measure impulse response of the channel; the swept frequency sounders, which operate in f-domain and measure the channel frequency response and use vector network analyser (VNA); and the sliding correlator sounders that applied spread spectrum techniques to measure multipath properties.

The VNA sounder is an expensive single equipment sounder compared to other types. The impulse-based sounder uses the chirp approach and employs floating-point gate arrays and direct frequency synthesizers, and hence, such a sounder system is very complex. Chirp-based sounder found use in measurement at 60 GHz. The unlicensed 60 GHZ band is widely used for Giga b/s data rate multimedia transmission and for unlicensed backhaul products [15]. Furthermore, impulse-based sounders are susceptible to interference. The sliding correlator [16–18] is inherently robust through the use of direct sequence spread spectrum (DSSS) signals, which provides improved dynamic range; bandwidth through temporal dilation, which is ideal for wideband measurements; and low peak power to perform the measurements of interference-sensitive channels. Sliding correlator sounders can be used for indoor, outdoor measurements, and also for mobile radio channels for the extraction of absolute delay and phase information.

Broadband sliding correlator channel sounders were employed in the 28 GHz and 38 GHz propagation measurement campaigns in [2]. Furthermore, data for both angle of arrival at receiver (AOA) and angle of departure at transmitter (AOD) are measured to determine multipath angular spread for AOA and AOD. During the measurement, 360° sweep was applied at the TX and RX antennas to determine the angles with the highest received power. The measurement shows an average path-loss exponent (*PLE*) $n = 1.65$ and 2.55 for LOS propagation and an average $n = 5.76$ for NLOS propagation.

The maximum coverage distance of BS and the TX-RX gain is depicted in Figure 6.3 for various PLE. The maximum coverage is calculated as follows: a total PL of 178 dB was measured. The measuring system requires approximately 10 dB SNR for a reliable reception. For two 24.5 dBi antennas, the actual measured PL is $178 - 2*24.5 + 10 = 119$ dB, which was used to calculate the coverage distances.

PL measurements at 38 GHz employ a transmit vertically polarized horn antenna with 25 dBi and two interchangeable vertically polarized receive horn antennas with 25 and 13.3 dBi gains. The measured PLE for the 25 dBi horn antennas was 2.3 for LOS and 3.86 for NLOS. Compared to measurements in urban environment at 28 GHz, where the LOS PLE and NLOS PLE were 2.55 and 5.76, respectively, it is clear that PLE at 38 GHz in the light urban environment is considerably lower.

Figure 6.4 illustrates the path loss scatterplot using 24.5 dBi TX and Rx antennas for LOS and NLOS environments. The omni-directional close-in reference path loss with respect to 1 m in LOS and NLOS measurements were carried out at 28 GHz. Nine AOA measurements were taken for three adjacent TX azimuth angles, separated by 5° with elevation down tilt of −10°. At each RX location of 10 antenna pointing combinations for nine RX azimuth sweeps where the RX antenna was rotated in 10° azimuthal increments for three pre-determined elevation planes of 5° and ±20°. The total path losses were obtained by summing all received powers for each unique azimuth and elevation pointing angle combination at each RX-TX location combination and removing the TX and RX antenna gains for each individual PDP recoding.

PLE n and standard deviations σ with respect to a 1 m free space reference distance obtained for 28 GHz were achieved 3.4 and 9.7, respectively. The PLE in LOS environments at 28 GHz was $n = 2.1$, which is relatively close to the expected theoretical free space PLE of $n = 2$.

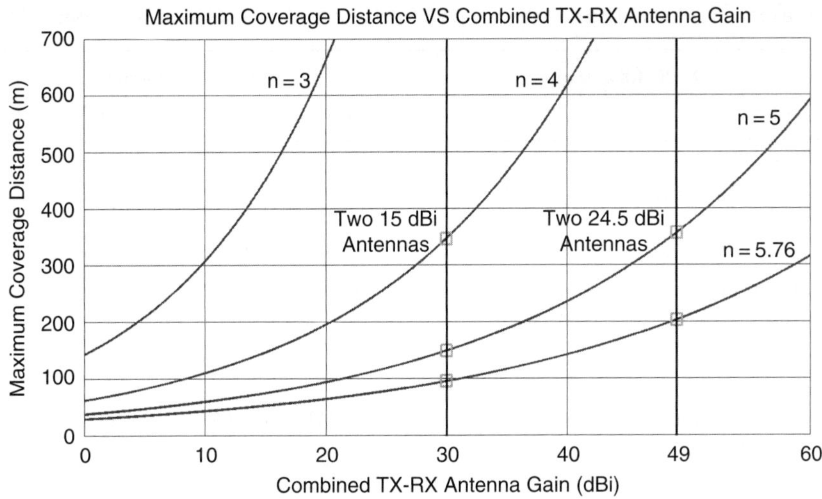

Figure 6.3 Maximum coverage distance at 28 GHz with 119 dB maximum path loss dynamic range without antenna gains and 10 dB SNR, as a function of path loss exponent n [2].

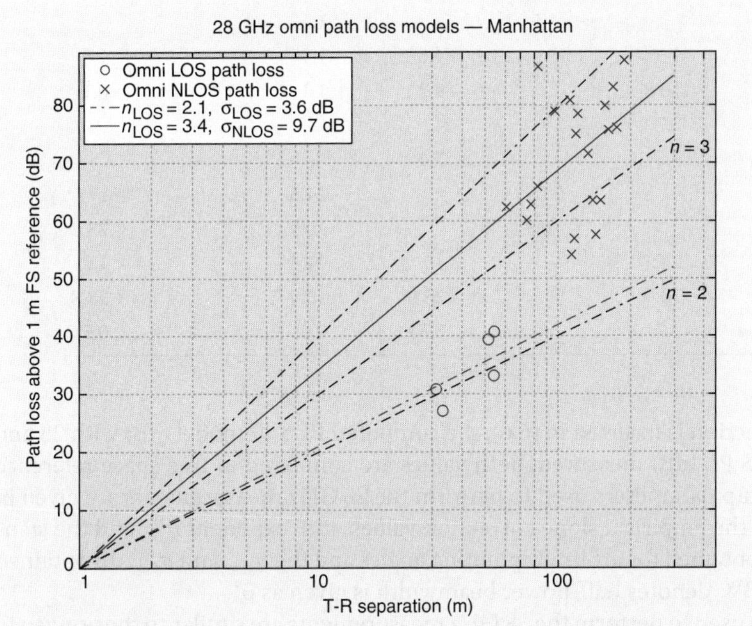

Figure 6.4 28 GHz LOS (circles) and NLOS (crosses): omnidirectional close-in reference distance Free Space (FS) path loss models with respect to 1 m for two 0 dBi isotropic TX and RX antennas [13].

The PLE and average and 99% RMS delay spread for 38 GHz are given in Table 6.2.

Measurement campaigns are carried out at 28 and 38 GHz to investigate signal scattering of multipath propagation due to a variety of objects in urban and suburban environments. The aim of the investigation is to modify the SUI model to take into account the scattering effects. In order to compare the propagation measurement made using steerable beam antennas that have antenna gains associated with the measurements and those measured with omni-directional SUI model, conversion factors are needed to make the two results with setups that have identical antenna gains. The SUI model in (6.6) should be modified by introducing slope correction

Table 6.2 Path loss (PL) exponent, PL standard deviation, and RMS delay spread for 38 GHz measurements [19].

	25 dBi RX antenna		13.3 dBi RX antenna	
	LOS	**NLOS**	**LOS**	**NLOS**
PLE	2.3 (clear LOS 1.89)	3.89 (best 3.2)	2.21 (Clear LOS 1.90)	3.18 (Best 2.56)
PL Standard Deviation dB	11.6 (Clear LOS 4.6)	13.4 (Best 11.7)	9.4 (Clear LOS 3.5)	11.0 (Best 8.4)
RMS Delay Spread (ns)	1.5 average 15.4 at 99%	14.3 average 133 at 99%	1.9 average 15.5 at 99%	13.7 average 117 at 99%

Table 6.3 Measurements support modified Stanford University Interim (SUI) PL model for 28GHz in NLOS environments and LOS [13].

	Frequency 28 GHz					
		NLOS			**LOS**	
TX antenna height m	7		17	17		17
RX antenna height m		1.5		1.5		
PLE n_{arb}	4.5		4.6	1.9		1.8
σ_{arb} dB	10.8		9.2	1.1		0.1
PLE n_{best}	3.7		4	—	—	—
σ_{best} dB	9.5		7.4	—	—	—
TX gain dBi		24.5		24.5		1
TX HPBW $\theta°$		10.9		10.9		28.8
RX gain dBi	24.5		15	24.5		15
RX HPBW $\theta°$	10.9		28.8	10.9		28.8
Slope correction factor α	0.71		0.88	0.95		0.9

factors. When the correction is inserted in (6.6), the computed PL SUI model (dB) with 28 and 38 GHz NLOS and LOS PL (dB) measured, both values are compared in free space reference distance of 1 m. The setup parameters used to perform the 28 GHz measurements are given in Table 6.3 together with the empirical slope correction values. PLE exponent n_{arb} and the standard deviation σ_{arb} dB obtained for arbitrary pointing angles and the n_{best} and σ_{best} dB obtained for best angles and HPBW denotes half-power beamwidth is given as θ°.

The setup parameters used to perform the 38 GHz measurements are similar to those given in Table 6.3. The slope corrections at 38 GHz are commonly smaller than that at 28 GHz for NLOS transmission but slope correction at 38 GHZ is larger or equal to that at 28 GHz for LOS. The slope correction values at 38 GHz are given in Table 6.4.

The asset of using SUI model is that it has additional parameters that allow frequency and height to be adjusted. So, the results in Tables 6.3 and 6.4 are computed using the measurements data at 28 and 38 GHz using the best minimum mean square (MMSE) line fit to measured data. The SUI model $PL_{SUI}(d)$ is modified at mmWave frequency bands and denoted $PL'_{SUI}(d)$. The modified model for mmWave transmission in NLOS environments is given as

$$PL'_{SUINLOS}(d) = \alpha_{NLOS}\,(PL_{SUI}(d) - PL_{SUI}(d_0)) + PL(d_0) + s \tag{6.8}$$

Table 6.4 Measurements supporting modified SUI PL model for 38GHz in NLOS environments and LOS environments [13].

	Frequency 38 GHz					
	NLOS			LOS		
TX antenna height m	23	8	36	23	8	36
RX antenna height m	1.5					
PLE n_{arb}	3.3, 27	3.8, 3.3	3.1, 2.7	2.0, 2.0	1.9, 2.0	1.9, 1.9
σ_{arb} dB	10.6, 8.1	11.1, 10.7	10.3, 8.0	2.3, 3.3	8.4, 4.3	3.7, 1.5
PLE n_{best}	2.7, 2.4	3.2, 2.6	2.6, 2.4	----, ----	----, ----	----, ----
σ_{best} dB	8.0, 6.0	10.3, 10.3	8.1, 5.1			
TX antenna gain dBi	25					
TX HPBW $\theta°$	7.8					
RX gain dBi	25, 13.3	25, 13.3	25, 13.3	25, 13.3	25, 13.3	25, 13.3
RX HPBW $\theta°$	7.8, 49.4	7.8, 49.4	7.8, 49.4	7.8, 49.4	7.8, 49.4	7.8, 49.4
Slope correction factor α	0.66, 0.54	0.62, 0.54	0.66, 0.58	1.0, 1.0	0.95, 1.0	0.95, 0.95

Substituting (6.7) into (6.8) we get

$$PL'_{SUI\,NLOS}(d) = 10n\,\alpha_{NLOS}.\log_{10}\left(\frac{d}{d_0}\right) + PL(d_0) + \Delta PL_{f_c} + \Delta PL_{h_{RX}} + s \tag{6.9}$$

Similarly, for a LOS environment, using Friis free space (FS) formula, the modified PL model is

$$PL'_{FS}(d) = \alpha_{LOS}\{PL_{FS}(d) - PL_{FS}(d_0)\} + PL(d_0) + s\text{ dB} \tag{6.10}$$

which can be simplified to

$$PL'(d) = \alpha_{LOS}PL_{FS}(d) + PL_{FS}(d_0)(1 - \alpha_{LOS}) + s\text{ dB} \tag{6.11}$$

where α_{NLOS} is the mean slope correction factor obtained as follows: the measurement data are plotted for MMSE best line fit and equated to data computed using the model in (6.7) (corrected to loss above $PL(d_0)$) and the mean slope correction is computed. It is worth noting that slope correction factors used in SUI model modification are dependent on TX and RX antennas gain, beamwidth, the variation of random variables such as scattering and shadowing processes and on the transmission environment such as NLOS and LOS and on the frequency of transmission.

6.3 Atmospheric Gaseous Losses

6.3.1 Introduction

Earth's atmosphere protects life by *absorbing* ultraviolet solar radiation. The atmospheric gases include two main gases, which account for 99% of the atmospheric gases, namely: oxygen 21% and nitrogen 78%. The remaining (1%) includes water vapour, carbon dioxide (0.037%), and organ (0.9%) and other rare gases. In addition to the path losses discussed in Section 6.2, the transmitted signal in space is exposed to an atmospheric attenuation caused principally by the atmospheric gases. At microwave frequency band ≤ 10 GHz, the attenuation is arguably predicted. However, for higher frequencies (i.e. the mmWave band), the attenuation not only increased but became dependent on the absorbing characteristics of the gaseous molecules.

Awareness of the atmospheric transmission losses knowledge is essential in the design of reliable mmWave communication systems in many fields such as remote sensing; radio astronomy and space and satellite communications and future mmWave cellular mobile communication systems.

Gaseous atmospheric attenuation is caused primarily by oxygen and water vapour molecules. The attenuation is down to the mechanical resonance of oxygen molecules in the vicinity of 60 and 118 GHz and is dependent on surrounding temperature, pressure, and for a given altitude may vary from place to place and time to time. Resonance of water vapour molecules causes peak absorption in the vicinity of 24 and 183 GHz. At these frequencies, peak absorption makes mmWave signal appropriate for short distance transmission. The water vapour content of the atmosphere varies naturally over a wide range of time and location. There are several atmospheric windows (i.e. frequencies that have minimum atmospheric loss) in the mmWave bands, namely at 35, 94,140, 220 GHz. Atmospheric gaseous access losses is loss of gases *above* transmission attenuation caused by free space.

In general, the average atmospheric gaseous total losses at mmWave band are much higher than at cm frequency bands, hence, the mmWave spectrum is considered by many researchers to be appropriate for short-distance communications, and this may thought to be a disadvantageous. However, cellular systems operating in mmWave bands can turn around this disadvantageous characteristic into beneficial opportunities. For one thing, mmWave transmission provides security for data transmission, and for another, it provides very efficient spectrum utilization through frequent band reuse.

6.3.2 Attenuation by Atmospheric Gases

The pioneering work by Van Vlec contributed to a better insight of the basic characteristics of mmWaves absorption by molecular of atmospheric gases. Van Vlec evaluated the experimental evidence available on the absorption of radio waves and predicted the attenuation due to oxygen and water vapour [20, 21]. Rosenkranz's work on the band loss model for the oxygen absorption significantly *contributed* to the *fundamental understanding* of the molecular absorption of radio waves [22]. The level of atmospheric gaseous absorption is determined by the sum of the contributions of the absorption lines and a continuous component contributes by frequencies located apart from absorption line peaks. The effects of minor gaseous components such as stratospheric ozone, variety of nitrogen, carbon, or sulphur oxides are usually neglected, since their attenuation is very small (approximately 0.2 dB/km) compared to oxygen and water vapour absorption [23].

Measurements of absorption spectrum of oxygen and water vapour have been conducted under simulated (laboratory) atmospheric conditions [24] to confirm theoretical investigations and empirical modelling. Generally such prediction models and techniques are shaped by the frequency, the temperature, the pressure, the humidity, the altitude, and the elevation angle. An EM wave is characterized by oscillation (variations) of the electric and the magnetic fields. The oscillating motion of the field vibrating at a particular point in space at a certain frequency excites similar vibrations at neighbouring points, and the wave is said to be travelling or propagating at the speed of light. There are many atmospheric gases molecules that can interact with the EM waves, but at mmWaves frequencies only the oxygen and the water vapour contribute significant absorptions. The permanent magnetic moment of the oxygen molecules interacting with the magnetic field of the EM wave, while the water vapour molecules interact with the electric field of the wave, produces a quantum level change in the rotational energy of the molecule, which absorbs energy from the EM wave [25]. At very low pressures, the change of

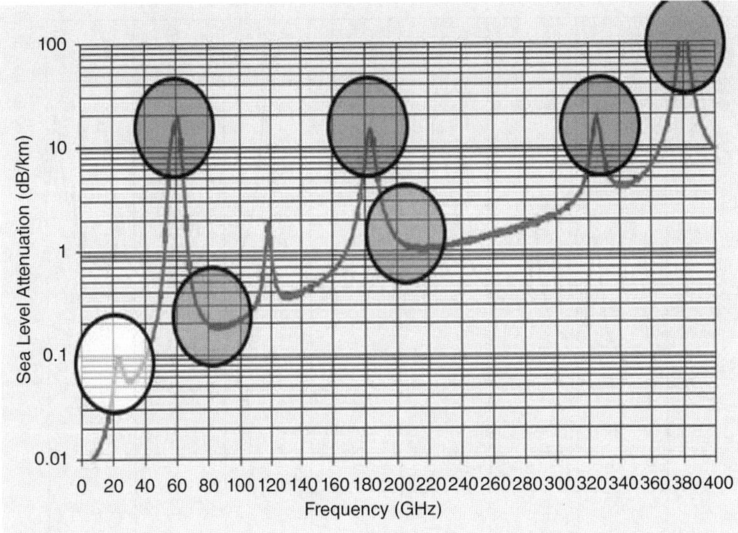

Figure 6.5 Atmospheric absorption across mmWave frequencies in dB/km [2].

the rotational energy of gas molecules is associated with a very narrow band of frequencies. At increased pressures, the energy is absorbed over a wider range.

Measurements of attenuation dB/km expected from the oxygen and the water vapour molecules are conducted [12, 24, 26] to reveal three absorption peaks centred at the frequencies of 22.2 GHz for water vapour maximum attenuation is 0.1 dB/km, and for oxygen the maximum attenuation is about 20 dB/km at 60 GHz and maximum attenuation of 1.7 dB/km at 118.8 reaffirm Van Vlec predictions. These measurements are depicted in Figure 6.5. However, small-cell coverage is within 200 m in radius. The frequencies that are likely to be used for 5G are 28 and 38 GHz. The attenuations caused by atmospheric gases (mainly oxygen) are 0.12 and 0.016 dB, respectively, over the small-cell coverage area. The rain attenuation contributes more attenuation at these frequencies, as we will find in Section 6.6.

Since the percentage of oxygen in the atmosphere is approximately constant with climate and altitude up to approximately 90 km, the oxygen absorption should not change radically with climate. However, changes in the atmospheric temperature and pressure profiles affect the molecules density, which eventually changes the molecules absorption. Furthermore, increasing the pressures, or decreasing temperatures increases the molecular density and hence increases the attenuation [27, 28].

The specific attenuation (dB/km) at frequencies up to 1000 GHz attributed to dry air and water vapour can be computed accurately at any value of pressure, temperature, and humidity by means of a summation of the individual resonance lines from gases and water vapour. The specific gaseous attenuation is characterized by a frequency-dependent complex refractivity N'' of the atmospheric gaseous refractive index N, defined as

$$N = N_0 + N' + jN'' \tag{6.12}$$

The real part $(N_0 + N')$ is associated with the wave propagation refraction and comprised of frequency-independent term N_0 and the frequency-dependent $N'(f)$. The imaginary part describes the radio radiation power absorption. Consequently, refractivity dictates the specific attenuation γ and wave phase β or delay rate τ.

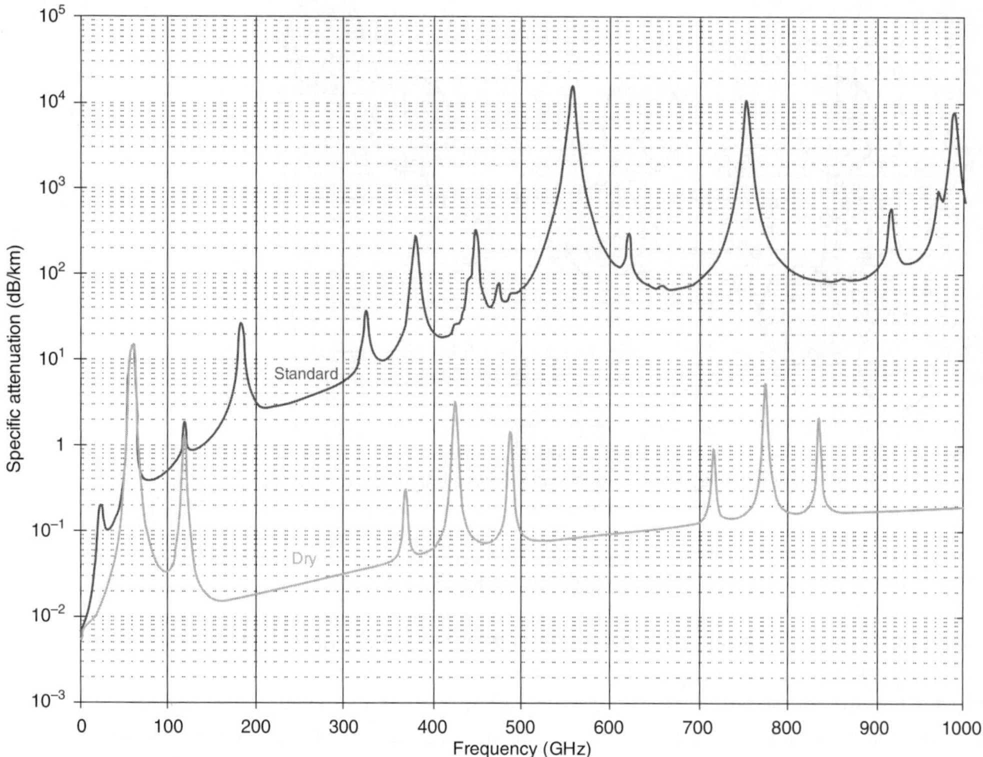

Figure 6.6 Specific attenuation due to atmospheric gases [29].

Figure 6.6 shows the specific attenuation using the International telecommunication Union (ITU) recommended model [29], calculated from 0 to 1000 GHz at 1 GHz interval for a standard atmosphere (pressure of 1013 hPa, temperature of 15 °C with oxygen and water attenuation) for the cases of a water-vapour density of 7.5 g/m^3 and a dry atmosphere.

Around 60 GHz, many oxygen-absorption lines merge together, at sea-level, to form a single, broad absorption band covering 50–70 GHz, as shown in Figure 6.7. This figure also shows the oxygen attenuation at higher altitudes, with the individual lines becoming resolved at lower pressures.

6.3.3 ITU Recommendations for Modelling Atmospheric Gaseous Attenuation

Recommendation ITU-R P.676-10 [29] described a process for accurately calculating the atmospheric gaseous attenuation and provides a reference standard for the atmospheres [30]. We now present the ITU-R recommended procedure for calculating the gaseous attenuation. Let us denote the specific attenuation dB/km by atmospheric gases as γ_g as the combination of specific attenuation of dry air γ_d dB/km and specific attenuation of water vapour γ_w dB/km. Mathematically, γ_g can be accurately calculated using line-by-line summation of gaseous attenuation described in ITU-R P.676-10 Annex 1 [29] of the recommendation and summarized as

$$\gamma_g = \gamma_d + \gamma_w = 0.182 f\, N''(f) \text{dB/km} \tag{6.13a}$$

$$\beta = 1.2008 f\, (N_0 + N') \text{ deg /km} \tag{6.13b}$$

$$\tau = 3.3356\, (N_0 + N') \text{ pico sec /km} \tag{6.13c}$$

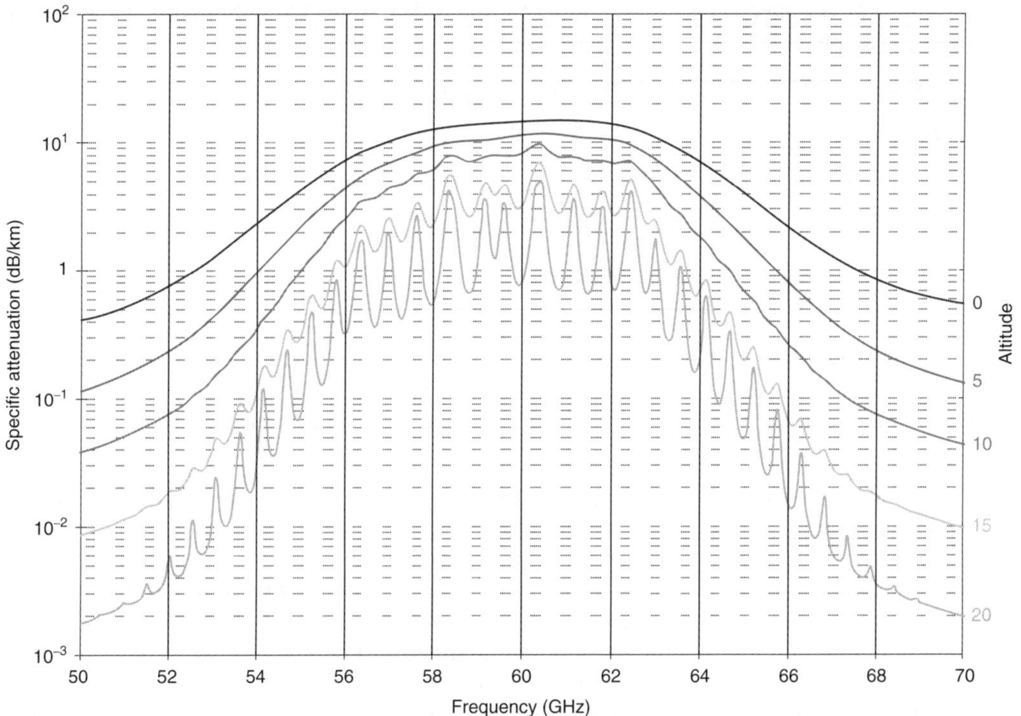

Figure 6.7 Attenuation in the range 50–70 GHz at the altitudes indicated [29].

where f is frequency in GHz, and $N''(f)$ is the imaginary part of refractivity of atmosphere. $N''(f)$ is responsible for atmospheric gaseous absorption given by

$$N''(f) = \sum_i S_i(f)\, F_i + N''_D(f) \tag{6.14}$$

where $S_i(f)$ absorption strength of the ith spectral line, F_i line shape factor, and the summation over all lines. Oxygen absorption lines in the vicinity of 60 GHz, lines from 1 to 37 should be included in the summation, and in the vicinity of 118 GHz the summation should begin at $i = 38$ rather than at $i = 1$). $N''_D(f)$ accounts for nonresonant Debye spectrum of oxygen below 10 GHz and a pressure-induced nitrogen attenuation above 100 GHz, which can be neglected for the band 10–100 GHz. The line strength is calculated using

$$S_i(f) = a_1 \text{ x } 10^{-7}\, p\, \theta^3\, e^{(1-\theta)a_2} \text{ for oxygen} \tag{6.15a}$$

$$S_i(f) = b_1 \text{x } 10^{-1}\, p_{wp}\, \theta^{3.5}\, e^{(1-\theta)b_2} \text{ for water vapour} \tag{6.15b}$$

where p is dry air pressure in hPa (1 hPa = 0.75 mmHg), p_{wp} water vapour partial pressure in hPa (total barometric pressure $p_{total} = p + p_{wp}$); $\theta = \frac{300}{T}$ and T temperature K; $p_{wp} = \frac{\rho T}{216.7}$ and ρ water vapour density.

The spectral line shape factor F_i is calculated using

$$F_i = \frac{f}{f_i}\left[\frac{\Delta f - \delta(f_i - f)}{(f_i - f)^2 + \Delta f^2} + \frac{\Delta f - \delta(f_i + f)}{(f_i + f)^2 + \Delta f^2}\right] \tag{6.16}$$

where f_i and Δf are spectral line frequency and line width, respectively. Spectral line width Δf is calculated using

$$\Delta f = a_3 \times 10^{-4}(p\,\theta^{(0.8-a_4)} + 1.1\,p_{wp}\theta) \text{ for oxygen} \quad (6.17a)$$

$$\Delta f = b_3 \times 10^{-4}(p\,\theta^{b_4} + b_5\,p_{wp}\theta^{b_6}) \text{ for water vapour} \quad (6.17b)$$

Spectral line width Δf calculated above is modified to $\Delta f'$ to account for Doppler broadening calculated using

$$\Delta f' = \sqrt{\Delta f^2 + 2.25 \text{ x } 10^{-6}} \text{ for oxygen} \quad (6.18a)$$

$$\Delta f' = 0.535\,\Delta f + \sqrt{0.217\,\Delta f^2 + \frac{2.1316 \text{ x } 10^{-12} f_i^2}{\theta}} \text{ for water vapour} \quad (6.18b)$$

The correction factor δ that is used to resolve the oxygen spectral line interference is calculated as

$$\delta = (a_5 + a_6\,\theta) \times 10^{-4}\,p_{total}\,\theta^{0.8} \text{ for oxygen} \quad (6.19a)$$

$$\delta = 0 \text{ for water vapour} \quad (6.19b)$$

The coefficients a_i and b_i are given in [29]. Atmospheric gases temperature, pressure calculation, and water-vapour pressure calculation as a function of altitude are outlined in Recommendation ITU-R P.835-5 [30]. The reference standard atmospheric temperature is divided into seven successive layers of altitudes showing linear variation with temperature, as depicted in Figure 6.8. In the next section, we present a detailed method to calculate the mean annual global reference atmosphere (temperature and pressure).

6.3.4 Temperature and Pressure

The temperature T at height h is given by

$$T(h) = T_i + L_i(h - H_i)\,\text{K} \quad (6.20)$$

where $T_i = T(H_i)$, L_i is temperature gradient at altitude H_i (height above sea level) and, their values are given in Table 6.5.

When the temperature gradient $L_i \neq 0$, the pressure is given by the equation:

$$P(h) = P_i\left[\frac{T_i}{T_i + L_i(h - H_i)}\right]^{\frac{-34.163}{L_i}} \text{hPa} \quad (6.21a)$$

and when the temperature gradient $L_i = 0$, the pressure is obtained by the equation:

$$P(h) = P_i\,e^{\left[\frac{-34.163(h-H_i)}{T_i}\right]} \text{hPa} \quad (6.21b)$$

The ground-level standard temperature and pressure are $T_0 = 288.15\,K$ and $P_0 = 1013.25$ hPa, respectively.

6.3.5 Water-Vapour Pressure

The distribution of water vapour in the atmosphere is generally extremely variable, but may be approximated by Eq. (6.22):

$$\rho(h) = \rho_0\,e^{\frac{-h}{h_0}} \quad \text{g/m}^3 \quad (6.22)$$

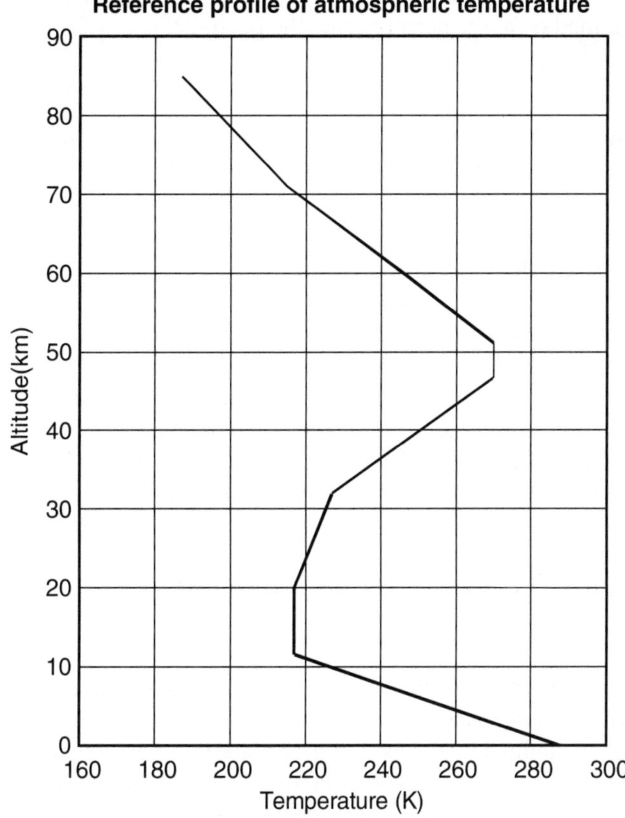

Figure 6.8 Reference profile of atmospheric temperature [30].

Table 6.5 Temperature (degrees K) versus altitude above sea level (in km) [30].

Subscript i	Altitude H_i km	Temperature gradient L_i K/km
0	0 (Sea level)	−6.5
1	11	0.0
2	20	+1.0
3	32	+2.8
4	47	0.0
5	51	−2.8
6	71	−2.0
7	85	—

where the scale height $h_0 = 2$ km, and the standard ground-level water-vapour density is $\rho_0 = 7.5$ g/m^3. Vapour pressure is obtained from the density using the Eq. (6.22).

$$e(h) = \frac{\rho(h)\,T(h)}{216.7} \quad \text{hPa} \tag{6.23}$$

Water-vapour density decreases exponentially with increasing altitude, up to an altitude where the mixing ratio $\frac{e(h)}{P(h)} = 2 \times 10^{-6}$. Above this altitude, the mixing ratio is assumed to be constant.

6.4 Dry Atmosphere for Attenuation Calculations

The profile of the density of atmospheric gases (other than water vapour), i.e. the *dry atmosphere,* may be found from the temperature and pressure profiles given in Section 6.3.4. For attenuation calculations, this density profile may be approximated by an exponential profile according to Eq. (6.22) with $h_0 = 6$ km. Temperature and pressure at different heights and water vapour pressure for low latitudes (smaller than 22°); mid latitudes (between 22° and 24°) and high latitudes (higher than 45°) for summer and winters seasons are given in [30].

6.5 Calculation of Atmospheric Gaseous Attenuation Using ITU-R Recommendations

Atmospheric gaseous specific attenuation can be calculated by two related methods: the computation-intensive line-by-line calculation described in ITU-R P.676-10 Annex-1 and the computationally less-intensive procedure defined in Annex-2 [29]. For line-by-line calculation, we use the procedure described in Section 6.3.3. Around 60 GHz, oxygen specific attenuation sharply peaked due to contributions from 37 spectral lines. For specific attenuation calculation in the vicinity of 118 GHz, the summation in (6.14) starts from spectral line $i = 38$. A sample of this calculation for the range 50–70 GHz is shown in Figure 6.7. The calculation method described in Annex-2 [29] is an accurate estimate of specific attenuation to within I dB. It requires less computation and is generated by a curve fitting to the line-by-line calculation for oxygen and the water vapour from sea level to altitude 10 km. Using the curve-fitting approximation method, the specific attenuation at frequencies up to 350 GHz is divided into seven bands for calculation purposes. The main equation used in the calculation is of the form

$$\varphi(.) = r_p^a\, r_t^b\, e^{c(1-r_p)+d(1-r_t)} \tag{6.24}$$

and

$$\varphi(.) = \varphi(r_p, r_t, a, b, c, d) \tag{6.25}$$

where $r_p = p_{\text{total}}/1013$ and p_{total} is total air pressure; $r_t = 288/(273 + \text{temperature}\ {}^\circ C)$; p pressure (hPa).

Parameters a, b, c, d are given in IT-R P.676-10 Annex- 2. The bands are:

$f \leq 54$ GHz, $54 < f \leq 60$ GHz, $60 < f \leq 62$ GHz, $62 < f \leq 66$ GHz, $66 < f \leq 120$ GHz, $120 < f \leq 350$ GHz.

As an example, let us consider the band $60 < f \leq 62$ GHz, with $r_p = \frac{P_{\text{total}}}{1013} = \frac{1023}{1013} = 1.0099$ and $r_t = \frac{288}{273+t} = \frac{288}{273+13} = 1.007$

The dry air attenuation γ_0 in dB/km is given by

$$\gamma_0 = \gamma_{60} + (\gamma_{62} - \gamma_{60})\frac{f-60}{2}, f \text{ in GHz} \tag{6.26}$$

Table 6.6 The dry air calculated attenuation γ_0 in dB/km for the mmWave band 60–62 GHz.

Frequency GHz	Total loss γ_0 dB/km
60.306056	15.2741
60.434778	15.2231
61.150562	14.9396
61.800158	14.6822
62.0000000	14.6030

$$\gamma_{60} = 15.0\,\varphi(r_p, r_t, 0.9003, 4.1335, 0.0427, 1.6088)$$

$$\gamma_{62} = 14.28\,\varphi(r_p, r_t, 0.9886, 3.4176, 0.1827, 1.3429)$$

Substituting γ_{60} and γ_{62} in (6.26), we get the values of γ_0 in Table 6.6.

Dry air nonresonant refractivity N''_D for Debye spectrum for oxygen *below* 10 GHz and the pressure-induced nitrogen absorption above 100 GHz contributes a small amount to the complex refractivity. For absorption outside this limit, we can calculate N''_Dcontribution using the following empirical formula [29]:

$$N''_D = f\,p\,\theta^2 \left[\frac{6.14 * 10^{-5}}{d\left(1 + \left(\frac{f}{d}\right)^2\right)} + \frac{1.4 * 10^{-12} * p * \theta^{1.5}}{1 + 1.9 * 10^{-5} * f^{1.5}} \right] \tag{6.27}$$

where d is the width parameter for the Debye spectrum given by

$$d = 5.6 * 10^{-4} * (p + p_{wp}) * \theta^{0.8} \tag{6.28}$$

6.6 Rain Attenuation at mmWave Frequency Bands

6.6.1 Introduction

In the previous sections, we considered the attenuation due to atmospheric gases (mainly oxygen) and water vapour (the gaseous *form* of *water*). In addition to the latter, water can exist in the atmosphere in several other phases, such rain (liquid), fog (suspended water droplets), clouds (tiny drops of water), snow, and hail (solid). Liquid water and ice are both heavier than air and the *air currents keep them* floating above the *ground*. Water transforms to ice crystals when the air temperature drops below zero and ice crystals are small enough to be supported by the air. Ice crystals clump together to form large particles that are too heavy to be suspended by the air and fall as rain, snow, or hail. Raindrops are large particles that cause the incident radio wave energy scattering.

Local rain can typically be classified into five types, namely: drizzle with rate (for example 1 mm/h); steady rain with nearly constant rainfall rate (around 5 mm/h) that does not depend on the point of origin of observations; heavy showers with rainfall rate of 20 mm/h; downpour (a heavy continuous fall of rain at rate of 100 mm/h; and cloudburst rain with an extreme amount of rain (250 mm/h) over short duration that is capable of causing flooding conditions.

It is well known that radio wave propagation at frequencies above 10 GHz is greatly affected by rain intensity. Raindrops are dielectrics that absorb radio energy causing wireless signal fading and communication system's long outage and hence loss of service availability. Modelling these fade events requires information on the local rainfall attenuation based on the local rain rate loss. Recent development of procedures permits a statistical prediction of the rain attenuation using local rainfall statistics. These methods can be used to take into account the hasty changes of the rain intensity and consequently the temporal fluctuations, and hence the statistical behaviour of the associated rain attenuation.

6.6.2 Research Development

Rain attenuation has been investigated through long-running measurements and through theoretical approaches to develop accurate models for predicting rain attenuation statistics. The contemporary models can be subdivided into two categories: Models that employ empirical algorithms, with certain assumptions, to derive local rainfall accumulative distribution function $P(R)$ based on local rainfall rate $R(P)$ with respect to certain probability level (typically 0.01%). These models implement power law relationship methods. Examples of the first category included in the ITU recommendations. Detailed recommendations of the ITU work in this field are summarized in Section 6.8. Another conversion factor modelled by Segal [31] with a power law known as conversion factor $\rho_T(p)$, modelled with a power law expresses the conversion in terms of equiprobable rainfall rates

$$\rho_T(p) = \frac{R_1(p)}{R_T(p)} \tag{6.29}$$

where T is integration time, R_1 and R_T are the rainfall rates values exceeded with same probability, p, for two integration times (e.g. 5 minutes and 10 minutes).

The ratio ρ_T may itself be expressed by p (this referred to as mode-1) or it can be function of R_T (mode-2), which can be expressed as

$$\Gamma_T(p) = \frac{p_1(R)}{p_T(R)} \tag{6.30}$$

where p_1 and p_T are the average probabilities that the rainfall rate, R, is exceeded for raingage integration times of 1 minutes and T minutes, respectively.

The two equations (6.29) and (6.30) are equally valid. However, Γ was found to vary over a much greater range than ρ at any given location. Measurement campaigns at 45 locations in Canada have been analysed to realize empirical conversion ratios suitable for 5- or 10-minute sampling times. Two methods are examined. In one the conversion of equiprobable 1-minute to T-minute rainfall rates is treated as a function of their probability of occurrence, while the other is treated as a function of T-min rainfall intensity and fitted in a power law relation. Convert 5- to 60-minute statistics into corresponding 1-minute cumulative distribution for mode conversion

$$\rho_T(p) = a * p^b \tag{6.31}$$

where a and b are regression coefficients given in [31] over the probability range $10^{-5} \leq p \leq 10^{-4}$. The power law relation for the other conversion model is

$$\rho_T = c * {R_T}^d \tag{6.32}$$

Where c and d are regression coefficients. The power law coefficients (a, b, c, d) are given in [31].

Bryant et al. [32] examined the dynamics of the horizontal and vertical shape of the rain. The dynamics are exactly associated to rain rate, rainfall volume, local terrain, and climate.

A breakpoint in the rain rate exceedance occurs close to 105 mm/h. At the breakpoint, rain can be divided into two large categories: below the breakpoint rain is uniformly distributed over the rainfall area, above the breakpoint the rain contains growingly horizontal irregularity with deep rain columns entrenched in a surroundings of less intense rain. The work is related to the prediction of satellite link attenuation based on the point rain rate, which is dependent on the rain dynamics. The rain attenuation exceedances were 12 dB. A rain cell is an area inside which the rain rate is equal to or higher than the threshold rain rate τ. So, the cell is continuous and along the counter that bounds it, the rain rate is at the threshold value [33]. The area where the rain rate is lower than the threshold rain rate is ignored. The description of rain field is characterised by the rain cell size distribution. As a characteristic measure, most meteorological researchers have considered the radius of the circular cell of equivalent area to rain field, and proposed an exponential form to represent the rain radius distribution as

$$N(r,\tau) = N_0(\tau)\exp[-\lambda(\tau)r] \tag{6.33}$$

where r is the radius of the equivalent circular cell area with the rain rate threshold τ; $N(r, \tau)$ is the number of cells having a radius r per unit area inside an observed area; $N_0(\tau)$ and $\lambda(\tau)$ are the two parameters of the distribution, that is, the intercept and the slope, respectively. It is assumed, a priori, that N_0 and λ depend on the rain rate threshold. Note that r and λ are in km and km^{-1}. The number density of the cells as a function of spatial coordinates is defined as number of cells observed in the surveyed area per unit range of definitive parameters.

The rain cells have rotational symmetry and exponential distribution of rain with different radii and peak rain rates. Basically, a rain cell is defined as any connected region of space that consists of points where the rain rate goes above a given intensity threshold. In any location, rain cells have the same exponential distribution of rain intensity and peak value / cell dimension relation but the difference in the rain field from one location to another location is the cells probability of occurrence.

6.7 The Physical Rain (EXCELL) Capsoni Model

This model was developed by Capsoni et al. [34] for predicting the rain degradation of radio wave propagation. The model assumes the rainfall distinctiveness of any location can be statistically characterised using an admixture of synthetic rain cells.

A large volume of radar-measured rain cells is used in the design and evaluation of the rain cell model according to cells exponential shape and both rotational and biaxial symmetry for the horizontal cross section. An algorithm was developed for using the model at any location but with requirements prerequisite the local cumulative distribution function (CDF) of point rain as an input to provide the number of cells per square km and per unit range of parameters. The mode allows the possibility of predicting the statistics of propagation parameters, such as attenuate and rain scattering, which are determined by the rain cell characteristics and their probability of occurrence.

Two models of a cell are dealt with: a monoaxial model and a biaxial model. The last model takes into account the geographical asymmetry of rain cells since this characteristic can play an important part in some applications. Frequently, the model cells are characterized by two or three parameters, so it is appropriate that the cells be defined by joint probability distributions, which give number of cells examined in a given area per unit range of definitive parameters. Two categories of cells are defined: the proposed model related cells referred to as *model cells* and observed cells known as *radar cells*.

The radar cell area A is given as

$$A = \iint_{cell} dx\, dy \tag{6.34}$$

and the average rain rate of the radar cell is

$$\overline{R} = \frac{Q}{A} \tag{6.35}$$

where Q is the cumulative quantity of rain given as

$$Q = \iint_{cell} R(x, y) dx\, dy \tag{6.36}$$

and $R(x, y)$ is the rain rate at point (x, y).

6.7.1 Model Cells

So as to realize an analytical interpretation of the profile of rain rate inside a cell, various shapes were studied and assessed. The fundamental prerequisite of the model was the comprehension to optimally replicate the point rain rate cumulative distribution P(R) created by the entire group of measured cells to empower the derivation of model with parameter values from an available meteorological data.

It emerged that the exponential shape for rain rate inside the cell was the one that fulfilled this concern. The results acquired for the two types of exponential models, the *biaxial* and the monoaxial one, are cited in this section [34]. Considering a reference system along the principal directions of the cells, the rain rate for the biaxial model is defined by the function

$$R(x, y) = R_M \exp\left(-\sqrt{\left(\frac{x}{\rho_x}\right)^2 + \left(\frac{y}{\rho_y}\right)^2}\right) \tag{6.37}$$

and the rain rate for the monoaxial model is defined by the function

$$R(x, y) = R_M \exp\left(-\sqrt{\left(\frac{x^2 + y^2}{\rho_0}\right)}\right) \tag{6.38}$$

In (6.37) and (6.38) R_M is the maximum rain rate, ρ_x, ρ_y and ρ_0 are the rain cell radii along the respective axes and the rain cell radius of the monoaxial model.

The postulation of a complete decay of R with increasing distance from the peak does not concur to the measurements as rain cells with more than one relative maximum are common; nevertheless, shapes as in (6.37) and (6.38) replicate adequately the statistical behaviour of the rain rate profile along a path. A comprehensive interpretation of the rain rate models require description of the values of the parameters R_M, ρ_x, ρ_y for the *biaxial model* and R_M and ρ_0 for the *monoaxial (in a single direction) model*. These parameters are determined on a cell-by-cell basis after imposing appropriate cell descriptions to be identical in the radar and in the model cell and then selecting the parameter values that best adhere to the key prerequisite concerning the $P(R)$.

The aim is to match the cell area A and the average rain rate $\overline{R}$ given by (6.34) and (6.35) of radar cell to the A and $\overline{R}$ of the model cell. This can be accomplished using the expressions

$$A = \pi \rho_x \rho_y \log^2\left(\frac{R_M}{5}\right) = \pi \rho_0^2 \log^2\left(\frac{R_M}{5}\right) \tag{6.39}$$

So, $\rho_0^2 = \rho_x \, \rho_y$, and

$$\overline{R} = 2R_M \frac{\left[1 - \left(\frac{5}{R_M}\right)\left(1 + \log\left(\frac{R_M}{5}\right)\right)\right]}{\log^2\left(\frac{R_M}{5}\right)}, R_M > 5 \tag{6.40}$$

Eqs. (6.39) and (6.40) accomplish the fitting of the biaxial model. The degree of divergence of an ellipse from a circle (i.e. cell ellipticity) is

$$\frac{\rho_x}{\rho_y} = \frac{\rho_{\min}}{\rho_{\max}} \tag{6.41}$$

6.7.2 Monoaxial Cell and Biaxial Cell Models

Monoaxial rain cells modelled by (6.38) are described by the parameters R_M and ρ_0. Accordingly, modelling the joint density function of R_M and ρ_0 provides the broad description of the cell model.

Research carried out on the subject recommends exponential models for both cumulative distribution and probability density of ρ_0 in addition to exponentially distributed cell radii as highlighted earlier.

Normalizing ρ_0 to the value $\overline{\rho}_0$ for which the probability decays by a factor of $1/e$, the *number density per unit range of* $\frac{\rho_0}{\overline{\rho}_0}$ *and* R_M is given by

$$N(\rho_0, R_M) = N_0(R_M) \exp\left(-\frac{\rho_0}{\overline{\rho}_0\,(R_M)}\right) \tag{6.42}$$

Let us presuppose that (6.42) supports any value of ρ_0 (from 0 to ∞), so $N_0(R_M)$ is the *total number of cells per unit* R_M *range* and $\overline{\rho}_0\,(R_M)$ becomes the conditional *average Monoaxial cell radius*. The experimental values are adequately expressed by the *double power law* form as

$$\overline{\rho}_0\,(R_M) = 1.7\left[\left(\frac{R_M}{6}\right)^{-10} + \left(\frac{R_M}{6}\right)^{-0.26}\right], \quad R_M > 5 \tag{6.43}$$

The density $N_0(R_M)$ can be modelled in the range $6 < R_M < 150$, mm/h as

$$N_0(R_M) = 8.3\, R_M^{-2.13} \tag{6.44}$$

A broadening of the monoaxial model statistics to the biaxial case expressed in (6.37) is attained by concluding the set of cell parameters R_M, ρ_0 with the axial ratio $\frac{\rho_x}{\rho_y}$ and by studying the two rain rate models jointly.

6.7.3 Fitting the Model to the Local Meteorological Data

The statistical characterization of the monoaxial cell model is defined by substituting (6.43) and (6.44) in (6.42) and multiplying by a constant that transforms the number density $N(\rho_0, R_M)$ to a number density $N^*(\rho_0, R_M)$ in number of cells per km^2. Further, $N_0(R_M)$ in (6.44) is unique to the measurement location and the augmentation to any other site can be made by assuming that the conditional probability of ρ_0 is still exponential, with the average value given by (6.43), so that $N^*(\rho_0, R_M)$ can be written as

$$N^*(\rho_0, R_M) = N_0^*(R_M) \exp\left[-\frac{\rho_0}{\overline{\rho}_0\,(R_M)}\right] \tag{6.45}$$

It is worth noting that $N^*(R_M)$ implies the spatial density of the cells with peak rain rate R_M, regardless of radius ρ_0.

The expression $N^*(\rho_0, R_M)$ in (6.45) can be used for formation of the point rain rate CDF through the expression

$$P(R) = \int_R^{\infty} \int_0^{\infty} S_0(R, R_M, \rho_0)\, N^*(\rho_0, R_M)\, d\left(\frac{\rho_0}{\overline{\rho}_0}\right) dR_M \tag{6.46}$$

For rain circular cell radius ρ, the cell area is $\pi\rho^2$. $S_0(R, R_M, \rho_0)$ is the cell area where the rain rate exceeds R and for radius ρ using (6.39).

$$S_0(R, R_M, \rho_0) = \pi\rho^2 = \pi\rho_0^2 \log^2\left(\frac{R_M}{R}\right) \tag{6.47}$$

Substituting (6.47) in (6.46) we get

$$P(R) = \int_R^{\infty} \int_0^{\infty} \pi\rho_0^2 \left(\log\left(\frac{R_M}{R}\right)\right)^2 N^*(\rho_0, R_M)\, d\left(\frac{\rho_0}{\overline{\rho}_0}\right) dR_M \tag{6.48}$$

Use (6.45) to substitute for $N^*(\rho_0, R_M)$ and integrate (6.48) with respect to $\frac{\rho_0}{\overline{\rho}_0}$ as follows:

$$\int_0^{\infty} \pi\rho_0^2\, N^*(\rho_0, R_M)\, d\left(\frac{\rho_0}{\overline{\rho}_0}\right) = \int_0^{\infty} \pi\rho_0^2\, N_0^*(R_M) \exp\left[-\frac{\rho_0}{\overline{\rho}_0}\right] d\left(\frac{\rho_0}{\overline{\rho}_0}\right)$$

$$= \pi\overline{\rho}_0^{-2} N_0^*(R_M) \int_0^{\infty} \rho_0^2 x^2\ \exp[-x]\, d(x)$$

So

$$\int_0^{\infty} \pi\rho_0^2\, N^*(\rho_0, R_M)\, d\left(\frac{\rho_0}{\overline{\rho}_0}\right) = 2\pi\overline{\rho}_0^{-2} N_0^*(R_M) \tag{6.49}$$

Recalling that

$$N_0^*(R_M)\, dR_M = N_0^*(\log R_M)\, d(\log R_M)$$

Accordingly, expression (6.46) simplified to

$$P(R) = 2\pi \int_{\log R}^{\infty} \left(\log\left(\frac{R_M}{\mathrm{R}}\right)\right)^2 \overline{\rho}_0^{-2}(R_M) N_0^*(\log R_M)\, d(\log R_M)$$

$$P(R) = 2\pi \int_{\log R}^{\infty} (\log(R_M) - \log(R))^2\, \overline{\rho}_0^{-2}(R_M) N_0^*(\log R_M)\, d(\log R_M)$$

where the density $N_0^*(R_M)$ has been replaced with $N_0^*(\log R_M)$ for analysis convenience.

Using the mathematical identity

$$f(t) = -\frac{1}{2}\left[\frac{d^3}{d\tilde{t}^3} \int_t^{\infty} (t - \tilde{t}) f(t) dt\right]_{\tilde{t}=t} \tag{6.50}$$

which holds for function that are $O\left(\frac{1}{t^2}\right)$ when$\rightarrow\infty$.

Let $f(t) = 2\pi 2\, \overline{\rho}_0^{-2}(R_M) N_0^*(\log R_M)$; $(t - \tilde{t}) = \log(R_M) - \log(R)$

$$2\pi\overline{\rho}_0^{-2}(R_M) N_0^*(\log R_M) = -\frac{1}{2}\frac{d^3(P(R)}{d(\log R)^3}\bigg|_{R=R_M}$$

Taking the third-order derivative to (6.48) with respect to log$R$, we get

$$4\pi\overline{\rho}_0^{-2}(R_M) N_0^*(\log R_M) = -\frac{d^3 P(R)}{d(\log R)^3}\bigg|_{R=R_M} \tag{6.51}$$

which gives the desired spatial density N_0^*.

6.7.4 Development of the Capsoni EXCELL Model

The first modification to Capsoni EXCELL model introduced the lowered EXCELL model in [35].

The EXCELL model already described inferred that cells spatially extend to ∞. The lowered EXCELL model introduced a cell-lowering factor R_{low} as illustrated in Figure 6.9 . The exponential shaped cells are defined by

$$R(\rho) = (R_M + R_{Low})\exp\left(-\frac{\rho}{\rho_0}\right) - R_{low}, \quad \rho \leq \rho_{\max} \tag{6.52a}$$

$$R(\rho) = 0 \qquad \rho > \rho_{\max} \tag{6.52b}$$

where ρ is the distance from cell centre where the peak rain rate R_M is located. The implication for R_{low} is to allow the rain rate profile $R(\rho)$ to attain to zero at

$$0 = (R_M + R_{Low})\exp\left(-\frac{\rho_{\max}}{\rho_0}\right) - R_{low} \tag{6.52c}$$

Simplifying (6.52c), we get

$$\rho_{\max} = \rho_0 \log\left(1 + \frac{R_M}{R_{low}}\right) \tag{6.53}$$

where ρ_0 is the monoaxial cell radius. Substituting ρ_0 for ρ in (6.52a) we get

$$R(\rho_0) = \frac{(R_M + R_{Low})}{e} - R_{low} \approx \frac{R_M}{e}, for\ R_{low} \leq 2\ \text{mm/h}$$

where $e = 2.718$.

The second modification appeared [36, 37] permits the retrieval of the CDF of point rain rate with one-minute integration time $P(R)_1$ from rain rate CDFs with longer integration time i.e. one-hour time average rain rate statistics $P(R)_{60}$. The method initiates the scaling factors between $P(R)_{60}$ and $P(R)_1$ as a function of the probability level and of the site physical coordinates. From the scaling factors, a scaling law is concocted.

The first step in the conversion law is to choose a first guess $P(R)_1$ using the method described in the ITU recommendation and apply the lowered EXCELL model with $P(R)_1$ as an input and determine the probability of occurrence of all the cells in the rain field. The conversion factors from one-hour to one-minute integration time is calculated as

$$\rho_{60}(P) = \frac{R_1(P)}{R_{60}(P)} \tag{6.54}$$

where P is the probability level for which the scaling factor is calculated, $R_1(P)$ and $R_{60}(P)$ are the rain rate values of $P(R)_1$ and $P(R)_{60}$, respectively, and the probability level is P . The conversion factors are applied to the conversion law as

$$\rho_{60}(P) = a\,P^b + c\,e^{dP} \tag{6.55}$$

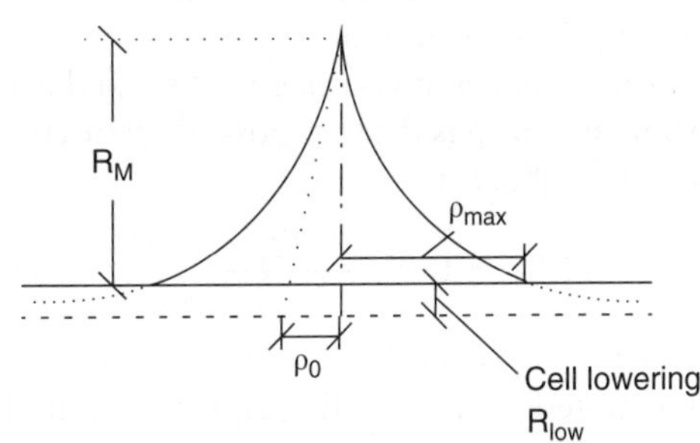

Figure 6.9 Rain cell model, including lowering through the factor R_{low} [35].

The first step in the conversion law is to choose a first guess $P(R)_1$ using the method described in the ITU recommendations.

The third modification is known as EXCELL Rainfall Statistics Conversion, or EXCELL RSC [38]. The proposed method has a strong physical basis and it is simple, accurate, and has global applicability. The conversion method is based on the simulated movement of rain cells over a computer-generated raingage (pluviometer) with a given integration time T that conversion velocity depends both on the type of precipitation and on the observation period. The analytical expression of the input $P(R)$ is a log-power law of the form

$$P(R) = P_0 \log^n \left(\frac{R_{low} + R_{asint}}{R_{low} + R} \right) \tag{6.56}$$

where P_0 decides the behaviour of the curve for $R \to 0$, R_{asint} is the asymptotical value of $P(R)$ directly related to the maximum measured point rain rate, n determines the shape of the curve, and R_{low} is chosen so as to achieve the best possible fit. R_{low} permits the probability level to take a finite value when $R \to 0$ as in the physical case. The four parameters in (6.56), namely: (P_0, n, R_{low}, R_{asint}) produce a good analytical fit of the measured $P(R)$ and allows easy computation.

A good index for the average rain cell speed at a given site is the wind velocity at the 600 mb altitude v_{600} measured at the altitude corresponding to approximately 4000–4200 m above the ground.

Convection as a process implies the travel of heat through air, water, gases, and liquids. Cloud system made of small droplets and ice crystals and larger rain drops and ice. Convective cloud systems comprise two distinct regions: convective region encompasses clouds with strong vertical motion and heavy rainfall, much like thunderstorms and stratiform region that is a broader area of clouds with weak vertical motion and light rainfall. These cloud particles intermingle differently in the two regions. Their interaction causes the release of heat. Accordingly, the separation into two regions affirms the vertical location and amount of heating in each region and the amount of rainfall in each region [39, 40]. Further, it seems the convective cells convert faster than stratiform cells. So, the rainfall peaks in stratiform rain rarely exceed 10 mm/h. Accordingly, the movement of convective rain can be converted at velocity $v_{conv}(T)$ [36]:

$$v_{conv}(T) = \frac{v_{600}}{k_1(T)} \tag{6.57}$$

The stratiform cells appear to covert at a speed a fraction of $v_{conv}(T)$:

$$v_{strat}(T) = k_2(T)\, v_{conv}(T) \tag{6.58}$$

where $k_1(T)$ (>1) and $k_2(T)$ (<1) are velocity reduction factors.

The raingage simulator described so far receives the local $P(R)_1$ as an input and output an estimates of the T-minute integrated rainfall CDF. However, the requirement to provide a method to predict $P(R)_1$ from a given measured $P(R)_T$, i.e. the algorithm is driven backward. This goal can be achieved by suggesting a first-guess $P(R)_1$ as an initial estimate of $P(R)_T$. The first guess $P(R)_1$ is replace by an iterative inversion method that depends only on measure $P(R)_T$ without any a priori assumption.

The optimization is designed to find the local $P(R)_1$ that, when applied as input to the raingage simulator, brings the best possible estimate of the measured $P(R)_T$. Let us specify the following error symbol ε as

$$\varepsilon(P)_T = 100 * \frac{R_e(P)_T - R_m(P)_T}{R_m(P)_T} \%\ \text{error} \tag{6.59}$$

where $R_e(P)_T$ and $R_m(P)_T$ are the rain rate values of the estimated and the measured T-minute integrated rainfall CDF, respectively, at the same probability level P. The objective of the optimization method is to minimize the error function given by the root mean square value (RMS) of $\varepsilon(P)_T$ and denoted ψ_T:

$$\psi_T = \sqrt{\frac{1}{N} P_i \sum_{i=1}^{N} [\varepsilon(P_i)_T]^2} \tag{6.60}$$

where P_i is the ith probability level decided by the ITU in the range from $10^{-3}\%$ to 1% and N is the total number of the available probability values.

The conversion of rainfall statistics from long to short integration time can be carried out by genetic algorithm that primarily comprises three steps [38]:

1. First, decide on initial possible solutions, i.e. N individuals, set of P_0, n, R_{asint}.
2. At each iteration, rank and choose N individuals according to ψ_T, the lower its value, the fittest the associated solution, the more probable its selection (fittest individuals can be chosen more than once).
3. At each iteration, group pair-wise and recombine chosen individuals $\frac{N}{2}$ pairs whose generation (N in total) expands the possible solutions.

Steps 2 and 3 are repeated until the algorithm convergence criteria are met (i.e. any more iterations make negligible changes).The application of the genetic algorithm to the rainfall conversion optimization problem has shown that the convergence can be reached in a few seconds and provides the requested set P_0, n, R_{asint} .

6.8 ITU Recommendations on Rainfall Rate Conversion

6.8.1 Introduction

The rain attenuation prediction methods have been developing since the 1980s. Raindrops radio energy attenuation in dB can be derived from the EM theory based on drop size and shape of raindrop size (or mass) distribution (also called raindrop spectra). Raindrop distributions can be represented by an exponential law function (*commonly known in atmospheric chemistry and physics as Marsh and Palmer [MP] distribution*), or as gamma distribution function. In fact the MP distribution can be considered as a special case for the gamma distribution. However, it is easier to consider raindrop attenuation using ITU-R recommendations.

The ITU recommends the use of rain rate CDF**s** with one-minute integration time denoted as $P(R)_1$ as an input to rain attenuation model since such integration interval assured an acceptable level of accuracy in rain attenuation variability. However, one-minute integration time is not the standard interval used in meteorology activities because there is no concern in fast changes of rain intensity but rather in reliable average quantities of cumulated rain. So, rain data are collected for meteorological centres cover long integration periods that are easily available worldwide. By contrast, rain attenuation for radio wave propagation is conducted on an instant basis at specific locations. By contrast, the ITU conversion method makes use of average coefficients determined from rainfall rate CDFs collected in various climatic regions and with different integration times.

Rain rate (empirical) conversion methods are widely used to derive $P(R)_1$ from the locally measured *rain rate CDFs* with longer integration time denoted as $P(R)_T$ where T is the integration time expressed in minutes. These methods have the advantage of confining as much as is feasible the local irregularities of precipitation (topographic effects and yearly variability), but they usually provide only regional coefficients, since they are developed from local data.

Rain rate prediction methods can be grouped into three disparate categories.

- Meteorology-based prediction methods with input generic climatic information taken from meteorological centres, such as the annual or monthly average rain rate, the number of rainy days per year, the peak annual rain rate, and so on.
- Analytical methods based on the assumptions that the rain rate CDF is maintained while changing the integration time. So, the one-minute and one-hour CDFs are different because the factors of the CDF equations are different. These methods predict how the factors of the distribution change according to the integration time.

Empirical methods that provide conversion factors between the known CDF and the one to be estimated as a function of the probability level using a fitting power-law on probability for the $P(R)_{60}$ to $P(R)_1$ conversion factors.

6.8.2 Recommendations ITU–R P.530-17 and ITU-R P.838-3

Recommendation ITU–R p. 530-17 [41] defines prediction methods for the propagation effects that should be aware of in the design of digital fixed LOS links in rainfall conditions. For rainfall attenuation and prediction of the rain rate $R_{0.01}$ exceeded for 0.01% of the time with an integration time of one minute and computed the specific attenuation γ_R dB/km for the frequency, polarization, and rain rate of interest using recommendation ITU-R P.838-3 [42], which defines the specific rain attenuation γ_R in dB/km rather than attenuation in dB as in the EM theory derivation. The value of γ_R depends on the rain rate mm/h (accumulated depth of rainfall per a given time) at the area of interest; polarization and frequency, f (GHz) of transmitted signal. The specific attenuation γ_R is calculated from the rain rate R in mm/h using the power-law expression:

$$\gamma_R = k\,R^{\alpha}\ \text{dB/km} \tag{6.61}$$

where the coefficients k and α are a function of the frequency in GHz in the range of 1 to 1000 GHz and determined from *curve-fitting to power-law coefficients derived from the scattering calculations*

$$\log_{10}k = \sum_{j=1}^{4} a_j \exp\left[-\left(\frac{\log_{10}f - b_j}{c_j}\right)^2\right] + m_k\log_{10}f + c_k \tag{6.62}$$

$$\alpha = \sum_{j=1}^{5} a_j \exp\left[-\left(\frac{\log_{10}f - b_j}{c_j}\right)^2\right] + m_\alpha\log_{10}f + c_\alpha \tag{6.63}$$

where the coefficients k and α can be for horizontal polarization (k_H, α_H) or for vertical polarization (k_V, α_V) . The values for the constants (a, b, c, and m) are given in Tables 6.7–6.10.

6.8.2.1 Linear and Circular Polarization

The coefficients α and k are given by

$$k = (k_H + k_V + (k_H - k_V)\cos^2\theta\,\cos 2\tau)/2 \tag{6.64}$$

$$\alpha = k_H\,\alpha_H + k_V\,\alpha_V + (k_H\,\alpha_H - k_V\,\alpha_V)\cos^2\theta\,\cos 2\tau)/2k \tag{6.65}$$

where θ is the path elevation angle and τ is the polarization tilt angle relative to the horizontal ($\tau = 45°$ for circular polarization). The coefficients k and α are given in Table 6.11 for the frequencies 1 to 30 GHz. Higher coefficient values at frequencies between 30 -50 GHz can be found in [42].

Using equations (6.64) and (6.65) and the data given in Table 6.11, the attenuation by rain in dB/km at 30 GHz with rain rate 5 mm/h and 150 mm/h is computed for circular polarization when the elevation angle is 45° and shown in Table 6.12.

Table 6.7 Constant rain loss coefficient k for horizontal polarization k_H [42].

		Coefficients for horizontal polarization k_H			
j	a_j	b_j	c_j	m_k	c_k
1	−5.33980	−0.10008	1.13098	−0.18961	0.71147
2	−0.35351	1.26970	0.45400		
3	−0.23789	0.86036	0.15354		
4	−0.94158	0.64552	0.16817		

Table 6.8 Constant rain loss coefficients α for horizontal polarization α_H [42].

		Coefficients for horizontal polarization α_H			
j	a_j	b_j	c_j	m_k	c_k
1	−0.14318	1.82442	−0.55187	0.67849	−1.95537
2	0.29591	0.77564	0.19822		
3	0.32177	0.63773	0.13164		
4	−5.37610	−0.96230	1.47828		
5	16.1721	−3.29980	3.43990		

Table 6.9 Constants rain loss coefficients α for horizontal polarization k_V [42].

		Coefficients for vertical polarization k_V			
j	a_j	b_j	c_j	m_k	c_k
1	−3.80595	0.56934	0.81061	−0.16398	0.63297
2	−3.44965	−0.22911	0.51059		
3	−0.39902	0.73042	0.11899		
4	0.50167	1.07319	0.27195		

Table 6.10 Constants rain loss coefficients α for vertical polarization α_V [42].

		Coefficients for vertical polarization α_V			
j	a_j	b_j	c_j	m_α	c_α
1	−0.07771	2.33840	−0.76284	−0.053739	0.83433
2	0.56727	0.95545	0.54039		
3	−0.20238	1.14520	0.26809		
4	−48.2991	0.791669	0.116226		
5	48.5833	0.791459	0.116479		

Table 6.11 Values of the coefficients k_H, α_H, k_V, and α_V at frequencies from 1 to 30 GHz [42].

Frequency GHz	k_H	α_H	k_V	α_V
1	0.0000259	0.9691	0.0000308	0.8592
1.5	0.0000443	1.0185	0.0000574	0.8957
2	0.0000847	1.0664	0.0000998	0.9490
2.5	0.0001321	1.1209	0.0001464	1.0085
3	0.0001390	1.2322	0.0001942	1.0688
3.5	0.0001155	1.4189	0.0002346	1.1387
4	0.0001071	1.6009	0.0002461	1.2476
4.5	0.0001340	1.6948	0.0002347	1.3987
5	0.0002162	1.6969	0.0002428	1.5317
5.5	0.0003909	1.6499	0.0003115	1.5882
6	0.0007056	1.5900	0.0004878	1.5728
7	0.001915	1.4810	0.001425	1.4745
8	0.004115	1.3905	0.003450	1.3797
9	0.007535	1.3155	0.006691	1.2895
10	0.01217	1.2571	0.01129	1.2156
11	0.01772	1.2140	0.01731	1.1617
12	0.02386	1.1825	0.02455	1.1216
13	0.03041	1.1586	0.03266	1.0901
14	0.03738	1.1396	0.04126	1.0646
15	0.04481	1.1233	0.05008	1.0440
16	0.05282	1.1086	0.05899	1.0273
17	0.06146	1.0949	0.06797	1.0137
18	0.07078	1.0818	0.07708	1.0025
19	0.08084	1.0691	0.08642	0.9930
20	0.09164	1.0568	0.09611	0.9847
21	0.1032	1.0447	0.1063	0.9771
22	0.1155	1.0329	0.1170	0.9700
23	0.1286	1.0214	0.1284	0.9630
24	0.1425	1.0101	0.1404	0.9561
25	0.1571	0.9991	0.1533	0.9491
26	0.1724	0.9884	0.1669	0.9421
27	0.1884	0.9780	0.1813	0.9349
28	0.2051	0.9679	0.1964	0.9277
29	0.2224	0.9580	0.2124	0.9203
30	0.2403	0.9485	0.2291	0.9129

Low rainfall rate attenuation at mmWave frequency band is less significant compared to high rainfall rate. The latter has to be accounted for and it is of the order of 2 dB per 100 m. A linear plot of rain specific attenuation at temperature of 20°C and for rain rate 2.5–150 mm/h. is given in [26] based on calculations made by Setzer [43] and depicted in Figure 6.10.

The plot shows the rain attenuation increases with frequency and levels off around 100 GHz and even decrease slightly with increasing frequency.

Table 6.12 Calculated attenuation by rain at 18 GHz and 30 GHz with rain rate 5 mm/h and 150 mm/h.

18 GHz		30 GHz	
Rain loss dB/km for rain rate 5 mm/h	**Rain loss dB/km for rain rate 150 mm/h**	**Rain loss dB/km for rain rate 5 mm/h**	**Rain loss dB/km for rain rate 150 mm/h**
0.4	13.58	1.05	24.93

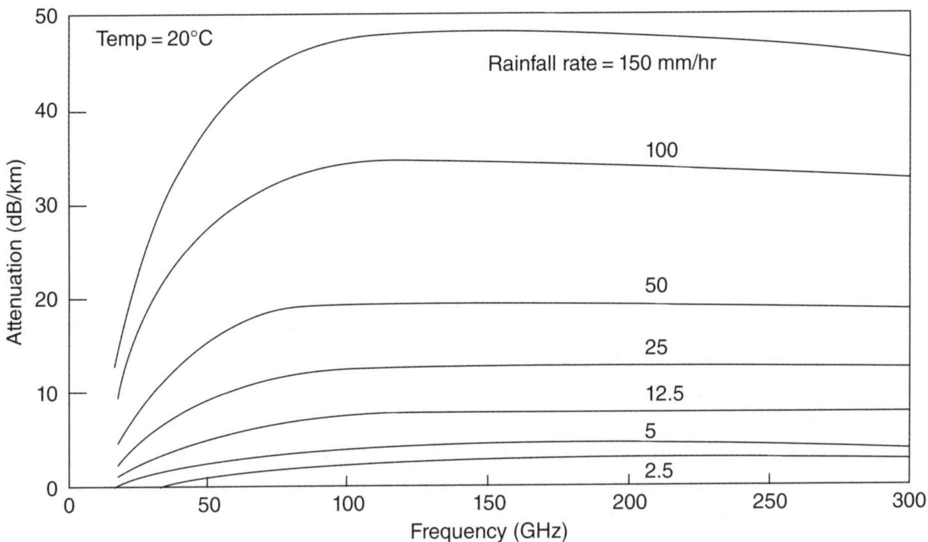

Figure 6.10 Rain attenuation (dB/km) for various rainfall rates at frequencies 1 to 300 GHz [26].

6.8.3 Recommendations ITU-R P.1144-6 and ITU-R P.837-7

Recommendation ITU-R P.1144-6 Annex1 provides a guide on how to determine an unknown position of a grid point $I(r, c)$ where r is a fractional row number and c is a fractional column number using bilinear interpolation, as illustrated in Figure 6.11.

The latitude grid *lat* defined by rows (Rs) and longitude grid by columns (Cs), respectively. The grid point is denoted as $I(R, C)$.For example, assume we are given the four grid points: $I(R, C)$; $I(R, C+1)$; $I(R+1, C)$; and $I(R+1, C+1)$. The grid point $I(r, c)$ is computed using bilinear interpolation as

$$\begin{aligned} I(r,c) = I(R,C) * [(R+1-r)(C+1-c)] + I(R+1,C) \\ * [(r-R)(C+1-c)] + I(R,C+1) * [(R+1-r)(c-C)] + I(R+1,C+1) \\ * [(r-R)(c-C))] \end{aligned} \tag{6.66a}$$

This could be more appropriately expressed as

$$I(r,c) = \left[I(R,C)\ \ I(R+1,C)\ \ I(R,C+1)\ \ I(R+1,C+1)\right] \begin{bmatrix} [(R+1-r)(C+1-c)] \\ [(r-R)(C+1-c)] \\ [(R+1-r)(c-C)] \\ [(r-R)(c-C))] \end{bmatrix} \tag{6.66b}$$

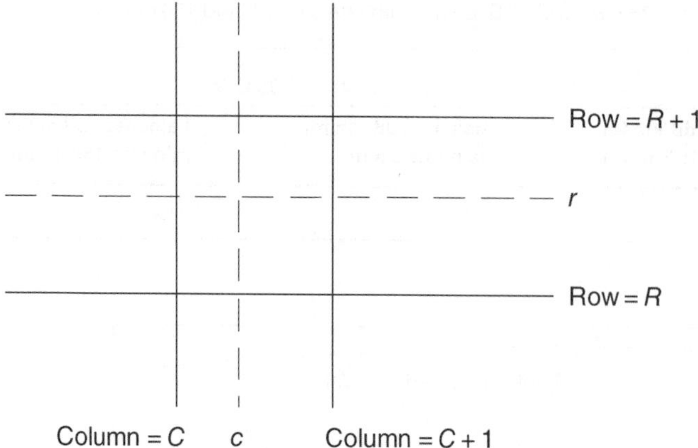

Figure 6.11 Sketch showing the location of area of interest for rain rate at grid point (r, c) [44].

See 6.A for a tutorial on bilinear interpolation.

Recommendation ITU-R P.837-7 [45], issued in June 2017 as an improvement to Recommendation ITU 837-6 [46], presents methods for determining rainfall statistics with one-minute integration time in two annexes. Annex-1 confers the case when reliable long-term rainfall rate data is *not available* at the point of interest, while Annex-2 considers the case when reliable long-term local rainfall rate data is *available.*

Annex-1 prediction method calculates the rainfall rate exceeded, for a desired average annual probability of exceedance and a given location, using digital maps of *monthly total rainfall* and *monthly mean surface temperature.* The monthly mean rainfall maps were derived from 50 years (1951–2000) of data over land and from 36 years (1979–2014) over water.

Let us denote the monthly mean total rainfall data as MT_{ii} *mm* is available as a digital map where *ii* is the month number. The latitude grid is from −90.125° N to +90.125° N organized in steps of 0.25°, and the longitude grid in steps of 0.25° from −180.125° E to +180.125° E. Define the following parameters: P for the desired annual probability of exceedance in %; *Lat* for the latitude of the desired location in degrees north; *Lon* for longitude of the desired location degrees east. The output of the prediction method R_p is the rainfall rate exceeded for the desired probability of exceedance in mm/h. The number of days N_{ii} in each month of the year defined in the prediction method is as follows:

Month	Jan	Feb	Mar	Apr	May	June	July	Aug	Sep	Oct	Nov	Dec
ii	01	02	03	04	05	06	07	08	09	10	11	12
N_{ii}	31	28.25	31	30	31	30	31	31	30	31	30	31

The monthly mean surface temperature is denoted as T_{ii} deg. K at the desired location *Lat, Lon* from reliable long-term local data. If reliable long-term local data is not available, T_{ii} deg. K at the desired location can be obtained from the digital maps of monthly mean surface temperature in Recommendation ITU-R P.1510, which is presented in Section 6.8.4.

For each month, determine MT_{ii} *mm* at the desired location from available reliable long-term local data. If such reliable data is not available, MT_{ii} *mm* can be determined as follows. Identify the four grid points Lat_1, Lon_1; Lat_2, Lon_2; Lat_3, Lon_3; and Lat_4, lon_4 surrounding the desired location *Lat, Lon.*

Now determine monthly mean total rainfall at the four grid points $MT_{ii,1}$; $MT_{ii,2}$; $MT_{ii,3}$; and $MT_{ii,4}$. We can use the bilinear interpolation described in (6.66) to determine the monthly mean total rainfall MT_{ii} at the desired location,*Lon*. Next we convert the temperature from $T_{ii}\, deg.K$ to $t_{ii}{}^{o}\, C$ and calculate r_{ii} as

$$r_{ii} = 0.5874 \exp(0.0883 t_{ii}) \quad for \;\; t_{ii} \geq 0^{o}\, C \;\; \text{mm/h}$$

$$r_{ii} = 0.5874 \qquad\qquad for \;\; t_{ii} < 0^{o}\, C \;\; \text{mm/h}$$

The monthly probability of rain is given as

$$P_{0_{ii}} = 100\, \frac{MT_{ii}}{24 \text{x} N_{ii} \text{x} r_{ii}} \%$$

For each month, if $P_{0_{ii}} > 70\%$, set it to $P_{0_{ii}} = 70\%$ and $r_{ii} = \frac{70}{100} \frac{MT_{ii}}{24 \text{x} N_{ii}}$. The annual probability of rain $P_{0_{ann}} = P(R > 0)$ is as follows:

$$P_{0_{ann}} = \frac{\sum_{ii=1}^{12} N_{ii} \text{ x } P_{0_{ii}}}{365.25} \%$$

A few repudiations in the prediction method are in order:

1. If the desired rainfall rate probability of exceedance $p > P_{0_{ann}}$, then set the rainfall rate at the desired rainfall rate of exceedance to zero mm/h.
2. If the desired rainfall rate probability of exceedance $p \leq P_{0_{ann}}$, adjust the rainfall rate R_{ref} until the absolute relative error between annual rainfall rate probability of exceedance $P(R > R_{ref})$ and the desired rainfall rate probability of exceedance p is less than 0.001%, i.e.

$$100 \left| \frac{P(R > R_{ref})}{p} - 1 \right| < 0.001$$

where

$$P(R > R_{ref}) = \frac{\sum_{ii=1}^{112} N_{ii} P_{ii}(R > R_{ref})}{365.25} \%$$

$$P_{ii}(R > R_{ref}) = P_{0_{ii}}\, Q\left(\frac{\log(R_{ref}) + 0.7938 - \log(r_{ii})}{1.26}\right) \%$$

and $Q(x)$ is the tail distribution function of standard normal distribution. At the end of the adjustment, set $R_p = R_{ref}$.

6.8.4 Recommendation ITU R P.1510-1

The annual and monthly mean surface temperature when no local data are available can be determined by the bilinear method described in Recommendation ITR P.1510-1 Annex-1 [47]. The monthly mean surface temperature data T_{ii} deg. K at 2 m above the surface of the earth are available as digital maps. The latitude grid is from $-90°\, N$ to $+90°\, N$ organized in steps of 0.75°, and the longitude grid in steps of 0.75° from $-180°\, E$ to $+180°\, E$. The annual mean surface temperature is also determined at 2 m above Earth's surface with identical *Lat, Lon* range. These digital maps are available by request to the ITU.

Similar to the monthly mean total rainfall data, choose four grid points, Lat_1, Lon_1; Lat_2, Lon_2;Lat_3, Lon_3; and Lat_4, lon_4 surrounding the desired location *Lat, Lon* and found out their monthly and annually mean surface temperature at 2 m above surface of the earth T_1, T_2, T_3, T_4. Subsequently, determine T at the desired location *Lat, Lon* by carrying out a bilinear interpolation method described above.

6.9 Attenuation from Snow and Hail

Snow can be viewed as a mixture of ice, water, and air in proportions that fluctuate depending on the type of snow. There are three types of snow namely: dry snow with equivalent rainfall rate of 0.1 mm/h; wet snow that is a mixture of rain and snow with equivalent rainfall rate of 5 mm/h, and moist snow with equivalent rainfall of 1 mm/h.

The dielectric constant of ice is much smaller than that of water, so for a given physical cross section, ice scattering is much smaller than those of liquid water drops [26]. In addition, ice particles absorb less power than a comparable water drop. So, for equivalent rates of precipitation, melted ice equivalent in mm/h, the attenuation due to snow and hail is expected to be considerably less than that produced by an equivalent rainfall. For mmWave frequencies (below 100 GHz), snow and hail attenuation is likely to be much less than rainfall attenuation.

Snow is challenging to characterize because it is a very complex material with a wide range of properties such as *surface roughness*; *density* and *particle distribution*; and *wetness*. These properties are extremely difficult to measure or mathematically formulate in an empirical model. The level of the difficulty depends on the condition of the snow. Mechanically, snow is a nonlinear viscoelastic material with a high degree of compressibility that can be modified by heat.

6.9.1 EM Propagation Properties Through Snow

EM propagation through snow is defined by the magnetic permeability μ and the relative permittivity also called the relative complex dielectric constant ε of the snow medium. For most types of snow, μ is very close to the magnetic permeability of free space μ_0 so the propagation through snow is entirely driven by ε that has the following complex form

$$\varepsilon = \varepsilon' - j\,\varepsilon'' \tag{6.67}$$

where ε' and ε'' are the real and imaginary terms, respectively. The dielectric behaviour of air is similar to that of free space, but the dielectric constants for water and snow are vastly different than free space, both can be described approximately to the first order by the Debye equation:

$$\varepsilon = \varepsilon_\infty + \frac{\varepsilon_0 - \varepsilon_\infty}{1 + j2\pi f\tau} \tag{6.68}$$

where ε_0 static dielectric constant, i.e. $\lim_{f\to 0}(\varepsilon)$ and $\varepsilon_\infty = \lim_{f\to\infty}(\varepsilon)$ where τ material relaxation time is in seconds:

f EM wave frequency in Hz

Debye Eq. (6.68) can be articulated in the complex form (6.67):

$$\varepsilon' = \varepsilon_\infty + \frac{\varepsilon_0 - \varepsilon_\infty}{1 + (2\pi f\tau)^2} \tag{6.69}$$

$$\varepsilon'' = \frac{2\pi f\tau(\varepsilon_0 - \varepsilon_\infty)}{1 + (2\pi f\tau)^2} \tag{6.70}$$

Using Debye equation and assuming $\varepsilon_0 = 87.9$ and $\varepsilon_\infty = 4.9$, the plot of the behaviour of the real, ε_w', and the imaginary ε_w'' terms of the water dielectric constant at 0 °C as a function of frequency is depicted in Figure 6.12.

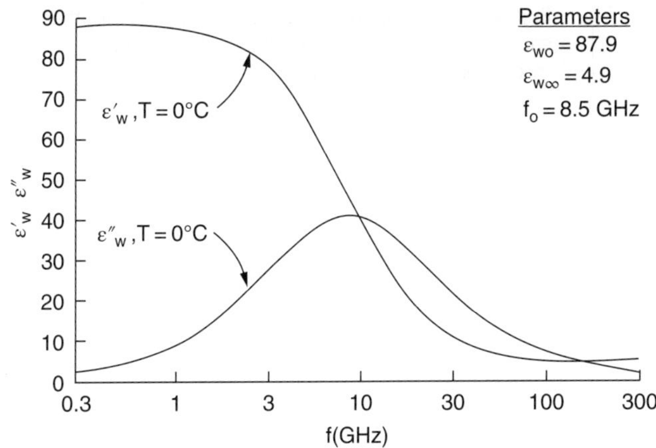

Figure 6.12 Complex dielectric constant of water at 0°C versus frequency in GHz [48].

The peak in ε_w'' occurs at 8.5 GHz, which is identified as the *relaxation frequency* f_r given as [48]

$$f_r = \frac{1}{2\pi\tau} \tag{6.71}$$

and τ is the relation time of the material in sec. The dielectric properties of ice over a large frequency range can also be interpreted by the Debye equation of the same form as for water. Nonetheless, relaxation frequency of water lies in the radio region, yet for ice, the relaxation frequency is reduced to kHz region and as for water, it is temperature dependent.

The penetration depth δ_p of EM wave in a medium is defined as the depth at which the average power travelling downward in the snow medium is equal to $\frac{1}{e}$ of the power at a point just beneath the surface of the medium where e = 2.718. For a medium, the uniform extinction coefficient k_e accounts for two components: an EM wave absorption defined by a coefficient k_a, and an EM wave scattering expressed by a coefficient k_s:

$$k_e = k_a + k_s \tag{6.72}$$

and the penetration depth is

$$\delta_p = \frac{1}{k_e} \tag{6.73}$$

Figure 6.13 shows the computed values of δ_p plotted as a function of fraction liquid water content within the snow sample, v_w, for three radio frequencies. For dry snow, $v_w = 0$.

Wet snow is inferred in [49] to contain either dry snow with $\varepsilon_{\text{ds}} = \varepsilon_{\text{ds}}' - j\,\varepsilon_{\text{ds}}''$ and water with $\varepsilon_{\text{ds}} = \varepsilon_w' - j\,\varepsilon_w''$ or air with $\varepsilon_a = 1$, ice with $\varepsilon_i = \varepsilon_i' - j\,\varepsilon_i''$ and water replacing the real and imaginary terms of the relative permittivity, ε, indicated in (6.69) and (6.70), respectively. A Debye-like semi-empirical model for wet snow is derived in [50] as

$$\varepsilon_{ws}' = A + \frac{Bv_w^x}{1 + \left(\frac{f}{f_r}\right)^2} \tag{6.74}$$

$$\varepsilon_{ws}'' = \frac{C\left(\frac{f}{f_r}\right)v_w^x}{1 + \left(\frac{f}{f_r}\right)^2} \tag{6.75}$$

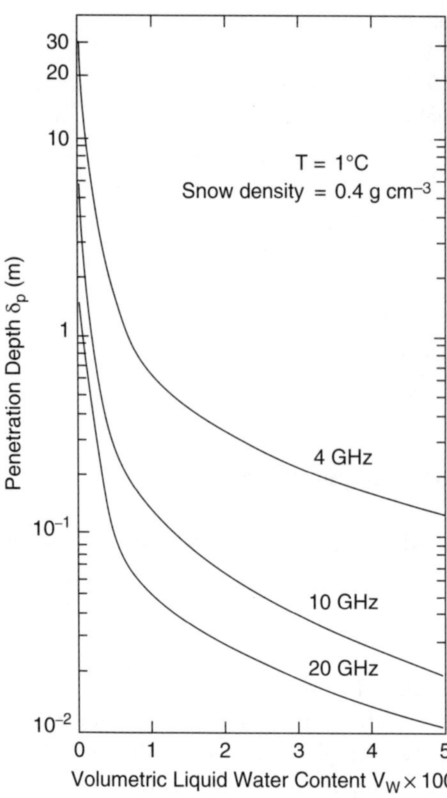

Figure 6.13 Computed values of penetration depth in snow as a function of snow liquid water content v_w for three radio frequencies [48].

where A, B, C, and x are all constants to be determined by fitting the model to the measured data in the frequency range 3–37 GHz range, and v_w defined above. This approach gives the following values to these constants:

$$A = 1.0 + 1.83\,\rho_{ds} + 0.02\,a_1 v_w^{1.015} + b_1 \tag{6.76}$$

$$B = 0.073\,a_1 \tag{6.77}$$

$$C = 0.073\,a_2 \tag{6.78}$$

$$x = 1.31 \tag{6.79}$$

$$f_r = 9.07\ \text{GHz} \tag{6.80}$$

and ρ_{ds} is the density of dry snow in gm. cm^{-3}, ρ_{ws} is the density of wet snow in gm/cm^{-3} . For the *Debye-like* model we have

$$a_1 = a_2 = 1.0 \text{ and } b_1 = 0 \tag{6.81}$$

The Debye-like model is simpler to use but its accuracy is poorer above 15 GHz. The accuracy may be improved by expressing the constants A, B, and C in (6.74) and (6.75) as frequency-dependent coefficients for $f > 15$ empirically defined functions as

$$a_1 = 0.78 + 0.03f - 0.58 * 10^{-3} f^2 \tag{6.82}$$

$$a_2 = 0.97 - 0.39 * 10^{-2} f + 0.39 * 10^{-3} f^2 \tag{6.83}$$

$$b_1 = 0.31 - 0.05f + 0.87 * 10^{-3} f^2 \tag{6.84}$$

where f is in GHz and $f > 15$ GHz . For<15 GHz, constant values become as in (6.81). The new empirical model is known as *modified Debye-like.* The modified Debye-like

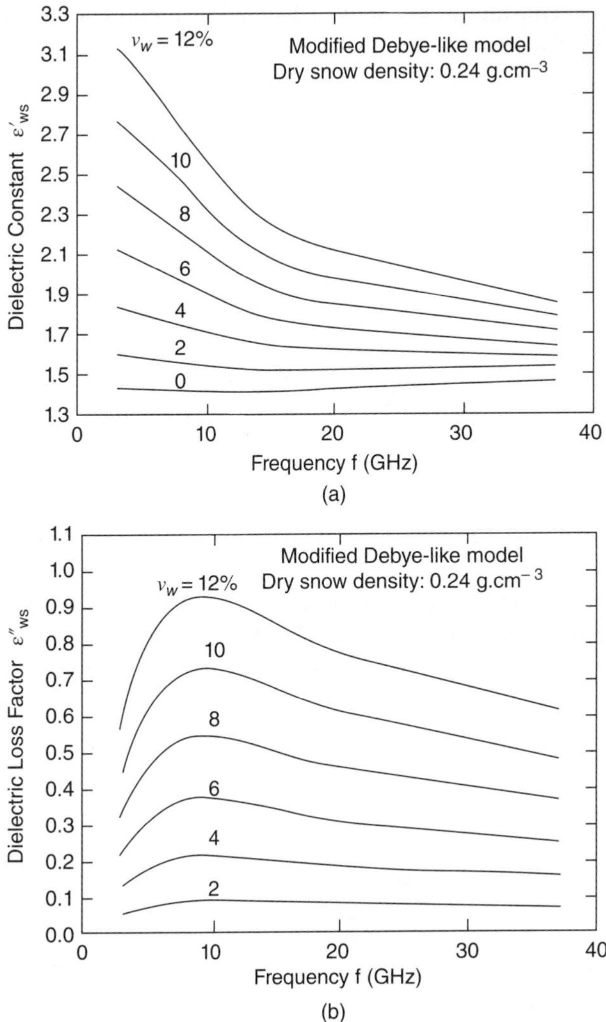

Figure 6.14 (a) Dielectric constant of snow, plotted as a function of frequency with liquid-water content as a parameter [50]. (b) Loss factor of snow, plotted as a function of frequency with liquid-water content as a parameter [50].

frequency-dependent coefficients in Eqs. (6.82)–(6.84) were used to calculate the values of ε'_{ws} and ε''_{ws} for wet snow and illustrated in Figure 6.14. Figure 6.14a shows the effect of liquid water content on the dielectric constant ε'_{ws} according to the modified Debye-like model with an average dry snow density $\rho_{ds} = 0.24$ g/m^3 in the frequency range 3–37 GHz. Figure 6.14b exhibits the loss factor of snow ε''_{ws} with water content as parameter. The loss factor reaches its maximum value at 9.0 GHz, *which* is typical water relaxation frequency at 0°C. The sharp increase in ε''_{ws} between 3 and 6 GHz is due to the large increase in the magnitude of ε''_{w} for water over this frequency range.

In order to obtain the density of dry snow ρ_{ds} (g. cm^{-3}) from the measured density of a wet snow sample, ρ_{ws} (g. cm^{-3}), and ν_w is the volume fraction of liquid water (the volumetric water content) within the snow sample. The following formula is used in [50]:

$$\rho_{ds} = \frac{\rho_{ws} - \nu_w}{1.0 - \nu_w} \tag{6.85}$$

Furthermore, the real part of dielectric constant of dry snow is defined by empirical model fitted to measured dielectric constant ε'_{ds} data taken at frequency range 4–18 GHz [49]:

$$\varepsilon'_{ds} = 1 + 1.91\, \rho_{ds} \;\; \rho_{ds} \leq 0.38\,\text{gm.cm}^{-3} \tag{6.86}$$

The variation of the real part of the dielectric constant of wet snow, ε'_{ws} with liquid water content, v_w, is denoted by the incremental dielectric constant, $\Delta\varepsilon'_{ws}$ is given as

$$\Delta\varepsilon'_{ws} = \varepsilon'_{ws} - \varepsilon'_{ds} \tag{6.87}$$

Substituting for ε'_{ds} (6.86) in (6.87), we get

$$\Delta\varepsilon'_{ws} = \varepsilon'_{ws} - 1.91\, \rho_{ds} - 1 \tag{6.88}$$

So the contribution due to liquid water is given by (6.88). Nonetheless, measurement carried out in [51] demonstrated that the increase in of the real term of the dielectric constant $\Delta\varepsilon'_{ws}$ caused by liquid water can be model as

$$\Delta\varepsilon'_{ws} = \varepsilon'_{ws} - 1.7\, \rho_{ds} - 0.7\rho_{ds} - 1 \tag{6.89}$$

Furthermore, for ε''_{ws}, liquid water dominates the attenuation, and there is no need to define an incremental quantity.

A comparison of model predictions two-phase Polder-Van Santen model, Debye-like model, modified Debye model with the measured values of $\Delta\varepsilon'_{ws}$ and ε''_{ws} at 37 GHz is depicted in Figure 6.15. The overall fit of the two-phase Polder-Van Santen model to the data is quite good

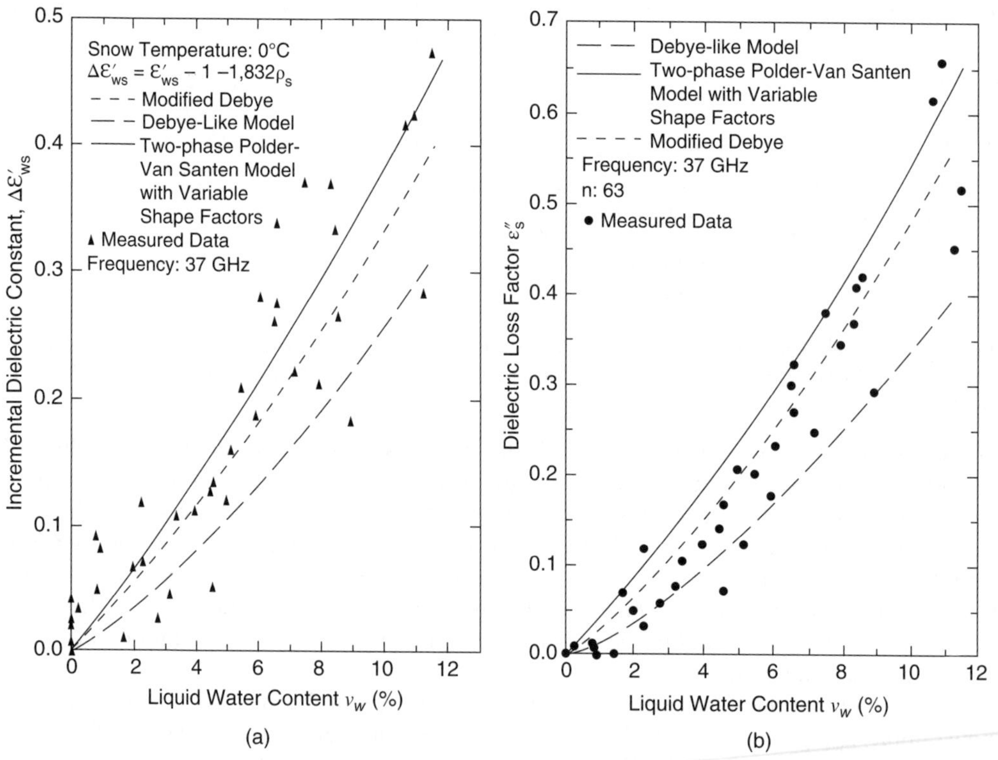

Figure 6.15 Comparison of model predictions with measured values of $\Delta\varepsilon'_{ws}$ and ε''_{ws} at 37 GHz. [50].

to measured $\Delta\varepsilon'_{ws}$, as seen in Figure 6.15a, while modified Debye model provides the best fit to measured ε''_{ws}, as seen in Figure 6.15b. For further details on the Polder-Van Santen model, please refer to [52].

6.9.2 Transmission Model for Ice Slab

Consider a slab of scatterers represented by a snow slab (i.e. dry snow) of thickness (d) with an incident plane EM wave as depicted in Figure 6.16.

At the air-snow interface, some of the incident EM wave is scattered, thus reducing the wave intensity by T, the Fresnel power transmission coefficient, and by S due to scattering by a rough surface. Thus, the EM wave at the output of the slab is $\mathrm{I}_0 TS$. A simple model for S has an exponential form as

$$S = \exp[-(2k_0 h_s)^2] \tag{6.90}$$

where $k_0 = \frac{2\pi}{\lambda}$ and h_s is the rms height of the surface of the slab determined by measurement. The total intensity of the EM wave at the output is

$$\mathrm{I_d} = \mathrm{I}_0 T \exp[-(2k_0 h_s)^2] = \mathrm{I_c} + \mathrm{I}_i \tag{6.91}$$

The latter two components at the output are: the coherent EM wave that would be attenuated by absorption and scattering and the other component is the incoherent that would be attenuated due to the scattering mostly in the forward direction. It was shown in [53] that for a small thickness snow slab, the loss factor L_{ice} (dB) is given as

$$L_{ice} \cong L_s + 4.34\, k_e d \text{ dB} \tag{6.92}$$

where

$$L_s\,(dB) = 34.72\, k_0^2 h_s^2 \tag{6.93}$$

The extinction coefficient k_e is the sum of absorption coefficient k_a and the forward scattering coefficient, k_s, that is

$$k_e = k_a + k_s \tag{6.94}$$

On the other hand, for large slab thickness, the snow slab attenuation can be approximated as

$$L_{ice} \cong L_s - 10\log_{10} q + 4.34\, k_a d \text{ dB} \tag{6.95}$$

where $0 \le q \le 1$ represents a fraction of the total scattered power picked up by the receiving antenna and k_a is the absorption coefficient.

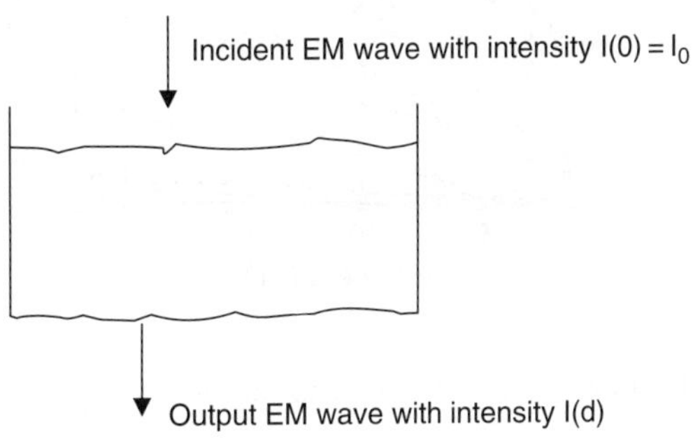

Figure 6.16 Sketch of transmission of an electromagnetic wave through an ice slab.

6.9.3 Empirical Model for Snow Attenuation

Dry snow is a mixture of air and fresh-water ice. The dielectric constant of air may be taken as unity so the dielectric constant for ice is similar to that for dry snow. Therefore, Eqs. (6.92)–(6.95) apply to dry snow. The first step in *deriving any empirical model* is to conduct a campaign to measure the transmission loss factor L_{ice} (dB) using snow samples (acquired from the natural snow) with various grain sizes and densities. Such measurements were carried out in 1985 in Helsinki. The snow aggregate grain size varies between 0.2 and1.6 mm; surface roughness between 0 and 4 mm; and density between 0.172 and 0.380 g/cm^3 were used in the measurement at 18, 35, 60, and 90 GHz. Snow samples have a wide ranging of surface roughness (0–4 mm), and grain density (0.172–0.380 g/cm^3). Furthermore, samples are also used in investigating the effects of temperature on snow loss factor. The loss factor (dB) measurements are depicted in Figure 6.17 showing loss factor (dB) versus snow thickness (0–20 cm) for the 23 snow grain sizes shown in Table 6.13. The measurements are repeated at frequencies 18, 35, 60, and 90 GHz.

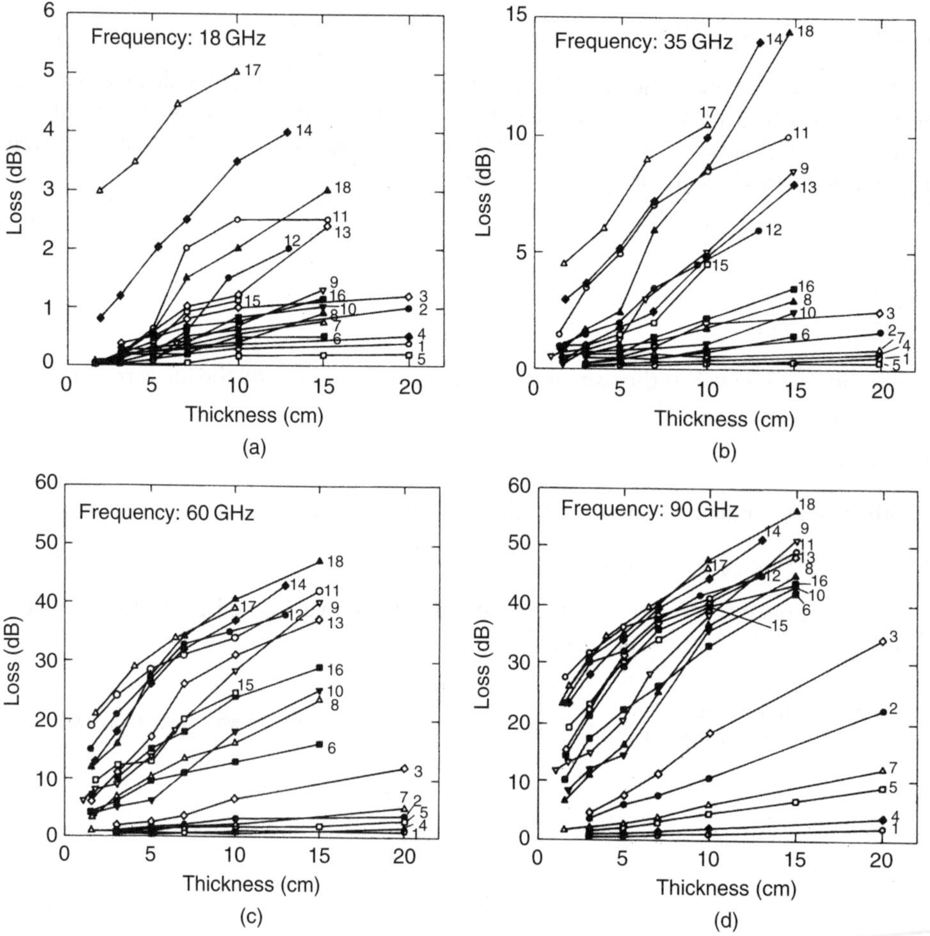

Figure 6.17 Measured transmission loss for snow samples [53].

Table 6.13 Properties of snow cover (both top and bottom layers) samples [53].

No.	Date (1985)	Depth in Snowpack	Observed Mean Grain Size (mm)	Surface Roughness (mm)	Melt-Freeze Cycles	Clustering	Density (g/cm^3)	Dielectric Constant at 10 GHz	Comments
1	Feb. 5	Top	0.2	0	None	None	0.172	1.31	One-hour old snow
2	Feb. 15	Near Top	0.5	0	None	None	0.194	1.34	Newly fallen snow
3	Feb. 16	Near Top	0.7	0	None	None	0.217	1.39	*Newly* fallen snow
4	Feb. 18	Top	0.2	0	None	None	0.322	1.58	Wind-driven 5-day-old snow
5	March 12	Top	0.3	0	None	None	0.277	1.52	
6	March 12	Middle	0.9	0	None	None	0.268	1.49	
7	March 18	Near Top	0.4	0	None	None	0.235	1.41	
8	March 21	Near Bottom	1.0	1	None	None	0.315	1.58	
9	March 29	Top	1.0	1	Few	None	0.385	1.75	Hard snow
10	March 29	Near Bottom	1.0	1	None	None	0.276	1.50	Separate grains
11	April 7	Top	1.3	2 to 3	Some	Some	0.307	1.61	
12	April 11	Top	1.2	3	Some	Some	0.304	1.61	
13	April 11	Near Bottom	1.3	1 to 3	None,	None	0.293	1.54	
14	April 13	Top	1.5	2 to 3	Some	Some	0.345	1.64	
15	April 13	Middle	1.1	1 to 2	Few	None	0.332	1.63	
16	April 16	Bottom	1.1	1 to 2	None	None	0.361	1.77	Separate grains; no continuous structure
17	April 17	Top	1.5	2 to 4	Some	Some	0.390	1.79	
18	April 30	Near Top	1.6	2 to 3	Several	Some	0.351	1.66	
19	March 18	Near Top	0.4	0	None	None	0.240	1.43	Acquired 1 m away from No. 7
20	March 29	Near Bottom	1.0	1	None	None	0.271	1.46	Acquired 1 m away from No. 10
21	April 7	Top	1.3	2 to 3	Some	Some	0.311	1.64	Acquired 1 m away from No. 11
22	April 13	Top	1.5	2 to 3	Some	Some	0.350	1.68	Acquired 1 m away from No. 14
23	April 17	Top	1.5	2 to 4	Some	Some	0.380	1.72	Acquired 1 m away from No. 17

Samples 1 to 18 were measured as a function of sample thickness and samples 19 to 23 as a function of temperature.

The following characteristics can be observed from the measured attenuation shown in Figure 6.17:

- The snow transmission loss increases rapidly with frequency: max attenuation at 18 GHz is about 5 dB (sample number 17), at 35 GHz nearly 15 dB (sample number 18), at 60 GHz about 50 dB (sample number 18), and at 90 GHz is very close to 60 dB (sample number 18).
- The snow transmission loss increases with snow grain size. Sample number 18 has the highest loss at frequency range 35–90 GHz and a mean grain size of 1.6 mm.
- The snow attenuation increases with snow thickness.
- Finally for a typical snow sample, the snow loss contained two identifiable regions, linear attenuation- thickness relation at small thickness and nonlinear as it grows larger.

The last observation has a pivotal role in deriving an empirical model based on the available experimental results. The behaviour of the loss factor (dB) versus snow thickness appears very much like the sketch in Figure 6.18.

Using (6.92) for small snow thickness, it is possible to determine L_s from the measured Loss (dB) as follows: extend the linear attenuation of the individual snow sample to meet the loss axis at $d = 0$ to give L_s *at* $d = 0$.

Furthermore, there are two slopes in Figure 6.18:

- Slope 1 is represented by (6.92).
- Slope 2 is represented in (6.95).

We determine slope 1 by differentiating the right hand side of equation (6.92) with respect to d to get

$$\text{Slope } 1 = 4.34\, k_e \tag{6.96a}$$

This can be obtained from the measured loss values and hence provide an estimate for k_e applying (6.96a). Similarly, by differentiating the right hand-side of equation (6.95) with respect to d we get

$$\text{Slope } 2 = 4.34\, k_a \tag{6.96b}$$

As before, we find slope 2 from the measurement to provide estimate k_a using (6.96b) and hence k_s.

However, It is clear that the values of L_s, k_e, and k_s exhibit a strong variation with snow type and snow grain size. The following empirical model for k_e was proposed in [53], in

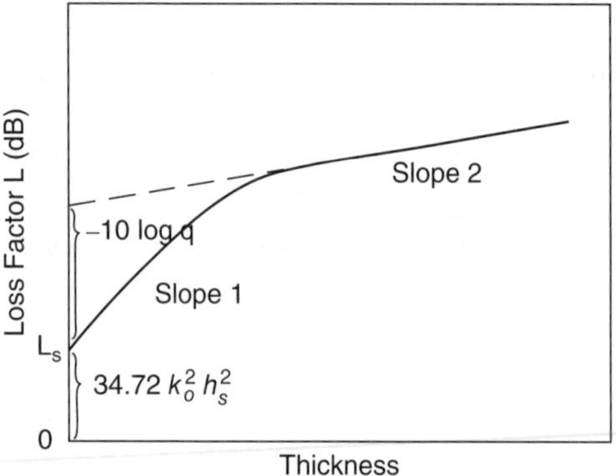

Figure 6.18 Plot of transmission loss versus snow thickness [53].

terms of frequency f in GHz and snow particle diameter d_0 in mm for the frequency band 18–60 GHz:

$$k_e = 0.0018 f^{2.8} d_0^2 \text{ dB/m} \tag{6.97}$$

where f is in GHz and d_0 is the observed snow particle diameter in mm.

Using (6.93) for the estimated value the surface loss L_s in dB, we can calculate the effective RMS height of the surface fluctuation as

$$h_s = \sqrt{\frac{L_s}{34.72}}\ \frac{1}{k_0} \approx 0.17\ \frac{\sqrt{L_s}}{k_0} \tag{6.98}$$

where L_s is surface loss in dB and k_0 is the wave number in free space.

6.9.4 Strong Fluctuation Theory

There are two analytical approaches for studying the EM wave scattering by snow constituent grains. The traditional approach based on the *Maxwells's equations* with random perturbations in dielectric properties and the other approach is based on the *radiation transfer*, which became popular in the past due to its simplicity. The *radiation transfer* mostly generally adopted the Mie scattering theory to determine the scattering coefficients. However, there are strong disagreements against its applicability to snow grains. Most important among them are that the ice grains containing snow are so closely packed that the scatterers in the vicinity of a particular grain are in the EM near field. Thus, the use of far field scattering coefficients is inappropriate; coupling between particles is not accounted for; and the application of Mie theory to find the scattering coefficients is not appropriate since the ice grains are not spherical. As an alternative to Mie scattering, Rayleigh scattering is sometimes assumed. However, Rayleigh is not an appropriate assumption above 20 GHz.

For these reasons, there is an alternative scattering theory, known as *strong fluctuation theory*, that permits better scattering assessments. The strong fluctuation theory in [54] is applied to the study of electromagnetic wave scattering from a layer of random discrete scatterers. A key application of this theory is *remote sensing* to provide useful information about the Earth's atmosphere, the land, and the oceans. Typical example of its application is the random mixture of snow, soil, and vegetation, each with a definite dielectric constant. One of its unique features is the determination of the dielectric characteristics of a random scattering medium in terms of the medium constituents and their correlation functions, with the basic assumption that no two constituents exist in the same part of the space exposed to the same emitted electric field.

When snow becomes moist, its water content is not distributed entirely at random. Part of the water forms a film around the snow grains so the snow grains are not in the same mean external field as the water, and hence a basic assumption of the theory is violated. To alleviate such a weakness, researchers developed a theory of mixtures, which assumes a transition at the surface of the particles to allow for a change in the effective field. However, the new theory is limited to low frequencies where snow grains size are much greater than the frequency wavelength, and so particle size effects may be ignored. The strong fluctuation theory is modified in [55] to account for a thin film surrounding at least some of the snow grains medium.

6.10 Snow Dielectric Constant Formulation Using Strong Fluctuation Theory

We now follow the derivation in [55] where the modified strong fluctuation is used to formulate the effective complex dielectric constant for dry snow. Measurements data show that the real dry snow dielectric constant is independent of frequency between 100 MHZ and 900 GHz but it is *temperature- dependent*. In the temperature range −50 to 0 °C, the dry snow real dielectric constant is given by

$$Re(\varepsilon_{ds}) = \varepsilon_{ds}{}' = \frac{3.099T - 992.65}{T - 318.896} \tag{6.99}$$

where T is temperature in Kelvin.

The imaginary part of dry snow ($\varepsilon_{ds}{}''$) depends on both the temperature and frequency. For frequencies above the dry snow relaxation frequency and based on a best available experimental data, the $\varepsilon_{ds}{}''$ may be formulated as

$$\varepsilon_{ds}{}'' = \frac{A(T)}{f} + B(T) * f \tag{6.100}$$

where the coefficients A and B are temperature-dependent only and f in GHz. For the best fit of the available data, A and B coefficients can be formulated as

$$A(T) = e^{\frac{\left[12.50 - \frac{3.77*10^3}{T}\right]}{T}} \tag{6.101}$$

$$B(T) = 10^{-4} * \varepsilon_{ds}{}'(273.41 - T)^{-1/2} \tag{6.102}$$

Beside the information on dielectric characteristics of snow, strong fluctuation theory requires information on the correlation function between constituents of the snow medium. The correlation function was assumed to be exponential with correlation length and well supported by experimental data. The correlation length l_{corr} for dry snow can be modelled as

$$l_{corr} = \frac{2}{3}(1 - v)d \tag{6.103}$$

where v and d are the volume fraction of ice in the snow and the mean ice grain diameter, respectively. In wet snow, the water is assumed to be in menisci (i.e. a crescent shaped) between the ice grains, while part forms a thin film that surrounds the ice grains.

The correlation functions corresponding to the air component and the component corresponding to the water film covered ice particles were assumed to have identical correlation lengths given by (6.103). The remainder of the liquid (which is not in the film around the ice grains), is distributed in the form of pendular water rings and has a correlation length given by

$$l_{corr} = c(1 - v_p)\left(\frac{v_p}{v_c}\right)^{1/2} d \tag{6.104}$$

Where v_p and v_c are volume fractions of pendular water and volume fraction of composite particles of ice grains and film water and c is a constant with a typical value range from 0.25 to 0.56. The fraction of the water in the film around ice grains was estimated to be

$$w_f = 0.261 - 0.724\, v_w \tag{6.105}$$

where v_w is the total volume fraction of free water in the snow. Therefore, v_p may be expressed as

$$v_p = (1 - w_f)v_w \tag{6.106}$$

6.11 Summary

It is widely known that communications systems operating at microwave frequencies and above are exposed to fades, long outage periods, and severe reduction in service availability.

The scenarios are due to our atmosphere and due to the atmospheric interaction with radio transmission. In this chapter, we directed our consideration to the effects of gases composed in the atmosphere and on the atmospheric impact that shaped our climate and weather patterns. Such atmospheric events as rain, snow, ice, etc. degraded the radio signal's reception quality by injecting losses on the propagated waves in radio and mmWave frequency bands. We researched the contemporary developed propagation models to evaluate the losses that are due to oxygen and water vapour, and compare these models with models recommended by the international standards organizations such ITU.

Contemporary wave propagation models developed by AT&T and modified by researchers at Stanford University were summarized in Section 6.2. Signal attenuation by atmospheric gases were derived and the corresponding ITU recommendations were explained in detail and the effects of such varying temperature and pressure were taken into account. All these topics were presented in Section 6.3. The calculation of atmospheric gaseous attenuation for dry atmosphere was considered in Section 6.4, and the relevant ITU recommendations were explained in Section 6.5.

The rain attenuation at mmWave frequency bands for horizontal and vertical linear and circular polarizations were covered in Section 6.6. Modelling the rainfall rate at any location and the conversion of rainfall rate from T- min integration time to the ITU standards one-minute integration time were viewed in details in Sections 6.7 and 6.8. The attenuation by snow and hail and EM propagation transmission models, together with an introduction to the strong fluctuation theory are presented in Section 6.9. Finally, the snow dielectric properties using the strong fluctuation theory are given in Section 6.10.

6.A Bilinear Interpolation

Linear interpolation is widely used in image wireless communication processing. Interpolation is basically a form of curve – fitting exploiting a linear polynomial to create new points within the range of the know points. Let us denote the known point by the coordinates (x_1, y_1) and (x_2, y_2), the linear interpolation is based on the line between these points as shown in Figure 6.A.1 and we wish to find y as function of x i.e. $y = f(x)$.

The slope equation of a straight line is

$$\frac{y_2 - y_1}{x_2 - x_1} = \frac{y - y_1}{x - x_1} \tag{6.A.1}$$

We can simplify (6.A.1) to,

$$y = y_1 - x_1\frac{y_2 - y_1}{x_2 - x_1} + \frac{y_2 - y_1}{x_2 - x_1}x \tag{6.A.2}$$

We can rewrite (6.A.2) as a first order polynomial equation as

$$y = f(x) = a + bx \tag{6.A.3}$$

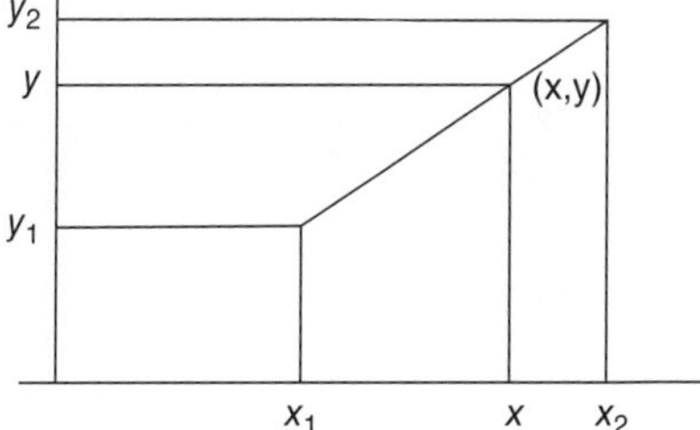

Figure 6.A.1 Interpolation of the position of a point at (x, y) using the coordinates of two surrounding points.

where

$$a = y_1 - x_1 b \tag{6.A.4a}$$

$$b = \frac{y_2 - y_1}{x_2 - x_1} \tag{6.A.4b}$$

Next we consider the bilinear interpolation using four known points surrounding the point of interest with the (x, y) coordinates. This scenario is depicted in Figure 6.A.2. The important concept is to perform linear interpolation first in one direction, and then again in the other direction. Whereas the interpolation in each direction is linear, the interpolation as a whole in nonlinear; in fact, it is quadratic. Suppose we want to determine the value of the unknown function f at (x, y), i.e. $f(x, y)$. Let us assume that we know the value of f at the four points denoted as Q_{11}; Q_{12}; Q_{21}; and Q_{22} corresponding to the coordinates (x_1, y_1); (x_1, y_2); (x_2, y_1); (x_2, y_2), respectively. The coordinates of the point of interest are

$$x_1 \le x \le x_2 \text{ and } y_1 \le y \le y_2$$

We first do a linear interpolation in the x direction to bring in the following equation:

$$f(x, y_1) \approx \frac{x_2 - x}{x_2 - x_1} Q_{11} + \frac{x - x_1}{x_2 - x_1} Q_{21} \tag{6.A.5}$$

$$f(x, y_2) \approx \frac{x_2 - x}{x_2 - x_1} Q_{12} + \frac{x - x_1}{x_2 - x_1} Q_{22} \tag{6.A.6}$$

We carry on the interpolating in the y direction to obtain the desired $f(x, y)$:

$$f(x, y) \approx \frac{y_2 - y}{y_2 - y_1} f(x, y_1) + \frac{y - y_1}{y_2 - y_1} f(x, y_2) \tag{6.A.7}$$

Use (6.A.3) and (6.A.4) to substitute for $f(x, y_1)$ and $f(x, y_2)$ in (6.A.7) to get

$$\begin{aligned} f(x, y) = {} & \frac{y_2 - y}{y_2 - y_1}\left[\frac{x_2 - x}{x_2 - x_1} Q_{11} + \frac{x - x_1}{x_2 - x_1} Q_{21}\right] \\ & + \frac{y - y_1}{y_2 - y_1}\left[\frac{x_2 - x}{x_2 - x_1} Q_{12} + \frac{x - x_1}{x_2 - x_1} Q_{22}\right] \end{aligned} \tag{6.A.8a}$$

$$\begin{aligned} f(y, x) = {} & \frac{(y_2 - y)(x_2 - x)}{(y_2 - y_1)(x_2 - x_1)} Q_{11} + \frac{y - y_1}{y_2 - y_1}\frac{x_2 - x}{x_2 - x_1} Q_{12} \\ & + \frac{y_2 - y}{y_2 - y_1}\frac{x - x_1}{x_2 - x_1} Q_{21} + \frac{y - y_1}{y_2 - y_1}\frac{x - x_1}{x_2 - x_1} Q_{22} \end{aligned} \tag{6.A.8b}$$

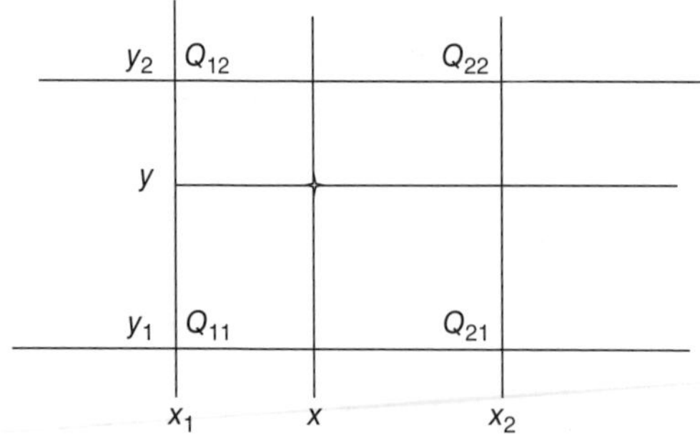

Figure 6.A.2 Interpolating the position of a point at (x, y) given the coordinates of four surrounding points.

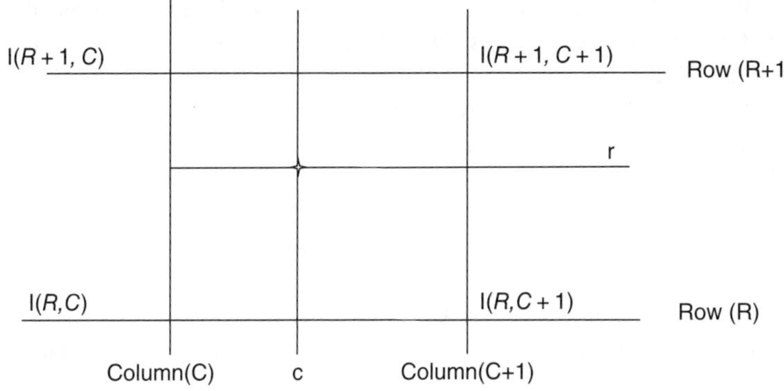

Figure 6.A.3 Interpolating the position of a point (*r*, *c*) given the positions of four surrounding points using Rows and columns coordinates.

The position of a point (r, c) using the positions of four surrounding point is depicted in Figure 6.A.3.

$$Q_{11} = \mathrm{I}(R, C), Q_{21} = \mathrm{I}(R, C+1), Q_{12} = \mathrm{I}(R+1, C), Q_{22} = \mathrm{I}(R+1, C+1)$$

$$f(y, x) = \left[\left[Q_{11}\ Q_{12}\ Q_{21}\ Q_{22}\right]\right] \begin{bmatrix} [(y_2 - y)(x_2 - x)] \\ [(y - y_1)(x_2 - x)] \\ [(y_2 - y)(x - x_1)] \\ [(y - y_1)(x - x_1))] \end{bmatrix} \tag{6.A.9}$$

Switching the (x, y) coordinates to rows and columns to determine the position of a point (r, c), we get

$$I(r, c) = \left[I(R, C)\ I(R+1, C)\ I(R, C+1)\ I(R+1, C+1)\right] \begin{bmatrix} [(R+1-r)(C+1-c)] \\ [(r-R)(C+1-c)] \\ [(R+1-r)(c-C)] \\ [(r-R)(c-C))] \end{bmatrix} \tag{6.A.10}$$

References

1 Rangan, S., Rappaport, T.S., and Erkip, E. (2014). Millimeter-wave cellular wireless networks: potentials and challenges. *Proceedings of the IEEE* 102 (3): 336–385.

2 Rappaport, T.S., Shu, S., Mayzus, R. et al. (2013). Millimeter wave mobile communications for 5G cellular: it will work! *IEEE Access* 1: 335–349.

3 Chao-Kai, W., Shi, J., Kai-Kit, W. et al. (2015). Channel estimation for massive MIMO using Gaussian-mixture Bayesian learning. *IEEE Transactions on Wireless Communications* 14 (3): 1356–1368.

4 Bai, T., Alkhateeb, A., and Heath, R.W. Jr., (2014). Coverage and capacity of millimeter-wave cellular networks. *IEEE Communications Magazine* 52 (9): 70–77.

5 Zhang, J.A., Huang, X., Dyadyuk, V., and Guo, Y.J. (2015). Massive hybrid antenna array for millimeter-wave cellular communications. *IEEE Wireless Communications* 22 (1): 79–87.

6 El Ayach, O., Heath, R.W., Jr., Abu-Surra, S., Rajagopal, S. and Pi, Z. (2012). 'Low complexity precoding for large millimeter wave MIMO systems', IEEE ICC -Signal Processing for Communications Symposium, 3724–3729.
7 Sayeed, A.M. and Raghavan, V. (2006). 'The ideal MIMO channel: maximizing capacity in sparse multipath with reconfigurable arrays', IEEE International Symposium on Information Theory (ISIT), 1036–1040.
8 Sayeed, A.M. and Raghavan, V. (2007). Maximizing MIMO capacity in sparse multipath with reconfigurable antenna arrays. *IEEE Journal of Selected Topics in Signal Processing* 1 (1): 156–166.
9 Erceg, V., Greenstein, L.J., Tjandra, S.Y. et al. (1999). An Empirically Based Path Loss Model for Wireless Channels in Suburban Environments. *IEEE Journal on Selected Areas in Communications* 17 (7): 1205–1211.
10 Hari, K.V.S. (2000). 'Interim Channel Models for G2 MMDS Fixed Wireless Applications', *IEEE 802.16 Broadband Wireless Access Working Group.*
11 Katev, P.D. (2012). 'Propagation models for WiMAX at 3.5 GHz', IEEE-Elektro Conference, 61–65.
12 Kim, T., Park, J., Seol, J-Y., Jeong, S., Cho, J. and Roh, W. (2013). 'Tens of Gbps support with mmWave beamforming systems for next generation communication', IEEE Globecom: Wireless Communications Symposium, 3685–3690.
13 Ahmed Iyanda Sulyman, A.I., Nassar, T.N., Samimi, M.K. et al. (2014). Radio propagation path loss models for 5G cellular networks in the 28 GHz and 38 GHz millimeter-wave bands. *IEEE Communications Magazine* 52 (9): 78–86.
14 Rappaport, T.S., Gutierrez, F. Jr.,, Ben-Dor, E. et al. (2013). Broadband millimeter-wave propagation measurements and models using adaptive-beam antennas for outdoor urban cellular communications. *IEEE Transactions on Antennas and Propagation* 61 (4): 1850–1859.
15 Smulders, P.F.M. and Correia, L.M. (1997). Characterization of propagation in 60 GHz radio channels. *IET Electronics & Communication Engineering Journal* 9 (2): 73–80.
16 Ben-Dor, E., Rappaport, T.S., Qiao, Y., and Lauffenburger, S.J. (2011). Millimeter-wave 60 GHz outdoor and vehicle AOA propagation measurements using a broadband channel sounder. *IEEE Globecom Proceedings* 1–6.
17 Pirkl, R. and Durgin, G.D. (2008). Optimum sliding correlator channel sounder design. *IEEE Transactions on Wireless Communications* 7 (9): 3488–3497.
18 Murdock, J.N., Ben-Dor, E., Qiao, Y., Tamir, J.I and Rappaport, T.S. (2012). 'A 38 GHz cellular outage study for an urban outdoor campus environment, IEEE Wireless Communications and Networking Conference: Mobile and Wireless Networks, 3085–3090.
19 Rappaport, T.S., Qiao, Y., Tamir, J.I., Murdock, J.N., Ben-Dor, E. (2012). 'Cellular broadband millimeter wave propagation and angle of arrival for adaptive beam steering systems', IEEE Radio and Wireless Symposium (RWS), 151–154.
20 Van Vleck, J.H. (1947). The absorption of microwaves by oxygen. *Physical Review* 71: 413–424.
21 Van Vleck, J.H. (1947). The absorption of microwaves by uncondensed water vapour. *Physical Review* 71: 425–433.
22 Rosenkranz, P.W. (1975). Shape of the 5 mm oxygen band in the atmosphere. *IEEE Transactions on Antennas and Propagation* 23 (4): 498–506.
23 Straiton, A.W. and Tolbert, C.W. (1960). Anomalies in the absorption of radio waves by atmospheric gases. *Proceedings of the IRE* 48 (5): 898–903.

24 Liebe, H.J., Gimmestan, G.G., and Hopponen, J.D. (1977). Atmospheric oxygen microwave spectrum-experiment versus theory. *IEEE Transactions on Antennas and Propagation* 25 (3): 327–335.

25 Ippolito, L.J. (1981). Radio propagation for space communications systems. *Proceedings of the IEEE* 69 (6): 697–727.

26 Dudzinsky, S.J. Jr., (1974). *Atmospheric Effects on Terrestrial Millimeter-Wave Communications.* Rand, California, USA: Defence Advanced research Projects Agency.

27 Reber, E.E., Mitchell, R.L., and Carter, C.J. (1970). Attenuation of the 5-mm wavelength band in a variable atmosphere. *IEEE Transactions on Antennas and Propagation* 18 (4): 472–479.

28 Liebe, H.J., Hufford, G.A. and Cotton, M.G. (1993) 'Propagation modelling of moist air and suspended water/ice particles at frequencies below 1000 GHz', AGARD Conference Proceedings 542, Atmospheric Propagation Effects through Natural and Man-Made Obscurants for Visible to MM-Wave Radiation, 31- 3-11.

29 Recommendation ITU-R P.676-10 (2013) 'Attenuation by atmospheric gases'.

30 Recommendation ITU-R P.835-5 (2012) 'Reference standard atmospheres'.

31 Segal, B. (1986). The influence of raingage integration time on measured rainfall-intensity distribution functions. *Journal of Atmospheric and Oceanic Technology* 3 (12): 662–671.

32 Bryant, G.H., Adimula, I., Riva, V., and Brussaard, G. (2001). Rain attenuation statistics from rain cell diameters and heights. *International Journal of Satellite Communications* 19 (3): 263–283.

33 Sauvageot, H. and Mesnard, F.D. (1999). The relation between the area-average rain rate and the rain cell size distribution parameters. *American Meteorological Society* 56: 57–70.

34 Capsoni, C., Fedi, F., Magistroni, C. et al. (1987). Data and theory for a new model of the horizontal structure of rain cells for propagation applications. *Radio Science* 22 (03): 395–404.

35 Capsoni, C., Luini, L., Paraboni, A., and Pawlina, A. (2006). Stratiform and Convection rain discrimination deduced from Local *P*(*R*). *IEEE Transactions on antennas and propagation* 54 (11): 3566–3569.

36 Capsoni, C. and Luini, L. (2008). 1-Min rain rate statistics predictions from 1-hour rain rate statistics measurements. *IEEE Transactions on Antennas and Propagation* 56 (03): 815–824.

37 Emiliani, L., Luini, L., and Capsoni, C. (2010). On the optimum estimation of 1-minute integrated rainfall statistics from data with longer integration time. In: *Proceedings of the Fourth European conference on Antennas and Propagation (EuAP)*, 1–5.

38 Capsoni, C. and Luini, L. (2009). A physically based method for then conversion of rainfall statistics from long tom short integration time. *IEEE Transactions on antennas and propagation* 57 (11): 3692–3696.

39 Lang, S., Tao, W.-K., Simpson, J., and Ferrier, B. (2003). Modelling of convective-stratiform precipitation processes: sensitivity to partitioning methods. *Journal of applied meteorology,* American meteorological society 42: 505–527.

40 Ahmed, F. and Schumacher, C. (2015). Convective and stratiform components of the precipitation-moisture relationship. *Geophysical Research Letters* 42: 10453–10462.

41 Recommendation ITU-R P.530-17, (12-2017), 'Propagation data and prediction methods required for the design of terrestrial line-of-sight systems'

42 Recommendation ITU-R P.838-3 (2005) 'Specific attenuation model for rain for use in prediction methods'.

43 Setzer, D.E. (1970). Computed transmission through rain at microwave and visible frequencies. *The Bell System Technical Journal* 49 (8): 1873–1892.

44 Recommendation ITU-R P.1144-6, (02-2012), 'Guide to the application of the propagation methods of radio communication' study group 3.

45 Recommendation ITU-R P.837-7 (06-2017) 'Characteristics of precipitation for propagation modelling.

46 Recommendation ITU-R P.837-6 (2012) 'Characteristics of precipitation for propagation modelling (cal. rain rate)'.

47 Recommendation ITU-R P.1510-1 (06/2017) Mean surface temperature.

48 Stiles, W.H. and Ulaby F.T. (1981) 'Dielectric properties of snow', Remote Sensing Laboratory, University of Kansas, Center of Research,' Inc., RSL Tech. Rep. 527-1.

49 Hallikainen, M., Ulaby F., and Abdel-Razik, M. (1982) 'Measurements of the dielectric properties of snow in the 4-18 GHz frequency range',12th European Microwave Conference, 151–156.

50 Hallikainen, M.T., Ulaby, F.T., and Abdelrazik, M. (1986). Dielectric properties of snow in the 3 - to 37 GHz range. *IEEE Transactions on Antennas an Propagation*, AP 34 (11): 1329–1340.

51 Tiuri, M.E., Sihvola, A.H., Nyfors, E.G., and Hallikaiken, M.T. (1984). The complex dielectric constant of snow at microwave frequencies. *IEEE Journal of Oceanic Engineering* OE-9 (5): 377–382.

52 Polder, D. and Van Santen, J.H. (1946). The effective permeability of mixtures of solids. *Physica* 12 (5): 257–271.

53 Hallikaiken, M.T., Ulaby, F.T., and Van Deventer, T.E. (1987). Extinction Behavior of Dry Snow in the 18- to 90-GHz Range. *IEEE Transactions on Geoscience and Remote Sensing* GE-25 (6): 737–745.

54 Tsang, L. and Kong, J.A. (1981). Scattering of electromagnetic waves from random media with strong permittivity fluctuations. *Radio Science* 16 (3): 303–320.

55 Stogryn, A. (1986). A study of the microwave brightness temperature of snow from the point of view of strong fluctuation theory. *IEEE Transactions on Geoscience and Remote Sensing* GE-24 (2): 220–231.

7

mmWave Propagation Modelling – Weather, Vegetation, and Building Material Losses

7.1 Introduction

When an electromagnetic (EM) wave propagates through the atmosphere, it is absorbed by the atmospheric gases and rain (as we explored in Chapter 6). The wave scattered is absorbed by atmospheric aerosols such as clouds, fog, and haze. The attenuation of the EM radiation due to these adverse weather conditions somewhat degrades the performance of radio communication systems. There has been a volume of work in open literature on the weather attenuation of radio frequency (RF) radiation in the band 1–100 GHz for civil and military applications due to clouds. Haze is transparent to RF radiation when its particles are much smaller than the RF radiation wavelength. The attenuation due to clouds at 18 °C for both dense clouds (2.5 g/m^3) and thin clouds (g/m^3) is negligible at 10 GHz. For dense clouds, the attenuation increases to about 1.5 dB/km at 35 GHz and to a little more than 10 dB/km at 94 GHz. Attenuation of fog at 18 °C, with inland and coastal visibility of 30 m, can also be ignored at 10 GHz, but at 35 GHz, the coastal fog attenuation is about 1.5 dB/km and increases to 10 dB/km at 94 GHz. The inland fog attenuation rates are less than coastal fog at 35 and 94 GHz frequencies.

At higher temperatures, the clouds will have lower attenuation at all frequencies. Similarly, better visibility fog will generate lower attenuation as well. Attenuation due to haze is very complex to define because of the diversity of particles in the haze atmosphere. The haze attenuation depends on these particles' nature, e.g. their refractive index. Luckily, the haze attenuation is negligible at mmWave frequencies. Future communication systems operating at mmWave must account for fog and cloud attenuation to achieve improved reliability, especially in coastal areas where moist fog and dense clouds can degrade the system performance [1].

Vegetation (e.g. forests) contributes sizable signal attenuation that affects system reliability and performance at high frequencies. Analysis of the vegetation attenuation is subject to the theory that vegetation is a random medium that allows multiple scattering effects and takes full account while interference effects are neglected. Transport theory, as it is called, splits the EM radiated field into two components: coherent component and incoherent component. The incoherent component is formulated by four key parameters that depend on the specific type of vegetation. The excess attenuation of vegetation is the sum of the contributions from propagation on top, sides and ground reflections and multiple scatterers and generally range between a 20 dB to as high as 50 dB taking into account the type of vegetation and the seasonal effects.

The other main contributors to radio radiation loss come from building materials. For example, a brick wall contributes 2–8 dB radiation loss, a concrete wall 10–15 dB, and even the human body contributes about 3 dB in radiation loss. Building material attenuation can seriously damage indoor radio communication, especially if the base station (BS) is located outside the building. Attenuation due to clouds, fog, vegetation, and building materials are clearly stated and theoretically analysed in this chapter.

5G Physical Layer Technologies, First Edition. Mosa Ali Abu-Rgheff.

7.2 Attenuation Due to Clouds and Fog

There are many types of visually identifiable clouds in the atmosphere. According to UK Met Office (with similar schemes used elsewhere), clouds can be classified appropriately to the altitude of cloud base: low; mid-level, and high altitude. Within each altitude class, there can be any of four basic types and combinations thereof. These types are cirrus (meaning hair like), stratus (meaning layer), cumulus (meaning pile), and nimbus (meaning rain producing). Often, several types of clouds will be present at different levels of the atmosphere at the same time. For practical purposes, rain attenuation is greater than that of the clouds, but clouds develop more often than rain.

Another categorization of the clouds is based on various phases of water, such as liquid clouds when clouds constituent particles are in the water state. Ice clouds constituent particles that are in the form of ice; and mixed-phase clouds exist when constituent particles are in both ice and liquid phases. Clouds are defined as an aggregate of very small water droplets (generally less than 0.01 cm in diameter), ice crystals of various shapes, or a mixture of both that are suspended in the air by turbulence.

According to the UK Meteorological Office, fog can be categorized with respect to visibility into three levels:

1. The international definition of fog is a visibility up to 1 km
2. Thick fog with visibility up to 200 m
3. Dense fog with visibility below 50 m

Thick fog comprises liquid water with a density of about 0.05 g/m^3 and the dense fog liquid water density is about 0.5 g/m^3. The clouds can be split into three levels: high clouds at a level 6 km or above and contain ice crystals of a variety shapes; medium clouds at level between 2 and 6 km; and low clouds below 2 km (see www.metoffice.gov.uk).

When an EM wave radiation interacts with the clouds water droplets, part of its energy is absorbed and the other part is scattered so the EM wave level is reduced. Consequently, the presence of hydrometeors in the atmosphere (e.g. fog, mist, clouds, sleet, rain, snow, and hail) present a very severe challenge to the signal propagation [2, 3]. For scattering consideration, we can think of the droplets as suspended spherical dielectric; however, scattering losses due to clouds is secondary to the absorption losses. Attenuation due to fog and clouds (as well as snow and rain presented in Chapter 6) can lead to the degradation of communication performance quality of service. It is vital to have reliable and realistic propagation models for predicting the attenuation by suspended water droplets (SWD) [4, 5]. These models are commonly based on a number of input parameters, such as frequency; temperature; water contents, and relative humidity. Clouds and fog attenuation are greatly influenced by their water content in g/m^3.

7.3 The Microphysical Modelling

Statistical functions are frequently used to model rain droplet size distribution [6]. Two statistical distribution functions are frequently used, namely: lognormal distribution function or modified gamma distribution function. Before we introduce the lognormal distribution, let us briefly review the normal (Gaussian) distributions. The normal distribution is based on the assumption that a given droplet of a diameter (d) *occurs randomly distributed*. This distribution function is relatively easy to use but is constrained to situations where droplets occur randomly. The normal distribution function $\mathcal{N}(d)$ expresses the number of droplets with a given random diameter d by the expression:

$$\mathcal{N}(d) = \frac{1}{\sigma\sqrt{2\pi}} e^{-\frac{(d-\overline{d})^2}{2\sigma^2}} \tag{7.1}$$

where σ is the standard deviation of the values of d, the mean value $\overline{d}$ and σ^2 is the variance of values of d. However, though attractive in its simplicity, it fails to fit the observed instantaneous sampling time of one-minute required by the International Telecommunication Union (ITU) recommendations.

In the lognormal distribution, the rain droplets diameter distribution is given by the logarithm of the normally distributed droplet diameter (d). The raindrop diameter distribution can be best described by a lognormal distribution. For raindrop diameter d the distribution $n(d)$ is [7]

$$n(d) = \frac{1}{d \log \sigma_w \sqrt{2\pi}} \exp\left[-\frac{1}{2} \left(\frac{\log \left({}^{d}/_{d_w} \right)}{\log \sigma_w} \right)^2 \right] \tag{7.2}$$

where d_w is the mean geometric diameter and σ_w is the geometric standard deviation derived in Appendix 7.A as

$$d_w = e^{\overline{\log d}} \tag{7.3}$$

$$\sigma_w = \exp \sqrt{\overline{\left(\log \frac{d_i}{d_w} \right)^2}} \tag{7.4}$$

In the lognormal distribution above, we assumed the raindrop number concentration (i.e. 0th moment of the distribution) is equal to unity.

Figure 7.1 presents a theoretical example of lognormal distribution for droplets diameters. The values of d_g and σ_g are 18.4746 and 2.6777, respectively. Similar lognormal distribution can be obtained for the radius of water cloud droplets instead of the diameter. The only difference is the geometric mean will be shifted by 0.69.

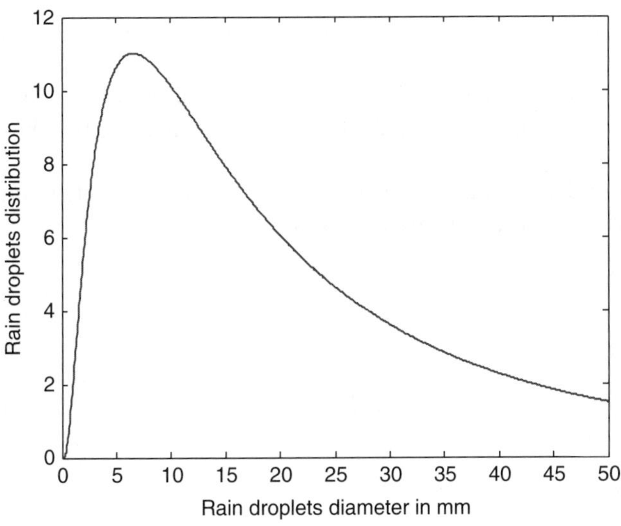

Figure 7.1 lognormal distributions for rain droplets diameters up to 50 mm.

In satellite communications, accurate models exist for certain phenomena, such as rain, clouds, in addition to low angle fading along the earth-satellite path. These effects are combined into a single model but since the occurrence probabilities of different impairments are not known, an empirical prediction model based on available cloud data and average properties of different cloud types was developed in [8]. The model was derived using the average cloud properties together with the assumption that the statistical distribution of cloud attenuation, following the lognormal probability law.

7.4 Modified Gamma Droplets Size Distribution

7.4.1 Analysis of the Size Distribution

The attenuation caused by clouds being made by the interaction of cloud water droplets with EM radiation. The loss of EM energy through a cloud volume is determined by the sum of absorption and scattering effects by the water droplets. Also these droplets can be considered as perfect dielectric spheres. Mie theory provides the essentials to accurately estimate (normally per m^2) the extinction coefficient σ_e; the scattering coefficient σ_s; and the absorption coefficient σ_a as

$$\sigma_e = \sigma_s + \sigma_a \tag{7.5}$$

These estimated coefficients (in dB/km) can be used to calculate the specific attenuation γ_{cloud} dB/km due to a population of water droplets spheres with various radii r (in μm) using the following integration:

$$\gamma_{cloud} = 4.343 * 10^3 * \int \sigma_e(r)\,\mathrm{n(r)\,dr} \tag{7.6}$$

where n(r) is the cloud water droplets radius distribution in m^{-3} mm^{-1}, and n(r) dr gives the density of cloud water droplets with radius r. The water droplets radius distribution is frequently modelled by a *modified gamma exponential* function [9, 10] of the form

$$\mathrm{n(r)} = a\,r^{\alpha}\exp(-b\,r^{\beta}),\quad 0 \le r \le \infty \tag{7.7}$$

where radius r expressed in μm, parameters α, β, a, b are all positive and real constants, and a is an integer the scaling factor that effects the total concentration of droplets. Parameter α is also called the shape parameter, b controls the positive slope of the size distribution, and β controls the negative slope. The number of particles per unit volume of air with droplets radius size between (r) and $(r + dr)$ is $\mathrm{n}(r)dr$.

The mode radius r_c(μm) of particle radius is defined as the radius of maximum frequency and is expressed in terms of the droplet size distribution parameters as

$$r_c = \left(\frac{\alpha}{b\beta}\right)^{\frac{1}{\beta}} \tag{7.8}$$

and

$$b = \frac{\alpha}{\beta r_c^{\beta}} \tag{7.9}$$

Substitute (7.9) in (7.7), we get

$$n(r) = a\,r^{\alpha}\exp\left[-\frac{\alpha}{\beta}\left(\frac{r}{r_c}\right)^{\beta}\right] \tag{7.10}$$

Table 7.1 Modified gamma model parameters for fog and clouds droplets size distributions [6].

Cloud type	α	b	a	w (g m^{-3})	r_c (μm)	N (cm^{-3})
Heavy advection fog	3	0.3	0.027	0.37	10.0	20
Moderate radiation fog	6	3.0	607.5	0.02	2.0	20
Cumulus	3	0.5	2.604	1.00	6.0	250
Stratus	2	0.6	27.0	0.29	3.33	250
Stratus/strato-cumulus	2	0.75	52.734	0.15	2.67	250
Alto-stratus	5	1.111	6.268	0.41	4.5	400
Nimbo-stratus	2	0.425	7.676	0.65	4.7	200
Cirrus	6	0.09375	2.21×10^{-12}	0.06405	64.0	0.025
Thin cirrus	6	1.5	0.011865	3.128×10^{-4}	4.0	0.5

The droplets size distribution at the mode radius expression in (7.10) becomes

$$n(r_c) = a r_c^{\alpha} \exp\left(-\frac{\alpha}{\beta}\right) \tag{7.11}$$

The derivation of expressions in (7.9) and (7.11) are given in Appendix 7.B.

The *liquid water content* (LWC) expressed in g m^{-3} of air corresponding to 1 km horizontal visibility can be computed as [11, 12].

$$LWC = \frac{4}{3}\, 10^{-6}\, N\, \pi\, \rho \int_0^{\infty} n(r) r^3\, dr \tag{7.12}$$

where ρ is the density of liquid water in 1 g cm^{-3}.

Substitute for $n(r)$ in (7.10) in (7.12), we get

$$LWC = \frac{4}{3}\, 10^{-6}\, aN\, \pi\, \rho \int_0^{\infty} r^{\alpha+3} \exp\left[-\frac{\alpha}{\beta}\left(\frac{r}{r_c}\right)^{\beta}\right] dr \tag{7.13}$$

Now

$$\int_0^{\infty} r^{\alpha+3} \exp\left[-\frac{\alpha}{\beta}\left(\frac{r}{r_c}\right)^{\beta}\right] dr = \int_0^{\infty} r^{\alpha+3} \exp\left[-\frac{\alpha}{\beta r_c^{\beta}} (r)^{\beta}\right] dr$$

It can be shown that

$$\int_0^{\infty} r^{\alpha+3} \exp\left[-\frac{\alpha}{\beta}\left(\frac{r}{r_c}\right)^{\beta}\right] dr = \frac{1}{\beta}\left(\frac{\alpha}{\beta r_c^{\beta}}\right)^{-\frac{\alpha+4}{\beta}} \Gamma\left(\frac{\alpha+4}{\beta}\right) \tag{7.14}$$

Substitute (7.14) in (7.13), we get the LWC as

$$LWC = \frac{4}{3}\, 10^{-6}\, aN\, \pi\, \rho\, \frac{1}{\beta}\left(\frac{\alpha}{\beta r_c^{\beta}}\right)^{-\frac{\alpha+4}{\beta}} \Gamma\left(\frac{\alpha+4}{\beta}\right) \tag{7.15}$$

The number density N i.e. total concentration of water droplets in cm^{-3} of air, is the cumulative number density of the water droplets radius (r) in *number* cm^{-3} expressed as

$$N = \int_0^{\infty} n(r).dr \qquad \text{number cm}^{-3} \tag{7.16}$$

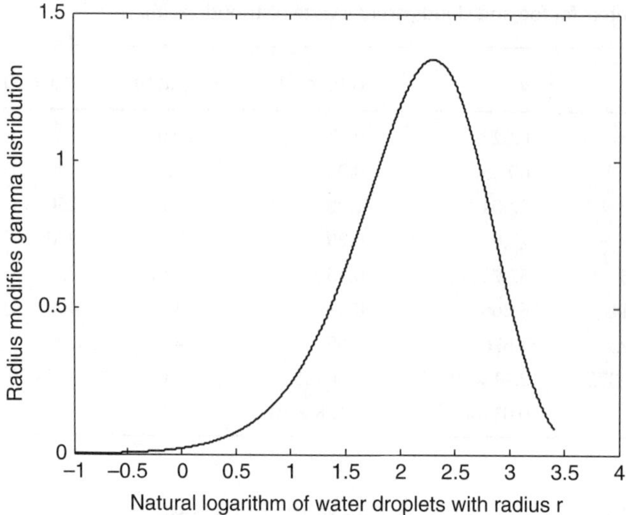

Figure 7.2 Modified gamma distribution of water droplets radius r expressed in μm.

After carrying out the integration of the integral in (7.7), we get the total particle density (N) is

$$N = \frac{a}{\beta} b^{-\frac{\alpha+1}{\gamma}} \Gamma\left(\frac{\alpha+1}{\beta}\right) \quad \text{particles} \quad \text{cm}^{-3} \tag{7.17}$$

Substitute for N given by (7.17) into (7.15), we get

$$LWC = \frac{4}{3} 10^{-6} \frac{a^2}{\beta^3} b^{-\frac{\alpha+1}{\beta}} \pi \rho \left(\frac{\alpha}{r_c{}^{\beta}}\right)^{-\frac{\alpha+4}{\beta}} \Gamma\left(\frac{\alpha+1}{\beta}\right) \Gamma\left(\frac{\alpha+4}{\beta}\right) \tag{7.18}$$

A plot for the droplets size distribution in (7.7) for heavy advection fog is depicted in Figure 7.2 where the calculated geometric mean is 3.0945.

The values for parameters a, b, α with $\beta = 1$ are given in Table 7.1.

7.4.2 Skewness and Kurtosis of Modified Gamma Distribution

Skewness is a measure of the asymmetry (off centre) of the distribution. A distribution is symmetric if it looks the same to the left and right of the centre point. The skewness can come as a negative or positive skewness depending on whether the distribution is skewed to the left or the right of the data average. A distribution has a positive skewness when a *longer tail* is presented to right of the mode (maximum of the statistical distribution) than to the left. Otherwise it has a negative skewness. Normal *distributions have* zero *skewness.*

The parameters that appear to be significant to the formation and growth of the fog and the clouds droplets are the *moment coefficient of skewness* denoted as a_3 which is a measure of the degree of asymmetry of the distribution curve and *Kurtosis coefficient*, which is a measure of whether the data distributions are peaked or flat topped relative to a normal distribution. *Kurtosis* β_2 is defined as standardized distribution 4th moment relative standard deviation to power 4 [13].

$$\beta_2 = \frac{\mathbb{E}(X-\mu)^4}{(\mathbb{E}(X-\mu)^2)^2)} = \frac{\mu_4}{\sigma^4} \tag{7.19}$$

where μ is the mean, μ_4 is the 4th moment about the mean, and σ is the standard deviation. The normal distribution has a kurtosis of 3, so $\beta_2 - 3$ is frequently used as reference to normal

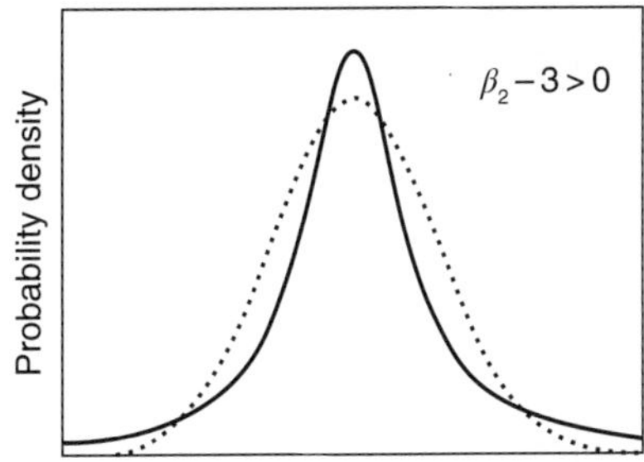

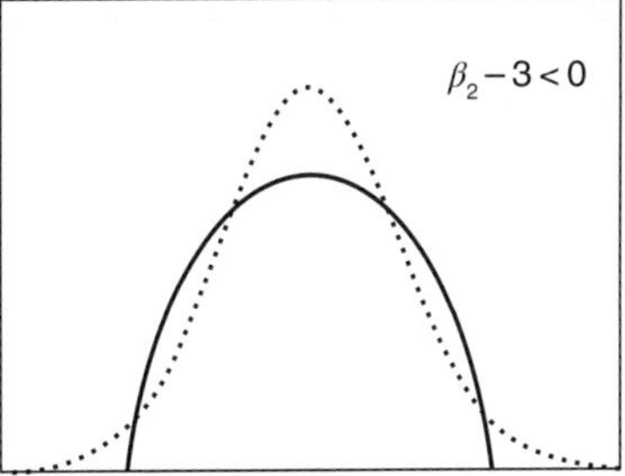

Figure 7.3 Illustration of kurtosis with positive and negative kurtosis relative to normal distribution shown as dotted lines [13].

Figure 7.4 Uniform distribution and the normal distribution, both with a variance of 1 [13].

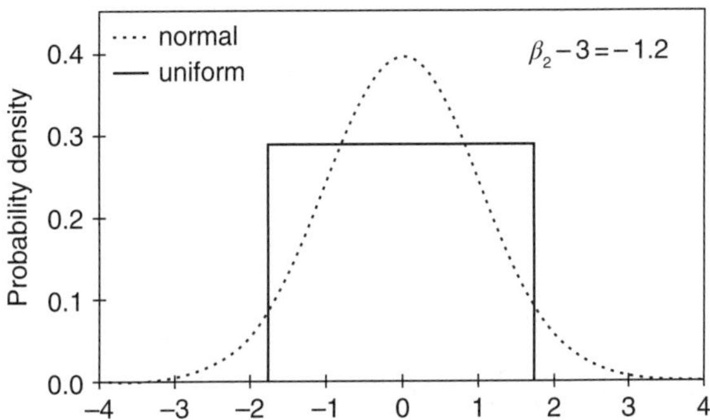

distribution, which has kurtosis of zero. An illustration of kurtosis with reference to normal distribution of kurtosis zero is shown in Figure 7.3. Kurtosis of uniform distribution compared to normal distribution is shown in Figure 7.4 when both distributions have a variance of 1. The uniform distribution seems to have negative kurtosis.

The uniform distribution in Figure 7.4 is plotted for the range $\pm\sqrt{3}$. Compared to the normal distribution, uniform distribution has light tails, a flat centre part, and considerable shoulders, so $\sqrt{3}-3 \approx -1.2$.

The modified gamma water droplets size distribution is asymmetric and appears more peaked in the centre and skewed in the tails. The skewness measure is related to the size distribution and the growth of the fog and cloud droplets. The moment coefficient of skewness (a_3) is mostly given by the ratio of the third moment about the mean to the 3/2 power of the second moment of the distribution curve. Kurtosis predicts the degree of peakedness distribution referenced to normal distribution. The skewness moment coefficients a_3 and a_4 are given in [14] as

$$a_3 = \frac{[\Gamma(\varrho_1)]^2\,\Gamma(\varrho_4) - 3\Gamma(\varrho_1)\,\Gamma(\varrho_2)\,\Gamma(\varrho_3) + 2[\Gamma(\varrho_2)]^3}{\{\Gamma(\varrho_1)\,\Gamma(\varrho_3) - [\Gamma(\varrho_2)]^2\}^{3/2}} \tag{7.20}$$

$$a_4 = \frac{[\Gamma(\varrho_1)]^3\,\Gamma(\varrho_5) - 4[\Gamma(\varrho_1)]^2\Gamma(\varrho_2)\Gamma(\varrho_4) + 6\Gamma(\varrho_1)[\Gamma(\varrho_2)]^2\,\Gamma(\varrho_3) - 3\,[\Gamma(\varrho_2)]^4}{\{\Gamma(\varrho_1)\,\Gamma(\varrho_3) - [\Gamma(\varrho_2)]^2\}^2} \tag{7.21}$$

where

$$\varrho_k = \frac{\alpha + k}{\beta} \tag{7.22}$$

Plots of skewness and kurtosis coefficients are shown in Figure 7.5. Skewness decreases as α and β increase and on the other hand, kurtosis exhibits a minimum of around 3 for all values of α.

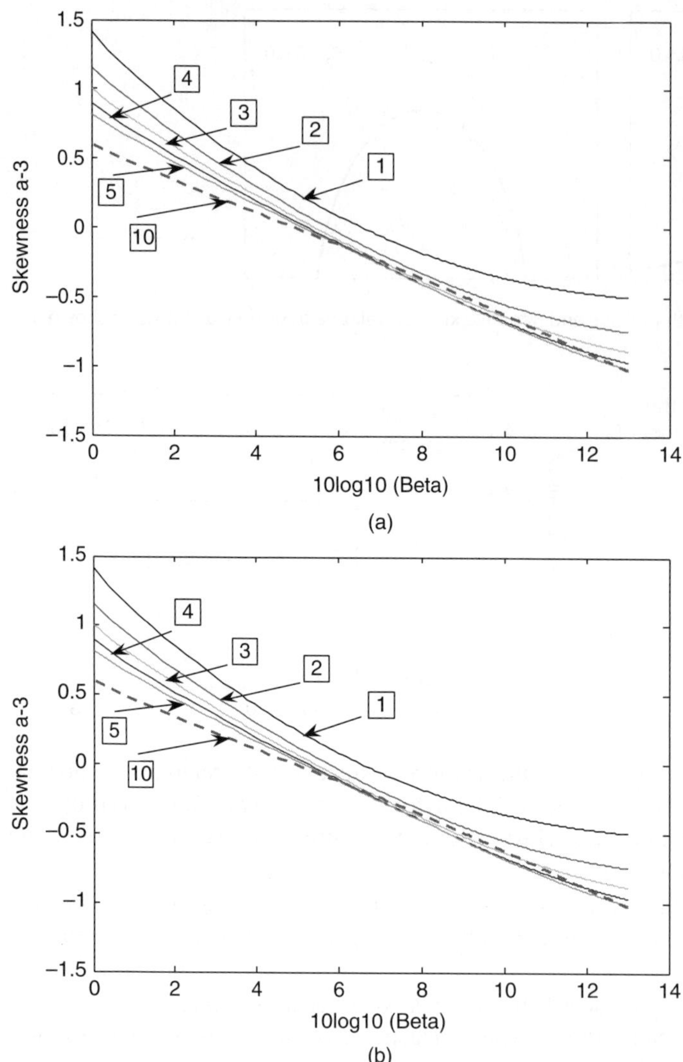

Figure 7.5 (a) Moment coefficients of skewness a_3 plotted as a function of the parameter 10 log 10 (β) for different integer values of the parameter α shown on the curves by arrows. (b) Moment coefficients of kurtosis β_2 plotted as a function of the parameter10 log 10 (β) for different integer values of the parameter α are shown on the curves by arrows [14].

The clouds specific attenuation γ_{cloud} in dB/km due to water droplets radius r μm is expressed in (7.6). The density of the water droplets within the radius r is defined by $(r)\ dr$. The attenuation (extinction) coefficient is denoted by $\sigma_e(r)$ given by the sum of σ_a and σ_s where a and s are absorption and scattering coefficients, respectively. Under the Rayleigh regime, the scattering coefficient is much less than the absorption coefficient (i.e. $\sigma_s \ll \sigma_a$). Thus, we can ignore wave scattering, and the total extinction is reduced to absorption effects only, given in [15] as

$$\sigma_e(r) = \sigma_a(r) = 8.18\, LWC\, Img(-K)/\lambda \tag{7.23}$$

where *LWC* is the *liquid water content* (LWC) expressed in g m^{-3} defined in (7.12), and K is expressed in terms of the complex refractive index n of water or ice:

$$K = \frac{n^2 - 1}{n^2 + 2} \tag{7.24a}$$

where

$$n = n_r + jn_i \tag{7.24b}$$

So, (7.23) can be simplified as

$$\sigma_e(r) = 8.18\, LWC(r)\, \frac{2([n_r n_i][1 - (n_r)^2 + (n_i)^2] - n_r n_i[2 + (n_r)^2 - (n_i)^2])}{[2 + (n_r)^2 - (n_i)^2]^2 + [2n_r n_i]^2} / \lambda \tag{7.25}$$

Substituting (7.25) into (7.6), we get the clouds specific attenuation:

$$\gamma_{cloud} = 35.5 * 10^3 \int_0^{\infty} LWC(r) \left[\frac{2([n_r n_i][1 - (n_r)^2 + (n_i)^2] - n_r n_i[2 + (n_r)^2 - (n_i)^2])}{[2 + (n_r)^2 - (n_i)^2]^2 + [2n_r n_i]^2} / \lambda\right] n(r)\, dr \text{dB/km} \tag{7.26}$$

Equation (7.26) is derived for the calculation of the cloud specific attenuation at 94 GHz and has to be checked for use under other frequencies.

7.5 Rayleigh and Mie Scattering Distributions

When particle size is much less than the wavelength of a radio wave, scattering takes place when the wave interacts with the particles and can be approximated by Rayleigh distribution where scattered energy resembles a radiation pattern like that of a dipole. If the particle size is of the order of a wavelength or greater, Mie scattering mechanism occurs and the radiation pattern contains a large forward lobe. The reasoning for this is simple: the choice between Rayleigh and Mie [16] is determined not only on the basis of particle size profile but also by the dielectric properties of the particle itself. The Rayleigh approximation can be used, provided that the following condition [9] is satisfied:

$$|n|\, \chi < 0.5 \tag{7.27}$$

where n is the complex refractive index of water droplets (in parts per million) and $\chi = \frac{2\pi r}{\lambda}$ is the Mie parameter. The frequency variation of $|n|\, \chi$ for different sizes of water cloud spheres, at 0 and 10 °C, are depicted in Figure 7.6, and the limit $|n|\, \chi < 0.5$ is shown on the plot when we can use the Rayleigh approximation. We will show in Appendix 7.C that the absolute refractive index can be expressed in terms of the water droplets dielectric constant and (7.27) becomes

$$\sqrt{n_r^2 + n_i^2} * \chi < 0.5 \tag{7.28}$$

which is similar to

$$|\varepsilon| * \chi < 0.5 \tag{7.29}$$

We may conclude from Figure 7.6 that for the water droplets size between 5 and 10 μm, Rayleigh approximation can be used up to at least 300 GHz. Above this frequency limit, the Mie theory has to be used to evaluate both scattering and absorption contributions of liquid water droplets.

The microphysical properties of ice-crystal clouds differ from those of water droplets clouds as they consist of ice crystals of various shapes [17] and their sizes are larger than the size of

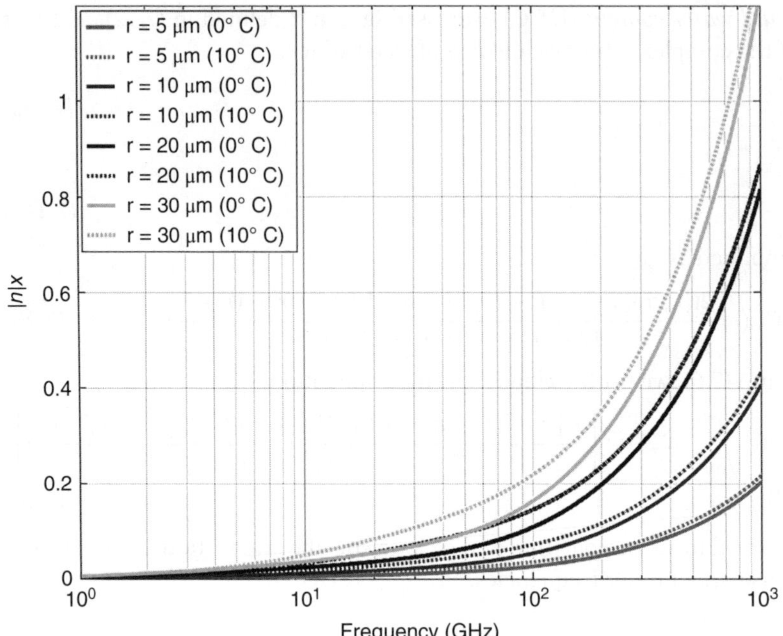

Figure 7.6 Frequency variation of $|n|\ \chi$ for different sizes of water cloud spheres, at 0 and 10 °C. Rayleigh approximation can be assumed if the condition $|n|\ \chi < 0.5$ is verified [9].

water droplets. Consequently, the characteristics of ice clouds differ from those of water clouds. Compared to attenuation caused by water clouds, the attenuation due to ice clouds is small. For example, measured for a rain rate of 1 mm/h is 0.70 dB/km, while the maximum attenuation that was measured in ice cloud is 0.28 dB/km.

7.6 ITU Empirical Model for Clouds and Fog Attenuation Calculation

Clouds (and fog) consist wholly of small water droplets and therefore we may express the attenuation of clouds (and fog) to radio wave propagation in terms of the total water content per unit volume. The recommendations ITU-R P.840-6 [18] hold for frequencies higher than 10 GHz but lower than 200 GHz, and express the specific attenuation γ_{cloud} within a cloud or fog as follows:

$$\gamma_{cloud} = K_c M \quad \text{dB/km} \tag{7.30}$$

where K_c and M are specific attenuation coefficient in (dB/km) per g/m^3 within the cloud and liquid water density in the cloud/fog in g/m^3, respectively. It is worth comparing how we express the cloud-specific attenuation in (7.6) and the ITU formulation (7.30). In (7.6) we express the specific attenuation mainly in terms of the droplets radius, while (7.30) expresses the specific attenuation mainly in terms of the content liquid water density and water dielectric properties. Furthermore, the complex refractive index and the complex permittivity in some open literature are expressed with a negative imaginary part conveying the fact that it accounts for loss. However, formulae derived, whether from negative or positive imaginary part, result in similar equations presented in this book.

The specific attenuation coefficient K_c depends on frequency, dielectric properties of water, and the temperature. K_c can be calculated for frequencies up to 1000 GHz, assuming a Rayleigh scattering of the droplets. Following is the double-Debye model for the complex dielectric permittivity $\varepsilon(f)$:

$$K_c = \frac{0.819 f}{\varepsilon_i(1+\eta^2)} \quad \text{dB/km per g/m}^3 \tag{7.31}$$

where f and ε_i are frequency in GHz and imaginary part of the complex dielectric permittivity ε, respectively, and

$$\eta = \frac{2+\varepsilon_r}{\varepsilon_i} \tag{7.32}$$

where ε_r is the real and ε_i is the imaginary parts of the $\varepsilon(f)$, both parts are frequency dependent.

The complex dielectric permittivity of the water droplets is given [4, 18, 19] as

$$\varepsilon(f) = \varepsilon_r(f) + j\varepsilon_r(f) \tag{7.33}$$

where

$$\varepsilon_r(f) = \frac{\varepsilon_0 - \varepsilon_1}{1+\left(\frac{f}{f_p}\right)^2} + \frac{\varepsilon_1 - \varepsilon_2}{1+\left(\frac{f}{f_s}\right)^2} + \varepsilon_2 \tag{7.34}$$

$$\varepsilon_i(f) = \frac{f(\varepsilon_0 - \varepsilon_1)}{f_p\left[1+\left(\frac{f}{f_p}\right)^2\right]} + \frac{f(\varepsilon_1 - \varepsilon_2)}{f_s\left[1+\left(\frac{f}{f_s}\right)^2\right]} \tag{7.35}$$

where the static dielectric constant ε_0 of pure water depends on temperature only, and can be as follows:

$$\varepsilon_0(T) = 77.66 + 103.3\,(\theta - 1) \tag{7.36}$$

$$\varepsilon_1 = 0.067\,\varepsilon_0 \tag{7.37}$$

$$\varepsilon_2 = 3.52 \tag{7.38}$$

$$\theta = \frac{300}{T} \tag{7.39}$$

where $T = temp°\text{C} + 273.15$ is the temperature in degrees K. At room temperature, $5.2 \le \varepsilon_1 \le 5.5$. The principal f_p and the secondary f_s Debye relaxation frequencies are

$$f_p = 20.20 - 146(\theta - 1) + 316(\theta - 1)^2 \;\text{ GHz} \tag{7.40}$$

$$f_s = 39.8 f_p \text{ GHz} \tag{7.41}$$

It is worth noting that the value for ε_1 in [18] is temperature dependent, while in [4] it is constant because it is computed at a given room temperature. The equations in this section followed those given in the ITU recommendations. Attenuation calculation of ice clouds is difficult since there may be several different ice crystal shapes observed in ice clouds. Considering ice water is the main contributor of the attenuation, the specific attenuation γ_{ic} per km [20] due to ice cloud can be derived from the ice water content (IWC):

$$\gamma_{ic} = 9.27\,(IWC)^{0.68} \tag{7.42}$$

where the IWC is expressed in g/m^3.

7.7 Building Material Attenuation

Indoor users of mobile communication, in most case, have to depend on outdoor-deployed macro/microcells infrastructures for services, meaning that indoor users must incur high losses originated by signal penetration through external building walls' materials to reach the indoor environment. Building penetration and reflection measurements are carried out through intensive campaigns, and the reported results are widely available in the literature.

7.7.1 Penetration Losses for Various Building Materials

A recent measurements campaign was carried out to confirm to the penetration losses through building materials and were collected at 28 GHz and reported in [21]. These materials include tinted glass, bricks, clear nontinted glass, and drywall. The setup used for the measurements is depicted in Figure 7.7. In the setup, the transmitter (TX) and the receiver (RX) antennas were placed 5 m away from each other and the free space attenuation reference at 5 m distance was first measured equal 75.3 dB, and then the TX and RX moved to a position on the opposite side of building test materials at equal heights of 1.5 m with antennas pointing at each other and both 2.5 m away from the test material. Both of the horn antennas had 24.5 dBi gains with 10° half power beamwidth. The penetration losses through various common building materials in access of free space path loss of 75.3 dB at 28 GHz are summarized in Table 7.2.

The penetration loss is presented in Table 7.2 at 28 GHz showed a high penetration loss from the exterior of urban buildings (tinted glass and bricks), reflecting the fact that penetration of mmWave from outdoor cell to indoor will be heavily attenuated, degrading the quality of

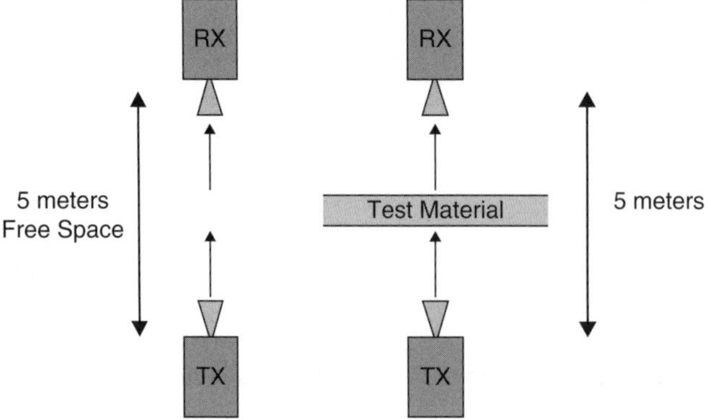

Figure 7.7 Test setup for penetration loss of various building materials at 28 GHz [21].

Table 7.2 Penetration losses through various common building materials at 28 GHz [21].

Environment	Material	Thickness (cm)	Penetration loss (dB)
Outdoor	Tinted glass	3.8	40.1
	Brick	185.4	28.3
Indoor	Tinted glass	<1.3	24.5
	Clear glass	<1.3	3.6
	Wall	38.1	6.8

reception from outdoor networks to indoor mobile units. On the other hand, general indoor materials such as dry wall and nontinted glass have relatively lower losses.

7.7.2 Penetration Losses for Indoor Obstructions in an Office Environment at 28 GHz

The penetration losses for indoor obstructions are conducted inside an office of 34 m by 43 m in [21]. A single TX was positioned in a convenient corner inside the office building and the TX was capable of three levels of transmitting power: – 8.6, 11.6, and 21.4 dBm. Eight RXs were placed at different positions inside the office at increasing distances from TX to include as many obstacles as possible. At certain locations, a weak signal was soon to be distinguished from noise but not enough to be acquired. In other locations, no signal could be detected, i.e. channel outage indicating no service is available.

At the RX sites, with TX-RX separation distances of 25.6 and 11.4 m, the measured penetration loss was 45.1 dB; however, the site with 25.6 m separation distance had obstructions of four walls and two cubicles, while the other site with 11.4 m separation distance had obstructions of three walls and one door. Furthermore, a single outage was experienced at the RX site with a separation distance of 35.8 m, which was expected due to the large separation distance and the inability of millimetre frequency signal to penetrate the metallic elevator bank (a system of elevators going to different floors).

The measured data were grouped into three subsections: signal acquired (i.e. signal power to noise power ratio [SNR] is sufficiently high to accurately acquire the signal); signal detected (SNR is high enough to distinguish signal from noise but not strong enough to acquire the signal); and no signal detected, indicating outage. The penetration losses for multiple indoor obstructions in an office environment at 28 GHz are presented in Table 7.3. It is clear from these data that penetration loss was not a strong dependent of TX-RX separation distance; e.g. TX-RX separation distances 11.4 and 25.6 m have identical measured penetration loss of 45.1 dB, even though the number of obstructions in both cases is not the same. Furthermore, when TX-RX is separated by 35.8 m and metallic elevators were included as obstacles, the system experienced signal outage.

7.7.3 The Penetration Loss for the Exterior of the House

While the measurement results presented in the previous section were very helpful in understanding the 5G transmission and penetration loss through a building, this does not model an

Table 7.3 Penetration losses for multiple indoor obstructions in an office environment at 28 GHz [21].

RX ID	TX-RX separation (m)	# of partitions				Transmitted power (dBm)	Received power – free space (dBm)	Received power-test material (dBm)	Penetration loss (dB)
		Wall	Door	Cubicles	Elevator				
1	4.7	2	0	0	0	−8.6	−34.4	−58.8	24.4
2	7.8	3	0	0	0	−8.6	−38.7	−79.8	41.1
3	11.4	3	1	0	0	11.6	−21.9	−67.1	45.1
5	25.6	4	0	2	0	21.4	−19.0	−64.1	45.1
4	30.1	3	2	0	0	21.4	−30.4	Signal detected	
6	30.7	4	0	2	0	21.4	−30.5		
7	32.2	5	2	2	0	21.4	−30.9		
8	35.8	5	0	2	1	21.4	−31.9	No signal detected	

indoor mobile communication using outdoor deployed BS. An interesting research result published recently [22] exploring the outdoor deployed BS transmitting at high frequencies (20, 30, 60 GHz) to indoor coverage of a single multi-story building. The propagation loss in this measurement from an outdoor BS to an indoor node (terminal) carries out the assessment of four possible candidate paths, one through each external wall in the building, to select the path that gives the lowest total path loss when summing up the outdoor loss, building penetration loss (BPL), and the indoor loss. The BPL depends critically on the type of building materials used. The 'older house'(s) used different material from the 'new house'(s), so it may be sensible to categorize the houses, for the penetration loss evaluation, as an 'old house' and a 'new house'. The 'old house' material corresponds to 70% concrete wall and 30% clear (standard) glass window, while the 'new house' composite corresponds to 70% infrared reflective (IRR) glass windows and 30% concrete walls.

7.8 Modelling the Penetration Loss for Building Materials

An empirical model based on available measured data has been developed in [22] for the frequency dependency of penetration loss of glass windows and concrete walls. The effective glass transmission loss L_{glass} was modelled as

$$L_{\text{glass}} = 0.1 f_{GHz} + 1 \text{ dB} \tag{7.43}$$

And for concrete wall loss $L_{concrete}$ as

$$L_{concrete} = 4 f_{GHz} + 5 \text{ dB} \tag{7.44}$$

where f_{GHz} is carrier frequency in GHz.

For example, for a standard glass window, the total transmission loss is by $2L_{\text{glass}}$ for two-glass windows and the more modern triple-pane IRR windows by $3L_{\text{glass}} + 20$ where the loss 20 dB accounts for the IRR layer loss. Furthermore, the angular wall loss is added to account for the incident angle θ. This loss is modelled as

$$L_{angle} = 20\,(1 - \cos\theta)^2 \text{ dB} \tag{7.45}$$

The maximum angular loss here is 20 dB. Equation (7.45) can be assumed for both line-of-sight (LOS) and non-line-of-sight (NLOS) propagation and the maximum angular loss can be as much as 20 dB.

7.9 Modelling the Penetration Loss for Indoor Environments

An indoor environment is commonly assumed to be open with standard glass or plaster walls along with indoor walls of an average distance of 4 m, and wall loss is a function of operating frequency. Two indoor wall loss models have been developed. Loss model-1 $L_{i-wall}{}^1$ assumes the wall is made of glass and so the wall loss is equal to that of standard glass layer given as

$$L_{i-wall}{}^1 = L_{\text{glass}} \text{ dB} \tag{7.46}$$

where L_{glass} is given by (7.43). Loss model-2, $L_{i-wall}{}^2$ is based on measured data and is given by

$$L_{i-wall}{}^2 = 0.2 f_{GHz} + 1.7 \text{ dB} \tag{7.47}$$

Body loss is caused by the user in a position that blocks the strongest signal path and may cause severe shadowing. Another factor that adds to the body loss is when the antenna is completely blocked by the user's hand. The addition loss depends on how the user is holding the antenna, and varies from 3 to 15 dB. The average body loss can be modelled as

$$L_{body} = \frac{f_{GHz}}{60} + 3\text{ dB} \tag{7.48}$$

where loss of 3 dB is due to possible hand/figure blocking of the antenna.

The outdoor-to-indoor coverage models in [22] are used to evaluate the performance of outdoor small cells aiming at a 21-story building at a distance of 35 m from the cell BS. The BS transmit power, bandwidth, and heights are 33 dBm, 100 MHz, 31.5 m, respectively. The carrier frequencies are 10, 30, and 60 GHz.

The key conclusions that can be drawn from the above deliberations are, subject to the assumptions made on the propagation model, antenna pattern, and bandwidth size, that the outdoor to indoor coverage with downlink (DL) throughput as high as 10 Mb/se coverage at frequencies in the mmWave band is challenging, especially for a new building model. However, with old building models, DL throughput of at least 10 Mb/s can be achieved at the same bands with low BS density. A high gain antenna, building type, exterior wall material, and interior office layout have significant influence on the DL throughput in the mmWave band.

7.10 Attenuation of Propagated Radio Waves in Vegetation

Attenuation in vegetation can be important for low transmit power mobile communications devices, such as those proposed for use in 5G. The vegetation presented in this section entails large clusters of trees or of isolated trees that affect the radio path. The development of a generic accurate model, either analytic or empirical, to account for attenuation due to the vegetation is hindered with difficulties due to the irregular vegetation environment, many types of species, vegetation densities, and foliage water content, thus requiring input parameters that are difficult to obtain. The required parameters are/or combinations of the following: height of the vegetation, leaf size and shape, trunk size, and canopy height. Indeed, some parameters are difficult to quantify for use in a practical (engineering) model.

Deterministic analytical models were developed based on sound theories, including the transport theory of millimetre wave propagation in woods and forests based on the principles of radiative energy transfer (RET), are presented in this section, together with their advantages and disadvantages. The models can be used to estimate the access (above free space loss) attenuation of vegetation. These models make use of input parameters that have to be specified at each frequency band by measurements for the specific geometries of the vegetation, nature of scatter functions, and for various tree species. The vegetation access attenuation can be modelled by accounting for the following radio propagation components:

- Ground-reflected component
- Top-diffracted component
- Side-diffracted component
- Through and scattered components of the radio beam passing through the foliage

7.10.1 Foliage Propagation Path Models

The vegetation loss prediction models can be categorized as a *horizontal propagation path* similar to the one shown in Figure 7.8 where the elevation angle is not greater than 3° and can

Figure 7.8 Illustrative diagram of the horizontal foliage path [23].

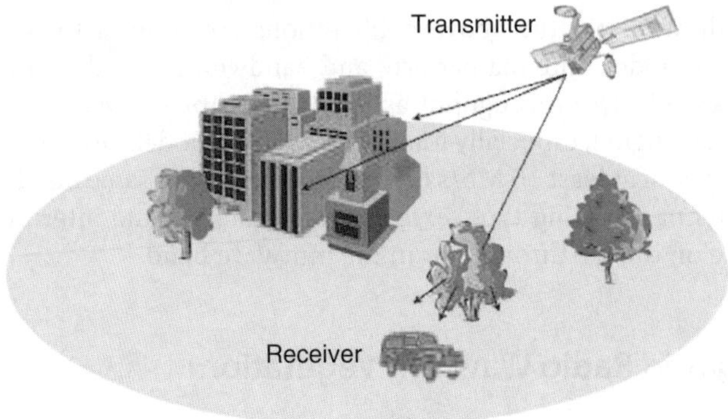

Figure 7.9 Illustrative diagram of the slant foliage path [23].

account for a short foliage path of one or two trees and a long foliage path of many trees. The other category is the slant path similar to one shown in Figure 7.9, where the elevation angle is normally greater than 10° and accounts for a short foliage path of few trees.

7.10.2 Review of Horizontal Empirical Models

Empirical models are easy to calculate since their mathematical expressions in general are simple. However, they usually do not provide any insight in the process of propagation within vegetation. They are formulated to a specific measured data set. Parameters used in these models are frequency, incident angles, path length through vegetation, and other parameters associated with the specific measurement environment. These parameters are usually computed through regression curves fitted to conform to the measurement data. On the other hand, analytical models (i.e. based on RET) give little account of the dynamic effects of the propagation channel and no account of the wideband effects of the vegetation medium.

7.10.3 Weissberger MED Vegetation Loss Model

There are several well-known empirical models, such as Weissberger Modified Exponential Decay model (MED) and the International Telecommunication Union-Radio (ITU-R) Model; COST 235. This section considers the Weissberger MED model [24], which is supported by measured data of the attenuation due to trees and underbrush that are located between a radio transmitter and receiver at the frequencies from 230 MHz to 35 GHz. The model is appropriate for environments where the propagation occurs through the body of trees rather than by propagation due to radio wave diffraction. The Weissberger MED path loss through dense, dry, in-leaf trees is expressed mathematically as

$$\alpha = 1.33 f^{0.284} d_f^{\ 0.588} \ \text{dB} \ \text{for} \ 14 \ \text{m} \le d_f \le 400 \ \text{m} \tag{7.49}$$

$$\alpha = 0.45 f^{0.284} d_f \ \text{dB} \quad \text{for} \ 0 \ \text{m} \le d_f \le 14 \ \text{m} \tag{7.50}$$

where α is the differential attenuation due to the trees in dB/m, f is the frequency in GHz, and d_f in m is depth of trees (foliage). The differential attenuation is obtained from the measurement using

$$\alpha = \frac{L_1 - L_2}{d_f} \tag{7.51}$$

where

L_1 = Total measured loss in dB
L_2 = Measured loss in dB on a comparable path *without trees*

Therefore, $\alpha \ d_f$ represents the additional dB of attenuation due to trees in the radio path. However, the Weissberger MED model offers a very poor fit to measured data at frequencies higher than 1 GHz. In addition, it takes no account of the vegetation environment, such as the density of trees and the moisture content in the leaves.

7.10.4 Recommendation ITU Vegetation Loss Model

The initial vegetation loss model was approved by the Consultative Committee on International Radio (CCIR) in 1986 [25]. CCIR became ITU-R in 1992. The model conforms to the Weissberger MED and the excess loss due to propagation through forest is modelled as

$$L = 1.5887 f^{0.3} d^{0.6} \ \text{dB for} \ d < 400 \ \text{m} \tag{7.52}$$

where f is the frequency in GHz, and d is the depth of trees in metres and applies for depths less than 400 m. For example, the foliage attenuation at 28 GHz and 38 GHz for a penetration of 10 m is about 17 dB and 18.8 dB, respectively. Clearly this amount of attenuation has a significant impact on the signal reception and system performance. The ITU model for the forest excess loss given in (7.53) is plotted in Figure 7.10. The excess loss increases with both frequency and penetration depth. For example, at f = 30 GHz, the excess attenuation at depth 5 m was 11.58 dB and increased to 22.38 dB at 15 m depth, i.e. the attenuation almost doubled.

7.10.5 The Maximum Attenuation (MA) Vegetation Loss Model

Recommendation ITU-R P.833-8 was formally announced in September 2013 and updated to Recommendation ITU-R P.833-9 in 2016. It presented several models that are suitable for a variety of vegetation types and various path geometries appropriate to estimate the attenuation of a radio wave passing through the vegetation. We consider the ITU-R attenuation model for a terrestrial path with a single terminal in *woodland* commonly known as *the maximum attenuation (MA) model*. The vegetation loss can be characterised by the specific attenuate rate (dB/m) caused by the scattering of radio energy out of the radio path; and the maximum additional attenuation due to other processes such as surface wave propagation over the top of the vegetation medium. The system model that we consider involves a receiver at a certain distance

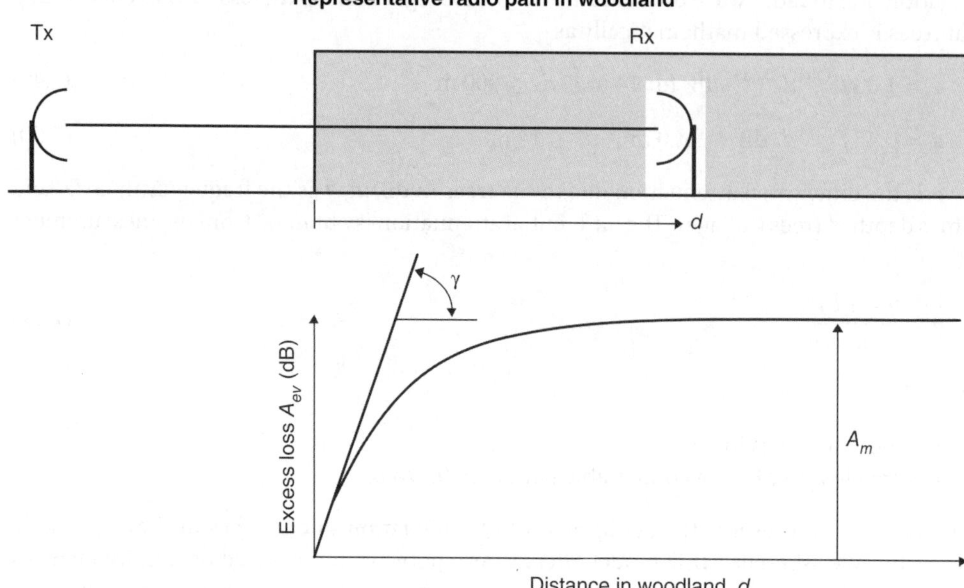

Figure 7.10 Illustration of the radio propagation path within woodland [26].

d within the wood and the transmitter located outside the wood. Denote the excess attenuation due to the vegetation as A_e given by [26] as

$$A_e = A_{v\,max}\left[1 - \exp(-\frac{d\gamma}{A_{v\,max}}\right] \tag{7.53a}$$

where d is the length of the radio path within the woodland in m; γ is the specific attenuation for very short vegetation paths (dB/m); and $A_{v\,max}$ is the MA in dB for a single terminal within a specific type of vegetation. The excess vegetation loss implies the extra loss on top of the free space loss. The ITU-R model for access loss is plotted in Figure 7.11.

Denote $x = \frac{d\gamma}{A_{v\,max}}$; then the exponential in (7.53) can be expressed as a series:

$$e^{-x} = 1 - \frac{x}{1!} + \frac{x^2}{2!} - \frac{x^3}{3!} + \ldots\ldots\ldots\ldots, x^2 < \infty \tag{7.53b}$$

For small x,

$$e^{-x} = \exp\left(\frac{-d\gamma}{A_{v\,max}}\right) \cong 1 - \frac{d\gamma}{A_{v\,max}}$$

For very short radio path, we have

$$A_e = A_{v\,max}\left(1 - \left(1 - \frac{d\gamma}{A_{v\,max}}\right)\right) = d\gamma \text{ dB} \tag{7.54a}$$

When $d = 1$ (m), γ dB/m is the attenuation for a short radio path in woodland. Typical values for γ in dB/m are empirically specified by the ITU after checking various measurements over the frequency range 30 MHz to about 30 GHz in woodland, as shown in Figure 7.11. It is worth noting that below about 1 GHz, vertically polarized signal experiences a higher attenuation than the horizontally polarization due to scattering from tree trunks. The $A_{v\,max}$ dB is frequency dependent:

$$A_{v\,max} = A_1 f^{\alpha} \tag{7.54b}$$

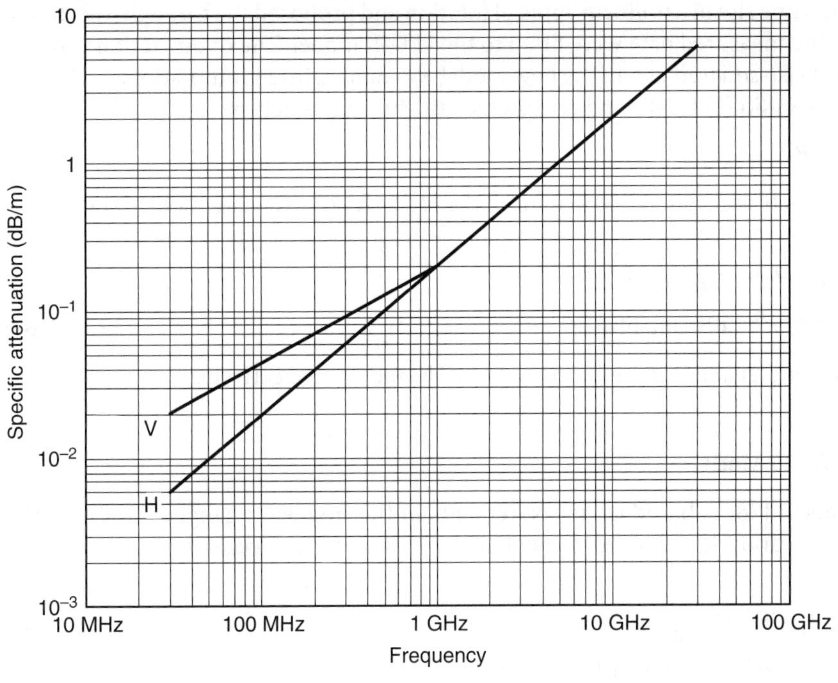

Figure 7.11 Specific attenuation due to woodland [26].

The model described in (7.54b) is only valid at frequencies below 3 GHz and its validity must be checked at higher frequencies.

7.10.6 The Modified and Fitted ITU-R (MITU-R) and (FITU-R) Vegetation Loss Models

The ITU-R model represented by Eq. (7.52) has to account for tree leaves. The modified International Telecommunication Union-Radio (MITU-R) model for attenuation due to vegetation (in excess to the free space) in dB is expressed as [27]

$$L = cd^n \tag{7.55}$$

where c is constant at a specific frequency, L is the loss in dB, and d is the foliage depth in m. The measurements were conducted at 11.2 GHz and the site consisted of a uniformly planted apple orchard. The new model was realized by optimizing the values of c and n to achieve a best fit with the measured data at 11.2 GHz. The in-leaf and out-of-leaf excess loss in (7.55) can be expressed as

$$L = 11.93d^{0.398} \text{ dB in-leaf} \tag{7.56}$$

and

$$L = 1.75d \quad d \leq 31 \text{ m} \;\; \text{dB out-of-leaf} \tag{7.57}$$

$$L = 28.1d^{0.17} \;\; d > 31 \text{ m} \;\; \text{dB out-of-leaf} \tag{7.58}$$

Further development on the ITU-R model was carried out using measurement data at 11.2 and 20 GHz in the same site of a uniform apple plantation and proposed the through-vegetation path loss model. The measured data were fitted to the ITU-R model. The optimization was carried out using the least-squared error. The new model was named fitted International Telecommunication Union-Radio (FITU-R) and is expressed by three parameters:

$$L = a f^b d^c \text{ dB} \tag{7.59}$$

Optimization of (a, b, and c) in (7.59) produced the FITU-R [27]. The latter added the following two formulae:

$$L = 5.7685 f^{0.39} d^{0.25} \text{ dB In-leaf} \tag{7.60}$$

$$L = 1.2829 f^{0.18} d^{0.59} \text{ dB Out-of-leaf} \tag{7.61}$$

where frequency f and distance d are expressed in GHz and m, respectively.

7.10.7 The COST235 Model

The COST235 model [28] is based on measurements in the frequency range (9.5–57.6 GHz) through small penetration ($d < 200$ m) through a grove of trees. The vegetation excess loss is expressed as

$$L = 14.6597 f^{-0.009} d^{0.26} \text{ dB In-leaf} \tag{7.62}$$

$$L = 6.6816 f^{-0.2} d^{0.5} \text{ dB Out-of-leaf} \tag{7.63}$$

where f is the frequency in GHz, and d is the depth of the trees in metres. For example, at 30 GHz and at 10 m inside the forest, the in-leaf loss and out-of-leaf loss is 25.87 and 10.7 dB, respectively.

Both COST235 and FITU-R models show received signal decay at a faster rate of propagation through a small number of trees obstructing the signal, relative to that obtained with a larger number of trees. This can be explained by the interplay of coherent components dominating the loss at short distances but strongly attenuated to the signal and the scattering component, which dominates larger paths but at less attenuation. Furthermore, foliage attenuation appeared to be higher for trees with leaves. This is because of higher absorption per unit volume, in addition to the fact that leaves dimensions and small branches are comparable to the wavelength at higher frequencies.

7.10.8 The Nonzero Gradient (NZG) Vegetation Loss Model

The maximum attenuation (MA) model defined by equation (7.53a) above was developed by including a final asymptotic attenuation rate together with the initial attenuation rate. The new model was called nonzero gradient (NZG) model in [29]. Mathematically, the NZG attenuation model can be expressed as

$$A = R_\infty d + k \left(1 - \exp\left[-\frac{(R_0 - R_\infty) d}{k}\right]\right) \tag{7.64}$$

where R_0, R_∞, and k are initial asymptotic attenuation rate in dB/m, final attenuation rate in dB/m, and offset in dB of asymptotic attenuation rate, respectively. The three-parameter model is attained by minimizing the sum of the least squares (LS) with the constraint that the final

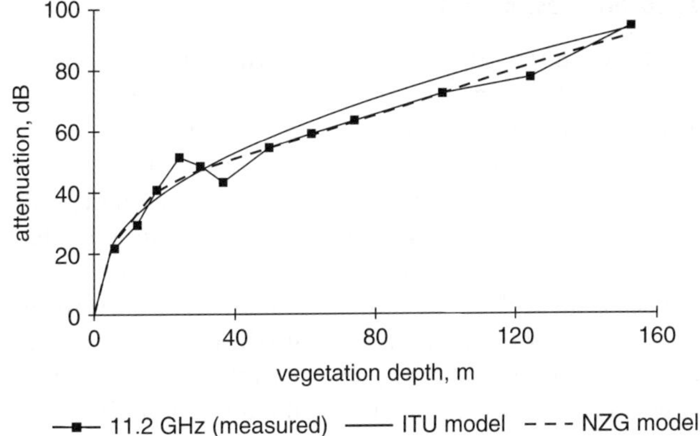

Figure 7.12 Comparison of measured attenuation rate with nonzero gradient model [29].

slope R_∞ is non-negative. The NZG model was tested using measured data at 38 GHz and the optimized fitted model conferred a final slope of zero, which reduced the NZG model to the MA model solution. Using vegetation loss measurement data at 11.2 GHz, the new model NZG is compared to the ITU model [25] as depicted in Figure 7.12. The three parameters used in the development of the NZG are given by Figure 7.12 as

$$R_0 = 4.84\ \mathrm{dB/m}, R_\infty = 0.35\ \mathrm{dB/m}, \text{and } k = 37.02\ \mathrm{dB}.$$

The NZG model was tested using loss data measured at 9.6, 28.8, and 57.6 GHz and on a copse of sycamore trees measured at 38 GHz and plotted in Figure 7.13.

Figure 7.13 presents a comparison of the NZG model fitted to all six of the data sets considered. The data plots for 9.6, 28.8, and 57.6 GHz seem to have the same initial and final slopes, which may relate to the measurement geometry and vegetation type, but their attenuation offset varies with frequency. The two 38 GHz measurement sets differ in their offset values by about 10 dB, which is equal to the widths of the illuminated vegetation areas for the two measurements.

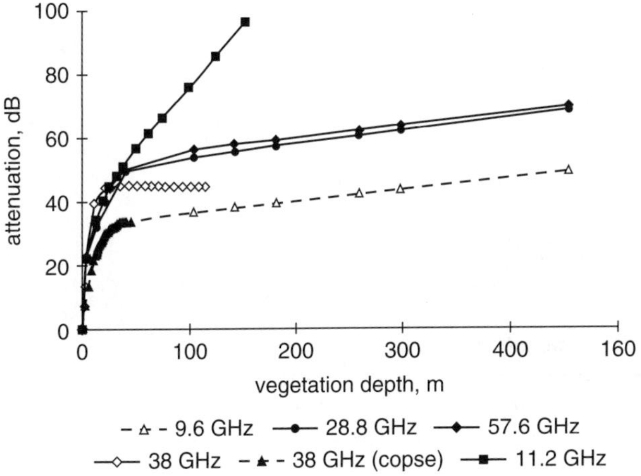

Figure 7.13 Best fit NZG model for five frequencies [29].

7.10.9 The Dual-Gradient (DG) Vegetation Loss Model

The NZG model was further developed to account for the site geometry, which the third parameter in the three-parameter model already mentioned. The vegetation site geometry includes the radio wave illumination of the vegetation medium and the antennas different beamwidths. The illumination width is the maximum effective coupling width between transmit and receive antenna beamwidths inside the vegetation medium. Vegetation measurement geometry with illumination of width W is shown in Figure 7.14. Here W is a measure of the maximum effective radio energy illumination width between the transmit and receive antennas that lies within the vegetation medium, θ_{tx}, θ_{rx} are angles representing the direction of energy departure and arrival, respectively, ω is the maximum height of the vegetation, r_1, r_2 are the distances of transmitter and receiver from both sides of the vegetation, and finally d is the width of the vegetation. The new model we are considering is generally known as the dual gradient (DG) model.

The best fit to the measurement data originated the DG model expression as

$$A = \frac{R_\infty}{f^a\, W^b}\, d + \frac{k}{W^c}\left(1 - \exp\left(-\frac{(R_0 - R_{R_\infty})W^c\, d}{k}\right)\right) \text{ dB} \tag{7.65}$$

where f is frequency of transmitted signal in GHz, and, b, c, k, R_0, and R_∞ are constants given in Table 7.4.

The maximum effective radio energy illumination width is given as

$$W = \min\begin{pmatrix} \dfrac{(r_1 + d + r_2)\tan(\theta_{tx})\,\tan(\theta_{rx})}{\tan(\theta_{tx}) + \tan(\theta_{rx})} \\ (r_1 + d)\tan(\theta_{tx}) \\ (d + r_2)\tan(\theta_{rx}) \\ \omega \end{pmatrix} \tag{7.66}$$

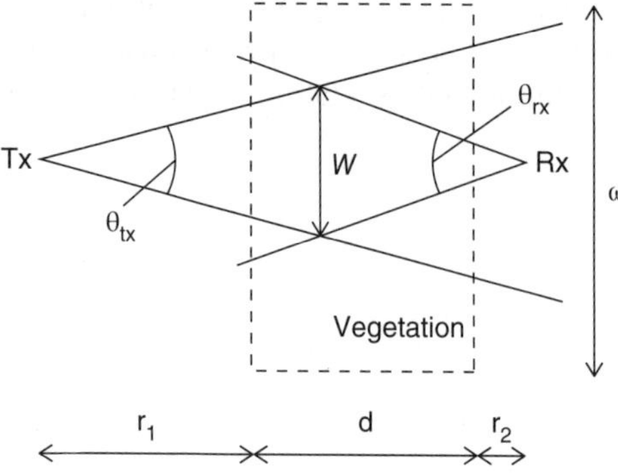

Figure 7.14 Vegetation measurement geometry [30].

Table 7.4 The DG model constant values for Eq. (7.65) [30].

Constant parameter	In leaf	Out of leaf
a	0.7	0.64
b	0.81	0.43
c	0.37	0.97
k	68.8	114.7
R_0	16.7	6.59
R_∞	8.77	3.89

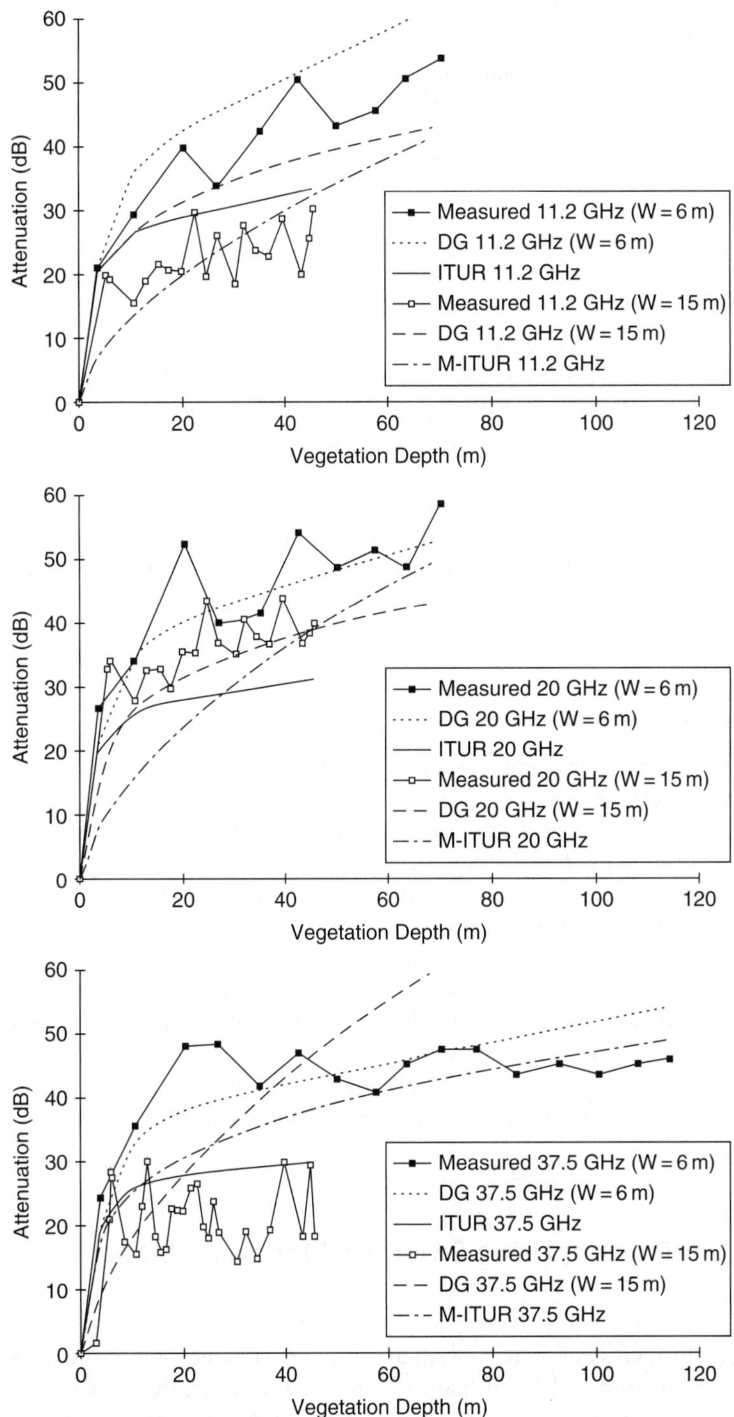

Figure 7.15 Comparison of vegetation attenuation models with measured data for three frequencies at two sites (Vegetation in leaf) [30].

Table 7.5 Average vegetation attenuation at mm wave frequencies [30].

Frequencies (GHz)	Small indoor trees attenuation (dB)		
	35	37.5	40
Norway spruce	9.0	9.0	9.8
Ficus	19.3	21.1	22.4
Leyland cypress	22.0	21.0	21.6

Two measurement sites at 11.2, 20, and 37.5 GHz are plotted as shown in Figure 7.15 and compared with other models discussed previously. For the best fit, the DG model error between measured and fitted data comprised a standard deviation $\sigma = 8.1$ dB for out-of-leaf case and $\sigma = 8.4$ dB for the in-leaf case, which is reduced by more than half the standard deviation produced by the ITU-R model. It should be noted that the inverse relationship with frequency f^a and with $0 < a < 1$ suggests a decreasing attenuation as the frequency increases, which is in contradiction with the ITU-R and other models described previously. For example, for a specific depth $d = 40$ m and $W = 6$ m, the DG model attenuation at 11.2, 19.6, and 37.5 GHz are roughly close to 50 dB, about 40, and less than 40 dB, respectively so the DG may need to be examined again.

7.10.10 Indoor Vegetation Attenuation Measurement

An indoor vegetation attenuation measurement was made at the Rutherford Appleton laboratory on three small trees in an anechoic chamber. The trees are a Norway spruce; a ficus; and a Leyland cypress. Each tree was placed on a turntable and the radio field was allowed to propagate through the tree as it was rotated at five degree intervals. The attenuation measurements were conducted at mmWave frequencies from 35 to 40 GHz. The average overall measurement attenuation for each of the three trees is given in Table 7.5.

There are two main observations from the indoor vegetation attenuation measurements at mmWave results: the attenuation from the Norway spruce over the band is the lowest and the highest is found with the Leyland cypress. This may be due to the fact that Leylandi has denser leaves, thus absorbing more energy and initiating more obstruction to the propagation path. The other observation is that the lowest increase in the attenuation over the 5 GHz band is less than 1 dB due to the Norway spruce and the highest increase is about 1.5 dB with the Leyland cypress.

7.11 Review of Vegetation Loss Using Empirical Models for Slant Propagation Path

When radio waves propagate through slant paths, initiated by mobile satellite systems, the trees introduce shadowing. The combination of the multipath and the shadowing are the main causes of signal fading. A typical scenario may be a direct signal received at the same time as multipath signals generated via scattering from a nearby tree(s). The received signals may add either constructively or destructively result in signal gain or signal fades.

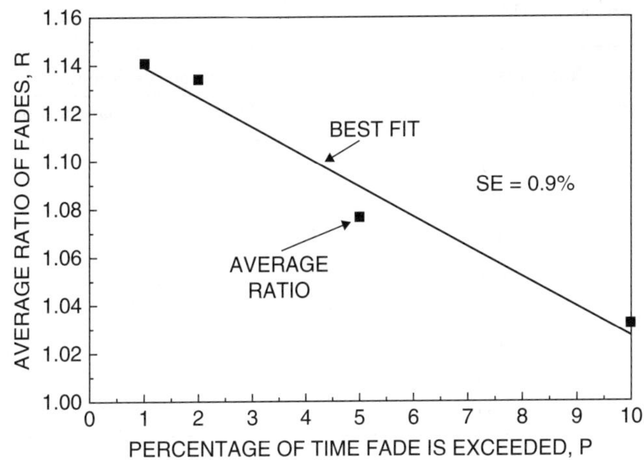

Figure 7.16 Average ratio for L-band to UHF fades versus percentage of time fade is exceeded [31].

Fading results in [31, 32] related to land mobile satellite communications at L-band (1502 MHz) and ultra-high frequency (UHF, 870 MHz) show that the fade distributions at 1% and 5% of the time the fades of 5.5 and 2.6 dB were, on average, exceeded at L-band and 4.8 and 2.4 dB were exceeded at UHF, respectively, for a path elevation angle of 45°. The fades were also found to be dependent on both *frequency* and *propagation path.* A least-squares fit to the average ratio of L-band fade $A(L)$ to UHF fade $A(U)$ versus the percentage of the time the individual fade is exceeded is depicted in Figure 7.16. The ratios of fades are approximately within the interval 1.02–1.14 corresponding to the respective 10–1% levels. The best fit line in this percentage range may be expressed as

$$R_f = \alpha + \beta P \tag{7.67}$$

where

$$R_f = \frac{A(L)}{A(U)} \tag{7.68}$$

and

$$\alpha = 1.152, \beta = -0.0123 \tag{7.69}$$

The fades correspond to % P and on average 1% level correspond to about 15 seconds. The standard error (SE) of the best fit was within 0.9% compared to average ratio points. Note SE is the standard error of the mean, which measures the accuracy with which the best fit compared to the average ratio.

An extensive land mobile satellite propagation measurements campaign was carried out at 19.6 GHz on a 30° to estimate the slant path attenuation from roadside trees and to compare these results with measurements at 1.6 GHz. Both sets of measurements were repeated on a pecan tree with and without leaves and on evergreen magnolia using similar geometry. The results at 19.6 GHz have shown that for the pecan with and without leaves, the fades exceed ~23 and 7 dB, respectively, as shown in Table 7.6. The cumulative distribution functions of the percentage of locations at which the fade in dB was exceeded were shown in Figure 7.17. These functions are comparing measurements that have been taken at 1.6 and 19.6 GHz and include clear LOS, pecan and magnolia with and without leaves.

Table 7.6 Median and 1% fades at L-band and K-band [33].

		Total fade		Attenuation coefficient	
		L-band (dB)	K-band (dB)	L-band (dB/m)	K-band (dB/m)
Clear L-O-S	Median		0.5		
	1%		2.6		
Bare pecan	Median	10.3	6.9	1.1	0.75
	1%	18.4	25.0	2.0	2.8
Pecan in leaf	Median	11.6	22.7	1.3	2.5
	1%	18.6	43	2.1	4.8
Magnolia	Median		19.6		4.4
	1%		39.6		8.8

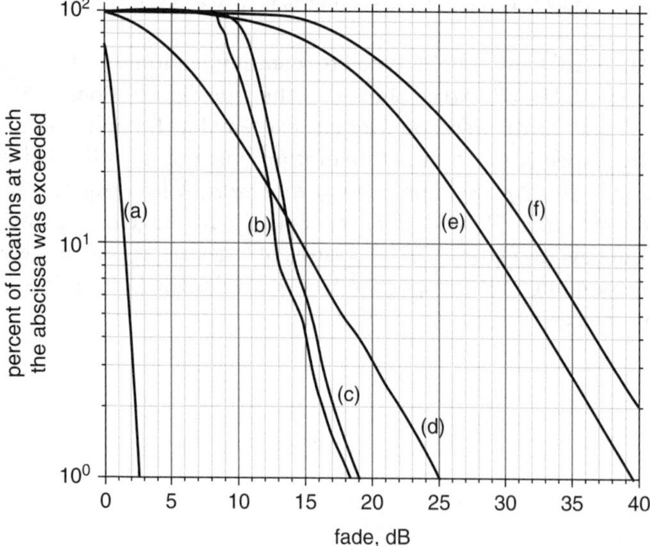

Figure 7.17 Cumulative distribution functions for clear LOS, as well as pecan and magnolia trees with and without foliage. (a) clear LOS (b) 1.6 GHz bare pecan (c) 1.6 GHz pecan in leaf (d) 19.6 GHz bare ecan (e) 19.6 GHz magnolia (f) 19.6 GHz pecan in leaf [33].

The distributions in Figure 7.17 show that all locations of pecan trees with leaves and without leaves have fades exceeding 10 dB at 1.6 GHz, while all locations of magnolia and pecan in leaves have fades exceeding 15 dB at 19.6 GHz. Furthermore, the presence of leaves increases the median at 1.6 GHz and 19.6 GHz fades by about 1 dB and 17 dB compared to the case with no leaves, respectively.

This difference in fades with the frequency may be attributed to the Fresnel zone size. The radio wave propagates along the diameter of the Fresnel zone. The Fresnel zone diameters are 2.7 m and 0.7 m at 1.6 GHz and 19.6 GHz, respectively. Assume the branches of trees without leaves are conducting mesh. Then as the Fresnel zone gets smaller (i.e.19.6 GHz), the presence of leaves introduces more energy absorbers along the wave propagation path and hence introduces more attenuation compared to the case of bare trees. When the Fresnel zone is large, the additional attenuation from the leaves is much smaller.

The effects of radio wave propagation through trees with and without leaves are overviewed in [23, 34, 35]. The attenuation contribution from Callery Pear tree with leaves, A(with leaves) at 870 MHz is related to the attenuation of bare tree A(bare tree) for elevation angle 15 – 40° as

$$\text{A(with leaves)} \approx 1.35\ \text{A(bare tree) dB} \tag{7.70}$$

The attenuation at 50° associated with transmission at 1.6 GHz through a canopy of red pine foliage is related to measurements at 870 MHz in this way:

$$\text{A(1.6 GHz)} \approx 1.36\ \text{A(870 MHz)} \tag{7.71}$$

These relationships are generalized in [33] as follows. Denote the median attenuations, $A(f_1)$ and $A(f_2)$ in dB of tree canopies with leaves at frequencies f_1 and f_2. Then their median attenuations are related as

$$A(f_2) = A(f_1)\exp\left\{b\left[\frac{1}{f_1^{0.5}} - \frac{1}{f_2^{0.5}}\right]\right\} \tag{7.72}$$

where $b = 1.173$.

Expression (7.72) has proven to be applicable in the frequency range 0.8–19.6 GHz and in agreement with the measured results within a maximum error of 0.2 dB.

7.12 Microphysical Modelling of Vegetation Attenuation

Microphysical modelling of vegetation loss mathematically describes the functions of a system involved in complex physical processes to define, simulate, or predict its perform*ance*. Compared with empirical/semi-empirical models, microphysical models *provide a physical insight into* the propagation of radio waves through vegetation. During the analytical development, *two* alternative vegetation mediums are the basis of two *different types of analytical* models. The first type of models defines the vegetation medium as a *homogeneous isotropic dielectric material* with constant permittivity and conductivity. While the second type designated the vegetation medium as a *mixture of trunks, branches, leaves, and air,* each material is represented by its complex permittivity, the effective volume it occupies, and by the statistical spatial orientation distribution of its elements.

In this section, we consider the propagation of radio waves through a multi-layered model of a forest and examine the different types of common waves that can exist and define the lateral waves that can be accommodated on top of a forest. Specifically, we consider the two-layered and the four-layered models to analyse the radio transmission loss.

A forest was first assumed to be a conductive slab. The basic geometry for a two-layered forest propagation model is illustrated in Figure 7.18 where the layers are the forest layer and the air layer, as identified in the figure. The model involves three interfaces, namely: ground–forest, forest–air, and air–ionosphere. The various waves produced by the radio transmission illuminating the forest are shown in Figure 7.19. These waves include a direct ray between transmitter and receiver within the forest; a reflected ray at the forest air interface; a reflected at the ionosphere (i.e. sky ray); and a wave propagated along the forest air interface (i.e. later ray).

The first interface (ground-forest) performs a minor contribution when the forest is adequately dense. In contrast, the other two interfaces accommodate strong fields in the form of lateral ray and sky ray, respectively. While the sky ray (wave) is propagated by being reflected by the Earth's ionosphere towards a certain receiver, the lateral ray is propagated by skimming across the treetops. Let the transmitter be located at a height z_0 above ground and consist of a

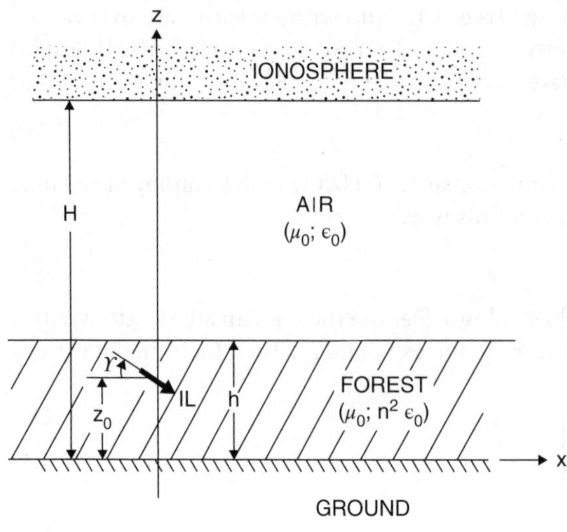

Figure 7.18 Basic geometry for the forest propagation model [36].

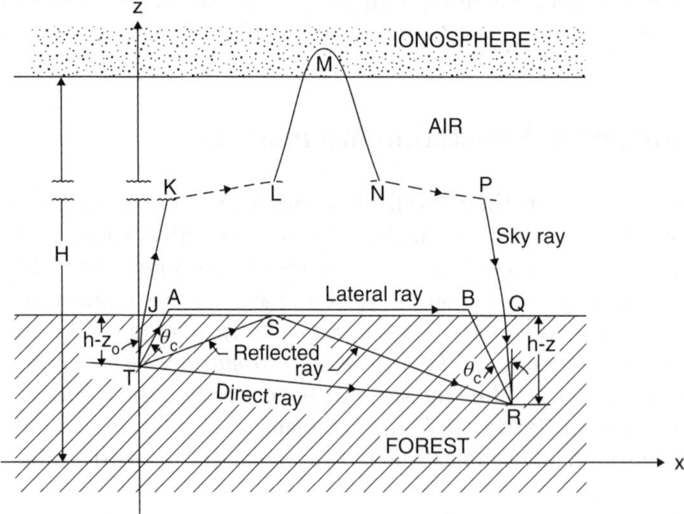

Figure 7.19 Various waves generated within the two-layered forest model [36].

small radiator inclined at an angle γ with respect to the x-axis. Furthermore, assume the forest average tree height is h and the ionosphere height H corresponds to the height of the wave reflection plane at any given frequency f. The lateral wave loss rate α_L in dB/m [36] is given by

$$\alpha_L = \frac{2\pi}{\lambda} Im(\sqrt{n^2 - 1}) \tag{7.73}$$

where n is the forest complex refractive index given by

$$n^2 = \varepsilon_1 - j\frac{\sigma_1}{2\pi f \varepsilon_0} \tag{7.74}$$

where ε_1 denotes the average relative permittivity, σ_1 is the average conductivity of the forest medium and ε_0 is the absolute permittivity. The theoretical and experimental values of the lateral wave attenuation rate in dB/m versus frequency in MHz are shown in Figure 7.20.

In Figure 7.20 the experimental points are obtained with horizontally polarized transmitted radio signals denoted as H points and with the vertically polarized signal denoted as V points.

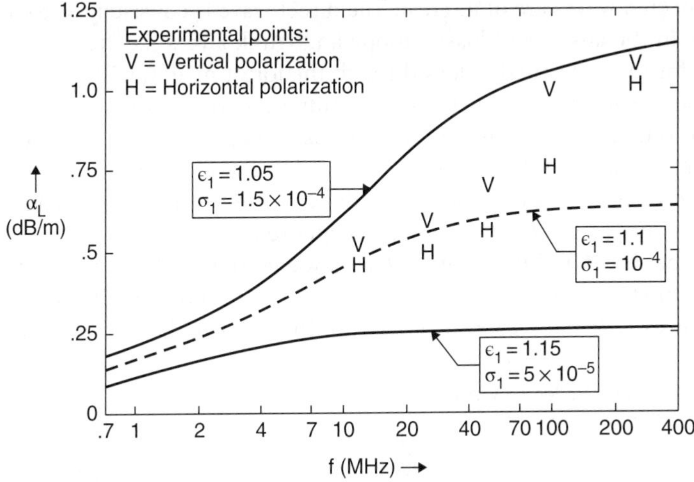

Figure 7.20 Theoretical and experimental values of the lateral- wave attenuation α_L dB/m versus frequency f in MHz [36].

The H and V points are obtained by field strength measurements at various antenna heights. Theoretical prediction results are obtained by sets of the refractive index n values. The two theoretical attenuation rate predictions by later-wave models differ by about 1 db/m at 400 MHz so accurate determination of both $\sigma_1, \in_1$ is necessary for determining the total attenuation rate of later-waves. Over practical distances greater than 1 km, the lateral wave propagates on the treetop line even if the vegetation extended over a terrain comprised hills or other obstructions. A sky wave is important only at large distances >10 km and low frequencies <10 MHz.

Analysis of the radio waves that are exited from a dipole antenna located inside a four-layered forest medium is illustrated in Figure 7.21. Layer 1 represents the air (free space); layer 2 represents the forest canopy; layer 3 illustrates the trunk layer of the forest; and layer 4 is the ground layer. The four-layered geometry has been commonly used for a typical forest model and generally applied in modelling the radio transmission loss. The model has shown that the transmission loss is an increasing function of the operating frequency. The electric fields are expressed in terms of direct waves; multiple reflected waves; and lateral waves. As the radio waves travel through the specified four layers, it appears that the lateral waves along the upper-side air-canopy interface play a dominant role.

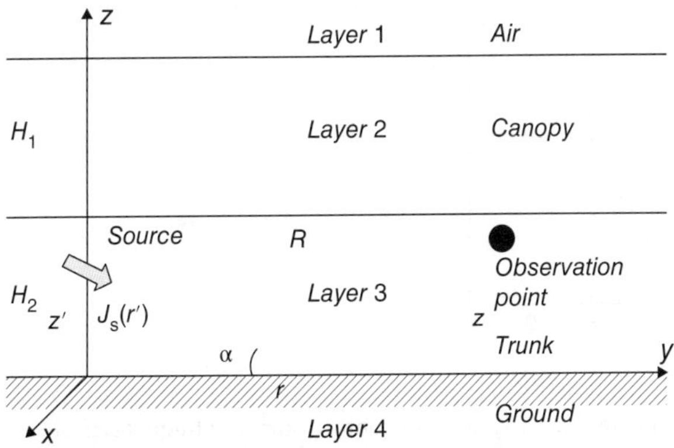

Figure 7.21 Geometry of a four-layered model of rain forest [37].

The direct wave and the reflected waves are of interest. The direct wave is consistently used to calculate the electric fields in the absence of lossy canopy and trunk layers (i.e. free space loss), and it is a key measure for the transmission loss through the forest medium. The direct wave becomes dominant when the distance between the transmitter and receiver is small. The reflected waves could exist for one-hop; two-hops and three-hops. For each of these scenarios the waves excited by the dipole could be reflected directly by either the first, second, or the third interface. These reflected waves modify the dominant direct wave. For near zone fields, the lateral waves propagating along the three interfaces can be ignored.

The transmission losses of the direct and the multi-reflected waves and the three interface lateral waves are computed within the range of 0.1–10 km at frequencies of 100 MHz, 500 MHz, and 3 GHz and plotted in Figures 7.22 and 7.23, respectively. In these cases, the transmission

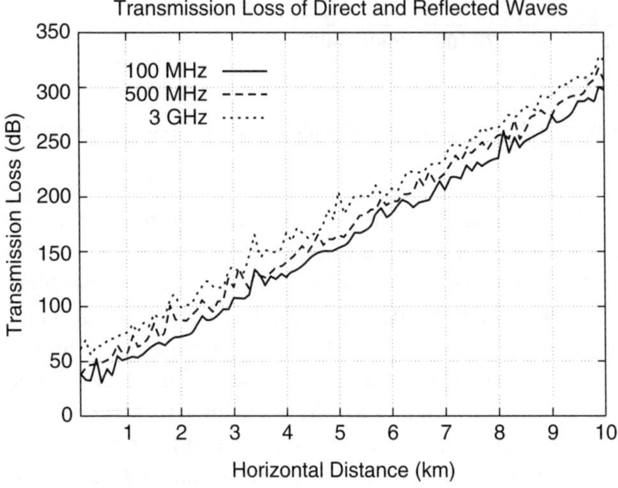

Figure 7.22 Transmission loss of direct and multi reflected waves against horizontal distance at frequencies of 100 MHz, 500 MHz, and 3 GHz [37].

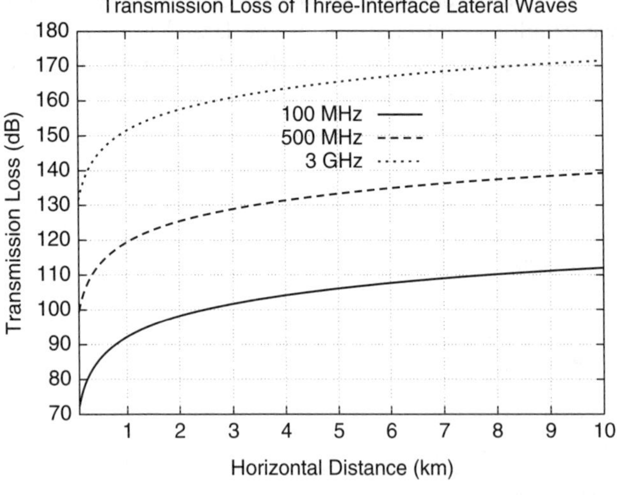

Figure 7.23 Transmission loss of three-interface lateral waves against horizontal distance at frequencies of 100 MHz, 500 MHz, and 3 GHz [37].

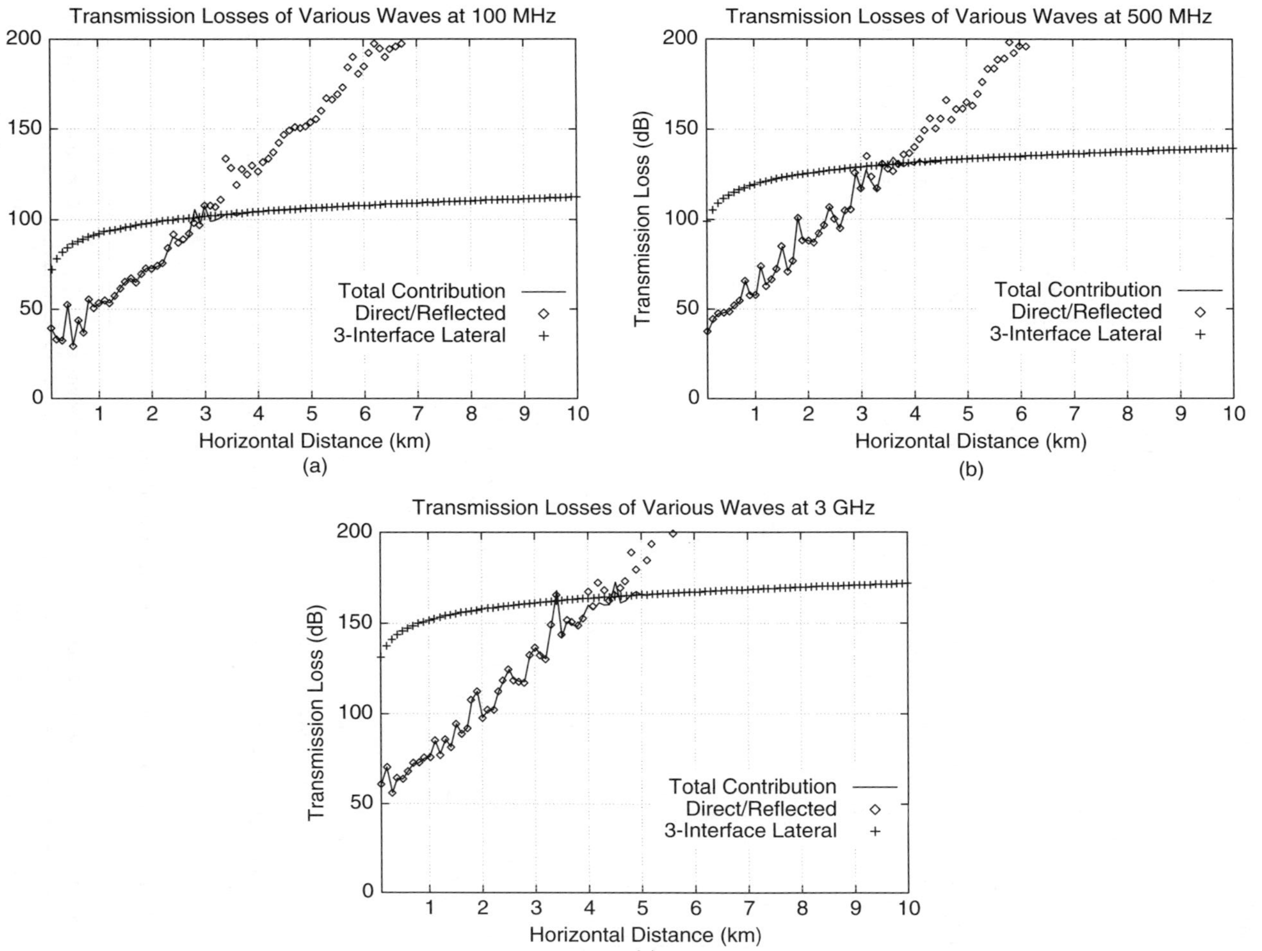

Figure 7.24 Comparison of transmission losses of: (a) the overall waves; (b) the direct and multi reflected waves; and (c) the three-interface lateral waves at frequencies of 100 MHz, 500 MHz, and 3 GHz [37].

loss increases gradually with the operating frequency at a known distance or rapidly with the horizontal distance at a given frequency. However, the transmission loss of the latter waves varies gently with the horizontal distant and rapidly with the operating frequency.

The contribution due to an individual group of waves to the transmission losses, including direct and multi-reflected waves and multi-interface lateral waves; together with the overall losses at frequencies of 0.1, 0.5, and 3 GHz, are shown in Figure 7.24. At short distances from the transmitter, the dominant waves in the total electric fields are due to the direct waves and multi-reflected waves. At 0.1 GHz, the near zone horizontal is about 3 km from the transmitter. At 0.5 GHz, it is about 4 km; and at 3 GHz it reaches 5 km. At the near zone horizontal distance, the loss contributions of the direct/reflected wave are almost equal to the total loss. On the other hand, the lateral waves along the upper side of the air canopy contribution become important at a large distance from the transmitter, i.e. larger than 3 km at 0.1 GHz; 4 km at 0.5 GHz; and 5 km at 3 GHz, respectively.

Transmission loss not only varies with distance but also with the operating frequencies, as illustrated in Figure 7.25. The transmission loss due to direct and reflected waves displays profound changes, unlike the lateral wave transmission loss, which faints with the frequency. The calculation of the forest transmission loss adopts the dielectric parameters given in Table 7.7.

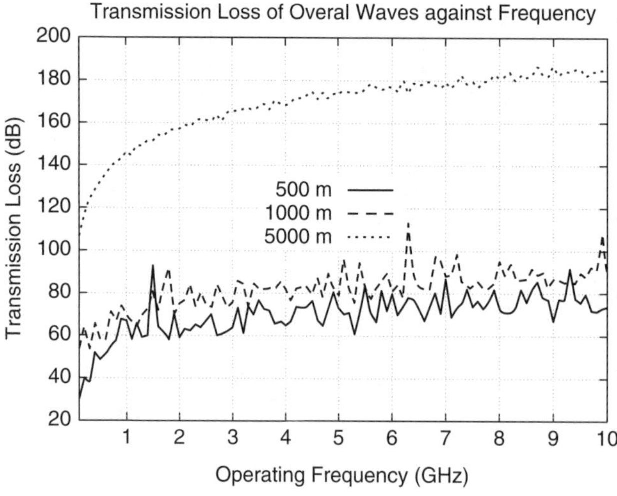

Figure 7.25 Transmission losses of the overall waves against operating frequency at a horizontal distance of 0.5, 1, and 5 km [37].

Table 7.7 Dielectric parameters of the typical four-layered forest model [37].

Layer number	Region extent	Permittivity ϵ_r	Conductivity σ (mS/m)
1	$H_1 + H_2 \leq z < \infty$	1	0
2	$H_1 \leq z \leq H_1 + H_2$	40	0.3
3	$0 \leq z \leq H_1$	35	0.1
4	$-\infty \leq z \leq 0$	50	100

H_1 and H_2 are chosen to be 10 m and 20 m, respectively.
mS/m is SI unit of conductivity in milli-Siemens per meter.

7.13 Attenuation in Vegetation Due to Diffraction

Diffraction is a phenomenon by which propagating waves enter into the shadow of an obstacle. One of the fundamental diffraction phenomena is that associated with the edge diffraction. Keller's geometrical theory of diffraction (GTD) [38] accounts for diffraction at the edges; corners; or vertices of boundary surfaces, making use of the general ray theory.

The concept of edge-diffracted rays making use of the general ray theory is introduced in [39]. However, GTD is only valid in deep shadows and fails in a shadow boundary region or partially shaded region. This limitation of GTD is removed by the uniform geometrical theory of diffraction (UTD) in [40]. The uniform asymptotic theory of edge diffraction in [41] can be applied for an edge in a perfectly conducting surface. UTD overcomes GTD limitations while retaining all the advantages of GTD. A UTD-based model for wooded crossroad configuration is presented in [42]. A ray-based transmit and receive antennas in the vertical polarization is used. A double-diffracted component is assumed over the canopy, and a direct transmission component is taken into account. The model has been validated with propagation measurements taken at 1.9 GHz.

Ray-based methods are formed on the geometrical optics and are applied to cases in which objects are of a much larger size than the wavelength, and in such cases the EM waves have flat wavefronts and thus behave similar to rays. At the locations where rays intersect an object, at least one new reflected, absorbed, diffracted, or scattered ray begins. As a result, a large number of rays arise throughout the environment. The computational load linearly increases with the number of reflections and transmissions and, therefore, ray launching and 'intelligent' ray tracing are the preferred techniques for rich scattering scenarios [43].

The analysis of vegetation obstacles in the radio wave propagation, in most cases, uses idealised forms of the obstacles, i.e. thick, smooth obstacles with a well-defined shape. Consequently, the results obtained using these forms must be regarded only as an approximate to the real things.

7.14 Recommendation ITU-R 526-7

The geometry of a single knife-edged obstacle is recommended in ITU-R 526-7 and illustrated in Figure 7.26. The geometrical parameters of the idealized case are shown here.

h	Height of the top of the obstacle above the straight line joining the two ends of the path. If the height is below this line, h is negative.
d_i	Distances of the two ends of the path from the top of the obstacle ($i = 1, 2$)
d	Length of the path
θ	Angle of diffraction (in rad.); its sign is the same as that of h. The angle θ is assumed to be less than about 0.2 rad., (1 rad. = 57.29558°; 0.2 rad. = 11.459 ≈12°).
α_i	Angles between the top of the obstacle and one end as seen from the other end. α_i have the same sign as h, where h, d_1, d_2, d, and λ in m.

These parameters are combined in a single (dimensionless) parameter v as

$$v(h) = h\sqrt{\frac{2}{\lambda}\left(\frac{1}{d_1} + \frac{1}{d_2}\right)} \tag{7.75}$$

where h and λ are in m, and d_1 and d_2 in km.

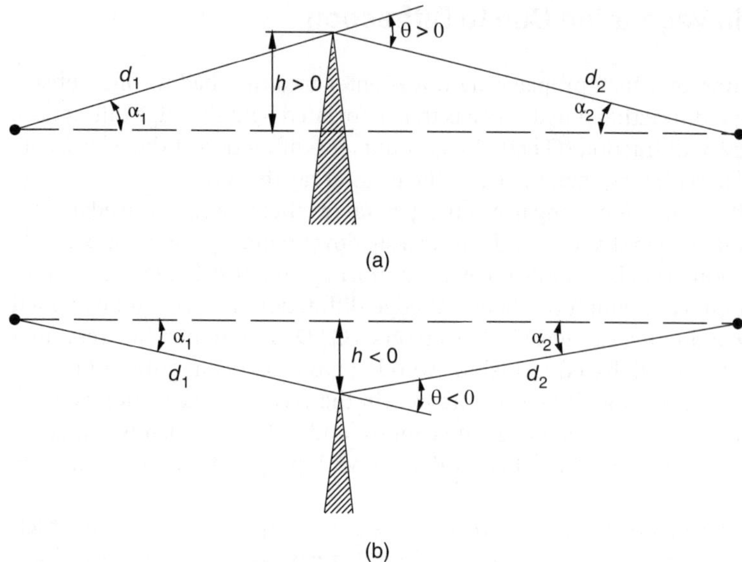

Figure 7.26 The geometry of a single knife-edge obstacle [44].

However, ν can assume other equivalent forms such as

$$\nu(\theta) = \theta\sqrt{\frac{2}{\lambda\left(\frac{1}{d_1}+\frac{1}{d_2}\right)}} \tag{7.76}$$

$$\nu(h,\theta) = \sqrt{\frac{2h\theta}{\lambda}} \tag{7.77}$$

$$\nu(\alpha_1,\alpha_2) = \sqrt{\frac{2d}{\lambda}\alpha_1\alpha_2} \tag{7.78}$$

For ν greater than -0.7, an approximate value for the knife-edge diffraction loss L_K can be obtained from the expression:

$$L_K(\nu) = 6.9 + 20\,log(\sqrt{(\nu-0.1)^2+1}+\nu-0.1)\text{ dB} \tag{7.79}$$

The knife-edge diffraction loss in dB is computed using formula (7.79) and plotted in Figure 7.27.

7.15 Propagation Modes Connected with the Vegetation Foliage

There are several possible modes of propagation related to vegetation, namely: the ground-reflected mode, the edge-diffracted mode, and the through foliage or scattered mode. The overall attenuation of the foliage should be inclusive of these modes. The more accurate inclusive attenuation model is based on the radiative energy transfer (RET) theory proposed by Johnson and Schwering in 1985. RET is briefly overviewed in Section 7.16. Measurement campaigns were carried out at many locations in the United Kingdom to measure the propagation

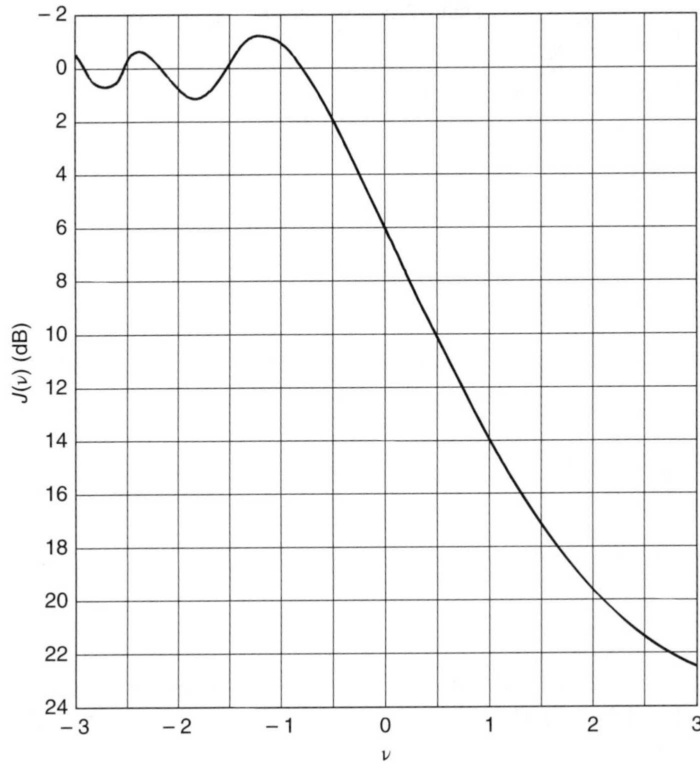

Figure 7.27 Knife-edge diffraction loss [44].

loss due to different species of trees; different formations of trees single trees, lines of trees, and dense wood and with different geometries for the aim of determining a generic loss model [45]. The findings of the measurements are outlined in the following items of this section.

7.15.1 Calculation of the Attenuation of the Top Diffracted Component

We consider that a single knife-edge diffraction method cannot sufficiently account for the signal path diffracted over vegetation. The geometry for double-isolated knife-edge diffraction is illustrated in Figure 7.28 and given by ITU-R P.526-7.

Total loss of the top diffraction is given by the sum of the two obstacles' individual losses. The first diffraction path is defined by distances a and b and the height h_{ab} produces a loss L_1 dB.

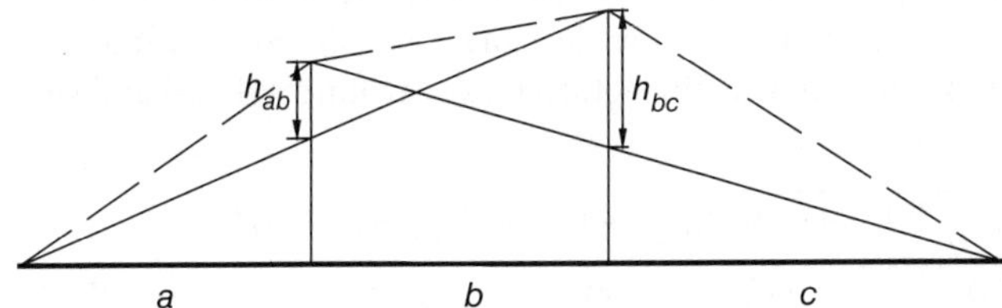

Figure 7.28 Method for double-isolated edges [44].

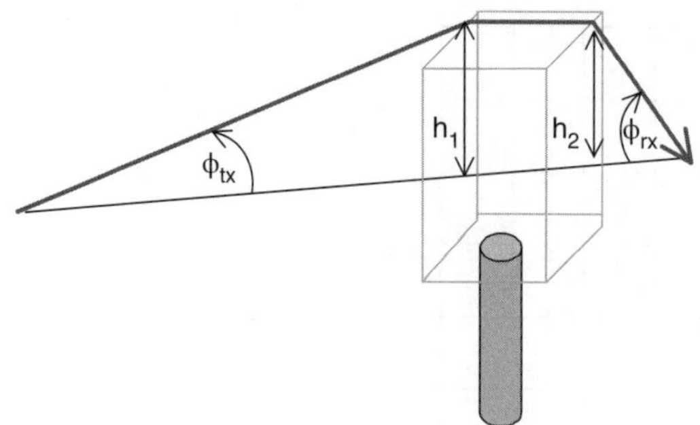

Figure 7.29 Geometry for top-diffracted wave [45].

The second diffraction path is defined by the distances b and c and the height h_{bc} and had a loss L_2 dB. A correction loss L_c is added to account for the separation b between the edges. L_c and is given as

$$L_c = 10\log_{10}\left(\frac{(a+b)(b+c)}{b(a+b+c)}\right) \tag{7.80}$$

Equation (7.80) is valid when diffraction loss at each edge exceeds about 15 dB. The total diffraction loss L is specified as

$$L = L_1 + L_2 + L_c$$

The attenuation experienced by the signal path diffracted over the top of vegetation may be treated as a double-isolated knife-edge diffraction for the geometry outlined in Figure 7.29. The total diffraction loss L_{top} dB is given by

$$L_{top} = L_1 + L_2 + L_c + G_{TX}(\varphi_{tx}) + G_{RX}(\varphi_{rx}) \tag{7.81}$$

where $G_{TX}(\varphi_{tx})$ and $G_{RX}(\varphi_{rx})$ [45] are the losses due to angles of the diffracted wave, leaving the transmit antenna and entering the receive antenna, respectively. Substituting for L_1 and L_2 using (7.79) and L_c using (7.80) we get

$$\begin{aligned} L_{top} &= 6.9 + 20\log_{10}(\sqrt{(v(h_1)-0.1)^2+1} + v(h_1) - 0.1) + G_{TX}(\varphi_{tx}) + 6.9 \\ &+ 20\log_{10}(\sqrt{(v(h_2)-0.1)^2+1} + v(h_2) - 0.1) + L_c + G_{RX}(\varphi_{rx}) \end{aligned} \tag{7.82}$$

The values for $G_{TX}(\varphi_{tx})$ and $G_{RX}(\varphi_{rx})$ losses can be read from the radiation patterns of transmit and receive antennas.

7.15.2 Attenuation Components Due to Side Diffraction

The diffraction access loss, L_{sidea} and L_{sideb}, in dB experienced by the signal diffracted wave around each side of the vegetation may be treated as double-isolated knife-edge diffraction. Considering the geometry defined in Figure 7.30 the diffraction losses around side (a) and side (b) are

$$\begin{aligned} L_{sidea} &= 6.9 + 20\log_{10}(\sqrt{(v(h_{1a})-0.1)^2+1} + v(h_{1a}) - 0.1) + G_{TX}(\theta_{txa}) + 6.9 \\ &+ 20\log_{10}(\sqrt{(v(h_{2a})-0.1)^2+1} + v(h_{2a}) - 0.1) + L_C + G_{RX}(\theta_{rxa}) \end{aligned} \tag{7.83}$$

$$\begin{aligned} L_{sideb} &= 6.9 + 20\log_{10}(\sqrt{(v(h_{1b})-0.1)^2+1} + v(h_{1b}) - 0.1) + G_{TX}(\theta_{txb}) + 6.9 \\ &+ 20\log_{10}(\sqrt{(v(h_{2b})-0.1)^2+1} + v(h_{2b}) - 0.1) + L_c + G_{RX}(\theta_{rxb}) \end{aligned} \tag{7.84}$$

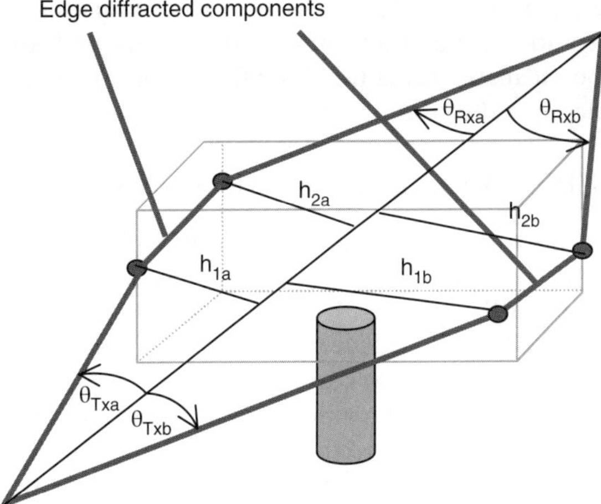

Figure 7.30 Geometry for edge-diffracted wave [46].

where $G_{TX}(\theta_{tx\,a,\,b})$ and $G_{RX}(\theta_{rx\,a,\,b})$ are the losses due to angles of the diffracted wave leaving the transmit antenna and coming into the receive antenna for sides a and b, respectively, and L_C is given by (7.80) with similar definitions for a, b and c for each side. The total access attenuation due to diffraction around the sides L_{side} in dB is given as

$$L_{side} = L_{sidea} + L_{sideb}\ \text{dB} \tag{7.85}$$

7.15.3 Attenuation of the Ground Reflection Component

The ground-reflected wave can be modelled as shown in Figure 7.31. The attenuation of the ground-reflected wave L_g dB at the receiver may be expressed in terms of the distances of the direct wave distance d_0 and the reflected waves distances d_1 and d_2 and the ground reflection coefficient δ at grazing angle θ_g. The ground reflection coefficient depends on the values for the ground permittivity and conductivity that are available in ITU-R P.527-4 [47] for frequencies up to 1000 GHz. The reflection coefficient magnitude and phase of a plane surface as function of grazing angle for vertical and horizontal polarization in the mmWave band are given in CCIR Report 1008-1 [48].

The attenuation experienced by the ground-reflected wave, L_g, in dB [46] is

$$L_g = L_{gveg}(d_{veg}) + 20\log_{10}\left(\frac{d_1 + d_2}{d_0}\right) - 20\log_{10}R_0 + G_{TX}(\vartheta_{tx}) + G_{RX}(\vartheta_{rx}) \tag{7.86}$$

where R_0 is the reflection coefficient, $G_{TX}(\vartheta_{tx})$ and $G_{RX}(\vartheta_{rx})$ are the losses due to angles of the reflected wave leaving the transmit antenna and coming into the receive antenna, respectively,

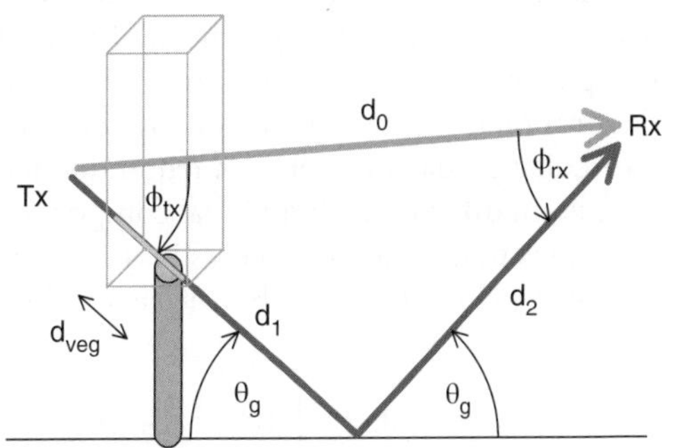

Figure 7.31 Geometry for ground-reflected wave [45].

and can be read from antenna radiation patterns, and $L_{gveg}(d_{veg})$ is the loss due to the propagation of the ground-reflected wave through the total vegetation depth, d_{veg}. This loss can be calculated using the through vegetation loss presented in Section 7.16, with the vegetation depth of the ground-reflected wave d_{veg} being substituted for z.

7.15.4 Attenuation of the 'Through' or Scattered Component

The scattering loss is ascertained by applying the RET theory so it is pertinent to be presented with the RET theory in one section i.e. Section 7.16. Let as denote the vegetation scattering loss as L_{scat} dB.

The scattering of wireless energy in a vegetation medium is dependent on many parameters, and these parameters are essential to the calculation of the scattering attenuation. Denote the spectral scattering coefficient by σ_s. These parameters include a number that are *species-specific,* namely: the ratio α of the forward scattered power to the total scattered power; the coefficient σ_τ, which is the sum of the absorption σ_a and the scattering σ_s coefficients i.e. $\sigma_\tau = \sigma_a + \sigma_s$; the albedo W; the typical leaf size of the vegetation; and the leaf area index LA I. The other parameters are antenna-specific: the beamwidth of the phase function β; the beamwidth of the receive antenna Δ_{γ_R}; and of course, the operating frequency in GHz. Measurement campaigns are carried out to measure these parameters for some native plants by the UK Radiocommunications Agency (RA), and the final report was published in 2002 [46]. The measured database submitted to the ITU-R as a proposed revision to the prediction vegetation attenuation above 1 GHz in recommendation ITU-R P.833-3.

Tables of the measured data can be found in [46] for operating frequency band 1.3–61.5 GHz as follows: α in Table 3-1; β in Table 3-2; W in Table 3-3; and σ_τ in Table 3-4 for operating frequency band 1.3–61.5 GHz together with the appropriate frequencies, leaf area indexes and leaf sizes, and the 3 dB beamwidth of the receiving antennas. Further measurement data can also be found in [49, 50].

In particular, the albedo W is defined as the fraction of the EM radiation reflected from the object surface. Parameters σ_τ and W are determined by measuring of the excess loss due to the vegetation medium. The parameters α and β were determined by measuring the scattering *phase function* of the vegetation medium. The phase function characterises the scattering behaviour of the medium at short distances after the air vegetation interface. In many cases the phase function measurement will therefore be conducted in conjunction with the measurements of the excess attenuation experienced in the vegetation medium as a function of depth inside the vegetation. An ideal measurement site will comprise a group of trees, where the receiver is located at various depths where possible.

7.15.5 Combination of the Individual Attenuation Components

Formulae that can be used to calculate the excess loss of components in dB were presented in Section 7.15. However, to get the total access loss, we cannot add up the access loss of individual components in dB as we commonly do when calculating the amplification of a cascaded amplifier or total transmission loss of a system. Here individual loss components have experienced a particular propagation phenomenon that has occurred. So what we need to do first is convert the loss in dB from an individual component to a *numerical value loss* and add all the numerical values to get the total access loss and finally convert the numerical total into dB. Thus, the total excess loss, L_{total}, in dB experienced by a signal propagating through trees can be given by (7.87):

$$L_{total} = -10\log_{10}\left\{10^{\left(\frac{-L_{top}}{10}\right)} + 10^{\left(\frac{-L_{side}}{10}\right)} + 10^{\left(\frac{-L_{g}}{10}\right)} + 10^{\left(\frac{-L_{scat}}{10}\right)}\right\} \text{ dB} \tag{7.87}$$

7.16 Radiative Energy Transfer (RET) Theory

7.16.1 Introduction

Energy transport between donor molecules to acceptors is possible when the emission and the absorption spectrums (partially) overlap. When an energy transport process is repeating itself so that most of the energy migrates to several molecules, it is called *energy transfer*. Classically, energy transfer is considered as an interaction of two oscillating molecules, the donor is initially in oscillation and the acceptor is initially at rest and during the oscillation, the energy is progressively transferred from the first molecule to the second. According to the quantum theory, the transfer occurs randomly and therefore the classical results can be seen as a *statistical average*. The power radiated by a donor is proportional to the fourth power of the frequency and the power absorbed by the acceptor is proportional to the acceptor's absorption cross section. RET theory enables the calculation of attenuation of a randomly distributed scattering medium such as vegetation loss [51–54].

The radiation transfer in emitting, absorbing, and scattering is concerned with the volumetric form of Kirchhoff's law. The local thermal equilibrium (LTE) is always applied in deriving the radiation intensity. The LTE conditions imply that any small volume element of a medium is at the LTE so that any point in the volume can be identified by a local temperature. This consideration also can be applied to the collisions of particles so that they result in a LTE at each point of the medium. Then the emission of the radiation by a small volume can be expressed in terms of Planck function of an ideal black body. This leads to the principles of the radiative energy transfer (also known as the transport theory).

A forest is represented as a *random medium* for the development of an mmWave propagation model. The medium can be thought as statistically homogeneous. In this Section the wave propagation in the forest medium is explored by applying the radiative energy transfer (RET). It is worth noting that the coherent length of mm Waves, is defined by $\left(\frac{c}{nB}\right)$ where c is speed of light, n medium refractive index, and B radiated signal bandwidth. Further experimental facts suggest that the coherence length of mm waves in vegetation is short so any interference during the wave propagation can be ignored. Scattering in a forest medium includes tree trunks; branches; and leaves which are longer than mmWave length and contribute strong forward scattering.

The different approaches of the RET theory compared to Maxwell theory can be comprehended with a preparation note on the foundation of the RET theory. After such note the use of specialized symbols and figures when needed to analyze the scattering loss related to EM waves propagated within a forest using the RET theory.

The EM field in a random scattering medium can be split into two parts: the direct path (coherent) and the non direct (incoherent) path. The direct path controls the short distances from the source but the non direct path becomes important at longer distances. The propagation properties to be determined include the distance dependence of direct path (coherent) and non direct path (incoherent) field components of mmWave beams transmitted through the forests/woods as a function of vegetation depth and density.

The key measure used in RET is the specific intensities of the coherent and incoherent field components. The total intensity of the radiation field is the sum of the two intensities. The key quantity used in the RET is the specific intensity defined as the *power per unit area* and *per unit solid angle* propagating at certain point into certain direction and at certain frequency.

A beam of EM radiation in certain direction in a given medium with intensity $I(\mathbf{r}, \mathbf{\Omega})$ propagates with absorption, emission, and scattering. Naturally, the radiation energy decreases due to the absorption. But the effect of scattering can be both positive and negative on the initial

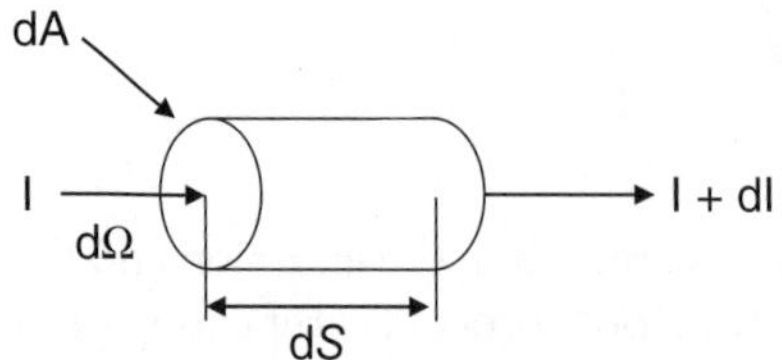

Figure 7.32 Schematic presentation of the geometry of the radiative transfer energy (RET).

radiation intensity. Consider such radiated beam is restricted by a solid angle dΩ and incident perpendicularly to the element of surface area dA of a flat object of width dS as shown in Figure 7.32.

As the incident radiation propagates in the substance, part of the radiation is absorbed by the substance. Denote the spectral absorption coefficient by $\sigma_a(\mathrm{r})$. The fraction of the incident radiation being absorbed by the substance per unit radiation path length has the dimension of m^{-1} if the path length is measured in metres. The absorption of the intensity of the incident radiation arrived at the object surface from the direction Ω per unit time in a *unit of volume* dA dS and in a unit frequency band is,

$$\sigma_a(\mathbf{r})\,\mathrm{I}(\mathbf{r},\boldsymbol{\Omega})\,\mathrm{d}\Omega \tag{7.88}$$

Total absorption from all directions within the solid angle can be determined by integrating (7.88) over all solid angles to give the absorption per unit time, in a unit of volume and in a unit frequency (W *per* m^3 Hz) as given by [53],

$$\sigma_a(\mathbf{r}) \int_0^{2\pi} \int_{\mu=-1}^{+1} \mathrm{I}_\mathrm{a}(\mathbf{r},\mu,\phi)\,\mathrm{d}\mu\,\mathrm{d}\phi \tag{7.89}$$

Consider the radiation beam again but this time to account for the scattering of the incident radiation. While part of the radiation is absorbed, the remainder of the incident radiation passes through the medium and scattered in all directions over the unit length of the propagation path. Denote the spectral scattering coefficient by σ_s. The scattering in all directions per unit time, in a unit of volume, and in a unit frequency band is,

$$\sigma_s(\mathbf{r})\,\mathrm{I}(\mathbf{r},\boldsymbol{\Omega})\mathrm{d}\Omega \tag{7.90a}$$

Equation (7.90) identifies the part of the energy which will be removed from the incident beam in a unit volume in the direction of Ω. The distribution of the scattering over all the directions can be described by a normalized phase function. The scattering of the radiation is given by [53]

$$\frac{1}{4\pi}\sigma_s(\mathbf{r})\,\mathrm{d}\Omega \iint \mathrm{I}(\mathbf{r},\boldsymbol{\Omega})\,p(\Omega)\mathrm{d}\Omega \tag{7.90b}$$

Forest can be described as a strongly scattering medium so when it is radiated by an EM beam, it creates a random distribution of discrete scatterers. A mathematical model of the propagation and scattering of the waves can be derived in two different approaches: Analytical theory approach based on the *Maxwell's equations*; and by the *transport theory*. In an analytical case using Maxwell approach, we start with Maxwell's equations to derive the characteristic nature of the medium including the statistical moments of the wave. In principle, this is the most fundamental approach, and many investigations have extensively exploited this approach. However, its drawback is the mathematical complexities involved that limit the depth of the finding usefulness.

Transport theory, on the other hand, does not start with Maxwell's equations. It deals directly with the transport of energy through dense medium. Since both Maxwell and transport theories deal with the same physical problem, one would expect there should be some relation between them. In fact there have been many attempts to link the two theories but with varying degrees

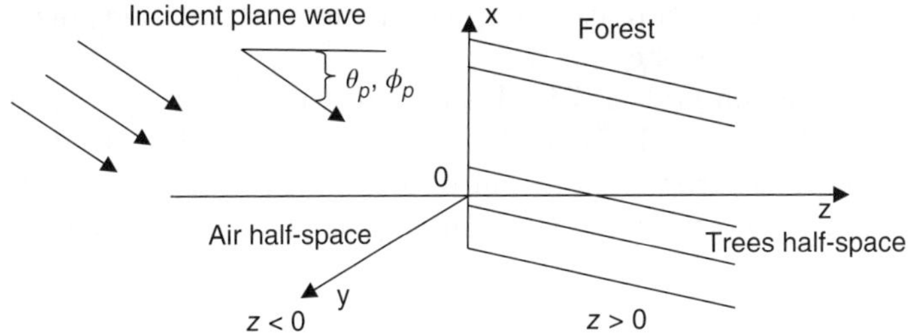

Figure 7.33 Forest half space ($z > 0$) illuminated by incident plane EM wave.

of success. But in spite of such heuristic development, the transport theory is used extensively to solve a large number of practical problems.

The RET theory allows an analytical calculation of the excess attenuation of a scattering medium, such as vegetation. It requires six input parameters, four of which are specific for the scattering medium and two are related to the geometry and system descriptors. The four medium-dependent parameters need to be established experimentally and are dependent on the type of vegetation, frequency, and state of foliation.

A forest is assumed to occupy the half-space region $z > 0$ as depicted in Figure 7.33, to be statistically homogeneous and to consist of a random distribution of particles, which scatter and absorb radiation. A plane wave from the air region ($z < 0$) is assumed to be normally incident onto the homogenous medium (forest). The site needs to be free of objects that can cause spurious reflected signals to interfere with the scattered signal from the trees to the receiver location. Phase function measurements combined with excess attenuation measurements, performed for a line of trees made up of the same tree species can jointly provide the RET parameters.

Both transmit and receive antennas are arranged to have a common volume irradiated in the scattering medium. Under these conditions the RET theory hypothesizes a phase function with Gaussian shaped forward lobe and an isotropic background. The two parameters α and β define the exact shape, which is dependent on factors such as the vegetation species and density and foliation state (in leaf; out of leaf). When the measurement location of the phase function moves inside the medium, the forward lobe broadens suggesting an optimum location within, which the measurement may be conducted. The beamwidth of the transmit antenna needs to provide volume for the measurement. On the other hand, the beamwidth must be narrow enough as not to permit diffraction and reflection signal components that may affect the measured phase function pattern. The receive antenna should have a very small (few degrees) beamwidth, in order to measure the relatively narrow beamwidth of the phase function itself.

7.16.2 RET Attenuation Prediction Model

The scattering phase function is assumed to consist of a narrow Gaussian forward lobe superimposed over an isotropic background,

$$p(\gamma) = \alpha\, q(\gamma) + (1 - \alpha) \tag{7.91a}$$

where,

$$q(\gamma) = \left(\frac{2}{\beta}\right)^2 e^{-\left(\frac{\gamma}{\beta}\right)^2} \qquad \Delta\gamma_s \ll \pi \tag{7.91b}$$

where, $\gamma = \cos^{-1}(\theta, \phi)$, α is the ratio of the forward scattered power to the total scattered power and $\beta = \Delta\gamma_s$ is the beamwidth of the forward lobe.

The forward lobe of the scatter function $p(\gamma)$ in (7.93a) is narrow Gaussian function, and the second term $(1 - \alpha)$ is isotropic background. The RET model equation can be solved by splitting the specific intensity I into two parts: the coherent incident intensity component I_c (direct component), which dominate at short distances into the vegetation and the incoherent (scattered) diffused intensity component I_d. Therefore

$$\mathrm{I}(z;\theta,\phi) = \mathrm{I}_c(z;\theta,\phi) + I_d(z;\theta,\phi) \tag{7.92}$$

where z; θ, ϕ represent the coordinate system, namely: z is the coordinate normal to air half space –forest half space interface, θ is the angle between positive z direction and scattering direction, and ϕ is the projection of θ in z plane, as shown in illustration in Figure 7.34.

The specific intensity I in the forest medium depends on the coordinate; angles; and ϕ only, and it is independent of the cross-sectional coordinates X and y.

We can split the diffuse intensity I_d into two parts:

$$\mathrm{I}_d(z;\theta,\phi) = \mathrm{I}_1(z;\theta,\phi) + \mathrm{I}_2(z;\theta) \tag{7.93}$$

where I_1 is determined by the forward lobe $q(\gamma)$ and I_2 is determined by the isotropic background. Finally, the specific intensity at a given point within the vegetation medium is

$$\mathrm{I} = \mathrm{I}_c + \mathrm{I}_1 + \mathrm{I}_2 \tag{7.94}$$

The receiving antenna is characterised by its power radiation pattern $G_R(\gamma_R)$, where γ_R is the angle for the direction of observation θ, ϕ includes with the pointing direction of the antenna axis θ_R, ϕ_R (main beam direction). Therefore

$$G_R(\gamma_R) = \left(\frac{2}{\beta}\right)^2 e^{-\left(\frac{\gamma_R}{\beta}\right)^2} \tag{7.95}$$

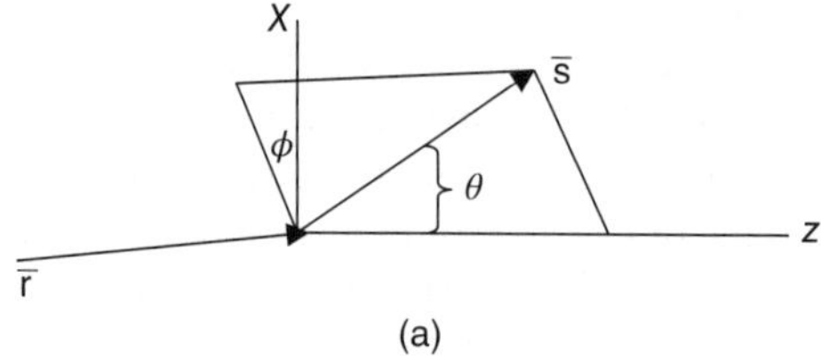

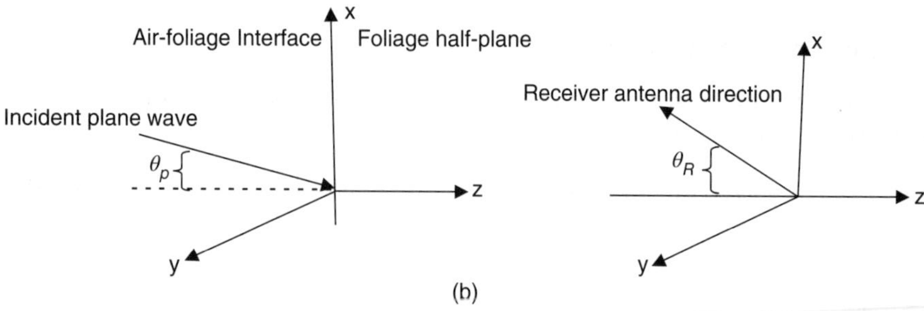

Figure 7.34 (a) The specific Intensity coordination system. (b) Incident angle and the receiver antenna angle.

$G_R(\gamma_R)$ is assumed to be normalizes such that

$$\int_0^{2\pi}\int_0^{\pi} G_R(\gamma_R)\ \sin\theta\ \mathrm{d}\theta\ \mathrm{d}\phi = 4\pi \tag{7.96}$$

To calculate the received power, we assume the transmit antenna is located at a sufficiently large distance from the edge of the forest and a highly directive receive antenna of narrow beamwidth is located inside the foliage area. The directivity gain of the receive antenna is given by $G_R(0)$. The received power is determined using the relation $A_{eff} = \frac{\lambda^2}{4\pi} G_R(\gamma_R)$ for the effective cross section. The available power at the output of the receiving antenna $P_R(\tau;\ \mu_R,\ \phi_R)$ is the product of the effective area of the receiving antenna in direction θ_R, ϕ_R and the flux density from that direction:

$$P_R(\tau;\mu_R,\phi_R) = \frac{\lambda^2}{4\pi}\int_0^{2\pi}\int_0^{\pi} G_R(\gamma_R).\ \mathrm{I}(\tau;\theta,\phi)\mathrm{d}\mu\ \mathrm{d}\phi \tag{7.97}$$

where λ is the wavelength at the operating frequency, and the normalized coordinates $\tau = (\sigma_a + \sigma_s)Z$, $\mu_R = \cos\theta_R$, and $\mu = \cos\theta$.

The received power P_R depends on the location and orientation of the antenna. The maximum received power P_{max} is received if the receive antenna is placed at the forest boundary $z = 0$ and is aligned with the incident radiation, i.e. $\theta_R = \theta_P$ and $\phi_R = \phi_P$. Therefore, using (7.95) with $\gamma_R = 0$, we have

$$P_{max} = P_R(0;\theta_P,\phi_P) = \frac{\lambda^2}{4\pi} G_R(0) = \frac{\lambda^2}{4\pi}\left(\frac{2}{\Delta_{\gamma_R}}\right)^2 S_P \tag{7.98}$$

where S_P is the Poynting vector of the incident wave. Johnson and Schwering [51] have derived the received power normalized to the maximum power is articulated in the next section.

7.16.2.1 Scattering Loss for *Slant Radiation*

$$\frac{P_R}{P_{max}} = e^{-\left(\frac{\gamma_{RP}}{\Delta_{\gamma_R}}\right)^2 - \tau}\ (\mathrm{I}_c) \tag{7.99a}$$

$$+\frac{(\Delta_{\gamma_R})^2}{4}\left(\left(\mathrm{e}^{-\frac{\hat{\tau}}{\mu_P}} - \mathrm{e}^{-\frac{\tau}{\mu_P}}\right)\overline{\mathrm{q}}_{\mathrm{M}}(\gamma_{RP}) + \mathrm{e}^{-\frac{\tau}{\mu_P}}\sum_{\mathrm{m}=1}^{\mathrm{M}}\frac{1}{\mathrm{m}!}(\alpha W\tau)^{\mathrm{m}}\left[\overline{\mathrm{q}}_{\mathrm{m}}(\gamma_{RP}) - \overline{\mathrm{q}}_{\mathrm{M}}(\gamma_{RP})\right]\right)\ (\mathrm{I}_1) \tag{7.99b}$$

$$+\frac{(\Delta_{\gamma_R})^2}{2}\left(-e^{-\frac{\hat{\tau}}{\mu_P}}\frac{F_j(\mu_R)}{P_j} + \sum_{k=\frac{N+1}{2}}^{N}\left[A_k e^{-\frac{\hat{\tau}}{s_k}}\sum_{n=0}^{N}\frac{F_n(\mu_R)}{1-\frac{\mu_n}{s_k}}\right]\right)\ (\mathrm{I}_2) \tag{7.99c}$$

where,

$$\overline{\mathrm{q}}_{\mathrm{m}}(\gamma_{RP}) = \frac{4}{(\Delta_{\gamma_R})^2 + m\beta^2}\exp\left[\frac{(\gamma_{RP})^2}{(\Delta_{\gamma_R})^2 + m\beta^2}\right] \tag{7.99d}$$

$$\gamma_{RP} = \cos^{-1}\left(\sqrt{(1-\mu_P^2)(\mu_R^2)} + (\mu_P)(\mu_R)\right) \tag{7.99e}$$

$$\mu_P = \cos\theta_P \tag{7.99f}$$

$$\mu_R = \cos\theta_R \tag{7.99g}$$

where P_r is the received power by the receiving antenna within its gain pattern, P_{max} is the received power in the absence of the vegetation, Δ_{γ_R} is the beamwidth of the receiving antenna, and m is a calculation index used in (7.99). Calculation of (7.99) does not change significantly for $m > 10$. In most cases M is taken equal 10, N is an odd number greater than 1. Large N will increase the computation time and smaller N will introduces error in the calculation. A compromised range of values were found to be in the limit $11 < N < 21$.

7.16.2.2 Scattering Loss for *Normal Radiation*

It is worth noting that when $\theta_P = \theta_R = 90°$, then $\mu_P = \mu_R = 1$ and $\gamma_{RP} = 0$ and $\overline{q}_M(\gamma_{RP}) = \overline{q}_M(0)$

$$\overline{q}_m = \frac{4}{(\Delta_{\gamma_R})^2 + m\beta^2} \tag{7.100}$$

Accordingly simplified (7.99) to

$$\frac{P_r}{P_{max}} = e^{-\tau} + \frac{(\Delta_{\gamma_R})^2}{4}(e^{-\hat{\tau}} - e^{-\tau})\,\overline{q}_M + e^{-\tau}\sum_{m=1}^{M}\frac{1}{m!}(\alpha W\tau)^m\,[\overline{q}_m - \overline{q}_M])$$
$$+\frac{(\Delta_{\gamma_R})^2}{2}\left(-e^{-\hat{\tau}}\frac{1}{P_N} + \sum_{k=\frac{N+1}{2}}^{N}\left[A_k e^{\frac{\hat{\tau}}{s_k}}\sum_{n=\frac{N+1}{2}}^{N}\frac{1}{1-\frac{\mu_n}{s_k}}\right]\right) \tag{7.101}$$

The following parameters are used in the computing of the excess attenuation L_s using (7.99) or (7.101):

$$\Delta_{\gamma_R} = 0.6\,\Delta_{\gamma_{3\,dB}} \tag{7.102a}$$

where Δ_{γ_R} is beamwidth of the receiving antenna in terms of the 3 dB beamwidth $\Delta_{\gamma_{3\,dB}}$, which should be known from the specifications of the receiving antenna.

$$\tau = (\sigma_a + \sigma_s)\,.\,z = \sigma_\tau\,.\,z \tag{7.102b}$$

where τ defines the optical density and the vegetation depth distance z in m.

$$\mu_n = -\cos\left(\frac{n\pi}{N}\right) \text{ for } n = 0, , \ldots \ldots , N-1 \tag{7.102c}$$

$$P_n = \sin^2\left(\frac{\pi}{2N}\right) \quad \text{for } n = 0, N$$

$$P_n = \sin\left(\frac{\pi}{N}\right)\sin\left(\frac{n\pi}{N}\right) \text{ for } n = 1, \ldots \ldots , N-1 \tag{7.102d}$$

$$\hat{\tau} = (1 - \alpha W)\tau \tag{7.102e}$$

$$\widehat{W} = \frac{(1-\alpha)\,W}{1-\alpha W} \tag{7.102f}$$

where $\widehat{W}$ is commonly known as a reduced albedo.

The excess attenuation due to scatter through the vegetation depth L_s in dB is given by

$$L_s = 10\log_{10}\frac{P_r}{P_{max}} \tag{7.103}$$

Computing the first three terms in (7.99)/(7.101) is straight forward using any one of a number of available computing platforms. However, the last term in (7.99)/(7.101) is a bit more complicated in the sense it requires the computing of two essential coefficients namely: the attenuation coefficients s_k and the amplitude coefficients A_k. We find the attenuation coefficients, s_k as the

$(N+1)$ solutions of the characteristic equation given in (7.104), and then we use the values of s_k coefficients to find the values of the amplitude coefficients A_k using a set of linear characteristic equations.

First, the attenuation coefficient s_k are solutions of the characteristic equation [51 equ.32],

$$\frac{\widehat{W}}{2}\sum_{n=0}^{N}\frac{P_n}{1-\frac{\mu_n}{S}}=1 \tag{7.104}$$

The attenuation coefficients s_k are solutions of the characteristic equation (7.104), which can be solved numerically to determine the s_k spectrum. Let us denote the left-hand side (LHS) of the characteristic equation as (S). It is helpful to understand the behaviour of the attenuation coefficients s_k by plotting the $L(S)$ versus values of S making sure $L(S)$ is equal to 1 for every attenuation coefficient s_k in the coefficients' spectrum. It is clear from (7.104) that when $s_k = \mu_n$ the $L(S)$ goes to infinity, i.e. there will be a pole at the points $s_k = \mu_n$ and there are $N+1$ of such poles. The plot of $L(S)$ versus S is shown in Figure 7.35. In between any adjacent poles, $L(S)$ varies from $-\infty$ to $+\infty$, except for the two poles closest to $s_k = 0$, the poles go towards $-\infty$. As $L(S)$ goes towards $+\infty$, it crosses $L(S) = 1$ and there are $N-1$ of such crosses between $-1 \leq s_k \leq 1$. Furthermore, $L(S)$ crosses $L(S) = 1$ twice when $|S| > 1$, hence we have $N+1$ solutions s_k to (7.104). The left-hand and the right-hand sides of the characteristic equations in (7.104) are symmetric in S. Numbering these solutions in increasing value order, the first $\frac{N-1}{2}$ solutions give $s_0, \ldots\ldots\ldots, s_{\frac{N-1}{2}}$ are negative, the remaining $\frac{N+1}{2}$ solutions give $s_{\frac{N+1}{2}}, \ldots\ldots\ldots, s_N$ are positive.

Because of the boundary conditions

$$A_k = 0 \quad for\, k = 0, 1, \ldots\ldots\ldots\ldots\ldots\ldots, \frac{N-1}{2} \tag{7.105}$$

The remaining amplitude coefficients $A_{\frac{N+1}{2}}, \ldots\ldots\ldots\ldots, A_N$ are determined by the following system of linear equations:

$$\sum_{k=\frac{N+1}{2}}^{N}\frac{A_k}{1-\frac{\mu_n}{S_k}}=\frac{\delta_{nN}}{P_N}\ \text{for}\ k,n=\frac{N+1}{2},\ldots\ldots\ldots\ldots\ldots, N \tag{7.106}$$

And

$$\delta_{nN} = \begin{bmatrix} 1 \ for\, n = N \\ 0 \ for\, n \neq N \end{bmatrix}$$

In the calculations, we only need to calculate the value of $\boldsymbol{P_n}$ at $\boldsymbol{n = N}$ since otherwise $\boldsymbol{\delta_{nN}}$. The set of equations in (7.107) have to be solved for coefficients $\boldsymbol{A_k}$. A step-by-step tutorial, on the calculation of the excess through (scatter) loss in vegetation, is given in Appendix 7.D. The excess attenuation due to scatter through the vegetation depth $\boldsymbol{L_s}$ in dB is given by

$$L_{scat}\, in\ \text{dB} = 10\log_{10}\frac{P_r}{P_{max}}$$

7.16.3 Determination of the Medium-Dependent Parameters from Measurement Data

The RET model parameters that represent the foliage medium are α, β, σ_τ, and W, as defined above. These parameters are dependent on the type of vegetation, frequency of the radiation, and state of the foliage (in-leaf/out of leaf). They must be determined by measurements carried out on the specific foliage medium in order to perform the calculation of the excess attenuation

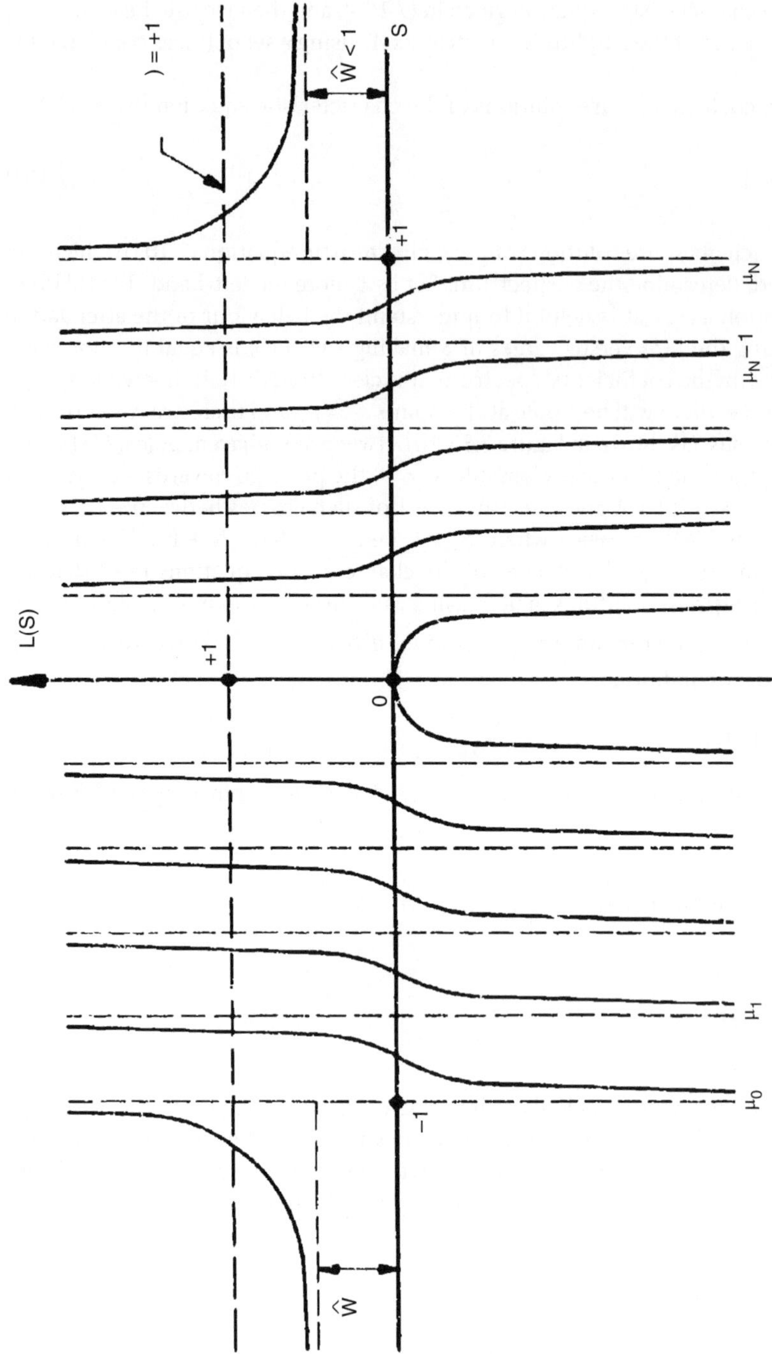

Figure 7.35 Solution of characteristic equation (7.104) [51].

using (7.101). The values for α and β can be decided by measuring the scatter function of the medium; and the values of σ_τ and W can be determined from the excess loss curve.

The scattering phase function of the medium comprises the scatter radiation pattern of the foliage medium. In the predominantly coherent region, the scatter function can be described by a narrow forward lobe and the isotropic is the backscatter. Measurement of the scatter function is, in general, conducted in conjunction with the excess attenuation in the medium. The important issue here is to make sure that the scatter medium resembles a homogenous infinite half space and is free from small particles of dust or drops of water. These particles can influence the scatter radiation pattern. The scatter pattern measurement has to be conducted over a range of angles large enough for the expected beamwidth β.

Three key techniques can be used to measure the scatter function, namely: the moving receiver technique; the moving transmitter technique; and the scanning technique. In the scanning technique [54], the transmitter is situated so that there is enough forefront clearance (i.e. 50 m) between the transmitter and the vegetation, and the receiver is placed behind the air/foliage interface at appropriate vegetation depth. The receiver antenna is rotated in discrete increments about its vertical axis at a fixed location inside the medium to perform the signal scan. The received signal amplitude (dB relative to free space) is determined when the transmitter and the receiver are aligned so the degradation in the phase function can be seen. The received signal normalized the maximum received signal and is plotted as a function of the scanning angle. A curve-fitting process is then applied on the measured normalized received signal amplitude in dB plotted versus the receive antenna angle of rotation increments to model the phase function for the vegetation using the scanning technique. Since α and β vary independently, therefore, they can be determined separately by an iterative algorithm.

The extinction cross section σ_τ determines the attenuation rate at short distances into the random medium where the coherent component dominates. Let P be the measured signal power inside the vegetation and P_{max} be the received power measured at the incident air-vegetation interface. At short distance z into the medium, we have

$$P = P_{max}\, e^{-\sigma_\tau z} \tag{7.107}$$

$$10 \log_{10}\left(\frac{P}{P_0}\right) = -4.3429\sigma_\tau z$$

The initial slope of measured attenuation dB/m of (I_c) component is

$$\frac{\Delta A}{\Delta z} = -4.3429\sigma_\tau$$

where

$$A = 10 \log_{10}\left(\frac{P}{P_0}\right)$$

Therefore

$$\sigma_\tau = \frac{\text{Slope}}{4.3429} \tag{7.108}$$

Therefore, an estimate value for (σ_τ) can be calculated using (7.108). However, this value has to be adjusted to be appropriately over the range of the measurement (not only I_c). To do this, an iterative curve-fitting procedure, together with minimum least errors fit, can be used to construct theoretical attenuation curves using RET access attenuation over a wide range of values for α, β, W, and a small number of different values for σ_τ close to the value obtained from the measurement to establish the best fit with measured data.

7.17 Summary

A brief introduction in Section 7.1 explains the importance of understanding the attenuation contribution due to some key topics covered such as clouds, fog, snow, and ice, vegetation. Understanding such natural phenomena requires theoretical analysis underpinned by on-site measurements. These are the aims of the chapter. Included in the topics is the impact of these phenomena on the quality and reliability of modern mobile communication systems. Section 7.2 introduced the many different types and phases of clouds that can be identified in the atmosphere and classified scientifically by specialists in research and meteorological centres. An important constituent of the clouds that is relevant to the attenuation is the water droplets or ice in the clouds since it absorbs and scatters radio waves. Similarly, fog comprises liquid water with its density measured in g/m^3. A microphysical model was developed for the clouds water droplets in Section 7.3. A modified gamma water droplet radius distribution was formulated as well as the water droplets absorption was modelled in Section 7.4. The scatter attenuation due to clouds/fog droplets was explained using Rayleigh/Mie scattering distributions in Section 7.5. The ITU empirical model for the attenuation due to clouds and fog was presented in Section 7.6.

Building material such as bricks, glass windows, plasterboard walls, metal frames and doors, concrete walls, etc. contribute a lot of attenuation at mmWave bands. Consequently, mathematical models to calculate the overall contributions of building material to the mmWave signals attenuation can be used to improve systems performance. These models were presented in Sections 7.7 and 7.8. The indoor environment is not immune to the degradation of the building materials and the radio wave penetration energy loss was investigated in Section 7.9.

The attenuation of propagated radio waves through vegetation was presented in Section 7.10. In this section, we presented an overview of the contemporary vegetation loss models that can be used over 1 GHz frequency bands and highlighted the strong and weak point of each. The models we explored include: Weissberger MED, modified, and fitted ITU-R (MITU-R) and (FITU-R), COST235, NZG, DG, as well as the ITU-R P.526-7, and the proposed revision to the prediction vegetation attenuation above 1 GHz in recommendation ITU-R P.833-4.

In Section 7.11 we checked the early work on vegetation loss using empirical models for slant propagation path initiated by mobile-satellite systems and examined the fades related to land mobile satellite communications at UHF, L-band and 20 GHz. The microphysical models of the vegetation loss were explored in Section 7.12. The vegetation loss due diffraction was explained in Section 7.13. An overview of the ITU-R P.526-7 was considered in Section 7.14.

Section 7.15 encompassed the work sponsored by the UK Radiocommunication Agency for the propagation loss through vegetation by carrying out a comprehensive campaign of measurements comprising different species of native trees, in-leaf and out-of-leaf, different formation of trees, single trees, lines of trees, dense wood, and with different geometries for the aim of determining a generic loss model. The RET theory was introduced in Section 7.16 and was used to derive the vegetation scattering loss model for slant and normal radiation in the mmWave frequency bands based on the UK RA database.

7.A Lognormal Distributed Random Numbers

A random variable is a lognormally distributed if a random variable logarithm is normally distributed. A lognormal distribution is a probability distribution with a normally distributed variable logarithm. Random variables must have positive values as logx exists only for positive values of x. logx is a natural (base e) logarithm (*ln*).

The probability density function of a lognormally distributed random positive variable x is defined by the mean μ and standard deviation σ as

$$\mathcal{N}(\log \mathrm{x}; \mu, \sigma) = \frac{1}{\sigma\sqrt{2\pi}} \exp\left[-\frac{(\log \mathrm{x} - \mu)^2}{2\sigma^2}\right] \quad \mathrm{x} > 0 \tag{7.A.1}$$

Consider a set of lognormal distributed variables $\mathbf{x} = [\mathrm{x}_1, \mathrm{x}_2, \ldots \ldots \ldots \ldots, \mathrm{x}_n]$ and denote the *geometric mean* of the set as μ. Then, by definition,

$$\overline{\mathrm{x}}_{ln} = (\Pi_{i=1}^{n} \mathrm{x}_i)^{\frac{1}{n}} = \sqrt[n]{\mathrm{x}_1 * \mathrm{x}_2 \ldots \ldots * \mathrm{x}_n} \tag{7.A.2}$$

Then,

$$\log \overline{\mathrm{x}}_{ln} = \frac{1}{n} \log(\Pi_{i=1}^{n} \mathrm{x}_i)$$

$$\log \overline{\mathrm{x}}_{ln} = \frac{1}{n}[\log x_1 + \log \mathrm{x}_2 + \ldots \ldots \ldots + \log \mathrm{x}_n] \tag{7.A.3}$$

Therefore, $\log \overline{\mathrm{x}}_{ln}$ is the *arithmetic mean* of the set lognormally distributed random variables $[\log \mathrm{x}_1, \log \mathrm{x}_2, \ldots \ldots \ldots \ldots \ldots, \log \mathrm{x}_\mathrm{n}]$ that is

$$\log \overline{\mathrm{x}}_{ln} = \overline{\log \mathbf{x}} \tag{7.A.4}$$

$$\overline{\mathrm{x}}_{ln} = e^{\overline{\log \mathbf{x}}} \tag{7.A.5}$$

The geometric standard deviation is

$$\log \sigma_{ln} = \left(\sqrt{\frac{\sum_{i=1}^{n} (\log \mathrm{x}_i - \log \overline{\mathrm{x}}_{ln})^2}{n}} \right) \tag{7.A.6}$$

This simplifies to

$$\sigma_{ln} = \exp \sqrt{\frac{\sum_{i=1}^{n} (\log \mathrm{x}_i - \log \overline{\mathrm{x}}_{ln})^2}{n}}) \tag{7.A.7}$$

$$\sigma_{ln} = \exp \sqrt{\frac{\sum_{i=1}^{n} \left(\log \frac{\mathrm{x}_i}{\overline{\mathrm{x}}_{ln}}\right)^2}{n}} \tag{7.A.8}$$

$$(\log \sigma_{ln})^2 = \frac{\sum_{i=1}^{n} \left(\log \frac{\mathrm{x}_i}{\overline{\mathrm{x}}_{ln}}\right)^2}{n} = \overline{\left(\log \frac{\mathrm{x}_i}{\overline{\mathrm{x}}_{ln}}\right)^2} \tag{7.A.9}$$

The parameter μ in the lognormal distribution is the mean of the distribution. The parameter σ is the standard deviation of the lognormal distribution.

Consider the values of the random variables **x** and denote the mean and standard deviation of the non-logarithmic values as m and s. The lognormal distribution mean μ can be calculated from mean m and standard deviation s of the non-logarithmic distribution as

$$\mu = \log\left(\frac{\mathrm{m}^2}{\sqrt{\mathrm{s}^2 + \mathrm{m}^2}}\right) \tag{7.A.10}$$

The lognormal distribution standard deviation σ can be calculated from mean m and standard deviation s of the non-logarithmic distribution as

$$\sigma = \sqrt{\log\left(1 + \frac{s^2}{m^2}\right)} \tag{7.A.11}$$

The expectation values of lognormal distribution parameters μ and σ are given by

$$\mathbb{E}[\mu] = \exp\left[\mu + \frac{\sigma^2}{2}\right] \tag{7.A.12}$$

$$\mathbb{E}[\sigma] = \sqrt{(\exp[\sigma^2] - 1)\,\exp[2\mu + \sigma^2]} \tag{7.A.13}$$

7.B Derivation of Cloud Water Droplets Mode Radius

The cloud water droplets radius r distributed by a *modified gamma exponential* function is given in (7.7) as

$$\text{n(r)} = a\,r^\alpha \exp(-b\,r^\beta),\quad 0 \le r \le \infty$$

$$\text{Let } y(r) = -b\,r^\beta \tag{7.B.1}$$

So,

$$n(r) = a\,r^\alpha\,e^y \tag{7.B.2}$$

Differentiate (7.B.1) with respect to r we get

$$\frac{dy}{dr} = -b\beta\,r^{\beta-1}$$

Differentiate (7.B.2) with respect to r we have,

$$\frac{dn}{dr} = a\left(r^\alpha * \frac{d}{dr}(e^y) + \alpha r^{\alpha-1} * e^y\right)$$

$$\frac{d}{dr}(e^y) = e^y * \frac{dy}{dr} = -b\beta r^{\beta-1} e^y$$

Thus

$$\frac{dn}{dr} = a(-r^\alpha * b\beta r^{\beta-1} e^y + \alpha r^{\alpha-1} * e^y) \tag{7.B.3}$$

At the max (7.B.3) becomes zero; i.e.

$$\frac{dn}{dr} = 0$$

$$a(-r^\alpha * b\beta r^{\beta-1} e^y + \alpha r^{\alpha-1} * e^y) = -ar^{\alpha-1} e^y (b\beta r\,r^{\beta-1} - \alpha) = 0$$

$$(b\beta\,r^\beta - \alpha) = 0$$

So,

$$b = \frac{\alpha}{\beta r_c^\beta} \tag{7.B.4}$$

Thus, the mode radius of the distribution is given by (7.B.4):

$$r_c = \left(\frac{\alpha}{b\beta}\right)^{\frac{1}{\beta}} \tag{7.B.5}$$

and

$$\alpha = b\beta\, r_c^{\beta} \tag{7.B.6}$$

Accordingly, $n(r_c)$ is obtained by substituting for r_c (7.B.5) and α (7.B.6) in

$$n(r) = a\, r^{\alpha} \exp(-b\, r^{\beta})$$

We get

$$n(r_c) = a\, {r_c}^{\alpha} \exp(-b\, {r_c}^{\beta}) = a\, {r_c}^{\alpha}\, \exp\left(-\frac{\alpha}{\beta}\right) \tag{7.B.7}$$

7.C The Complex Relative Permittivity and the Complex Relative Refractive Index Relationship

The refractive index n of an EM radiation is given by

$$n = \sqrt{\varepsilon\mu} \tag{7.C.1}$$

where ε is the medium relative permittivity and μ is its relative permeability. The refractive index is used predominantly in optics while the relative permittivity and permeability are used broadly in Maxwell's equation in electronics. Most naturally occurring mediums are nonmagnetic, such as air. That is, μ is very close to 1; therefore,

$$n \sim \sqrt{\varepsilon} \tag{7.C.2}$$

The complex relative permittivity, in this case, is made up of two quantities, a real part ε_r and imaginary part ε_i. Thus, the complex refractive index n is also with real part n_r and imaginary part n_i (n_i is also called the *extinction coefficient*):

$$\varepsilon = \varepsilon_r + j\varepsilon_i = n^2 = (n_r + jn_i)^2 \tag{7.C.3}$$

$$\varepsilon = n_r^2 - n_i^2 + j2\, n_r n_i \tag{7.C.4}$$

$$\varepsilon_r = n_r^2 - n_i^2 \tag{7.C.5}$$

$$\varepsilon_i = 2\, n_r n_i \tag{7.C.6}$$

$$\varepsilon_r^2 + \varepsilon_i^2 = (n_r^2 + n_i^2)^2$$

$$n_r^2 + n_i^2 = \sqrt{\varepsilon_r^2 + \varepsilon_i^2} \tag{7.C.7}$$

Adding (7.B.5) and (7.B.7), we have

$$n_r^2 = \frac{1}{2}\left(\sqrt{\varepsilon_r^2 + \varepsilon_i^2} + \varepsilon_r\right)$$

And now with (7.B.5) and (7.B.7), we have

$$n_r^2 = \frac{1}{2}\left(\sqrt{\varepsilon_r^2 + \varepsilon_i^2} - \varepsilon_r\right)$$

So, the complex refractive index components are y:

$$n_r = \sqrt{\frac{1}{2}\left(\sqrt{\varepsilon_r^2 + \varepsilon_i^2} + \varepsilon_r\right)} \tag{7.C.8}$$

$$n_i = \sqrt{\frac{1}{2}\left(\sqrt{\varepsilon_r^2 + \varepsilon_i^2} - \varepsilon_r\right)} \tag{7.C.9}$$

7.D Step-by-Step Tutorial to Calculate the Excess Through (Scatter) Loss in Vegetation

In this appendix, we provide a step by step calculation of all the terms used in the excess scatter loss suggested in (7.101) according to the RET theory. We consider a theoretical example with several arbitrary parameters taken from the literature. Although these methods are providing deep insight into the theoretical analysis, they are not generally used in practical systems, as they are time consuming. The results obtained in the tutorial are mainly designed to complement the tutorial and are not to be used for any other application. Nevertheless, the aim of the tutorial is to present well known methods so appropriate software codes/computing platforms can be used to obtain the final results.

The excess scatter loss ratio $\frac{P_r}{P_{max}}$ is given in Section 7.16.2.2:

$$\frac{P_r}{P_{max}} = e^{-\tau} + \frac{(\Delta_{\gamma_R})^2}{4}(e^{-\hat{\tau}} - e^{-\tau})\overline{q}_M + e^{-\tau}\sum_{m=1}^{M}\frac{1}{m!}(\alpha W\tau)^m\,[\overline{q}_m - \overline{q}_M])$$
$$+ \frac{(\Delta_{\gamma_R})^2}{2}\left(-e^{-\hat{\tau}}\frac{1}{P_N} + \sum_{k=\frac{N+1}{2}}^{N}\left[A_k e^{\frac{\hat{\tau}}{s_k}}\sum_{n=0}^{N}\frac{1}{1-\frac{\mu_N}{s_k}}\right]\right)$$

For clarity of the tutorial, we present the computing of the above equation into two parts:

$$\left(\frac{P_r}{P_{max}}\right)_{I_1} = e^{-\tau} + \frac{(\Delta_{\gamma_R})^2}{4}(e^{-\hat{\tau}} - e^{-\tau})\overline{q}_M + e^{-\tau}\sum_{m=1}^{M}\frac{1}{m!}(\alpha W\tau)^m\,[\overline{q}_m - \overline{q}_M]) \tag{7.D.1}$$

$$\left(\frac{P_r}{P_{max}}\right)_{I_2} = \frac{(\Delta_{\gamma_R})^2}{2}\left(-e^{-\hat{\tau}}\frac{1}{P_N} + \sum_{k=\frac{N+1}{2}}^{N}\left[A_k e^{\frac{\hat{\tau}}{s_k}}\sum_{n=\frac{N+1}{2}}^{N}\frac{1}{1-\frac{\mu_n}{s_k}}\right]\right) \tag{7.D.2}$$

In presenting such tutorials, we consider the following measured parameters taken from [46] for London Plane in leaf *Platanus hispanica Muenchh:* =1.93:

Leaf size 0.25; *Fliage depth* $z = 10$ m; $\alpha = 0.7$; $\beta = 6° = 0.10471976$ rad ; $W = 0.95$; $\sigma_\tau = 0.75$, frequency of measurement 11 GHz.

Furthermore, a 3-dB beamwidth $\Delta_{\gamma_{3dB-R}}$ of 3.2° is assumed for the receiving antenna. Then a beamwidth at the receiving antenna is $\Delta_{\gamma_R} = 0.6 * \Delta_{\gamma_{3dB-R}}$.

1. Calculation of $\left(\frac{P_r}{P_{max}}\right)_{I_1}$

   The optical density is

   $\tau = \sigma_\tau . z = 7.5$; the beamwidth at the receiving antenna is $\Delta_{\gamma_R} = 0.6 * \Delta_{\gamma_{3dB-R}} = 1.92^{\circ} = 0.0335$ rad; $\hat{\tau} = (1 - \alpha\ W)\tau$; the beamwidth of the forward lobe β is, $\beta = 6° = 0.1047$ rad; $\overline{q}_m = \frac{4}{(\Delta_{\gamma_R})^2 + m\beta^2}$ $m = 1, 2, \ldots\ldots .M$; $\widehat{W} = \frac{(1-\alpha)\ W}{1-\alpha W} = 0.8507$

m	1	2	3	4	5	6	7	8	9	10
$\bar{q}_m$	330.578	173.913	117.647	88.888	71.556	59.790	51.347	45.045	40.080	36.133

$\hat{\tau} = 0.251$; $\tau = \sigma_\tau . z = 7.5$; $\Delta_{\gamma_R} = 0.0335$; $\alpha = 0.7$; $W = 0.95$. Inserting the values of these parameters in (7.D.1), we have $\left(\frac{P_r}{P_{max}}\right)_{I_1} = 39.0208$ dB.

$\frac{\widehat{W}}{2} \sum_{n=0}^{N-1} \frac{P_n}{1-\frac{\mu_n}{s}} = 1$; $P_n = \sin^2\left(\frac{\pi}{2N}\right)$ for $n = 0, N$; $P_n = \sin\left(\frac{\pi}{N}\right) \sin\left(\frac{n\pi}{N}\right)$ for $n = 1, \ldots \ldots, N-1$.

n	0	1	2	3	4	5	6	7	8
P_n	0.0085	0.0338	0.0664	0.0967	0.1238	0.1466	0.1645	0.1767	0.1830
n	9	10	11	12	13	14	15	16	17
P_n	0.1830	0.1767	0.1645	0.1466	0.1238	0.0967	0.0664	0.0338	0.0085

$\mu_n = -\cos\left(\frac{n\pi}{N}\right)$ for $= 0, , \ldots \ldots, N-1$,

n	0	1	2	3	4	5	6	7	8
μ_n	−1	−0.983	−0.9325	−0.8502	−0.7390	−0.6026	−0.4457	−0.2737	−0.0923

n	9	10	11	12	13	14	15	16
μ_n	0.0923	0.2737	0.4457	0.6026	0.7390	0.8502	0.9325	1

2. Calculation of $\left(\frac{P_r}{P_{max}}\right)_{I_2}$

$$P_n = \sin^2\left(\frac{\pi}{2N}\right) \quad \text{for } n = 0, N \tag{7.D.3a}$$

$$P_n = \sin\left(\frac{\pi}{N}\right) \sin\left(\frac{n\pi}{N}\right) \quad \text{for } n = 1, \ldots \ldots, N-1 \tag{7.D.3b}$$

Considering $11 < N < 21$ we have chosen N is to be 17.

$$P_N = 0.0085$$

$$\mu_n = -\cos\left(\frac{n\pi}{N}\right) \tag{7.D.4}$$

μ_0	μ_1	μ_2	μ_3	μ_4	μ_5	μ_6	μ_7	μ_8	μ_9	μ_{10}	μ_{11}	μ_{12}	μ_{13}	μ_{14}	μ_{15}	μ_{16}
−1	−0.983	−0.932	−0.850	−0.739	−0.602	−0.445	−0.273	−0.092	−0.092	0.273	0.445	0.602	0.739	0.850	0.932	0.983

$\hat{\tau} = 0.251$; $\Delta_{\gamma_R} = 0.6 * \Delta_{\gamma_{3dB-R}} = 1.92^\circ = 0.0335$

The essential parameters in the second term $\left(\frac{P_r}{P_{max}}\right)_{I_2}$ are the attenuation coefficients s_k and the amplitude coefficients A_k, which we determine now. The attenuation coefficients s_k has $(N+1)$ (not all used in the calculation) values, which should be determined first so we can use them to find the values of the amplitude coefficients A_k. The attenuation coefficients, s_k , are solutions of the characteristic equations [51]:

$$Ls = \frac{\widehat{W}}{2} \sum_{n=0}^{N} \frac{P_n}{1 - \frac{\mu_n}{s}} = 1 \tag{7.D.5}$$

Accordingly, (7.D.3) should be true for every attenuation coefficient s_k of S. Between $s_k = \pm 1$, Ls varies from $-\infty$ *to* $+\infty$ $(N+1)$ times (except at $s_k = \pm 0$ where Ls varies from 0 *to* $-\infty$) and in between these two limits, Ls intersect $Ls = 1$ level generating $N-1$ attenuation coefficients s_k symmetrical about the origin. The other two coefficients are $|s_k| > 1$. Recall that when $\mu_n = s_k$, the loss tends to infinite which is called a pole. Foe finding a large number of the attenuation coefficients applying a search engine is the best option. However, to calculate a few s_k coefficients, there are a number of computing platforms that can be used, such Matlab, to try to search close to the poles. And look for $Ls = 1$. For the assumed parameters above, the attenuation coefficients are listed below.

k	9	10	11	12	13	14	15	16
s_k	0.10011	0.29644	0.481286	0.64772	0.78875	0.8984	0.9831	1.5572

The amplitude coefficients A_k are determined by a set of linear equations given by (7.D.6),

$$\sum_{k=\frac{N+1}{2}}^{N}\left[A_k \sum_{n=\frac{N+1}{2}}^{N} \frac{1}{1-\frac{\mu_n}{s_k}}\right] = \frac{\delta_{nN}}{P_N} \text{ for } n = \frac{N+1}{2}, \ldots\ldots\ldots\ldots\ldots., N \tag{7.D.6}$$

The important point to observe is that each A_k is multiplied by the summation $\frac{1}{1-\frac{\mu_n}{s_k}}$ for $n = \frac{N+1}{2}, \ldots\ldots\ldots\ldots\ldots., N$ generating a set of $\frac{N-1}{2}$ linear equations,

$$\begin{aligned}
&A_k\frac{1}{1-\frac{\mu_n}{s_k}} + A_{k+1}\frac{1}{1-\frac{\mu_n}{s_{k+1}}} + A_{k+2}\frac{1}{1-\frac{\mu_n}{s_{k+2}}} + A_{k+3}\frac{1}{1-\frac{\mu_n}{s_{k+3}}} + A_{k+4}\frac{1}{1-\frac{\mu_n}{s_{k+4}}}\\
&\quad + \ldots\ldots\ldots\ldots + A_N\frac{1}{1-\frac{\mu_n}{s_N}} = 0\\
&A_k\frac{1}{1-\frac{\mu_{n+1}}{s_k}} + A_{k+1}\frac{1}{1-\frac{\mu_{n+1}}{s_{k+1}}} + A_{k+2}\frac{1}{1-\frac{\mu_{n+1}}{s_{k+2}}} + A_{k+3}\frac{1}{1-\frac{\mu_n}{s_{k+3}}} + A_{k+4}\frac{1}{1-\frac{\mu_{n+1}}{s_{k+4}}}\\
&\quad + \ldots\ldots\ldots\ldots + A_N\frac{1}{1-\frac{\mu_{n+1}}{s_N}} = 0\\
&A_k\frac{1}{1-\frac{\mu_N}{s_k}} + A_{k+1}\frac{1}{1-\frac{\mu_N}{s_{k+1}}} + A_{k+2}\frac{1}{1-\frac{\mu_N}{s_{k+2}}} + A_{k+3}\frac{1}{1-\frac{\mu_N}{s_{k+3}}} + A_{k+4}\frac{1}{1-\frac{\mu_N}{s_{k+4}}}\\
&\quad + \ldots\ldots\ldots\ldots + A_N\frac{1}{1-\frac{\mu_N}{s_N}} = \frac{1}{P_N}
\end{aligned} \tag{7.D.7}$$

The solution to the set of equations provide the values of A_k. There are several different methods that you choose from depending on the problem at hand but the most prominent of them are the direct elimination method and equation in matrix form is the Gaussian elimination method.

References

1 Chen, C.C. (1975) 'Attenuation of Electromagnetic Radiation by Haze, Interim Fog, Clouds, and Rain', A Report prepared for US AIR FORCE Project RAND, 29 pages.

2 Ulaby, F.T., Moore, R.K., and Fung, A.K. (1981). Microwave remote sensing: Active and passive microwave. In: *Volume 1: Remote Sensing Fundamentals and Radiometry*. Artech House. ISBN: 0-89006-192-0.

3 Awan, M.S., Capsoni, C., Leitgeb, E. et al. (2008). FSO-relevant new measurement results under moderate continental fog conditions at Graz and Milan. In: *Advanced Satellite Mobile Systems*, 116–121.

4 Liebe, H., Manabe, T., and Hufford, G.A. (1989). Millimeter wave attenuation and delay rates due to fog/cloud conditions. *IEEE Transactions on Antennas and Propagation* 37 (12): 1617–1623.

5 Luini, L. and Capsoni, C. (2013) 'Performance evaluation of satellite communication systems operating in the Q/V/W bands', *Research Report to European Office of Aerospace Research and Development*, APO AE 09421 - 4515.

6 Shettle, E.P. (1989) 'Models of Aerosols, clouds and precipitation for Atmospheric Propagation Studies'. In: *Nato Conference on Atmospheric Propagation in the UV, visible, IR and MM-Wave Region and Related Systems Aspects*, Proceedings N o.454, Denmark, Copenhagen.

7 Geoffroy, O., Siebesma, A.P., and Burnet, F. (2014). Characteristics of the raindrop distributions in RICO shallow cumulus. *Atmospheric Chemistry and Physics Discussions* 14 (1): 677–705.

8 Dissanayake, A., Allnutt, J., and Haidara, F. (1997). A prediction model that combines rain attenuation and other propagation impairments along earth-satellite paths. *IEEE Transactions on Antennas and Propagation* 45 (10): 1546–1558.

9 Siles, G.A., Riera, J.M., and García-del-Pino, P. (2015). Atmospheric attenuation in wireless communication systems at millimeter and THz frequencies. *IEEE Antennas and Propagation Magazine* 57 (1): 48–61.

10 Li, Y., Hoogeboom, P., and Russchenberg, H. (2014). 'Radar observations and modelling of fog at 35 GHz'. In: *European Conference on Antennas and Propagation (EuAP)*, 1053–1057.

11 Tomasi, C. and Tampieri, F. (1976). Features of the proportionality coefficient in the relationship between visibility and liquid water content in haze and fog. *Atmosphere - Ocean* 14 (2): 61–76.

12 Grainger, R.G. (2015). *Some Useful Formulae for Aerosol Size Distributions and Optical Properties*. Oxford, UK: Department of Physics, University of Oxford.

13 DeCarlo, L.T. (1997). On the meaning and use of kurtosis. *Psychological Methods* 2 (3): 292–307.

14 Tampieri, F. and Tomasi, C. (1977). Size distributions of tropospheric particles in terms of the modified gamma function and relationships between skewness and mode radius, *Wiley online, Meteorology,. Tellus* 29 (1): 66–74.

15 Lhermitte, R.M. (1988). Cloud and precipitation remote sensing at 94 GHz. *IEEE Transactions on Geoscience and Remote Sensing* 26 (3): 207–216.

16 Awan, M.S., Leitgeb, E., Sheikh-Muhammad, S., et al. (2008) 'Distribution function for continental and maritime fog environments for optical wireless communication'. In: *IEEE International Symposium on Communication Systems Networks and Digital Signal Processing (CSNDSP)*, 260–264.

17 Wang, J., Ge, J., Wei, M. and Yu, W. (2013) 'Influence of scattering properties due to complex refractive index of ice'. In: *IEEE International Conference on Information Science and Technology*, 997–999.

18 Recommendation ITU-R P.840-6 (09-2013) Attenuation due to clouds and fog.

19 Liebe, H.J., Hufford, G.A., and Manabe, T. (1991). A model for the complex permittivity of water at frequencies below 1 THz. *International Journal of Infrared and Millimeter Waves* 12 (7): 659–675.

20 Capsoni, C., Luini, L. and Nebuloni, R. (2012) 'Prediction of cloud attenuation on earth-space optical links', 6th European Conference on Antennas and Propagation, 326–329.

21 Zhao, H., Mayzus, R., Sun, S., et al. (2013) '28 GHz millimeter wave cellular communication measurements for reflection and penetration loss in and around buildings in New York City', IEEE International Conference on Communications-Wireless Communications Symposium, 5163–5167.

22 Semaan, E., Harrysson, F., Furuskar, A. and Asplund, H. (2014) 'Outdoor–to–indoor coverage in high frequency bands', IEEE Globecom Workshop – Mobile Communications in High Frequency Bands, 393–398.

23 Meng, Y.S. and Lee, Y.H. (2010). Investigations of foliage effect on modern wireless communication systems: a review. *Progress in Electromagnetics Research* 105: 313–332.

24 Weissberger, M.A. (1981). *An Initial Critical Summary of Models for Predicting the Attenuation of Radio Waves by Foliage*. Annapolis, MD: Electromagnetic Compatibility Analysis Center, CAC-TR-81-101.

25 CCIR (1986). Influences of terrain irregularities and vegetation on troposphere propagation, CCIR Report, 235–236, Geneva.

26 Recommendation ITU-R P.833-8 (09/2013) 'Attenuation in vegetation', P Series, Radio Propagation.

27 Al-Nuaimi, M.O. and Stephens, R.B.L. (1998). Measurements and prediction model optimization for signal attenuation in vegetation media at centimetre wave frequencies. *IEE Proceedings-Microwave Antennas and Propagation* 145 (3): 201–206.

28 COST235 (1996) 'Radio propagation effects on next-generation fixed- service terrestrial telecommunication systems', Final Report, Luxembourg.

29 Seville, A. and Craig, K.H. (1995). Semi-empirical model for millimetre-wave vegetation attenuation rates. *Electronics Letters* 31 (17): 1507–1508.

30 Seville, A. (1997) 'Vegetation attenuation-modelling and measurements at millimetric frequencies', IEE International Conference on Antennas and Propagation, 2.5–2.8.

31 Vogel, W.J. and Goldhirsh, J. (1988). Fade measurements at L-band and UHF in mountainous terrain for land mobile satellite systems. *IEEE Transactions on Antennas and Propagation* 36 (1): 104–113.

32 Goldhirsh, J. and Vogel, W.J. (1989). Mobile satellite system fade statistics for shadowing and multipath from roadside trees at UHF and L-band. *IEEE Transactions on Antennas and Propagation* 37 (4): 489–498.

33 Vogel, W.J. and Goldhirsh, J. (1993). Earth -satellite tree attenuation at 20 GHz: foliage effects. *IET Electronics Letters* 29 (18): 1640–1641.

34 Goldhirsh, J. and Vogel, W.J. (1992) 'Propagation Effects for Land Mobile Satellite Systems: Overview of Experimental and Modeling Results', NASA Reference Publication 1274.

35 Sofos, T. and Constantinou, P. (2004). Propagation model for vegetation effects in terrestrial satellite mobile systems. *IEEE Transactions on Antennas and propagation* 52 (7): 1917–1920.

36 Tamir, T. (1967). On radio-wave propagation in forest environments. *IEEE Transactions on Antennas and Propagation* 15 (6): 806–817.

37 Li, L., Yeo, T.-S., Kooi, P.-S., and Leong, M.-S. (1988). Radio wave propagation along mixed paths through a four-layered model of rain forest: an analytic approach. *IEEE Transactions on Antennas and Propagation* 46 (7): 1098–1111.

38 Keller, J.B. (1962). Geometrical theory of diffraction. *Journal of the Optical Society of America* 52 (2): 116–130.

39 Pathak, P.H., Carluccio, G., and Albani, M. (2013). The uniform geometrical theory of diffraction and some of its applications. *IEEE Antennas and Propagation Magazine* 55 (4): 42–69.

40 Liang, M.C., Chuang, C.W., and Pathak, P.H. (1996). A generalized uniform geometrical theory of diffraction ray solution for the diffraction by a wedge with convex faces. *Radio Science* 31 (4): 679–691.

41 Lewis, R.M. and Boersma, J. (1969). Uniform asymptotic theory of edge diffraction. *Journal of Mathematical Physics* 10 (12): 2291–2305.

42 Matschak, R., Linot, B. and Sizun, H. (1999) 'Model for wave propagation in presence of vegetation based on the UTD associating transmitted and lateral waves', IEE National Conference on Antennas and Propagation, 120–123.

43 Valcarce, A., de la Roche, G., Wagen, J.-F., and Gorce, J.-M. (2011). A new trend in propagation prediction. *IEEE Vehicular Technology Magazine* 6 (2): 73–81.

44 RECOMMENDATION ITU-R P.526-7- Propagation by diffraction, 2001.

45 Richter, J., Caldeirinha, R.F.S., Al-Nuaimi, M.O. et al. (2005). A generic narrowband model for radio wave propagation through vegetation. In: *IEEE Vehicular Technology Conference*, vol. 1, 39–43.

46 Rogers, C., Saville, A., Richter, J. et al. (2002). *A Generic Model of 1-60 GHz Radio Propagation through Vegetation - Final Report*. UK: Radiocommunications Agency (*Ofcom*).

47 Recommendation ITU-R P. 527-4-Electrical characteristics of the surface of the earth 06/2017

48 ITU Report 1008-1 (1990) 'Reflection from the surface of the earth', 75–82.

49 St Michael, H., Al-Nuaimi, M.O., and Caldeirinha, R. (2001) 'Radiative energy transfer prediction of excess attenuation of microwave radio signals in a regularly planted orchard', IEEE High Frequency Postgraduate Colloquium, 130–135.

50 Al-Nuaimi, M.O. and Hammoudeh, A. (1993) 'Theoretical and experimental study of attenuation and scatter of microwave signals by trees', IEEE International Conference on Antennas and Propagation, 808–811.

51 Johnson, R.A. and Schwering, F. (1985) 'A Transport Theory of Millimeter Wave Propagation in Woods and Forests', Research and Development Technical Report CECOM-TR-85-1, U S Army Communications-Electronics Command USA.

52 Sharkov, E. (2003). *Passive Microwave Remote Sensing of the Earth*, Chapter 9. Springer-Verlag. ISBN: 3-540-43946-3, 357-395.

53 Ishimoru, A. (1991). Wave propagation and scattering in random media and rough surfaces. *Proceedings of the IEEE* 79 (I0): 1539–1366.

54 Richter, J., Al-Nuaimi, M., and Caldeirinha, R. (2002) 'Phase function measurement for modelling radiowave attenuation and scatter in vegetation based on the theory of radiative energy transfer', IEEE International Symposium on Personal, Indoor, and Mobile Radio Communications, 1, 146–150.

8

Wireless Channel Modelling and Array Mutual Coupling

8.1 Key Parameters in Wireless Channel Modelling

We open this chapter with a discussion of the key parameters in wireless channel modeling [1].

8.1.1 Doppler Spread

In general, relative motion of the receiver with respect to the transmitter changes the frequency f_c of the transmitted radio wave by Doppler shift f_d and the frequency of the received signal $f_i(\alpha)$ is given by:

$$f_i(\alpha) = f_c + \frac{v}{c} f_c \cos\alpha \tag{8.1}$$

where f_c is the operating frequency, v is the speed of the receiver (in a vehicle for example) and c is the speed of light (3×10^8 m/s) and α is the angle between the direction of the signal received and the direction of the vehicle. Since $-1 \leq \cos\alpha \leq +1$, the received energy will be spread into the range $f_c \pm \frac{v}{c} f_c$. The maximum Doppler frequency shift f_d, also known as *maximum fade rate,* is given by

$$f_d = \frac{v}{c} f_c \tag{8.2}$$

The effects of receiver motion on the signal reception can be acknowledged in view of a vehicle travelling at speed v and receiving a single unmodulated carrier transmitted by a vertically polarized omni-directional antenna. In an urban environment, the receiver acquires a large number of independent scattered N waves, as shown in Figure 8.1 where $i = 1, 2, \ldots, N$. Each of these waves has random amplitude A_i with uniformly distributed phase θ_i and an incident angle α_i to the direction of vehicle travel.

Thus, the ith received wave $r_i(t)$ can be expressed as

$$r_i(t) = A_i \cos(\omega_i(\alpha_i)t + \theta_i) \tag{8.3}$$

The resulting received signal $r_i(t)$ is the sum on N waves as

$$r(t) = \sum_{i=0}^{N-1} A_i \cos(\omega_i(\alpha_i)t + \theta_i) \tag{8.4}$$

Let us denote the statistical distribution of α by $p(\alpha)$. Exploiting an isotropic antenna, the received power in a differential angle $d\alpha$ is given by P. $p(\alpha)d\alpha$ where P is the average received power by an isotropic antenna.

5G Physical Layer Technologies, First Edition. Mosa Ali Abu-Rgheff.

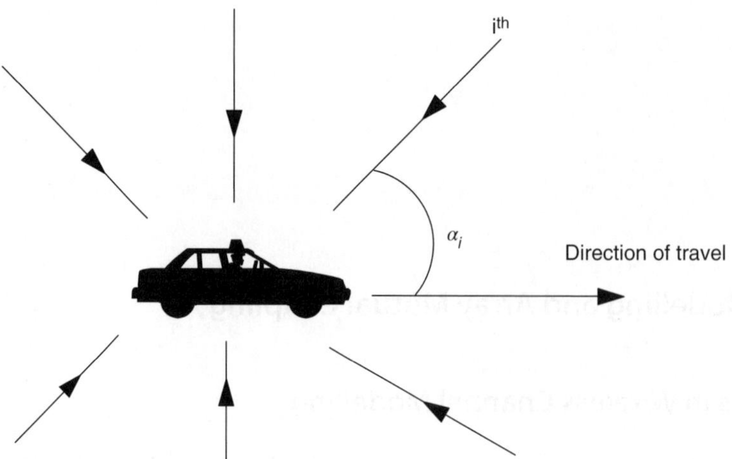

Figure 8.1 Reception of scattered waves by a mobile station [1].

Consider an ideal transmission scenario where a variation of received power is mainly due to Doppler effects so the received power within the differential frequency df is given by $S(f)$ * df where $S(f)$ is the Doppler power spectral density (PSD). Consequently, we have:

$$\mathrm{P}[\mathrm{g}(\alpha)\,\mathrm{p}(\alpha) + \mathrm{g}(-\alpha)\mathrm{p}(-\alpha)]\,|\mathrm{d}\alpha| = S(f) * |\mathrm{d}f| \tag{8.5}$$

where g(α) is the antenna gain at angle α. Now $f = f_c + f_d \cos\alpha$ so

$$|\mathrm{d}f| = f_\mathrm{d}\,\sin\alpha\,|\mathrm{d}\alpha| = f_\mathrm{d}\sqrt{1-\left(\frac{f-f_c}{f_\mathrm{d}}\right)^2}\,|\mathrm{d}\alpha| \tag{8.6}$$

Therefore,

$$S(f) = \frac{\mathrm{P}[\mathrm{g}(\alpha)\mathrm{p}(\alpha) + (-\alpha)\mathrm{p}(-\alpha)]}{f_\mathrm{d}\sqrt{1-\left(\frac{f-f_c}{f_\mathrm{d}}\right)^2}} \quad \text{for } f - f_c < f_\mathrm{d}$$
$$= 0 \quad \text{otherwise} \tag{8.7}$$

An omni-directional receiving antenna with constant gain g so that g(α) = g(−α) = g and assuming the incident angle is uniformly distributed between $-\pi \leq \alpha \leq \pi$, so that $\mathrm{p}(\alpha) = \frac{1}{2\pi}$. Substituting the value of $\mathrm{p}(\alpha)$ in Eq. (8.7), we get the Doppler PSD as

$$S(f) = \frac{\mathrm{P}\cdot\mathrm{g}}{\pi\cdot f_\mathrm{d}\sqrt{1-\left(\frac{f-f_c}{f_\mathrm{d}}\right)^2}} \quad \text{for } f - f_c < f_\mathrm{d}$$
$$= 0 \quad \text{otherwise} \tag{8.8}$$

$S(f)$ as given by (8.8) is plotted in Figure 8.2, which shows the classical Doppler PSD of Clarke-Jakes.

8.1.2 Coherence Time

Channel coherence time T_c is determined by the Doppler effects. Coherence time is the duration over which the impulse response of the wireless channel is considered to be not varying.

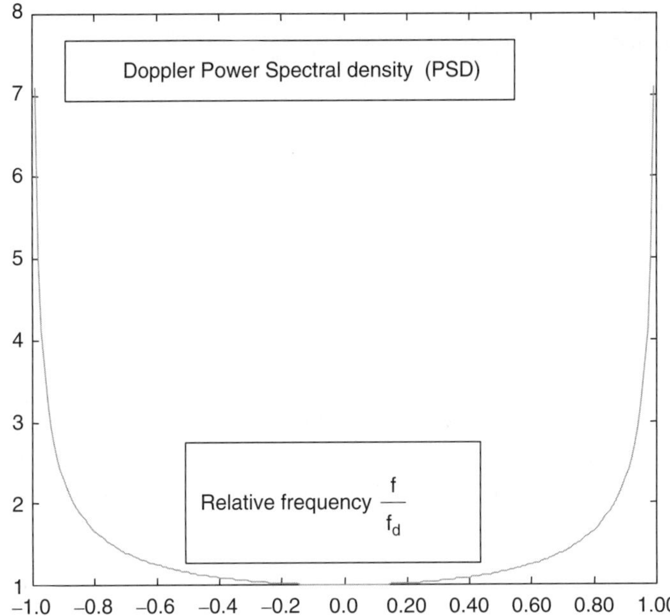

Figure 8.2 Doppler power spectral density [1].

Consider a signal $x(t)$ sent through a channel with impulse response $h(t)$. Let $x(t)$ sent at time t_1. Then the received signal $y(t)$ is

$$y_{t_1}(t) = x(t - t_1) * h_{t_1}(t)$$

where the symbol $x * h$ denotes convolution of x and h. Now we send $x(t)$ through the same channel at time t_2; then the received signal is

$$y_{t_2}(t) = x(t - t_2) * h_{t_2}(t)$$

Now if $h_{t_2}(t) - h_{t_1}(t)$ is very small, then coherence time T_c is given by

$$T_c = t_2 - t_1 \tag{8.9}$$

The coherence time and the maximum difference between Doppler shifts D_s (*Doppler spread*) are related as

$$T_c = \frac{1}{k\,D_s} \tag{8.10}$$

where k is a constant. The coherence time is determined by the Doppler spread and the relationship is reciprocal, so the larger the Doppler spread, the smaller the coherence time. The relation is difficult to determine precisely, since the largest Doppler shifts may belong to signals that are too weak to make any difference.

8.1.3 Delay Spread

In a multipath radio environment, a single transmitted impulse results in multiple received impulses with different amplitude and arriving at different times, as shown in Figure 8.3. The difference in propagation time between the longest path and the shortest path for paths with significant energy is called multipath delay spread T_d. Mathematically, T_d can be expressed as

$$T_d := max|\tau_i(t) - \tau_j(t)| \tag{8.11}$$

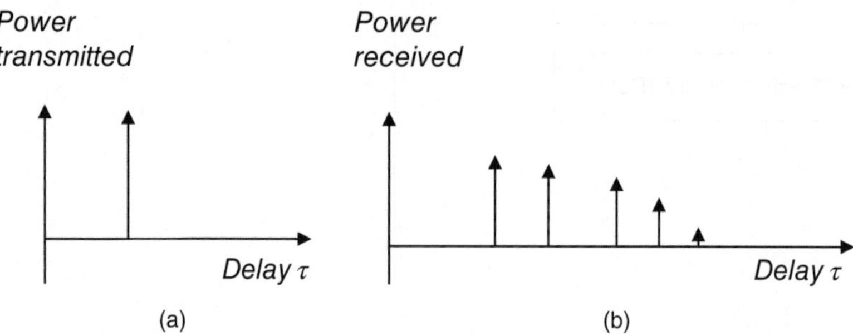

Figure 8.3 Multipath radio environment (a) transmit pulse (b) received pulses.

where $\tau_i(t)$ and $\tau_j(t)$ are delays for i^{th} and j^{th} paths with significant energy. The power delay profile is a plot of the received power versus the arrival times of the different paths between the transmitter and the receiver. The times are measured relative to the arrival time of the line-of-sight (LOS) path. Let the received power at delay τ_k be denoted by $P(\tau_k)$; then the average delay $\overline{T}_d$ is

$$\overline{T}_d = \frac{\sum_k P(\tau_k)\,\tau_k}{\sum_k P(\tau_k)} \tag{8.12}$$

Let $\overline{T}_d^2$ be the average delay square, then the root mean square (RMS) delay spread $\widehat{T}_d$ is

$$\widehat{T}_d = \sqrt{\overline{T}_d^2 - (\overline{T}_d)^2} \tag{8.13}$$

where

$$\overline{T}_d^2 = \frac{\sum_k P(\tau_k)\,\tau_k^2}{\sum_k P(\tau_k)} \tag{8.14}$$

8.1.4 Coherence Bandwidth

The coherence bandwidth, B_c, is defined as a statistical measure of the range of frequencies over which the radio channel passes all spectral components of the transmitted radio signal with approximately equal gain and linear phase. This means that the signal components within the B_c have strong amplitude correlation. As an approximation, the coherence bandwidth is

$$B_c \approx \frac{1}{T_d} \tag{8.15}$$

For 0.9 correlation between signal components within B_c, then

$$B_c \approx \frac{1}{50\,\widehat{T}_d} \tag{8.16a}$$

And for 0.5 correlation:

$$B_c \approx \frac{0.276}{\widehat{T}_d} \tag{8.16b}$$

However, the commonly used value is

$$B_c \approx \frac{0.2}{\widehat{T}_d} \tag{8.17}$$

8.2 Signal Fading

The wireless channels are generally classified on the rate of change of the signal amplitude, i.e. fast or slow. The wireless channel response varies randomly with time. In addition, a sudden time-variant change of the response causes the radio signals to loss the amplitude or phase or both. This phenomenon is called *signal fading*. The wireless channel can be modelled by key parameters such as Doppler spread, coherence time delay spread, and coherence bandwidth, as discussed above. However, these parameters most casually vary significantly with a radio environment such as terrain (flat versus mountainous), radio coverage (rural versus urban), or indoor versus outdoor wireless communications. In addition, different kinds of fading can occur, but mostly they are grouped into two prominent types. Their classification is based on time/distance scale such as large-time scale fading (i.e. path loss) and small-time scale fading (multipath fading).

Small-time scale fading is a stochastic process that causes great variation over small distances (within a wavelength). It is caused by moving scatterers and multipath. Multipath propagation causes signal fading that varies with frequency since each frequency arrives at the receiver through a different radio path. When a radio signal is made of a wide band of frequencies that is transmitted simultaneously, each frequency varies according to a specific fading mechanism, resulting in frequency selective fading, which causes severe distortion. The envelope of multipath signal is random variable with Rayleigh distribution. When the received signal encloses a strong LOS component in addition to the multipath components, the envelope varies randomly with Rician distribution. Statistic measurements also show a good fit to Nakagami-m and Weibull distributions.

Large time-scale fading is a deterministic process, and its variations take place over larger areas, caused by terrain, building, and foliage obstructions. The large-scale fading due to various obstacles causes the shadowing phenomenon and is commonly expressed with a log-normal distribution.

8.2.1 Small-Scale Fading Channels

There are two types of small time-scale fading: *slow fading* and *fast fading*. The terms *slow* and *fast* fading refer to the rate of change of the signal magnitude and phase is imposed by the channel.

8.2.1.1 Slow Fading

Slow fading occurs when the symbol duration is smaller than the coherence time. Coherence time is defined in Section 8.1.2. In other words, the amplitude and phase change imposed by the channel can be considered roughly constant over the symbol period. Slow fading can be caused by shadowing due to a large obstruction such as a hill or large building obscuring the main signal path between the transmitter and the receiver. The amplitude change caused by shadowing is often modelled using a log-normal distribution.

8.2.1.2 Fast Fading

Fast fading occurs when the symbol duration is larger than the coherence time. In other words, channel impulse response changes rapidly within the symbol duration. High Doppler rate changes the fast fading channel into ergodic channel averaging out the randomness of channel gain allowing long-term constant bit rates to be supported like additive white Gaussian noise (AWGN) channels. Ergodic channel gain time average is equal to the ensemble average, and the randomness of the channel gain can be averaged out (removed) over time. This is in contrast with a non-ergodic channel, where channel gain is random but not changing with time, so that its randomness cannot be removed over time and consequently, constant bit rates cannot be supported.

8.2.1.3 Frequency Selective Fading

When the bandwidth of the radio signal is larger than B_c, the channel is called *frequency-selective,* and causes severe distortion to certain signal frequencies. In such a case, components with larger frequencies than B_c are subjected to uncorrelated fading while components with frequencies less than B_c are affected by flat fading. The channel is flat or *frequency nonselective* when the signal bandwidth is smaller or equal B_c.

8.2.2 Large-Scale Fading Channels

The large-scale fading channel where the signal $m(t)$ is a time-variant random variable multiplier of the channel. The signal $m(t)$ can be characterized by its mean (deterministic part) and a statistical deviation from the mean. The mean of $m(t)$ is determined from the received Power law decay. For free space (stationary transmitter – receiver), the power decay is determined by the path loss which follows inverse-square law with respect to transmitter – receiver separation. For mobile transmitter/receiver), the power decay is inversely proportional to the nth power of transmit-receive separation distance, where n takes on integer values from 2 to 5.

8.2.3 Statistics of Wireless Channel

The radio signal transported over a mobile channel can be described by a random process with certain statistics that are closely related to the terrain configuration. While the base station (BS) antenna is commonly clear of its neighbour's coverage, the mobile unit's antenna may be within the coverage of more than one BS. The mobile will receive a number of independent reflected waves, each with random phase and amplitude and combined at the receiver. The total received signal undergoes short-term fading, due to multipath propagation. The received signal can be described by quadrature components that are uncorrelated Gaussian processes with zero mean and variance σ^2. The amplitude, r of the received signal $x(t)$ can be described by a Rayleigh probability density function (PDF) $p_R(r)$ as:

$$\begin{aligned} pR(r) &= \frac{r}{2\pi\sigma^2}\exp\left(-\frac{r^2}{2\sigma^2}\right) \quad \text{for } r \geq 0 \text{ and } -\pi \leq \theta \leq \pi \\ &= 0 \quad \text{Otherwise} \end{aligned} \tag{8.18}$$

where θ is the signal phase with uniform distribution.

When the received signal is made up of a LOS component of amplitude A, plus multiple reflected waves, the envelope of the received signal has a Rician PDF expressed as Eq. (8.19):

$$\begin{aligned} price(r) &= \frac{r}{\sigma^2}\exp\left(-\frac{r^2+A^2}{2\sigma^2}\right) I_0\left(r\frac{A}{\sigma^2}\right) \quad \text{for } r \geq 0 \\ &= 0 \quad \text{Otherwise} \end{aligned} \tag{8.19}$$

where $I_0(.)$ represents Bessel's function of zero order.

It is worth noting that if the multipath propagation is eradicated, the Rician random envelope transmitted in a Gaussian channel is modified to a Gaussian envelope describing the envelope of the direct (unfaded) wave. Large obstacles like hills and buildings outdoors, walls and furniture indoors, often block the propagation of radio waves. The received signal variation due to these obstacles is called *shadow fading*. Measurements of the received power taken at a specified location between transmitter and receiver have shown the received power to be a random variable with certain distribution. Let the received power be P_r watts; then received power in dBm $= P'_r = \log_{10}\left(\frac{P_r}{1000}\right)$. The random variable P'_r is known to have *log-normal distribution*. Let $\widehat{P}_r$ be the average power in dBm taken at the same location, and let the standard deviation of the shadowing process be σ_{shadow} in dB. Then the PDF of the shadow fading is given by

$$p_{shadow}(P'_r) = \frac{1}{\sqrt{2\pi\sigma^2_{shadow}}} \exp\left(-\frac{(P'_r - \widehat{P}_r)^2}{2\sigma^2_{shadow}}\right) \tag{8.20}$$

The shadowing effects are usually included into the path loss estimates by adding zero mean Gaussian random variable with variance σ^2_{shadow}. The values for σ_{shadow} in dB depend on the type of the obstruction and are often estimated by empirical methods. Commonly accepted values for σ_{shadow} are between 6 and 8 dB. For fixed transmitted power, the shadow fading changes the received power as the mobile unit moves around the obstacle's area.

8.3 MIMO Channel Models [2]

Consider a multiple-input, multiple-output (MIMO) system comprising transmitter with M antennas and a receiver provided with N antennas. The wireless channel matrix $\mathbf{H} \in \mathbb{C}^{N\times M}$ with elements h_{ij} defining the channel gain between the jth transmit antenna and the ith receive antenna. The received vector $\mathbf{y} \in \mathbb{C}^{N\times 1}$ is delivered as

$$\mathbf{y} = \mathbf{H}\,\mathbf{x} + \mathbf{n} \tag{8.21}$$

where $\mathbf{x} \in \mathbb{C}^{M\times 1}$ is the transmit vector and $\mathbf{n} \in \mathbb{C}^{N\times 1}$ is the additive white noise with zero-mean and normalized covariance matrix given by an identity matrix in most scenarios. A mathematical model defined in Eq. (8.21) is a description of a MIMO system we will refer to extensively in the analysis that follows. During the sequels, we will also define a complex Gaussian distribution of vector **x** being a circularly symmetric if for any $\alpha \in [0, 2\pi]$, the distribution of **x** will be the same as the original distribution, i.e. $e^{i\alpha}\,\mathbf{x}$.

8.3.1 MIMO Channel Model Based on Perfect CSIT or CSIR

In this case, the channel state information (CSI) is perfectly known at the transmitter or at the receiver. The channel state is defined by a matrix **H** and the channel realization distributions are perfectly and instantaneously known at the transmitter or the receiver, respectively.

8.3.2 MIMO Channel Model Based on Perfect CSIR and CDIT

In this model, the channel state information at the receiver (CSIR) and its cumulative distribution function at the transmitter are perfectly known. Assume a *low-rate noise free* feedback link has been established between the receiver and the transmitter and a perfect CSI knowledge at the receiver enables the receiver to accurately tracks the CSI and relays back CSI estimates to the transmitter using the feedback link. The important factor in this link is the channel distribution which is time-varying and the channel realizations at various instants are independently and identically distributed (*i. i. d*). So, the channel statistics are changing with time as a result of the mobility pattern of the transmitter, receiver, and the scattering in the surrounding environment. In the short-term, the channel realizations have a nonzero mean and a set of correlations described by the propagation environment while in the long-term; the channel realizations have zero mean and are uncorrelated due to the averaging over time.

This channel model is based on the condition of the receiver ability to accurately track the CSI and feedback the CSI estimates to the transmitter. The viability of the model depends greatly on the accuracy of the channel estimates, and of course these estimates incur certain delay associated with them. Given the channel distribution, we consider three most prominent cases for modelling **H** namely: the zero-mean spatially white (ZMSW) model; the channel mean information (CMI) model; and the channel covariance information (CCI) model. So, in the ZMSW model, the channel has zero mean and the covariance is modelled as white. The channel matrix elements are *i. i. d* random variables. The ZMSW model is suitable for long-term average distribution of the channel realizations. The channel mean is nonzero in the CMI model but the covariance is white as in the ZMSW model. The problem of the receiver outdated measurement arriving at the transmitter due to delay over the feedback link produces imperfect estimate at the transmitter and a number of attempts are made to rectify the problem with varying successes. The CMI model includes a constant factor as a scaling factor α to account for the delay that may cause for the imperfect estimate arriving at the transmitter.

Under the CCI model, the channel is assumed to be varying rapidly to track its mean, so the mean is set to zero and the channel covariance is nonwhite. Denote the channel matrix $\mathbf{H}_w \in \mathbb{C}^{N \times M}$ with zero mean, unit variance, and circularly symmetric Gaussian channel. Mathematically, we can define these models as

$$\text{ZMSW model}: \mathbb{E}[\mathbf{H}] = 0, \mathbf{H} = \mathbf{H}_w \tag{8.22a}$$

$$\text{CMI model}: \mathbb{E}[\mathbf{H}] = \mathbf{H}_\mathrm{m}, \mathbf{H} = \mathbf{H}_\mathrm{m} + \sqrt{\alpha}\mathbf{H}_w \tag{8.22b}$$

$$\text{CCI model}: \mathbb{E}[\mathbf{H}] = 0, \mathbf{H} = \sqrt{\mathbf{R}_r}\,\mathbf{H}_w\,\sqrt{\mathbf{R}_t} \tag{8.22c}$$

where $\mathbf{R}_r$, $\mathbf{R}_t$ are receive antenna and transmit antenna correlation matrices, respectively. The CCI is verified by measurement to be an appropriately accurate interpretation for the fade distribution in most viable cellular systems. Indeed, the CSI estimates can be quantized at the receiver and fed back to the transmitter. Such a model is called quantized channel state information (QCI) model and is suitable for use in low fading environments. Using B bits for the QCI model can predetermine a set of $\mathrm{N} = 2^B$ vectors to represent the channel at the transmitter. The QCI channel model seems to be practical, since many wireless systems have a low-rate feedback link.

8.3.3 MIMO Channel Model Based on Perfect CDIT and CDIR

In the previous deliberations, we assumed a perfect CSIR. However, for rapidly changing mobile channels, such an implicit assumption may not be realistic. In such an environment, it may be possible to estimate the distribution of the channel rather than the channel itself, since in general the distribution varies at a slower rate to the channel variation. Therefore, the estimated distribution can be made available to the transmitter using the low-rate feedback link. The downside of this model is that in practice the receiver estimates the distribution from the received signal, which is degraded by noise and possibly interference and therefore will not have a perfect distribution estimate.

8.4 Massive MIMO Channel Models

In general, there are two categories of channel models that are widely used in the performance evaluation of MIMO systems. They are the correlation-based stochastic (CBS) models and the geometry-based stochastic (GBS) models [3]. The latter is appropriate for applications that use geometry-oriented massive MIMO systems and can accurately predict the realistic characteristic of the channel, even though at a higher computing cost. The CBS models can further be divided into three types: the i.i.d Rayleigh model where channel realizations are i.i.d complex Gaussian variables; the correlation model that involves correlation between transmit antennas and/or receive antennas; and the mutual coupling model that takes into account the antenna impedance, load impedance, and mutual impedance to characterize the antenna coupling. The GBS models can be divided into two types: the 2D model in connection with 2D linear antenna array where wave propagates in 2D plane, the 3D model covers spherical and cylindrical antenna arrays, and the wave propagates in 3D plane. Massive MIMO systems are likely to adopt the time-division duplexing (TDD) transmission protocol. The following analysis focuses on single antenna users for uplink transmission only since the downlink (DL) CSI can be obtained from the uplink models due to TDD reciprocity property.

The uplink multiple-input multiple-output (UL MIMO) system consists of K single antenna users transmitting simultaneously to a BS that is equipped with receive M_B antennas. The uplink channel operating in a fading environment can be modelled [3] as

$$\mathbf{H} = \mathbf{G}\sqrt{\mathbf{D}} \tag{8.23}$$

where $\mathbf{D} = \mathrm{diag}\,\{\beta_1, \beta_2, \ldots, k, \ldots\ldots, \beta_K\}$ is $K \times K$ Large-scale fading (attenuation and shadow) diagonal matrix and $\mathbf{H} \in \mathbb{C}^{\mathrm{M}_B \times K}$ matrix with fading coefficients.

8.4.1 i.i.d. Rayleigh Channel Model

In this model, we assume the transmit antennas and receive antennas are uncorrelated and uncoupled so the fading channel matrix $\mathbf{H} \in \mathbb{C}^{\mathrm{M}_B \times K} = [\mathbf{h}_1, \mathbf{h}_2, \ldots, \mathbf{h}_K]$ where the channel elements are i.i.d Gaussian random variables, $\mathrm{h}_{n,k} \sim \mathbb{C}\mathrm{N}(0, 1)$ where $n = 1, 2, \ldots \ldots, \mathrm{M}_B$ and $k = 1, \ldots, K$. In a massive MIMO system, the Rayleigh channel can be approximated as in [4]:

$$\mathbf{H}^H\, \mathbf{H} = \sqrt{\mathbf{D}}\mathbf{G}^H\mathbf{G}\sqrt{\mathbf{D}} = \mathbf{D}\, \mathbf{G}^H\mathbf{G} \tag{8.24}$$

where $\mathbf{D}$ is diagonal matrix given as $\mathbf{D} = \mathrm{diag}\{\beta_1, \beta_2, \ldots\ldots\ldots, \beta_K\}$ and β is constant vector.

An important advantage in using massive MIMO for multi-users uplink channel is that the individual user channels are orthogonal, i.e.

$$\frac{1}{\mathrm{M}_B}\,\mathbf{h}_i^H\,\mathbf{h}_j \approx \begin{cases} 0, & i \neq j \\ 1, & i = j \end{cases} \tag{8.25}$$

Another phenomenon that often occurs in massive MIMO system is *channel hardening* [5]. Channel hardening is a condition that occurs when receiver and transmit antennas N, M, respectively, are both very large but $\frac{\mathrm{M}}{\mathrm{N}}$ is kept constant. Then the off-diagonal terms of the $\mathbf{H}^H\mathbf{H}$ matrix become increasingly weaker compared to the diagonal terms as the size of the channel matrix $\mathbf{H}$ increases – i.e.

$$\frac{\mathbf{H}^H\mathbf{H}}{\mathrm{M}_B} \to \mathbf{I}_{\mathrm{M}_B} \tag{8.26}$$

It is important to note that the user channels are orthogonal, as depicted in (8.25), and thus mitigating the inter-user/inter-cell interference, which improves the system capacity. Channel hardening mitigates the impact of fast fading on the scheduling gain [6], and thus simplifies the design of scheduling schemes.

8.5 Correlation Inspired Channel Models

8.5.1 Introduction

Analysis in this section consider a narrowband MIMO system with M transmit antennas and N receive antennas. The received signal vector $\mathbf{y}$ corresponding to one symbol interval is the N-dimensional vector given by

$$\mathbf{y} = \sqrt{\frac{\rho}{\mathrm{M}}}\mathbf{H}\,\mathbf{x} + \mathbf{n} \tag{8.27}$$

where $\mathbf{x}$ is the M-dimensional transmit vector, $\mathbf{H} \in \mathbb{C}^{\mathrm{N}\times\mathrm{M}}$ is the channel matrix, and $\mathbf{n} \sim \mathcal{CN}(\mathbf{0}, \mathbf{I})$ is complex AWGN. We assume an average input power constraint of $\mathbb{E}[\mathbf{x}^H\,\mathbf{x}] \le \mathrm{M}$ so that ρ represents the effective signal-to-noise ratio (SNR) at the transmitter. Furthermore, we assume that the entries of $\mathbf{H}$ are $\mathcal{CN}(0, 1)$ random variables, and that the channel changes will not fluctuate its statistical properties such as the theoretical mean and variance with time from symbol to symbol. In general, the matrix $\mathbf{H}$ will consist of correlated entries as a result of the physical characteristics of the scattering environment.

The MIMO system in uncorrelated Rayleigh environments can provide huge Shannon capacities. However, in a realistic environment, antenna correlations exist due to either insufficient antenna spacing or poor scattering conditions. Consider the MIMO system modelled as in (8.21). The channel matrix $\mathbf{H}$ is random matrix with complex elements $\{\mathrm{h}_{i,j}\}$ describing the gain of the wireless propagation channel between the jth transmitting antenna and the i^{th}receiving antenna. We denote the jth column of $\mathbf{H}$ by $\mathbf{h}_j$ corresponding to the jth transmitted signal. For uncorrelated MIMO Rayleigh-fading channels, the entries of $\mathbf{H}$ are i.i.d. Gaussian random variables with zero-mean, independent real and imaginary parts with equal variances. When a correlation among the receiving antennas exists, the columns of $\mathbf{H}$ are independent random vectors, but the elements of each column are correlated with the same mean and covariance matrix. Similarly, a correlation among the transmitting antennas can be considered in the same definitions. Nonetheless, in this case the rows of $\mathbf{H}$ are independent, but the elements of each row are correlated with a given covariance matrix (the same for all rows). The channel $\mathbf{H}$ is subject to flat fading when its delay spread is negligible compared to the inverse of the bandwidth.

For the case of Rayleigh fading, this implies that $\mathbb{E}\{\mathbf{h}_j\} = 0$ and a correlation matrix is defined as $\Sigma = \mathbb{E}\{\mathbf{h}_j\,\mathbf{h}_j^{\mathrm{T}}\}$ for $j = 1, 2, \ldots \ldots \mathrm{M}$. Without loss of generality, we can normalize the diagonal elements of Σ to one, i.e. $\mathbb{E}\{|\mathrm{h}_{jj}|^2\} = 1$.

Most practical MIMO channels are exposed to not only spatial correlations but also temporal correlations. The latter occurs between the channel matrix realizations at different time instants. We focus on the spatial correlations only and its effects on the uplink capacity. We don't need to investigate both uplink and downlink channels since the downlink propagation channel is the conjugate transpose of the uplink propagation channel.

The channel, $\mathbf{h}_k$ is $\mathrm{M}\times 1$ vector channel between the BS and k^{th} user where

$$\mathbf{h}_k = \sqrt{\beta_k}\;\mathbf{g}_k \qquad k = 1, 2\ldots\ldots, K \tag{8.28}$$

where β_k is the large scale fading accounting for both attenuation and shadow fading expressed as

$$\beta_k = \frac{\alpha_k}{r_k^{\varkappa}} \tag{8.29}$$

where α_k is log-normal random variable representing shadow fading with standard deviation σ_{shadow}, r_k is the distance between k^{th} user and the BS, and $\varkappa$ is the path loss exponent. The small-scale fading vector $\boldsymbol{g}_k \sim \mathbb{C}\mathcal{N}(0, \mathbf{I}_{\mathrm{M}})$ is complex Gaussian variables with zero mean and unit variance. We assume $\boldsymbol{g}_k$ to be statistically independent across the users. The spatially correlated channel is modelled [7] and the analysis of the correlated channel models is presented in Section 9.5.2.

The capacity performance of a MIMO system under uncorrelated Rayleigh fading channel and the correlated Rayleigh fading are compared in [8]. A realistic DL scenario is simulated instead of the uplink reciprocity. The simulation results are presented for the DL channel with a MIMO system comprising a BS equipped with up to 400 transmit antennas serving 10 single antenna users using multiuser eigen beamforming (BF) and regularized zero-forcing (RZF) precoding. The distance-based path-loss exponent is 3.7, but shadowing effects are not included. Figure 8.4 depicts the achievable rate per user in b/s/Hz versus the number of the BS antennas. The ultimate theoretical average rates for an unlimited number of antennas are 7.2 and 7.08 b/s/Hz for BF and RFZ precoding, respectively. Figure 8.4 shows the shapes of the curves for both precoders to be similar for correlated and uncorrelated antennas but not identical, and it becomes clear that correlation matrices seriously degrade the MIMO system capacity performance.

The uncorrelated channel with RZF precoding achieves about 5.8 b/s/Hz per user (0.82% of the ultimate rate) compared to the per user rate of the BF precoded of 4.57 b/s/Hz (0.63%), and this high performance is due to the reduction of the multiuser interference by RZF. However, the correlated channel with RZF precoding achieves about 4.86 b/s/Hz per user (0.69% of the ultimate rate) compared to the per user rate of the BF precoded of 3.14 b/s/Hz (0.44% of the ultimate rate).

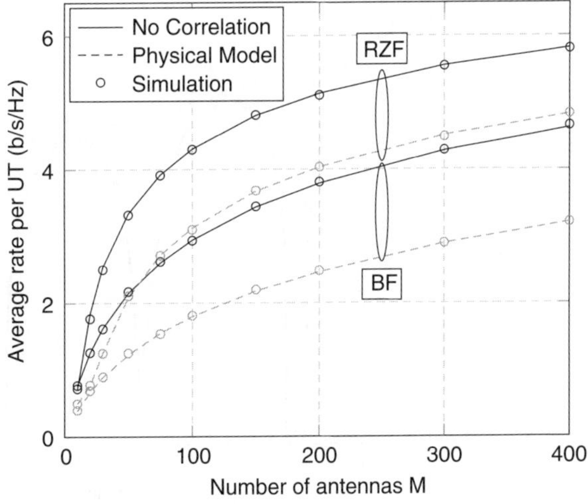

Figure 8.4 Average per-user rate with BF and RZF precoding versus the number of antennas M. Solid and dashed lines depict the asymptotic approximations and markers for the simulation results [8].

The impact of spatial correlation on the performance of MIMO systems can be gauged by looking at the eigenvalues distribution of the channel covariance matrix [9]. The eigenvalues will be almost identical in the weak correlation and only few dominant in the strong correlation. Consequently, in a highly correlated system, the channel is disclosed to finite eigen subspace. In the uncorrelated scenario, the eigenvectors are important as they point to the direction of the largest variance *orthogonal* to other eigenvectors. BS in an urban environment is placed higher than the surrounding area and exposed to little scattering and thus BS antennas experience strong spatial correlation. On the other hand, mobile users exposed to rich scattering have weak antenna correlation.

The antennas correlation is computed using the radiation patterns of the antennas [10], as well as a complicated integration operation over the whole space, meaning high computing complexity. A simpler-to-calculate and easier-to-measure technique for the antennas correlation is proposed in [11] based on the antennas scattering parameters, commonly known as *S-parameters*. These parameters can be measured quite easily using laboratory equipment such as vector network analysers (see 8.A). The S-parameters scheme is further developed in [12, 13]. In [11], and an exact expression between two adjacent antennas is derived without the knowledge of their radiation patterns. Instead, two antennas are driven by two generators at the same frequency, and the total power radiated by both antennas is mathematically derived by considering the total radiated electric field in the far-field region using spherical coordinates. The envelope correlation $Corr_{env}$, in terms of the S-parameters of the two antennas is given as

$$Corr_{env} = \frac{|S_{11}^* S_{12} + S_{21}^* S_{22}|^2}{\{1 - (|S_{11}|^2 + |S_{21}|^2)\}\{1 - (|S_{22}|^2 + |S_{12}|^2)\}} \tag{8.30}$$

Figure 8.5 depicts the envelope correlation in dB versus the separation between antennas in wavelength, for a two-antenna system formed by two collinear half-wave dipoles comparing

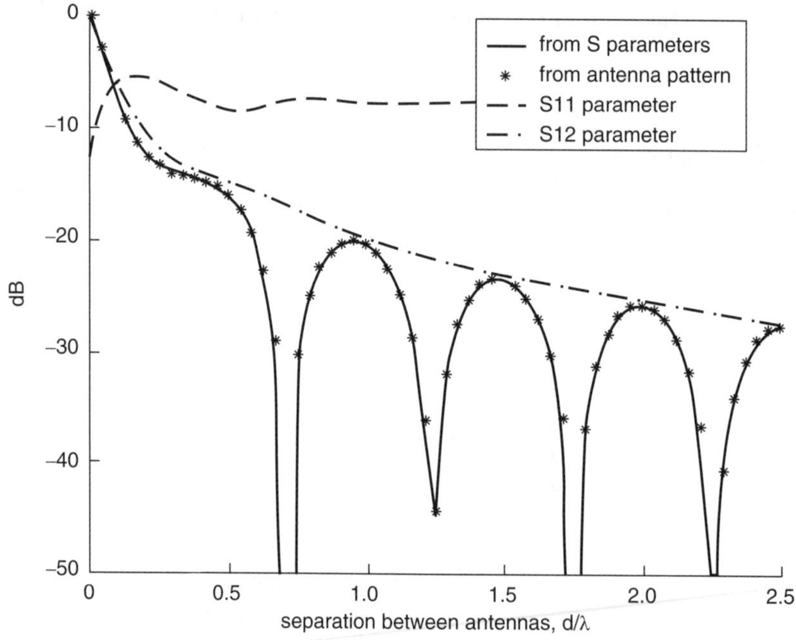

Figure 8.5 Envelope correlation and S-parameters for two collinear half-wave dipoles against their separation [11].

the antenna radiation pattern and S-parameters methods. It is clearly seen that the envelope correlation decreases rapidly with increasing the separation of the antennas and that both methods provide exactly the same result but with the S-parameters have the advantages of easily measured and the envelope correlation simply computed.

8.5.2 Formation of Kronecker Channel Model

The idealised statistical model of a spatial MIMO channel corresponds to a rich scattering environment but does not experience correlation and thus the elements of **H** are i.i.d. complex Gaussian random variables. However, the elements of **H** are correlated in realistic environments since in practice the rich scattering environments do not always occur. Under perfect conditions, it is well known that the theoretic capacity of a MIMO system increases linearly with the minimum of transmit and receive antennas. However, various measurements show that practical MIMO channels have a significantly lower capacity, as we previously suggested. This reduction of capacity is due to the spatial correlation at either the receive side or the transmit side of radio link or at both sides simultaneously [12]. The random channel matrix $\mathbf{H} \in \mathbb{C}^{N \times M}$ is subjected to small-scale fading. The link's ends change their roles with respect to transmitting and receiving the radio signal. For clarity, we denote the link ends with the labels '*rx*' for receiving side and '*tx*' for transmitting side. Both link ends can be receiver or transmitter, depending on the link direction.

The signal vector **y** at the N receive antennas is given by (8.27) and repeated here:

$$\mathbf{y} = \sqrt{\frac{\rho}{M}}\mathbf{Hx} + \mathbf{n}$$

The total mean transmit power P_H is constraint to $P_H \triangleq \mathbb{E}_H\{\mathrm{tr}(\mathbf{HH}^H)\}$. The expectation operator is performed with respect to different channel realizations of one and the same scenarios. A scenario is defined as a region in time and space for which the channel statistics are approximately constant.

In order to characterise a correlated spatial MIMO channel, a full correlation matrix that specifies the N M × N M mutual correlation values between all channel matrix elements is required. In accordance to the literature, we define the correlation matrix [13] as

$$\mathbf{R_H} \triangleq \mathbb{E}_\mathbf{H}\,\{\mathrm{vec}(\mathbf{H})\mathrm{vec}(\mathbf{H})^H\} \tag{8.31}$$

where vec(**H**) is an operator that stacks columns of **H** one under another in a single long column. Let us denote the matrix square roots as $\sqrt{(.)}$ and expresses the (N M × 1) matrix $\mathbf{R_H}$ by

$$\mathbf{R_H} = \sqrt{\mathbf{R_H}}.(\sqrt{\mathbf{R_H}})^H \tag{8.32}$$

For the Kronecker model, we examine each side correlation matrix independently so each correlation matrix expresses the spatial correlation of one link end only. The link ends correlation matrices, i.e. the receive and transmit correlation matrices, are given as

$$\mathbf{R}_r = \mathbb{E}\{\mathbf{H}\,\mathbf{H}^H\} \tag{8.33a}$$

$$\mathbf{R}_t = \mathbb{E}\{(\mathbf{H}^H\,\mathbf{H})^T\} \tag{8.33b}$$

The normalized spatial correlation matrix of the MIMO radio channel is the Kronecker product of spatial correlation matrix at the transmitter, and the receiver is given in [14] as

$$\mathbf{R_H} = \frac{1}{\mathrm{tr}\{\mathbf{R}_r\}}\,(\mathbf{R}_t \otimes \mathbf{R}_{rx}) \tag{8.34}$$

Consequently, the correlated Rayleigh fading channel matrix $\widetilde{\mathbf{H}} \in \mathbb{C}^{N \times M}$ is given by

$$\mathrm{vec}(\widetilde{\mathbf{H}}) = \sqrt{\mathbf{R_H}}\,\mathrm{vec}(\mathbf{G}) \tag{8.35}$$

where $\mathbf{G}$ is an N x M matrix consisting of i.i.d zero-mean complex Gaussian entries with unit variance. Substituting (8.34) for $\mathbf{R_H}$ in (8.35) we get

$$\text{vec}(\widetilde{\mathbf{H}}) = \frac{1}{\sqrt{\text{tr}\{\mathbf{R}_r\}}}(\sqrt{\mathbf{R}_t} \otimes \sqrt{\mathbf{R}_r})\text{vec}(\mathbf{G}) \tag{8.36}$$

Using the identity, $[\mathbf{B}^T \otimes \mathbf{A}]\text{vec}(\mathbf{D}) = \text{vec}(\mathbf{ADB})$ to facilitate the spatial MIMO channel Kronecker model, $\mathbf{H}_{kron}$ as

$$\widetilde{\mathbf{H}} = \mathbf{H}_{kron} = \frac{1}{P_H}\sqrt{\mathbf{R}_r}\,\mathbf{G}\,(\sqrt{\mathbf{R}_t})^T \tag{8.37}$$

The channel model expressed in (8.37) facilitates the analysis and simulation of the correlated MIMO systems which has made the Kronecker channel popular. However, evidence based on measurements taken over different environments, frequencies, and number of researchers indicates that when high correlation endures at either or both of the MIMO link, the Kronecker model predicts a sum capacity lower than that measured (at times as low as 20%) [13, 14]. The reduction is mainly due to rendering the multipath structure incorrectly.

8.6 Weichselberger Channel Model

8.6.1 Introduction

Thus far, the correlation of a MIMO system at the ends of a link is considered as two separate independent matrices. However, the one-sided correlation matrices lack the credibility in evaluating the achievable capacity of a practical MIMO system. The two ends of a MIMO channel are fully associated with the spatial structure of the MIMO channel. So, the two link ends of a MIMO channel cannot be considered as independent [15]. Accordingly, the one-sided correlation matrices have to be parameterized by the signal statistics of the other link end. That is a stochastic characterization of the link end correlation matrix given by (8.33) has to include extra parameter to account for the statistical properties of the signal transmitted from the other end. Let the subscripts r and t indicate to which side of the link an identity is associated.

Assume the MIMO channel matrix $\mathbf{H}$ of size N × M denoting the receive antennas and transmit antennas respectively. The spatial characterization of a general MIMO channel is described by the correlation matrix $\mathbf{R_H}$. The correlation between all channel matrix elements defined by $\mathbf{R_H}$ can be expressed as in (8.31),

$$\mathbf{R_H} = \mathbb{E}_{\mathbf{H}}\{\text{vec}(\mathbf{H})\text{vec}(\mathbf{H})^H\} \tag{8.38}$$

We define the mth column of the channel matrix by $\mathbf{h}_{col,m}$, vec($\mathbf{H}$) for the channel matrix $\mathbf{H}$ can be expressed as

$$\text{vec}(\mathbf{H}) = [\,\mathbf{h}^T_{col,1}\ \mathbf{h}^T_{col,2} \cdots\cdots \mathbf{h}^T_{col,m} \cdots\cdots\cdot\cdot \mathbf{h}^T_{col,\text{M}}\,]^{\text{T}} \tag{8.39}$$

According to Eq. (8.31) $\mathbf{R_H}$ is N M × N M matrix and can be composed as a block matrix of M × M blocks, each of size N × N as expressed as

$$\mathbf{R_H} = \begin{bmatrix} \mathbf{R}_{r,1,1} & \mathbf{R}_{r,1,2} & - - - - - - & \mathbf{R}_{r,1,\text{M}} \\ \mathbf{R}_{r,2,1} & \mathbf{R}_{r,2,2} & - - - - - - & \mathbf{R}_{r,2,\text{M}} \\ - & - & - - - - - - & - \\ - & - & - - - - - - & - \\ - & - & - - - - - - & - \\ \mathbf{R}_{r,\text{M},1} & \mathbf{R}_{r,\text{M},2} & - - - - - - & \mathbf{R}_{r,\text{M},\text{M}} \end{bmatrix} \tag{8.40a}$$

A single block out of the M × M blocks defined by $\mathbf{R}_{r,m_1,m_2}$ is given by

$$\mathbf{R}_{r,m_1,m_2} \triangleq \mathbb{E}_{\mathbf{H}}\{\mathbf{h}_{col,m_1}\, \mathbf{h}^H_{col,m_2}\} \tag{8.40b}$$

A diagonal block $\mathbf{R}_{r,\,m,\,m}$ seen as the spatial correlation matrix at the receive end caused by the single transmit antenna m, i.e. the auto-correlation matrix of the vector $\mathbf{h}_{col,\,m}$. The off-diagonal blocks $\mathbf{R}_{r,m_1,m_2}$ represent spatial cross-correlation matrices at the receive end initiated by channel vectors $\mathbf{h}_{col,m_1}$ and $\mathbf{h}_{col,m_2}$, one channel vector is developed by transmitting from antenna m_1, the other one by transmitting from antenna m_2.

For a transmit beamforming with fixed transmit weights $\boldsymbol{w}_t$, whose elements are denoted as $w_{tx,\,m}$, the correlation matrix at the receive link end is

$$\mathbf{R}_{r,\boldsymbol{w}_t} \triangleq \mathbb{E}_{\mathbf{H}}\{(\mathbf{H}\,\boldsymbol{w}_t)\,(\mathbf{H}\,\boldsymbol{w}_t)^H\} \tag{8.41a}$$

$$\mathbf{R}_{r,\boldsymbol{w}_t} = \sum_{m_1}^{M}\sum_{m_2}^{M} w_{t,m_1}\, w^*_{t,m_2}\, \mathbf{R}_{r,m_1,m_2} \tag{8.41b}$$

where t and r indicate the side of the transmit side and receive side of link respectively.

When multiple signal streams are transmitted over the MIMO channel or the spatial transmit weights are time varying, the statistical charaterisation of the transmit signal is determined by the signal covariance matrix $\mathbf{Q}_{\text{sig}}$. Using $\mathbf{Q}_{\text{sig}}$ of the transmitted signal with eigenvectors $\boldsymbol{u}_{\text{sig},\,m}$ and eigenvalues $\lambda_{\text{sig},\,m}$, the parameterized correlation matrix at the receive link end can be expressed as

$$\mathbf{R}_{r,\mathbf{Q}_{\text{sig}}} \triangleq \mathbb{E}_{\mathbf{H}}\{\mathbf{H}\,\mathbf{Q}_{\text{sig}}\,\mathbf{H}^H\} = \sum_{m=1}^{M} \lambda_{\text{sig},m}\, \mathbf{R}_{r,\boldsymbol{u}_{\text{sig},m}} \tag{8.42}$$

We should recall that random variables such as the signal streams, the covariance is closely related to their correlation coefficients and zero covariance implies the variables are uncorrelated. In fact the correlation coefficient is equal to scaled down covariance. Let transmit and receive link ends have signal covariance matrices $\boldsymbol{Q}_r$ and $\boldsymbol{Q}_t$, respectively. Therefore, the parameterized receive and transmit correlation matrices are given as

$$\mathbf{R}_{r,\boldsymbol{Q}_t} \triangleq \mathbb{E}_H\,\{\mathbf{H}\boldsymbol{Q}_t\mathbf{H}^H\} \tag{8.43a}$$

$$\mathbf{R}_{t,\boldsymbol{Q}_r} \triangleq \mathbb{E}_H\{\mathbf{H}^H\,\boldsymbol{Q}_r\mathbf{H}\}^T = \mathbb{E}_H\{\mathbf{H}^T\boldsymbol{Q}_r\mathbf{H}^*\} \tag{8.43b}$$

The unparameterized one-sided correlation matrices given by (8.33) assumes the signal covariance of the other end is spatially white MIMO channel, i.e. the MIMO signals are completely uncorrelated as given in (8.33) and repeated here:

$$\mathbf{R}_r = \mathbb{E}\{\mathbf{H}\,\mathbf{H}^H\} = \sum_{m=1}^{\mathrm{M}} \mathbf{R}_{r,m,m} \tag{8.44a}$$

$$\mathbf{R}_t = \mathbb{E}\{(\mathbf{H}^H\,\mathbf{H})^T\} = \sum_{n=1}^{\mathrm{N}} \mathbf{R}_{r,n,n} \tag{8.44b}$$

The total mean energy $E_{\mathbf{H}}$ of the MIMO channel $\mathbf{H}$ is defined by the trace of the correlation matrices:

$$\mathbb{E}_{\mathbf{H}} \triangleq \mathrm{tr}(\mathbf{R}_r) = \mathrm{tr}(\mathbf{R}_t) = \sum_{n=1}^{\mathrm{N}}\sum_{m=1}^{\mathrm{M}} \mathbb{E}_{\mathbf{H}}\{|\mathrm{h}_{n,m}|^2\} \tag{8.45}$$

In Weichselberger channel model [16, 17], it was assumed that the *auto-and cross-correlation of a link end* have the same *eigenbasis* (i.e. comprises the same orthonormal eigenvectors). The transmit eigenbasis mirrors the spatial structure of the scatterers that are concerning the receive array. On the other hand, the eigenvalues affirm the way the scatterers are illuminated by the radio waves propagating from the transmitter. However, radiating in a certain direction may illuminate only certain scatterers so the physical implication is that the eigenvalues may differ for each correlation matrix. Mathematically, the assumption can be articulated when we apply the eigendecomposition to the correlation matrices at the receive and transmit link ends to get,

$$\mathbf{R}_{r,m_1,m_2} = \mathbf{U}_r\, \mathbf{\Lambda}_{r,m_1,m_2}\, \mathbf{U}_r^H \tag{8.46}$$

$$\mathbf{R}_{t,n_1,n_2} = \mathbf{U}_t\, \mathbf{\Lambda}_{t,n_1,n_2}\, \mathbf{U}_t^H \tag{8.47}$$

where $\mathbf{U}_r$ denotes the eigenbasis, and $\mathbf{\Lambda}_{r,m_1,m_2}$ is a diagonal matrix whose elements are the eigenvalues of the correlation matrix. The $\mathbf{U}_r$ *does not change with* the indices m_1 and m_2; it is the same for all correlation matrices on the receive link end. In general, however, the *eigenvalues do differ* for each correlation matrix. Due to the assumption stated above, all one-sided correlation matrices at the receiver have the same common eigenbasis

$$\mathbf{R}_{r,m_1,m_2} = \mathbf{U}_r\, \mathbf{\Lambda}_{r,m_1,m_2}\, \mathbf{U}_r^H \tag{8.48}$$

$$\mathbf{R}_{r,\mathbf{w}_t} = \mathbf{U}_r\, \mathbf{\Lambda}_{r,\mathbf{w}_t}\, \mathbf{U}_r^H \tag{8.49}$$

$$\mathbf{R}_{r,\mathbf{Q}_{\text{sig}}} = \mathbf{U}_r\, \mathbf{\Lambda}_{r,\mathbf{Q}_{\text{sig}}}\, \mathbf{U}_r^H \tag{8.50}$$

8.6.2 Formulation of Weichselberger Channel Model

The Weichselberger model develops the joint correlation properties of both wireless link ends. The development confirms the dependencies between the direction of arrivals (DoAs) and the direction of departures (DoDs). The model comprised three key elements that are illustrated in Figure 8.6: the spatial eigenbasis at side A is denoted as $\mathbf{U}_\text{A}$ on the receive end, i.e. ($\mathbf{U}_r$); the spatial eigenbasis at side B is denoted as $\mathbf{U}_B$ one the transmit end, i.e. $\mathbf{U}_t$; and the power-coupling matrix $\mathbf{\Omega}$ whose structure is greatly dependent of the wireless environment. Each pair of receive-transmit eigenmodes (transmission links) connects a single-input single output (SISO) channel and its average power specifies an element of the matrix. The SISO channels are completely uncorrelated.

Let us define eigenbasis channel matrix $\mathbf{H}_\text{eig}$ given by the transform of the channel matrix $\mathbf{H}$ with common eigenbasis at both link ends as

$$\mathbf{H}_\text{eig} \triangleq \mathbf{U}_r^H\, \mathbf{H}\, \mathbf{U}_t^* \tag{8.51a}$$

The entries of the matrix $\mathbf{H}_\text{eig}$ are given as

$$[\mathbf{H}_\text{eig}]_{n,m} = \boldsymbol{u}_{r,n}^H\, \mathbf{H}\, \boldsymbol{u}_{t,m}^* \tag{8.51b}$$

Equation (8.51b) expresses the complex amplitude of the single SISO channel between the mth transmit eigenvector and the nth receive eigenvector. These channels, as we previously highlighted, are completely *uncorrelated*, which is due to the properties of the eigendecomposition

Figure 8.6 Parameters required for the Weichselberger channel model [16].

(i.e. linearly independent eigenvectors). Using (8.51a),

$$\begin{aligned}\mathbb{E}_{\mathbf{H}}\{\mathbf{H}_{\text{eig}}\,\mathbf{Q}_{\text{t}}'\,\mathbf{H}_{\text{eig}}^{\text{H}}\} &= \mathbb{E}_{\mathbf{H}}\{\mathbf{U}_r^H\,\mathbf{H}\,\mathbf{U}_t^*\,\mathbf{Q}_t'\,\mathbf{U}_t^T\,\mathbf{H}^H\mathbf{U}_r\}\\ &= \mathbb{E}_{\mathbf{H}}\{\mathbf{U}_r^H\,\mathbf{H}\,\mathbf{Q}_t\,\mathbf{H}^H\mathbf{U}_r\}\end{aligned} \tag{8.52}$$

where

$$\mathbf{Q}_t = \mathbf{U}_t^*\,\mathbf{Q}_t'\,\mathbf{U}_t^T \tag{8.53}$$

Therefore, we have

$$\mathbb{E}_{\mathbf{H}}\{\mathbf{H}_{\text{eig}}\,\mathbf{Q}_{\text{t}}'\,\mathbf{H}_{\text{eig}}^{\text{H}}\} = \mathbf{U}_r^H\,\mathbb{E}_{\mathbf{H}}\{\mathbf{H}\,\mathbf{Q}_t\,\mathbf{H}^H\}\mathbf{U}_r$$

Applying (8.50) to the middle term we get

$$\mathbb{E}_{\mathbf{H}}\{\mathbf{H}_{\text{eig}}\,\mathbf{Q}_{\text{t}}'\,\mathbf{H}_{\text{eig}}^{\text{H}}\} = \mathbf{\Lambda}_{r,\mathbf{Q}_t} \tag{8.54}$$

Using a similar analysis again, we can show that

$$\mathbb{E}_{\mathbf{H}}\{\mathbf{H}_{\text{eig}}^T\,\mathbf{Q}_r'\,\mathbf{H}_{\text{eig}}^*\} = \mathbf{\Lambda}_{t,\mathbf{Q}_r} \tag{8.55}$$

where

$$\mathbf{Q}_r = \mathbf{U}_r^*\,\mathbf{Q}_r'\,\mathbf{U}_r^T \tag{8.56}$$

where $\mathbf{Q}_t$ and $\mathbf{Q}_r$ are signal covariance matrices at the transmit end and receive link end, and $\mathbf{\Lambda}_{r,\mathbf{Q}_t}$ and $\mathbf{\Lambda}_{t,\mathbf{Q}_r}$ are diagonal matrices. Equations (8.54) and (8.55) hold correct for all $\mathbf{Q}_t$ and $\mathbf{Q}_r$ *if and only if* all elements of $\mathbf{H}_{\text{eig}}$ are mutually uncorrelated. The power coupling matrix $\mathbf{\Omega}$ specifies the average energy coupled between an eigenvector at the transmit side and an eigenvector at the receive side. An estimate of the coupling matrix $\mathbf{\Omega}$ of the Weichselberger spatial channel model is

$$\hat{\mathbf{\Omega}} = \mathbb{E}_{\mathbf{H}}\{(\mathbf{U}_{\text{A}}^H\,\mathbf{H}\,\mathbf{U}_{\text{B}}^*)\odot(\mathbf{U}_{\text{A}}^H\,\mathbf{H}\,\mathbf{U}_B^*)^*\} \tag{8.57}$$

The coefficients of the coupling matrix $\mathbf{\Omega}$ define the mean amount of energy that is coupled from the mth eigenvector of side-A to the nth eigenvector of side B (or vice versa). The expected energy propagating between the mth transmit eigenmode and the nth receive eigenmode is

$$[\hat{\mathbf{\Omega}}]_{m,n} = \mathbb{E}_{\mathbf{H}}\{|\boldsymbol{u}_{\text{A},m}^H\,\mathbf{H}\,\boldsymbol{u}_{\text{B},m}^*|^2\} \tag{8.58}$$

Let us introduce $\mathbf{G}$ again as $\text{N}\times\text{M}$ matrix consisting of i.i.d zero-mean complex Gaussian elements $g_{m,n}$ and express $\mathbf{H}_{weic}$ as a sum of all eigenmodes. We can stack the channel matrix $\mathbf{H}_{weic}$:

$$\text{vec}(\mathbf{H}_{weic}) = \sum_{n=1}^{\text{N}}\sum_{m=1}^{\text{M}}\text{vec}\left(\sqrt{\text{w}_{m,n}}\,g_{m,n}\left(\boldsymbol{u}_{\text{A},m}\,\boldsymbol{u}_{\text{B},n}^T\right)\right) \tag{8.59}$$

The correlation matrix of $\mathbf{H}_{weic}$ can be calculated as

$$\mathbf{R}_{weic} = \sum_{n=1}^{\text{N}}\sum_{m=1}^{\text{M}}\text{w}_{m,n}\,(\boldsymbol{u}_{\text{B},n}\otimes\boldsymbol{u}_{\text{A},m})(\boldsymbol{u}_{\text{B},n}\otimes\boldsymbol{u}_{\text{A},m})^H \tag{8.60}$$

where $\text{w}_{m,n}$ denotes the eigenvalues.

Accordingly, we have:

$$\mathbf{H}_{weic} = \mathbf{U}_r\,(\mathbf{\Omega}^{\overset{0.5}{\leftrightarrow}}\odot\mathbf{G})\,\mathbf{U}_t^T \tag{8.61}$$

where $\odot$ notation of element-wise multiplication, $\mathbf{G}$ is a random matrix with i.i.d. zero-mean complex-normal entries with unit variance, and $\mathbf{\Omega}^{\overset{0.5}{\leftrightarrow}}$ is defined as the element-wise square root of $\mathbf{\Omega}$. The configuration of the coupling matrix $\mathbf{\Omega}$ echoes the spatial arrangement of scattering objects and affects the system capacity. Examples of such configurations are related to the physical radio environments as illustrated in Figure 8.7. We define $\omega_{m,\,n}$ as the average power of the single SISO channel between each nth eigenmode of the receive end (side A) and each

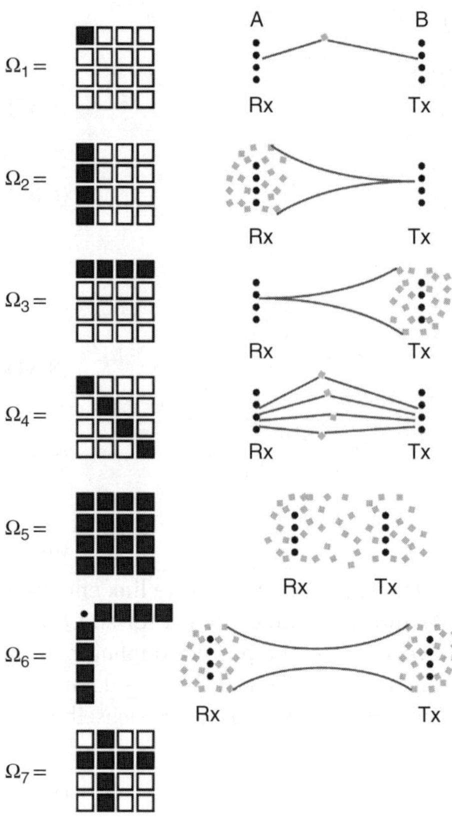

Figure 8.7 Various structures of the coupling matrix $\mathbf{\Omega}$ (grey squares: significant power; blank squares: no power). Black dots: antenna array; gray squares: scatterers [16].

m^{th} eigenmode of the transmit end (side B) linking the correlation of both ends. The left-hand side (LHS) of Figure 8.7 displays the eigenmodes as directions. Although this is not a complete representation, it simplifies the representation for deep understanding. We assume the number of eigenmodes present in the channel considered equals the number of resolvable multipath components. Example $\mathbf{\Omega}_1$ has only one nonzero element in the coupling matrix. So, there is a single resolvable multipath element. Whether or not a single path is a LOS or scattered path cannot be determined by $\mathbf{\Omega}$. On the other hand, $\mathbf{\Omega}_2$has a full column of the coupling matrix. Nonetheless, the transmit link end sees this behaviour as similar to $\mathbf{\Omega}_1$, i.e. it sees that a single spatial channel receive link end, at least N independent multipath components arrive at the Rx antenna, causing a spatial diversity of order N. The third example $\mathbf{\Omega}_3$ is the same as $\mathbf{\Omega}_2$ with sides A and B alternated. The channels thus far cannot support spatial multiplexing as collapsed to single input, single output (SISO), single input, multiple output (SIMO), and multiple input single output (MISO).

$\mathbf{\Omega}_4$ is a diagonal coupling matrix. This example is equivalent to any channel that connects a single entry on each raw and column of the coupling matrix, i.e. each transmit eigenmode is linked to a single receive eigenmode, possible LOS, single scatterer, or multiple scatterers. The number of independent multipaths and thus the order of diversity is limited to min(N, M).

Example $\mathbf{\Omega}_5$ involves the full coupling matrix. Each transmit eigenmode is connected to each Rx eigenmode. A possible physical environment leading to $\mathbf{\Omega}_5$ is a rich scattering cluster containing both link ends.

Example $\mathbf{\Omega}_6$ is a spectial case of Example 5 and is identified with the Kronecker model. Research has shown the Kronecker model bears a deficiency in the degrees of freedom of the MIMO channels – its N M elements of Kronecker coupling matrix are determined by means of N + M eigenvalues. The model cannot generate entirely diagonally coupling matrices. Indeed, it is most likely that the elements of the correlation matrix are uniformly distributed as possible, which drives the wrong capacity estimates.

Example $\mathbf{\Omega}_7$ is an infrequent case of a cross-like coupling matrix that leads to a cross-like correlation matrices at both ends with rank limited to 2. The channel realization is cross-like and is similar to the keyhole channel. Keyhole effect results from some propagation scenarios where the MIMO channel capacity can be low compared to the SISO capacity, even though the signals at the MIMO antennas are uncorrelated. Ideal keyhole properties are: the capacity is low, the rank of the MIMO channel matrix is nearly one, and the correlation between the antenna elements is low [18]. The keyhole channel [19] is represented by two vectors: the receive column vector is of size N × 1 and transmit row vector is of size 1 × M. The difference between the two is that the channel correlation matrix is given by the knowledge of the MIMO channel while the keyhole phenomenon does not appear in the channel keyhole correlation matrix [20].

8.7 Virtual Channel Representation

In realistic environments, the statistics of $\mathbf{H}_{VCR}$ are dictated by the scattering and array characteristics, such as angular spreads of scattering clusters and antenna spacing. The virtual channel is explored in [21–23] to obtain the statistical structure of correlated fading channels by considering the clustered scattering environments.

Consider uniform linear arrays of antennas at both the transmitter and receiver operating in the far-field so the channel matrix can then be described using the array steering and the vectors representing the signal response given by

$$\mathbf{a}_t(\theta_t) = \frac{1}{\sqrt{\mathrm{M}}}[1, e^{-j2\pi\theta_t}, \ldots\ldots\ldots\ldots, e^{-j2\pi(\mathrm{M}-1)\theta_t}]^T \tag{8.62a}$$

$$\mathbf{a}_r(\theta_r) = \frac{1}{\sqrt{\mathrm{N}}}[1, e^{-j2\pi\theta_r}, \ldots\ldots\ldots\ldots, e^{-j2\pi(\mathrm{N}-1)\theta_r}]^T \tag{8.62b}$$

where θ represents the angle variable and is related to the physical angle (measured with respect to the horizontal axis) as $\theta = \mathrm{d}\frac{\sin\phi}{\lambda} = \alpha \sin\phi$, $\phi = \sin^{-1}\frac{\theta}{\alpha}$, where λ is the wavelength of propagation and d is the antenna spacing. The vector $\mathbf{a}_r(\theta_r)$ describes the signal response at the receiver array attributed to a point source in the direction θ_r. Likewise, $\mathbf{a}_t(\theta_t)$ expresses the array weights desired to transmit a beam focused in the direction of θ_t. The receive array of N elements accumulates a few signals from directions within the range of θ_r and the transmit array of M elements set energy at angles within the range of θ_t.

For $-\frac{\pi}{2} \le \phi \le \frac{\pi}{2}$, i.e. $-\alpha \le \theta \le \alpha$ both the steering and response vectors in (8.62) are periodic in θ with period 1. Uniform sampling of θ in the main period ($\theta \in [-0.5, 0.5)$) is a common range for virtual spatial angles, such that:

$$\theta_{r,n} = \frac{n-1}{\mathrm{N}} - 0.5, \theta_{t,m} = \frac{m-1}{\mathrm{M}} - 0.5 \tag{8.63}$$

A widely accepted *virtual channel representation of a* physical scattering environment is illustrated in Figure 8.8. The effective representation relates to beamforming in fixed directions defined by the resolution of the arrays. Each scattering path is accompanied with a fading gain β_l and a unique pair of transmit and receive angles ϕ_t, ϕ_r corresponding to scatterers distributed within the angular spreads. For L-path environment, the physical channel $\mathbf{H}$ model

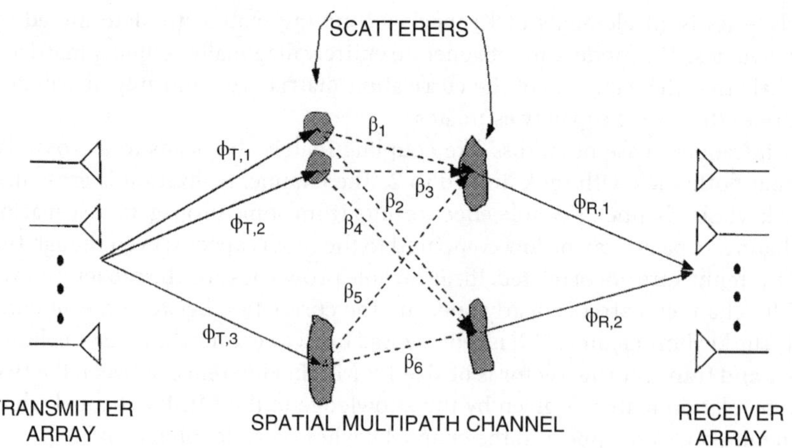

Figure 8.8 Illustrating physical scattering environment [21].

can be expressed mathematical as

$$\mathbf{H} = \sum_{l=1}^{L} \beta_l \, \mathbf{a}_r(\theta_{r,l}) \, \mathbf{a}_t^H(\theta_{t,l}) \tag{8.64}$$

$\theta_{r,\,l}$ represents the angle of arrival (AoA), and $\theta_{t,\,l}$ represents the angle of departure (AoD) associated with the lth path. Conversely, the virtual channel representation exploits the signal space to determine the effect of the scattering environment. The virtual channel representation can be expressed as

$$\mathbf{H} = \sum_{n=1}^{\mathrm{N}} \sum_{m=1}^{\mathrm{M}} [\mathbf{H}_{\mathrm{V}}]_{n,m} \, \mathbf{a}_r(\theta_{r,n}) \;\; \mathbf{a}_t^H(\theta_{t,m}) \tag{8.65a}$$

$$\mathbf{H} = \mathbf{A}_r(\boldsymbol{\theta}_r) \, \mathbf{H}_{\mathrm{V}} \, \mathbf{A}_t^H(\boldsymbol{\theta}_t) \tag{8.65b}$$

where $\mathbf{A}_r(\boldsymbol{\theta}_r)$ and $\mathbf{A}_t(\boldsymbol{\theta}_t)$ are given as.

$$\mathbf{A}_r(\boldsymbol{\theta}_r) \in \mathcal{C}^{\mathrm{N \times N}} = [\mathbf{a}_r(\theta_{r,1}).,\, \ldots \ldots \ldots,\, \mathbf{a}_r(\theta_{r,\mathrm{N}})] \tag{8.66a}$$

$$\mathbf{A}_t(\boldsymbol{\theta}_t) \in \mathcal{C}^{\mathrm{M \times M}} = [\mathbf{a}_t(\theta_{t,1}).,\, \ldots \ldots \ldots,\, \mathbf{a}_t(\theta_{t,\mathrm{M}})] \tag{8.66b}$$

where $\boldsymbol{a}_r(\theta_{r,\,n})$ and $\boldsymbol{\theta}_t(\theta_{t,\,m})$ are given as

$$\boldsymbol{a}_r(\theta_{r,n}) = \frac{1}{\sqrt{\mathrm{N}}}[1, e^{-j2\pi\theta_{r,n}}, \ldots \ldots \ldots \ldots, e^{-j2\pi(\mathrm{N}-1)\theta_{r,n}}\,]^T \tag{8.67a}$$

$$\mathbf{a}_t(\theta_{t,m}) = \frac{1}{\sqrt{\mathrm{M}}}\,[1, e^{-j2\pi\theta_{t},m}, \ldots \ldots \ldots \ldots, e^{-j2\pi(\mathrm{M}-1)\theta_{t,m}}\,]^T \tag{8.67b}$$

In addition, $\mathbf{A}_r$ and $\mathbf{A}_t$ are unitary matrices, i.e.

$$\mathbf{A}_r\,\mathbf{A}_r^H = \mathbf{I}_{\mathrm{N}} \in \boldsymbol{\mathcal{R}}^{\mathrm{N \times N}} \text{ and } \mathbf{A}_t\,\mathbf{A}_t^H = \mathbf{I}_{\mathrm{M}} \in \boldsymbol{\mathcal{R}}^{\mathrm{M \times M}}. \tag{8.68}$$

Applying vec(.) operator to both sides of (8.65) to find vec($\mathbf{H}$) as

$$\mathrm{vec}(\mathbf{H}) = \mathrm{vec}(\mathbf{A}_r(\boldsymbol{\theta}_r)\,\mathbf{H}_{\mathrm{V}}\,\mathbf{A}_t^H(\boldsymbol{\theta}_t)) \tag{8.69}$$

We use the identity $[\mathbf{B}^T \otimes \mathbf{A}]\mathrm{vec}(\mathbf{D}) = \mathrm{vec}(\mathbf{ADB})$ to simplify (8.69) as follows: Substitute $\mathbf{A}_r(\boldsymbol{\theta}_r)$ as $\mathbf{A}$, $\mathbf{H}_V$ as $\mathbf{D}$, and $\mathbf{A}_t^H(\boldsymbol{\theta}_t) = \mathbf{B}$ in (8.69) to get,

$$\mathrm{vec}(\mathbf{A}_r^H\,\mathbf{H}_{\mathrm{V}}\,\mathbf{A}_t) = [\mathbf{A}_t^*(\boldsymbol{\theta}_t) \otimes \mathbf{A}_r(\boldsymbol{\theta}_r)]\,\mathrm{vec}(\mathbf{H}_V)$$

Therefore,

$$\mathrm{vec}(\mathbf{H}) = [\mathbf{A}_t^*(\boldsymbol{\theta}_t) \otimes \mathbf{A}_r(\boldsymbol{\theta}_r)]\,\mathrm{vec}(\mathbf{H}_V) \tag{8.70a}$$

Equation (8.73a) can also be written as a summation over (N, M),

$$\text{vec}(\mathbf{H}) = \sum_{n}^{N} \sum_{m}^{M} [\mathbf{H}_V]_{n,m} (\mathbf{a}_t^*(\theta_{t,m}) \otimes \mathbf{a}_r(\theta_{r,n})) \tag{8.70b}$$

The matrix $\mathbf{H}_V$ presents a spontaneous representation for an actual propagation environment in terms of different clusters corresponding to the nonvanishing elements of $\mathbf{H}_V$, and these nonvanishing elements are usually assumed to be uncorrelated.

Let the correlation matrix of vec(**H**) be denoted as **R** where:

$$\begin{aligned} \mathbf{R} &= \mathbb{E}(\text{vec}(\mathbf{H})\text{vec}(\mathbf{H})^H) \\ &= \mathbb{E}([\mathbf{A}_t^* \otimes \mathbf{A}_r]\text{vec}(\mathbf{H}_V)\text{vec}(\mathbf{H}_V)^H [\mathbf{A}_t^T \otimes \mathbf{A}_r^H]) \end{aligned}$$

$$\mathbf{R} = ([\mathbf{A}_t^* \otimes \mathbf{A}_r]\mathbf{R}_v[\mathbf{A}_t^T \otimes \mathbf{A}_r^H]) \tag{8.71a}$$

where $\mathbf{R}_v = \mathbb{E}(\text{vec}(\mathbf{H}_V)\text{vec}(\mathbf{H}_V)^H)$.

We can express (8.71) as a summation in M, N as

$$\mathbf{R} \approx \sum_{n=1}^{N} \sum_{m=1}^{M} \mathbb{E}|[\mathbf{H}_V]_{n,m}|^2 \, [\mathbf{a}_t^*(\theta_{t,m}) \otimes \mathbf{a}_r(\theta_{r,n})] \cdot [\mathbf{a}_t^T(\theta_{t,m}) \otimes \mathbf{a}_r^H(\theta_{r,n})] \tag{8.71b}$$

Owing to the sparse structure of $\mathbf{H}_V$, its elements are generally uncorrelated, which forces $\mathbf{R}_v$ to be almost diagonal. In most scenarios, we can safely assume $\mathbf{R}_v$ to be exactly diagonal. Additionally, $\mathbf{R}_v$ diagonal matrix may contain some zero elements corresponding to the vanishing elements in $\mathbf{H}_V$ as a result of the clustered scattering. It is worth bearing in mind that **R** and $\mathbf{R}_v$ are unitarily equivalent since the Kronecker product of two unitary matrices is also unitary.

The average energy of the virtual SISO link between each eigenmode of the receive side and the eigenmode of the transmit side are the element of the diagonal coupling matrix $\boldsymbol{\Omega}$ linking the correlation properties of both ends. Since $\mathbf{A}_r$ and $\mathbf{A}_t$ are unitary matrices, using (8.68b) we can expresses $\mathbf{H}_V$ as

$$\mathbf{H}_V = \mathbf{A}_r^H(\theta_r)\mathbf{H}\mathbf{A}_t(\theta_t) \tag{8.72}$$

The diagonal elements of matrix $\boldsymbol{\Omega}$ are given by $|[\mathbf{H}_V]_{n,m}|^2$, i.e.

$$[\boldsymbol{\Omega}]_{n,m} = \mathbb{E}|[\mathbf{H}_V]_{n,m}|^2 = \mathbb{E}|[\mathbf{A}_r^H \mathbf{H}\mathbf{A}_t]_{n,m}|^2 \tag{8.73}$$

Each element of $\boldsymbol{\Omega}$ is multiplied by the corresponding gain of **G**. Accordingly, the virtual channel representation can be articulated as

$$\mathbf{H}_{\text{virt}} = \mathbf{A}_r \, (\boldsymbol{\Omega}_{\text{virt}}^{\overset{0.5}{\leftrightarrow}} \odot \mathbf{G}) \, \boldsymbol{\Lambda}_t^H \tag{8.74}$$

where $\boldsymbol{\Omega}_{\text{virt}}^{\overset{0.5}{\leftrightarrow}}$ is defined as the element-wise square root of $\boldsymbol{\Omega}_{\text{virt}}$.

G is an independent and identically distributed (i.i.d.) random matrix with zero mean and unit variance elements.

8.8 Mutual Coupling in Wireless Antenna Systems

This section discusses several aspects related to mutual coupling in wireless antenna systems, as noted in [24, 25]

8.8.1 Array Mutual Coupling

The electromagnetic (EM) action by the array elements on each other is referred to as array mutual coupling. The action takes the form of energy absorption by one antenna element when a nearby element is operating. Energy absorption by mutual coupling reduces antenna efficiency and its performance. Mutual coupling problems in applications that employ antenna arrays attract interest not only from antenna engineers and researchers but also from researchers in various disciplines, such as wireless communications and biomedical imaging, including magnetic resonance imaging (MRI) systems that operate antenna arrays.

Unlike a single antenna system that delivers information to a single receiver, an MIMO system provides information to various receivers. Such an operation requires the channels between different transmit and receive antennas to be independent and their coefficients identically distributed (i.i.d.). The latter property can be achieved by spacing the antennas as far apart as possible. In practice, due to constraints on the physical dimensions of available spaces, the distances between the multiple antennas are usually small and antenna elements interact with each other (i.e. they are mutually coupled). Consequently, to improve the MIMO system efficiency, we need to find ways to decouple the transmit and/or the receive antenna arrays. Mutual coupling between two antennas is characterized by mutual impedances whose definition is taken from that used originally in circuit analysis, i.e. the Z parameters in network analysis [26, 27]. The operation of the transmit antennas is dissimilar to receive antennas; therefore, mutual coupling manifested differently in transmitting and receiving antenna arrays [28, 29].

8.8.2 Mutual Coupling of Antenna Arrays Operating in Transmit and Receive Modes

Let antenna n be excited by source n while antenna m is passive. The generated power travels along path (0) towards antenna n and radiates to the free space, shown as paths (1), and some of the power is received by antenna m along path (2), and hence a current is induced in antenna m and the current generates energy that radiates in free space along path (3), but some energy is dissipated in antenna m along path (4). In addition, it is feasible that some of the energy along path (3) is received by antenna n along path (5). This process will continue indefinitely. A similar process takes place in any element in array n when excited. However, when both antennas m and n are excited at the same time, the total field will be the sum of the radiated field and the re-scattered field from both antennas. The coupling paths of the antenna array with elements m and n in the transmit mode are shown in Figure 8.9.

Now let us consider the antenna array in the receive mode as depicted in Figure 8.10. The incident wave from free space (0) induces a current in antenna m and radiates some of the generated energy into free space (2), and the rest of the energy is dissipated by the antenna (1). If the antenna m is not matched to the load (Z), part of the energy (1) will be reflected back and radiated to free space.

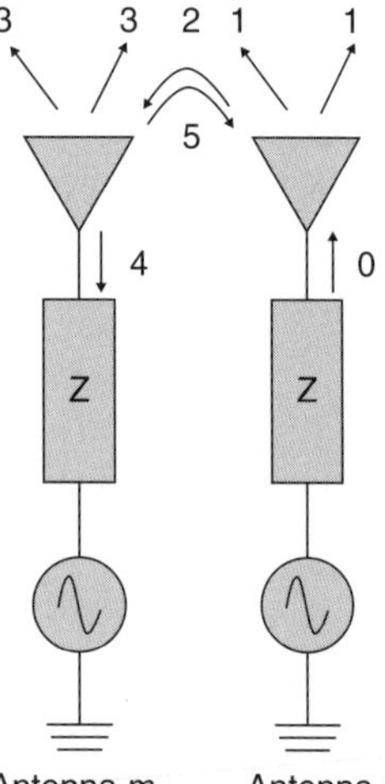

Figure 8.9 Coupling paths for the antennas m and n for an antenna array and n in transmitting mode [28, 29].

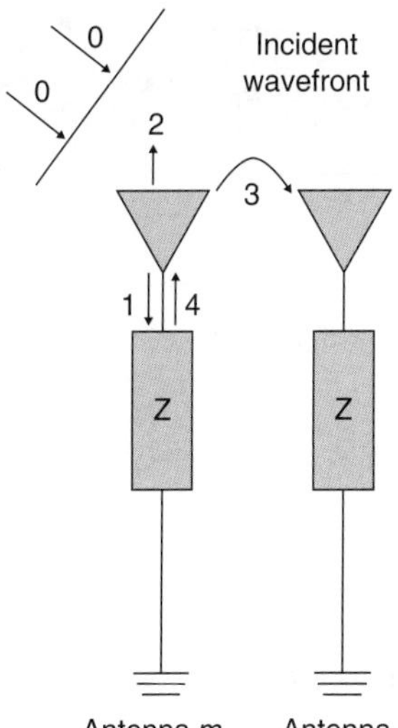

Figure 8.10 Coupling paths for the antennas *m* and *n* for an antenna array in receiving mode [28, 29].

Some of the free space energy (1) is picked up by the antenna *n* (3). The latter induces current in antenna *n* and the energy due to this current is radiated by antenna *n*. A similar process will take place if any other antenna in array *m* strikes by the incident wave. If both antennas *m* and *n* are in the receive mode simultaneously, then the total field will be the sum of the radiated field and the re-scattered field from both antennas, as in the case of the transmit mode.

8.8.3 BS Antennas Mutual Coupling in MIMO Systems

Antennas mutual coupling generally use impedance matrices to represent the antenna array. Nonetheless, we use scattering parameter (S-parameter) matrices instead, since it simplifies the analysis and measurements [30]. The S-parameter expresses an inward-propagating and an outward-propagating wave with complex envelopes denoted as complex vectors $\boldsymbol{a}$, $\boldsymbol{b}$, respectively, are related by an S-parameter matrix $\boldsymbol{S}$:

$$\boldsymbol{b} = \boldsymbol{S}\,\boldsymbol{a} \tag{8.75a}$$

where $\boldsymbol{S}$ is the S-parameters matrix for multi-port network is defined (see 8.A S-parameters) as

$$S = \begin{pmatrix} S_{SS} & S_{SR} \\ S_{RS} & S_{RR} \end{pmatrix} \tag{8.75b}$$

Consider a receive network consisting of N antenna array that converts the EM received wave into a source vector $\boldsymbol{b}_0$ so if the source is terminated with its characteristic impedance, the total power accumulated by the load is

$$\|\boldsymbol{b}_0\|^2 = \boldsymbol{b}_0^H\,\boldsymbol{b}_0 \tag{8.76}$$

The receive array that generates $\boldsymbol{b}_0$ wave is represented by $\boldsymbol{S}_{RR}$ matrix such that

$$\boldsymbol{b}_R = \boldsymbol{b}_0 + \boldsymbol{S}_{RR}\,\boldsymbol{a}_R \tag{8.77}$$

The array matrix is followed at reference plane 1 by a matching network with S-parameter matrix $\boldsymbol{S}_M$ to maximize the power transferred from the array to the load of resistance $Z_0 = 50\ \Omega$ at reference plane 2, as shown in Figure 8.11.

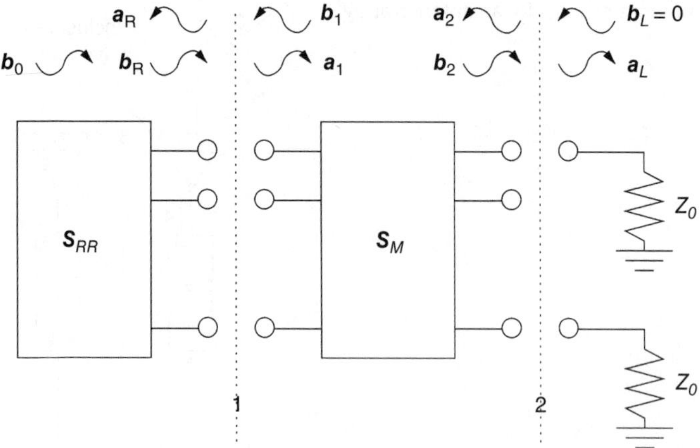

Figure 8.11 Network model for the receive subsystem [31].

8.8.4 Total Power Collected by the Receiving Array

The matching circuit with S-parameters $\boldsymbol{S}_M$ is ideally made of passive reactive components so it is lossless to avoid unnecessary loss of power and also is reciprocal. If it is lossless, then we have,

$$\boldsymbol{S}_M^H \boldsymbol{S}_M = \mathbf{I} \text{ (i.e.identity matrix)}$$

and when it is reciprocal,

$$\boldsymbol{S}_M = \boldsymbol{S}_M^T$$

The matching network is required to maximum power to resistance Z_0. We partition the matching network matrix as

$$\boldsymbol{S}_M = \begin{bmatrix} \boldsymbol{S}_{11} & \boldsymbol{S}_{12} \\ \boldsymbol{S}_{21} & \boldsymbol{S}_{22} \end{bmatrix} \tag{8.78}$$

where the subscripts 1 and 2 refer to input and output ports respectively. The matrix is partitioned into input $\boldsymbol{S}_{11}$, output $\boldsymbol{S}_{22}$ blocks and input-output $\boldsymbol{S}_{12}$ and output-input $\boldsymbol{S}_{21}$ blocks as well as in [31, 32]. The outward-propagating and the inward-propagating waves of $\boldsymbol{S}_M$ are related as

$$\begin{pmatrix} \boldsymbol{b}_1 \\ \boldsymbol{b}_2 \end{pmatrix} = \begin{bmatrix} \boldsymbol{S}_{11} & \boldsymbol{S}_{12} \\ \boldsymbol{S}_{21} & \boldsymbol{S}_{22} \end{bmatrix} \begin{pmatrix} \boldsymbol{a}_1 \\ \boldsymbol{a}_2 \end{pmatrix} \tag{8.79}$$

Since $\boldsymbol{b}_L = \mathbf{0}$ so $\boldsymbol{a}_2 = \mathbf{0}$, and

$$\boldsymbol{b}_1 = \boldsymbol{S}_{11}\, \boldsymbol{a}_1 \text{ and } \boldsymbol{b}_2 = \boldsymbol{S}_{21} \boldsymbol{a}_1. \tag{8.80}$$

So, we can replace the entirety to the right of reference plane 1 in Figure. 8.11 (i.e. the matching network) as $\boldsymbol{S}_{11}$. The source block can now be modelled as multi-input ports equal to the output ports, as shown in Figure 8.12 where the S-parameter matrix $\boldsymbol{S}$ is given by (8.75b).

The outward-propagating wave and the inward-propagating wave vectors in model in Figure 8.12 are related as

$$\begin{pmatrix} \boldsymbol{b}_S \\ \boldsymbol{b}_R \end{pmatrix} = \begin{bmatrix} \boldsymbol{S}_{SS} & \boldsymbol{S}_{SR} \\ \boldsymbol{S}_{RS} & \boldsymbol{S}_{RR} \end{bmatrix} \begin{pmatrix} \boldsymbol{a}_s \\ \boldsymbol{a}_R \end{pmatrix} \tag{8.81}$$

$$\boldsymbol{b}_R = \boldsymbol{S}_{RS}\, \boldsymbol{a}_s + \boldsymbol{S}_{RR} \boldsymbol{a}_R \tag{8.82}$$

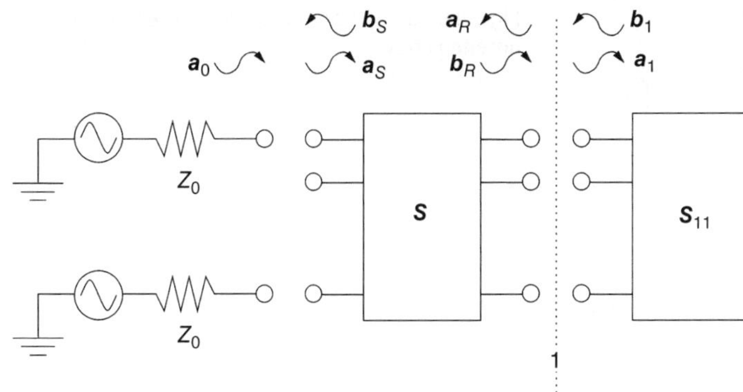

Figure 8.12 Network model for the equivalent receive impedance matching block [31].

From (8.77), we have

$$b_R = b_0 + S_{RR}\, a_R$$

To make the model in Figure 8.12 equivalent to the model in Figure 8.11, we set (8.77) equal (8.82):

$b_0 = S_{RS}\, a_0$ and $a_s = a_0$, accordingly we have

$$b_R = a_1 = b_0 + S_{RR}\, a_R \tag{8.83a}$$

and

$$a_R = b_1 \tag{8.83b}$$

Using (8.80) to substitute for $b_1 = S_{11}\, a_1$, therefore,

$$a_1 = b_0 + S_{RR}\, S_{11}\, a_1 \tag{8.84}$$

Multiply both sides of (8.84) by a_1^{-1}:

$$a_1 = (\mathbf{I} - S_{RR}\, S_{11})^{-1} b_0 \tag{8.85}$$

From (8.80), $b_2 = S_{21} a_1$ and after substituting for a_1, we get

$$b_2 = S_{21}(\mathbf{I} - S_{RR}\, S_{11})^{-1} b_0 \tag{8.86}$$

The total power collected by receive array P(S) is

$$\mathrm{P}(S) = b_2^H\, b_2 = [S_{21}(\mathbf{I} - S_{RR}\, S_{11})^{-1} b_0]^H [S_{21}(\mathbf{I} - S_{RR}\, S_{11})^{-1} b_0]$$

$$\mathrm{P}(S) = b_0^H\, [(\mathbf{I} - S_{RR}\, S_{11})^H]^{-1} S_{21}^H\, S_{21}(\mathbf{I} - S_{RR}\, S_{11})^{-1} b_0 \tag{8.87a}$$

Denote $W(S_{11})$ as

$$W(S_{11}) = [(\mathbf{I} - S_{RR}\, S_{11})^H]^{-1} S_{21}^H\, S_{21}(\mathbf{I} - S_{RR}\, S_{11})^{-1} \tag{8.87b}$$

Therefore, we can rewrite (8.87a) as

$$\mathrm{P}(S) = b_0^H\, W(S_{11})\, b_0 \tag{8.87c}$$

For an arbitrary b_0, we will show in 8.B the power collected by the receive array is maximum when

$$S_{11} = S_{RR}^H \tag{8.87d}$$

A passive and lossless network as a prerequisite ensures that

$$S_{11}^H\, S_{11} + S_{21}^H\, S_{21} = \mathbf{I} \tag{8.88a}$$

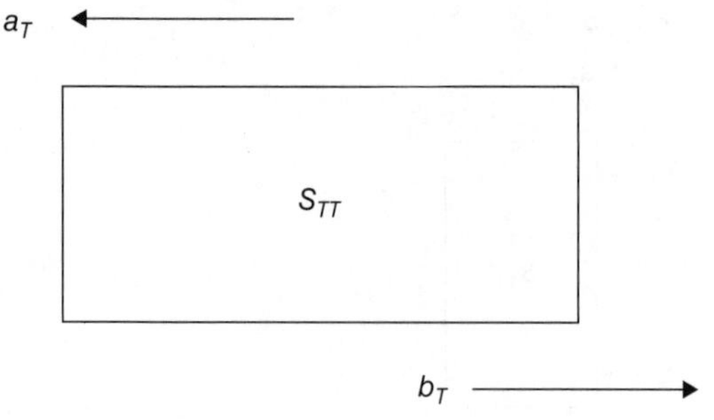

Figure 8.13 S blocks representing the transmit antenna array.

So,

$$S_{11}^H S_{11} = \mathbf{I} - S_{21}^H S_{21} \tag{8.88b}$$

Substituting (8.87d) and (8.88b) in (8.87b) we get

$$W(S_{11}) = [(\mathbf{I} - S_{RR} S_{11})^H]^{-1} S_{21}^H S_{21} (\mathbf{I} - \mathbf{I} + S_{21}^H S_{21})^{-1}$$
$$W(S_{11}) = [(\mathbf{I} - S_{RR} S_{11})^H]^{-1} S_{21}^H S_{21} (S_{21}^H S_{21})^{-1}$$
$$W(S_{11}) = [\mathbf{I} - S_{RR} S_{RR}^H]^{-1} \tag{8.89}$$

Therefore, the maximum power collected by the receive array is

$$\mathrm{P}(S) = b_0^H (\mathbf{I} - S_{RR} S_{RR}^H)^{-1} b_0 \tag{8.90}$$

8.9 Mutual Coupling Constrained on Transmit Radiated Power

Conventionally, the antennas mutual coupling is not taken into account when considering the radiated power. Consider a transmit antenna array with M elements and S-parameters block S_{TT} shown in Figure 8.13.

From Figure 8.13, we have

$$b_T = S_{TT} a_T \tag{8.91}$$

The instantaneous transmit power $\mathrm{P}_\mathrm{T}^\mathrm{Inst}$ entering the network for lossless array is

$$\|a_T\|^2 - \|b_T\|^2 = \|a_T\|^2 - \|S_{TT} a_T\|^2$$
$$\|a_T\|^2 - \|b_T\|^2 = a_T^H a_T - (a_T^H S_{TT}^H)(S_{TT} a_T)$$
$$= a_T^H (\mathbf{I} - S_{TT}^H S_{TT}) a_T$$
$$= \mathbf{x}^\mathrm{H} A \mathbf{x} \tag{8.92}$$

where $\mathbf{x}$ and A are called the *transmit signal vector* and *coherence matrix, respectively*. Assuming zero mean signals, then the average radiated power P_T is

$$\mathrm{P}_\mathrm{T} = \mathrm{tr}(\mathbf{K}_\mathbf{x} A) \tag{8.93}$$

where $\mathbf{K}_\mathbf{x} = \mathbb{E}\{\mathbf{x}\,\mathbf{x}^\mathrm{H}\}$.

8.10 Analysis Voltage Induced at the Receive Antenna Port

A MIMOM Communication system model in S-parameter block representations is shown in Figure 8.14. A MIMO system comprised M, N transmit antenna and receive antenna arrays, respectively. The wave vectors a_T and b_T at transmit array and the wave vectors a_R and b_R waves

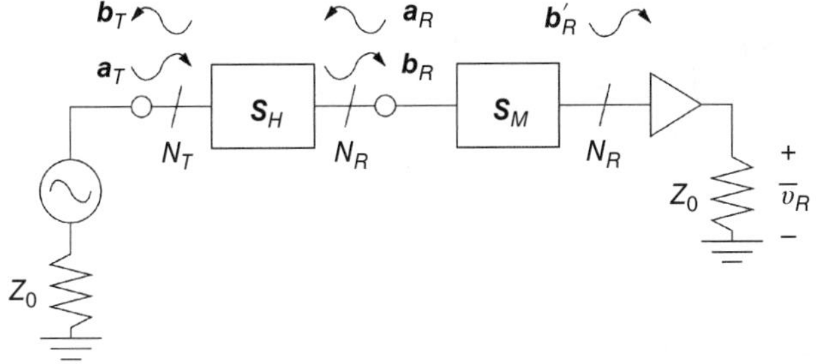

Figure 8.14 S blocks representation for the MIMO system [31].

at the receive antenna array are defined in Figures 8.13 and 8.12, respectively. The combined transmit array and channel $\boldsymbol{S}$-parameter matrix is denoted as $\boldsymbol{S}_H$. A unit gain network match is used to match the output impedance to reference impedance Z_0 and AWGN transmission is assumed. Each receive antenna port is connected to Z_0, the voltage across which is $\boldsymbol{v}_R$. At the output of the matched network, only the outward waves exist. The aim of the analysis is to find an S-parameter expression for the *induced voltage* across the load connected to the receive antenna port.

The antenna arrays are operating in a linear scattering medium so the combined transmit antenna/channel matrix $\boldsymbol{S}_H$ is given as

$$\begin{bmatrix} \boldsymbol{b}_T \\ \boldsymbol{b}_R \end{bmatrix} = \underbrace{\begin{bmatrix} \boldsymbol{S}_{TT} & \boldsymbol{S}_{TR} \\ \boldsymbol{S}_{RT} & \boldsymbol{S}_{RR} \end{bmatrix}}_{\boldsymbol{S}_H} \begin{bmatrix} \boldsymbol{a}_T \\ \boldsymbol{a}_R \end{bmatrix} \tag{8.94}$$

In the $\boldsymbol{S}_H$ matrix, $\boldsymbol{S}_{TR}$ represents the power reflected from the receive array and picked-up by the transmit array. Assuming transmit and receive arrays are a large distant apart, then $\boldsymbol{S}_{TR} \cong \boldsymbol{0}$.

Accordingly, the voltage induced at the receive antenna port, $\boldsymbol{v}_R$ and the corresponding current, $\boldsymbol{i}_R$ are defined by

$$\boldsymbol{v}_R := \sqrt{Z_0}\,(\boldsymbol{a} - \boldsymbol{b}) \tag{8.95}$$

$$\boldsymbol{i}_R = \frac{\boldsymbol{v}_R}{Z_0} = \frac{(\boldsymbol{a} - \boldsymbol{b})}{\sqrt{Z_0}} \tag{8.96}$$

where $\boldsymbol{a},\boldsymbol{b}$ are inward-propagating and outward-propagating waves, respectively (relative to the load). In our system $\boldsymbol{a} = \boldsymbol{b}'_R$ and the outward-propagating wave is $\boldsymbol{b} = \boldsymbol{0}$. Assume the matching network to be lossless and draw the matching network with inward-propagating and outward-propagating waves from Figures 8.11 and 8.14, as displayed below:

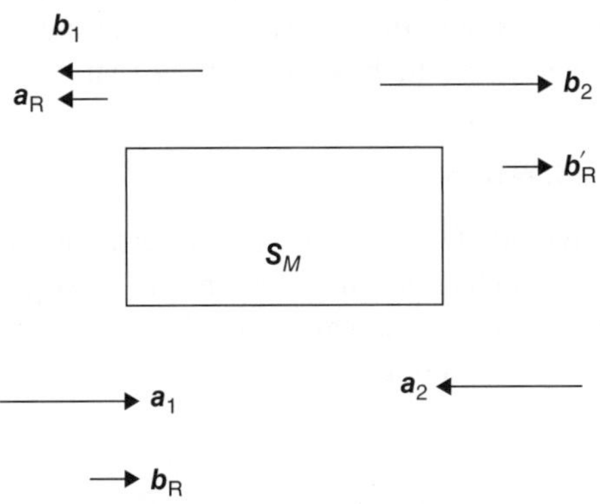

$$\begin{pmatrix} b_1 \sim & a_R \\ b_2 \sim & b'_R \end{pmatrix} = \begin{bmatrix} S_{11} & S_{12} \\ S_{21} & S_{22} \end{bmatrix} \begin{pmatrix} a_1 \sim b_R \\ a_2 = \mathbf{0} \end{pmatrix}$$

$$b'_R = S_{21} b_R$$

and

$$a_R = S_{11} b_R$$

Considering the S_H matrix as expressed in (8.94), we have,

$$b_R = S_{RT} a_T + S_{RR}\, a_R \tag{8.97}$$

Substituting for a_R,

$$b_R = S_{RT} a_T + S_{RR}\, S_{11} b_R$$

Multiply both sides by b_R^{-1} we get

$$\mathbf{I} = S_{RT} a_T b_R^{-1} + S_{RR}\, S_{11}$$

$$\mathbf{I} - S_{RR}\, S_{11} = S_{RT} a_T b_R^{-1}$$

Multiply both sides by $(\mathbf{I} - S_{RR}\, S_{11})^{-1}$, and the by b_R:

$$b_R = (\mathbf{I} - S_{RR}\, S_{11})^{-1} S_{RT} a_T \tag{8.98a}$$

And b'_R is given as

$$b'_R = a = S_{21} b_R = S_{21} (\mathbf{I} - S_{RR}\, S_{11})^{-1} S_{RT} a_T \tag{8.98b}$$

Therefore, in the noiseless case, the voltage created at the receive port is related to the received signal according to the following:

$$\underset{Y}{\underleftrightarrow{v_R}} = \underset{\mathbf{H}(S_M)}{\underleftrightarrow{\sqrt{Z_0}\, S_{21} (\mathbf{I} - S_{RR}\, S_{11})^{-1}}} \underset{X}{\underleftrightarrow{S_{RT} a_T}} \tag{8.99}$$

where X and Y are inward-propagating wave and voltage across Z_0 vector, respectively, and the effective $\mathrm{N} \times \mathrm{M}$ channel matrix $\mathbf{H}(S_M)$ that depends on the matching network S_M.

8.11 MIMO Channel Capacity of Mutually Coupled Wireless Systems

Consider the MIMO system depicted in Figure 8.14, for analysing the MIMO channel capacity for mutually coupled array. In this case, there are two practical sources of noise: receiver thermal noise and noise due to interference, which is much larger than the receiver thermal noise.

8.11.1 Interference Consideration

The channel experiences two kind of noise: noise from interference and thermal noise. Since the interference is expected to be much larger than the thermal noise generated by the user receiver front end, we assume that thermal noise can be neglected. Define the forward-propagating noise wave on the ith receive antenna port as $b_{RN,\,i}$ given by

$$b_{RN,i} := \sqrt{Z_0}\, V_n \tag{8.100}$$

where V_n is the effective noise voltage. After passing through the matching network and for maximum power transfer using (8.88b) and (8.90), the channel noise forward wave becomes

$$\boldsymbol{b}_{RN} = \sqrt{Z_0}\,(\mathbf{I} - \boldsymbol{S}_{RR}\,\boldsymbol{S}_{11})^{-1}\boldsymbol{V} \tag{8.101}$$

where $\boldsymbol{V}$ is noise vector for all ports of the network. Superimposing the signal vector and channel noise vector, we get

$$\boldsymbol{v}_R = \sqrt{Z_0}\,\boldsymbol{S}_{21}(\mathbf{I} - \boldsymbol{S}_{RR}\,\boldsymbol{S}_{11})^{-1}\boldsymbol{S}_{RT}\boldsymbol{a}_T + \sqrt{Z_0}(\mathbf{I} - \boldsymbol{S}_{RR}\,\boldsymbol{S}_{11})^{-1}\boldsymbol{V} \tag{8.102a}$$

Equation (8.102a) can be simplified to

$$\underbrace{\boldsymbol{v}_R}_{\mathbf{y}} = \underbrace{\boldsymbol{S}_{21}(\mathbf{I} - \boldsymbol{S}_{RR}\,\boldsymbol{S}_{11})^{-1}}_{\boldsymbol{P}}\left(\underbrace{\sqrt{Z_0}\boldsymbol{S}_{RT}}_{\mathbf{H}}\underbrace{\boldsymbol{a}_T}_{\mathbf{x}} + \boldsymbol{V}\right) \tag{8.102b}$$

Assume complex Gaussian signal signalling and Gaussian distributed interference (noise). The MIMO channel capacity is given in terms of output signal $\mathbf{y}$ and the input signal $\mathbf{x}$ vectors. The MIMO channel capacity C of the mutually coupled array is defined and given by the maximum mutual information $I(\mathbf{y};\mathbf{x})$ between output signal $\mathbf{y}$ and input signal $\mathbf{x}$ vectors.

In addition, we are dealing with a continuous random variable $\mathbf{x}$, where the differential entropy $\mathrm{h}(\mathbf{X})$ is defined [33, 34] as

$$\mathrm{h}(\mathbf{x}) = \int_{-\infty}^{\infty} f_{\mathbf{x}}(\mathbf{x})\log_2\left[\frac{1}{f_{\mathbf{x}}(\mathbf{x})}\right]\mathrm{d}\mathbf{x} \tag{8.103}$$

Assume a complex Gaussian signal vector, $\boldsymbol{a}_T$, and a complex channel noise voltage $\boldsymbol{V}$. The mutual information $I(\mathbf{y};\mathbf{X})$ is

$$I(\mathbf{y};\mathbf{x}) = \mathrm{h}(\mathbf{y}) - \mathrm{h}(\mathbf{y}\mid\mathbf{x}) \tag{8.104}$$

where h(.) is differential entropy. Substituting (8.103) for the differential entropy terms in (8.104), we get

$$I(\mathbf{y};\mathbf{x}) = \mathrm{h}(\boldsymbol{P}\mathbf{H}\mathbf{x} + \boldsymbol{P}\boldsymbol{V}) - \mathrm{h}(\boldsymbol{P}\boldsymbol{V}) \tag{8.105}$$

$$I(\mathbf{y};\mathbf{x}) = \log_2\frac{|\boldsymbol{P}\,\boldsymbol{S}_{RT}\,\boldsymbol{K}_{\mathrm{x}}\,\boldsymbol{S}_{RT}^H\,\boldsymbol{P}^H + \boldsymbol{P}\,\boldsymbol{K}_V\,\boldsymbol{P}^H|}{|\boldsymbol{P}\,\boldsymbol{K}_V\,\boldsymbol{P}^H|} \tag{8.106}$$

where $\boldsymbol{K}_{\mathrm{x}} = \mathbb{E}(\mathbf{x}\,\mathbf{x}^H)$, and $\boldsymbol{K}_V = \mathbb{E}(\boldsymbol{V}\,\boldsymbol{V}^H)$. Assume $\boldsymbol{P}$ is nonsingular, and since both signal and noise pass through the same matching network, we can remove the matching element in this case without changing the mutual information, that is $\boldsymbol{S}_{11} = \mathbf{0}$ and $\boldsymbol{S}_{21} = \mathbf{I}$. Therefore $(\boldsymbol{P})$ simplifies to $\boldsymbol{P} = \mathbf{I}$, and if we assume $\boldsymbol{V}$ to be Gaussian noise with co-variances $\boldsymbol{K}_V = \sigma_V^2\,\mathbf{I}$, then Eq. (8.106) becomes

$$I(\mathbf{y};\mathbf{x}) = \log_2\frac{|\,\boldsymbol{S}_{RT}\,\boldsymbol{K}_{\mathrm{x}}\,\boldsymbol{S}_{RT}^H + \boldsymbol{K}_V\,|}{|\,\boldsymbol{K}_V\,|} \tag{8.107}$$

Substituting for $\boldsymbol{K}_V$ in (8.107), we get

$$I(\mathbf{y};\mathbf{x}) = \log_2\left|\frac{\boldsymbol{S}_{RT}\,\boldsymbol{K}_{\mathrm{x}}\,\boldsymbol{S}_{RT}^H}{\sigma_V^2} + \mathbf{I}\right| \tag{8.108}$$

8.11.2 Users Receiver Noise Consideration

In the absence of multiuser interference, the main source of noise is thermal, created by the receiver frontend (receive antennas and low signal amplifier), and we proceed to develop the expression for the mutual information under AWGN channel. We note that the AWGN does not pass through the matching network in the case we are considering now. We can use the expression developed in (8.102b) relating the output signal vector (y) and the input signal vector ($\mathbf{x}$),

$$\underbrace{\bar{\boldsymbol{v}}_R}_{\mathbf{y}} = \underbrace{\sqrt{Z_0}\boldsymbol{S}_{21}(\mathbf{I}-\boldsymbol{S}_{RR}\,\boldsymbol{S}_{11})^{-1}\boldsymbol{S}_{RT}}_{\mathbf{H}(\boldsymbol{S}_M)}\,\underbrace{\boldsymbol{a}_T}_{\mathbf{x}} + \boldsymbol{V}_{awgn} \tag{8.109}$$

8.11.3 Formulation of MIMO Channel Capacity

The mutual information is formulated in the same process as in (8.107),

$$I(\mathbf{y};\mathbf{x}) = \log_2\left|\frac{\mathbf{H}(\boldsymbol{S}_M)\,\boldsymbol{K}_\mathrm{x}\,\mathbf{H}(\boldsymbol{S}_M)^H}{\sigma^2_{awgn}}+\mathbf{I}\right| \tag{8.110}$$

We can simplify (8.110) as follows:

$$\mathbf{H}(\boldsymbol{S}_M)\,\boldsymbol{K}_\mathrm{x}\,\mathbf{H}(\boldsymbol{S}_M)^H = Z_0\boldsymbol{S}_{21}(\mathbf{I}-\boldsymbol{S}_{RR}\,\boldsymbol{S}_{11})^{-1}\boldsymbol{S}_{RT}\boldsymbol{K}_\mathrm{x}\,\boldsymbol{S}^H_{RT}((\mathbf{I}-\boldsymbol{S}_{RR}\,\boldsymbol{S}_{11})^{-1})^H\,\boldsymbol{S}^H_{21} \tag{8.111}$$

and if the matching network is to be lossless, then, $\boldsymbol{S}^H_{11}\,\boldsymbol{S}_{11}+\boldsymbol{S}^H_{21}\,\boldsymbol{S}_{21}=\mathbf{I}$ .Therefore,

$$\boldsymbol{S}^H_{21}\,\boldsymbol{S}_{21} = \mathbf{I}-\boldsymbol{S}^H_{11}\,\boldsymbol{S}_{11} \tag{8.112a}$$

Define $\boldsymbol{M} := \boldsymbol{S}_{RT}\boldsymbol{K}_\mathrm{x}\,\boldsymbol{S}^H_{RT}$ and use (8.112a), then (8.111) becomes

$$\mathbf{H}(\boldsymbol{S}_M)\,\boldsymbol{K}_\mathrm{x}\,\mathbf{H}(\boldsymbol{S}_M)^H = Z_0\boldsymbol{S}_{21}(\mathbf{I}-\boldsymbol{S}_{RR}\,\boldsymbol{S}_{11})^{-1}\boldsymbol{M}((\mathbf{I}-\boldsymbol{S}_{RR}\,\boldsymbol{S}_{11})^{-1})^H\,\boldsymbol{S}^H_{21}$$

For maximum power transfer:

$$\boldsymbol{S}^H_{RR} = \boldsymbol{S}_{11} \tag{8.112b}$$

Substituting (8.112a) and (8.112b) in (8.111) we have

$$\mathbf{H}(\boldsymbol{S}_M)\,\boldsymbol{K}_\mathrm{x}\,\mathbf{H}(\boldsymbol{S}_M)^H = (\mathbf{I}-\boldsymbol{S}_{RR}\,\boldsymbol{S}_{11})^{-1}\,\boldsymbol{M} \tag{8.113}$$

Define

$$\boldsymbol{W}(\boldsymbol{S}_{11}) := (\mathbf{I}-\boldsymbol{S}_{RR}\,\boldsymbol{S}_{11})^{-1} \tag{8.114}$$

Hence, Eq. (8.113) becomes

$$\mathbf{H}(\boldsymbol{S}_M)\,\boldsymbol{K}_\mathrm{x}\,\mathbf{H}(\boldsymbol{S}_M)^H = \boldsymbol{W}(\boldsymbol{S}_{11})\,\boldsymbol{M} \tag{8.115}$$

The mutual information is formulated by substituting (8.115) into (8.110):

$$\boldsymbol{I}(\mathbf{y};\mathbf{x}) = \log_2\left|\frac{\boldsymbol{W}(\boldsymbol{S}_{11})\,\boldsymbol{M}}{\sigma^2_{awgn}}+\mathbf{I}\right| \tag{8.116}$$

It is clear that $\boldsymbol{M}$ is a Hermitian matrix (i.e. $\boldsymbol{M}=\boldsymbol{M}^H$) and positive semi-definitive. In addition, $\boldsymbol{M}$ is a square normal matrix, so we can apply the eigenvalue decomposition (EVD) of $\boldsymbol{M}$ to get

$$\boldsymbol{M} = \xi_M\,\boldsymbol{\Lambda}_M\,\xi^H_M = \left(\xi_M\boldsymbol{\Lambda}^{\frac{1}{2}}_M\right)\left(\xi_M\boldsymbol{\Lambda}^{\frac{1}{2}}_M\right)^H \tag{8.117}$$

Denote $\boldsymbol{M}^{\frac{1}{2}} = \xi_M\boldsymbol{\Lambda}^{\frac{1}{2}}_M = \boldsymbol{M}'$so Eq. (8.116) can be written as

$$\boldsymbol{I}(\mathbf{y};\mathbf{x}) = \log_2\left|\frac{(\boldsymbol{M}')^H\,\boldsymbol{W}(\boldsymbol{S}_{11})\,\boldsymbol{M}'}{\sigma^2_{awgn}}+\mathbf{I}\right| \tag{8.118}$$

Maximizing the mutual information requires maximizing the RHS of (8.118) over all $\boldsymbol{S}_{11}$ and $\boldsymbol{K}_\mathrm{x}$ using conjugate matching (i.e. $\boldsymbol{S}_{11}=\boldsymbol{S}^H_{RR}$) for fixed but arbitrary $\boldsymbol{K}_\mathrm{x}$.

Substitute for $\boldsymbol{S}_{11}$ in $\boldsymbol{W}(\boldsymbol{S}_{11})$ (Eq. (8.114)):

$$\boldsymbol{W}(\boldsymbol{S}_{RR}^{H}) = (\mathbf{I} - \boldsymbol{S}_{RR}\,\boldsymbol{S}_{RR}^{H})^{-1} \tag{8.119}$$

which gives the maximum mutual information as

$$\max_{\boldsymbol{S}_{RR}^{H},\boldsymbol{K}_X} \boldsymbol{I}(\mathbf{y};\mathbf{x}) = \log_2 \left| \frac{Z_0\,(\mathbf{I} - \boldsymbol{S}_{RR}\,\boldsymbol{S}_{RR}^{H})^{-1}\,\boldsymbol{S}_{RT}\boldsymbol{K}_{\mathrm{x}}\,\boldsymbol{S}_{RT}^{H}}{\sigma_{awgn}^{2}} + \mathbf{I} \right| \tag{8.120}$$

Since the matrix $(\mathbf{I} - \boldsymbol{S}_{RR}\,\boldsymbol{S}_{RR}^{H})$ is Hermitian and positive definitive, we can use EVD:

$$Z_0\,(\mathbf{I} - \boldsymbol{S}_{RR}\,\boldsymbol{S}_{RR}^{H})^{-1} = \xi'\,\boldsymbol{\Lambda}\,\xi'^{H}$$

Define $\boldsymbol{Q} := \boldsymbol{\Lambda}^{\frac{1}{2}}\,\xi'^{H}\,\boldsymbol{S}_{RT}$. We can write (8.75) as

$$\max_{\boldsymbol{S}_{RR}^{H},\boldsymbol{K}_X} \boldsymbol{I}(\mathbf{y};\mathbf{x}) = \log_2 \left| \frac{\boldsymbol{Q}\,\boldsymbol{K}_{\mathrm{x}}\,\boldsymbol{Q}^{H}}{\sigma_{awgn}^{2}} + \mathbf{I} \right| \tag{8.121}$$

Correlation can be used to assess the MIMO system capacity since high correlation between channel matrix elements reduces the channel capacity with low correlation offers high capacity. We explore the correlation of two-element array with matching networks. The matching networks construct *open circuit, self-impedance matched load* z_{11}^{*}, and *multiport Conjugate (MP conj.) matched* terminations. The induced voltage correlation coefficient is given as $\mathbb{E}[v_i v_i^{*}]$ where $i = 1, 2$ for the two ports of the antennas matching network.

The incident wireless signal of plane waves with azimuth angles of arrival that are uniformly distributed on $[0, 2\pi]$. The magnitude correlation coefficients are plotted versus antenna spacing in Figure 8.15.

The matching network offers the *conjugate matching* (*i.e.* $\boldsymbol{S}_{11} = \boldsymbol{S}_{RR}^{H}$) an optimal output zero correlation (i.e. decorrelation) for non-zero spacing symmetric dipoles, albeit a nonstable output. Clearly, perfect decorrelation occurs when matching network changes the radiation patterns of each antenna element so they are orthogonal over $[0, 2\pi]$ in azimuth. In addition, perfect decorrelation leads to unequal power at the matching network outputs and does not necessarily improve MIMO performance.

When correlation amplitudes are compared to correlation due to Jakes' model (mutual coupling in jakes' model is ignored), the self-matching antenna termination generates lower correlation amplitudes, and the open circuit antenna termination produces the highest correlation amplitudes for all antenna spacing. Thus, the curves show that correlation is load dependent.

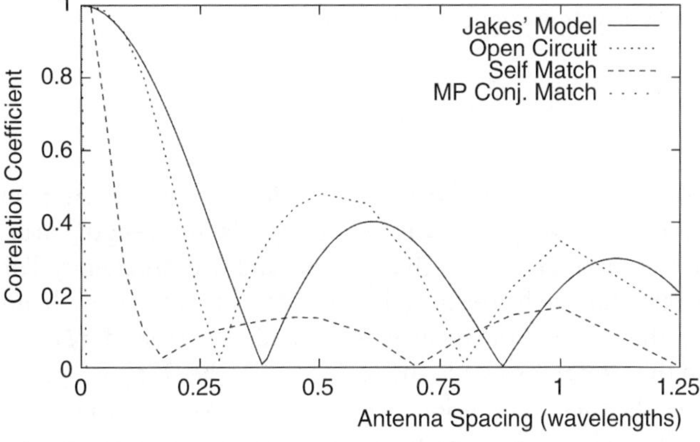

Figure 8.15 Correlation coefficient versus dipole antenna spacing [31].

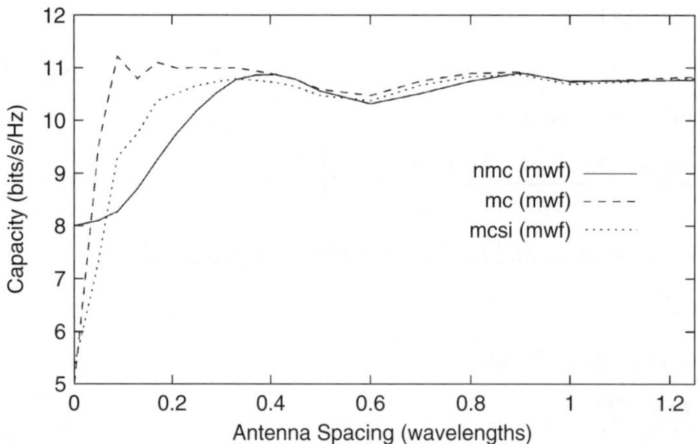

Figure 8.16 Mean capacity versus transmit and receive antenna spacing for coupling assumptions [31].

MIMO channel capacity is considered with a system consisting of a two transmit and two receive dipole antennas and a channel with four wave propagation paths and are used to simulate the effects of mutual coupling on the channel capacity. For each channel realization, the receive signal was averaged to fix the SNR at 20 dB and the mean capacity was computed over 7000 realizations for each antenna spacing. The MIMO system capacity simulations are aimed at the effects of the mutual coupling due to the combined effect of transmit and receive antennas. We activate both transmit and receive coupling to evaluate their effects on the mean capacity, and the results as shown in Figure 8.16. In this case, the spacing of both transmit and receive antennas are varied to equal spacing and the capacity was computed when the spacing is increased and the antennas are separated over 1λ, i.e. antennas elements are separated by a space long enough to generate no mutual coupling (*nmc*). Then the capacity was simulated for smaller spacing to introduce mutual coupling (*mc*) and explore its effects on the capacity. The channel capacity is also simulated for mutual coupling antennas with self-impedance matching (*mcsi*). The channel capacity is determined for the case of a modified water-filling scheme (*mwf*) power allocation solution with antenna elements mutual coupling.

ideal (*nmc*), (*mc*) with optimal matching at transmit and receive, and with suboptimal self-impedance matching (*mcsi*), all using (*mwf*) solution. For spacing $0.1\ \lambda \leq d \leq 0.3\lambda$, (*mc*) offers more capacity than (*nmc*). For spacing $\geq 1\lambda$, the coupling effect vanishes and the mean capacity of (*mc*), (*nmc*) and *mcsi* tends to 10.75 bits/sec/Hz. However, for spacing $d < 0.1\lambda$, both (*mc*) and (*mcsi*) actually degrade the system capacity (Figure 8.16).

8.12 Summary

The physical parameters of the wireless channel are important for understanding the channel and its model issues. A key characteristic of the mobile wireless channel is its strength variations with the movements of transmitter or receiver or both over time and over frequency. For example, a single tone transmitted from a BS will be received by a hand-held receiver as a band of frequencies of width depending on the speed of the receive device and its surrounding environment. This phenomenon is called the Doppler spread. The reciprocal of the Doppler spread determines the coherence time of the channel. In addition, the coherence BW represents the band of frequencies over which the channel passes all frequency components with equal gain and linear phase. These parameters were presented in detail in Section 8.1.

The wireless channels are often categorized on the rate of change of the signal amplitude carried by the channel, i.e. fast or slow. The changes are named as fast fading and slow fading. The fading can be categorized based on time/distance scale such as large-scale fading and small-scale fading. When the bandwidth of the radio signal is larger than B_c, the channel is called *frequency-selective*, causing severe distortion to the signal components with frequencies larger than B_c Signal fading and the wireless channel statistics were described in Section 8.2.

The traditional MIMO wireless channel modelling is based on the knowledge of perfect CSI at transmitter/receiver or both. Models can be based on perfect statistical knowledge as provided in Section 8.3. Section 8.4 dealt with massive MIMO channel modelling of the Rayleigh channels where the channel elements are independent and identically distributed zero mean complex Gaussian variables. The transmitter and receiver antennas are uncoupled and the MIMO streams at both link ends and are uncorrelated.

Under perfect conditions, it is well known the theoretic capacity of a MIMO system increases linearly with the minimum of transmit and receive antennas. However, various measurements show that practical MIMO channels show a significantly lower capacity as we suggested. This reduction of capacity is due to the spatial correlation at either the receive side or the transmit side of radio link or at both sides simultaneously.

We examined the case when a correlation exists at both link ends; however, the link ends correlations are not joined. This category of correlation inspired channel models is well-known, as Kronecker channel models explored in Section 8.5. Nevertheless, measurement evidence shows the Kronecker channel model predicts a sum capacity lower than that measured and the reduction is mainly due to adapting the multipath structure incorrectly caused by assuming link end correlations are independent.

In Section 8.6, we investigated the case when the two ends of a MIMO channel are fully associated with the spatial structure of the MIMO channel. This category of MIMO channel models are generally known as Weichselberger channel models. We analyse such MIMO channels to identify three key parameters that characterized such models, namely: The spatial eigenbasis at one side of the model, the spatial eigenbasis at the other side, and the energy coupling matrix in between.

A third MIMO channel model category is based on the physical scattering environment known as the virtual channel representation. We explored the MIMO channel representation to obtain the statistical structure of correlated fading initiated by the clustered scattering environments. The effective representation relates to beamforming in fixed directions defined by the resolution of the arrays. The virtual channel representation is explored in detail to derive the correlation matrix, as presented in Section 8.7

Mutual coupling causes undesirable influences on the performance of a wireless communication system. Most importantly, it degrades the system capacity. Specifically, the EM energy that is transmitted from a transceiver unit to a distance receiver is absorbed by receiving antenna of the unit itself. Mutual coupling between array elements also affects the reciprocity property in MIMO system. We have examined the antenna mutual coupling in both transmit and receive modes, and use the scattering parameters (S-parameters) to determine the mutual coupling in a BS MIMO system. In addition, we have extended our analysis to determine the maximum power collected at the MIMO receive array. These issues were discussed in details in Section 8.8.

In Section 8.9, we have taken account of the array coupling on the transmit power and determine, using of S-parameters, the instantaneous transmit power using mutually coupled transmit array. The voltage induced at a receive antenna port in the MIMO system was analysed in Section 8.10 and the expression of the induced voltage is used in determining the MIMO channel capacity of mutually coupled array systems allowing for the interference and user receiver noise, as described in Section 8.11.

8.A S-Parameters

At low frequencies, various techniques are available to represent the electrical behaviour of a linear multiport device (or a network). The most common approach adopts parameters H, Y, and Z. These parameters depend on measuring the total voltage and/or current, (as a function of frequency), at the input or output ports of the device. Yet, at RF frequency, it is not appropriate to measure voltages and currents. Instead, we think of a voltage as a travelling wave. As the wave is applied to the input of the device, there will be scattering (i.e. part of the wave travels through the device and the remainder is reflected. Therefore, it is sensible to think of parameters that replicate the scattering method; hence, the birth of the scattering concept in Bell Labs in 1965 to mathematically represent devices at RF frequency. The method explored the scattering approach and adopted the scattering parameters generally known as S-parameters. S-parameters are complex numbers, having real and imaginary parts or magnitude and phase parts of the incident signal wave. The S-parameters of a network are usually displayed in **S**-matrix, which can accurately describe the characteristics of a complicated circuit inside a black box. The S-parameters are initially used to construct a mathematical model for microwave networks. For lossless network, the **S**-matrix must satisfy

$$\mathbf{S}^H \mathbf{S} = \mathbf{I} \tag{8.A.1}$$

where **I** is a unit matrix. The **S**-matrix for an N-port network contains N^2 S-parameters, each represents a possible input-output path. Examples of the S-matrices for one-, two-, and three-port networks are given below:

$$\text{one} - \text{port}\ (S_{11})$$

$$\text{two} - \text{port}\begin{pmatrix} S_{11} & S_{12} \\ S_{21} & S_{22} \end{pmatrix}$$

$$\text{three} - \text{port}\begin{pmatrix} S_{11} & S_{12} & S_{13} \\ S_{21} & S_{22} & S_{23} \\ S_{31} & S_{32} & S_{33} \end{pmatrix}$$

Consider the two-port network shown below:

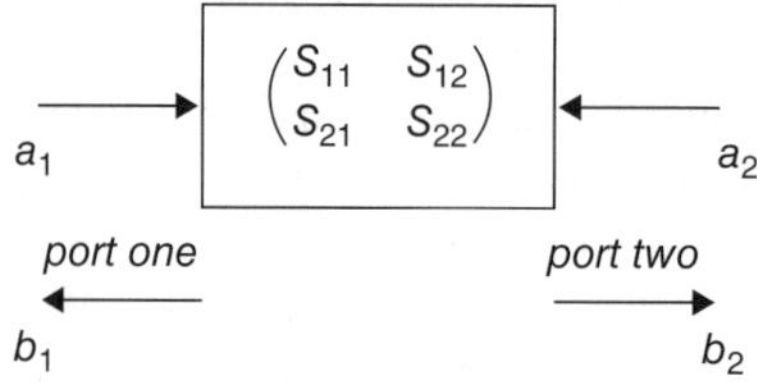

The S-parameters in two-port network

For lossless two-port network (8.A.1) can be expressed as

$$|S_{11}|^2 + |S_{22}|^2 = 1 \tag{8.A.2a}$$

$$|S_{12}|^2 + |S_{21}|^2 = 1 \tag{8.A.2b}$$

$$S_{11}^* S_{12} + S_{21}^* S_{22} = 0 \tag{8.A.2c}$$

Let's examine the two-port network. Each port has two nodes, with arrows in opposite directions. The signal at each network port represents two waves travelling in opposite directions. These waves are conventionally denoted as a_i and b_i representing the incident and reflected waves at the ith port. The magnitude of each wave is normalized to specific reference impedance z_0. We now define the four S-parameters of the two ports:

$S_{11} \triangleq \frac{b_1}{a_1}\Big| a_2 = 0$ Input reflection coefficient with device output matched

$S_{21} \triangleq \frac{b_2}{a_1}\Big| a_2 = 0$ Forward transmission coefficient with device out matched

$S_{22} \triangleq \frac{b_2}{a_2}\Big| a_1 = 0$ Output reflection coefficient with device input matched

$S_{12} \triangleq \frac{b_1}{a_2}\Big| a_1 = 0$ Reverse transmission coefficient with device input matched

The two-port network can be represented in a matrix form:

$$\begin{pmatrix} b_1 \\ b_2 \end{pmatrix} = \begin{pmatrix} S_{11} & S_{12} \\ S_{21} & S_{22} \end{pmatrix} \begin{pmatrix} a_1 \\ a_2 \end{pmatrix} \tag{8.A.3}$$

The S-parameter in dB is given by the following formula:

$$S_{ij}(dB) = 20\log_{10}|S_{ij}| \tag{8.A.4}$$

Under matched conditions, four figure of merits can be used to characterize the DUT, namely:

$$\text{Insertion loss dB} \triangleq -20\log_{10}|S_{21}| \tag{8.A.5a}$$

$$\text{Input return loss dB} \triangleq -20\log_{10}|S_{11}| \tag{8.A.5b}$$

$$\text{Output return loss dB} \triangleq -20\log_{10}|S_{22}| \tag{8.A.5c}$$

$$\text{Voltage standing wave ratio (VSWR)} \triangleq \frac{1+|S_{kk}|}{1-|S_{kk}|}\quad k = 1, 2 \tag{8.A.6}$$

The S-parameters can be measured using network analyser (an important equipment in any RF laboratory) as shown below:

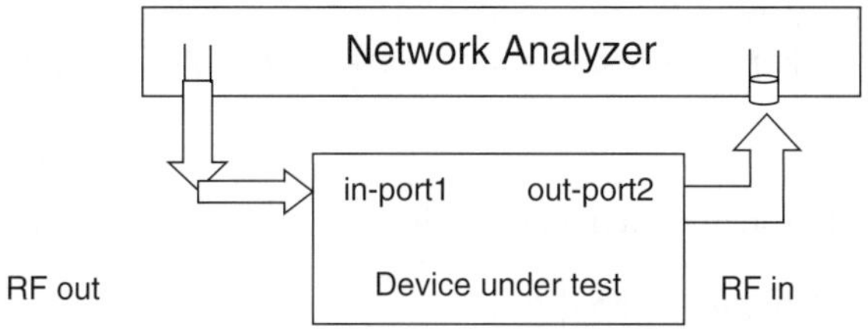

The network analyser has to be calibrated before measurement to improve measurement accuracy. Calibration procedure is usually provided by the manufacturer. Once the analyser is calibrated, $\log_{10}S_{11}$ and $\log_{10}S_{21}$ can be taken over the band of frequency with the device connected. Similarly, $\log_{10}S_{22}$ and $\log_{10}S_{12}$ can be taken over the same band by reversing the connection of the device (i.e. Analyser RF Out connected to device output and Analyser RF In to device input). Finally, we present the graphical equivalent to the two-port network in terms of the S-parameters:

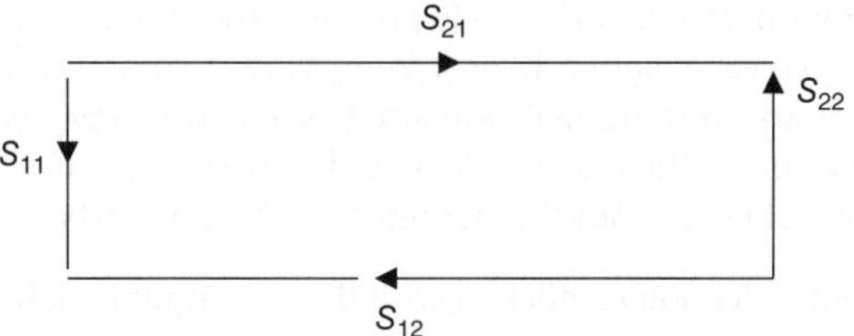

Graphical representation of S-parameters of a linear two-port network.

After this brief about S-parameters, we now investigate its application to quantify the correlation MIMO antennas system. Suppose we want to evaluate the correlation between adjacent antennas:

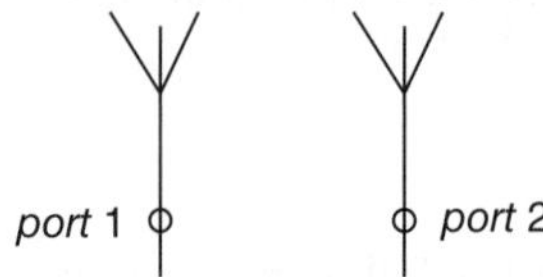

S-parameters in two adjacent antennas

The power is applied to the two antennas through port 1 and port 2. When we applied the power to port 1, we terminate port 2 with matched impedance. Then S_{11} would be the ratio of the reflected power to the input power of antenna 1; S_{22} would be the ratio of the reflected power to input power to antenna 2 when port 1 is terminated with a matched impedance; S_{12} is the power from antenna 2 that is picked up by antenna 1; and S_{21} is the power from antenna 1 that is picked up by antenna 2. Furthermore, S_{ij}(dB) is the power (dB) received by i^{th} antenna from the power input to j^{th} antenna. For example, suppose 1 W (i.e. 0 dB) is delivered to j^{th} antenna and we measure $S_{ij} = -10$ dB. Then the power received by i^{th} antenna is 0.1 W.

8.B Power Collected by the Receive Array Is Maximum When $S_{11} = S_{RR}^H$

Consider the network model in Figure 8.12, the power available at the output of the source block $\boldsymbol{S}_{RR}$ is fixed for a given $\boldsymbol{b}_0$. We wish to choose a matching block $\mathbf{S}_M$ to maximize deliver power delivered to the load Z_0. For a lossless block network as we noted in Appendix 8.A we have

$$\mathbf{S}^H\,\mathbf{S} = \mathbf{I}$$

The power available to the matched load is

$$\boldsymbol{a}_0^H\,\boldsymbol{a}_0 = (\mathbf{S}_{RS}^{-1}\,\boldsymbol{b}_0)^H(\mathbf{S}_{RS}^{-1}\,\boldsymbol{b}_0) = \|\mathbf{S}_{RS}^{-1}\,\boldsymbol{b}_0\|^2 \tag{8.B.1}$$

We now want to analyse the condition that maximizes the power collected by the receive array and ensures $\boldsymbol{b}_s = \mathbf{0}$. Considering the S-parameters matching network block in Figure 8.12 we have (8.81) as

$$\begin{pmatrix}\boldsymbol{b}_S\\ \boldsymbol{b}_R\end{pmatrix} = \begin{bmatrix}\mathbf{S}_{SS} & \mathbf{S}_{SR}\\ \mathbf{S}_{RS} & \mathbf{S}_{RR}\end{bmatrix}\begin{pmatrix}\boldsymbol{a}_S\\ \boldsymbol{a}_R\end{pmatrix}$$

$$\boldsymbol{b}_S = \mathbf{S}_{SS}\,\boldsymbol{a}_s + \mathbf{S}_{SR}\,\boldsymbol{a}_R \tag{8.B.2}$$

From the equivalence of the models in Figures 8.11 and 8.12 we get

$$\boldsymbol{b}_0 = \mathbf{S}_{RS}\,\boldsymbol{a}_0 \tag{8.B.3a}$$

$$\boldsymbol{a}_S = \boldsymbol{a}_0, \text{ i.e.} \boldsymbol{b}_0 = \mathbf{S}_{RS}\,\boldsymbol{a}_S \tag{8.B.3b}$$

So, from (8.B.3a) and (8.B.3b) we get

$$\boldsymbol{a}_S = \mathbf{S}_{RS}^{-1}\,\boldsymbol{b}_0 \tag{8.B.4}$$

From Figure 8.12 we have

$$\boldsymbol{a}_R = \boldsymbol{b}_1 \tag{8.B.5a}$$

And from (8.80):

$$\boldsymbol{b}_1 = \mathbf{S}_{11}\,\boldsymbol{a}_1 \tag{8.B.5b}$$

Combining (8.B.5a) and (8.B.5b), we get

$$\boldsymbol{a}_R = \mathbf{S}_{11}\,\boldsymbol{a}_1 \tag{8.B.5c}$$

Furthermore, (8.85) gives

$$\boldsymbol{a}_1 = (\mathbf{I} - \boldsymbol{S}_{RR}\,\mathbf{S}_{11})^{-1}\boldsymbol{b}_0 \tag{8.B.5d}$$

Substituting (8.B.5d) in (8.B.5c), we get

$$\boldsymbol{a}_R = \mathbf{S}_{11}\,(\mathbf{I} - \mathbf{S}_{RR}\,\mathbf{S}_{11})^{-1}\boldsymbol{b}_0 \tag{8.B.6}$$

Substituting (8.B.4) and (8.B.6) in (8.B.2), we get

$$\boldsymbol{b}_S = [\mathbf{S}_{SS}\,\mathbf{S}_{RS}^{-1} + \mathbf{S}_{SR}\,(\mathbf{I} - \mathbf{S}_{RR}\,\mathbf{S}_{11})^{-1}\mathbf{S}_{11}]\,\boldsymbol{b}_0 \tag{8.B.7}$$

We desire the **S** matrix of the matching network to be lossless; then $\mathbf{S}^H\,\mathbf{S} = \mathbf{I}$. The matching network **S** matrix is given by (8.78b)

$$\begin{pmatrix} \mathbf{S}_{SS} & \mathbf{S}_{SR} \\ \boldsymbol{S}_{RS} & \mathbf{S}_{RR} \end{pmatrix} \tag{8.B.8}$$

We apply singular value decomposition (SVD) operators on $\mathbf{S}_{RR}$ and $\mathbf{S}_{RR}$ to get

$$\mathbf{S}_{RR} = \mathbf{U}_{RR}\,\boldsymbol{\Lambda}_{RR}^{\frac{1}{2}}\,\mathbf{V}_{RR}^{H} \tag{8.B.9a}$$

$$\mathbf{S}_{SS} = \mathbf{U}_{SS}\,\boldsymbol{\Lambda}_{RR}^{\frac{1}{2}}\,\mathbf{V}_{SS}^{H} \tag{8.B.9b}$$

where $\mathbf{U}_{SS}$ and $\mathbf{V}_{SS}$ are arbitrary unitary matrices. In addition, we also have the following:

$$\mathbf{S}_{SR} = \mathbf{U}_{SS}\,\mathbf{D}\,(\mathbf{I} - \boldsymbol{\Lambda}_{RR})^{\frac{1}{2}}\,\mathbf{V}_{RR}^{H} \tag{8.B.9c}$$

$$\mathbf{S}_{\mathrm{RS}} = -\mathbf{U}_{RR}\,\mathbf{D}^{H}\,(\mathbf{I} - \boldsymbol{\Lambda}_{RR})^{\frac{1}{2}}\,\mathbf{V}_{SS}^{H} \tag{8.B.9d}$$

where **D** is a complex diagonal matrix with unit magnitude elements and arbitrary phase. If $\boldsymbol{S}$ matrix of the matching network is reciprocal then, $\boldsymbol{S} = \boldsymbol{S}^T$ and we choose the block networks such that

$$\mathbf{U}_{RR} = \mathbf{V}_{RR}^{*}, \mathbf{U}_{SS} = \mathbf{V}_{SS}^{*} \text{ and } = \mathrm{j}\,\mathbf{I}. \tag{8.B.10}$$

Let us assume that $\boldsymbol{S}_{11} = \boldsymbol{S}_{RR}^{H}$, and analyse the consequence of this assumption on $\boldsymbol{b}_S$.

$$\mathbf{S}_{11} = \mathbf{S}_{RR}^{H} = \mathbf{V}_{RR}\,\boldsymbol{\Lambda}_{RR}^{\frac{1}{2}}\,\mathbf{U}_{RR}^{H} \tag{8.B.11}$$

Substituting (8.B.11) in (8.B.7), we get,

$$\boldsymbol{b}_s = [\mathbf{S}_{SS}\,\mathbf{S}_{RS}^{-1} + \mathbf{S}_{SR}\,(\mathbf{I} - \mathbf{S}_{11}\,\mathbf{S}_{RR})^{-1}\,\mathbf{S}_{11}]\,\boldsymbol{b}_0$$

$$\boldsymbol{b}_S = \Bigg[-\mathbf{U}_{SS}\,\boldsymbol{\Lambda}_{RR}^{\frac{1}{2}}\,\mathbf{V}_{SS}^{H}\,(\mathbf{V}_{SS}^{H})^{-1}\,\mathbf{D}^{-1}(\mathbf{I} - \boldsymbol{\Lambda}_{RR})^{\frac{-1}{2}}\,\mathbf{U}_{RR}^{-1} + \mathbf{U}_{SS}\,\mathbf{D}\,(\mathbf{I} - \boldsymbol{\Lambda}_{RR})^{\frac{1}{2}}\,\mathbf{V}_{RR}^{H}\,(\mathbf{I} - \mathbf{S}_{RR}\,\mathbf{S}_{RR}^{H})^{-1}\mathbf{V}_{RR}\,\boldsymbol{\Lambda}_{RR}^{\frac{1}{2}}\,\mathbf{U}_{RR}^{H}\Bigg]\,\boldsymbol{b}_0 \tag{8.B.12}$$

since $\mathbf{U}_{RR}$ is unitary matrix $\mathbf{U}_{RR}^{-1} = \mathbf{U}_{RR}^{H}$. Further, seeing that we assumed $\mathbf{D} = \mathrm{j}\,\mathbf{I}$ then, $\mathbf{D}^{-1} = \mathbf{D}$, since $\boldsymbol{V}_{ss}^{H}$ is unitary matrix $(\mathbf{V}_{SS}^{H})^{-1} = \mathbf{V}_{SS}$, and (8.B.12) can be simplified to

$$\boldsymbol{b}_S = \Bigg[-\mathbf{U}_{SS}\,\boldsymbol{\Lambda}_{RR}^{\frac{1}{2}}\,\mathbf{V}_{SS}^{H}\,\mathbf{V}_{SS}\,\mathbf{D}(\mathbf{I} - \boldsymbol{\Lambda}_{RR})^{\frac{-1}{2}}\,\mathbf{U}_{RR}^{H} + \mathbf{U}_{SS}\,\mathbf{D}\,(\mathbf{I} - \boldsymbol{\Lambda}_{RR})^{\frac{1}{2}}\,\mathbf{V}_{RR}^{H}\,(\mathbf{I} - \mathbf{S}_{RR}\,\mathbf{S}_{RR}^{H})^{-1}\mathbf{V}_{RR}\,\boldsymbol{\Lambda}_{RR}^{\frac{1}{2}}\,\mathbf{U}_{RR}^{H}\Bigg]\,\boldsymbol{b}_0 \tag{8.B.13}$$

Also,

$$(\mathbf{I} - \mathbf{S}_{11}\boldsymbol{S}_{RR})^{-1} = \left(\mathbf{I} - \mathbf{V}_{RR}\,\boldsymbol{\Lambda}_{RR}^{\frac{1}{2}}\,\mathbf{U}_{RR}^{H}\,\mathbf{U}_{RR}\boldsymbol{\Lambda}_{RR}^{\frac{1}{2}}\,\mathbf{V}_{RR}^{H}\right)^{-1} = (\mathbf{I} - \boldsymbol{\Lambda}_{RR})^{-1} \tag{8.B.14}$$

Substituting (8.B.14) in (8.B.13), we get

$$\boldsymbol{b}_S = \Bigg[-\mathbf{U}_{SS}\,\boldsymbol{\Lambda}_{RR}^{\frac{1}{2}}\,\mathbf{V}_{SS}^{H}\,\mathbf{V}_{SS}\,\mathbf{D}(\mathbf{I} - \boldsymbol{\Lambda}_{RR})^{\frac{-1}{2}}\,\mathbf{U}_{RR}^{H} + \mathbf{U}_{SS}\,\mathbf{D}\,(\mathbf{I} - \boldsymbol{\Lambda}_{RR})^{\frac{1}{2}}\,\mathbf{V}_{RR}^{H}\,(\mathbf{I} - \boldsymbol{\Lambda}_{RR})^{-1}\mathbf{V}_{RR}\,\boldsymbol{\Lambda}_{RR}^{\frac{1}{2}}\,\mathbf{U}_{RR}^{H}\Bigg]\,\boldsymbol{b}_0 \tag{8.B.15}$$

We can write (8.B.15) as

$$\boldsymbol{b}_s = [\mathbf{U}_{SS}\,\mathbf{M}\,\mathbf{U}_{RR}^{H}\,] \tag{8.B.16}$$

where

$$\mathbf{M} = -\boldsymbol{\Lambda}_{RR}^{\frac{1}{2}}\,\mathbf{V}_{SS}^{H}\,\mathbf{V}_{SS}\,\mathbf{D}(\mathbf{I} - \boldsymbol{\Lambda}_{RR})^{\frac{-1}{2}} + \mathbf{D}\,(\mathbf{I} - \boldsymbol{\Lambda}_{RR})^{\frac{1}{2}}.\mathbf{V}_{RR}^{H}\,(\mathbf{I} - \boldsymbol{\Lambda}_{RR})^{-1}\mathbf{V}_{RR}\,\boldsymbol{\Lambda}_{RR}^{\frac{1}{2}} \tag{8.B.17}$$

Or equivalently,

$$\mathbf{M} = -\boldsymbol{\Lambda}_{RR}^{\frac{1}{2}}\,\mathbf{V}_{SS}^{H}\,\mathbf{V}_{SS}\,\mathbf{D}(\mathbf{I} - \boldsymbol{\Lambda}_{RR})^{\frac{-1}{2}} + \mathbf{D}\,(\mathbf{I} - \boldsymbol{\Lambda}_{RR})^{\frac{1}{2}}.\mathbf{V}_{RR}^{H}\mathbf{V}_{RR}\,(\mathbf{I} - \boldsymbol{\Lambda}_{RR})^{-1}\mathbf{V}_{RR}^{H}\mathbf{V}_{RR}\,\boldsymbol{\Lambda}_{RR}^{\frac{1}{2}} \tag{8.B.18}$$

After cancelling the unitary matrices in (8.B.18), we get

$$\mathbf{M} = -\boldsymbol{\Lambda}_{RR}^{\frac{1}{2}}\,\mathbf{D}(\mathbf{I} - \boldsymbol{\Lambda}_{RR})^{\frac{-1}{2}} + \mathbf{D}\,(\mathbf{I} - \boldsymbol{\Lambda}_{RR})^{\frac{-1}{2}}\,\boldsymbol{\Lambda}_{RR}^{\frac{1}{2}} = \mathbf{0} \tag{8.B.19}$$

Thus, $\boldsymbol{b}_S$ is the zero vector for any $\boldsymbol{b}_0$, and hence the assumption, that $\mathbf{S}_{11} = \mathbf{S}_{RR}^{H}$ ensures all the available power is dissipated into the load, maximizes the collected receive power.

References

1 Abu-Rgheff, M.A. (2007). *Introduction to CDMA Wireless Communications*. Academic Press. ISBN: 978-0-75-065252-0.

2 Biglieri, E., Calderbank, R., Constantinides, A. et al. (2007). *MIMO Wireless Communications*. Cambridge, England, UK: Cambridge University Press. ISBN: 13 978-0-521-87328-4.

3 Zheng, K., Ou, S., and Yin, X. (2014). Massive MIMO channel models: a survey. *International Journal of Antennas and Propagation,* Hindawi Publishing Corporation, Article ID: 848071, 10 pages.
4 Marzetta, T.L. (2010). Noncooperative cellular wireless with unlimited numbers of base station antennas. *IEEE Transaction on Wireless Communications* 9 (11): 3590–3600.
5 Lakshmi, N.T. and Chockalingam, A. (2014). Channel hardening – exploiting message passing (CHEMP) receiver in large – scale MIMO systems. *IEEE Journal of Selected Topics in Signal Processing* 8 (5): 847–860.
6 Hochwald, B.M., Marzetta, T.L., and Tarokh, V. (2004). Multiple-antenna channel hardening and its implications for rate feedback and scheduling. *IEEE Transactions on Information Theory* 50 (9): 1893–1909.
7 Shin, H. and Lee, J.H. (2003). Capacity of multiple-antenna fading channels: spatial fading correlation, double scattering, and keyhole. *IEEE Transactions on Information Theory* 49 (10): 2636–2647.
8 Hoydis, J., Brink, S.T., and Debbah, M. (2013). Massive MIMO in the UL/DL of cellular networks: how many antennas do we need? *IEEE Journal on Selected Areas in Communications* 31 (2): 160–171.
9 Bjornson, E., Ottersten, B. and Jorswieck, E. (2009) 'On the impact of spatial correlation and precoder design on the performance of MIMO systems with space-time coding'. In: *IEEE International Conference on Acoustics, Speech and Signal Processing* (ICASSP), 2741–2744
10 Vaughan, R.G. and Andersex, J.B. (1987). Antenna diversity in mobile communications. *IEEE Transactions on Vehicular Technology* VT-36 (4): 149–172.
11 Blanch, S., Romeu, J., and Corbella, I. (2003). Exact representation of antenna system diversity performance from input parameter description. *Electronics Letters* 39 (9): 705–707.
12 Shiu, D.-S., Foschini, G.J., Gans, M.J., and Kahn, J.M. (2000). Fading correlation and its effect on the capacity of multielement antenna systems. *IEEE Transactions on Communications* 48 (3): 502–513.
13 Orcelik, H., Herdin, M., Weichselberger, W. et al. (2003). Deficiencies of 'Kronecker' MIMO radio channel model. *Electronics Letters* 39 (16): 1209–1210.
14 Oestges, C. (2006) 'Validity of the Kronecker model for MIMO correlated channels'. In: *IEEE Vehicular Technology Conference,* 6, 2818–2822.
15 Weichselberger, W., Herdin, M., and Bonek, E. (2003). A novel stochastic MIMO channel model and its physical interpretation. In: *IEEE International Symposium on Wireless Personal Multimedia Communications,* 1–6.
16 Weichselberger, W., Herdin, M., Ozcelik, H., and Bonek, E. (2006). A stochastic MIMO channel model with joint correlation of both link ends. *IEEE Transactions on Wireless Communications* 5 (1): 90–100.
17 Wood, L. and Hodgkiss, W.S. (2008). Understanding the Weichselberger model: a detailed investigation. In: *IEEE Military Communications Conference,* 1–7.
18 Almers, P. and Tufvesson, F. (2003). *Keyhole Effects in MIMO Wireless Channels – Measurements and Theory.* Cambridge, MA, USA: Mitsubishi Electric Research Laboratories.
19 Sanayei, S. and Nosratinia, A. (2007). Antenna selection in keyhole channels. *IEEE Transactions on Communications* 55 (3): 404–408.
20 Chizhik, D., Foschini, G.J., Gans, M.J., and Valenzuela, R.A. Keyholes, correlations, and capacities of multielement transmit and receive antennas. *IEEE Transactions on Wireless Communications* 1 (2): 361–368.
21 Sayeed, A.M. (2002). Deconstruction multiantenna fading channels. *IEEE Transactions on Signal Processing* 50 (10): 2563–2579.

22 Hong, Z., Liu, K., Heath, R.W., and Sayeed, A.M. (2003). Spatial multiplexing in correlated fading via the virtual channel representation. *IEEE Journal on Selected Areas in Communications* 21 (5): 856–866.

23 Tong, H. and Zekavat, S.A. (2006). Spatially correlated MIMO channel-generation via virtual channel representation. *IEEE Communications Letters* 10 (5): 332–334.

24 Chae, S.H., Oh, S.-K., and Park, S.-O. (2007). Analysis of mutual coupling, correlations, and TARC in WiBro MIMO array antenna. *IEEE Antennas and Wireless Propagation Letters* 6: 122–125.

25 Wang, X., Nguyen, H.D. and Hui, H.T. (2011) 'Correlation coefficient expression by S-parameters for two omni-directional MIMO antennas', *IEEE International Symposium on Antennas and Propagation and National Radio Science meeting* (AP-S/URSI), 301–304.

26 Gupta, I.J. and Ksienski, A.A. (1983). Effect of mutual coupling on the performance of adaptive arrays. *IEEE Transactions on Antennas and Propagation* 31 (9): 785–791.

27 Hui, H.T. (2004). A new definition of mutual impedance for application in dipole receiving antenna arrays. *IEEE Antennas and Wireless Propagation Letters* 3: 364–367.

28 Lui, H.-S., Hui, H.T., and Leong, M.S. (2009). A note on the mutual-coupling problems in transmitting and receiving antenna arrays. *IEEE Antennas and Propagation Magazine* 51 (5): 171–176.

29 Balanis, C.A. (2005). *Antenna Theory: Analysis and Design*, 3e. Wiley.

30 Pozer, D.M. (1998). *Microwave Engineering* Chapter 4, 2e. Wiley.

31 Wallace, J.W. and Jensen, M.A. (2004). Mutual coupling in MIMO wireless systems: a rigorous network theory analysis. *IEEE Transactions on Wireless Communications* 3 (4): 1317–1325.

32 Hui, H.T. (2007). Decoupling methods for the mutual coupling effect in antenna arrays: a review. *Recent Patents on Engineering* 1: 187–193.

33 Viterbi, A.J. and Omura, J.K. (2009). *Principles of Digital Communication and Coding.* Dover publications Inc.

34 Haykin, S. (1988). *Digital Communications.* Willey.

Further Reading

Liang, C.-H., Shi, Y., and Su, T. (2010). S parameter theory of lossless block Network. *Progress in Electromagnetics Research, PIER* 104: 253–266.

Frickey, D.A. (1994). Conversions between S, 2, Y, h, ABCD, and T parameters which are valid for complex source and load impedances. *IEEE Transactions on Microwave Theory and Techniques* 42 (2): 205–211.

Marks, R.B. and Williams, D.F. (1995). Comments on conversions between S, Z, Y, h, ABCD, and T parameters which are valid for complex source and load impedances. *IEEE Transactions on Microwave Theory and Techniques* 43 (4): 914–915.

Liang, C.-H., Shi, Y., and Su, T. (2010). S parameter theory of lossless block network. *Progress in Electromagnetics Research, PIER* 104: 253–266.

Gupta, G. and Chaturvedi, A.K. (2013). Conditional entropy based user selection for multiuser MIMO systems. *IEEE Communications Letters* 17 (8): 1628–1631.

Telatar, E. (2008). Capacity of multiple-antenna Gaussian channels. *European Transactions on Telecommunications* 10 (6): 585–595. *Invited paper.*

Cichocki, A., Cruces, S. and Amari, S-I. (2014) 'Log-Determinant Divergences Revisited: Alpha–Beta and Gamma Log-Det Divergences', available @ arXiv:1412.7146v2

9

Massive Array Configurations and 3D Channel Modelling

9.1 Massive Antenna Array Configurations at BS

Base station antenna array (BSAA) greatly enhances the performance of multiple-input, multiple-output (MIMO) wireless communication. Elements of the BSAA are commonly assumed to be identical. An array is a set of antennas operating together to perform a certain radiation pattern. A number of factors control the performance of the array, such as the geometric configurations (i.e. linear, square, circular, spherical, etc.), the spacing between consecutive elements and whether such spacing is uniform or nonuniform, the radiation pattern of the individual elements, and the amplitude and phase of the excitation signal. Examples of possible BSAA configurations are shown in Figure 9.1. Considering an electromagnetic (EM) field arriving at the array, the ratio of the E-field strength to induced voltage across the antenna terminals, $\frac{E}{V}$ is called the *array factor* (AF). The total far-field received by the base station (BS) array is equal to the field of a single element, positioned at the origin, multiplied by the AF. In the following, we provide a brief summary of the relevant factors related to the array configurations, which are likely to be useful in the design and analysis of some configurations that may be employed in 5G systems. The simplest configuration indubitably is that of a linear (1-D) array, which we deal with first.

9.2 Uniform Linear Arrays

In a uniform linear array with identical elements, positioned along a straight line, with equal element separations is shown in Figure 9.2a. The expression for the AF is a function of the geometry of the array and the phase of the excitation. The array characteristic can be controlled by changing the elements separation and/or the excitation phase of the elements. The normalized AF of N-element linear array AF_{LA} with uniform antenna spacing and amplitude is given by

$$AF_{LA}(\Psi) = \frac{1}{\Gamma_{\text{LA}}}\left[\frac{\sin\left(\frac{N}{2}\Psi\right)}{\sin\left(\frac{1}{2}\Psi\right)}\right] \tag{9.1}$$

where Γ_{LA} is a constant to make the largest value of $|AF_{LA}(\Psi)|$ equal to one. Note that Γ_{LA} is the normalization constant and it is not necessarily equal to N; $\Psi = kd\cos\theta + \beta$; k is the wave-number defined as $k := \frac{2\pi}{\lambda}$ in radians per unit distance; d is the element spacing; θ is the angle measured from the dipoles axis; and β is the phase between the elements. The maximum amplitude of AF is given by (9.1).

5G Physical Layer Technologies, First Edition. Mosa Ali Abu-Rgheff.

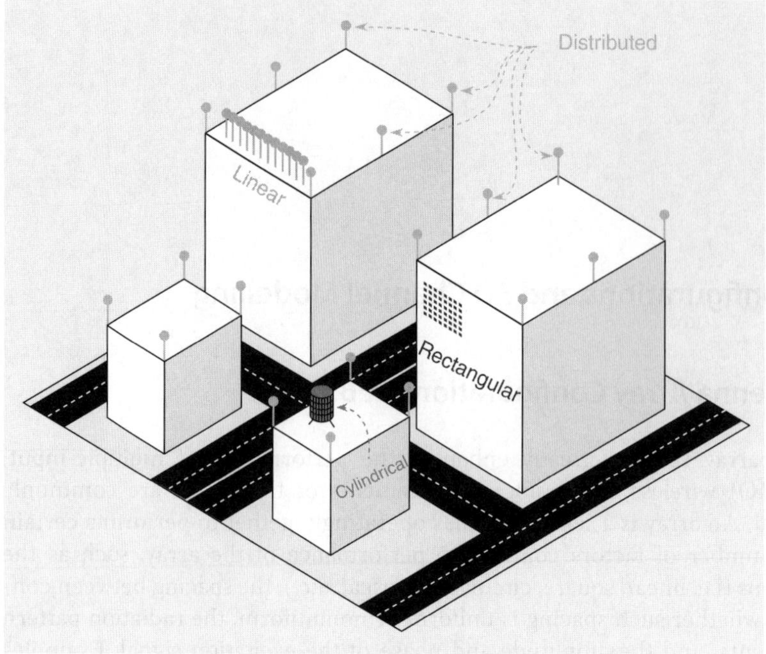

Figure 9.1 Some possible massive MIMO BS antenna configurations [1].

9.3 Rectangular Planar Arrays

Rectangular planar arrays are characterised by their directional beams, low side lobes symmetrical patterns, and much higher directivities compared to linear arrays.

A rectangular array with M elements along the x-axis and N element along the y-axis is depicted in Figure 9.2b. The array has uniform spacing on the 2D and uniform excitation amplitude of all the elements. The normalised AF of the M × N array denoted as AF_{RA} is given as

$$AF_{RA} = \frac{1}{\Gamma_{\mathrm{RA}}} \left(\frac{\sin\left(\frac{M}{2}\Psi_x\right)}{\sin\left(\frac{\Psi_x}{2}\right)} \right) \left(\frac{\sin\left(\frac{N}{2}\Psi_y\right)}{\sin\left(\frac{\Psi_y}{2}\right)} \right) \tag{9.2}$$

where Γ_{RA} is the normalisation constant chosen to make the maximum amplitude given by (9.2) as one and

$$\Psi_x = k\, d_x \sin\theta \cos\phi + \beta_x \tag{9.3}$$

$$\Psi_y = k\, d_y \sin\theta \cos\phi + \beta_y \tag{9.4}$$

Therefore, the AF of the rectangular planar array is the product of the AFs of the linear arrays in x and y directions.

9.4 Circular Arrays

The geometry of the circular array is shown in Figure 9.3 where the elements are positioned on the perimeter of a circle of radius a. The AF of the circular array, AF_C consisting of N isotropic

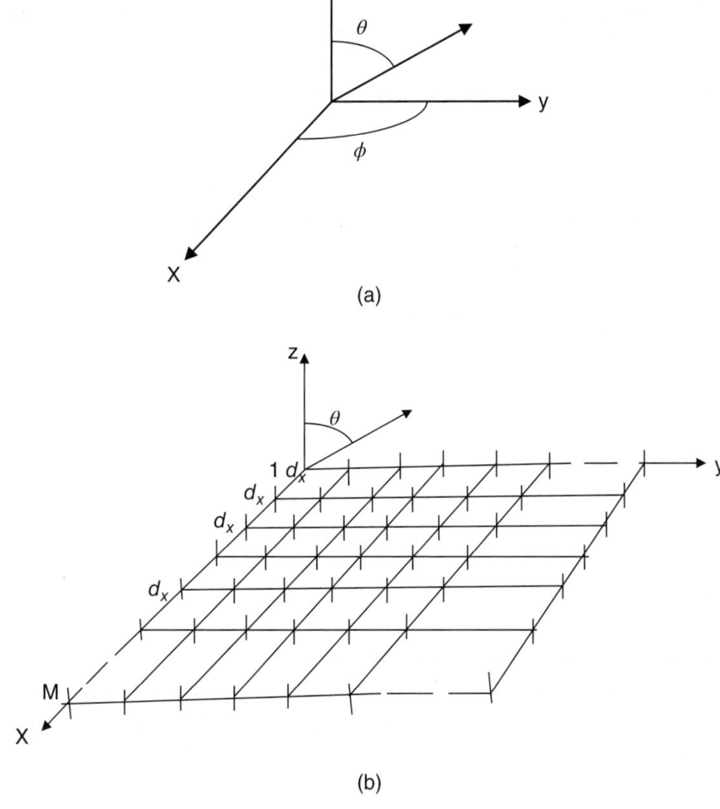

Figure 9.2 Linear and rectangular array. (a) Linear array. (b) Rectangular array [2].

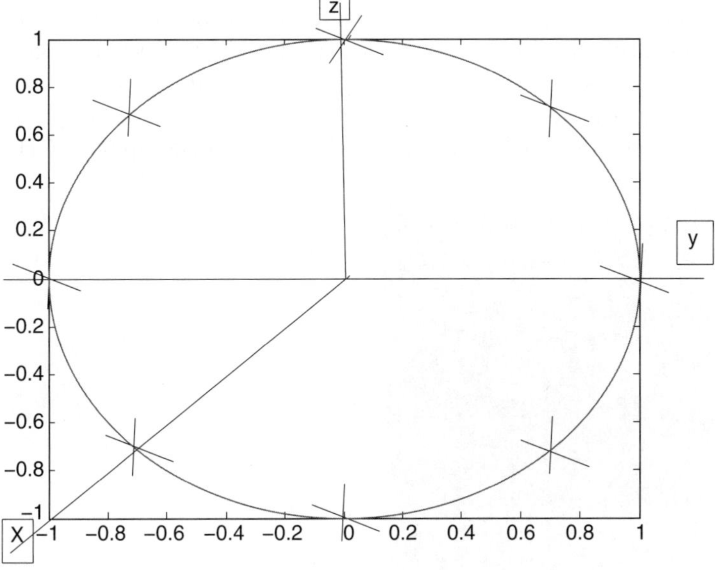

Figure 9.3 BS N-element circular array [2].

(omni-directional) elements that are equally spaced can be composed as

$$AF_C(\theta,\phi) = \sum_{n=1}^{N} I_n\, e^{j[ka \sin\theta \cos(\phi-\phi_n)+\alpha_n]} \tag{9.5}$$

The excitation coefficient of the nth element e_n can be expressed as

$$e_n = I_n\, e^{j\,\alpha_n} \tag{9.6}$$

where I_n is the excitation amplitude of the nth element and α_n is its phase relative to a reference element of zero phase. The maximum of the AF occurs when the phase is given as

$$ka \sin\theta \cos(\phi - \phi_n) + \alpha_n] = 2\pi m, \qquad m = 0, \pm 1, \dots \tag{9.7}$$

The main lobe maximum occurs at $m = 0$ in the direction (θ_0, ϕ_0) such that

$$\alpha_n = -ka \sin\theta_0 \cos(\phi_0 - \phi_n) \qquad n = 1, 2, \dots\dots, N \tag{9.8}$$

Substituting (9.8) for α_n into (9.5), we get

$$AF_C(\theta,\phi) = \sum_{n=1}^{N} I_n\, e^{jka[\sin\theta \cos(\phi-\phi_n) - \sin\theta_0 \cos(\phi_0-\phi_n)]}. \tag{9.9}$$

9.5 Cylindrical Arrays

Cylindrical arrays are configured in 3-D (x, y, z) so at the first instance we may think of the 2D circular array in the (x, y) plan with the array moving up and down the z-axis to map the 2D array into a cylindrical array. However, such cylindrical arrays are very challenging to analyse and to configure. Instead, the most practical cylindrical arrays design is based on *wraparound* arrays on the cylindrical planes.

Microstrip antennas in [3] and [4] are suited to conformal array applications. Depending on the radiation characteristics of the cylindrical circumferential arrays, polarized microstrip patches of a nearly square radiator can be used. Figure 9.4 depicts four such elements. The curvature seems to introduce only small variations on the final return loss of the array [5]. The advantages of microstrip arrays encompass low cost to fabricate; lightweight, easy to configure a large array spaced $\frac{\lambda}{2}$ element; and easy to form a curved surface on cylindrical and spherical surfaces i.e. conformal structures. However, it also has some disadvantages, such as low power handling, and even though microstrip technology can be used for frequencies up to 50 GHz,

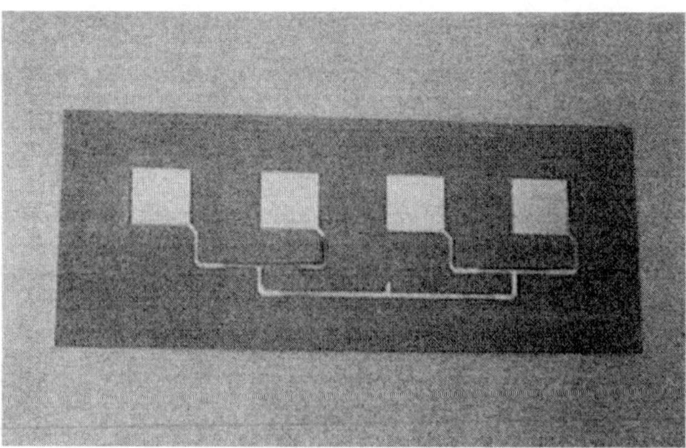

Figure 9.4 Four elements microstrip array [5].

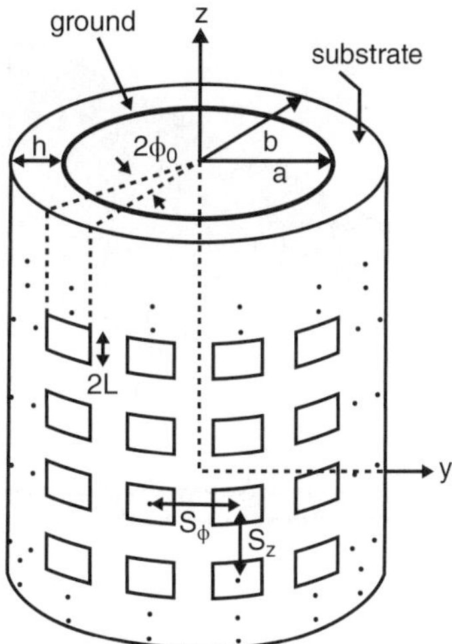

Figure 9.5 Geometry of microstrip antenna array mounted on a cylindrical body [7].

elements fabricated have a limited bandwidth. A helpful reference on the microstrip antennas is [6]. The geometry of a microstrip antenna array mounted on a cylindrical body is shown in Figure 9.5. A 4×4 microstrip designed with azimuthal array to operate at 8.15 GHz contributes a measured axial gain of 16.7 dB, measured azimuthal gain of 17.5 dB, the axial and azimuth insertion loss of 0.2, and 0.3 dB, respectively, described in [8].

9.6 Spherical Antenna Arrays

The conventional configuration of spherical arrays comprised a large number of radiating elements placed on a spherical surface. For a given beam direction, a sector of the array is excited and the active sector is defined by a cone angle, with the cone axis coinciding with the beam direction. Spherical arrays are highly symmetrical devices. The symmetry is also applied to the array elements.

The mmWave array package with beam steering capability for the 5G mobile terminal (MT) applications is proposed in [9]. The array can achieve 3D scanning coverage of the space with high gain beams. Three identical subarrays are tightly arranged along the edge of the mobile phone printed circuit board (PCB) for the proposed array. A PCB array is a multiple of a single PCB to make a larger array of connected boards. The desired coverage can be found by switching the feeding to one of the elements. Each subarray consists of eight patch antenna elements, and the centre-to-centre distance is about half a wavelength at the operating frequency. Figure 9.6 shows a schematic diagram of the proposed 5G antenna. Figure 9.6a shows a side view of the mobile phone PCB with subarray B. Figure 9.6b shows the location of the subarray A, subarray B, and subarray C. Figure 9.6c shows the top layer of subarray B and Figure 9.6d shows the ground layer of typical subarray.

Bias network can be employed to provide DC bias to the device used in the phased array whilst blocking the DC to the radio ports. Bias networks are now available for use in frequencies up to mmWave bands. Conventional phased array architecture is illustrated in Figure 9.7. The feed network is a key element in phased array for beam steering. Selection of a specific subarray can be achieved by using a mmWave switch. A proposed design uses three 1×8 uniform linear array elements where each element is excited by signals with equal amplitude.

A spherical array with 60 subarrays is described in [10, 11]. Each subarray is further divided into four triangular regions fitted with four identical isosceles triangles. Each tile is fitted with three radiating elements so the total elements of the array are 720. A schematic diagram of the

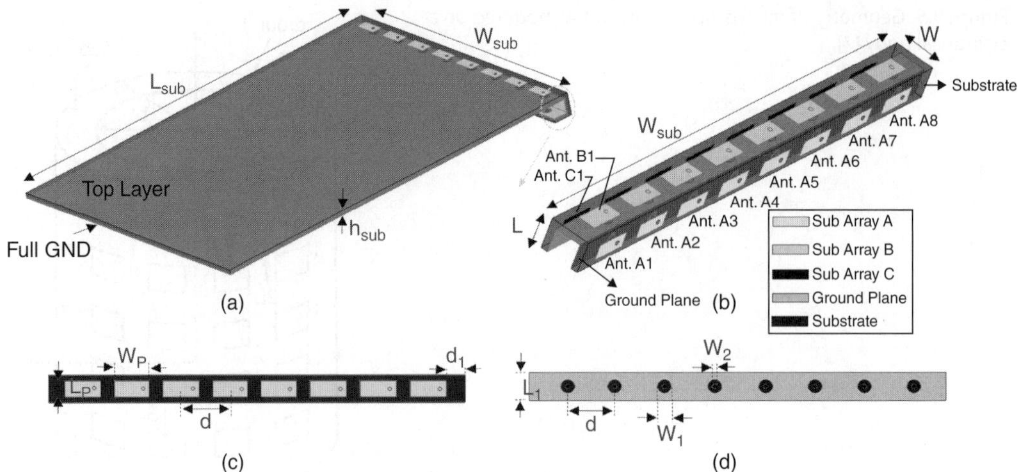

Figure 9.6 (a) Side view of a proposed 5G antenna with full ground plane, (b) antenna package configuration with three subarrays, (c) top layer of each subarray, and (d) bottom layer of each subarray [9].

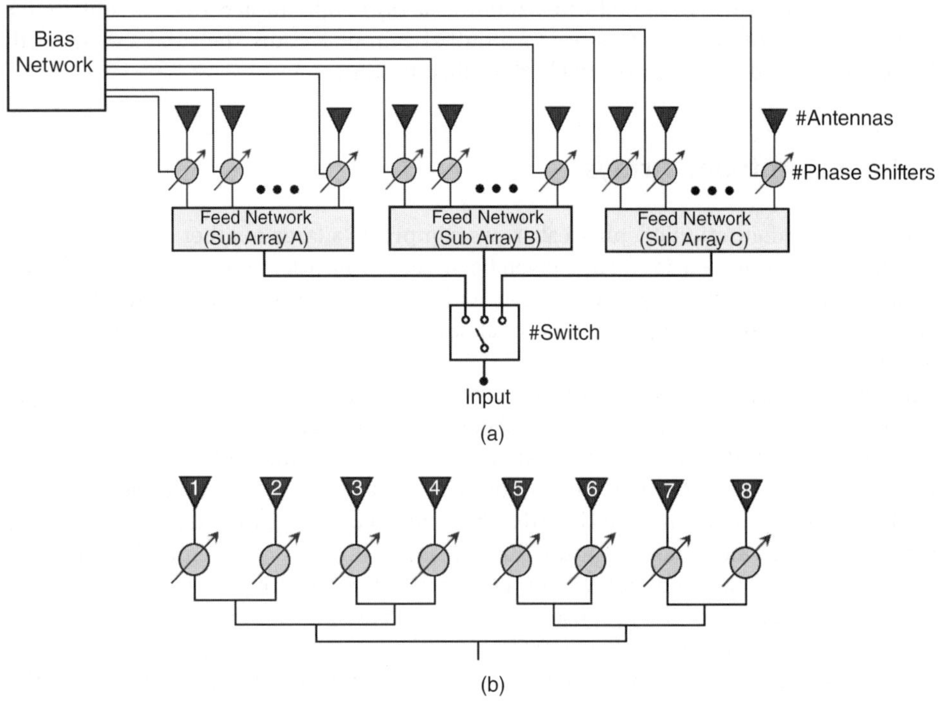

Figure 9.7 (a) Phased array architecture (b) feed network for a linear phased-array antenna [9].

proposed array with the radiating elements grouped in subarrays is shown in Figure 9.8. Each element is configured as a single-feed proximity coupled ring antenna, as shown in Figure 9.9. The proximity coupled ring element comprises two layers: the bottom layer accommodates the feeding line and the top layer a circular ring patch radiator. The ground plane is located underneath the bottom layer. The circular ring is held up by foam material. Figure 9.10 shows a

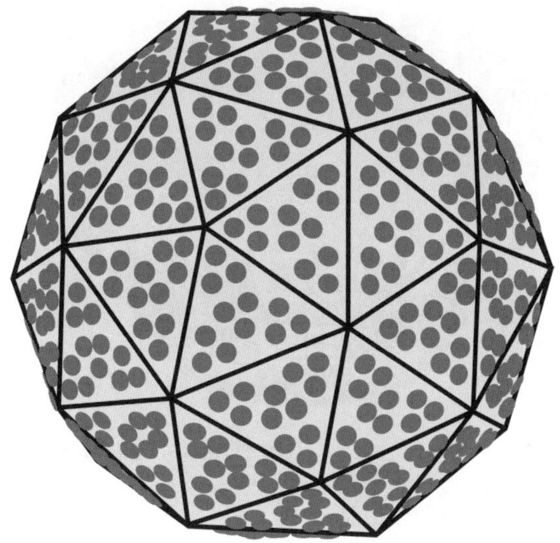

Figure 9.8 Schematic diagram of LISA, with the radiating elements grouped in subarrays [10].

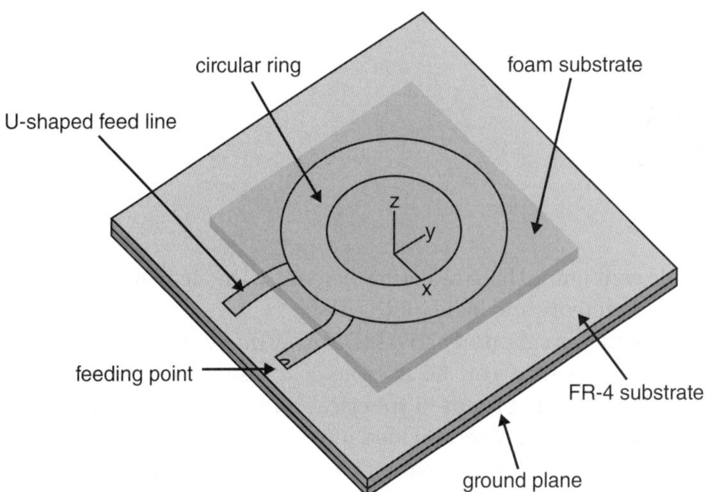

Figure 9.9 A single radiating element. There are 720 of these radiators on the spherical surface [10].

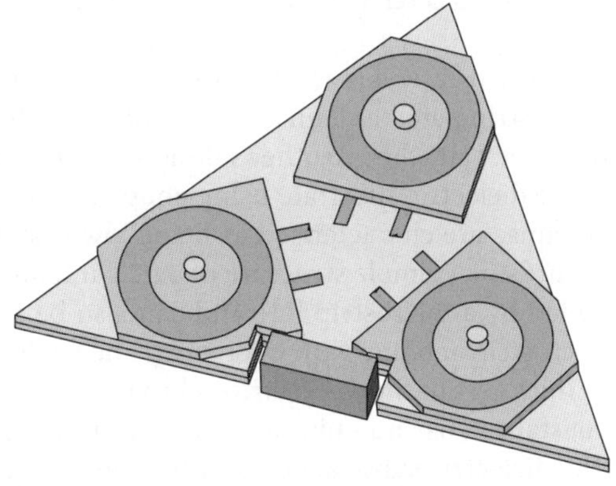

Figure 9.10 LISA consists of 240 tiles, each with three radiating elements [10].

Figure 9.11 Spherical array antenna [12].

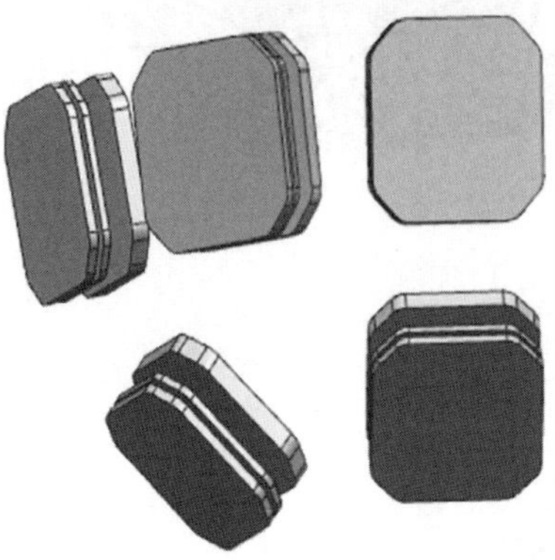

Figure 9.12 One of the subarrays [12].

schematic diagram of a typical tile containing three radiating elements. The elements are placed in sequential rotation to improve directivity and bandwidth (BW).

A 16-element spherical array for the direction of arrival (DOA) estimation in high altitude platforms is proposed in [12]. Each element can be used as a transmit port or a receive port. Figure 9.11 depicts the structure of the proposed spherical array and Figure 9.12 shows schematic diagram of a typical subarray. The array is divided into eight subarrays so that beam scanning covers the whole surrounding area. Each subarray covers 30°–60° of elevation angle and 15° – 45° of azimuth angle.

9.7 Microstrip Patch Antennas

Microstrip antennas are deployed in various applications, including wireless communications. Microstrip patch antennas consist of a radiating patch on one side a dielectric substrate with thin metal attached to other side acting as the ground. The patch is configured on the other side of the dielectric substrate using photo-etching optical corrosion to take a certain shape. Patch antennas are characterised as being low cost, lightweight, easy to fabricate, and conformable to surface. A simple structure of patch antenna is shown in Figure 9.13. The patch antenna can be fed by a microstrip line and can also be fed by a coaxial probe. A top and side view of a microstrip antenna with microstrip line feeding is shown in Figure 9.14.

Thickness of the microstrip is broadly $\ll\lambda_0$ where λ_0 is free space wavelength. The dielectric constant ε_r has no units, as it is the ratio of permittivity of the substance to that of free space. The dielectric substrate ε_r is in the range of 2–12. A dielectric constant in the lower range is

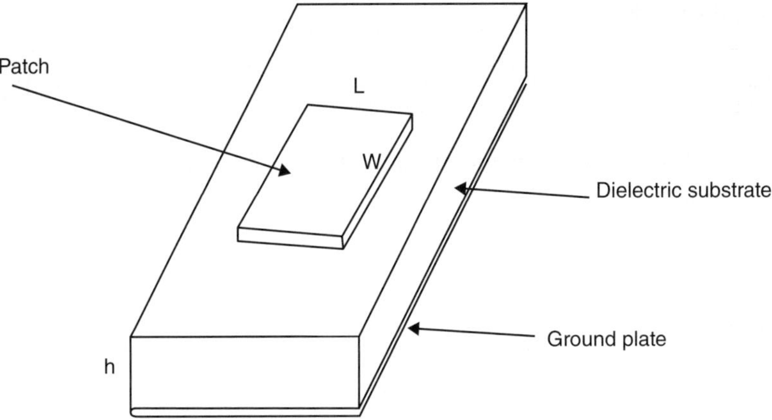

Figure 9.13 Basic structure of microstrip antenna [13].

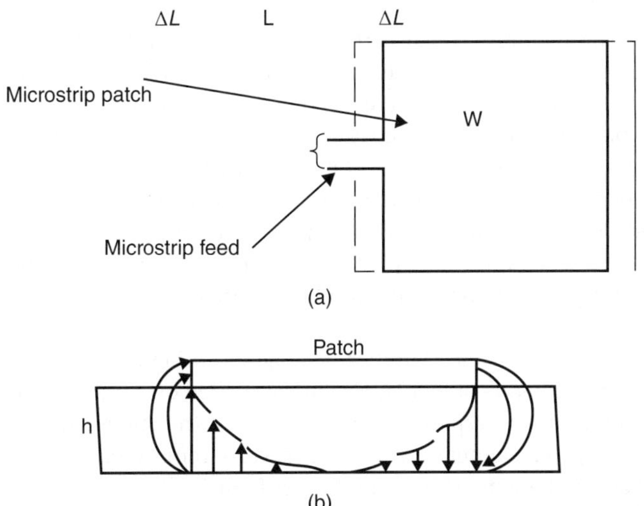

Figure 9.14 (a) Basic structure and dimension of microstrip patch. (b) The E-field of a microstrip patch.

desirable for high-quality antenna performance in terms of radiation, larger BW, and better efficiency. Since the length L and width W of the patch are finite, the fields at the two radiation ends of the microstrip antenna undergo fringing, as shown in Figure 9.14b.

The key design parameters in patch antenna are as follows [2]:

Effective dielectric constant ε_{reff},

$$\varepsilon_{reff} = \frac{\varepsilon_r + 1}{2} + \frac{\varepsilon_r - 1}{2}\left[1 + 12\,\frac{h}{W}\right]^{-\frac{1}{2}} \quad for\ \frac{W}{h} > 1 \tag{9.10}$$

where ε_r is the dielectric constant (relative permittivity). The length of the patch L is calculated as

$$L = \frac{c}{2f_r\sqrt{\varepsilon_{reff}}} - 2\,\Delta L \tag{9.11}$$

where c free space speed of light, 3×10^6 m/sec, f_r and ΔL are the resonance frequency (centre frequency) of the patch, and the patch length extension on each end is due to the fringing effect, respectively. The width of the patch W is given as

$$W = \frac{c}{2f_r}\sqrt{\frac{2}{\varepsilon_{reff}+1}} \tag{9.12}$$

The frequency of the operation of the patch antenna is determined by L, and its centre frequency is approximately given as

$$f_r = \frac{c}{2\sqrt{\varepsilon_{reff}}}\frac{1}{L+2\Delta L} \approx \frac{c}{2L\sqrt{\varepsilon_{reff}}} \tag{9.13}$$

where we assume $L \gg 2\Delta L$.

The length correction (ΔL) due to E-wave fringing is given as

$$\frac{\Delta L}{\text{h}} = 0.412\,\frac{(\varepsilon_{reff+0.3})\left(\frac{W}{h}+0.264\right)}{(\varepsilon_{reff-0.3})\left(\frac{W}{h}+0.8\right)} \tag{9.14}$$

Conformal arrays can be fitted to nonplanar surface parts on such as aircrafts, vehicles, or ships and are considered an attractive replacement to planar arrays, which have some drawbacks. The main advantages of conformal arrays are a large field of view, low-level observation, and improved aerodynamics. Conformal arrays (cylindrical or spherical) are most suitable where the number of antennas is required to be large in applications such as massive MIMO systems. However, deployment of conformal array technology in commercial applications is still not widespread. The modelling, analysis, and production of conformal arrays are much more complex, compared with linear or planar arrays. In addition, conformal arrays fabrication technology, assembly, and feeding of mounted conformal arrays on curved surfaces are challenging issues.

A spherical antenna array mounted in a anechoic chamber is depicted in Figure 9.15. The microstrip patch elements are designed to operate at 8–12 GHz and fabricated on a planar substrate and then cut and fixed into the cavities prepared on the surface of the sphere with a diameter of 300 mm. There are 95 triple-patch elements arranged in seven circumference rings. The minimum distance between adjacent elements should not be less than 1.5 λ at the operating frequency. Another example of 3×3 triple microstrip patch designed by the same author to function at a centre frequency of 9.5 GHz is shown in Figure 9.16.

Figure 9.15 Spherical antenna array mounted in anechoic chamber [14].

Figure 9.16 Array consisting of 3 × 3 triple patch subarrays [15].

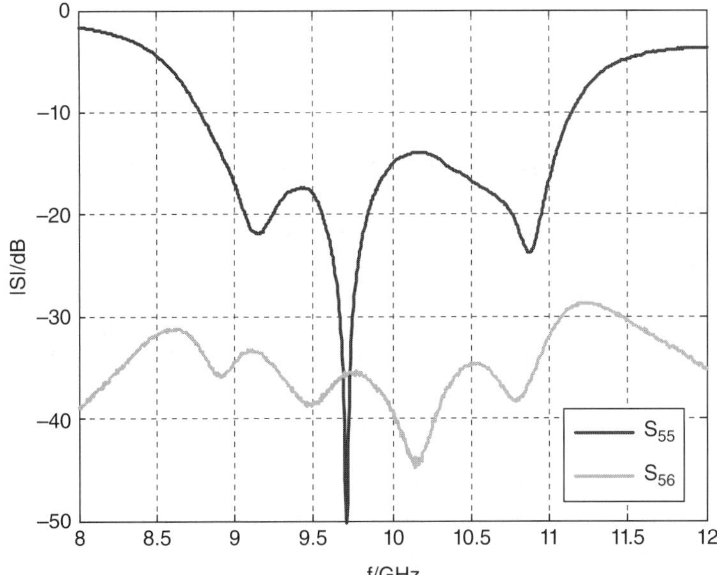

Figure 9.17 Measured reflection coefficient and mutual coupling of centre element [15].

Tests carried out on the 3 × 3 triple patch subarrays depicted in Figure 9.16 have shown that the measured amplitude of the reflection coefficient S_{11} of the centre element denoted as S_{55} implying S_{11} for the centre element number 5 and the mutual coupling between centre element, and its neighbour S_{12} denoted as S_{65} hinting S_{21} for the mutual coupling between the centre element number 5 and its neighbour element number 6 at frequencies between 8 and 12 GHz are shown in Figure 9.17. It can be seen that the −10 dB measured reflection coefficient bandwidth (calculated using S_{11}) is approximately 2.5 GHz while the mutual coupling is well below −30 dB over this band. The measurements were conducted inside an anechoic chamber.

Nonplanar surfaces such as aircraft surfaces, high-speed trains, and high-speed vehicles require microstrip antennas that are conformable to the shape of the surface and that also avoid high fuel consumption because they refrain from exerting extra drag. In most cases,

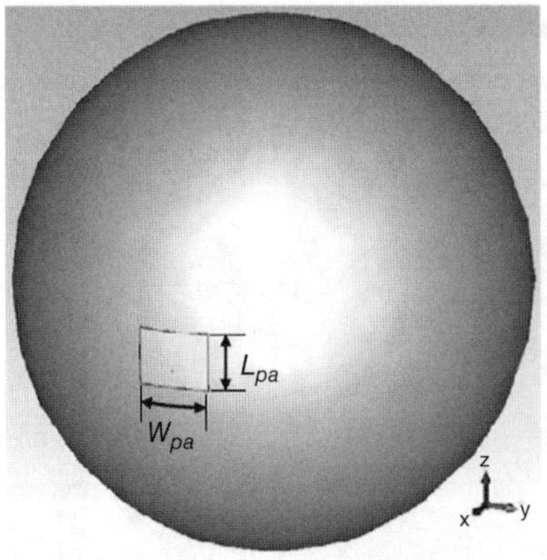

Figure 9.18 Patch microstrip antenna conformal to spherical surface [16].

microstrip patch antennas have to be conformal on a spherical structure with patch length L_{PA} and patch width W_{PA}, as shown in Figure 9.18.

9.8 EU WINNER Projects [17]

The European Union (EU) World Wireless Initiative New Radio (WINNER) projects are endeavoured to investigate the radio propagation in scenarios such as office indoor, indoor-to-indoor, indoor-to-outdoor, outdoor-to-outdoor, suburban macrocell, urban microcell, and rural mobile networks built on a single radio access technology with improved abilities. WINNER I project work conducted during 2004–2005 and WINNER II is extension of the WINNER I project ended in 2007. WINNER I channel models were based on channel measurements performed at 2 and 5 GHz bands during the project. WINNER II project work resumed the channel modelling work of WINNER I and extended the model to a geometry-based stochastic channel modelling approach features, frequency range (2–6 GHz), and for the same scenarios.

The channel models investigated in WINNER are antenna independent, i.e. different antenna configurations and different element patterns were investigated. The channel parameters were defined stochastically, according to statistical distributions derived from channel measurement. The distributions were defined for, e.g. delay spread (DS), delay values, angle spread, shadow fading (SF), and cross-polarization ratio. A number of measurement campaigns provided the environments for the parameterisation of the propagation scenarios for both line-of-sight (LOS) and non-LOS (NLOS) conditions. The system considered by WINNER projects comprised multiple BSs, multiple relay stations, and multiple mobile stations (MSs), as depicted in Figure 9.19. In WINNER channel models were assumed independent of distance, which is probably not strictly valid, it is used for simplicity of the model. Furthermore, the angles of arrival and departure were specified in 2D, i.e. only azimuth angles are considered. However, for the indoor and outdoor-to-indoor cases, the angles to be considered are azimuth and elevation, and the channel models should be constructed as 3D.

The small-scale parameters like delays, powers, and direction of arrival and departure (DoA, DoD) are randomly drawn from appropriate random distributed functions. In a spatial channel model (SCM), the performance of the single link is defined by small-scale parameters of multipath components (MCPs) between BS and MS. WINNER project findings were adopted by 3GPP3 Release 6, which is examined in the next section.

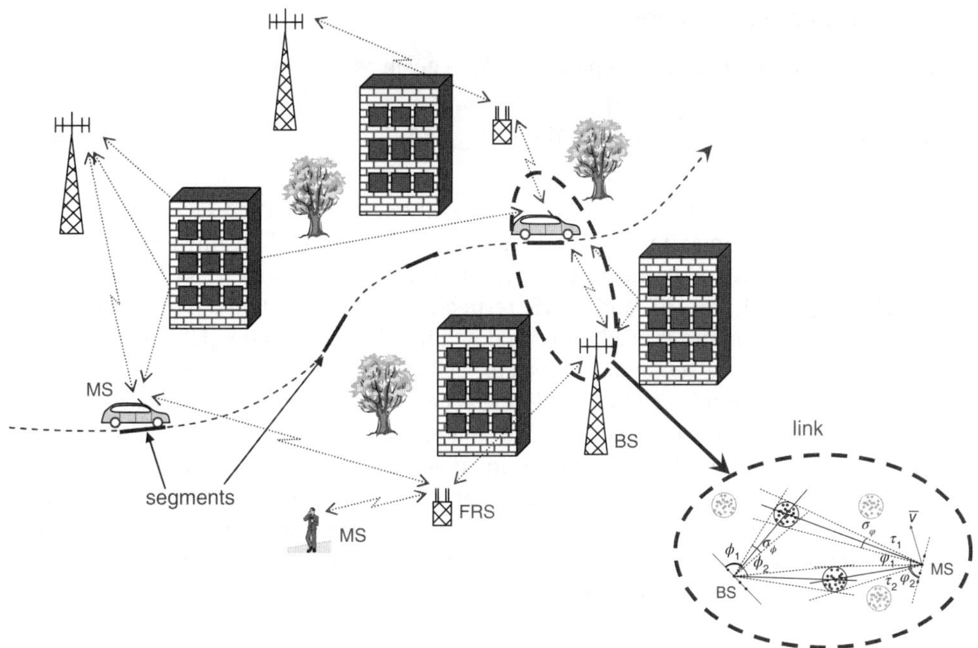

Figure 9.19 System level approach, several drops [17].

9.9 Spatial MIMO Channel Model in 3GPP Release 6 [18]

The 2D SCMs are defined in x, y Cartesian coordinate plane in multipath environments. Paths are assumed to be independent and each resolvable path is characterised by its own spatial parameters such as DS, AS, angle of arrival, and power azimuth spectrum (PAS). Assumptions declared in the deliberation apply to the BS and the MS spatial parameters for *link-level channel model.*

The scattering during the wave propagation generates clusters of subpaths. For example, Release 6 defined a wireless propagation of 6 paths, each comprising 20 subpaths. The spatial channels models are generally stochastic models, and two or three levels of randomness are used when theoretically generating a model. Macro suburban/urban and urban microcell parameters such as the *mean* of the azimuth spread at BS and MS are fixed, while the azimuth spread at BS and MS are random variables with *lognormal distributions* defined in terms of measured mean and standard deviation values.

The angle of departure (AOD) per path at the BS and the angle of arrival (AOA) at MS were described as random variables with uniform distribution. The SF is defined as random lognormal distributed variables with zero mean and measured standard deviation [19]. The propagation loss is defined linearly in terms of the distance.

A list of the symbols used in the following analysis is given by 3GPP Release 6:

σ_{AS} Angle spread = Azimuth spread
σ_{DS} delay spread
σ_{SF} lognormal SF random variable
σ_{SH} log normal SF constant
$\mathcal{N}(a, b)$ represents a random normal (Gaussian) distribution with mean a and variance b

9.9.1 BS and MS Antenna Patterns

The BS three-sector antenna radiation pattern $A(\theta)$ is given as

$$A(\theta) = -\min\left[12\left(\frac{\theta}{\theta_{3dB}}\right)^2, A_m\right] \quad -180 \leq \theta \leq 180 \tag{9.15}$$

where θ is defined as the angle between the direction of interest and the *boresight* of the antenna (i.e. axis of maximum gain/maximum radiated power), θ_{3dB} is the 3 dB beamwidth in degrees, and A_m is the maximum attenuation. For three-sector cell, θ_{3dB} is 70° and $A_m = 20$ dB while for six-sector cell, θ_{3dB} and A_m are 35° and 23 dB, respectively. The BS antenna pattern described in (9.15) covers each sector and is used for both downlink (DL) and uplink (UL). The radiation patterns are plotted for three-sector and six-sector cells and shown in Figure 9.20.

The gain for the three-sector cell 70° is 14 dBi. When the beamwidth is reduced by half to 35°, the corresponding gain is increased by 3 dB higher, resulting in 17 dBi. The BS antennas are designed with large inter-element spacing for diversity gain. However, for beamforming applications, small spacing is required, so different antenna designs may have to be considered, which give a different antenna pattern. For each MS antenna element, the radiation is assumed to be omni-directional with an antenna gain of −1 dBi. On the user's side, an MS antenna gain is determined using angle of arrival (AOA).

9.9.2 Per-Path BS and MS Angle Spread (AS)

The BS per-path AS is described as the root mean square (RMS) of angles with which an arriving path's signal is received by the BS array. The individual path powers are defined in the time-based channel model. For the 2D SCM, 3GPP Release 6 specified two values of BS angle spread (AS), each associated with a corresponding AoD: AS= 2° at AoD 50° and AS = 5° at AoD 20°. It is worth noting that the *link performance* with the two AS values will be different since the BS antenna gain for the two corresponding AoDs is different.

The MS per-path AS is interpreted as the RMS of angles of an incident path's signal at the MS array.

Two values of the path's AS for the MS are quoted in 3GPP Release 6: AS = 104° for a uniform PAS over 360°, and AS = 35° for a Laplacian PAS. A tutorial on Laplacian distribution is presented in 9.A.

9.9.3 Per-Path BS and MS Power Azimuth Spectrum

The paths signal arriving at the BS is expected to have a PAS with a Laplacian distribution. For an AoD, $\overline{\theta}$, and RMS angle-spread σ_{BS}, then the BS per path PAS estimate at an angle θ is given by

$$P_{BS}(\theta, \sigma_{BS}, \overline{\theta}) = C_0^{BS} \exp\left[\frac{-\sqrt{2}\,[\theta - \overline{\theta}]}{\sigma_{BS}}\right] G_{BS}(\theta) \tag{9.16}$$

where C_0^{BS} is a normalization constant, G_{BS} is the BS antenna gain, and both angles θ and $\overline{\theta}$ are given with respect to the boresight of the antenna elements assuming that all antenna elements' orientations are aligned. The average received power P is the average received power, and the BS average gain is given by

$$G_{BS} = 10^{\frac{A_{BS}(\theta)}{10}} \tag{9.17}$$

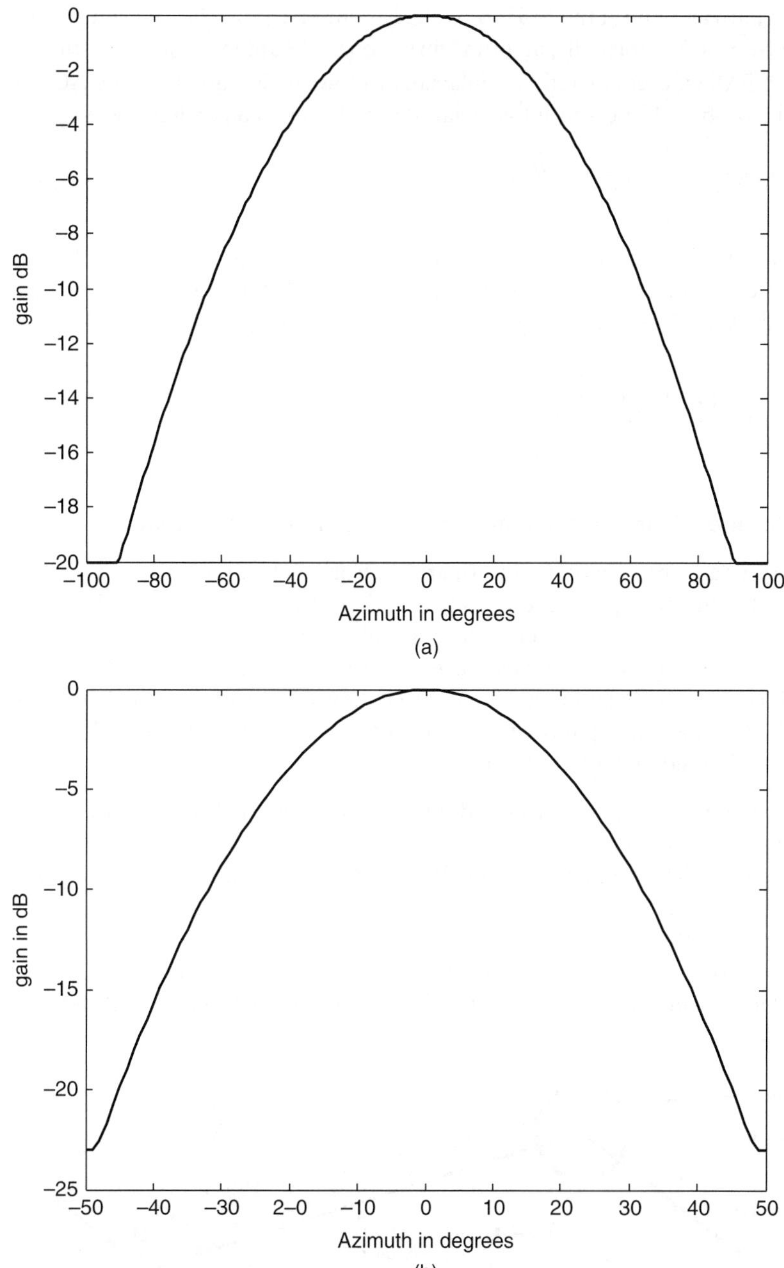

Figure 9.20 (a) Antenna radiation pattern for three-sector cells. (b) Antenna radiation Pattern for six-sector cells [18].

where $A_{BS}(\theta)$ is the BS antenna radiation pattern given by (9.15), and C_0^{BS} is a normalization constant given as

$$(C_0^{BS})^{-1} = \int_{-\pi+\bar{\theta}}^{\pi+\bar{\theta}} \exp\left[\frac{-\sqrt{2}\,[\theta - \bar{\theta}]}{\sigma_{BS}}\right] G_{BS}(\theta)\, d\theta \tag{9.18}$$

The per-path PAS of a path arriving at the MS is modelled as either a Laplacian distribution or a uniform distribution over 360°. Normally, an omni-directional MS antenna gain is assumed, so the received per-path PAS will remain either Laplacian or uniform. For an incoming AOA $\overline{\theta}$ and RMS angle-spread σ_{MS}, the MS per-path Laplacian PAS value at an angle θ is given by

$$P_{MS}(\theta, \sigma_{MS}, \overline{\theta}) = C_0^{MS} \exp\left[\frac{-\sqrt{2}\,[\theta - \overline{\theta}]}{\sigma_{MS}}\right] \tag{9.19}$$

Comparing (9.19) with (9.18), we noticed the MS antenna gain is missing, i.e. because the MS antenna pattern is omni-directional (i.e. gain is zero dB). It is assumed that all antenna elements' orientations are aligned. Also, P is the average received power and C_0 is the normalization constant:

$$(C_0^{MS})^{-1} = \int_{-\pi+\overline{\theta}}^{\pi+\overline{\theta}} \exp\left[\frac{-\sqrt{2}\,[\theta - \overline{\theta}]}{\sigma_{MS}}\right] d\theta \tag{9.20}$$

9.9.4 Definitions of BS and MS Angle Parameters for a Scattering Environment

Consider the BS and MS arrangement where a multipath environment exists between the ends of the transmission link. Let the multipath received signal at the MS consists of N, $n = 1, \ldots.. N$ paths defined by powers and delays and be chosen randomly according to the channel generation procedure. The scattering process takes place at various obstructs during EM wave propagation, each path consists of M subpaths. Figure 9.21 shows the scattering environment between BS array and MS with angle parameters used in the model. The BS and MS angle parameters are listed and defined in 3GPP Release 6 as follows:

Ω_{BS} BS antenna array orientation is defined as the difference between the broadside of the BS array and the absolute north reference direction.
θ_{BS} Line-of-sight (LOS) AOD direction between the BS and MS, with respect to the broadside of the BS array.
$\delta_{n,\,AoD}$ AOD for the nth path with respect to the LOS AOD θ_0
$\Delta_{n,\,m,\,AoD}$ Offset for the mth ($m = 1 \ldots M$) subpath of the nth path with respect to $\delta_{n,\,AoD}$
$\theta_{n,\,m,\,AoD}$ Absolute AoD for the mth ($m = 1 \ldots M$) subpath of the nth path at the BS with respect to the BS broadside.

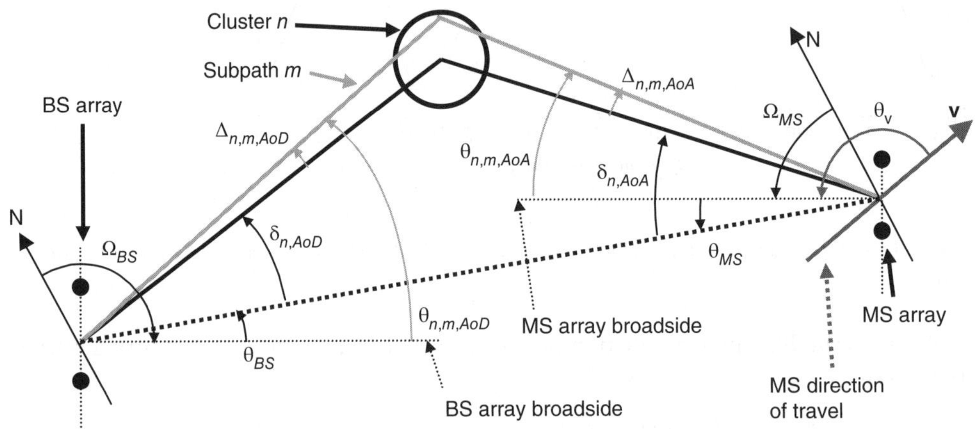

Figure 9.21 The scattering environment between BS and MS with angle parameters [18].

Ω_{MS} MS antenna array orientation defined as the difference between the broadside of the MS array and the absolute North reference direction.
θ_{MS} Angle between the BS-MS LOS and the MS broadside.
$\delta_{n,\,AoA}$ AoA for the nth path with respect to the LOS AoA $\theta_{0.\,MS}$
$\Delta_{n,\,m,\,AoA}$ Offset for the mth subpath of the nth path with respect to $\delta_{n,\,AoA}$
$\theta_{n,\,m,\,AoA}$ Absolute AoA for the mth subpath of the nth path at the MS with respect to the BS broadside
V MS velocity vector
θ_v Angle of the velocity vector with respect to the MS broadside: $\theta_v = \arg(V)$

The angles that are measured in a clockwise direction are assumed to be negative in value.

9.10 The Scattering Environments

Three categories of environments are scrutinized by 3GPP Release 6, namely: suburban macrocell, urban macrocell, and urban microcell [18]. The fast-fading per-path is changing in time, although parameters, including AS, DS, lognormal shadowing, and MS location will remain fixed during the period of its evaluation. The following are general assumptions that are independent of environment:

1. *UL-DL reciprocity*. The AOD/AOA values are identical between the UL and DL.
2. For frequency-division duplex (FDD) systems, random subpath phases between UL, DL are uncorrelated since they operate at different frequencies. However, for time-division duplex (TDD) systems, the phases will be fully correlated.
3. Shadowing among different mobiles is uncorrelated except when mobiles are very close to each other. This assumption is widely used to simplify the model analysis.
4. The SCM is assumed to work with any type of antenna configuration (whose size is smaller than the shadowing coherence distance). Reference antenna configurations are uniform linear array with 0.5λ or more interelement spacing commonly used.
5. The composite Azimuth angle spread (AS), DS, and lognormal SF, which may be correlated parameters depending on the channel scenarios, are applicable to all sectors/antennas of a given BS.
   Subpath phases are random variables between coverage sectors. The AS is composed of multipaths, each path has numerous subpaths, and each subpath has a precise angle of departure that corresponds to an antenna gain from each BS antenna. The SF is a common parameter among all the BS antennas or sectors.
6. The elevation spread is not modelled in the 2D model.
7. The generation of the channel coefficients assumes linear arrays.

9.11 Large-Scale Parameters (LSPs)

Thus far, we have investigated small-scale parameters like delays, powers, directions of departure and arrival, and their values are picked randomly from tabulated distribution functions. Random large-scale parameters (LSPs) are second moments like SF, (i.e. power fluctuation over large area), standard deviation, DS and distribution, angle of departure spread and distribution, angle of arrival spread and distribution, and Rician K-factor.

These LSPs are obtained through averaging over a long channel segment (i.e. over distance of some tens of wavelengths λ). The LSPs have the following support parameters: scaling parameter for delay distribution, cross-polarization power ratios, number of clusters, cluster AS of departure, cluster AS of arrival, per cluster shadowing, auto-correlations of the LSPs, cross-correlations of the LSPs, and number of rays per cluster. LSPs are drawn randomly from random distribution functions based on measured mean values and standard deviation values.

A spatial link shows inter-dependence of LSPs described by correlation observed in measured data. Such dependence is specified through cross correlation value defined for LSPs as:

$$\rho_{xy} = \frac{C_{xy}}{\sqrt{C_{xx} C_{yy}}} \tag{9.21}$$

where C_{xy} is the cross-covariance of LSPs values in x and y. At system level, two types of correlations could be defined:

1. Between MSs being connected to the same BS
2. Correlations of links from the same MS to multiple BSs as shown in Figure 9.22. These correlations are mostly caused by scattering in nearby links in a similar environment. The cross correlation coefficient describes the LSPs dependence is exponentially decaying with distances. Accordingly, LSPs of two MSs links to same BS undergo correlations that are exponential since correlations are proportional to their relative distances. Further, correlation coefficients matrices of neighbouring links of MSs are not independent over distances so that:

$$\rho_{xy}(d_{MS}) = \frac{C_{xy}(d_{MS})}{\sqrt{C_{xx} C_{yy}}} \tag{9.22}$$

The SF for links from one MS to different BSs shows constant correlation coefficient equal to 0.5. Additionally, correlation properties of links from the same MS to multiple BSs were examined in [17]. The results obtained from different measurement scenarios are inconsistent, showing rather high correlation for one measurement and quite low for another. Their results could not show clear correlation behaviour between different BSs.

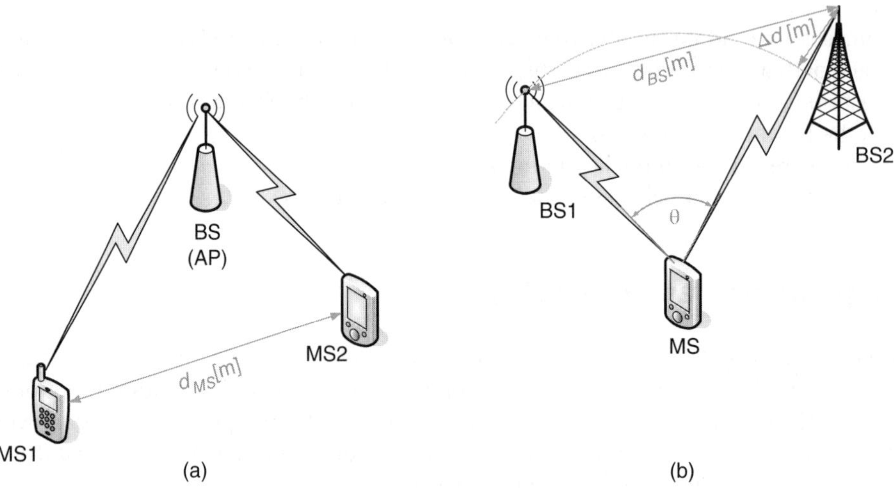

Figure 9.22 Links towards common BS exhibits inter-correlations: (a) fixed common station, (b) mobile common station [17].

9.11.1 Correlation Between Channel Parameters in 3GPP Release 6

The DS and SF were investigated by Greenstein et al. [19]. Their results show that DS and SF are indeed lognormal distributed at any given distance. In addition, they can be assumed to be correlated. However, the median of the DS distributions grows as some power of the distance. The variation of the DS distribution around the median can be negatively correlated with the SF. When there is a significant negative correlation between lognormal SF and DS, this indicates that for a larger SF, the DS is reduced, and for a smaller SF, the DS is increased.

The Azimuth spread is also lognormal distributed, and correlated to the DS and SF. The correlation functions of these parameters are normally quite high; hence, a spatial channel has to be modelled, such that it can reproduce this correlation behaviour along with the expected probability and range of each parameter.

Suppose we want to generate the values of DS, AS, and SF for the bth BS where $b = 1, \ldots, B$ with respect of a given MS. Let us denote these values as $\sigma_{DS,b}$, $\sigma_{AS,b}$, and $\sigma_{SF,b}$. These values are a function of the respective correlation Gaussian random variables α_b, β_b, and γ_b. The latter random variables are, in turn, respectively generated from independent Gaussian random distribution.

To guarantee that the full correlation matrix is *positive semi-definite*, the intra-site correlations between SF, DS, and AS are

$\rho_{\alpha\beta}$ = Correlation between DS and AS = +0.5
$\rho_{\gamma\beta}$ = Correlation between SF and AS = −0.6
$\rho_{\gamma\alpha}$ = Correlation between SF and DS = −0.6

The resulting intra-site correlation matrix, is given by

$$\mathbf{A} = \begin{bmatrix} 1 & \rho_{\alpha\beta} & \rho_{\gamma\alpha} \\ \rho_{\alpha\beta} & 1 & \rho_{\gamma\beta} \\ \rho_{\gamma\alpha} & \rho_{\gamma\beta} & 1 \end{bmatrix} = \begin{bmatrix} 1 & +0.5 & -0.6 \\ +0.5 & 1 & -0.6 \\ -0.6 & -0.6 & 1 \end{bmatrix} \tag{9.23}$$

In addition to *intra-site* correlations given by $\mathbf{A}$, there are also correlations between these parameters between different sites. These *inter-site* correlations are given by $\mathbf{B}$

$$\mathbf{B} = \begin{bmatrix} 0 & 0 & 0 \\ 0 & 0 & 0 \\ 0 & 0 & \zeta \end{bmatrix} = \begin{bmatrix} 0 & 0 & 0 \\ 0 & 0 & 0 \\ 0 & 0 & 0.5 \end{bmatrix} \tag{9.24}$$

Essentially, only inter-site correlations between SF are included. ζ is the SF correlation, which is $\zeta = 0.5$.

9.11.2 Generation of Values of DS, AS, SF

The values $\sigma_{DS,b}$, $\sigma_{AS,b}$, and $\sigma_{SF,b}$ for the bth BS regarding a given MS are functions of the respective correlated *Gaussian random variables* α_b, β_b, *and* γ_b, which are, in turn, respectively, generated from independent Gaussian distributions w_{b1}, w_{b2}, and w_{b3} as well as three global (applicable to all BSs) independent Gaussian random variables ξ_1, ξ_2, ξ_3. The variables α_b, β_b, and γ_b are then given by

$$\begin{bmatrix} \alpha_b \\ \beta_b \\ \gamma_b \end{bmatrix} = \begin{bmatrix} c_{11} & c_{12} & c_{13} \\ c_{21} & c_{22} & c_{23} \\ c_{31} & c_{32} & c_{33} \end{bmatrix}^{\frac{1}{2}} \begin{bmatrix} \xi_1 \\ \xi_2 \\ \xi_3 \end{bmatrix} \tag{9.25}$$

Thus

$$\mathbf{A}-\mathbf{B}=\begin{bmatrix} 1 & \rho_{\alpha\beta} & \rho_{\gamma\alpha} \\ \rho_{\alpha\beta} & 1 & \rho_{\gamma\beta} \\ \rho_{\gamma\alpha} & \rho_{\gamma\beta} & 1-\zeta \end{bmatrix}$$

where the matrix $\mathbf{C}$ with elements c_{ij} is given by

$$\mathbf{C}^{\frac{1}{2}}=(\mathbf{A}-\mathbf{B})^{\frac{1}{2}}=\begin{bmatrix} 1 & \rho_{\alpha\beta} & \rho_{\gamma\alpha} \\ \rho_{\alpha\beta} & 1 & \rho_{\gamma\beta} \\ \rho_{\gamma\alpha} & \rho_{\gamma\beta} & 1-\zeta \end{bmatrix}^{\frac{1}{2}} \tag{9.26}$$

Using A and B in Eqs. (9.23) and (9.24) gives the matrix $\mathbf{C}$ with elements c_{ij}. The square root of matrix $\mathbf{C}$ can be acquired using Cholesky decomposition as

$$\sqrt{\mathbf{C}}=\begin{pmatrix} 1 & 0.5 & -0.6 \\ 0 & 0.8660 & -0.3464 \\ 0 & 0 & 0.1414 \end{pmatrix}$$

Equation (9.25) can be simplified as

$$\begin{bmatrix} \alpha_b \\ \beta_b \\ \gamma_b \end{bmatrix}=\begin{bmatrix} 1 & 0.5 & -0.6 \\ 0 & 0.8660 & -0.3464 \\ 0 & 0 & 0.1414 \end{bmatrix}\begin{bmatrix} \xi_1 \\ \xi_2 \\ \xi_3 \end{bmatrix} \tag{9.27}$$

The random variables ξ_1, ξ_2, ξ_3 are generated using three independent Gaussian distributions. The distribution of DS for the bth BS is given by

$$\sigma_{DS,b}=10^{\varepsilon_{DS}\alpha_b+\mu_{DS}} \tag{9.28}$$

where α_b is generated in (9.27), μ_{DS} are determined by measurements as the logarithmic mean of the distribution of DS as,

$$\mu_{DS}=\mathbb{E}[\log_{10}(\sigma_{DS,b})] \tag{9.29}$$

The ε_{DS} is the logarithmic standard deviation of the distribution of DS as

$$\varepsilon_{DS}=\sqrt{\mathbb{E}[\log_{10}(\sigma_{DS,b})]^2-\mu_{DS}^2} \tag{9.30}$$

Similarly the distribution of AS is given by

$$\sigma_{AS,n}=10^{\varepsilon_{AS}\beta_b+\mu_{AS}} \tag{9.31}$$

where β_b is generated in (9.27), ε_{AS} and μ_{AS} are determined by measurements. The logarithmic mean of the distribution of AS is computed as

$$\mu_{AS}=\mathbb{E}[\log_{10}(\sigma_{AS,b})] \tag{9.32}$$

The logarithmic standard deviation of the distribution of AS is computed as

$$\varepsilon_{AS}=\sqrt{\mathbb{E}[\log_{10}(\sigma_{AS,b})]^2-\mu_{AS}^2} \tag{9.33}$$

Finally, the distribution for SF is given by

$$\sigma_{SF,n}=10^{\frac{\sigma_{SH}\gamma_b}{10}} \tag{9.34}$$

where γ_b is given in (9.27), and σ_{SH} is the SF standard deviation given in dB determined by measurement. The expected values for σ_{SH} are 8 and 10 dB for the macrocell and microcell cases, respectively. Note that the dB value for SF is simply given as $\sigma_{SH}\gamma_b$.

9.12 2D Spatial Channel Models (SCMs)

9.12.1 Spatial Channel Models with No Antennas Polarization

The development of the SCMs in 3GPP Release 6 is based on the statistical relations between angles, powers, as well as power profiles. As we emphasized previously, the models considered are for *urban* and *suburban macrocell* (distance between BS-to-BS = 3 km) and *urban microcell* (distance between BS to BS = 1 km), site to site correlation = 0.5, number of paths investigated is 6 and bandwidth less than 5 MHz.

9.12.2 Path Loss (PL)

The macrocell path-loss (PL) according to COST231 model is determined as

$$PL\,(dB) = (44.9 - 6.55\log_{10}h_{BS})\log_{10}\frac{d}{1000} + 45.5 + (35.46 - 1.1h_{MS})\log_{10}f_c - 13.82\log_{10}h_{BS} + 0.7h_{MS} + C \tag{9.35}$$

where h_{BS} and h_{MS} are BS and MS antennas heights in metres, d is distance between the BS and MS in metres (m), f_c is carrier frequency in MHz, and C is constant $C = 0\,dB$, $3\,dB$ for suburban and urban macrocell environments, respectively. Assuming f_c = 1900 MHz, $h_{BS} = 32\ m$ and $h_{MS} = 1.5\ m$, respectively, Eq. (9.35) is simplified to the following empirical equations:

$$PL\,(dB) \cong 31.5 + 35\log_{10} d \text{ for suburban macrocell environment} \tag{9.36}$$

$$PL\,(dB) \cong 34.5 + 35\log_{10} d \text{ for urban macrocell environment} \tag{9.37}$$

where d has to be with the urban and suburban macrocell ranges. For example, for d = 3 km macrocell, the *PL* computed by (9.35) is 150.648 dB and by (9.36) is 150.425 dB.

The microcell path-loss model based on the COST231 comprises the following parameters: h_{BS} = 12.5 m, h_{MS} = 1.5 m, building height = 12 m, building-to-building distance = 50 m, street width 25 m, and orientation for all paths is 30°. The NLOS model is expressed as

$$PL\,(dB) = -55.9 + 38\log_{10}d + \left(24.5 + \frac{1.5f_c}{925}\right)\log_{10}f_c \tag{9.38}$$

and for micro-cell LOS,

$$PL\,(dB) = -35.4 + 26\log_{10}d + 20\log_{10}f_c \tag{9.39}$$

Assume f_c = 1900 MHz, with microcell parameters, Eq. (9.38) is simplified by an empirical formula:

$$PL\,(dB) = 34.53 + 38\log_{10}d \tag{9.40}$$

In a similar manner, (9.39) can be simplified to

$$PL\,(dB) = 30.18 + 26\log_{10}d \tag{9.41}$$

For microcell NLOS d = 100 m, computed *PL* using (9.38) is 104.527 dB and using by (9.40) is 104.525.

9.12.3 2D Channel Coefficients

Consider a cellular system with linear BS array of M elements and MS array of N elements and a propagation environment consisting of J MCPs, each with I subpaths, the 2D channel coefficients are described by N × M matrix of complex amplitudes. The jth path component channel matrix is denoted as $\mathbf{H}_j(t)$ for $j = 1, \ldots \ldots \ldots, J$ and the $(n, m)^{th}$ component of $\mathbf{H}_j(t)$ is, $h_{n,m,j}(t)$ given by

$$h_{n,m,j}(t) = \sqrt{\frac{P_j \sigma_{SF}}{I}} \sum_{i=1}^{I} X1 * X2 * X3 \tag{9.42a}$$

$$X1 = \sqrt{G_{BS}(\theta_{j,i,AoD})}\, e^{j[k\, d_m \sin\theta_{j,i,AoD} + \Phi_{j,i}]} \tag{9.42b}$$

$$X2 = \sqrt{G_{MS}(\theta_{j,i,AoA})}\, e^{j[k\, d_n \sin\theta_{j,i,AoA}]} \tag{9.42c}$$

$$X3 = e^{jk\|V\| \cos(\theta_{j,i,AoA} - \theta_v)t} \tag{9.42d}$$

where $\Phi_{j,i}$ is the phase of the ith subpath of the jth path, $\|V\|$ is the magnitude of the MS velocity vector, θ_v is the angle of the MS velocity vector, and P_j is the power of the jth path based on a power-delay envelope, $\theta_{j,i,AoD}$ is the AoD for the ith subpath of the jth path, $\theta_{j,i,AoA}$ is the AoA for the ith subpath of the jth path, $G_{BS}(\theta_{j,i,AoD})$ is the BS antenna gain of each array element, $G_{MS}(\theta_{j,i,AoA})$ is the MS antenna gain of each array element, d_m is the distance in metres from BS antenna element m from the reference $m = 1$, and d_n is the distance in metres from MS antenna element n from the reference $n = 1$.

9.12.4 Generating Channel Parameters for Urban, Suburban Macrocell, and Urban Microcell Environments

In this section, we evaluate urban and suburban macrocells, with noted differences from urban microcells.

Step 1. Designate channel environment as either an urban macrocell or suburban macrocell.

Step 2. Ascertain various distance, angles, and orientation parameters in the cell layout: d is distance between MS and BS and compute the path loss, LOS directions relating to the BS and MS i.e. θ_{BS}, θ_{MS}, MS antenna array orientations $\boldsymbol{\Omega}_{MS}$ are *i. i. d.* picked from a uniform $0° - 360°$ distribution, $\|V\|$ is the magnitude of the MS velocity vector picked from the MS velocity distribution, and θ_v chosen from a uniform $0° - 360°$ distribution.

Step 3. Determine the *DS*, *AS*, *SF*.

These random variables σ_{DS}, σ_{AS}, σ_{SF}, respectively, are determined by measurements at the carrier frequency and given by (9.28), (9.31), and (9.34), respectively. For suburban macrocell and urban macrocell $\sigma_{SF} = 8\,dB$. Unlike macrocells, the urban microcell individual multipaths are independently shadowed so microcell *SF* for each path is generated using (9.34).

Step 4. Determine paths random delays.

For macrocell environment, the random delays for each of the J MCPs $\tau'_1, \ldots \ldots \ldots, \tau'_J$ are generated conforming to

$$\tau'_j = -r_{DS}\, \sigma_{DS} \log z_j \quad j = 1, \ldots \ldots \ldots, J$$

where z_j are *i. i. d* random variables with uniform distribution with values between 0 and 1, r_{DS} is obtained by measurements and σ_{DS} is generated using (9.28) above. These variables are ordered so that $\tau'_J > \tau'_{J-1} > \ldots\ldots\ldots > \tau'_1$ and the minimum is subtracted from all delays, giving the delay of the jth path a value of $\tau_j = \tau'_j - \tau'_1$. The delay τ_j is quantized in time.

The random delays for the microcell are *i. i. d* drawn from a uniform distribution from 0 to 1.2 μ *sec* and are quantized in time. The minimum of these delays is subtracted for all delays so that the first delay is zero and delays are quantized in time. For LOS, the delay of the direct component is set to zero.

For urban microcell environment, the random delays for each of the J MCPs τ_j $j = 1, \ldots, J$ are random variables picked from a uniform distribution from 0 to 1.2 μs.

Step 5. Determine random average power for each path of the multipath. Let the random average power of the jth path be P'_j. For suburban and urban macrocells, P'_j is given by

$$P'_j = e^{\frac{(1-r_{DS})(\tau'_j-\tau'_1)}{r_{DS}\sigma_{DS}}} 10^{-\frac{\xi_j}{10}}$$

where $j = 1, \ldots\ldots\ldots\ldots, 6$ and ξ_j are *i. i. d* Gaussian random variables with standard deviation 3 dB, and τ'_j is the unquantized channel delay. The randomizing process produced by ξ_j is to vary the powers as experienced in the actual channel, r_{DS} is the normalizing factor determined by measurements. The average powers for each of the J multi-paths sum up to give the power of the multipath, P_j, normalized (so the total power of all paths equal to one) is given as

$$P_j = \frac{P'_j}{\sum_{l=1}^{J} P'_l} \tag{9.43}$$

The urban microcell powers for each path are exponentially decaying in time with the addition of a lognormal randomness, which is independent of the path delay and is given as

$$P'_j = 10^{-\left(\tau_j + \frac{z_j}{10}\right)}$$

where τ_j are unquantized delays in μ *secs* and z_j is *i. i. d* zero mean Gaussian random variables with standard deviation 3 dB. The average powers P_j are given above:

The LOS power of the direct component, P_D and the ratio of direct power to scattered power K are related as

$$K = \frac{P_D}{1 - P_D} \tag{9.44}$$

and the normalized path powers are given by

$$P_j = \frac{P'_j}{(K+1)\sum_{l=1}^{J} P'_l}$$

Step 6. Determine AODs for each of the J MCPs.

Suburban and urban macrocell AODs random variables with Gaussian distribution, the AODs for the jth path can be generated using the zero mean *Gaussian distribution function* δ'_j defined as

$$\delta'_j \sim \mathcal{N}(0, \sigma^2_{AoD})$$

where $\sigma_{AoD} = r_{AS}\,\sigma_{AS}$, and r_{AS} is tabulated in Release 6 and σ_{AS} is determined in step 3, then variables δ'_n are ordered in increasing value, $|\delta'_1| < |\delta'_2| < \ldots.. < |\delta'_J|$. The AODs in degrees is given by $\delta_{j,AoD}$ assigned to the ordered variables as

$$\delta_{j,AoD} = \delta'_j \text{ for } j = 1, 2, \ldots, J.$$

For urban microcell, the AoDs (with respect to the LOS direction) are *i. i. d* random variables picked from a *uniform distribution* over –40° – +40°,

$$\delta_{j,AoD} \sim U(-40, +40)$$

The jth path AODs, $\delta_{j,\,AoD}$ is associated with paths power P_j. In LOS model, the AODs of the direct component is set to the LOS direction.

Step 7. Connect the multipath delays with AODs.

Connect the jth delay generated in Step 4 with j^{th} AOD $\delta_{j,\,AoD}$ generated in Step 6.

Step 8. Ascertain the powers, phases, and offset (drifting) of AODs of the 20 subpaths for each of the M paths at the BS.

Suburban and urban macrocells, as well as urban microcell subpaths associated with the jth path, have the same power given by $(\frac{P_j}{I})$ where P_j is given in Step 5 and their *i. i. d* phases $\Phi_{j,\,i}$ are picked from uniform 0°–360° distribution. The relative phase offsets of the i^{th}path ($i = 1, \ldots \ldots, I$) subpath of each path, is a fixed value given in Release 6.

Step 9 Determine *the* AOAs for each of the MPCs.

Suburban and urban macrocell AoAs are *i. i. d* Gaussian random variables given by

$$\delta_{j,AoA} \sim \mathcal{N}(0, \sigma^2_{j,AoA})$$

where $\sigma_{j,\,AoA}$ is defined by the following model:

$$\sigma_{j,AoA} = 104.12(1 - e^{-0.2175|10\log_{10}P_j|})$$

where P_j is the normalized power of the jth path generated in Step 5.

For urban microcell environments, $\sigma_{j,\,AoA}$ is defined by the following model:

$$\sigma_{n,AoA} = 104.12(1 - e^{-0.265|10\log_{10}P_j|})$$

For the LOS model, AOA of the direct component is set to the LOS direction.

Step 10. Determine the offset AOA at the user equipment (UE) of the 20 subpaths for each path at the MS.

The relative offset AOAs of subpath $\Delta_{j.\,i.\,AoA}$ is a fixed value, as for the AOD offsets in Step 8 given in Release 6.

Step 11. Associate the BS and the MS paths and subpaths.

The jth BS path (defined by its delay τ_j, power P_j, and AOD $\delta_{j,\,AOD}$) is associated with the jth MS path (defined by its offset $\Delta_{j.\,i.\,AOA}$). Then, for the jth path, randomly pair each of the *J BS* subpaths defined by its offset $\Delta_{j.\,i.\,AOD}$ with a MS subpath defined by its offset $\Delta_{j.\,i.\,AOA}$. Each subpath pair is connected so the phases are defined by $\Phi_{j,\,i}$ *as in Step 8.*

Step 12. Choose the BS and MS antenna gains for subpaths associated with subpath AODs and AOAs.

Suburban and urban macrocell, for the jth path, the AOD of the ith subpath with respect of the MS antenna array broadside is $\theta_{j,\,i,\,AOD}$, is defined as

$$\theta_{j,i,AOD} = \theta_{BS} + \delta_{j,AOD} + \Delta_{j.i.AOD}$$

where θ_{BS} is LOS AOD direction between the BS and MS, with respect to the broadside of the BS array, $\delta_{j,\,AoD}$ is AOD for the jth path with respect to the LOS AOD θ_0, and the $\Delta_{j.\,i.\,AOD}$ is offset for the ith subpath of the jth path with respect to $\delta_{j,\,AOD}$.

Similarly, the AOA of the ith subpath for the jth path with respect of the MS antenna array broadside $\theta_{j.\,i.\,AOA}$ is defined as

$$\theta_{j,i,AOA} = \theta_{MS} + \delta_{j,AOA} + \Delta_{j.i.AOA}$$

The BS and MS antenna gains depend on subpath AODs and AOAs, expressed as

$G_{BS}(\theta_{j,i,AOD})$ and $G_{MS}(\theta_{j,i,AOA})$.

Step 13. Exploit the path loss according to the BS-to-MS distance from 9.12.2.

9.13 2D Spatial Channel Models (SCMs) with Antenna Polarization

In the 2D SCMs presented in Section 9.12.2, antenna polarization is not included in the channel coefficients matrix. However, because of the limited space available at MS, polarized antennas are key to the implementation of future multiple antennas on handheld mobile devices. Indeed, polarization can reduce the number of antennas to half for the same throughput. In the following, we modified our SCM model representation to account for horizontal and vertical polarized antennas at BS and MS. Consequently, considering polarized antennas, we only need $\frac{M}{2}$ - element BS arrays instead of M and similarly $\frac{N}{2}$ -element MS arrays instead of N. That is, we consider cross-polarized antennas with the vertical (v) polarized and the horizontal (h) polarized antennas employ $\frac{M}{2}$ elements at BS and $\frac{N}{2}$ elements at MS. In addition, we model the additional processes related to the polarization itself as follows:

Step 14. Generate additional cross-polarized sub-paths.

We denote the paths and subpaths as in Section 9.12.2. For each of the *J* paths, we generate an additional *I* subpaths at MS and *I* subpaths at BS to represent the signal that flows into the cross -polarized antenna.

Step 15. Arrange subpath AoDs and AoAs.

We arrange the AoD and the AoA of each subpath to be equal to the corresponding subpath of the co-polarized antenna so that orthogonal subrays arrival/departure at common angles.

Step 16. Create phase offsets for the cross-polarized elements.

Defined $\Phi_{j,i}^{(x,y)}$ to be the phase offset of the ith subpath of the jth path between the x component (e.g. either the horizontal h or vertical v) of the BS element and the y component (e.g. either the h or v) of the MS element. We now use this definition to introduce four types of cross polarizations (XPs):

$\Phi_{j,i}^{(x,x)}$ set to be $\Phi_{j,i}$ generated for unpolarized arrays, and generate $\Phi_{j,i}^{(x,y)}, \Phi_{j,i}^{(y,x)}, \Phi_{j,i}^{(y,y)}$ as *i. i. d* random variables picked from a uniform 0 to 360°distribution. We use x and y for co-polarized and cross polarized antennas.

Step 17. Break down each of the co-polarized and cross-polarized rays into horizontal and vertical components.

Step 18. Determine cross-polarization discrimination ratio (*XPD*).

The cross-polarization discrimination ratio (*XPD*) is defined as

$$XPD = \frac{P_1}{P_2}$$

where P_2 is the coupled power of each subpath in the h direction and is set relative to the P_1 of each subpath in the v direction as per an *XPD* ratio. A single *XPD* ratio of each path is assumed to apply to all constituent subpaths of the named path. Each jth path undergoes an independent realization of *XPD*, which is drawn for a random distribution as follows:

For urban *macrocell environment* we set

$$P_2 = P_1 - A - B * \mathcal{N}(0,1) \tag{9.45}$$

where A = 0.34 * (mean relative path power in dB) + 7.2 dB and B = 5.5 dB is the standard deviation of the *XPD* variation.

For urban *microcell environment*, we define P_2 in a similar linear model to the urban macrocell but A = 8 dB and B = 8 dB is the standard deviation of the *XPD* variation. The value $\mathcal{N}(0,1)$ is a zero mean Gaussian random number with unit variance and is held constant for all subpaths of a given path.

The coupled power of the opposite polarization (i.e. horizontal to vertical) is the same. At the receive antennas each of the v and h are decomposed into co-polarized components and summed up.

The propagation characteristics of v to v paths are assumed to be equivalent to those of h to h paths. The fading between orthogonal polarization has been noticed by measurements to be independent, and therefore, the subrays phases can be chosen randomly.

9.13.1 2D Spatial Channel Model (SCMs) with Polarized Antennas

The SCM $h_{n,m,j}(t)$ is represented by a matrix defining the propagation of the horizontal h and vertical v amplitude of each subpath and their anticipated mixing, so the channel realization is defined as

$$h_{n,m,j}(t) = \sqrt{\frac{P_j \sigma_{SF}}{I}} \sum_{i=1}^{I} X1 * X2 * X3 \tag{9.46a}$$

where

$$X1 = \begin{bmatrix} \chi_{BS}^{(v)}(\theta_{j,i,AOD}) \\ \chi_{BS}^{(h)}(\theta_{j,i,AOD}) \end{bmatrix}^T \tag{9.46b}$$

$$X2 = \begin{bmatrix} e^{j\Phi_{j,i}^{(v,v)}} & \sqrt{r_{j1}}\, e^{j\Phi_{j,i}^{(v,h)}} \\ \sqrt{r_{j2}}\, e^{j\Phi_{j,i}^{(h,v)}} & e^{j\Phi_{j,i}^{(h,h)}} \end{bmatrix} \tag{9.46c}$$

$$X3 = \begin{bmatrix} \chi_{MS}^{(v)}(\theta_{j,i,AOA}) \\ \chi_{MS}^{(h)}(\theta_{j,i,AOA}) \end{bmatrix} e^{jkd_m \sin\theta_{j,i,AoD}}\, e^{jkd_n \sin\theta_{j,i,AoA}}\, e^{jk\|\mathrm{v}\| \cos(\theta_{j,i,AoA}-\theta_{\mathrm{v}})t} \tag{9.46d}$$

where $X2$ corresponds to the scattering phases and amplitudes of a plane wave with a given angle and polarization, $\chi_{BS}^{(v)}(\theta_{j,i,AoD})$ and $\chi_{BS}^{(h)}(\theta_{j,i,AoD})$ are the BS antenna complex response for the v and h components, respectively, $\chi_{MS}^{(v)}(\theta_{j,i,AoA})$ and $\chi_{MS}^{(h)}(\theta_{j,i,AoA})$ are the MS antenna complex response for the v and h components, respectively, $|\chi^{(.)}(.)|^2$ is the antenna gain, r_{j1} and r_{j2} are *i. i. d* random variables expresses the power ratio of waves of the jth path leaving the BS in the v direction and arriving at the MS in the h direction (i.e. $v \to h$) to those leaving in the v direction and arriving in the v direction (i.e. $v \to v$).

Similarly, leaving in h and arriving in v to those $v \to v$, respectively, $\Phi_{i,j}^{(x,y)}$ is phase offset of ith subpath of the jth path between x component (h *or* v) of the BS antenna element and the y component (h *or* v) of the MS antenna element.

The BS distance d_m is the distance in metres for BS antenna element m from the reference ($m = 1$) antenna. For the reference antenna $m = 1$ and $d_1 = 0$. The MS distance d_n is the distance in metres from MS antenna element n from the reference ($n = 1$) antenna. For the reference antenna $n = 1$, $d_1 = 0$. The wave-number $k = \frac{2\pi}{\lambda}$ where λ is the carrier wavelength in metres.

The above channel models in Sections 9.12.2 and 9.13.2 correspond to transmission from BS to MS (i.e DL channel). UL SCMs from MS to BS can easily be made by an appropriate UL-DL reciprocity and the fact that AoD/AoA values are identical between UL and DL.

9.14 3D Channel Models in 3GPP Release 14

This section discusses 3D channel models, including coordinate systems and line-of-sight (LOS), as further described in [20–22]

9.14.1 Coordinate Systems

A coordinate system used in map-based 3D channel model is described by the x, y, z axes with spherical angles, zenith angle θ, and azimuth angle ϕ as illustrated in Figure 9.23. In addition, the model calls for the contribution from the digital map to take into consideration the impacts from environmental structure such as trees, buildings, and building materials. The model can be used in the frequency band 0.5 GHz to 100 GHZ and doesn't need to be calibrated and can be used as per enterprise basis. Note that the zenith angle is measured from the z-axis in spherical coordinates and azimuth angle is measured from x axis in x, y plane in spherical coordinates.

9.14.2 Local and Global Coordinate Systems

Antenna array for a BS or a user terminal (UT) is generally defined in a local coordinate system (LCS). Such a system is used as reference to describe the far-field radiation power pattern and polarization of each element in the array. On the other hand, a network comprising multiple BSs and UTs are normally described in a different coordinate commonly known as global coordinate system (GCS). Accordingly, the engagement of individual BS /UT with the network call for LCS to be translated to wide network coordinates, i.e. GCS.

Consider a network with a GCS defined with coordinates (x, y, z, θ, ϕ) where θ defines the zenith angle and ϕ is the azimuth angle in a Cartesian coordinate system and unit vectors $\hat{\theta}, \hat{\phi}$. Similarly, we denote the LCS with primed-coordinates $(x', y', z', \theta', \phi')$ with unit vectors $\hat{\theta}', \hat{\phi}'$. Let us denote an antenna element radiation power pattern as $A(\theta', \phi')$ in the LCS and $A(\theta, \phi)$ denotes the same antenna element in GCS. Clearly the two are related as

$$A(\theta, \phi) = A'(\theta', \phi') \tag{9.47}$$

Figure 9.23 Definition of spherical angles and spherical unit vectors in a Cartesian coordinate system [20].

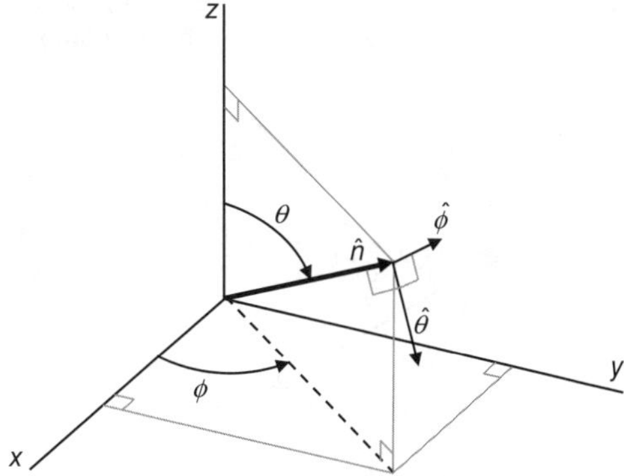

Let us denote the polarized field components in LCS and in GCS as $F'_{\theta'}(\theta', \phi'), F'_{\phi'}(\theta', \phi')$ and $F_\theta(\theta, \phi), F_\phi(\theta, \phi)$ and are related by

$$\begin{pmatrix} F_\theta(\theta, \phi) \\ F_\phi(\theta, \phi) \end{pmatrix} = \begin{pmatrix} \hat{\theta}(\theta, \phi)^T \boldsymbol{R} \widehat{\theta}'(\theta', \phi') & \hat{\theta}(\theta, \phi)^T \boldsymbol{R} \widehat{\boldsymbol{\phi}}'(\theta', \phi') \\ \widehat{\boldsymbol{\phi}}(\theta, \phi)^T \boldsymbol{R} \widehat{\theta}'(\theta', \phi') & \widehat{\boldsymbol{\phi}}(\theta, \phi)^T \boldsymbol{R} \widehat{\boldsymbol{\phi}}'(\theta', \phi') \end{pmatrix} \begin{pmatrix} F'_{\theta'}(\theta', \phi') \\ F'_{\phi'}(\theta', \phi') \end{pmatrix} \tag{9.48}$$

where $\widehat{\boldsymbol{\theta}}$ and $\widehat{\boldsymbol{\phi}}$ are the spherical unit vectors of the GCS and likewise $\widehat{\boldsymbol{\theta}}'$ and $\widehat{\boldsymbol{\phi}}'$ of the LCS respectively, and $\boldsymbol{R}$ is the rotation matrix transforming the LCS unit vectors to the GCS reference frame. The unit vectors $\hat{\boldsymbol{\theta}}$ and $\hat{\boldsymbol{\phi}}$ are given by

$$\hat{\boldsymbol{\theta}} = \begin{pmatrix} \cos\theta \cos\phi \\ \cos\theta \sin\phi \\ -\sin\theta \end{pmatrix} \tag{9.49a}$$

and

$$\widehat{\boldsymbol{\phi}} = \begin{pmatrix} -\sin\phi \\ \cos\phi \\ 0 \end{pmatrix} \tag{9.49b}$$

In addition, $\widehat{\boldsymbol{\theta}}$ and $\widehat{\boldsymbol{\phi}}$ are an orthogonal pair and $\widehat{\boldsymbol{\theta}}'$ and $\widehat{\boldsymbol{\phi}}'$ is another orthogonal pair as illustrated in Figure 9.24. The angular displacement ψ between the two pairs of GCS and LCS as function of their orientation (α, β, γ) are bearing angle, downtilt angle, slant angle, respectively, and the spherical position (θ, ϕ) is given by

$$\psi = \arccos\left(\frac{A_\psi - B_\psi \cos\theta}{\sqrt{1 - (C_\psi + D_\psi \sin\theta)^2}} \right) \tag{9.50a}$$

where

$$A_\psi = \cos\beta \cos\gamma \sin\theta \tag{9.50b}$$

$$B_\psi = \sin\beta \cos\gamma \cos(\phi - \alpha) - \sin\gamma \sin(\phi - \alpha) \tag{9.50c}$$

$$C_\psi = \cos\beta \cos\gamma \cos\theta \tag{9.50d}$$

$$D_\psi = \sin\beta \cos\gamma \cos(\phi - \alpha) - \sin\gamma \sin(\phi - \alpha) \tag{9.50e}$$

To attain the transformation of the coordinate system and the radiation pattern, we need to define the rotation matrix that can be used to transform a point (x, y, z) in the GCS into a point

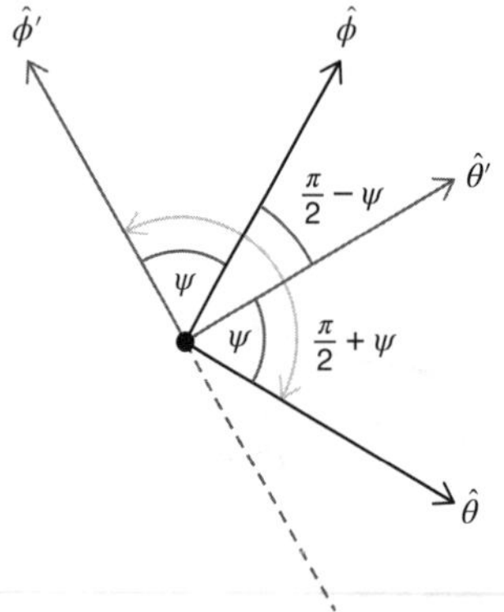

Figure 9.24 Rotation of the spherical basis vectors by an angle ψ due to the orientation of the LCS with respect to the GCS [20].

(x', y', z') in the LCS. The required matrix is actually the product *of three rotation matrices by angles* α, β, γ, respectively, are defined as

$$\boldsymbol{R}_Z(\alpha) = \begin{pmatrix} \cos\alpha & -\sin\alpha & 0 \\ \sin\alpha & \cos\alpha & 0 \\ 0 & 0 & 1 \end{pmatrix} \tag{9.51a}$$

$$\boldsymbol{R}_Y(\beta) = \begin{pmatrix} \cos\beta & 0 & \sin\beta \\ 0 & 1 & 0 \\ -\sin\beta & 0 & \cos\beta \end{pmatrix} \tag{9.51b}$$

$$\boldsymbol{R}_X(\gamma) = \begin{pmatrix} 1 & 0 & 0 \\ 0 & \cos\gamma & -\sin\gamma \\ 0 & \sin\gamma & \cos\gamma \end{pmatrix} \tag{9.52c}$$

The composite rotation matrix $\boldsymbol{R}$ is given by

$$\boldsymbol{R} = \boldsymbol{R}_Z(\alpha) * \boldsymbol{R}_Y(\beta) * \boldsymbol{R}_X(\gamma) \tag{9.53}$$

$$\boldsymbol{R} = \begin{pmatrix} \cos\alpha\cos\beta & \cos\alpha\sin\beta\sin\gamma - \sin\alpha\cos\gamma & \cos\alpha\sin\beta\cos\gamma + \sin\alpha\sin\gamma \\ \sin\alpha\cos\beta & \sin\alpha\sin\beta\sin\gamma + \cos\alpha\cos\gamma & \sin\alpha\sin\beta\cos\gamma - \cos\alpha\sin\gamma \\ -\sin\beta & \cos\alpha\sin\gamma & \cos\alpha\cos\gamma \end{pmatrix}$$

$\boldsymbol{R}$ is an orthogonal matrix. For transformation a point (x', y', z') in the LCS into a point (x, y, z) in the GCS, we use the inverse of the composite matrix R^{-1}, which is equal transpose of $\boldsymbol{R}$ matrix.

$$R^{-1} = \begin{pmatrix} \cos\alpha\cos\beta & \sin\alpha\cos\beta & -\sin\beta \\ \cos\alpha\sin\beta\sin\gamma - \sin\alpha\cos\gamma & \sin\alpha\sin\beta\sin\gamma + \cos\alpha\cos\gamma & \cos\beta\sin\gamma \\ \cos\alpha\sin\beta\cos\gamma + \sin\alpha\sin\gamma & \sin\alpha\sin\beta\sin\gamma - \cos\alpha\gamma & \cos\beta\cos\gamma \end{pmatrix} \tag{9.54}$$

Let us consider a point (x, y, z) on a unit sphere defined by the spherical coordinates (ρ, θ, ϕ) and we assume the sphere is a unit radius i.e. $\rho = 1$, as per Release 14. It can easily be shown that the zenith angle θ is given as $\theta = \cos^{-1}(\hat{\rho} \cdot \hat{z})$ measured from the z-axis.

The azimuth angle ϕ measured from x-axis in x–y plane is given as

$$\phi = \arg(\hat{x}.\hat{\rho} + j\hat{y}.\hat{\rho}) \tag{9.55a}$$

Now consider the case where a point $(1, \theta, \phi)$ is located in GCS and we want to find the corresponding position in the LCS. For this scenario, we use $R^{-1}\hat{\rho}$ to compute the local θ', ϕ' where $\hat{\rho}$ is given by

$$\hat{\rho} = \begin{bmatrix} x \\ y \\ z \end{bmatrix} = \begin{bmatrix} \sin\theta\cos\phi \\ \sin\theta\sin\phi \\ \cos\theta \end{bmatrix} \tag{9.55b}$$

After trigonometric simplifications, it can be shown that

$$\theta'(\alpha, \beta, \gamma, \theta, \phi) = \cos^{-1}\left\{\begin{bmatrix} 0 \\ 0 \\ 1 \end{bmatrix}^T R^{-1}\hat{\rho}\right\} = \cos^{-1}\{\mathrm{A} + \mathrm{B}\,\sin\theta\} \tag{9.56a}$$

where

$$A = \cos\beta \cos\gamma \cos\theta \tag{9.56b}$$

$$B = \sin\beta \cos\gamma \cos(\phi - \alpha) - \sin\gamma \sin(\phi - \alpha) \tag{9.56c}$$

The azimuth angle $\phi'(\alpha, \beta, \gamma, \theta, \phi)$ is given as

$$\phi'(\alpha,\beta,\gamma,\theta,\phi) = \arg\left\{\begin{bmatrix}1\\ j\\ 0\end{bmatrix}^T R^{-1}\hat{\rho}\right\} = \arg\{C + j(D + E\sin\theta)\} \tag{9.57a}$$

$$C = \cos\beta \sin\theta \cos(\phi - \alpha) - \sin\beta \cos\theta \tag{9.57b}$$

$$D = \sin\beta \sin\gamma(\phi - \alpha) \tag{9.57c}$$

$$E = (\sin\beta \sin\gamma \cos(\phi - \alpha) + \cos\gamma \sin(\phi - \alpha)) \tag{9.57d}$$

Spherical coordinates are presented in 9.B.

9.14.3 Scenarios Descriptions

The scenarios descriptions for the models constituted of various transmission environments. In urban microcell street canyon and urban macrocell, the models consider outdoor/indoor LOS /NLOS environments for both microcell and macrocell. The BS antenna heights for microcell and the macrocell are 10 and 25 m. UT height is 1.5 m and UT mobility on the horizontal plan is 3 km/h for both microcells and macrocells. The minimum distance between BS–UT in 2D is 10 m in the microcell and 35 m in macrocell. Note street canyon is a transmission environment comparable to a street bordered by high buildings. In the indoor-office scenarios, the BS antenna and UT heights are 3 m (i.e. ceiling) and 1 m and comprise LOS/NLOS environments. In the rural schemes a larger and constant coverage is required that suggests wide area coverage supporting high-speed vehicles. The schemes for rural macrocell are characterised by the BS antenna height of 35 m but UT height is still at 1.5 m. the contribution of indoor and vehicles outdoor is about 50/50. The carrier frequency is up to 7GHz and reception environments can be LOS and NLOS a minimum distance (2D) between BS and UT is 35 m.

In Section 9.9, we examined the horizontal part of the radiation power pattern associated with the 2D channel model. The antenna modelling for 3D channel models includes a vertical part of the radiation power pattern. In 3GPP release 14, the BS AA is formed by a 2D uniform rectangular panel antenna array. The panel arrays are placed equidistantly spaced in y- and the z-directions. The BS panel antenna arrays comprised M_g panels in each column and N_g panels in each row so there are M_g N_g panels in total. Antenna panels are uniformly spaced in the horizontal direction, with a spacing of $d_{g,H}$ and in the vertical direction with a spacing of $d_{g,V}$. On each panel, antenna elements are placed in M rows in the vertical direction and N columns in the horizontal direction, M is the number of antenna elements with the same polarization in each column. The antenna elements on each panel are uniformly spaced in the horizontal direction with a spacing $d_{g,H}$ and in the vertical direction with a spacing of $d_{g,V}$. Each antenna panel is either a single-polarized (co-polarized) $P = 1$ or dual-polarized (crossed polarized) $P = 2$. The rectangular panel antenna can be defined by M_g, N_g, M, N, P. However, the model represents a compromise between practicality and precision, as it does not include the mutual coupling effect as well as different propagation effects of horizontally and vertically polarized waves as illustrated in Figure 9.25.

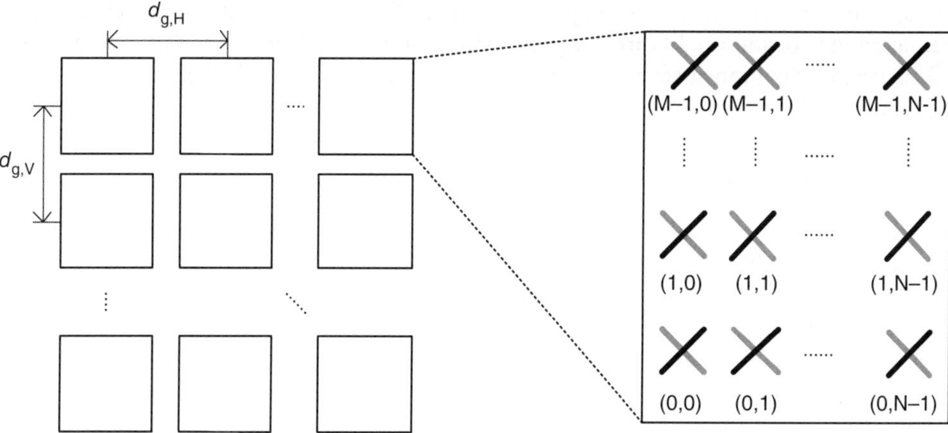

Figure 9.25 Cross-polarized panel array antenna model [20].

9.14.4 Antenna Modelling

The antenna array radiation power pattern in Section 9.9 given by (9.15) is developed for 3D channel as defined in Table 9.1:

The radiation power pattern and the antenna field pattern are related by

$$A''_{\text{tot}}(\theta'', \phi'') = |F''_{\theta''}(\theta'', \phi'')|^2 + |F''_{\phi''}(\theta'', \phi'')|^2 \tag{9.58}$$

where $F''_{\theta''}(\theta'', \phi'')$ and $F''_{\phi''}(\theta'', \phi'')$ are antenna element field patterns in the direction of the spherical unit vector $\hat{\boldsymbol{\theta}}$ and unit vector $\hat{\boldsymbol{\phi}}$ in a unpolarised scenario.

It is important to note that the antenna elements are equidistantly spaced in y- and z-directions. BS array elements electric beam steering (also known as electrical tilting) with a fixed phase shifts between its M elements to obtain a beam tilt angle θ_{etilt} applies complex weights to each antenna element in the vertical direction. Accordingly, the complex weights for the mth antenna element can be expressed as

$$w_m = \frac{1}{\sqrt{\text{M}}} \exp\left(-j\frac{2\pi}{\lambda}(m-1)d_{g.V} \cos\theta_{etilt}\right) \tag{9.59}$$

where $m = 1, \ldots \ldots \ldots \ldots \ldots, $ M and $d_{g.V}$ is the vertical element spacing. A conventional approach often used in uniform antenna array when determining the array radiation power

Table 9.1 Radiation power pattern of a single antenna element [20].

Parameter	Expression
Vertical radiation power pattern (dB)	$A''_{\text{V}}(\theta'', \phi'' = 0°)\text{dB} = -\min\left\{12\left(\frac{\theta''-90^o}{65^0}\right)^2, 30\text{ dB}\right\}\ \theta'' \in [0^o, 180^o]$
Horizontal radiation power pattern (dB)	$A''_{\text{H}}(\theta'' = 90^o, \phi'')\text{dB} = -\min\left\{12\left(\frac{\phi''}{65^0}\right)^2, 30\text{ dB}\right\}\ \phi'' \in [-180^o, 180^o]$
3D radiation power pattern (dB)	$A''_{\text{tot}}(\theta'', \phi'') = -\min\{-(A''_{\text{dB}}(\theta'', \phi'' = 0^o) + A''_{\text{dB}}(\theta'' = 90^o, \phi'')), 30\text{dB}\}$
Maximum directional gain of an antenna element $G_{\text{E, max}}$	8 dBi

pattern is to use the array factor (AF) applied to the radiation pattern of a single element. In 3D channel model, beam weights are assigned to the channel coefficients for each antenna element. The weighted composite channel coefficients for the ith sector of the BS for the jth cluster $[\mathbf{H}_{i,j}^{C}]_{n,m}$ is given as [23]:

$$[\mathbf{H}_{i,j}^{C}]_{n,m} = \sum_{n\in \mathrm{N}} w_n \sum_{m\in \mathrm{M}} w_m \mathrm{H}_{i,n,m,j}(t) \tag{9.60}$$

where $i = 0$ for the serving sector and $i = \{1, 2, \ldots.\}$ denotes the interfering sectors, $n \in \mathrm{N}$ refers to the sets of receive antenna elements and $m \in \mathrm{M}$ the transmit antenna elements and w_u, w_s are complex weights and account for the phase shifters for the beam electric tilting at the UT and BS, respectively.

In 3D channel mode, two antenna polarization models are known: a constant polarization model and a slanted polarization model. However, it was accepted that little difference is observed in the system performance between the two polarization models. The choice from the two models would be based on the expected antenna radiation pattern of the BSs (evolved Node B, eNBs).

Consider a polarized antenna array according to [20], when a constant polarization model is established a polarization, the power split and the polarisation are an angle-independent in both azimuth and elevation in an LCS. In other words, equal power split in the vertical and horizontal direction for all UT locations. For linear polarization model, the antenna element field pattern in the elevation and azimuth can be expressed as

$$\begin{pmatrix} F'_{\theta'}(\theta', \phi') \\ F'_{\phi'}(\theta', \phi') \end{pmatrix} = \begin{pmatrix} A'_{tot}(\theta', \phi') \cos \varphi \\ A'_{tot}(\theta', \phi') \sin \varphi \end{pmatrix} \tag{9.61}$$

where φ is the polarization slant angle, and $A'_{tot}(\theta', \phi')$ is the 3D antenna element power radiation pattern as a function of elevation and azimuth angles in the LCS and $\varphi = 0$ implies vertically polarized antenna element.

The slanted polarization model attains an equal power split in vertical and horizontal directions in the direction of antenna boresight but in general, the power split ratio depends on the UE location in both azimuth and elevation dimensions. A polarisation with slant angle $+45^o/-45^o$ represents cross-polarisation. Let us consider the polarisation slant angle ϕ and refer to the polarized field component in the LCS of the UT by $F'_{\theta'}(\theta', \phi')$ and $F'_{\phi'}(\theta', \phi')$ as before. Then the slanted polarization modifies the field components by ψ (angular displacement between two pairs of unit vectors defining the UT location) to

$$\begin{pmatrix} F'_{\theta'}(\theta', \phi') \\ F'_{\phi'}(\theta', \phi') \end{pmatrix} = \begin{pmatrix} \cos \psi & -\sin \psi \\ \sin \psi & \cos \psi \end{pmatrix} \begin{pmatrix} F''_{\theta''}(\theta'', \phi'') \\ F''_{\phi''}(\theta'', \phi'') \end{pmatrix} \tag{9.62a}$$

where ψ the angular displacement is given by

$$\psi = \arcsin\left(\frac{\sin \varphi \cos \phi'}{\sqrt{1 - (\cos \varphi \cos \theta' - \sin \varphi \sin \phi' \sin \theta')^2}} \right) \tag{9.62b}$$

9.14.5 Probability of LOS

3GPP and International Telecommunication Union (ITU) methods preceding the Release 14 consider LOS probability mainly at street levels and UEs height are not completely included. The effect of UE varying heights (typically in the range of 1.5–22.5 m) on LOS probability was examined in 3D scenarios. It may be perceived that modelling LOS probability requires the inclusion of precise building dimensions by applying ray tracing techniques. The LOS probability for 3D urban macro is developed as the sum of two probabilities: type-1 and type-2 probabilities.

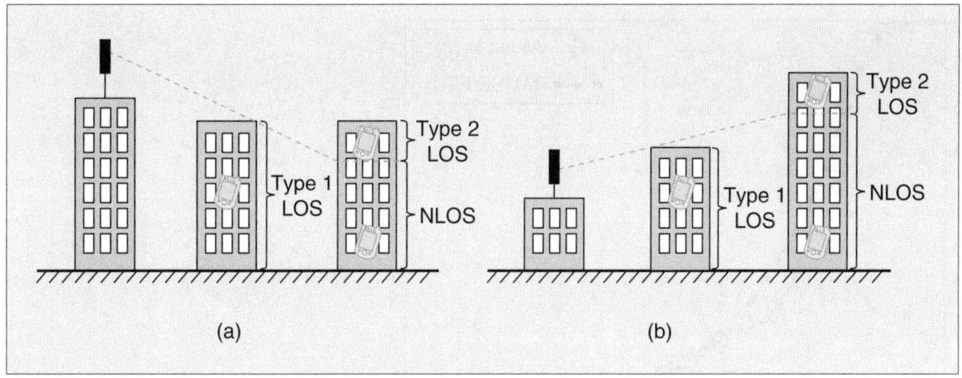

Figure 9.26 Type 1 and type 2 LOS probabilities: (a) 3D-UMa scenario; (b) 3D-UMi scenario [24].

The idea of breaking down of the LOS probabilities into two components is LOS probability modelling simplification.

Type-1 LOS relies on the horizontal distance between the BS and the UE as defined in International Telecommunication Union-Radio (ITU-R) M.2135-1 (2009). A user on a higher floor of a building is hypothesized to be in type-2 LOS if on the first floor (assumed to be type-1 LOS) of the same building can never be a LOS state. Figure 9.26 illustrates the two LOS probabilities for urban macro and urban micro scenarios in Figure 9.26a, b, respectively.

The probability of LOS can be defined as the likelihood that radiation from a given transmitter (TX) is travelling onwards on a straight and unobstructed propagation path in an urban environment destined to a certain RX, i.e. zero reflection on the path. Likewise, NLOS (non-LOS) probability can be delineated as the chance the TX radiation is likely to be interpreted by at least a single obstruction and travels onwards on its obstructed path to reach RX (i.e. travels along scattering environment). These two probabilities considerably influenced by the TX and RX location environment. The most popular method for attaining the LOS probabilities is ray tracing techniques. Such methods indicatively call for all buildings bordering the TXs to be modelled using simple geometric shapes such as cubes exploiting 3D database created by employing special algorithms with digital maps.

For every pair of TX-RX locations, a simple test is acted upon to decide if any of the buildings blocked the direct communications line (i.e. LOS) between the TX and RX. Distances flanked by the TX and Rx ranged from 10–200 m. Four BS locations are chosen to act as TX locations in the test and for RX of 1.5 m. Smaller obstacle objects such as trees, lampposts, vehicles can be ignored as they introduce little effects.

9.14.6 Estimate of the LOS Probability Using Ray Tracing

Prior to the ray tracing method, the LOS probability is pertained to actual measured RX locations. In ray tracing, this implied ascertaining the obstacles in the coverage area to identify LOS connections as far as 100 m or more (depending on the type of BS be it for macro or microcell) away from the BS and in incremental TX-RX separation, ensuring the probability of LOS remains constant with a value of 100%. LOS probability curves obtained for a number of TX (BS) sites and using the 3D site database for the specific location, LOS probability is plotted versus the TX–RX separations for appropriate number of sites (four sites or more to get a reasonably accurate estimate of the LOS probability for each T–R separation) and minimum mean square error (MMSE) fit curve is established defining the mean probability of LOS and shown in Figure 9.27

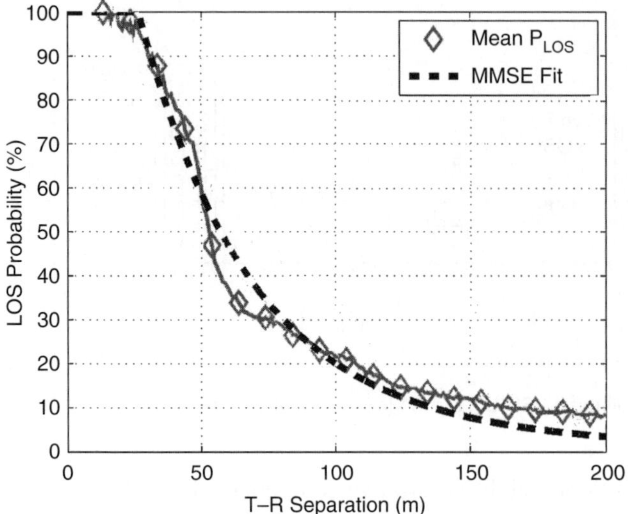

Figure 9.27 LOS probability curves from ray-tracing as a function of T-R separation distance [25].

The analytical expression from the MMSE fit curve can be expressed as

$$p_{\mathrm{LOS}}(d) = \left[\min\left(\frac{d_{\mathrm{BP}}}{d}, 1\right)\left(1 - \exp\left(-\frac{d}{\alpha}\right) + \exp\left(-\frac{d}{\alpha}\right)\right)\right]^2 \tag{9.63}$$

where d_{BP} is the breakpoint distance and at larger separation beyond such point the LOS probability is less than 100%, α m is a decay used in MMSE, which gives d_{BP}, α as 27 and 71 m, respectively.

9.14.7 LOS Probability in 3GPP Release 14

3GPP adopted the entirely stochastic approach of modelling as used in SCM and in WINNER II that does not depend on building/street dimensions definitively. The advantage of such an approach is to enable partially rehashing the existing modelling parameters from 2D stochastic models. Further, it also supports combining measurement results from multiple sources and reduces system simulations complexity.

For probability and path loss, 3GPP Release 14 defines the horizontal distance between BS and outdoor UE as d_{2D} and the heights of the BS and UTs as h_{BS} and h_{UT}, respectively, for urban macrocell and 3D distance between BS antenna and UT antenna as d_{3D} as illustrated in Figure 9.28. For indoor UT, the distance between outdoor BS antenna and the indoor UT antenna is the sum of the outdoor term d_{3D-out} and an indoor term d_{3D-in}, i.e. $(d_{3D-out} + d_{3D-in})$. The previous horizontal distance d_{2D} modified to $(d_{2D-out} + d_{2D-in})$, as shown in Figure 9.28. It is worth bearing in mind that d_{2D-in} defined the location of the UT inside the building. So, we can show that

$$d_{3D\text{-}out} + d_{3D\text{-}in} = \sqrt{(d_{2D\text{-}out} + d_{2D\text{-}in})^2 + (h_{BS} - h_{UT})^2} \tag{9.64a}$$

and

$$d_{3D\text{-}in} = \sqrt{(d_{2D\text{-}in})^2 + (h_{BS} - h_{UT})^2} \tag{9.64b}$$

The LOS probability models for RMa; UMa; and UMi–street canyon from [20] are shown in Table 9.2.

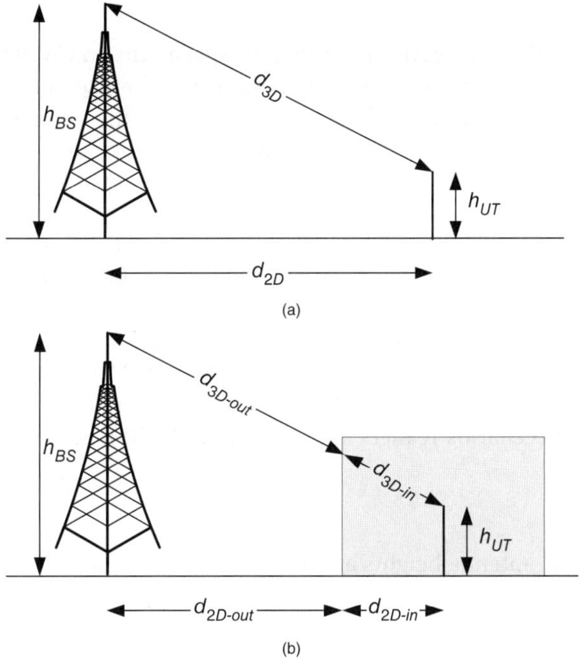

Figure 9.28 (a) Definition of d2D and d3D for outdoor UTs, (b) Definition of *d2D-out*, *d2D-in* and *d3D-out*, *d3D-in* for indoor UTs [20].

Table 9.2 The LOS probability models for RMa; UMa; and UMi – street canyon [20].

Scheme type	Probability of LOS
RMa	$p_{\mathrm{LOS}} = \left\{ \begin{array}{cc} 1 & , d_{\mathrm{2D-out}} \leq 10\,\mathrm{m} \\ \exp\left(-\frac{d_{\mathrm{2D-out}}-10}{1000}\right) & , 10\,\mathrm{m} < d_{\mathrm{2D-out}} \end{array} \right\}$
UMa	$p_{\mathrm{LOS}} = \left\{ \begin{array}{cc} 1 & , d_{\mathrm{2D-out}} \leq 18\,\mathrm{m} \\ A_{\mathrm{UMa}} * B_{\mathrm{UMa}} & , 18\,\mathrm{m} < d_{\mathrm{2D-out}} \end{array} \right\}$
UMi – street canyon	$p_{\mathrm{LOS}} = \left\{ \begin{array}{cc} 1 & , d_{\mathrm{2D-out}} \leq 18\,\mathrm{m} \\ \frac{18}{d_{\mathrm{2D-out}}} + \exp\left(-\frac{d_{\mathrm{2D-out}}}{36}\right)\left(1-\frac{18}{d_{\mathrm{2D-out}}}\right) & , 18\,\mathrm{m} < d_{\mathrm{2D-out}} \end{array} \right\}$

where

$$A_{\mathrm{UMa}} = \frac{18}{d_{\mathrm{2D-out}}} + \exp\left(-\frac{d_{\mathrm{2D-out}}}{63}\right)\left(1-\frac{18}{d_{\mathrm{2D-out}}}\right)$$

$$\mathrm{B}_{\mathrm{UMa}} = \left(1 + C_{\mathrm{UMa}}\frac{5}{4}\left(\frac{d_{2D-out}}{100}\right)^3 \exp\left(\frac{\mathrm{d}_{\mathrm{2D-out}}}{150}\right)\right)$$

$$C_{\mathrm{UMa}} = \left\{ \begin{array}{cc} 0 & , h_{\mathrm{UT}} \leq 13\,\mathrm{m} \\ \left(\frac{h_{\mathrm{UT}}-13}{10}\right)^{\frac{3}{2}} & , 13 < h_{\mathrm{UT}} \leq 23\,\mathrm{m}. \end{array} \right\}$$

9.14.8 Path Loss

Path loss is mainly due to energy loss as the EM wave propagates in space. The mmWave large-scale signal fading arises from free space path loss, vegetation loss, atmospheric gaseous loss, weather loss (rain, snow, fog, and clouds), and building penetration loss, the latter is due to building materials. The path-loss models are presented in Chapters 6 and 7, and the blockage loss and blockages modelling using random shape theory (RST) are presented in this chapter.

Accordingly, path loss is made up of several components, depending on the UT outdoor/indoor. Path loss generally is known as outdoor-2-indoor path loss, when EM wave is propagating from outdoor BS to an indoor UT (UT in office, UT in multi-story building, etc.). In most cases, the total path loss is the sum of the two. Furthermore, the path loss also depends on whether wave propagation is over a link that is clear LOS or non-line-of-sight (NLOS).

3GPP Release 14 recommends the use of two models for calculating the path loss for UMa are as follows, and for other environment scenarios readers are referred to 3GPP Release 14 in [20].

9.14.8.1 UMacell Path Loss

The path losses are expressed for the actual antenna heights at the BS and UT are $h_{BS} = 25$ m and $1.5 \leq h_{UT} \leq 22.5$, respectively.

9.14.8.2 LOS Channel Environment

Standard deviation of SF RMS spread $\sigma_{SF} = 4$, the path loss models are given by

$$PL_1^{UMa-LOS} = 28 + 22\log_{10}(d_{3D}) + 20\log_{10}(f_c) \tag{9.65a}$$

$$PL_2^{UMa-LOS} = 28 + 40\log_{10}(d_{3D}) + 20\log_{10}(f_c) - 9\log_{10}((d'_{BP})^2 + (h_{BS} - h_{UT})^2) \tag{9.65b}$$

where the effective breakpoint distance d'_{BP} is given as $d'_{BP} = 4\frac{h'_{BS}\,h'_{UT}}{\lambda}$, λ is the wavelength at the operating frequency in free space, and the effective antenna height at the BS and UT h'_{BS} and h'_{UT} are given as follows: $h'_{BS} = h_{BS} - h_E$ and $h'_{UT} = h_{UT} - h_E$, h_E is the effective terrain height, and the operating frequency $0.5 < f_c < 100$ in GHz, h'_{BS}, h'_{UT} are the heights of the antenna's centre of the radiation above ground for BS and UT, respectively, and $PL_1^{UMa-LOS}$ and $PL_2^{UMa-LOS}$ are valid for $10\text{m} \leq d_{2D} \leq d'_{BP}$ and $d'_{BP} \leq d_{2D} \leq 5\text{km}$, respectively, d_{2D} is the distance between two points in a plane with Cartesian coordinates x_1, y_1 and x_2, y_2 is $d_{2D} = \sqrt{(x_2 - x_1)^2 + (y_2 - y_1)^2}$. Similarly in 3D the two points Cartesian coordinates are x_1, y_1, z_1 and x_2, y_2, z_2 and $d_{3D} = \sqrt{(x_2 - x_1)^2 + (y_2 - y_1)^2 + (z_2 - z_1)^2}$.

9.14.8.3 Non-Line-of-Sight (NLOS)

Standard deviation of SF spread $\sigma_{SF} = 6$, the UMa path loss under NLOS $PL^{UMa-NLOS}$ is given by

$$PL^{UMa-NLOS} = \max(PL^{UMa-LOS}, PL')$$

where PL' is given as

$$PL' = 13.54 + 39.08\log_{10}(d_{3D}) + 20\log_{10}(f_c) - 0.6(h_{UT} - 1.5)$$

9.14.9 Fast-Fading Model for 3D Channels

Unlike the flat-fading Rayleigh channel that causes slow channel coefficients variation, the channel coefficients in fast-fading undergo faster time-varying fluctuations [20]. The fast-fading phenomenon is instigated by the amalgamation of multipath and enhanced by UT mobility. SF develops when an obstacle appeared between the UT and the transmitted signal. The shadowed signal strength is subjected to a significant reduction and modelled as log-normal distributed with two parameters: the standard deviation σ_{SF} dB characterises the width of the distribution and a correlation distance which defines how fast the SF varies when the UT moves through the environment. The RMS DS describes the delay time range of a multipath channel assumed to be lognormal distributed and defined by mean μ_{DS} and standard deviation σ_{DS} . For example, consider an UMa (NLOS) with mean $\mu_{DS} = -6.28 - 0.204 \log 10(f_c)$ where f_c in GHz. If the UMa cell is operating at 30 GHz $\mu_{DS} = -6.6$ and $\sigma_{DS} = 2.55$. The Rician K-factor is the ratio of the LOS path power to the power in the other scattered paths. It assumed that the Rician K-factor to be lognormal distributed. The angular spread determines the distributions of the departure and arrival angles of each of the MPCs in 3D space as seen by the TX and RX, respectively. Each path ascribed an azimuth angle in the horizontal plane and zenith angle in the vertical plane. There are four angular spreads per path. The cross polarization expressed how the polarization changes for a given MPC. The Initial polarization of a path is defined by the transmit antenna. In contrast, the NLOS signal components experience diffraction and scattering before reaching the RX.

Consider a cellular DL for investigating the fast fading. We must define two kinds of angles: the departure angles at the eNB (BS) side and the arrival angle at the UT side. The channel coefficients of the DL is ascertained by a compound channel response of the MPCs. Each MPC is identified by a path delay, a path power, and random phases ushered in during propagation. In addition, consider path angles: azimuth angles of departure and arrival (AOD, AOA), and zenith angles of departure and arrival (ZAOD, ZAOA), respectively. The outdoor LOS and NLOS environment is illustrated in Figure 9.29.

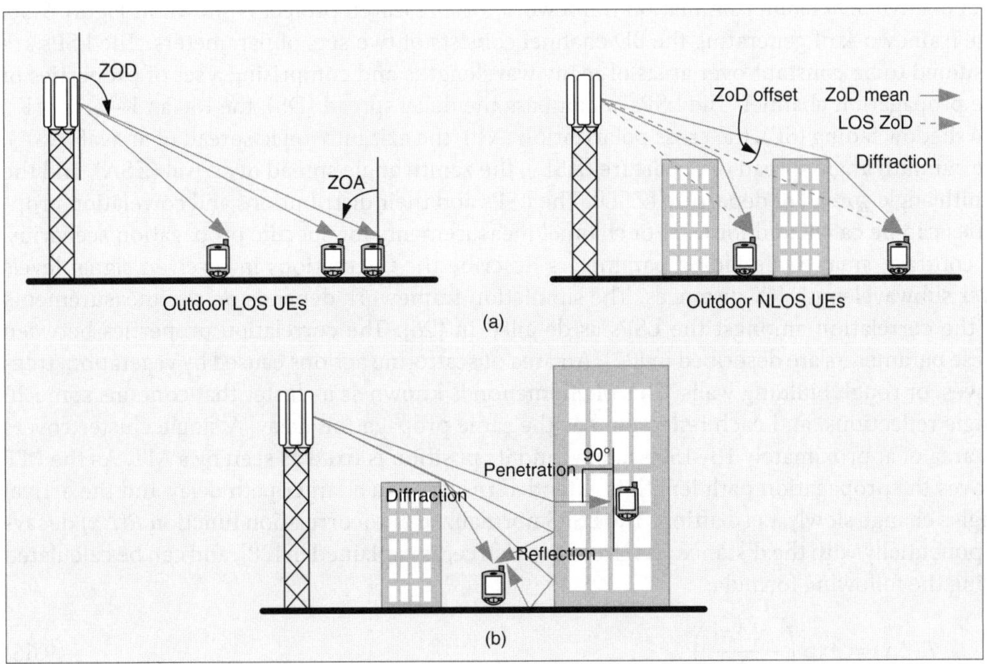

Figure 9.29 Propagation mechanisms (a) ZOD in outdoor LOS and NLOS conditions; and (b) ZOA in indoor/outdoor NLOS conditions [24].

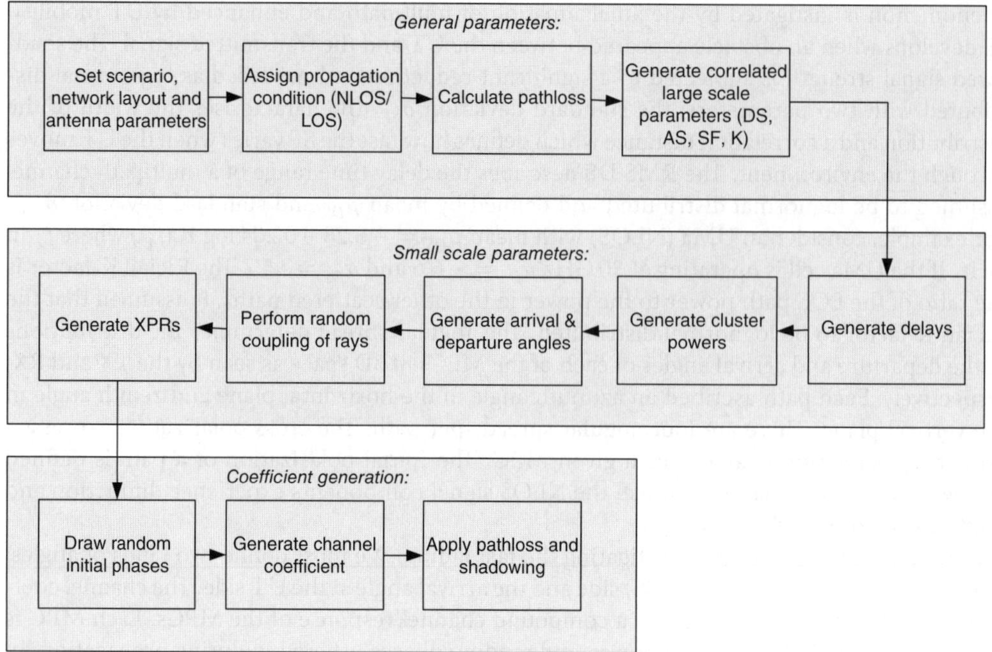

Figure 9.30 3GPP 3D channel coefficients generating [20].

9.14.10 Large-Scale Parameters

An effective 3D channel simulation framework for a research project is shown in Figure 9.30. The framework of generating the 3D channel consists of two sets of parameters. The LSPs are assumed to be constant over areas of many wavelengths and comprising a set of properties of the propagation channel. The LSPs encompass the delay spread (DS), the Rician K-factor (K), the shadow fading (SF), the cross polarization (XP), the azimuth angle spread of arrival (ASA), the azimuth angle spread of departure (ASD), the zenith angle spread of arrival (ZSA), and the zenith angle spread of departure (ZSD). The LSPs and their distributions and correlation properties can be calculated for a set of channel measurements for specific propagation scenarios. In contrast, small-scale fading parameters describe the fluctuations in received signal levels over subwavelength RX distances. The simulation framework depends on site measurements of the correlation amongst the LSPs as detailed in [26]. The correlation properties between these parameters are described in [27]. An area of scattering actions caused by vegetation, trees leaves, or rough building walls. This phenomenon is known as a cluster that contains some 20 single reflections, and each reflection has the same propagation delay. A single cluster covers an area of approximately 10–15 m radius, and its position is fixed as seen by a MT. As the MT moves the propagation path length is extended resulting in a larger path delay and the arrival angles change slowly, i.e. drifting. The LSPS normalized autocorrelation function $R(\Delta x)$ decays exponentially with the distance $|\Delta x|$ from the source, as explained in [28], and can be calculated using the following formula:

$$R(\Delta x) = \exp\left(-\frac{|\Delta x|}{d_{cor}}\right) \tag{9.66}$$

Table 9.3 Auto-correlation distances in the horizontal plane for UMA; RMa; and UMi – street canyon [20].

LSPs	Correlation distances in the horizontal plane [m]					
	Urban Macrocell (UMa)		Rural Macrocell (RMa)		UMi – street canyon	
	LOS	NLOS	LOS	NLOS	LOS	NLOS
SF	37	50	37	120	10	13
K	12	N/A	40	N/A	15	N/A
DS	30	40	50	36	7	10
ASD	18	50	25	30	8	10
ASA	15	50	35	40	8	9
ZSD	15	50	15	50	12	10
ZSA	15	50	15	50	12	10

where d_{cor} is the correlation distance defined as the distance that causes the correlation to decay to a value equal $e^{-1} = 0.37$. The correlation distance for various channel environments are shown in Table 9.3.

In a simulation platform, the channel coefficients need to be computed for each sampling point on the time-frequency grid. 3GPP channel model coefficients depend on the UT location in 3D space and for a massive MIMO channel, the complexity is huge, so computationally intensive tasks particularly may have to be computed off-line or on demand.

The step-wise procedure for generating the 3D channel coefficients includes 12 steps, each denoted as 'Step n' where $n = 1, ..12$ is illustrated in Figure 9.30. The channel model generation comprises two main parts: the first part deals with the general parameters of the channel model covered in Steps 1–4, and the second part considers the small-scale parameters in the model covered by Steps 5–9. Step 10 determines the blocked clusters and Steps 11–12 are involved in channel coefficients generation. The small-scale parameters in Steps 5–9 include the following steps:

Step 5. Generate delays for random clusters.
Step 6. Generate powers for random clusters.
Step 7. Generate arrival angles and departure angles for both azimuth and elevation, for each random cluster.
Step 8. Merge deterministic clusters and random clusters.
Step 9. Generate ray delays and ray angle offsets inside each cluster.

The first action to take is to ascertain the intended channel environment scenario in Step 1. The contemporary channel environment scenarios in [20] include the following channel environments: UMi-street canyon, UMa, RMa, outdoor to indoor (O2I), and indoor-office. Once the environment is determined, the propagation condition has to be assigned either LOS or NLOS suitable for the chosen environment scenario in Step 1. Step 2 choice of the propagation conditions is dependent on the probability of LOS/NLOS, which is dependent on the UT height and its distance to the macrocell / microcell BS. Step 3 computes the pathloss using the formulas in 9.14.8.

Step 4 handles the LSPs, which are relatively constant for several metres. When one BS communicates with two UTs located in the same vicinity or if two BSs communicate with a single UT, correlation between the two links LSPs prevails. Accordingly, in a system of a number of

Table 9.4 Cross-correlation values ρ between each two LSPs in 3D channel model.

$$X=\begin{pmatrix} 1 & \rho_{SF.KF} & \rho_{SF.DS} & \rho_{SF.ASD} & \rho_{SF.ASA} & \rho_{SF.ZSD} & \rho_{SF.ZSA} \\ \rho_{KF.SF} & 1 & \rho_{KF.DS} & \rho_{KF.ASD} & \rho_{KF.ASA} & \rho_{KF.ZSD} & \rho_{KF.ZSA} \\ \rho_{DS.SF} & \rho_{DS.KF} & 1 & \rho_{DS.ASD} & \rho_{DS.ASA} & \rho_{DS.ZSD} & \rho_{DS.ZSA} \\ \rho_{ASD.SF} & \rho_{ASD.KF} & \rho_{ASD.DS} & 1 & \rho_{ASD.ASA} & \rho_{ASD.ZSD} & \rho_{ASD.ZSA} \\ \rho_{ASA.SF} & \rho_{ASA.KF} & \rho_{ASA.DS} & \rho_{ASA.ASD} & 1 & \rho_{ASA.ZSD} & \rho_{ASA.ZSA} \\ \rho_{ZSD.SF} & \rho_{ZSD.KF} & \rho_{ZSD.DS} & \rho_{ZSD.ASD} & \rho_{ZSD.ASA} & 1 & \rho_{ZSD.ZSA} \\ \rho_{ZSA.SF} & \rho_{ZSA.KF} & \rho_{ZSA.DS} & \rho_{ZSA.ASD} & \rho_{ZSA.ASA} & \rho_{ZSA.ZSD} & 1 \end{pmatrix}$$

BSs serving a number of UTs, correlation between channels' LSPs of different links may also occur. This fact has been experimentally noticed in [29] and [30], where the autocorrelation of LSPS are discovered to be exponentially decaying with correlation distance of many metres. High correlation of the LSPs between UTs is likely when UTs move closer in a confined area. It is also expected that this may be the circumstance for multiple BSs that are closely located or in similar environments.

When UT travels along a certain path or if UTs are closely spaced together, their LSPs are correlated. Typically, LSPs are assumed to be lognormal distributed. The autocorrelation of the LSPs expressed in (9.66) is distance dependent. Additionally, the LSPs are cross correlated. A 7×7 matrix $\boldsymbol{X}$ containing the cross-correlation values ρ between each two LSPs is shown in Table 9.4

The square root of the cross-correlation matrix i.e. $\boldsymbol{X}^{\frac{1}{2}}$ can be determined by performing Cholesky decomposition of matrix $\boldsymbol{X}$. In order to do the decomposition, the matrix has to be symmetric. The correlation matrix $\boldsymbol{X}$ is symmetric but has to be positive definite also. For testing the positive definiteness of matrix $\boldsymbol{X}$ the following simple methods can be used: the eigenvalues of matrix X have to be positive and the all determinants of k x k submatrices starting from the upper-left of matrix $\boldsymbol{X}$, i.e. [1]; $[2\times2]$, …,$[7\times7]$ have to be positive. We now present examples of Cholesky decomposition of matrix $\boldsymbol{X}$ with cross correlation data taken from [20].

$$X=\begin{pmatrix} 1 & 0 & -0.4 & -0.5 & -0.5 & 0 & -0.8 \\ 0 & 1 & -0.4 & 0 & -0.2 & 0 & 0 \\ -0.4 & -0.4 & 1 & 0.4 & 0.8 & -0.2 & 0 \\ -0.5 & 0 & 0.4 & 1 & 0 & 0.5 & 0 \\ -0.5 & -0.2 & 0.8 & 0 & 1 & -0.3 & 0.4 \\ 0 & 0 & -0.2 & 0.5 & -0.3 & 1 & 0 \\ -0.8 & 0 & 0 & 0 & 0.4 & 0 & 1 \end{pmatrix}$$

The eigenvalues can be found using Matlab software platform as:

Eig($\boldsymbol{X}$):
0.0026
0.0083
0.4378
0.8415
1.3903
1.6510
2.6686

It appears that all eigenvalues are positive. Now we test the determinants of submatrices of $\boldsymbol{X}$:

$$[1] = 1; [2 \times 2] = 1; [3 \times 3] = 0.68; [4 \times 4] = 0.47; [5 \times 5] = 0.036; [6 \times 6] = 0.0028; [7 \times 7] = 0.48\text{x}10^{-4}$$

All sub-submatrices determinants are positive. We can declare matrix $\boldsymbol{X}$ positive definite, and we can perform the Cholesky decomposition as

$$\sqrt{X} = \begin{matrix} 1.0000 & 0 & -0.4000 & -0.5000 & -0.5000 & 0 & -0.8000 \\ 0 & 1.0000 & -0.4000 & 0 & -0.2000 & 0 & 0 \\ 0 & 0 & 0.8246 & 0.2425 & 0.6306 & -0.2425 & -0.3881 \\ 0 & 0 & 0 & 0.8314 & -0.4847 & 0.6722 & -0.3679 \\ 0 & 0 & 0 & 0 & 0.2783 & 0.6422 & 0.2385 \\ 0 & 0 & 0 & 0 & 0 & 0.2774 & -0.0000 \\ 0 & 0 & 0 & 0 & 0 & 0 & 0.1309 \end{matrix} \qquad (9.67a)$$

Assume we have K UTs connected through correlated links to the same BS site and denote the kth UT is located at x_k, y_k where $k = 1, \ldots \ldots \ldots\ldots, K$ and let be M LSPs per UT link, 3GPP require M to be 7. The 2D auto-correlation is distance-dependent generated to shape the UT correlation map. We generate homogeneous grid locations according to the coordination of the K UTs.

The correlation maps are generated for a fixed sampling grid. A random normal distributed sequence of values is filtered using FIR filter. The correlation map is represented by a matrix $\mathbf{B}$ with entries $B_{y,x}$ (y is the row index and x is the column index). The first value $B_{1,\,1}$ corresponds to the top left corner of the map. The FIR filter coefficients are calculated from correlation distance in metres d_{cor}. The distance dependent autocorrelation coefficient follows an exponential function as expressed in (9.66) where$|\Delta x|$ is replaced with d and lettered as

$$\rho(d) = \exp\left(\frac{d}{d_{cor}}\right)$$

where d is the distance between two positions [31]. Two sets of filter coefficients are used: one set of coefficients to be used for the horizontal and vertical directions and the other set for the diagonal directions [32]. These two sets are generated from (X) by substituting distance d with the relative distance d_{px} of two adjacent values. The impulse response of the two FIR filter are given by

$$a_k = \frac{1}{\sqrt{d_{cor}}} \exp\left(-\frac{k\, d_{\text{px}}}{d_{cor}}\right) \qquad (9.67b)$$

$$b_k = \frac{1}{\sqrt{d_{cor}}} \exp\left(-\frac{k\,\sqrt{2}\, d_{\text{px}}}{d_{cor}}\right) \qquad (9.67c)$$

where k is the running filter coefficient index. We assume the exponential function decays for a maximum distance of 4 d_{cor} and normalized with $\sqrt{d_{cor}}$.

The correlation map is initialised with random normal distributed numbers denoted as $B^{[1]}_{y,x}$ expressed as

$$B^{[1]}_{y,x} = X \text{ with } X \sim \mathcal{N}(0,1) \qquad (9.68a)$$

We filtered (9.68a) at point x in vertical direction from top to bottom to get

$$B^{[2]}_{y,x} = \sum_{k=0}^{\lfloor 4d_{cor}/d_{px} \rfloor} a_k\, B_{y-k,x} \tag{9.68b}$$

where $k = 0, 1\ldots\ldots$,'to the largest integer less than or equal to $4d_{cor}/d_{px}$', and filtered (M) in horizontal direction left to right to get

$$B^{[3]}_{y,x} = \sum_{k=0}^{\lfloor 4d_{cor}/d_{px} \rfloor} a_k\, B_{y,x-k} \tag{9.68c}$$

We filter the diagonal of the map using filter (9.67c) first from top left to bottom right to get

$$B^{[4]}_{y,x} = \sum_{k=0}^{\lfloor 4d_{cor}/d_{px} \rfloor} b_k\, B_{y-k,x-k} \tag{9.68d}$$

then from bottom left to top right to get

$$B^{[5]}_{y,x} = \sum_{k=0}^{\lfloor 4d_{cor}/d_{px} \rfloor} b_k\, B_{y+k,x-k} \tag{9.68e}$$

After filtering the correlated random numbers $\xi_M(x_k, y_k)$ at the K grid nodes defining the K UTs locations are multiplied with $\sqrt{X_{MxM}}$ to generate the independently correlated LPS of the K links

$$s_M(x_k, y_k) = \sqrt{X_{MxM}}\, \xi_M(x_k, y_k) = [s_{SF}\ s_K\ s_{DS}\ s_{ASD}\ s_{ASA}\ \ s_{ZSD}\ s_{ZSA}]^T \tag{9.69}$$

9.14.11 Small-Scale Parameters

The small-scale parameters in Steps 5–9 include the following.

Step 5. Generate clusters delay.

Cluster delays τ_l' are taken out randomly from exponential delay distribution calculated as

$$\tau_l' = -r_\tau\, \mathrm{DS} \ln(X_l)$$

where r_τ is a fixed delay scaling factor, DS is the delays mean DS generally expressed in $\log_{10}$DS and its values depend on the channel environments, i.e. UMa LOS/NLOS, UMi LOS/NLO, etc., and X_l is a uniform distribution within the range 0 to 1 and cluster index $l = 1, \ldots\ldots, L$. We normalise the delays and let τ_l denote the normalized delays and then sort the normalized delays to an increasing order:

$$\tau_l = sort(\tau_l' - \min(\tau_l'))$$

Measured instantaneous power delay profile in [33] shows the received power peaks at a number of relative delays. The received power gap between the strongest peak and the mid-sized peak could be as high as 10 dB. In addition there are many minor peaks that greatly reduce the RMS DS. Accordingly, for LOS condition an additional scaling of delays is essential to offset the LOS peak increase to the DS. The scaling factor (constant) C_τ is dependent on the Rician K-factor as

$$C_\tau = 0.7705 - 0.0433K + 0.0002K^2 + 0.000017K^3$$

where Rician K-factor is generated in Step 4. So, the scaled delays for LOS are given as

$$\tau_l^{LOS} = \frac{\tau_l}{C_\tau}$$

Step 6. Generate powers for random clusters.

The cluster powers assignment depends on the delay τ_l. Assuming a single slop exponential delay profile, the lth cluster power is calculated as

$$P_l' = e^{c*10^{-0.1\ Z_l}}$$

where c is a constant given as

$$c = -\tau_l \frac{r_\tau - 1}{r_\tau \mathrm{DS}}$$

and $Z_l \sim \mathcal{N}(0, \xi^2)$ is per cluster SF in dB and, $\xi = 3$ dB. Normalize the cluster powers so that the sum is equal to one using the following expression:

$$P_l = \frac{P_l'}{\sum\limits_{l=1}^{\mathrm{L}} P_l'}$$

For LOS case, an additional component is added to the first cluster so the power of a single LOS ray is given as

$$P_{1,LOS} = \frac{K_R}{K_R + 1}$$

The cluster powers are not normalized but the total clusters power is given as

$$P_l = \frac{1}{K_R + 1} \frac{P_l'}{\sum\limits_{l=1}^{\mathrm{L}} P_l'} + \delta(l-1) P_{1,LOS}$$

where K_R is the Rician K- factor converted to linear scale. We assign power of each ray within a cluster as

$$\frac{P_l}{number\ of\ rays\ per\ cluster}$$

We can delete clusters with less than −25 dB power compared to the maximum cluster power without changing the scaling factor.

Step 7. Generate arrival angles and departure angles for both azimuth and elevation for each random cluster.

The PAS of all clusters is modelled as *wrapped Gaussian (WG)*. For details about wrapped Gaussian distribution, see 9.C. Azimuth angles of arrival and departure are distributed as wrapped Gaussian. The azimuth angle of arrival and departure *AOA and AOD* are determined by using the inverse Gaussian with input P_l and RMS angle spread ASA. *AOA* in the lth cluster, $\phi_{l,AOA}'$ is given as

$$\phi_{l,AOA}' = \frac{2\left(\frac{ASA}{1.4}\right)}{C_\phi} \sqrt{-\ln\left(\frac{P_l}{\max(P_l)}\right)}$$

where *ASA* is the azimuth angle spread of arrival and C_ϕ is a constant scaling factor: its values are given as

$$C_\phi = \begin{Bmatrix} C_\phi^{NLOS}.(1.1035 - 0.028\,K - 0.002K^2 + 0.000K^3), & for\ LOS \\ C_\phi^{NLOS}, & for\ NLOS \end{Bmatrix}$$

Number of clusters	4	5	8	10	11	12	14	15	16	19	20
C_{ϕ}^{NLOS}	0.779	0.860	1.018	1.090	1.123	1.146	1.190	1.211	1.226	1.273	1.289

For the LOS case, additional scaling of the angles has to be used to offset the LOS peak anomaly.

Consider assigning +/− angles by multiplying with a random variable X_l with a uniform distributed variable chosen from the set $\{1, -1\}$ and add component a random variable Y_l picked from a Gaussian distribution $Y_n^{AOA} \sim \mathcal{N}\left(0, \left(\frac{ASA}{7}\right)^2\right)$ to generate random angles variation as

$$\phi_{l,AOA} = X_l\ \phi'_{l,AOA} + Y_l^{AOA} + \phi_{LOS}^{AOA}$$

where ϕ_{LOS}^{AOA} is the AoA in the LOS direction. Let us consider the LOS scenario in more details. In the deliberation, we assume the first cluster is in the LOS direction i.e. ϕ_{LOS}^{AOA} then,

$$\phi_{l,AOA} = (X_l\ \phi'_{l,AOA} + Y_l^{AOA}) - (X_1\ \phi'_{1,AOA} + Y_1^{AOA} - \phi_{LOS}^{AOA}$$

We add ray offset angles within a cluster to generate the azimuth angle of arrival of the uth ray in the lth cluster $\phi_{l,\,u,\,AOA}$ as

$$\phi_{l,u,AOA} = \phi_{l,AOA} + c_{ASA}\ \alpha_u$$

where c_{ASA} is cluster RMS azimuth spread of arrival and departure, and α_m is the offset angle for the uth ray. Ray offset angles are given in Table 9.5. The generation of AOD ($\phi_{l,\,u,\,AOD}$) imitates AOA procedure albeit with different values for the parameters. The power angular spectrum in the zenith direction for all clusters is assumed to be Laplacian distributed. (See 9.A for further details about the Laplace random variables distribution.) Zenith angles ZOA and ZOD are generated in similar procedure as that used for AOA albeit assigning different parameters values. Parameters for generation AOD, ZOA; and AOD are given in [20].

Table 9.5 Ray offset angles within a cluster for RMS angle spread normalized to 1 [20].

Ray number u	**Offset angles α_u**
1,2	$\pm$ 0.0447
3,4	$\pm$ 0.1413
5,6	$\pm$ 0.2492
7,8	$\pm$ 0.3715
9,10	$\pm$ 0.5129
11,12	$\pm$ 0.6797
13,14	$\pm$ 0.8844
15,16	$\pm$ 1.1481
17,18	$\pm$ 1.5195
19,20	$\pm$ 2.1551

Step 8. Coupling of rays within a cluster for both azimuth and elevation.
Pair randomly azimuth angles AOD angles $\phi_{l,\,u,\,AOD}$ to AOA angles $\phi_{l,\,u,\,AOA}$ within a cluster l or within subcluster for the case of two strong clusters. Similarly, pair randomly zenith angles $\phi_{l,\,u,\,ZOD}$ with zenith angles $\phi_{l,\,u,\,ZOA}$. Further pair randomly azimuth angles of departure AOD $\phi_{l,\,u,\,AOD}$ with zenith angles of departure $\phi_{l,\,u,\,ZOD}$ within a cluster l or within subclusters for the case of two strong clusters.

Step 9. Generate the cross polarization power ratio (XPR).
Create the XPR κ for each ray u of each cluster . XPR κ is lognormal distributed. Pick XPR values as

$$\kappa_{l,u} = 10^{0.1X_{l,u}}$$

where $X_{l,u} \sim \mathcal{N}(\mu_{XPR}, \sigma^2_{XPR})$ is normal distributed with mean μ_{XPR} and variance σ^2_{XPR}, independently picked $X_{l,u}$ for each ray and each cluster.

Step 10. Pick initial random phases.
Choose random initial phases $\{\Phi^{\theta\theta}_{l,u}, \Phi^{\theta\phi}_{l,u}, \Phi^{\phi\theta}_{l,u}, \Phi^{\phi\phi}_{l,u}\}$ using uniform distribution within $(-\pi, \pi)$ for each ray and each cluster and for four different polarization $(\theta\theta, \theta\phi, \phi\theta, \phi\phi)$.

Step 11. Generate channel coefficients for each cluster l and each RX n and TX m.
The generation of channel large and small parameters are helpful for simulation. However, they have to be added as extra modelling modules. Additional simulation modules include UT mobility, multifrequency simulations, very large arrays, and large BW simulations and blockage effects, atmospheric gaseous loss, rain, and snow loss to mention but a few.

9.14.11.1 Channel Coefficients for NLOS Channel Environment

The method used by 3GPP is to split the clusters into two groups: strong clusters encompassing cluster 1 and 2; and weak clusters for the rest L-2, i.e. $l = 3, 4 \ldots \ldots, L$.

The channel coefficients generation comprises L-2 weak clusters, i.e. $l = 3, 4 \ldots \ldots, L$ and the channel coefficients for these clusters are given as

$$\mathrm{H}^{NLOS}_{n,m,l}(t) = \sqrt{\frac{P_l}{\mathrm{L}}} \sum_{u=1}^{U} H^{NLOS}_{n,m,l,u}(t) \tag{9.70a}$$

where $H^{NLOS}_{n,m,l,u}(t)$ is given by

$$\begin{aligned} H^{NLOS}_{n,m,l,u}(t) = &\begin{bmatrix} \mathrm{F}_{rx,n,\theta}(\theta_{l,u,ZOA}, & \phi_{l,u,AOA}) \\ \mathrm{F}_{rx,n,\phi}(\theta_{l,u,ZOA}, & \phi_{l,u,AOA}) \end{bmatrix}^T \\ &\times \begin{bmatrix} \exp(j\Phi^{\theta\theta}_{l,u}) & \sqrt{\kappa^{-1}_{l,u}}\exp(j\Phi^{\theta\phi}_{l,u}) \\ \sqrt{\kappa^{-1}_{l,u}}\exp(j\Phi^{\phi\theta}_{l,u}) & \exp(j\Phi^{ŒŒ}_{l,u}) \end{bmatrix} \\ &\times \begin{bmatrix} \mathrm{F}_{tx,m,\theta}(\theta_{l,u,ZOD}, & \phi_{l,u,AOD}) \\ \mathrm{F}_{tx,m,\phi}(\theta_{l,u,ZOD}, & \phi_{l,u,AOD}) \end{bmatrix} \exp\left(\frac{j2\pi(\hat{r}^T_{rx,l,u}.\bar{d}_{rx,n})}{\lambda_0}\right) \exp\left(\frac{j2\pi(\hat{r}^T_{tx,l,u}.\bar{d}_{tx,m})}{\lambda_0}\right) \\ &\times \exp\left(j2\pi\frac{\hat{r}^T_{rx,l,u}.\bar{v}}{\lambda_0}t\right) \end{aligned} \tag{9.70b}$$

where $F_{rx,n,\theta}$ and $F_{rx,n,\phi}$ are the radiation patterns of receive antenna n describes in the zenith direction of spherical unit vector $\hat{\theta}$ and azimuth direction of spherical unit vector $\hat{\phi}$ as described in 9.14.1. The spherical unit vector $\boldsymbol{r}_{rx,l,u}$ pointing in the direction of arrival of elevation angle and azimuth angle for the uth ray of the lth cluster is given as

$$\boldsymbol{r}_{rx,l,u} = \begin{bmatrix} \sin\theta_{l,u,ZOA} & \cos\phi_{l,u,AOA} \\ \sin\theta_{l,u,ZOA} & \sin\phi_{l,u,AOA} \\ \multicolumn{2}{c}{\theta_{l,u,ZOA}} \end{bmatrix} \tag{9.70c}$$

The spherical unit vector $\boldsymbol{r}_{tx,l,u}$ with elevation departure angle $\theta_{n,m}$ and azimuth departure angle $\phi_{n,m}$ are given as

$$\boldsymbol{r}_{tx,l,u} = \begin{bmatrix} \sin\theta_{l,u,ZOD} & \cos\phi_{l,u,AOD} \\ \sin\theta_{l,u,ZOD} & \sin\phi_{l,u,AOD} \\ \multicolumn{2}{c}{\theta_{l,u,ZOD}} \end{bmatrix} \tag{9.70d}$$

where we use l, u to denote a cluster and a ray within cluster l, respectively, $\bar{d}_{tx,m}$ and $\bar{d}_{rx,n}$ are standing for the location vector of transmit antenna element m and receive antenna element n. The UT velocity vector $\bar{v}$ with speed v, navigation angles are azimuth angle ϕ_v and elevation angle θ_v. The Doppler frequency $v_{l,u}$ is given as

$$v_{l,u} = \frac{\hat{r}^T_{rx,l,u}.\bar{v}}{\lambda_0} \tag{9.70e}$$

where

$$\bar{v} = v.\left[\sin\theta_v\cos\phi_v \quad \sin\theta_v\sin\phi_v \quad \cos\theta_v\right]^T$$

Additionally, the strongest clusters cluster 1 and cluster 2, it was assumed that the rays are spread to three subclusters $i = 1, 2, 3$ per cluster with fixed delay offset, and the channel coefficients are a function of time and three delays per cluster as

$$H^{NLOS}_{n,m}(\tau, t) = \sum_{l=1}^{2}\sum^{3}\sum_{u} H^{NLOS}_{n,m,l,u}(t)\,\delta(\tau - \tau_{l,}) + H^{NLOS}_{n,m,l}(t)\,\delta(\tau - \tau_l) \tag{9.71}$$

where $H^{NLOS}_{n,m,l}(t)\,\delta(\tau - \tau_l)$ is given in (9.70a) and $H^{NLOS}_{n,m,l,u}(t)$ is given in (9.70b).

9.14.11.2 Channel Coefficients for LOS Channel Environment

The LOS conditions, the channel coefficients are given as

$$H^{LOS}_{n,m,1}(t) = \begin{bmatrix} F_{rx,n,\theta}(\theta_{LOS,ZOA}, & \phi_{LOS,AOA}) \\ F_{rx,n,\phi}(\theta_{LOS,ZOA}, & \phi_{LOS,AOA}) \end{bmatrix}^T \begin{bmatrix} 1 & 0 \\ 0 & -1 \end{bmatrix} \times \begin{bmatrix} F_{tx,m,\theta}(\theta_{LOS,ZOD}, & \phi_{l,u,AOD}) \\ F_{tx,m,\phi}(\theta_{LOS,ZOD}, & \phi_{l,u,AOD}) \end{bmatrix}$$

$$\times \exp\left(\frac{-j2\pi d_{3D}}{\lambda_0}\right)\exp\left(\frac{j2\pi(\boldsymbol{r}^T_{rx,LOS}.\bar{d}_{rx,n})}{\lambda_0}\right)\exp\left(\frac{j2\pi(\boldsymbol{r}^T_{tx,LOS}.\bar{d}_{tx,m})}{\lambda_0}\right)\exp\left(j2\pi\frac{\boldsymbol{r}^T_{rx,LOS}.\bar{v}}{\lambda_0}t\right) \tag{9.72a}$$

Table 9.6 Oxygen loss for frequencies 0–100 GHz [20].

Freq. [GHz]	0–52	53	54	55	56	57	58	59	60	61	62	63	64	65	66	67	68–100
$\alpha(f_c)$ dB/km	0	1	2.2	4	6.6	9.7	12.6	14.6	15	14.6	14.3	10.5	6.8	3.9	1.9	1	0

The LOS impulse response to is given by the LOS channel coefficients in (9.72a) to the NLOS channel impulse response in (9.71) and scaling each term of the sum by the appropriate K-factor as $\mathrm{K_R}$,

$$\mathrm{H}_{n,m}^{LOS}(t) = \sqrt{\frac{1}{\mathrm{K_R}+1}}\,\mathrm{H}_{n,m}^{NLOS}(\tau,t) + \sqrt{\frac{\mathrm{K_R}}{\mathrm{K_R}+1}}\,\mathrm{H}_{n,m,1}^{LOS}(t)\,\delta(\tau-\tau_1) \tag{9.72b}$$

Step 12. Apply path loss and shadowing for the channel coefficients.

In addition to path-loss calculation formulae in 9.14.8, 3GPP added the atmospheric gaseous loss, specifically oxygen absorption and blockage loss.

9.14.11.3 Oxygen Absorption

The physics of oxygen loss are conferred in more details in Chapter 6. Oxygen loss was defined by 3GPP for cluster l at frequency f_c $OL_l(f_c)$ was modelled as

$$OL_l(f_c) = \frac{\alpha(f_c)}{1000}(d_{3D} + c.(\tau_l + \tau_\Delta)) \quad \mathrm{dB} \tag{9.73}$$

where $\alpha(f_c)$ is loss in dB /km specified in Table 9.6, c is speed of light, τ_l is the lth cluster delay, and τ_Δ is 0 for LOS, and $\min(\tau_l')$ otherwise where $\min(\tau_l')$ is generated in Step 5. The values of rate of loss $\alpha(f_c)$ are given in Table 9.6.

9.14.11.4 Blockage Loss

3GPP defined two types of blockers: human and vehicular blocking. 3GPP recommended two alternative models to determine the blockage loss. Model A follows a stochastic approach and is appropriate for use when computationally efficient modelling is required, while model B adopts a geometric method for capturing the human and vehicular blocking and is suitable for use when realistic modelling is necessary. Model A is considered next.

Model A considered 2D angular blocking regions created around the UT in respect of centre angle, azimuth, and elevation span. These regions are: one is the self-blocking region associated with the human body blocking, and four others non-self-blocking regions including indoor (office) and outdoor UMa, RMa, and UMi environments. The self-blocking and the non-self-blocking regions are defined in respect of elevation and azimuth angles and their uniform distributions. The self-blocking loss varied between 0 dB and a maximum of 30 dB. The non-self-blocking loss is given as

$$L_{dB} - 20\log_{10}[1 - (F_{A1} + F_{A2})(F_{Z1} + F_{Z2})] \tag{9.74a}$$

where

$$F_x = \frac{\tan^{-1}\left(\pm\frac{\pi}{2}\sqrt{\frac{\pi}{\lambda}r\left(\frac{1}{\cos x} - 1\right)}\right)}{\pi} \tag{9.74b}$$

where r is the distance between the UT and the blocker, the four regions are represented by the four angle of arrival expressed as $\mathrm{x} = A1|A2|Z1|\ Z2|$, and A is azimuth angle of arrival, Z is the zenith angle of arrive as

$$A1|A2| = \phi_{AOA} \quad - \left(\phi_k \pm \frac{x_k}{2}\right)$$

$$Z1|Z2| = \theta_{ZOA} \quad - \left(\theta_k \pm \frac{y_k}{2}\right) \tag{9.74c}$$

where $k = 1, 2,\ldots, 4$ is the blocker index, x_k represents azimuth uniform distribution, and y_k zenith uniform distribution. Values for the parameters to computer the blockage loss are given in [20] and $r = 2$ m indoor high and $r = 10$ m for UMa, RMa, and UMi environments.

9.15 Blockage Modelling

9.15.1 Blockages Modelling Using Random Shape Theory

The most important features of mmWave signals you have to remember are: the mmWave signals are more sensitive to blockage such as high buildings than signals in a lower frequency band [31, 34–36]. Materials such as concrete have a high penetration loss, which makes the prediction of the cellular network performance in urban areas very challenging. The blockages have knock-on effects on network functions like user-BS association and user mobility. For example, a user connects with a farther away BS if it provides LOS reception rather than a closer BS that is blocked.

Conventionally, the effects of blockage are included into the shadowing and scattering that are modelled as lognormal distributed variables. Unfortunately, such a model is not viable at mmWave band, for it does not echo the distance-dependence of blockage effects, For example, the longer the link, the more buildings in urban areas intersect the link causing severe shadowing losses.

There are two favourable methods to model blockages in wireless propagation. One approach is using ray tracing to add terrain information (i.e. location and size of the blockages, BS location) to generate the received signal strength for site-specific simulations. Indeed, such an approach requires complex computation but gives accurate measurements for the specific site. The second approach is to use the stochastic method to model the random blockages with few parameters. Each blockage is defined with some probability, and it is assumed that signals are reflected when they touch the blockages with no power losses.

In this section, we consider a mathematical framework to model blockages with random sizes, locations, and orientations using the RST. A RST is a branch of stochastic geometry that examines the random distribution of shapes in space. We use RST to model buildings using a line segment with random lengths and orientations, located consistent with a spatial *Poisson point process* (PPP). Buildings are considered as large-scale blockages, obstructions by the human body that can be modelled in a similar way. Body blockage is notably important in mmWave transmission so is the foliage effect on mmWave signal propagation.

Let us consider a set of random objects S on $\mathbb{R}^n$ that are closed and bordered; i.e. have a finite area and perimeter. These objects could be a combination of line segments, rectangles, circles, or any other shape on $\mathbb{R}^2$ or a collection of cubes in $\mathbb{R}^3$. For each element $s\varepsilon S$, a centre point is identified that has a well-defined point but is not necessarily associated with a centre of gravity. Furthermore, nonsymmetric objects have a specific orientation in space, defined by a directional unit vector. In cellular network analysis, these random objects in S are subjected to blockages such as buildings, walls, and concrete enclosures.

A *random object process* (ROP) is assembled from randomly sampling objects from S and places their corresponding centre points at points of a random point process. The orientation

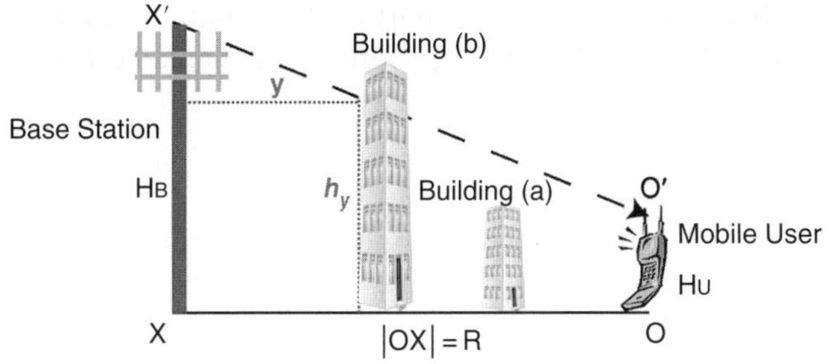

Figure 9.31 The base station located at **X** has a height of H_B, while the user has a height of H_U. [35].

of each object is specified according to certain probability distribution. Analysis of ROP is challenging especially when the sampling, location, and orientation of the objects are interrelated.

The simplest ROP $\mathscr{P}$ is PPP. An ROP with PPP distributed locations is known as the Boolean scheme. The Boolean scheme of objects S has to have the following features: (i) object centre points assemble a PPP; (ii) For all objects $\in S$, the attributes of each object such orientation, shape, and volume are independent determined; (iii) Any specific object, its orientation, shape and volume and also location are also statistically. Boolean scheme associates blockages with rectangles of centres denoted as $\{C_k\}$ to put together a homogenous PPP of density. The lengths, widths, and orientations of the rectangles are denoted as L_k, W_k, Θ_k, respectively. So, the Boolean scheme of blockages is effectively characterized by quadruple $\{C_k, L_k, W_k, \Theta_\text{k}\}$. Rectangles L_k, W_k are assumed to be i.i.d. distributed as per certain probability $f_L(x)$ and $f_W(x)$, respectively, and the orientation is uniformly distributed within $[0, 2\pi]$. Typically line segments are used to describe building, i.e. $W = 0$.

Consider a BS at X with height H_B and UT at O with a height H_U and the distance between the BS and UT is OX = R as depicted in Figure 9.28. Let the physical propagation path be represented by $\text{O}'\ X'$ in $\mathbb{R}^3$. Clearly, any building cross OX such as building (a) can block the signal propagation between BS and UT. A building (a) intersecting OX at point h_y away from X will block $\text{O}'\ X'$ if and only if its height h is greater than h_y where h_y can be computed as

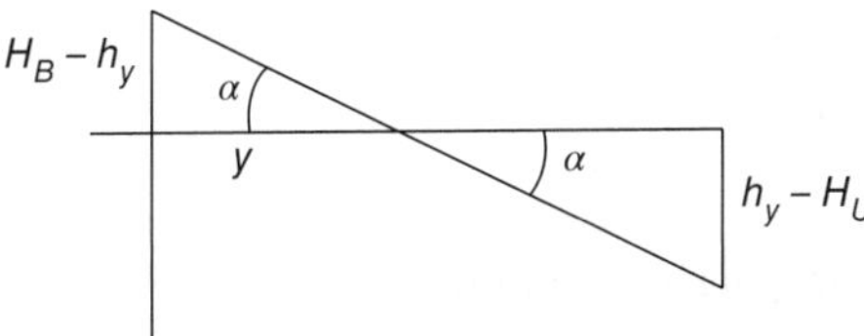

Take tan α and simplify the expression to get

$$h_y = \frac{y H_U + (R - y)H_B}{R} \tag{9.75}$$

Let K be a random Poisson variable denoting the *total number of blockages* crossing the direct line of propagation, then according to [35], the mean of K is

$$\mathbb{E}[K] = \frac{2\lambda(\mathbb{E}[W] + \mathbb{E}[L])}{\pi} R + \lambda\mathbb{E}[L]\mathbb{E}[W] = \beta R + \chi \tag{9.76a}$$

Accordingly, the average number of blockages on link is proportional to length of the link, which seems sensible; i.e. the longer the link, the more blockages intersect the link. Further, if $W = 0$ then,

$$\mathbb{E}[K] = \frac{2\lambda\mathbb{E}[L]}{\pi} R \tag{9.76b}$$

According to 3GPP, LOS probability decays exponentially as distance increases. So, the probability that a link of length R allows LOS propagation (i.e. no blockages intersect the link $K = 0$) is

$$\text{pr}(K = 0) = \exp(-(\beta R + \chi) \tag{9.77a}$$

The probability that a location in $\mathbb{R}^2$ is contained by blockages ($R = 0$) is

$$\text{pr}(K \neq 0) = 1 - \text{pr}(K = 0) = 1 - \exp(-\chi) \tag{9.77b}$$

Let the height of the kth blockage be H_k and we assume it is i.i.d. with certain probability function $f_H(x)$ and independent of $\{C_k, L_k, \Theta_k\}$. The building blocks and the direct propagation path as depicted in (Figure 9.31).

9.15.2 Analysis Using Random Shape Theory to Model Buildings

In this section, we examine model buildings located in urban wireless environment by applying a Boolean scheme. Buildings are presented as a set S of rectangles with random areas and random orientations.

Let us consider the power P_i received by UT from a distance BS located at $\boldsymbol{X}_i$ in the presence of buildings (line segments) with a small-scale fading g_i, a random variable with mean one, on a link $\mathbf{OX}_i$ from the origin to BS at $\boldsymbol{X}_i$. The exponent of path loss is α. Further, we designate K_i as the number of obstacles (buildings) on the ith link $\mathbf{OX}_i$ and denote γ_{ik} as the ratio of building penetration power losses created by the kth $(0 < k \leq K_i)$ blockage on $\mathbf{OX}_i$ where γ_{ik} holds value in [0, 1] for the blockages attenuate the received signal power when we ignore reflections. Accordingly, the received power is

$$P_i = \frac{A_{ant} g_i S_i}{R_i^{\alpha}} \tag{9.78a}$$

where $S_i = \Pi_{k=0}^{K_i} \gamma_{ik}$ and R_i^{α} expresses the distance-dependent path loss in the ith link and A_{ant} is a constant related to antenna gains at transmitted power and path-loss reference distance. The average received power $\mathbb{E}[P_i]$ is

$$\mathbb{E}[P_i] = \frac{A_{ant}\mathbb{E}[g_i]\mathbb{E}[S_i]}{R_i^{\alpha}} \tag{9.78b}$$

We know $\mathbb{E}[g_i] = 1$ and $\mathbb{E}[S_i]$ derived in [35] as

$$\mathbb{E}[S_i] = \exp(-(\beta R + \chi)(1 - \mathbb{E}[\gamma_{ik}]))$$

Thus

$$\mathbb{E}[P_i] \approx \frac{A_{ant} \quad \exp(-(\beta R + \chi)(1 - \mathbb{E}[\gamma_{ik}]))}{R_i^{\alpha}} \tag{9.78c}$$

9.15.3 Distance to Closest BS with Building Blockage

When a cellular network is blocked with impenetrable blockages, the blockages divide the network coverage into isolated areas, and only locations within the same area can communicate wirelessly. So it is important to explore the connectivity of impenetrable network to find how many unblocked BSs are available for a typical UT and what is the distribution of the distance to the nearest BS.

Consider two locations $\boldsymbol{X}, \boldsymbol{Y} \varepsilon \mathbb{R}^2$ and the number of blockages on the direct link connecting $\boldsymbol{X}$ and $\boldsymbol{Y}$ be K_{XY} . So $\boldsymbol{X}$ is visible by $\boldsymbol{Y}$ if and only if $K_{XY} = 0$ (i.e. there are no blockages between them). The visible area of location $\boldsymbol{X}$ is a set of locations that can connect to $\boldsymbol{X}$ with LOS link. Denote the visible area of $\boldsymbol{X}$ as Q_X defined such that a UT at location $\boldsymbol{X}$ can only connect to BSs in Q_X and if location $\boldsymbol{X}$ is blocked completely, then $Q_X = \emptyset$ is an empty set.

The average size of the visibility region in a cellular network with impenetrable blockages $\mathbb{E}[V(Q_X)]$ is given by

$$\mathbb{E}[V(Q_X)] = \frac{2\pi \exp[-\chi]}{\beta^2} \tag{9.79}$$

where $\chi = \lambda\, \mathbb{E}[L]\mathbb{E}[W]$ where L is length and W is width of the blockage. However, if $W \equiv 0$ when we use line segments to describe buildings then $\chi = 0$. Hence,

$$\mathbb{E}[V(Q_X)] = \frac{2\pi}{\beta^2} \tag{9.80}$$

An important metric to assess the network connectivity is the distribution of the distance of the nearest BS to a UT to achieve LOS reception (i.e. there is no blockage crossing the direct propagation path). Assume a UT is located at the origin and the distance to the nearest visible BS located at $R_0 > x$. The distribution of the distance to the nearest BS is

$$\Pr(R_0 > x) = \exp(-2\pi\mu\, U(x)) \tag{9.81a}$$

where μ is density of the BSs and $U(x)$ is defined as

$$U(x) = \frac{\exp(\chi)}{\beta^2}\,[1 - (1 + \beta x)\exp(-\beta x)] \tag{9.81b}$$

Furthermore, the probability density function (PDF) of R_0 is

$$f_{R_0}(x) = 2\pi\mu\, x \exp[-(\beta x + \chi + 2\pi\, \mu U(x))] \tag{9.82}$$

The probability $\Pr_c(x)$ that a UT at the origin connects with at least one BS is defined as

$$\Pr_c(x) := \Pr_c(R_0 < \infty) = 1 - \exp\left(\frac{2\pi\mu}{\beta^2}\exp(\chi)\right) \tag{9.83}$$

9.16 Summary

In this chapter, we have dealt with various configurations of antenna arrays and explored the development of channel modelling by 3GPP covering 2D SCMs WINNER channel models and the recent 3D channel models in 3GPP Release 14.

Section 9.1 introduced the antenna configurations by identifying a number of factors that control the performance of an array, such as the geometric configurations (i.e. linear, square, circular, spherical, etc.), the spacing between consecutive elements, and whether such spacing is uniform or non-uniform. An antenna array comprises multiple identical antennas, the radiation pattern of the antenna element is essentially used to define the array factor (AF). The AF of arrays was derived for 1D uniform linear arrays in Section 9.2, and for 2D rectangular planar arrays in Section 9.3, for circular arrays in Section 9.4, and for cylindrical arrays in Section 9.5. We examined the 3D spherical arrays as possible candidates for mmWave arrays for

5G applications in Section 9.6. The microstrip patch arrays design and physical dimensions were explored together with a number of patch microstrip arrays, and a patch microstrip antenna conformal to spherical surface was also presented in Section 9.7.

The EU WINNER project was briefly introduced in Section 9.8. The 2D spatial MIMO channel model in 3GPP Release 6 includes: BS and MS antenna patterns; BS and MS angle spread (AS) per path; BS and MS PAS per path; and BS and MS angle parameters definition for scattering environment. All these topics were discussed in more details in Section 9.9.

The scattering environments were considered in Section 9.10. The 2D SCM large scale parameters (LSPs) such as DS, AS, SF, and auto and cross correlation between the LSPs, were examined in detail in Section 9.11. The SCMs of the 2D channel coefficients are generated for urban, suburban macrocell and urban microcells, as described in Section 9.12.2. 2D SCMs with antenna polarization were considered in Section 9.13.2.

3GPP Release 14 3D channel models were investigated in depth, including the antenna array radiation power pattern for azimuth and elevation dimensions; probability of LOS and its estimate using ray tracing method; path loss for LOS and NLOS channel environments. Fast-fading and large-scale parameters and channel coefficients generations were dealt with more throughly in Section 9.14.3.

The RST is used to explain the blockage modelling including: ROP, random theory to model buildings, and estimates of the distance to closest BS in environments with building blockage were all examined in detail in Section 9.15.

9.A Laplace Random Variables Distribution

Laplace distribution has many important applications in mathematics, physics, various biological processes, engineering, and statistics. Some of the popular applications include finance, speech recognition and image processing, video coding, Kalman filtering, and in wireless indoor communications to model the indoor angle of arrival/departure and time of arrival. Laplace distribution, like normal (Gaussian) distribution, has an exponential PDF. However, normal distribution is used to analyse symmetrical data when it has a short tail, while Laplace distribution is peaky and is used when the data has a long tail (sec Figure 9.A.1).

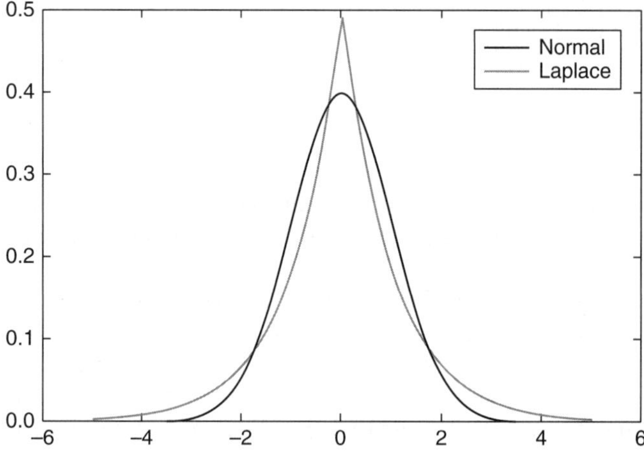

Figure 9.A.1 Comparison of Laplace and Gaussian (Normal) PDFs.

The continuous Laplace probability distribution, denoted by $\mathcal{L}(a, b)$, is defined by the probability density function $f_{a,\ b}(x)$ as

$$f_{a,\ b}(x) := \frac{1}{2b} \exp\left(\frac{-|x-a|}{b}\right), \forall x \in \mathbb{R} \tag{9.A.1}$$

where $a \in (-\infty, \infty)$ is called the location parameter and $b > 0$ scale parameter. The mean and variance of Laplace distribution are a and $2b^2$, respectively. Laplace distribution is a symmetric distribution whose tails fall gently. Compare (10.D.1) with the corresponding formulae to normal (Gaussian) distribution $\mathcal{N}(\mu, \sigma^2)$ defined as

$$f_{\mu,\ \sigma}(x) := \frac{1}{\sigma\sqrt{2\pi}} \exp\left(\frac{-(x-\mu)^2}{2\sigma^2}\right), \forall x \in \mathbb{R} \tag{9.A.2}$$

where μ is the mean value of the data and σ^2 is the variance. Equations (9.1) and (9.2) are plotted in Figure 9.31.

9.B Spherical Coordinates

Consider a point (x, y, z) on a sphere defined by the spherical coordinates (ρ, θ, ϕ) where ρ is the distance from the origin to the point (x, y, z), i.e. radius of the sphere as illustrated in Figure 9.B.1.

$$\hat{\rho} = \begin{pmatrix} x\,\hat{\imath} \\ y\hat{\jmath} \\ z\hat{k} \end{pmatrix} = \begin{pmatrix} \rho\ \sin\theta\cos\phi \\ \rho\ \sin\theta\sin\phi \\ \rho\cos\theta \end{pmatrix} \tag{9.B.1}$$

where the Cartesian unit vectors for (x, y, z) are $\hat{\imath} = (1,0,0), \hat{\jmath} = (0,1,0)$, $\hat{k} = (0,0,1)$, respectively, and $\hat{\rho} = x\,\hat{\imath} + y\hat{\jmath} + z\hat{k}$.

We also have

$$|\rho|^2 = x^2 + y^2 + z^2 \tag{9.B.2}$$

where $\rho \geq 0$.

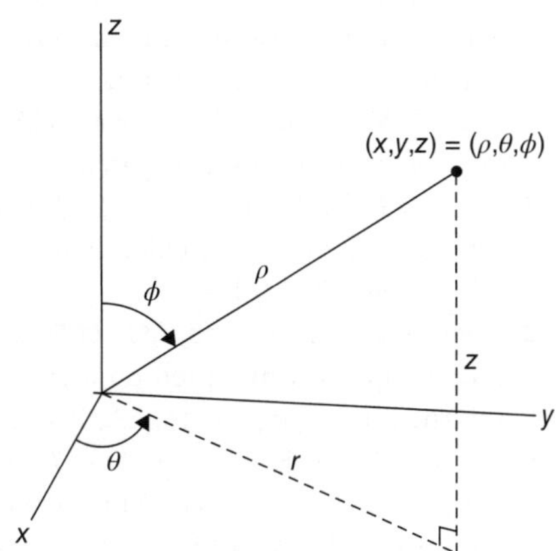

Figure 9.B.1 Illustration of spherical coordinates.

9.C Wrapped Gaussian Distribution

The wrapped Gaussian (WG) distribution is a widely used distribution in circular statistics ([37], [38], [39]). Traditionally, a uniform distribution for AoA at UT of MPCs has been assumed. However, it has been established by measurements that nonisotropic scattering in many transmission environments corresponds to *nonuniform* distribution for AOA at the UT. Further, the assumption of a uniform distribution of the AOA seems to bring small errors in its first moment (the mean value) of the received signal but a significant error on the second moment (correlation; variances). Wrapped Gaussian is used for nonuniform distributed AOA at the UT. Other applications include directional statistics, robot applications, signal processing such as filtering cases that involve a high level of noise, speech processing, and constrained tracking.

The fundamental concept underlying WG distribution is a one-dimensional Gaussian distribution wrapped around a unit circle. Consider a Gaussian-distributed random variable, x ~ $\mathcal{N}(\mu, \sigma^2)$ and wrap it around an interval of 2 π. Then we get a new random variable $X \in [\pi, \pi]$ and the distribution of X is called wrapped Gaussian (normal) distribution referred to it as $X \sim WN(\mu, \sigma^2)$. Mathematically, the PDF of X is given as

$$\mathrm{WN}(\theta) = \frac{1}{\sigma\sqrt{2\pi}} \sum_{k=-\infty}^{\infty} \exp\left(\frac{\theta - \mu + 2\pi k}{2\sigma^2}\right) \tag{9.C.1}$$

where $\sigma > 0$ and $\mu \in [-\pi, \pi]$. The wrapped distribution $\mathrm{WN}(\theta)$ is similar to the Gaussian distribution in that the sum of two wrapped Gaussian distributed variables is itself a wrapped Gaussian distributed random variable after mod $- 2\pi$ operation. However, one important feature in Gaussian distribution is missing in counterpart-wrapped Gaussian, which is the product of two wrapped Gaussian densities and is not itself wrapped Gaussian density.

A number of circular distributions were prevailed beside WG distribution, such as the von Mises distribution derived to avoid the inconvenient and challenging computations. The same property is maintained in Bingham distribution but unlike von Mises, Bingham distribution density has π symmetry.

References

1 Larsson, E.G., Edfors, O., Tufvesson, F., and Marzetta, T.L. (2014). Massive MIMO for next generation wireless systems. *IEEE Communications Magazine* 52 (2): 186–195.
2 Balanis, C.A. (2005). *Antenna Theory*. Hoboken, NJ: John Wiley & Sons.
3 Silva, C.M., Lumini, F., Lacava, J.C.S., and Richards, F.P. (1991). Analysis of cylindrical arrays of microstrip rectangular patches. *Electronics Letters* 27 (9): 778–780.
4 Li, P., Luk, K.M., and Lau, K.L. (2003). An omnidirectional high gain microstrip antenna array mounted on a circular cylinder. In: *IEEE Antennas and Propagation Symposium Digest*, 698–701. IEEE.
5 Heckler, M.V.T., Lacava, J.C. da S., and Cividanes, L. (2005) 'Design of a circularly polarized microstrip array mounted on a cylindrical surface', IEEE Antennas and Propagation Society International Symposium, 2A, 266–269.
6 Kumar, G. and Ray, K.P. (2003). *Broadband Microstrip Antennas*. Artech House.
7 Gui-Bin Hsieh, G.-B. and Kin-Lu Wong, K.-L. (1999) 'Radiation characteristics of cylindrical microstrip arrays', IEEE Antennas and Propagation Society International Symposium, 4, 2760–2763.

8 Ashkenazy, J., Shtrikman, S., and Treves, D. (1988). Conformal microstrip arrays on cylinders. *IEE Proceedings H (Microwaves, Antennas and Propagation)* 135 (2): 132–134.
9 Ojaroudiparchin, N., Shen, M., Zhang, S., and Pedersen, G.F. (2016). A switchable 3D coverage-phased array antenna package for 5G mobile terminals. *IEEE Antennas and Wireless Propagation Letters* 15: 1747–1750.
10 De Witte, E., Marantis, L., Tong, K-F., Brennan, P., and Griffiths, H. (2006) 'Design and development of a spherical array antenna, European Conference on Antennas and Propagation, 1–5.
11 Tomasic, B., Turtle, J., and Liu, S. (2005) 'Spherical arrays – design considerations', International Conference on Applied Electromagnetics and Communications, 1–8.
12 Xie, Y., Peng, C., Jiang, X., and Ouyang, S. (2014) 'Hardware design and implementation of DOA estimation algorithms for spherical array antennas', IEEE International Conference on Signal Processing, Communications and Computing (ICSPCC), 219–223.
13 Sood, D., Singh, G., Tripathi, C.C. et al. (2008). Design fabrication and characterization of microstrip square patch antenna array for X-band applications. *Indian Journal of Pure and Applied Physics* 46: 593–597.
14 Knott, P. (2007) Design and experimental results of a spherical antenna array for a conformal array demonstrator, International ITG Conference on Antennas (INICA), 120–123
15 Knott, P. (2006) 'Design of a triple patch antenna element for double curved conformal antenna arrays', European Conference on Antennas and Propagation, 1–4.
16 Sharma, A. and Gupta, S.D. (2015) 'Design and analysis of rectangular microstrip patch antenna conformal on spherical surface', IEEE International Conference on Signal Processing and Communications (ICSC), 366–369.
17 European Conference of Postal and Telecommunications Administrations (CEPT), 2008. Wireless World Initiative for Next Radio (WINNER), IST-4-027756 WINNER II, D1.1.2 V1.2, *Channel Models*, updated April 2.
18 ETSI 3GPP 3GPP TR 25.996 V6.1.0 Spatial Channel Model for Multiple Input- Multiple Out Simulations (Release 6) September 2003.
19 Greenstein, L.J., Erceg, V., Yeh, Y.S., and Clark, M.V. (1997). A new path-gain/delay-spread propagation model for digital cellular channels. *IEEE Transactions on Vehicular Technology* 46 (2): 477–485.
20 ETSI 3GPP. 2018. 3GPP TR 38.901 version 14.3,0 Release 14- 5G Study on channel model for frequencies from 0.5 to 100 GHz.
21 Akdeniz, M.R., Liu, Y., Samimi, M.K. et al. (2014). Millimeter wave channel modelling and cellular capacity evaluation. *IEEE Journal on Selected Areas in Communications* 32 (6): 1164–1179.
22 ETSI 3GPP 3GPP TR 36.873 v12.2.0 (2015) 'Study on 3D Channel Model for LTE (Release 12) September.
23 Adema, F., Taranetz, M., and Rupp, M. (2016). 3GPP 3D MIMO channel model: a holistic implementation guideline for open source simulation tools. *EURASIP Journal on Wireless Communications and Networking* 2016 (55): 1–14.
24 Mondal, B., Thomas, T.A., Visotsky, E. et al. (2015). 3D channel model in 3GPP. *IEEE Communications Magazine* 53 (3): 16–23.
25 Samimi, M.K., Rappaport, T.S., and MacCartney, G.R. Jr. (2015). Probabilitistic omnidirectional path loss models for millimeter-wave outdoor communications. *IEEE Wireless Communications Letters* 4 (4): 357–360.
26 Kalden, N., Zetterberg, P., Ottersten, B., and Garcia, L. (2007). Inter and intra site correlation of large scale parameters from macro cellular measurements at 1800 MHz. *EURASIP Journal on Wireless Communications and Networking* 2007 (3): 1–9.

27 Jaeckel, S., Liang, L., Jungnickel, V., Thiele, L., Jandura, C., Sommerkorn, G., and Schneider, C., (2009) 'Correlation properties of large and small-scale parameters from multicell channel measurements', EU Conference on Antennas and Propagation, Berlin, Germany, 3406–3410.

28 European Conference of Postal and Telecommunications Administrations (CEPT) EU project WINNER II Channel modes, 2007 and updated 2008.

29 Jaeckle, S., Jiang, L., Jungnickel, V., Thiele, L., Jandura, C., Sommerkorn, G, and Schneider, C. (2009) 'Correlation properties of large- and small parameters from multicell channel measurements', EU Conference on Antenna and Propagation , 3406–3410, Berlin, Germany, (invited paper).

30 Jaldén, N., Zetterberg, P., Ottersten, B., and Garcia, L. (2007). Inter- and intrasite correlations of large-scale parameters from macrocellular measurements at 1800 MHz. *EURASIP Journal on Wireless Communications and Networking* 2007 (3, Art. No. 25757): 1–12.

31 Cowan, R. (1989). Objects arranged randomly in space – an accessible theory. *Advances in Applied Probability* 21 (3): 543–569.

32 Jaeckel, S., Raschkowski, L., Boner, K., Thiele, L., Burkhardt, F., and Ebrlein, E (2017) 'Quasi Deterministic Radio Channel generator', User Manual and Documentation, Fraunhifer Heinrich Hertz Institute, Technical Report v2.0.0.

33 Asplund, H., Larsson, K., and Okvist, P. (2008). How typical is the 'typical urban' channel model? In: *IEEE Vehicular Technology Conference*, 340–343.

34 Bai, T., Vaze, R., and Heath, R.W. (2012) 'Using random shape theory to model blockage in random cellular networks', IEEE International Conference on Signal Processing and Communications (SPCOM), 1–5.

35 Bai, T., Vaze, R., and Heath, T.W. (2014). Analysis of blockage effects on urban cellular networks. *IEEE Transactions on Wireless Communications* 13 (9): 5070–5083.

36 Taranetz, M. and Muller, M.K. (2016). A survey on modelling interference and blockage in urban heterogeneous cellular networks. *EURASIP Journal on Wireless Communications and Networking* 2016 (252): 1–20.

37 Abdi, A., Barger, J., and Kaveh, M. (2002). A parametric model for the distribution of the angle of arrival and the associated correlation function and power spectrum at the mobile station. *IEEE Transactions on Vehicular Technology* 51 (3): 425–434.

38 Gilitschenski, I., Kurz, G., Hanebeck, U.D., and Siegwart, R. (2016) 'Optimal quantization of circular distributions', International Conference on Information Fusion, Heidelberg, Germany, 1–8.

39 Kurz, G. and Hanebeck, U.D. (2015) 'Parameter estimation for the bivariate wrapped normal distribution', IEEE Annual Conference on Decision and Control (CDC), Osaka, Japan, 1192–1198.

10

Massive MIMO Channel Estimation Schemes

10.1 Introduction

10.1.1 Cellular MIMO Channels

A cellular network consists of N_c cells and is commonly structured in clusters, each cluster comprises a number of cells (4-cell cluster/7-cell cluster/13-cell cluster, etc.). Each cell is provided with a fixed BS that is equipped with an array of antennas to serve a community of subscribed users and each user is supported with one or more antennas. The subscribed user is able to move within the network with seamless communication and unaware of the cell boundary. The most common form of shape used in the theoretical analysis for a radio cell is a hexagon but in reality, a radio cell has a no uniform shape and fundamentally the cell borders are dependent on the antenna radiation pattern.

The BSs are connected together by a fibre or microwave link and to a mobile switching centre. A basic multi-user cellular system is depicted in schematic form in Figure 10.1. There are two basic multiple-input, multiple-output (MIMO) channel models namely: The MIMO multiple-access channel (MAC) from MS transmitter towards the BS receiver, and the MIMO broadcast channel (BC) from the BS transmitter towards MS receivers.

Full reuse of frequency bands or pilot sequences across neighbouring cells develops severe intercell interference that degrades services offered by the network. Several solutions aimed at mitigating intercell interference have emerged in recent years; among them is the use of MIMO processing at BSs to optimize the beamforming vectors so that cooperating BSs can perform jointly. However, the coordinated beamforming scheme calls for a fast exchange of the channel state information at the transmitters (CSIT) on a low-latency basis, which is a challenge to achieve particularly in massive MIMO scheme.

Another solution presented in [1, 2] is by simply increasing the number of antennas at the BSs (i.e. massive MIMO). The idea behind this concept is simple: as the number of antennas at the BS increases, the desired channel's vector tends to be orthogonal to the interfering channel vectors and therefore it is possible to avoid the interference from all other users by aligning the beamforming vector with the desired channel vector. In general, channel state information (CSI) is acquired using finite pilot sequences in the presence of intercell interference. Such scenarios imply a possibility that pilot sequences from neighbouring cells contaminate each other.

Massive MIMO BSs employ a very large number of low-cost antennas to serve a relatively small number of users on the same time-frequency resource block. Such a system with linear beamforming and random MS scheduling drives the uncorrelated interference impairments to zero and achieves very high spectral efficiency. The only remaining impairment is due to pilot contamination. Various schemes are proposed to mitigate the pilot contamination interference

5G Physical Layer Technologies, First Edition. Mosa Ali Abu-Rgheff.

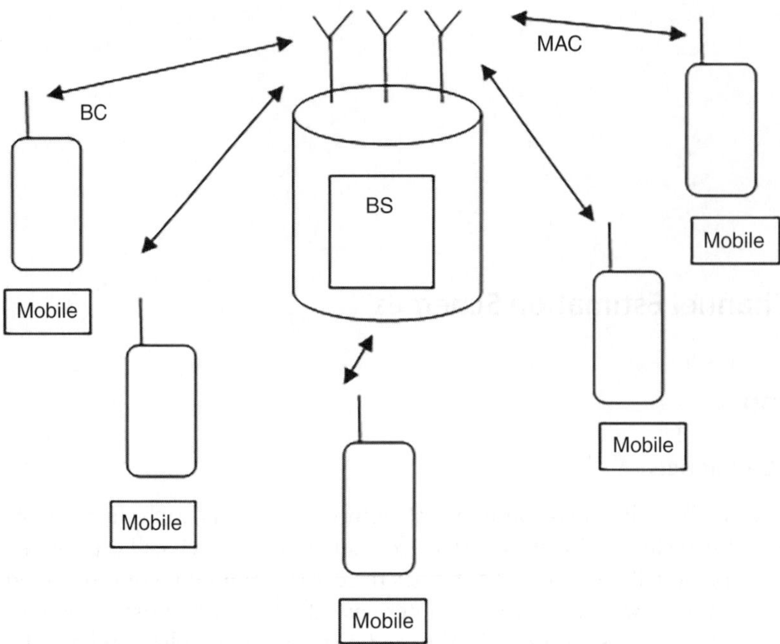

Figure 10.1 Schematic diagram showing the MAC and BC channels in a cellular system.

such as linear multi-user minimum mean square error (MMSE) precoding; linear combination of messages of MSs sharing the same pilot sequences; low rate coordination between BSs: and time-shifted pilot-based schemes by rearranging the uplink pilot transmission order for different cells to avoid simultaneous transmission in adjacent cells.

Knowledge of the channel matrix is helpful in receive and transmit system processing. Furthermore, Frobenius-squared norm (FSN) of the channel matrix provides important knowledge of channel gain that can be exploited for rate adaptation and scheduling. The FSN can be determined from the *estimated channel matrix*, or *determined using an* mmse *estimator*. Under (Rayleigh and Rician) fading environment, the former approach develops poor estimation performance at most signal power to interference power plus noise power ratio (SINR) compared to the latter. The nature of the MIMO channel, and indeed wireless channels in general, are stochastic that can benefit from Bayesian estimation, that is modelling the current channel state based on a known multivariate PDF function.

BSs must have accurate CSIs at the transmitter (CSIT) to carry out the transmit signals beamforming, that is an estimate of the DL channel. In frequency-division duplex (FDD) systems, uplink and downlink are operating at different frequencies in two separated bands and consequently, uplink-downlink channel reciprocity does not hold. DL training is used to estimate the CSI that is fed back on the uplink. This procedure (closed-loop feedback) may work when the number of antennas at BS is small. However, when the number of BS antennas is very large, the bandwidth required for CSI feedback on the uplink become excessive. In time-division duplexing (TDD), since uplink and downlink transmissions take place in the same channel coherence bandwidth, CSIT can be acquired directly from the uplink pilot symbols. Consequently, the 5G network is more likely to adapt TDD operation. We assume that the channel reciprocity between the propagation uplink (UL) and downlink (DL) applied; i.e. the DL channel vectors are the conjugate transpose of the UL ones, and so estimating the UL is sufficient to generate the DL channel.

10.2 Massive MIMO Channels Definition

In a typical single cell BS, the input data to the BS transmitter passed through multistage digital signal processing stages commonly used in any digital communication system such as channel coding, data interleaving, signal mapping (modulation), and possibly filtering followed by a MIMO precoding. The transmit signal can be launched from the transmit antennas and when it arrives at the receive antennas, the receiver acquires the signals at each receive antenna output and reverses the transmitter processing operations (i.e. space-time de-precoding, decoding, demodulation etc.). Due to space restrictions, we focus only on the MIMO system since the early stages are treated in many textbooks available on the market.

In this chapter we consider the MIMO channels, particularly when the MIMO size becomes massive. Two associated parameters with MIMO system design namely: the multiplexing gain and the diversity gain. The *maximum multiplexing gain* that can be achieved over a MIMO channel is defined as the asymptotic slope of the channel outage capacity for a fixed frame error rate (FER) and given signal power to noise power ratio (SNR). For optimally designed MIMO transmit-receive system, the *maximum multiplexing gain* is given by the minimum of (M,N) bit/sec/Hz and double for every 3 dB increase in SNR for a fixed FER. The *maximum diversity gain* over the MIMO channel is given by the negative asymptotic slope of FER (y-axis) plotted versus SNR (x-axis) both in log scales for fixed transmission rate. The *maximum diversity gain* over a MIMO channel for optimum designed transceiver is (NM) so that for a fixed transmission rate, an increase of 3 dB in SNR reduces the FER by a factor of 2^{-NM}.

The general MIMO channel within each cell can be represented by matrix $\mathbf{H} \in \mathbb{C}^{N \times M}$ where h_{nm} is the single-input, single-output (SISO) channel between the mth transmit and the nth receive antenna. The mth column expresses the special signature of the mth transmit antenna across the M transmit antennas array. The nth row represents the signals received from the N antennas array at a receiver. The SISO individual channel gains making the MIMO channel generally modelled as a *zero mean circularly symmetric complex Gaussian random variables* with Rayleigh distributed random variable amplitudes and exponentially distributed power gains. The correlation between the individual channels is a function of the scattering in the medium and the antenna arrays configuration at BS.

10.2.1 Massive MIMO UL Definition

Consider a cellular network with N_c noncooperative hexagonal cells working in TDD-based operation, each cell has a BS with M antennas, and we assume that M can be increased without a limit, i.e. $M \rightarrow \infty$. The M-component propagation vector between any MS and any BS grows as M and the inner product of any two different propagation vectors grows slowly and eventually disappears.

We assume that each cell serves K single antenna users that are located randomly within the cell with uniform distribution. At the centre of each cell is a BS with M omni-directional antennas. The hexagonal cell' radius is denoted as r_c where users are uniformly distributed over

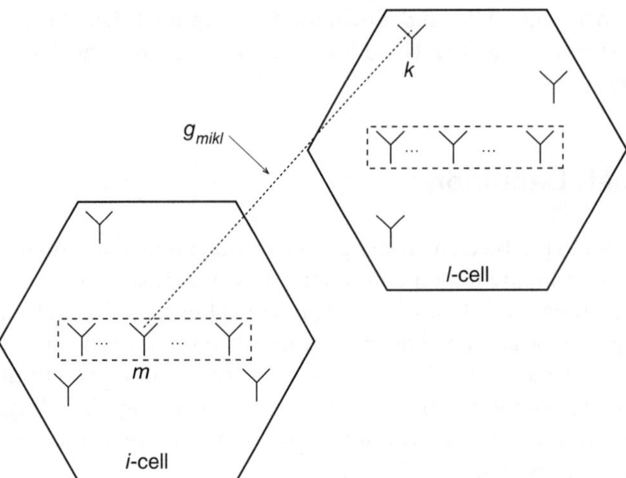

Figure 10.2 The propagation coefficient between the kth MS in the lth cell, and the mth BS antenna of the ith cell, is denoted by g_{mikl} [1].

their home cell except for a central disk r_h at the edge of the cell to ensure that MSs propagation model adopted is valid.

In a TDD-based operation, mutually orthogonal pilot sequences of length τ are used by *single-antenna* users in the same cell. However, when identical pilot sequences are used in more than one cell, intercell interference is initiated by the so-called pilot contamination. To facilitate the notation, we assume the first cell is the target cell unless otherwise stated. For ease of explanation, we consider the situation where a *single user per cell* transmits its pilot sequence to its serving BS.

Consider the transmission over the *uplink flat-fading channels* between single-antenna terminals in one cell and a BS in another cell as illustrated in Figure 10.2. Denote the complex propagation coefficients $\mathbf{h}_{ikl} \in \mathbb{C}^{M\times 1}$ vector channel between the mth BS antenna in the ith cell and the kth MS in the lth cell.

$$\mathbf{h}_{mikl} = \sqrt{\beta_{jkl}}\ \mathbf{g}_{mikl} \qquad k = 1, 2 \ldots \ldots, K, m = 1, \ldots .., \mathrm{M}, and \quad i, l = 1, \ldots .N_c \tag{10.1}$$

where β_{jkl} is the large-scale fading acting for both attenuation and shadow fading expressed as

$$\beta_{ikl} = \frac{\alpha_{ikl}}{r_{ikl}^{\varkappa}} \tag{10.2}$$

and α_{ikl} is lognormal random variable representing shadow fading with standard deviation σ_{shadow}, r_{ikl} is the distance between k^{th} user of the lth cell and the ith BS antenna, and $\varkappa$ is the attenuation power exponent. The small-scale fading vector $\mathbf{g}_{mikl} \sim \mathcal{CN}(0, \mathbf{I}_{\mathrm{M}})$ has complex Gaussian variables with zero mean and unit variance. We assume $\mathbf{g}_{mjkl}$ to be statistically independent across the users.

Denote $\mathbf{H}_{il} \in \mathbb{C}^{\mathrm{M}\times K}$ as a complex matrix between the K users in the lth cell and the M BS antennas of the ith cell where

$$\mathbf{H}_{il} = \sqrt{\beta_{jl}}\ \boldsymbol{G}_{il} \tag{10.3}$$

where $[\boldsymbol{G}_{il}]_{mk} \in \mathbf{g}_{mikl}$ small scale fading vectors between the kth MS in the l^{th}cell, and the mth BS antenna of the ith cell, $\boldsymbol{\beta}_{il}$ is a vector whose elements assembled a $K \times K$ diagonal matrix.

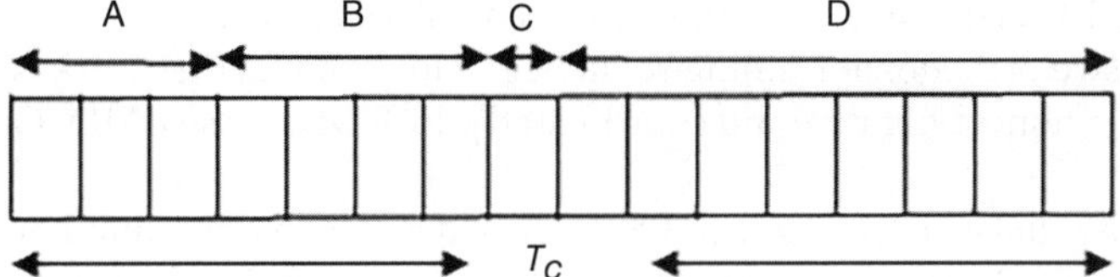

Figure 10.3 TDD transmission protocol (Aligned pilot transmission). Each block in the figure represents a single symbol duration so (in this example) each coherence interval has a length of 16 symbol durations. *A: Uplink data transmission; B: Uplink pilot transmission; C: Uplink channel vectors estimation; D: Downlink (BC) data transmission.*

10.3 Time-Division Duplexing (TDD) Transmission Protocol

Consider UL pilots that are transmitted using the TDD scheme to obtain the UL channel vector estimates. From the reciprocity concept, the corresponding DL vector estimates are expressed by conjugate transpose of the UL estimates vector within the coherence interval of the channel T_C [3–5]. The latter is defined as the time duration over which the channel response is invariant. We assume the spatial channel vectors remain constant over a coherence block of T channel uses given by

$$T = W_c T_c \tag{10.4}$$

where W_c the coherence bandwidth. The coherence interval is comprised of four segments to accommodate four phases of the TDD transmission, as shown in Figure 10.3. phase1, each user transmits uplink data necessary for the network to grant access to transmit to a specific BS in a specific cell, A (3) symbol periods; phase 2 each user transmits pilot sequence of length K to its home BS for B, a (4) symbol periods; phase 3 home BS uses the received pilot sequences to estimate the uplink channel vectors, C a single symbol period for the users to detect the data received on the uplink phase; Finally, phase 4 the home BS uses the conjugate transpose of the uplink channel vectors as downlink channel vectors to transmit downlink data, D (8) symbol periods to its mobile users. The traditional TDD protocol is depicted in Figure 10.3.

In massive MIMO systems, as $M \to \infty$, a single pilot sequence is reused by a user in more than a single cell. This generates interference commonly known as pilot contamination [6] and TDD transmission does not null the contamination.

10.4 Massive MIMO Channel Estimation in Noncooperative TDD Networks

Channel estimation in massive MIMO networks is corrupted by the pilot contamination effects creating a significant degrade to the estimates. This problem can be eased through coordination between cells, using a low-rate communication channel, during the channel estimation phase. The aim of the coordination is to make use of the second-order statistical information of the user channels creating selectivity across the interfering users' multipath even with strong correlated pilot sequences. Consequently, the massive MIMO technique tends to reduce or even completely remove the pilot contamination effect for certain channel covariance. A pilot contamination reduction/elimination was proposed in [8]. The scheme is based on using an orthogonal variable spreading factor (OVSF) code and a set of Zadoff-Chu (ZC) sequences for UL pilot training. The set of ZC sequences is multiplied (element-wise) with its OVCF code row at the BS to create massive orthogonal training sequences. However, further work is needed to investigate the scheme in noncooperative and coordinated pilot TDD massive MIMO systems.

We examine the performance of cellular networks with BSs provided with a massive number of antennas using noncooperative TDD transmission protocol. A precise analysis of accurate SINR is then derived for both UL and DL. In addition, a scheme to avoid pilot contamination and hence considerably reduce the inter-cell interference proposed by the authors of [7] which will be presented in the following deliberations.

10.4.1 Uplink Pilots' Transmission Using the Aligned Pilot Scheme

We use the MIMO UL system's parameters defined in 10.2.1 and we assume the same set of K orthogonal pilot sequences each of length K symbols are sent simultaneously by all MSs in the system. The kth MSs in every cell employs the same pilot sequence $\boldsymbol{\psi}_k = \psi_{k1}, \dots \dots, \psi_{kK}$ and $|\boldsymbol{\psi}_{kj}| = 1$, $|\boldsymbol{\psi}_{k'}^H \; \boldsymbol{\psi}_k| = K\delta_{k,k'}$ where $\boldsymbol{\psi}_{k'}$ and $\boldsymbol{\psi}_k$ are referring to the kth pilot sequence when applied in two different cells.

Denote ρ_{kl} as pilot power. At the pilot transmission stage, pilots are transmitted simultaneously by all users to respective BSs (*aligned pilot transmission scheme*). The UL transmitted signal of the kth user in the lth cell $\mathrm{x}_k^{i^{th}\,BS}$ is

$$\mathrm{x}_k^{i^{th}\,BS} = \sqrt{\rho_{kl}}\, \psi_k \tag{10.5}$$

where ρ_{kl} is the pilot transmit power for k^{th} user in the lth cell.

Define β_{ikl} and $\mathbf{g}_{ikl}$ as large-scale fading coefficients (lognormal distribution and power decaying) and the small scale fading vectors, $\mathbf{g}_{ikl} \sim \mathbb{C}\mathcal{N}(0, \mathbf{I})$ between the ith BS and the kth user of the lth cell respectively. The ith BS receives the pilot signal from N_c cells each with K users. Each user transmits only a single pilot sequence:

$$\mathbf{Y}^{i^{th}\,BS} = \sum_{l=1}^{N_c} \sum_{k=1}^{K} \sqrt{\rho_{kl}\beta_{ikl}}\, \mathbf{g}_{ikl}\, \boldsymbol{\psi}_{k'}^H + \mathbf{N}_i \tag{10.6}$$

where $\mathbf{N}_i \in \mathbb{C}^{M \times K}$ is the BS receiver additive noise with zero mean and variance of one, i.e. entries of $\mathbf{N}_i$ are i.i.d are random variables with $\mathrm{C}\mathcal{N}(0, 1)$. Note that C stands for a complex number and $\mathbb{C}$ stands for a set of complex numbers. The ith BS estimates the channel vectors $\widehat{\mathbf{h}}_{ik'i}$ for MSs located in the same cell as [7]:

$$\widehat{\mathbf{h}}_{ik'i} = \frac{\boldsymbol{y}^{i^{th}\,BS}\;\; \boldsymbol{\psi}_{k'}^H}{K} \tag{10.7}$$

Substituting (10.6) in (10.7) we get

$$\widehat{\mathbf{h}}_{ik'i} = \sqrt{\rho_{k'i}\beta_{ik'i}}\;\; \mathbf{g}_{ik'i} + \sum_{l=1, l\neq i}^{N_c} \sqrt{\rho_{k'l}\beta_{ik'l}}\;\; \mathbf{g}_{ik'l} + \mathbf{n}_i' \tag{10.8}$$

where

$$\mathbf{n}_i' = \mathbf{n}_i \frac{\boldsymbol{\psi}_{k'}^H}{K} \sim \mathbb{C}\mathcal{N}\left(0, \frac{1}{K}\mathbf{I}_{\mathrm{M}}\right) \tag{10.9}$$

It is interesting to note that the second term in (10.8) is an estimate error due to pilot *contamination* from the same pilot sequence used in neighbouring cells ($l = 1 \dots N_c$, $l \neq i$) and the last term error is due to filtered Gaussian noise.

The ith BS computes the normalized beamforming vector $\mathbf{w}_{k'i}$ to its k'^{th} user as

$$\mathbf{w}_{k'i} = \frac{\widehat{\mathbf{h}}_{ik'i}}{\|\widehat{\mathbf{h}}_{ik'i}\|} = \frac{\widehat{\mathbf{h}}_{ik'i}}{\alpha_{k'i}\sqrt{\mathrm{M}}} \tag{10.10}$$

where the normalization scalar factor $\alpha_{k'i} = \frac{\|\widehat{\mathbf{h}}_{ik'i}\|}{\sqrt{\mathrm{M}}}$. We now compute the beamforming vector $\mathbf{w}_{k'i}$ when $\mathrm{M} \to \infty$. It was shown in 10.A that as $\mathrm{M} \to \infty$, the approximate value $\alpha^2_{k'i}$ is

$$\lim_{\mathrm{M}\to\infty} \alpha^2_{k'i} \approx \sum_{l=1}^{N_c} \rho_{k'l}\beta_{ik'l} + \frac{1}{K} \tag{10.11}$$

The UL channel vectors are estimated at the BS using the received pilot sequences, and each BS transmits the DL data to its corresponding users.

10.4.2 SINR for Uplink Data Transmission

In order to investigate the SINR for UL data transmission, let use consider K users transmitting their data towards their respective BSs. The received signal at respective BS is a combination of the desired signal plus interference. The ith BS receives the signal $y^{i^{th}\,BS}$

$$y^{i^{th}\,BS} = \sum_{l=1}^{N_c}\sum_{k=1}^{K} \sqrt{P^U_{kl}\,\beta_{ikl}}\,\mathbf{g}_{ikl}\,q_{kl} + v_i \tag{10.12}$$

where P^U_{kl} is the total uplink transmit power, $\mathbf{g}_{ikl}$ is small-scale fading vector between the ith BS and the kth user of the lth cell, q_{kl} is *uplink transmitted signal* sent by the kth user of the lth BS, the same pilot sequence is used by $k'^{\,th}$ user of the ith BS and v_i is the additive white noise of the ith BS receiver. Let us assume the ith BS receiver uses MF equalizer by multiplying the received signal (10.12) with $(\mathbf{h}_{ik'i})^H$ when decoding signal sent by its $k'^{\,th}$ user [7]:

$$\widehat{q}_{k'i} = (\mathbf{h}_{ik'i})^H\, y^{i^{th}\,BS} \tag{10.13}$$

$$\widehat{q}_{k'i} = (\widehat{\mathbf{h}}_{ik'i})^H\, y^{i^{th}\,BS} = \sum_{l_1=1}^{N_c}\sum_{l_2=1}^{N_c}\sum_{k=1}^{K} \sqrt{\rho_{k'l_1}\beta_{ik'l_1}P^U_{kl_2}\beta_{ikl_2}}\,\mathbf{g}^H_{ik'l_1}\mathbf{g}_{ikl_2}q_{kl_2}$$

$$+\sum_{l}^{N_c}\sqrt{P^U_{kl}\beta_{ikl}}\;\boldsymbol{n}'^H\,\mathbf{g}_{ikl} + \sum_{l}^{N_c}\sqrt{\rho_{k'l}\beta_{ik'l}}\;\mathbf{g}^H_{ik'l}\,v_i + \boldsymbol{n}'^H_i v_i \tag{10.14}$$

Applying (10.8) and (10.12) in (10.14) we get

$$\hat{q}_{k'i} = \underbrace{\sqrt{\rho_{k'i}\,P^U_{k'i}}\,\beta_{ik'i}\,\|\mathbf{g}_{ik'i}\|^2\,q_{k'i}}_{\text{desired signal}} + \underbrace{\textstyle\sum_{l=1,l\neq i}^{N_c}\sqrt{\rho_{k'l}\,P^U_{k'l}}\,\beta_{ik'l}\,\|\mathbf{g}_{ik'l}\|^2\,q_{k'l}}_{\text{interference}} + O(\mathrm{M}) \tag{10.15}$$

where $O(\mathrm{M})$ is the small error that tends to zero when $\mathrm{M} \to \infty$.

The first term in (10.16) is the desired signal and the second term is the interference due to pilot sequence contamination. Therefore, the uplink SINR $SINR^{UL}_{k'i}$ for the $k'^{\,th}$ user in the ith cell is

$$\mathrm{SINR}^{UL}_{k'i} = \frac{\rho_{k'i}P^U_{k'i}\beta^2_{ik'i}}{\sum_{l=1,l\neq i}^{N_c}\rho_{k'l}P^U_{k'l}\beta^2_{ik'l}} \tag{10.16}$$

10.4.3 SINR for Downlink Data Transmission

Once the BS received the UL pilot sequences, it estimates the UL channel vectors. Each BS generated the corresponding DL channel (based on the reciprocity principle) computes the beamforming vector $\mathbf{w}_{k'i}$ using (10.10) to transmit its data to its respective users. The k'^{th} MS of the ith cell receives a signal given by

$$y_{k'i} = \left(\sum_{l=1}^{N_c} \sum_{k=1}^{K} \sqrt{P_{kl}\beta_{lk'i}} \ (\mathbf{g}_{lk'i})^H \mathbf{w}_{kl}\, s_{kl} \right) + v_{k'i} \tag{10.17}$$

where $v_{k'i}$ is the additive white noise with unit variance, P_{kl} power transmitted by l^{th} BS to send data to their MSs in the lth cell, $\beta_{lk'i}$, $\mathbf{g}_{lk'i}$, and $\boldsymbol{w}_{kl}$ are as defined in 10.4.2. The k'^{th} user in the ith cell uses match filtering (MF) equalizer $(\mathbf{h}_{lk'i})^H$ to decode its personal data.

Consider the double summation on the right-hand side of (10.17) and denote each term in the double summation by ϑ_{kl}. Let S_{kl} be the variance of the received signal, then

$$S_{kl} = \mathbb{E}[|\vartheta_{kl}|^2] = \mathbb{E}[P_{kl}\,\beta_{lk'i}|(\mathbf{g}_{lk'i})^H\,\mathbf{w}_{kl}|^2\,|s_{kl}\,|^2] \tag{10.18}$$

Substitute for $\boldsymbol{w}_{kl}$ in (10.18) using (10.10) we get

$$S_{kl} = \frac{P_{kl}\,\beta_{lk'i}}{\alpha_{kl}^2\,\mathrm{M}} \mathbb{E}[|(\mathbf{g}_{lk'i})^H\,\mathbf{g}_{lkl}\,|^2\,|s_{kl}|^2]$$

which can be expressed as

$$S_{kl} = \frac{P_{kl}\,\beta_{lk'i}}{\alpha_{kl}^2\,\mathrm{M}} \left| \sum_{l_1}^{N_c} \sqrt{\rho_{kl_1}\beta_{lkl_1}}\ \mathbf{g}_{l\,k'i}^H\,\mathbf{g}_{lkl_1} + \mathbf{g}_{l\,k'i}^H\,\mathbf{n}_i' \right|^2 \tag{10.19}$$

Now we use [7 Lemma 1], which states the fact that for two independent complex vectors $\mathbf{x,y} \in \mathbb{C}^{\mathrm{M\,x\,1}}$ with Gaussian distribution $\mathbb{C}\mathcal{N}(\mathbf{0}, c\,\mathbf{I})$ then

$$\lim_{\mathrm{M}\to\infty} \left(\frac{\mathbf{x}^H\,\mathbf{y}}{\mathrm{M}} \right) \xrightarrow{a.s} 0 \tag{10.20}$$

And

$$\lim_{\mathrm{M}\to\infty} \left(\frac{\mathbf{x}^H\,\mathbf{x}}{\mathrm{M}} \right) \xrightarrow{a.s} c \tag{10.21}$$

It can be seen that (10.20) is a consequence of the correlation between two independent random variables, while (10.21) is the result of a trace of a variable. When we apply eqs. (10.20)–(10.21) and consider the two scenarios for (10.19) we get

(a) when $k = k'$ in (10.19), we get [7]

$$\lim_{\mathrm{M}\to\infty} \left(\frac{\sum_{l_1}^{N_c} \sqrt{\rho_{kl_1}\beta_{lkl_1}}\ \mathbf{g}_{l\,k'i}^H\,\mathbf{g}_{lkl_1} + \mathbf{g}_{l\,k'i}^H\,\mathbf{n}_i'}{\mathrm{M}} \right) \xrightarrow{a.s} \sqrt{\rho_{k'i}\beta_{lk'i}} \tag{10.22}$$

Therefore, we have

$$\lim_{\mathrm{M}\to\infty} \left(\frac{S_{k'l}}{\mathrm{M}} \right) \xrightarrow{a.s} \frac{P_{k'l}\,\rho_{k'i}\beta_{lk'i}^2}{\alpha_{k'l}^2} \tag{10.23}$$

(b) when $k \neq k'$ in (10.19), according to (10.20), we get

$$\lim_{M \to \infty} \left(\frac{S_{kl}}{M} \right) = 0 \quad (10.24)$$

Careful examination of (10.17) reveals that the desired signal received is $s_{k'l}$ of the k'^{th} user in the ith cell contains components from k^{th} users in the neighbouring l^{th}cells using the same pilot sequence as the desired user signal and creating pilot contamination interference that does not vanish as $M \to \infty$, but the variance of the additive noise is null in the asymptotic region. The cause for the contamination interference is that the beamforming vectors applied by these users contain components directed close to the k'^{th} user in the ith cell hence causing the interference.

The DL SINR of the k'^{th} user in the ith cell is given by

$$SINR_{k'i}^{DL} = \frac{\frac{P_{k'i}\beta_{ik'i}^2}{\alpha_{k'i}^2}}{\sum_{l=1,l\neq i}^{N_c} \frac{P_{k'l}\beta^2{}_{lk'i}}{\alpha_{k'l}^2}} \quad (10.25)$$

where

$$\alpha_{k'l}^2 = \sum_{l=1}^{N_c} \rho_{k'l}\beta_{ik'l} + \frac{1}{K} \quad (10.26)$$

We now compare the SINRS in (10.25) and (10.16) in the analysis to examine the quality of DL and UL data transmission in noncooperative TDD systems. DL SINR in (10.25), pilot powers appear to divide the DL transmit power, while on the opposite it multiplies the UL transmit power in (10.16). Therefore, any changes in pilot power to MSs will influence the DL and UL SINRS diversely. Accordingly, the optimization of pilot powers for both transmission directions concurrently is difficult. However, adjustment in pilot power can be attained by transmit link power allocation. Therefore, there would be no loss experienced if we assign equal and constant pilot power $\rho_{kl} = \rho$ to all MSs over all network cells. Thence, the UL SINR in (10.16) becomes

$$\mathrm{SINR}_{k'i}^{UL} = \frac{P_{k'i}^{U}\beta_{ik'i}^2}{\sum_{l=1,l\neq i}^{N_c} P_{k'l}^{U}\beta_{ik'l}^2} \quad (10.27)$$

Note that due to fixing the pilot power, the normalization factor $\alpha_{k'l}^2$ becomes independent of pilot transmit power. Additionally, the pilot contamination interference effects transmission over DL and UL diversely. In the DL, each MS collects contamination interference from neighbouring BS. This is because the contaminated UL and hence the DL channel estimate used in the beamforming is in error as well. In the UL, the contamination interference is due to users in other cells using the same pilot sequences. The level of the interference depends on the distances between each BS and the users from other cells.

10.4.4 Massive MIMO Channels Estimation Using Time-Shifted Pilot Scheme (TSPS)

So far, we learn that transmission of pilot sequences concurrently contaminates the UL channel estimates even at $M \to \infty$. The time-shifted pilot transmission in TDD spread out the transmission using fixed time shift. In this scheme, we divide the system cells into groups such as $\mathcal{g}_1$, $\mathcal{g}_2, \ldots, \mathcal{g}_\phi, \ldots, \mathcal{g}_c$ and each group comprises a number of cells. The communication protocol during each coherence interval is partitioned into two stages. First, users from cells in the group

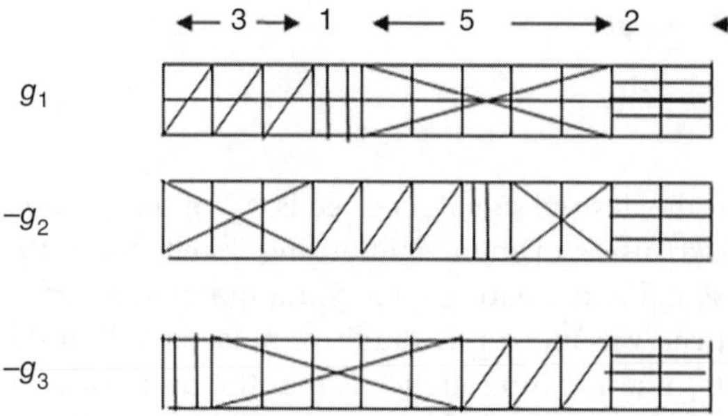

Figure 10.4 Time-shifted pilots scheme with number *Pilot sequences* = 3, *processing pilots* = 1, *DL data* = 5 and *UL data* = 2. Numbers are in symbol durations [7].

(say $\mathcal{g}_\phi$) transmit their pilot sequences simultaneously while users from all other groups receive their DL data. Once group $\mathcal{g}_\phi$ complete their pilot transmission, users of $\mathcal{g}_\phi$ start receiving their DL data and at the same time users from a different group transmit their pilot sequences. This cycle is repeated until all users in all groups transmit their pilot sequences. Pursuing the time-shifted pilot scheme (TSPS) reduces pilot contamination considerably. The remnant of the interference is due to MSs in the group only as the analysis implies. The time-shifted scheme is depicted in Figure 10.4 for three cell groups ($\mathcal{g}_1, \mathcal{g}_2, \mathcal{g}_3$).

The reciprocity between uplink and downlink channels is also valid as in the aligned pilots' scheme. It is worth noting that coherence interval includes a combined time allocated to uplink pilot transmission UL data, DL data transmission, and channel vector estimates processing. However, time allocation follows a tradeoff between data rates, number of users in the system, and coherence time interval.

In the time-shifted scheme, users in all cells belonging to, say, group $\mathcal{g}_\phi$ transmit their pilot sequences simultaneously over UL. At the same time, other cell groups transmit their data over DL to their respective users. Time-shifted pilot sequences transmission in both directions at the same time is of concern. On one hand, the BSs in group $\mathcal{g}_\phi$ received not only pilot signals but also DL signals transmitted by different BSs from all groups except $\mathcal{g}_\phi$ and, on the other hand, BS powers transmitted on the DL are typically more powerful than the pilot powers. We now investigate the degradation (if any) incurred on the UL channel estimates by the DL transmission. Let the ith cell be in group $\mathcal{g}_\phi$. During the pilot sequences transmission phase, this BS receives

$$\mathbf{y}^{i^{th}BS} = \sum_{j\in\mathcal{g}_\phi}\sum_{k=1}^{K}\sqrt{\rho\beta_{ikj}}\,\mathbf{g}_{ikj}\boldsymbol{\psi}_k + \sum_{l\notin\mathcal{g}_\phi}\sum_{k=1}^{K}\sqrt{P_{kl}c_{il}}\,\mathbf{g}_{il}\mathbf{w}_{kl}s_{kl} + \mathbf{n}_i \tag{10.28}$$

where ρ is pilot transmit power and β_{ikj}, $\mathbf{g}_{ikj}$, and P_{kl} are as defined in 10.2.1, $\mathbf{g}_{il}\in\mathbb{C}^{\mathrm{MxM}}$ is the channel matrix between antennas of the ith and l^{th} BSs, $\boldsymbol{w}_{kl}$ and $\boldsymbol{s}_{kl}$ are the beamforming vector for the kth users in the lth cells, and $K\times 1$ vector of the signal to the kth user in the kth cell, respectively, and c_{il} is long-scale fading between the ith and l^{th} BSs. We assume that both $\boldsymbol{g}_{il}$ and c_{il} are constant within the coherence interval and have $i.\ i.\ d$ Gaussian distribution and lognormal distribution, respectively.

The ith BS estimates the UL channel vector by multiplying $\boldsymbol{y}^{i^{th}BS}$ with MF equalizer $\boldsymbol{\psi}_{k'}^{H}$

$$\widehat{\mathbf{h}}_{ik'i} = \frac{\boldsymbol{y}^{i^{th}BS}\psi_{k'}^{H}}{K} = \sum_{j\in\mathcal{g}_\phi}\sqrt{\rho\beta_{ik'j}}\,\mathbf{g}_{ik'j} + \frac{1}{K}\sum_{l\,\notin\,\mathcal{g}_\phi}\sum_{k=1}^{K}\sqrt{P_{kl}c_{il}}\,\mathbf{g}_{il}\boldsymbol{w}_{kl}\boldsymbol{s}_{kl}\boldsymbol{\psi}_{k'}^{H} + \mathbf{n}_i' \tag{10.29}$$

The beamforming vector $\boldsymbol{w}_{k'i}$ is defined as in (10.10):

$$\mathbf{w}_{k'i} := \frac{\widehat{\mathbf{h}}_{ik'i}}{\|\widehat{\mathbf{h}}_{ik'i}\|} = \frac{\widehat{\mathbf{h}}_{ik'i}}{\alpha_{k'i}\sqrt{\mathrm{M}}} \tag{10.30}$$

where $\alpha_{k'i} = \frac{\|\widehat{\mathbf{h}}_{ik'i}\|}{\sqrt{\mathrm{M}}}$ is an asymptotic normalization factor.

$$\lim_{\mathrm{M}\to\infty} \alpha_{k'i}^2 = \lim_{\mathrm{M}} \frac{1}{\mathrm{M}}\left(\sum_{j\in\mathcal{G}_\phi} \rho\, \beta_{ik'j}\, \|\mathbf{g}_{ik'j}\|^2 + \frac{1}{K}\sum_{l\notin\mathcal{G}_\phi}\sum_{k=1}^{K} P_{kl}c_{il}\|\,\mathbf{h}_{il}\,\mathbf{w}_{kl}\|^2\|\boldsymbol{s}_{kl}\boldsymbol{\psi}_{k'}^H\|^2 + \|\mathbf{n}_i'\|^2 \right) \tag{10.31}$$

It is clear from (10.31) that $\alpha_{k'i}$ is influenced by a number of factors, including the DL signals, BS transmit powers of neighbouring BSs, and beamforming vectors. The ith BS can easily determine $\alpha_{k'i}$ once it computes the channel vector estimates. Derivation of theoretical analysis of the beamforming vectors are derived in 10.B.

Applying a similar procedure to the one used previously, it is shown in [7] that the expression for the SINR for the k'^{th} user of the ith BS downlink, $SINR_{k'i}^{DL}$ is

$$SINR_{k'i}^{DL} = \frac{\dfrac{P_{k'i}\,\beta_{ik'i}^2}{\alpha_{k'i}^2}}{\sum_{l\in\mathcal{G}_\phi, l\neq i}\dfrac{P_{k'l}\beta_{lk'i}^2}{\alpha_{k'l}^2}} \tag{10.32}$$

and the expression for the SINR for the k'^{th} user of the ith BS uplink, $SINR_{k'i}^{UL}$ is

$$SINR_{k'i}^{UL} = \frac{P_{k'i}^{U}\beta_{ik'i}^2}{\sum_{j\in\mathcal{G}_\phi, j\neq i} P_{k'j}^{U}\beta_{ik'j}^2} \tag{10.33}$$

Comparing the time-shift pilot scheme SINRs for the k'^{th} user of the ith BS UL in (10.33) and of the ith BS DL in (10.32) with the respective SINRs for the aligned pilots scheme in (10.16) and (10.25) we conclude that the time-shifted pilot scheme guarantees that only cells in the same group interfere with each other and that the scheme completely nulls interference from adjacent group cells provided the pilots do not overlap in time. Consequently, time-shifted pilot scheme promises significant gains in data transmission rates. Furthermore, using a time-shifted pilot scheme and ensuring orthogonal pilot sequences guarantees cancellation of pilot contamination interference.

A comparison of user above rate for aligned and time-shifted pilots for the DL and UL is shown in Figures 10.5 and 10.6, respectively, where r is pilot sequence reuse factor. While significant gains in transmission rates are possible with time-shifted pilot schemes over the aligned pilots, it has to bear in mind that these gains are dependent of many system parameters such power allocations, pilots re-user factor r time-sharing pilot for multi-user (MU) systems, and channel vector estimates. Both Figures 10.5 and 10.6 show a fraction of users using TSPS with rates above that obtained by using the aligned pilot scheme, both with equal power allocation for DL and UL. It is clear that the rate in TSPS is much higher than the corresponding rate in aligned pilot scheme, which is due to the huge reduction in the contamination interference for other groups.

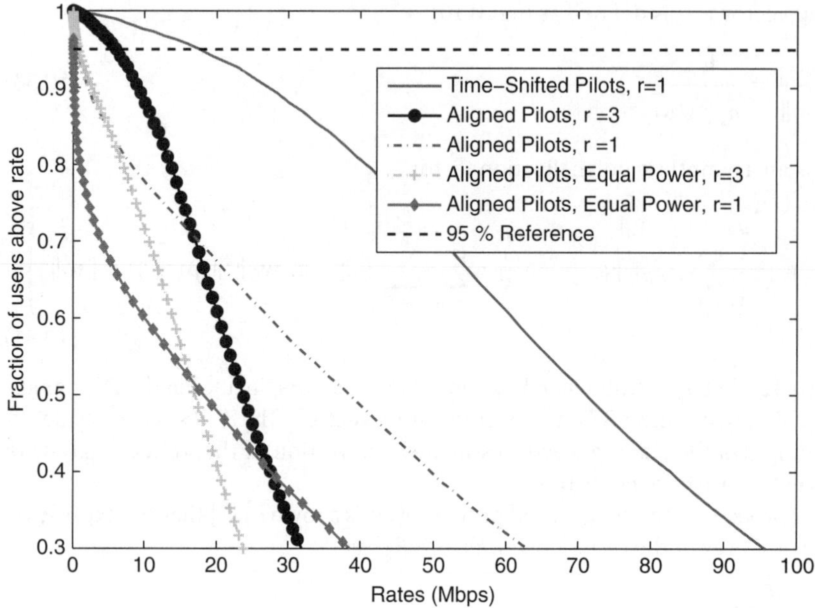

Figure 10.5 Fraction of the users above rate for the aligned and time-shifted pilots for the downlink [7].

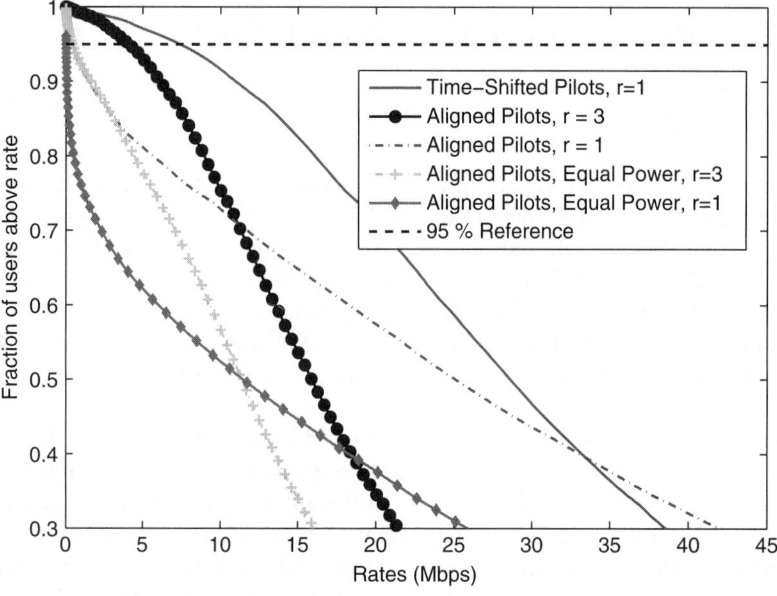

Figure 10.6 Fraction of the users above rate for the aligned and time-shifted pilots for the uplink [7].

10.5 Channel Estimation Using Coordinated Cells in MIMO System

So far in attempt to reduce the contamination interference we have examined pilot transmission in cellular network where cells and users are not cooperating to reduce the interference. The intercell interference is mainly duc to rcusing pilot sequences in neighbouring cells. We then examined the scenario where time-shifted pilot sequences are transmitted in organized groupings of the network cells and we find the contamination interference reduced to the group intra-users' interference.

In this section, we introduce a new pilot transmission scheme based on a coordinated pilot protocol. The aim of the coordination mode is to optimize the channel covariance matrices so that ideally the desired and the interference covariance matrices (a total of $K\,N_c$) occupy dissimilar spaces. To analyse the new scheme, we consider a cellular network that consists of N_c cells and select an arbitrary pilot sequence to assign to one (out of K) user in each of the N_c time-synchronized cells, i.e. N_c users are assigned an identical pilot sequences. But before we analyse the coordinated pilot assignment, we deal with the covariance-based channel estimation.

10.5.1 Bayesian Estimation of Uplink for All Users [2]

We assume the network of N_c cells to be time-synchronized so that the pilots nurture the worst contamination scenario since absence of synchronization is likely to statistically reduce the correlation of the pilots. Further, all the BSs are equipped with M antennas and we assume the first cell is the target cell unless otherwise stated. Pilots of length τ are transmitted by single antenna-users and pilots in the same cell are mutually orthogonal so their intra-interference is negligible. In addition, pilots are reused from cell to cell are nonorthogonal causing intercell interference from $N_c - 1$ interfering cells.

The channel vectors covariance matrices are related to the distribution (i.e. mean and spread) of the multipath angle of arrival (AOA) at the BS. Because of the commonly elevated positions of the BSs, rays arrive on the BS antennas will have a finite AOA spread. Normally, the mean AOA is dependent on user location.

Two Bayesian channel estimators are presented. In the first, all channels are estimated at the target BS (desired user and interfering users) and in the second, only the desired channel $\mathbf{h}_1$ is estimated. The channel vector between the user in the lth and the target BS is denoted as $\mathbf{h}_l$.

During the pilot phase, the $\mathrm{M} \times \tau$ signal received $\mathbf{Y}$ at the target BS (i.e. first cell) is given by

$$\mathbf{Y} = \sum_{l=1}^{N_c} \mathbf{h}_l\, \mathbf{s}_l^T + \mathbf{N} \tag{10.34}$$

where $\mathbf{s}_l = [s_{l1}, s_{l2}, \dots \dots \dots \dots, s_{l\tau}]^T$ is pilot sequence used by the lth cell user, τ is number of symbols in the pilot sequence, and $\mathbf{N}$ is complex $\mathrm{M} \times \tau$ additive white Gaussian noise (AWGN) with zero mean and element-wise variance σ_n^2. The powers of the pilot sequences are anticipated to be equal, i.e.

$$|s_{l1}|^2 + |s_{l2}|^2 + \dots \dots \dots \dots + |s_{l\tau}|^2 = \tau$$

where $l(=1, 2, \dots \dots\dots, N_c)$ is cell index. The pilot symbols are normalized so that

$$|s_{l1}|^2 = |s_{l2}|^2 = \dots \dots \dots \dots = |s_{l\tau}|^2 = 1$$

Let us denote the channel covariance matrix at the receive end in the lth cell as $(\mathrm{M} \times \mathrm{M})$ matrix $\mathbf{R}_l$ as

$$\mathbf{R}_l = \mathbb{E}\{\mathbf{h}_l\, \mathbf{h}_l^H\}$$

where $\mathbf{h}_l$ is channel vector between a user in the lth cell and the target BS. We denote the desired user channel vector as $\mathbf{h}_1$ at the same time as $\mathbf{h}_l$, $l > 1$ are interference channels.

The channel estimation method exploits the variance of the channel vectors to get further information about the distribution (mean and spread) of the multipath AOA at the target BS.

The focus is on the optimal pilot sequence design that enables a correct channel estimate [2, 8, 10].

Now we generalize the model given in (10.34) for all N_c cells so we have N_c models in (10.34) and use the vec(.) operator to express the received signals and noises vectors for the N_c cells as

$$\mathbf{y} = \widetilde{\mathbf{S}}\mathbf{h} + \mathbf{n} \tag{10.35}$$

Note vectorization denoted as vec(.) of a matrix is a linear transformation operator that converts the matrix into a single column vector. For example, $m \times n$ matrix **A** can be expressed as vec(**A**) the $mn \times 1$ column vector attained by laying the columns of the matrix **A** on top of one another.

The received signal vectors for the N_c cells are represented as $\mathbf{y} = \text{vec}(\mathbf{Y})$. Similarly, for noise $\mathbf{n} = \text{vec}(\mathbf{N})$ and $\mathbf{h} \in \mathbb{C}^{N_c\mathrm{M} \times 1}$. The pilot matrix $\widetilde{\mathbf{S}}$ is defined as

$$\widetilde{\mathbf{S}} \triangleq [\,\mathbf{s}_1 \oplus \mathbf{I}_\mathrm{M} \ldots\ldots\ldots\ldots\ldots\mathbf{s}_{N_c} \oplus \mathbf{I}_\mathrm{M}] \tag{10.36}$$

where $\mathbf{s}_i \oplus \mathbf{I}_\mathrm{M}$ represents the Kronecker product of matrices $\mathbf{s}_i$ and $\mathbf{I}_\mathrm{M}$. Note that if **A** and **B** are $m \times n$ and $l \times k$ matrices, then Kronecker product is

$$A \oplus B = \begin{bmatrix} a_{11}B & a_{12}B & & a_{1n}B \\ a_{21}B & & & \\ & & & \\ a_{m1}B & a_{m2}B & & a_{mn}B \end{bmatrix}$$

MATLAB has a built-in *function kron* that can be called as i.e. K = kron(**A**, **B**).

We consider **h** and **y** as two dependent random variables. Applying Bayes' rule, the conditional probability of **h** given the received pilots signal **y** is as follows:

$$\text{pr}(\mathbf{h} \mid \mathbf{y}) = \frac{\text{pr}(\mathbf{h})\text{pr}(\mathbf{y} \mid \mathbf{h})}{\text{pr}(\mathbf{y})} \tag{10.37}$$

We use the multivariate complex Gaussian PDF to formulate the joint PDF of the random variable **h**. Assuming that the **h** columns to be $(\mathbf{h}_1, \ldots\ldots\ldots, \mathbf{h}_l, \ldots\ldots.., \mathbf{h}_{N_c})$ are mutually independent, then the PDF of $\mathbf{h}_l$ is

$$\text{pr}(\mathbf{h}_l) = \frac{\exp(-\mathbf{h}_l^H\,\mathbf{R}_l^{-1}\,\mathbf{h}_l)}{\pi^{N_c\mathrm{M}}(\det \mathbf{R}_l)^\mathrm{M}}$$

Now

$$\text{pr}(\mathbf{h}) = \text{pr}(\mathbf{h}_1) * \text{pr}(\mathbf{h}_2) * \ldots\ldots\ldots.. * \text{pr}(\mathbf{h}_{N_c})$$

Therefore, we have

$$\text{pr}(\mathbf{h}) = \frac{\exp\left(-\sum_{l=1}^{N_c} \mathbf{h}_l^H\,\mathbf{R}_l^{-1}\,\mathbf{h}_l\right)}{\pi^{N_c\mathrm{M}}(\det \mathbf{R}_1 \ldots\ldots\ldots\ldots \det \mathbf{R}_{N_c})^\mathrm{M}} \tag{10.38}$$

where $\mathbf{R}_l \triangleq \mathbb{E}\{\mathbf{h}_l\,\mathbf{h}_l^H\}$is the covariance matrix, assuming $\mathbf{R}_l$ is invertible. We obtain the conditional distribution of **y** given the channel **h** using (10.35). For the complex Gaussian multivariate process, the conditional probability of received complex signal $M \times \tau$ matrix **y**, conditioned to complex channel **h** is

$$\text{pr}(\mathbf{y} \mid \mathbf{h}) = \frac{\exp(-(\mathbf{y} - \widetilde{\mathbf{S}}\mathbf{h})^H(\mathbf{y} - \widetilde{\mathbf{S}}\mathbf{h})/\sigma_n^2)}{(\pi\sigma_n^2)^{\mathrm{M}\tau}} \tag{10.39}$$

Substituting (10.38) and (10.39) into (10.37), we get

$$\mathrm{pr}(\,\mathbf{h} \mid \mathbf{y}) = \frac{\exp\left(-\left(\mathbf{h}^H R^{-1}\,\mathbf{h} + \frac{(\mathbf{y}-\widetilde{\mathbf{S}}\mathbf{h})^H(\mathbf{y}-\widetilde{\mathbf{S}}\mathbf{h})}{\sigma_n^2}\right)\right)}{\mathrm{pr}(\mathbf{y})(\pi\sigma_n^2)^{\mathrm{M}\tau}\quad \pi^{N_c\mathrm{M}}(\det \mathbf{R})^{\mathrm{M}}} \tag{10.40}$$

where

$$\mathbf{R} \triangleq \mathrm{diag}\,(\mathbf{R}_1, \mathbf{R}_2, \ldots\ldots\ldots., \mathbf{R}_{N_C})$$

Define $l(\mathbf{h})$, A, B as follows: [2]

$$l(\mathbf{h}) \triangleq \left(\mathbf{h}^H\,\mathbf{R}^{-1}\,\mathbf{h} + \frac{(\mathbf{y}-\widetilde{\mathbf{S}}\mathbf{h})^H(\mathbf{y}-\widetilde{\mathbf{S}}\mathbf{h})}{\sigma_n^2}\right)$$

$$\mathrm{A} \triangleq \mathrm{pr}(\boldsymbol{y})(\pi\sigma_n^2)^{\mathrm{M}\tau}$$

$$\mathrm{B} \triangleq \pi^{N_c\mathrm{M}}(\det \mathbf{R})^{\mathrm{M}}$$

Then we can write (10.40) as

$$\mathrm{pr}(\,\mathbf{h} \mid \mathbf{y}) = \frac{\exp(-l(\mathbf{h}))}{\mathrm{AB}}$$

The most probable channel estimate $\hat{\mathbf{h}}$ that maximizes pr($\mathbf{h} \mid \mathbf{y}$) for a given observation $\mathbf{y}$ is obtained by Maximum a posterior probability (MAP) decision rule with soft decision decoding:

$$\hat{\mathbf{h}} = \arg \max_{\mathbf{h}\varepsilon\mathbb{C}^{N_c\mathrm{M}\ \mathrm{x1}}} \mathrm{pr}(\,\mathbf{h} \mid \mathbf{y}) \tag{10.41a}$$

Substituting (10.40) for pr($\mathbf{h} \mid \mathbf{y}$) in (10.41a) we get

$$\hat{\mathbf{h}} = \arg \min_{\mathbf{h}\varepsilon\mathbb{C}^{N_c\mathrm{Mx1}}} \left(\mathbf{h}^H\,\mathbf{R}^{-1}\,\mathbf{h} + \frac{(\mathbf{y}-\widetilde{\mathbf{S}}\mathbf{h})^H(\mathbf{y}-\widetilde{\mathbf{S}}\mathbf{h})}{\sigma_n^2}\right) = \arg \min_{\mathbf{h}\varepsilon\mathbb{C}^{N_c\mathrm{Mx1}}} l(\mathbf{h}) \tag{10.41b}$$

We examine the channel estimate that makes $l(\mathbf{h})$ minimum by differentiating the expression in (10.41b) with respect to $\mathbf{h}^*$ and set the derivative to zero:

$$\frac{\partial l(\mathbf{h})}{\partial \mathbf{h}^*} = 0$$

$$\alpha = (\mathbf{h}^*)^T\,\mathbf{R}^{-1}\,\mathbf{h}$$

$$\frac{\partial \alpha}{\partial \boldsymbol{h}^*} = \mathbf{R}^{-1}\mathbf{h}$$

$$\beta = \frac{1}{\sigma_n^2}\,[((\mathbf{y}-\widetilde{\mathbf{S}}\mathbf{h})^*)^T(\mathbf{y}-\widetilde{\mathbf{S}}\mathbf{h})]$$

$$\frac{\partial \beta}{\partial \boldsymbol{h}^*} = \frac{-(\widetilde{\mathbf{S}})^H\mathbf{y} + (\widetilde{S}\,)^H\ \widetilde{S}\,\boldsymbol{h}}{\sigma_n^2}$$

$$\frac{\partial l(\mathbf{h})}{\partial \mathbf{h}^*} = \sigma_n^2[\mathbf{R}^{-1}\mathbf{h}] - (\widetilde{\mathbf{S}})^H\mathbf{y} + (\widetilde{S}\,)^T\ \widetilde{S}\,\mathbf{h} = 0$$

$$\hat{\mathbf{h}} = \{\sigma_n^2\ \mathbf{R}^{-1} + (\widetilde{S}\,)^T\ \widetilde{S}\}^{-1}(\widetilde{\mathbf{S}})^T\mathbf{y}$$

Let us introduce matrix $\mathbf{T}$ and let it denote the inverted part of $\hat{\mathbf{h}}$:

$$\mathbf{T} = \{\sigma_n^2\ \mathbf{R}^{-1} + (\widetilde{S}\,)^T\ \widetilde{S}\}^{-1}$$

Accordingly,

$$\hat{\mathbf{h}} = \mathbf{T}(\widetilde{\mathbf{S}})^T \mathbf{y}$$

Let us simplify

$$\mathbf{T} = \mathbf{R}^{-1}\{\sigma_n^2\ \mathbf{R}^{-1} + (\widetilde{S})^T\ \widetilde{S}\}^{-1}\mathbf{R} = \{\sigma_n^2 \mathbf{I}_{N_c M} + \mathbf{R}(\widetilde{S})^T\ \widetilde{S}\}^{-1}\mathbf{R}$$

which simplifies the expression for channel estimate $\hat{\mathbf{h}}$ to

$$\hat{\mathbf{h}}_{\text{Bayesian}} = (\sigma_n^2\ \mathbf{I}_{N_c \mathrm{M}} + \mathbf{R}\,\widetilde{\mathbf{S}}^H\,\widetilde{S})^{-1}\,\mathbf{R}\,\widetilde{\mathbf{S}}^H\ \mathbf{y} \tag{10.41c}$$

We may simplify (10.41c) further using the following matrix inversion equation

$$(\mathbf{I} + \mathbf{AB})^{-1}\mathbf{A} = \mathbf{A}(\mathbf{I} + \mathbf{BA})^{-1} \tag{10.42}$$

Apply (10.42) by defining

$$\mathbf{A} \triangleq \mathbf{R}\,\widetilde{\mathbf{S}}^{\mathrm{H}}, \mathbf{B} \triangleq \widetilde{\mathbf{S}}$$

Then, the expression (10.41c) becomes

$$\hat{\mathbf{h}}_{\text{MMSE}} = \mathbf{R}\,\widetilde{\mathbf{S}}^H(\sigma_n^2\ \mathbf{I}_{\tau\mathrm{M}} + \widetilde{\mathbf{S}}\mathbf{R}\,\widetilde{\mathbf{S}}^H\ )^{-1}\ \mathbf{y} \tag{10.43}$$

The channel estimate expressed in (10.43) is similar to the MMSE estimate expression of the MIMO channel in [9]. Accordingly, the Bayesian and the MMSE channel estimates are identical.

10.5.2 Bayesian Desired Channel Estimation with Full Pilot Reuse

The analysis in 10.5.1, we estimated the desired together with the interfering channels, simultaneously. Such an approach would be helpful in designing zero-forcing receivers. However, matched filter (MF) receivers are considered more appropriate to use in the massive MIMO systems for their simplicity and cost-effectiveness. However, the MF scheme requires the knowledge of the desired channel only i.e., without the interference channels. So now we consider the estimation of the desired channel only for matching processing. In addition, we will analyse the worst-case scenario where a unique pilot sequence s is reused in all the N_c network cells, i.e. $s_1 = \dots\dots\dots\dots\dots = s_{N_c} = s$. As in section 10.5.1, $s = [s_1\ s_2 \dots\ \dots\ \dots s_\tau]^T$. We also define a training matrix $\bar{S}$ for the MF receiver as

$$\bar{S} := s \otimes \mathbf{I}_{\mathrm{M}} \tag{10.44}$$

so that

$$\bar{S}^H\,\bar{S} = \tau\,\mathbf{I}_{\mathrm{M}} \tag{10.45}$$

The target BS received vector is

$$y = \bar{S}\sum_{l=1}^{N_c} \mathbf{h}_l + n \tag{10.46}$$

Using (10.43), the expression of the MMSE estimator for $\hat{h}_1$ is

$$\hat{\mathbf{h}}_1 = \mathbf{R}_1\bar{\mathbf{S}}^H\left(\left(\bar{\mathbf{S}}\sum_l^{N_c}\mathbf{R}_l\right)\bar{\mathbf{S}}^H + \sigma_n^2\,\mathbf{I}_{\tau\mathrm{M}}\right)^{-1}\mathbf{y} \tag{10.47a}$$

We apply the matrix inversion identity (10.42) again to (48) and denote

$$\mathbf{A} = \bar{\mathbf{S}}^H, \mathbf{B} = \frac{\bar{\mathbf{S}}\sum_l^{N_c}\mathbf{R}_l}{\sigma_n^2}$$

Consider the inverted term in (10.47a) and equated to $\mathbf{A}(\mathbf{I}+\mathbf{BA})^{-1}$,

$$\bar{\mathbf{S}}^H\left(\left(\bar{\mathbf{S}}\sum_l^{N_c}\mathbf{R}_l\right)\bar{\mathbf{S}}^H+\sigma_n^2\,\mathbf{I}_{\tau M}\right)^{-1}$$

$$=\sigma_n^2\,\bar{\mathbf{S}}^H\left(\mathbf{I}_{\tau M}+\frac{\left(\bar{\mathbf{S}}\sum_l^{N_c}\mathbf{R}_l\right)}{\sigma_n^2}\bar{\mathbf{S}}^H\right)^{-1}$$

Now apply matrix inversion equation $\mathbf{A}(\mathbf{I}+\mathbf{BA})^{-1}=(\mathbf{I}+\mathbf{AB})^{-1}\,\mathbf{A}$,

$$\mathbf{A}(\mathbf{I}+\mathbf{BA})^{-1}\equiv\bar{\mathbf{S}}^H\left(\mathbf{I}_{\tau M}+\frac{\left(\bar{\mathbf{S}}\sum_l^{N_c}\mathbf{R}_l\right)}{\sigma_n^2}\bar{\mathbf{S}}^H\right)^{-1}$$

$$=\widehat{\mathbf{h}}_1=\mathbf{R}_1\left(\sigma_n^2\,\mathbf{I}_{\tau M}+\bar{\mathbf{S}}^H\bar{\mathbf{S}}\sum_l^{N_c}\mathbf{R}_l\right)^{-1}\bar{\mathbf{S}}^H\mathbf{y}$$

Substituting τ for $\bar{\mathbf{S}}^H\bar{\mathbf{S}}$, we get

$$\widehat{\mathbf{h}}_1=\mathbf{R}_1\left(\sigma_n^2\,\mathbf{I}_{\tau M}+\tau\sum_l^{N_c}\mathbf{R}_l\right)^{-1}\bar{\mathbf{S}}^H\mathbf{y} \tag{10.47b}$$

Given that the same pilot sequence is reused in all the N_c cells of the network, it is appropriate to explore the degradation caused by the pilot contamination, which will peak in the scenario we are considering. In particular, we investigate how the adoption of the covariance matrices reduces the pilot contamination interference.

Our analysis of the contamination interference is aimed at the determination of the mean square error (MSE) $\mathcal{M}$ of the channel estimates. The MSE $\mathcal{M}$ for the channel estimate of (10.41c) is defined as

$$\mathcal{M}\triangleq\mathbb{E}\{\|\widehat{\mathbf{h}}-\mathbf{h}\|_F^2\} \tag{10.48a}$$

and is derived in 10.C as

$$\mathcal{M}=\mathrm{tr}\left\{\mathbf{R}\left(\mathbf{I}_{N_c M}+\widetilde{\mathbf{S}}^H\widetilde{\mathbf{S}}\frac{\mathbf{R}}{\sigma_n^2}\right)^{-1}\right\} \tag{10.48b}$$

and for a single user with identical pilots used in all N_c cell

$$\mathcal{M}_1\triangleq\mathbb{E}\{\|\widehat{\mathbf{h}}_1-\mathbf{h}_1\|_F^2\} \tag{10.49a}$$

and is derived in 10.C as

$$\mathcal{M}_1=\mathrm{tr}\left\{\mathbf{R}_1-\mathbf{R}_1^2\left(\frac{\sigma_n^2}{\tau}\,\mathbf{I}_M+\sum_{l=1}^{N_c}\mathbf{R}_l\right)^{-1}\right\} \tag{10.49b}$$

Using (10.47b) the single user channel estimate $\widehat{\mathbf{h}}_1$ with multi-user interference free scenario can easily be shown to be

$$\widehat{\mathbf{h}}_1^{no\ int}=\mathbf{R}_1(\tau\mathbf{R}_1+\sigma_n^2\,\mathbf{I}_M)^{-1}\widetilde{\mathbf{S}}^H(\widetilde{\mathbf{S}}\mathbf{h}_1+\mathbf{n}) \tag{10.50}$$

The corresponding MSE for interference free desired user only uplink is

$$\mathcal{M}_1^{no\ int} = \text{tr}\ \{\mathbf{R}_1\left(\mathbf{I}_M + \frac{\tau}{\sigma_n^2}\mathbf{R}_1\right)^{-1} \tag{10.51}$$

10.6 Bayesian Estimation of UL in a Massive MIMO System

In this section, we explore the Bayesian estimators that are used in a system of large antenna number M (i.e. massive MIMO system) at the BS consisting of a uniform linear antennas array with spacing between antennas d set at $\frac{\lambda}{2}$ or less. Furthermore, we assume that the wireless environment is made up of very large number of i.i.d paths n such that the i^{th}channel is given by

$$\mathbf{h}_i = \frac{1}{\sqrt{n}}\sum_{n=1}^{n} \mathbf{a}(\theta_{in})\alpha_{in} \tag{10.52}$$

where $\alpha_{in} \sim \mathcal{CN}(0, \delta_i^2)$ Gaussian random variable with zero mean and variance (channel attenuation) δ_i^2, $\mathbf{a}(\theta)$ antennas steering vector and $\theta_{in} \in [0, \pi]$ is a random AOA for the ith channel and scattered path n. The antennas steering vector is defined as

$$\mathbf{a}(\theta) := \begin{bmatrix} 1 \\ e^{-j2\pi\frac{d}{\lambda}\cos\theta} \\ e^{-j2\pi\frac{2d}{\lambda}\cos\theta} \\ --- \\ --- \\ --- \\ e^{-j2\pi\frac{(M-1)d}{\lambda}\cos\theta} \end{bmatrix} \tag{10.53}$$

The AOA encompasses a uniform distribution limited to $[0, \pi]$ since $[-\pi, 0]$ can be obtained by replacing θ by $-\theta$ giving the same steering vector. We assume that the AOA spread and the users' locations are such that multipath of the desired user is restricted to a space where interfering paths are highly unlikely to exist. Let us consider the structure that enables such assumption to be possible.

Suppose the multipath AOA θ that corresponds to channel $\mathbf{h}_j$ for $j = 1, \dots \dots, N_c$ is distributed according to PDF pr(θ), is restricted to $\theta \in [\theta_j^{min}, \theta_j^{max}]$ in the range $[0, \pi]$ such that $\text{pr}_j(\theta) = 0$ when $\theta \notin [\theta_j^{min}, \theta_j^{max}]$. If the intervals $N_c - 1$ $[\theta_i^{min}, \theta_i^{max}]$ where $i = 2, \dots \dots, N_c$ are not overlapping with the desired channel's AOA $[\theta_1^{min}, \theta_1^{max}]$ in target first cell, the estimate of desired the channel in (10.47b) will tend to the channel estimate free of interference given in (10.50), mathematically this can be expressed as

$$\lim_{M\to\infty} \widehat{\mathbf{h}}_1 \to \widehat{\mathbf{h}}_1^{no\ int} \tag{10.54}$$

We assume that the desired user location and AOA are confined in space region where interfering users are highly unlikely to exist. The ith channel covariance matrix $\mathbf{R}_i$ is expressed as

$$\mathbf{R}_i = \delta^2\ \mathbb{E}\{\mathbf{a}(\theta_i)\ \mathbf{a}(\theta_i)^H \tag{10.55}$$

According to [2, Lamma 2], the rank of channel covariance matrix $\mathbf{R}_i$ is

$$\frac{rank(\mathbf{R}_i)}{M} \le d_i \quad as\ M \to \infty \tag{10.56}$$

where d_i is defined as

$$d_i := (\cos\theta_i^{\min} - \cos\theta_i^{\max})\,\frac{d}{\lambda} \tag{10.57}$$

As M increases towards infinity, there exists a null space, null($\mathbf{R}_i$) of dimension $(1-d_i)$M. Furthermore, multipath components with AOA outside the AOA range of the desired user tends to fall in the null space of its covariance matrix when $M \to \infty$.

10.6.1 Rule of Coordinated Pilot Allocation

The covariance-based channel estimation is easily affected by the subspaces of the desired and interference channels covariance subspaces to overlap. We have shown in section 10.6 that when the signal and interference occupy distinguishable subspaces, pilot contamination tends to null as $M \to \infty$. For this scenario, we present a suitable coordination protocol for allocating pilot sequences to users in the N_c cells to eliminate the contamination interference. The task of the coordination is to develop the use of the channel covariance matrices to satisfy the nonoverlapping AOA constraint.

We describe the coordinated pilot assignment by considering the network of N_c cells. We assume that each cell serves K users and an arbitrary pilot sequence $\boldsymbol{s}$ is assigned to one (out of K) user in each of the N_c cells. Let us denote the set of the users as $\mathcal{G} \triangleq \{1, \ldots. \mathcal{K}_l \ldots \ldots, K\}$, then $\mathcal{K}_l \in \mathcal{G}$ is an indicator for the user in the lth cell, which is assigned the pilot sequence $\boldsymbol{s}$ and let u denote the set of users for the pilot sequence $\boldsymbol{s}$ assignment. In addition, we use the MSE in (10.49b) as a performance metric to be minimized in order to find the best user set for the assignment. Define the network utility function for a given user set u as

$$\mathrm{F}(u) \triangleq \sum_{j=1}^{|u|} \frac{\mathcal{M}_j(u)}{\mathrm{tr}\{\mathbf{R}_{jj}(u)\}} \tag{10.58}$$

where $|u|$ is the cardinal number of the set u (i.e. the number of distinct users in the set u), $\mathcal{M}_j(u)$ is the estimated MSE for the desired channel in the jth BS, where cell j is the target cell when computing $\mathcal{M}_j$, and $\mathbf{R}_{jj}(u)$ is the covariance matrix of the desired channel at the jth cell.

10.6.2 Evaluation of the Coordinated Pilot Assignment Protocol

Fundamental to the coordinated pilot assignment is to utilize the covariance information at all cells, a total of KN_c^2 covariance matrices to *minimize the sum* MSE *metric*. Hence, N_c users are assigned an identical pilot sequence when the corresponding N_c^2 covariance matrices exhibit the most orthogonal signal subspaces. The network utility function expressed in (10.58) can achieve this goal, since at high SNRs, users with covariance matrices showing maximum signal subspace orthogonality that minimizes the utility function. A greedy algorithm can be used for the utility function minimum points.

The coordination scheme can be explained as follows: a given BS assigns a given pilot to the user whose spatial feature dissimilar to those for the interfering users assigned the same pilot. Clearly, the performance improves with increased number of users, as it is highly likely to find users with distinct spatial statistics.

The coordinated pilot protocol has to ensure fairness among users to prevent high-SNR users from monopolizing the pilot assignment. One way to impose fairness is to consider that cell users are distributed uniformly (including users at the cell edge) with the same distance from their BSs. Alternatively, high-SNR users can be assigned together from a separate pilot pool.

Two distributions can be specified for the AOA, namely: Gaussian distribution and uniform distribution. In the Gaussian, the channel coefficients between u^{th} set user in the lth cell, $\mathbf{h}_{lu}$ and the AOA of all paths are i.i.d Gaussian random variables with mean $\overline{\theta}_{lu}$ and variance σ^2. Both desired and interference channels have the same distribution and variance. It is worth noting that Gaussian distribution does not provide nonoverlapping AOA, but our aim here is to compare the channel estimates performance of the AOA distribution of the Gaussian and the bounded uniform distributions. In the uniform AOA distribution, the AOA is uniformly distributed over $[\overline{\theta}_{lu} - \theta_\Delta,\ \overline{\theta}_{lu} + \theta_\Delta]$ where θ_Δ is an interval around the mean value $\overline{\theta}_{lu}$. We consider two performance metrics to evaluate the channel estimation scheme.

The first metric is the normalized channel estimate error *err* defined as

$$\text{err} \triangleq 10\log_{10}\left(\frac{\sum_{j=1}^{N_c} \|\widehat{\mathbf{h}}_{jj} - \mathbf{h}_{jj}\|_F^2}{\sum_{j=1}^{N_c} \|\mathbf{h}_{jj}\|_F^2}\right) \tag{10.59}$$

where the $\mathbf{h}_{jj}$ and $\widehat{\mathbf{h}}_{jj}$ are the desired channel at the jth BS and its estimate respectively. The second metric is the *per-cell rate for DL* obtained using maximal ratio combining MRC beamformer based on the channel estimate, with beamformer weights vector of the jth BS $\mathbf{w}_j^{MRC} = \widehat{\mathbf{h}}_{jj}$. The per-cell rate c in bits per sec per Hz is defined as

$$C \triangleq \frac{\sum_{j=1}^{N_c} \log_2(1 + \text{SINR}_j)}{N_c} \tag{10.60}$$

where SINR_j is the received signal to noise plus interference power ratio for the scheduled user in the jth cell.

Figure 10.7 illustrates the per-cell rate vs. BS antenna number, M, for AOA Gaussian distributed with standard deviation, $\sigma = 10$ degrees and two-cell network, while Figure 10.8 shows the per-cell rate for AOA Gaussian distributed and $0 \leq \sigma \leq \frac{\pi}{2}$ in a seven-cell network. In both figures, the coordinated pilot assignment-based estimation produces a higher per-cell rate compared with conventional channel LS estimation. Figure 10.7 shows the coordinated assignment with AOA Gaussian distributed attained about 17 bits/sec/Hz compared to about 6 bits/sec/Hz for M = 50 achieved with the conventional LS estimation. However, Figure 10.8 shows per-cell rate vs. standard deviation for 7-cell network with M = 10. As the AOA standard deviation increases, the per-cell rate decreases rapidly and for standard deviation 45° the rates are about 4.5 bits/sec/Hz accomplished by coordinated pilot assignment compared to about 2.75 bits/sec/Hz by conventional LS respectively.

A 2-cell network, the MSE is plotted against the BS antenna number with AOA Gaussian distributed, standard deviation 10 degrees, is shown in Figure 10.9. The MSE (denoted CPA in Figure 10.9) decreases with an increasing M. For M = 50 the coordinated pilot assignment MSE was close to −32.5 dB compared with the LS estimation MSE about −10 dB.

Figure 10.10 presents the 2-cell network channel estimates MSE versus M for AOA Uniformly distributed, with interval around the mean, 10 degrees. Again the MSE decreases with an increasing M. For example for M = 50 the MSE was nearly −40 dB for coordinated pilot assignment compared with −10 dB for LS estimation.

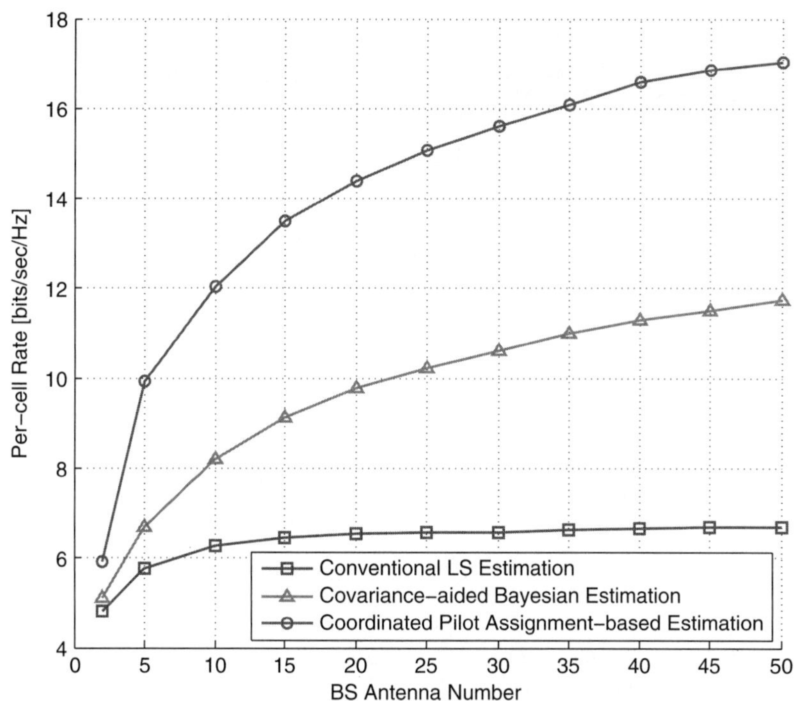

Figure 10.7 Per-cell rate vs. BS antenna number, two-cell network, Gaussian distributed AOAs with $\sigma = 10°$ [2].

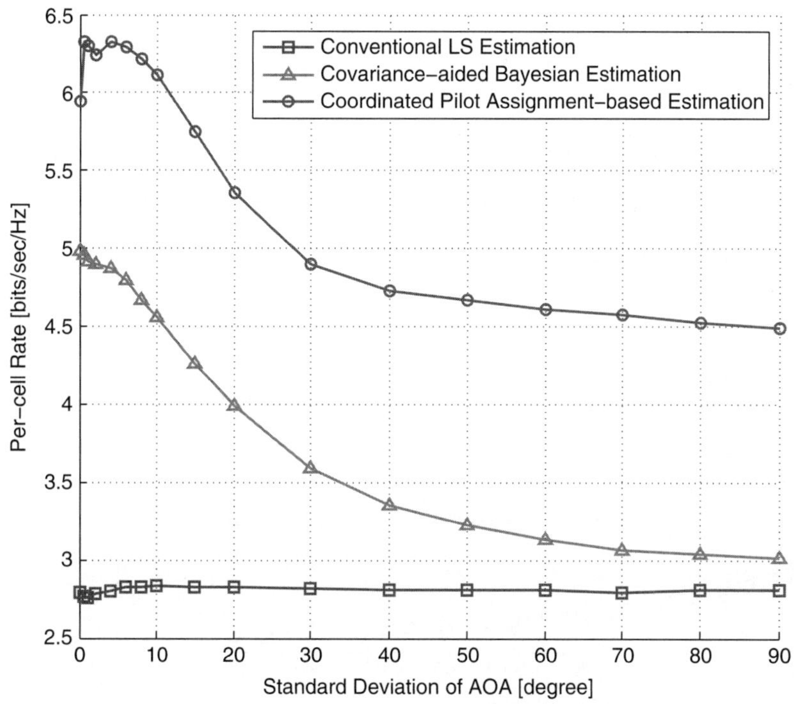

Figure 10.8 Per-cell rate vs. standard deviation of AOA (Gaussian distribution) with M = 10, 7-cell network [2].

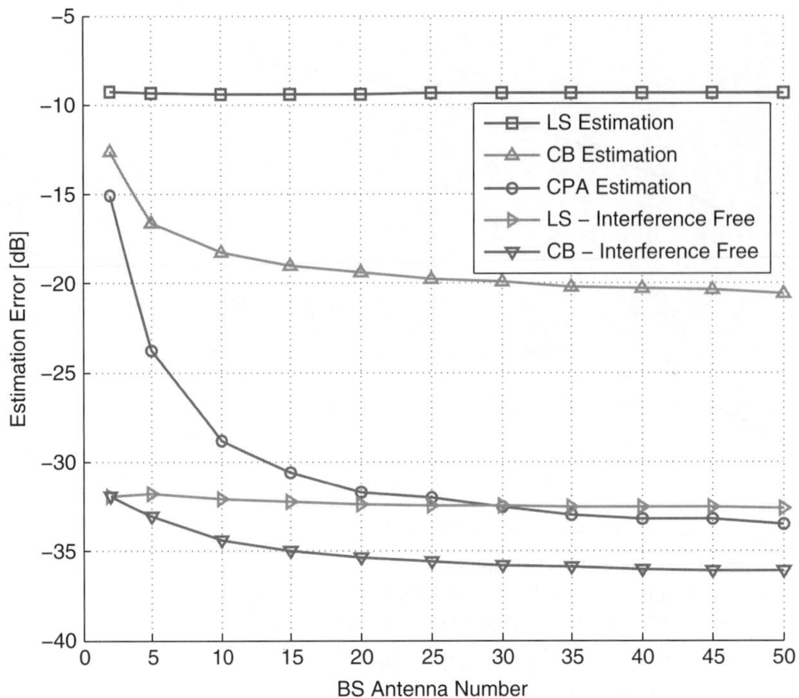

Figure 10.9 Estimation MSE vs. BS antenna number, Gaussian distributed AOAs with $\sigma = 10$, two-cell network [2].

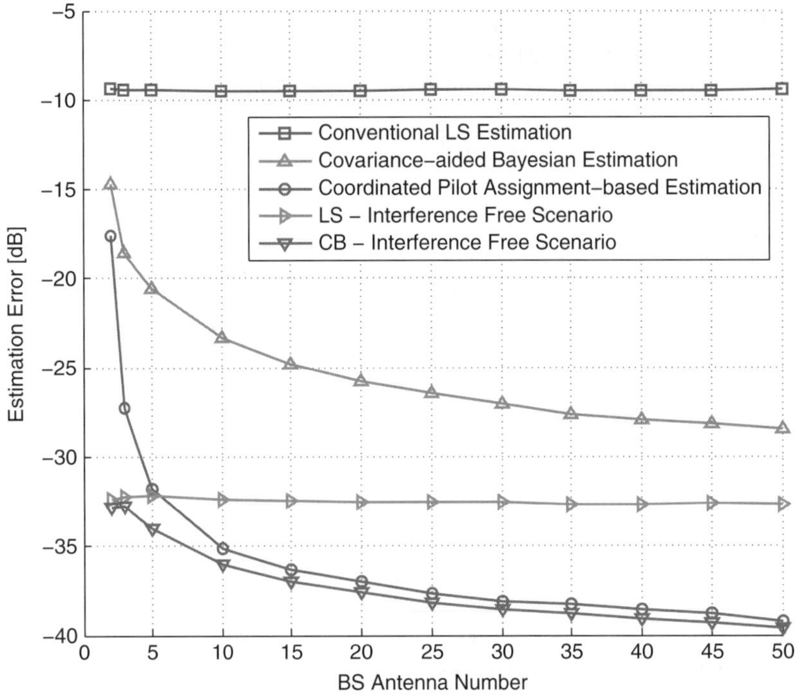

Figure 10.10 Estimation MSE vs. antenna number, uniformly distributed AOAs with $\theta\Delta = 10$, two-cell network [2].

10.7 Arbitrary Correlated Rician Fading Channel

Initial assessment of the MIMO system performance was carried out by Foschini, Telatar, and others [32–33] revealed that the system can provide considerable capacity gain compared to the capacity single antenna system. The evaluation is based on the assumption of either a rich Rayleigh scattering environment or uncorrelated independent and identically Rayleigh fading channels. However, recent works have shown that the rich scattering assumption is often inappropriate. Authors of [10–13] thoroughly investigated the correlated Rician channels. In addition, much of the contemporary available experimental evidence appeared to indicate that the i.i.d. assumption is unrealistic in practice. These facts are possibly due, for example, to insufficient angular spread induced by the scattering environment, or when there are dominant (possibly LOS) non-fading paths between the transmitter and receiver. In such situations, a spatially correlated Rician (i.e. Rayleigh + dominant LOS) channel model is more appropriate.

In fact, current wireless MIMO channel models are an evolution of the initial spatially uncorrelated Rayleigh fading MIMO channel: the evolution process kept the circularly symmetric complex Gaussian distribution but assume nonzero mean and several types of correlation structures (e.g. *nonzero-mean uncorrelated, zero-mean separately correlated (Kronecker)* at the receiver or the transmitter or both, *nonzero-mean separately correlated*). The evolved model is known as *arbitrary correlated Rician fading channel* [11–13]. The transmitter is assumed to know the channel distribution, i.e. channel matrix $\mathbf{H}$ complex Gaussian distributed with mean $\mathbf{H}_{\mathrm{m}}$ and transmit covariance $\mathbf{R}_{\mathrm{t}}$ is

$$\mathbf{H} = \mathbf{H}_{\mathrm{m}} + \mathbf{H}_{\mathrm{w}}\sqrt{\mathbf{R}_{\mathrm{t}}} \tag{10.61}$$

where $\mathbf{H}_{\mathrm{w}}$ has i.i.d zero-mean unit-variance Gaussian elements, the mean $\mathbf{H}_{\mathrm{m}}$, an arbitrary complex matrix, and the covariance $\mathbf{R}_t$ is a positive definite matrix. Variation of this model can exist in a practical system such as zero-mean correlated channel (i.e. $\mathbf{H}_{\mathrm{m}} = \mathbf{0}$ and $\mathbf{R}_t$ is arbitrary) and a non-zero-mean uncorrelated channel ($\mathbf{H}_{\mathrm{m}}$ is arbitrary and $\mathbf{R}_{\mathrm{t}} = \mathbf{I}$).

10.7.1 Estimation of Correlated Rician Channels Using MMSE Approach

We analyse a flat-fading generic MIMO system comprising an array of M transmit antennas and an array of N receive antennas and estimate the flat-fading channel. The aim of the analysis is to determine an optimal pilot sequence that minimizes the MSE in the Rician MMSE channel estimates. The transmitted and received signals at channel's use (t) such that $\mathbf{x}(t) \in \mathbb{C}^{\mathrm{M}}$ and $\mathbf{y}(t) \in \mathbb{C}^{\mathrm{N}}$ respectively are related as

$$\mathbf{y}(t) = \mathbf{H}\,\mathbf{x}(t) + \mathbf{n}(t) \tag{10.62}$$

where $\mathbf{n}(t) \in \mathbb{C}^{\mathrm{N}}$ represents the Gaussian *disruption,* including background *noise and interference* from nearby systems. The channel matrix $\mathbf{H} \in \mathbb{C}^{\mathrm{N\times M}}$ is modelled as Rician fading, with mean $\mathbf{H}_m \in \mathbb{C}^{\mathrm{M\times N}}$ is characterized by the column stacking of the channel matrix and its positive definite covariance is (MN × MN) matrix $\mathbf{R}$.Thus

$$\mathrm{vec}(\mathbf{H}) \in \mathcal{CN}(\mathrm{vec}(\mathbf{H}_m), \mathbf{R}) \tag{10.63}$$

The channel statistics are assumed to be known to the receiver, but in order to estimate the received data, the CSI $\mathbf{H}$ needs to be estimated and to obtain the estimate, the transmitter commonly sends a sequence of training vectors that span the transmit antennas. Consider training matrix sequences of arbitrary length B and denoted as $\mathbf{S} \in \mathbb{C}^{\mathrm{M}\times B}$. The training matrix power constraint should satisfy $\mathrm{tr}(\mathbf{S}^H\,\mathbf{S}) = \mathcal{P}$. The training matrix maximal rank rank(S)

is defined by the maximal number of spatial channels excited by the training sequence, i.e. rank(S) $\triangleq$ min(M, B).

$$\text{Training sequence matrix } \mathbf{S} = \begin{bmatrix} s_{1,1} & ------ & s_{1,B} \\ - & ------ & - \\ s_{M,1} & ------ & s_{M,B} \end{bmatrix} \tag{10.64}$$

During the channel's uses $t = 1, \ldots\ldots, B$, training symbols **S** are transmitted so Eq. (10.62) becomes

$$\mathbf{Y} \in \mathbb{C}^{N \times B} = [\mathbf{y}(1) \quad - \quad - \quad - \quad \mathbf{y}(B)] = \mathbf{H}\,\mathbf{S} + \mathbf{N} \tag{10.65}$$

The Gaussian disturbance matrix vec(**N**) = [**n**(1), … … … … …, **n**(B)] $\varepsilon\ \mathbb{C}^{N \times B}$.

$$\mathbf{N} \in \mathbb{C}^{N \times B} = [\mathbf{n}(1), \ldots\ldots\ldots\ldots\ldots, \mathbf{n}(B)] \tag{10.66}$$

where

$$\text{vec}(\mathbf{N}) \in \mathcal{CN}(\text{vec}(\overline{\mathbf{N}}), \mathbf{R_N}) \tag{10.67}$$

Adopting the vec(.) of a product of matrices,

$$\text{vec}(\mathbf{ABC}) = (\mathbf{C}^T \oplus \mathbf{A})\,\text{vec}(\mathbf{B})$$

becomes,

$$\text{vec}(\mathbf{Y}) = \widetilde{\mathbf{S}}\ \text{vec}(\mathbf{H}) + \text{vec}(\mathbf{N}) \tag{10.68}$$

and vec($\overline{\mathbf{N}}$) is the disturbance (noise plus interference) mean and $\mathbf{R_N}$ is a positive definite covariance of the disturbance matrix (BN × BN). It is assumed that the channel **H** estimates will remain unchanged during the whole training sequence transmission and independent of previous channel estimates.

The pilot matrix $\widetilde{\mathbf{S}}$ is defined as

$$\widetilde{\mathbf{S}} \triangleq [\,S^T \oplus \mathbf{I}_{B \times M}]$$

using Kronecker product of matrices S and $\mathbf{I}_{B \times M}$.

In this section, we consider MMSE estimation of the Rician channel matrix from the observations taken at the receiver during a training transmission. In Section 10.5.1, we had shown the Bayesian and the MMSE channel estimates are identical. The received signal vectors for the N_c cells are represented by vector vec(**Y**). Similarly, for noise $\mathbf{n} \in C(\overline{\mathbf{n}}, \mathbf{R_n})$ = vec(**N**) and vec($\mathbf{H}_{Ric}$) $\in \mathbb{C}^{N_c M \times 1}$.

The Rician fading channel matrix vec($\widehat{\mathbf{H}}_{Ric}$) can be derived in a method similar to the one used in Section 10.5.1 when accounting for Rician channel mean vec($\mathbf{H}_m$). The mean vec($\mathbf{H}_m$) introduces a bias in the estimate vec($\widehat{\mathbf{H}}_{Ric}$) and alters the value of the received vector vec(**Y**). In addition, the variance of the Rician channel disturbance comprises Gaussian noise plus interference and is denoted by $\mathbf{R_n}$. Accordingly, the MMSE *channel estimate* (10.43) is now modified to give the Rician channel estimate as

$$\text{vec}(\widehat{\mathbf{H}}_{Ric}) = \text{vec}(\mathbf{H}_m) + \mathbf{R}\,\widetilde{\mathbf{S}}^H(\mathbf{R_n} + \widetilde{\mathbf{S}}\mathbf{R}\,\widetilde{\mathbf{S}}^H\,)^{-1}\ (\text{vec}(\mathbf{Y}) - \widetilde{\mathbf{S}}\,\text{vec}(\mathbf{H}_m) - \text{vec}(\overline{\mathbf{N}}) \tag{10.69}$$

The MSE $\mathcal{M}_{Ric}$ of the Rician channel with the channel disturbance (noise plus interference) can be derived utilising the method used to obtain expression (10.49) except for the term to account for the noise is replaced with expression that accounts for the noise plus interference. Consequently, the $\mathcal{M}_{Ric}$ is [10]

$$\mathcal{M}_{Ric} = \text{tr}\{(\,\mathbf{R}^{-1} + \widetilde{\mathbf{S}}^H\,\mathbf{R_n}^{-1}\,\widetilde{\mathbf{S}}\ )^{-1}\} \tag{10.70}$$

The MIMO channel covariance matrix $\mathbf{R}$ is usually described by the Kronecker product of two spatial covariance matrices, one at the transmitter by (M × M) matrix denoted $\mathbf{R}_t$ and the other at the receiver side by (N × N) matrix, denoted $\mathbf{R}_r$,

$$\mathbf{R} = \mathbf{R}_t^T \oplus \mathbf{R}_r \tag{10.71}$$

The covariance matrix of the sum of the Gaussian noise and the interference $\mathbf{R_n}$ can be represented as a Kronecker product of two spatial covariance matrices as well: the spatial matrix $(B \times B)$ matrix is denoted $\mathbf{R}_{\mathbf{n},q}$ and at the receive side the (N × N) matrix is denoted $\mathbf{R}_{\mathbf{n},r}$,

$$\mathbf{R_n} = \mathbf{R}_{\mathbf{n},q}^T \oplus \mathbf{R}_{\mathbf{n},r} \tag{10.72}$$

We expect that $\mathbf{R}_r$ and $\mathbf{R}_{\mathbf{n},r}$ have indistinguishable eigenvectors so that the disturbance (interference and noise) is either spatially uncorrelated or shares the spatial structure of the channel and such interference is arriving from the same spatial direction. In essence, this means that we can describe a number of scenarios for the (interference) such as: noise limited, interference limited, noise and temporally uncorrelated interference, and noise and spatially uncorrelated interference.

10.7.2 Pilot Sequence Optimization for Channel Matrix Estimation

Next we examine the design of the pilot sequence $\mathbf{S}$ that achieves the optimization of the MMSE channel estimator in (10.69). The optimization problem can be expressed as

$$\min_{\mathbf{S}} \operatorname{tr}((\mathbf{R}^{-1} + \widetilde{\mathbf{S}}^H \mathbf{R_n}^{-1} \widetilde{\mathbf{S}})^{-1})$$

Substituting $\widetilde{\mathbf{S}} = \mathbf{S}^T \otimes \mathbf{I}$ we get

$$\min_{\mathbf{S}} \operatorname{tr}((\mathbf{R}^{-1} + (\mathbf{S}^T \otimes \mathbf{I})^H \mathbf{R_n}^{-1} (\mathbf{S}^T \otimes \mathbf{I}))^{-1}) \tag{10.73a}$$

The optimization is subject to:

$$\operatorname{tr}(\mathbf{S}^H \mathbf{S}) \leq \mathcal{P} \tag{10.73b}$$

Considering the MSE of the Rician channel in (10.70), it is clear that the Rician channel MSE depends on the covariance matrices of the channel $\mathbf{R}$ and the (interference and noise) $\mathbf{R_n}$ and it is unaltered by their mean values. Consequently, we design the pilot matrix $\mathbf{S}$ to optimize the performance of these covariance matrices. Thus, more power should be allocated to estimate the channel in a strong direction i.e. when the eigenvalues are large reducing the estimate errors. The preferred approach to optimize the pilot matrix is to use structured covariance matrices for the channel $\mathbf{R}$ and the noise $\mathbf{R_n}$, since the pilot matrix $\mathbf{S}$ may inherit this structure. In addition, pilot matrix $\mathbf{S}$ only affects the channel matrix from the transmit side. Accordingly, we consider the covariance matrices that can be split between the transmit side and the receive side. This model is known as a Kronecker-structure and can be used for a number of correlated channels but not in an environment of weak scattering cases. In general, we use the Kronecker model as an appropriate approximation in the analysis. We use the following eigenvalues decompositions for the spatial covariance matrices in (10.71):

$$\mathbf{R}_t = \mathbf{U}_t \boldsymbol{\Lambda}_t \mathbf{U}_t^H, \mathbf{R}_r = \mathbf{U}_r \boldsymbol{\Lambda}_r \mathbf{U}_r^H \tag{10.74}$$

The covariance matrix is the sum of the Gaussian noise and the interference, $\mathbf{R_n}$ in (10.72),

$$\mathbf{R}_{\mathbf{n},q} = \mathbf{V}_{\mathbf{n},q} \boldsymbol{\Sigma}_{\mathbf{n},q} \mathbf{V}_{\mathbf{n},q}^H, \mathbf{R}_{n,r} = \mathbf{U}_{\mathbf{n},r} \boldsymbol{\Sigma}_{\mathbf{n},r} \mathbf{U}_{\mathbf{n},r}^H \tag{10.75}$$

where $\boldsymbol{\Lambda}_t$ and $\boldsymbol{\Sigma}_{\mathbf{n},q}$ are diagonal matrices with diagonal eigenvalues λ^t and $\sigma^{\mathbf{n},q}$,

$\mathbf{\Lambda}_t = \text{diag}(\lambda_1^t, \ldots\ldots\ldots, \lambda_M^t)$ and $\mathbf{\Sigma}_{\mathbf{n},q} = \text{diag}(\sigma_1^{\mathbf{n},q}, \ldots\ldots\ldots\ldots, \sigma_M^{\mathbf{n},q})$ are ordered in decreasingly and increasingly, respectively. The matrices $\mathbf{\Lambda}_r$ and $\mathbf{\Sigma}_{\mathbf{n},r}$ are diagonal matrices with diagonal elements, which are the eigenvalues λ^r and $\sigma^{\mathbf{n},r}$:

$$\mathbf{\Lambda}_r = \text{diag}(\lambda_1^r, \ldots\ldots\ldots, \lambda_N^r) \tag{10.76}$$

$$\mathbf{\Sigma}_{\mathbf{n},q} = \text{diag}(\sigma_1^{\mathbf{n},r}, \ldots\ldots\ldots\ldots, \sigma_N^{\mathbf{n},r}) \tag{10.77}$$

and arbitrarily ordered. The Kronecker model provides the solution to the optimization problem expressed in (10.73) based on the singular decomposition of the pilot matrix $\mathbf{S}$ as

$$\mathbf{S} = \mathbf{U}_t \mathbf{D} \mathbf{V}_{\mathbf{n},q}^H \tag{10.78}$$

where diagonal matrix $\mathbf{D} \in \mathbb{C}^{M \times B}$ on its principle diagonal elements are the pilot powers $(\sqrt{p_1}, \ldots\ldots, \sqrt{p_m})$. Note that because we assumed the pilot matrix $\mathbf{S}$ is not necessarily a square matrix, eigenvalue decomposition cannot be used.

The MSE of such a pilot matrix is convex with respect to the pilots' powers $p_1, \ldots\ \ldots\ \ldots, p_m$ and the pilots' powers should be ordered such that $\frac{s_j}{\sigma_j^{\mathbf{n},q}}$ decreases with j the same order as λ_j^t. Define the Lagrange multiplier $\alpha > 0$ [10] to satisfy the power constraint $\sum_{j=1}^{m} p_j = \mathcal{P}$; then

$$\alpha \triangleq \sum_{l=1}^{N} \frac{(\lambda_j^t\ \lambda_l^r)^2\, \sigma_j^{\mathbf{n},q}\, \sigma_l^{\mathbf{n},r}}{(\sigma_j^{\mathbf{n},q}\, \sigma_l^{\mathbf{n},r} + s_j\, \lambda_j^t\ \lambda_l^r)^2} \quad \textit{for all } j \tag{10.79}$$

The pilot power allocation that minimizes MSE is attained when

$$\alpha < \sum_{l=1}^{m} \frac{(\lambda_j^t\ \lambda_l^r)^2}{(\sigma_j^{N,q}\, \sigma_l^{N,r})} \tag{10.80a}$$

$$\text{and}\ p_j = 0 \text{ otherwise} \tag{10.80b}$$

The interpretation of (10.78) is that the optimum pilot matrix $\mathbf{S} = \mathbf{U}_t \mathbf{D} \mathbf{V}_{\mathbf{n},q}^H$ structure is based on the eigenvectors of the channel at the transmitter side ($\mathbf{U}_t$), eigenvectors of the noise plus interference ($\mathbf{V}_{\mathbf{n},q}^H$) at the transmit side, and pilot power allocation diagonal matrix ($\mathbf{D}$). An Eigen direction is a subspace spanned by a single eigenvector. In the MSE minimization, we assign the jth strongest channel Eigen direction (vector) to the jth weakest Eigen direction $\mathbf{V}_{\mathbf{n},q}^H$ i.e. opposite order of magnitude. That is, the strongest channel direction is estimated when (noise plus interference) is as weak as possible (and vice versa). Consequently, high pilot power is allocated to the m strongest Eigen direction of channel, and the lower pilot power is allocated in a single direction when a combination of strong channel gain and weak noise plus interference exists.

10.7.3 Optimal Length of Pilot Sequences

Analysis conducted so far has assumed an arbitrary pilot sequence length B. In fact, in spatially uncorrelated and noise-limited systems, the optimum pilot sequence length is exactly equal to M. Under various other system statistics (i.e. correlated system), the optimum length can be selected based on rank($\mathbf{S}$). As we highlighted previously, for low $\mathcal{P}$ all power is allocated in a single Eigen direction, i.e. rank($\mathbf{S}$) = 1. Water-filling scheme allocates more pilot power to the strong Eigen directions than the weak and only a subset $\widetilde{m}$ of $(p_1, \ldots\ \ldots\ \ldots, p_m)$ with cardinality $\widetilde{m} \leq m$ will receive any power. In this case, rank($\mathbf{S}$) = $\widetilde{m}$. In other words, the pilot power $\mathcal{P}$ is distributed over the $\widetilde{m}$ best channel uses out of B uses and $B - \widetilde{m}$ pilots have no allocated power. Thus, unless noise and interference change dramatically over time, it makes sense to select $B = \widetilde{m}$. In general, one would expect that the pilot matrix $\mathbf{S}$ would be rank deficient when training power $\mathcal{P}$ is limited and strong eigenvalues exist in either channel or noise plus interference covariance. Furthermore, in correlated disturbance, the length of the optimal pilot sequence can be reduced towards rank(**S**) with only small degradation in MSE but with a possible improved overall throughput.

10.8 Massive MIMO Antennas Calibration

Beamforming (BF) is an attractive transmission technology used for steering a signal towards a specific receiver to maximize the SNR for the considered receiver. In order to perform the transmit beam, estimates of the CSI at the transmitter are required. In real-time transmission, both UL and DL are used. However, MIMO system estimates of UL or DL have to be calibrated because of several discrepancies between the radio frequency circuits in the transmit side and the receive side. Such discrepancies greatly influence the channel reciprocity property. In TDD reciprocity means that we can define the DL estimates from estimates of the UL assuming that the RF circuits of the transceiver are identical. In a practical system such an assumption does not hold. That is why the channel reciprocity (i.e. UL and DL are reciprocal) is imperfect.

At the heart of the channel reciprocity is not only used for the multi-user beamforming (MUBF) but also applied in the MU precoding coefficients for DL transmission. The solution to the problem can be provided mainly at the BS side. The BS sounds its antennas one by one and at the same time receives information with other BS antennas. Using this scenario and assuming M antennas BS, there will be M(M − 1) signals available for use in the calibration scheme. It must be emphasized that the propagation channels between the BS and MSs are reciprocal but the transceiver RF chains at BS and MSs are not reciprocal. The objective of the calibration is to estimate the difference between the UL and DL components of the transceiver hardware chains. This estimation process is called *reciprocity calibration.*

Since the RF transceivers at BS and user device are not necessarily identical, the actual UL and actual DL are not reciprocal. So, in order for the BS to get the actual DL CSI using the UL channel pilot method, the reciprocity issue has to be resolved and since antenna gain is likely to be constant over hours or even days, it seem reasonable to calibrate the antennas to resolve the nonreciprocity problem. In addition, the antenna calibration needs to be performed between long intervals. The hardware calibration methods explicitly calibrate the BS antennas are proposed in [20, 21, 23] which offer a number of calibration methods that appeared in the literature. We classify the calibration methods in two categories and briefly define their advantages/disadvantages.

The first category is the hardware-based RF circuit calibration procedure that exploits extra equipment at the BS such as switches and direction couplers to connect each antenna transmit RF circuit to the receive RF. In addition, the transceiver of each antenna is connected (by switches and attenuators) with one of the other antennas. In this category, the BS selects one antenna as a reference antenna and measures the calibration coefficient between the reference antenna and the other antennas. RF circuits of the other antennas and self-transmitted signals are used as calibration signals.

The second category of calibration methods doesn't employ extra hardware circuits but use the space signals in the calibration. Examples from this category are based on exchanging the calibration signals between the transmitter and receiver and are known as the total least squares-based (TLS) method described in [22]. This method generates considerable feedback and as such is ineffectual for massive MIMO system.

Another realizable calibration method is the on-the-air (OTA) method mostly used in single cell. The procedure needs no extra new hardware in the calibration process as explained in [24] and it is initially proposed for use in the coordinated multi-point (CoMP) systems. In this method, one or a number of users with good-quality channel(s) are called supporters chosen to estimate and feedback the DL CSI to the BS. The key issue in this method is how to select the appropriate MS for calibration. The accuracy of OTA method depends on the channel quality between the BS and the supporter who feedbacks the DL CSI. The on-the-air method is impractical for the massive MIMO systems because of the high feedback overhead.

Calibration procedures that exclude users from participating in the calibration procedure are called self-calibration methods i.e. calibration is achieved by the BS only. Example for self-calibration method is the Argos method, analysed next.

Mutual coupling between array elements was presented in Chapter 9 Section 9.5. When the distance between two antennas becomes less than half a wavelength, the mutual coupling effects become apparent. The inherent coupling in an antenna array can be utilised in the calibration method described in 10.8.2. It is worth noting that we only need to carry out calibration at the transmitter since the RF mismatch at the user's device produces negligible degradation of the system performance.

10.8.1 Argos Method

High-capacity gain can be achieved through using massive MIMO system with MUBF. However, in order to achieve target capacity, numerous challenges have been resolved. As an example, it is beneficial to have the BS architecture hierarchical and modular. The channel calibration procedure has to be implemented so that many users can be served at the same time [17, 18]. MUBF system simultaneously realises the directional transmission and precoded multiple data streams, each intended to a specific receiver, as shown in Figure 10.11. The Argos system [14] is based on modular design, so scaling up the MIMO system can easily be accomplished by adding extra modules (i.e. Radio modules called WARP platform [19]). In addition, the entire computation can be divided among the modules so that computation time can be reduced.

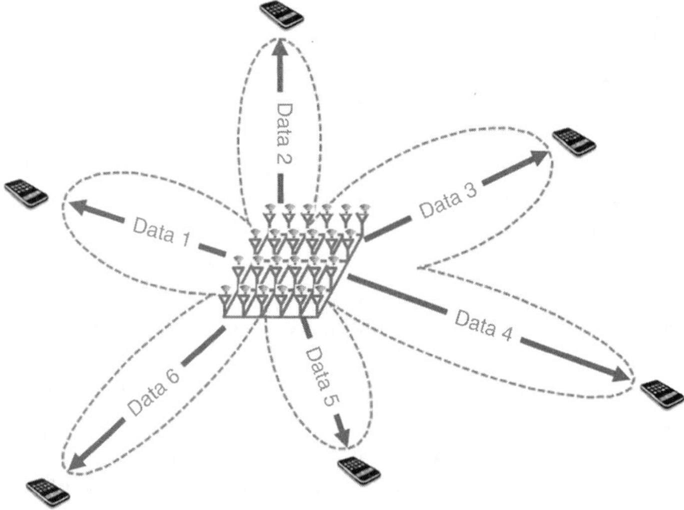

Figure 10.11 Multi-user beamforming transmitting independent precoded streams of data to multiple terminals simultaneously.

Argos relative calibration method is based on an earlier scheme proposed by Guillaud, Slock, and Knopp (GSK) in 2005. The GSK model was considered in depth in Chapter 5. The GSK technique is capable of estimating the DL accurately; the computing complexity increases with the numbers of BS antennas and users' terminals. Given the DL estimate should be acquired in duration much less than the channel coherence, it is clear that such a scheme is not viable at mmWave band and notably for a massive number of antennas.

Argos is a sample-based method intended to be used in massive MIMO systems with small cells. Contrary to the GSK model, the additive noise is not included in the Argos consideration, which implies the measured channel estimate contains error terms and Argos channels' estimates are not optimized with respect to noise level. In addition, the delay τ employed in (GSK) method is ignored in the Argos system model. Argos method focuses on the RF part of the transceivers.

We modify Figure 5.8 of the reciprocity model in Chapter 5 as illustrated in Figure 10.12 to analyse the Argos method. In Chapter 5, we characterized the estimated UL $\widehat{h}_{km}$ and DL $\widehat{g}_{mk}$ as the product of three terms: the RF *frequency response of the transmit hardware*; the wireless *propagation channel C*; and the frequency response of the receive hardware, respectively. The estimated DL channel is

$$\widehat{g}_{mk} = T_m\, C^T\, R_k \tag{10.81}$$

and the estimated UL channel

$$\widehat{h}_{km} = R_m.C\, T_k \tag{10.82}$$

According to the EM theory, the UL and DL propagation channels are reciprocal (i.e. C and C^T are reciprocal) within the channel coherence time. The *RF circuits* at transmit and receive side of the link include the antennas, RF mixers, A/D converters, RF power and low noise amplifiers etc.

Consider BS equipped with M antenna serving K single antenna devices. To empower the BS to formulate the MUBF, it has to acquire the DL CSI from each BS antenna to each user k, i.e. $\widehat{g}_{mk}$ where $m = 1, 2, \ldots \ldots, M$ and $k = 1, 2, \ldots\ldots, K$. The UL CSI is denoted as $\widehat{h}_{km}$.

Argos defines a *calibration coefficient* $b_{i\to j}$ between the mth BS antenna and the kth user transceiver radios as [16]:

$$b_{mk} := \frac{\widehat{g}_{mk}}{\widehat{h}_{km}} \tag{10.83}$$

where

$$\widehat{g}_{mk} = T_m\, C^T\, R_k$$

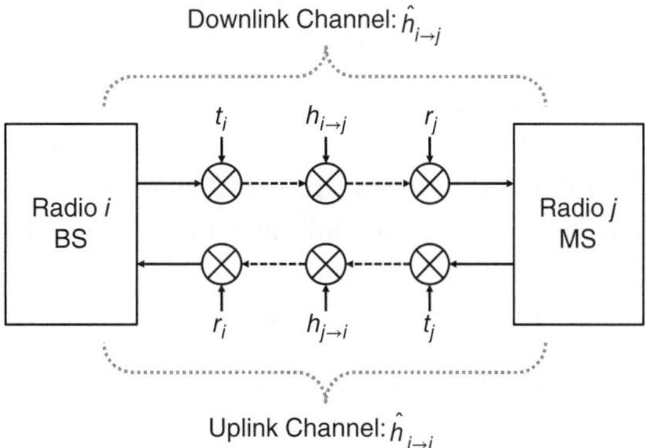

Figure 10.12 Transmit and receive hardware model [16].

$$\widehat{h}_{km} = R_m C\, T_k$$

Substituting (10.81) and (10.82) into (10.83), we get

$$b_{mk} := \frac{T_m\, C^T R_k}{R_m C\, T_k} = \frac{T_m\, R_k}{R_m\, T_k} = \frac{1}{b_{km}} \tag{10.84}$$

where we assumed the corresponding propagation channels are reciprocal and measured within the coherence interval, so:

$$C = C^T \tag{10.85}$$

The implication of (10.83) is that if we know calibration coefficient b_{mk} between the two transceiver radios, and one channel estimate, we can find the corresponding reciprocal channel estimate, i.e.

$$\widehat{g}_{mk} = b_{mk}\ \widehat{h}_{km} \tag{10.86}$$

It is worth noting that the Argos method is based on per estimate calibration. Let us apply (10.86) to estimate the DL. That is we need to estimate $\widehat{g}_{mk}$ from the UL $\widehat{h}_{km}$ but to do this, we need to know the values of b_{mk} for all user antenna–BS antenna pair for $k = 1, \ldots.$ K and $m = 1, \ldots.$ M. Effort as this would be impractical in real massive MIMO system, because it has to be completed within a time less than channel coherence intervals. Furthermore, there could be a synchronization implication. Meaning, the hardware transmit / receive channels may drift apart over time. Accordingly, the channel calibration must occur intermittently. Hence, the only feasible practical solution considering massive number of antennas is the internal or self-calibration, commonly known as relative calibration. That is, we determine the BS antennas calibration coefficients relative to one of the BS antennas called the reference antenna.

Let us choose BS antenna 1 as a reference. That is, we calibrate the BS antennas relative to antenna 1. In another words, we find the calibration b_{m1} for $m = 2, 3, \ldots \ldots$, M, using (10.86). These b_{m1} values are stable over a long interval since BS antennas share the same clock.

The new scenario works on the fact that if we know the calibration coefficient between two transceivers i and j and a reference transceiver y then we can derive the calibration coefficient between the two transceivers:

$$\frac{b_{ij}}{b_{iy}} = \frac{\frac{T_i\ R_j}{R_i.T_j}}{\frac{T_i\ R_y}{R_i.T_y}} = \frac{T_y\, R_j}{T_j R_y} = b_{yj} \tag{10.87}$$

The conclusion we can draw from (10.87) is that if we know the calibration coefficient between the reference (antenna 1) and BS antenna m and between the reference and terminal k, then we can compute the calibration coefficient between BS antenna m and terminal k, i.e. substitute $i = 1,\ y = m,\ j = k$ in (10.87) we get

$$\frac{b_{1k}}{b_{1m}} = b_{mk} \tag{10.88}$$

In addition, if we have one UL channel estimate $\widehat{h}_{km}$, we can find the corresponding reciprocal DL channel estimate. Multiply both sides of (10.88) with $\widehat{h}_{km}$,

$$\widehat{h}_{km}\frac{b_{1k}}{b_{1m}} = b_{mk}\,\widehat{h}_{km} = \widehat{g}_{mk} \tag{10.89}$$

So, all we have to do is to multiply the ratio of the LHS of (10.88) with UL channel estimated to get the corresponding DL estimate. However, to fully implement (10.89), we have to find

b_{1k} for each and every user that according to (10.83) requires heavy computation and feedback overhead even for a moderate number of users and completely reduces the channel capacity. Nonetheless, to implement (10.89), it is unnecessary to find the absolutely accurate estimate of the DL $\widehat{g}_{mk}$. When using linear precoding, MUBF can be established using *relatively accurate estimates.* In other words, as long as each BS antenna's DL estimate deviate from the *actual estimate* by a fixed multiplication factor for all antennas, the MUBF keep its same pattern. We can set

$$b_{1k} = 1 \tag{10.90}$$

Apply assumption (10.90) to rewrite (10.89)

$$\widehat{g}'_{mk} = \frac{\widehat{h}_{km}}{b_{1m}} = \widehat{h}_{km}.b_{m1} \tag{10.91}$$

where the relative DL= $\widehat{g}'_{mk}$ is taking into account $b_{1k} = 1$. Now we determine the relationship between b_{1m} and b_{m1} as follows:

$$b_{mk} := \frac{\widehat{h}_{mk}}{\widehat{h}_{km}}, b_{1m} = \frac{\widehat{h}_{1m}}{\widehat{h}_{m1}}, \text{ and } b_{m1} := \frac{\widehat{h}_{m1}}{\widehat{h}_{1m}}$$

Therefore, we get

$$\frac{1}{b_{1m}} = \frac{\widehat{h}_{m1}}{\widehat{h}_{1m}} = b_{m1}$$

where we have

$$\frac{1}{b_{1\to m}} = b_{m\to 1} \tag{10.92}$$

Applying assumption (10.90) and using (10.92), we have to rewrite (10.89) as

$$\widehat{g}'_{mk} = \frac{\widehat{h}_{km}}{b_{1m}} = \widehat{h}_{km}\, b_{m1} \tag{10.93}$$

The relative estimate scheme will drastically reduce the computation. Estimation of the relative DL in (10.93) can be implemented in three steps:

1. Obtain the internal BS antennas calibration, b_{1m} by sending pilots to and from each BS antenna m and reference BS antenna 1.
2. Determine the UL $\widehat{h}_{km}$ by sending K orthogonal pilots from each user $k = 1, \ldots..$ K to each BS antenna i.e. $m = 1, \ldots..$ M.
3. Determine $\widehat{g}'_{mk}$ from (10.93) and then use $\widehat{g}'_{mk}$ to compute the MUBF weights, which are available locally to avoid transport overhead.

However, the Argos relative channel calibration method is very sensitive to the placement of the reference BS antenna, and the system performance is likely to be unstable unless all the other BS antennas have very good SNR to the reference antenna.

10.8.2 Mutual Coupling Calibration Antennas Method

A channel reciprocity calibration method based on the mutual coupling between adjacent BS antennas, referred to as mutual coupling calibration method, is reported in [23]. We have examined the mutual coupling in Sections 9.5 and 9.6. The proposed approach applies the outcome of the mutual coupling to characterise the channel between co-located BS's antennas. The calibration method can be enforced to calibrate the BS antennas without feedback overhead and without involving the MSs or any external hardware. We analyse the DL transmission to evaluate the effect of the RF mismatches at both BS and end users.

To help us analyse and evaluate the new calibration method, we consider a system model consisting of a BS equipped with M serving K single antenna users where M ≫ K such is the case for the massive MIMO system. The system model used for the analysis is illustrated in Figure 10.13. Each antenna at the BS and user's device has RF transmit and RF receive circuits that are supporting the BS / user antennas for transmission and reception. Let us denote the *complex matrices* representing the *frequency response (complex gain)* of transmit and receive RF circuits (hardware) at the BS, $\boldsymbol{C}_{BS,\,t}$ and $\boldsymbol{C}_{BS,\,r}$, respectively. Similar corresponding matrices at the user's device (UE) are $\boldsymbol{C}_{UE,\,t}$ and $\boldsymbol{C}_{UE,\,r}$ respectively. The matrices are diagonal and defined as

$$\boldsymbol{C}_{BS,t} := \text{diag}\,(t_{BS,1}, \quad t_{BS,m} \quad t_{BS,\text{M}}) \tag{10.94}$$

$$\boldsymbol{C}_{BS,r} := \text{diag}\left(r_{BS,1}, \quad r_{BS,m} \quad r_{BS,\text{M}}\right) \tag{10.95}$$

$$\boldsymbol{C}_{UE,t} := \text{diag}\left(t_{UE,1}, \quad t_{UE,k} \quad t_{UE,K}\right) \tag{10.96}$$

$$\boldsymbol{C}_{UE,r} := \text{diag}\left(r_{UE,1}, \quad r_{UE,k} \quad r_{UE,K}\right) \tag{10.97}$$

where $t_{BS,\,m}, r_{BS,\,m}$ and $t_{UE,\,k}, r_{UE,\,k}$ are RF complex gains expressed by

$$t(r)_{BS,m} = |t(r)_{BS,m}|\, e^{i\phi^{t(r)}_{\text{BS,m}}},\ t(r)_{UE,k} = |t(r)_{UE,k}|\, e^{i\phi^{t(r)}_{\text{UE,k}}} \tag{10.98}$$

and $m = 1, 2, \ldots\ \ldots\ \ldots\ \ldots\ \ldots\ \ldots, \text{M}$, $k = 1, 2, \ldots\ \ldots\ \ldots\ \ldots\ \ldots\ \ldots, K$, and $t(r)_{BS,\,m}$, $t(r)_{UE,\,k}$ are the complex gains at transmitter(receiver) for BS and user terminal antennas, respectively. The complex gains are assumed to have lognormal distributed amplitudes with zero mean and variance $\sigma^2_{BS,t(r)}$ and , $\sigma^2_{UE,t(r)}$, and uniform distributed phases $\mathcal{U}$ as

$$\log|t(r)_{BS,m}| \sim \mathcal{N}(0,\ \sigma^2_{BS,t(r)}),\ \phi^{t(r)}_{BS,m} \sim \mathcal{U}(-\theta_{BS,t(r)}, \theta_{BS,t(r)}) \tag{10.99}$$

$$\log|t(r)_{UE,k}| \sim \mathcal{N}(0,\ \sigma^2_{UE,t(r)}),\ \phi^{t(r)}_{UE,k} \sim \mathcal{U}(-\theta_{UE,t(r)}, \theta_{UE,t(r)}) \tag{10.100}$$

where logx expresses the natural algorithm (lnx).

Consider the *actual DL wireless channel* $\boldsymbol{G}_{DL}$, which consists of: the propagation channel $\mathbf{H}$ the complex gain matrices of the transmit and receive RF circuits, $\boldsymbol{C}_{BS,\,t}$, $\boldsymbol{C}_{BS,\,r}$ and the transmit

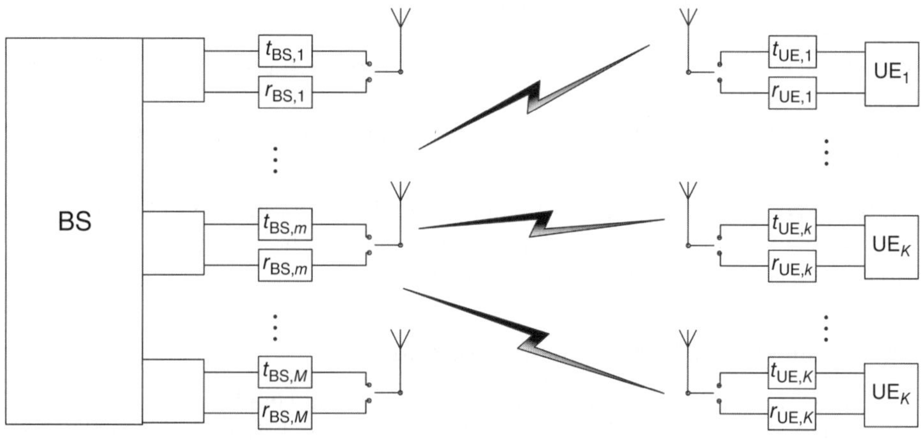

Figure 10.13 System model for the analysis of mutual coupling calibration [23]

and receive *spatial correlation* matrices at the BS, $\mathbf{R}_{BS,\,t}$, $\mathbf{R}_{BS,\,r}$ and using similar formulation as in the Argos method, i.e.

$$\mathbf{G}_{DL} = C_{UE,r}\mathbf{H}\sqrt{\mathbf{R}_{BS,t}}\,C_{BS,t} \tag{10.101}$$

The spatial correlation at the UE is not considered since each UE has only one antenna and the separation between arbitrary two UEs is likely to be very long. The actual UL wireless channel $\mathbf{G}_{UL}$ is given by

$$\mathbf{G}_{UL} = C_{BS,r}\sqrt{\mathbf{R}_{BS,r}}\,\mathbf{H}^T\,C_{UE,t} \tag{10.102}$$

where the DL channel matrix $\mathbf{H} \in \mathbb{C}^{K\times M}$ is the small-signal channel matrix and each element, $\mathrm{h}_{k,\,m}$, is circular symmetric complex Gaussian variable with zero mean and variance of unity, i.e. $\mathrm{h}_{k,m} \sim \mathcal{N}(0,1)$. We assumed the UL/DL propagation channels are reciprocal. Hence, the BS transmit / receive antennas correlation matrices are reciprocal as well, i.e. $\mathbf{R}_{BS,t} = \mathbf{R}_{BS,r}^T$ and

$$\mathbf{H}\sqrt{\mathbf{R}_{BS,t}} = \mathbf{H}(\sqrt{\mathbf{R}_{BS,r}})^T = [\sqrt{\mathbf{R}_{BS,r}}\,H^T]^T \tag{10.103}$$

DL in (10.101) can be simplified with (10.103) to

$$\mathbf{G}_{DL} = C_{UE,r}[\sqrt{\mathbf{R}_{BS,r}}\,\mathbf{H}^T]^T\,C_{BS,t} \tag{10.104}$$

Considering (10.104) and (10.102), we see that the *propagation channel reciprocity* still holds but the propagation channel is multiplied by the transpose of the spatial correlation at the BS receive antennas. According to [26, 27], each entry to the spatial correlation matrix $\mathbf{R}_{BS,\,r}$ between the BS array elements, with *limited spacing between adjacent antennas,* is a function of antenna spacing in wavelength, array geometry, and the angular energy distribution [28]. The angular spread has a range from close to 0.0 (directional case) to 1.0 (omni-directional). It is justifiable to assume the spatial correlation matrix $\mathbf{R}_{BS,\,r}$ to be established according to the BS antenna array structure. Therefore, as in the analysis of the Argos method, we will not take the spatial correlation into account in the analysis [23]. Therefore, (10.102) and (10.104) become

$$\mathbf{G}_{UL} = C_{BS,r}\;\mathbf{H}^T C_{UE,t} \tag{10.105}$$

$$G_{DL} = C_{UE,r}\mathbf{H}\,C_{BS,t} \tag{10.106}$$

Comparing (10.105) and (10.106) with the Argos method expression (10.89) and (10.91) we make the following remarks: mutual coupling method analysis is matrix-based, which is appropriately applied to massive MIMO, unlike the analysis of the Argos method, which is sample-based. In addition, the DL/UL channel matrix multiplication follows transmit to receive direction, while in [23] the direction is receive to transmit direction but this may not change the final results.

As we have highlighted in the introduction of 10.8, the aim of antenna calibration is to eliminate the nonreciprocity provoked by the RF hardware mismatches at the BS and user terminal transceiver. We now introduce a calibration matrix, C_{cal} and postulate that

$$C_{BS,t}C_{cal}C_{BS,r}^{*} = \alpha_{cal}\,\mathbf{I}_M \tag{10.107}$$

where $C_{BS,r}^{*}$ is the conjugate of $C_{BS,\,r}$ α_{cal} is a scalar multiplier to eliminate the effect of the RF hardware mismatch defined as the calibration coefficient. Consider three adjacent antennas, $A_m, A_{\mathcal{V}}, and\;\; A_{\mathcal{V}}$ in a hexagonal array at the BS, as illustrated in Figure 10.14. The distance between any adjacent antennas is $\frac{\lambda}{2}$.

The array configuration in Figure 10.14 makes the distances between any two of the antennas equal and thus the factors of the mutual coupling $Z_{m,\mathcal{V}}, Z_{\mathcal{V},m}, Z_{m,\mathcal{V}}, and\; Z_{\mathcal{V},m}$ are equal (say Z_C)

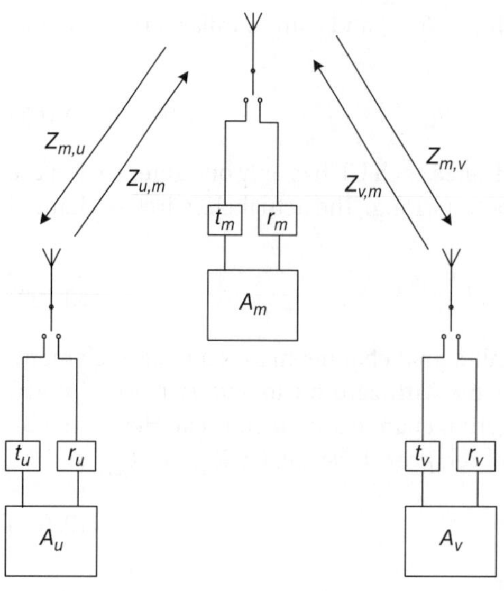

Figure 10.14 Three adjacent antennas at the BS forming equilateral triangle [23]

because the distances between any pair of these three pairs are equal. Denote the three transmit - receive RF circuits gains pairs as (t_m, r_m), $(t_{\mathcal{V}}, r_{\mathcal{V}})$, and $(t_{\mathcal{U}}, r_{\mathcal{U}})$, respectively. Using the Argos method definition [16], the calibration coefficients $\alpha_{\mathcal{U},\mathcal{V}}$, $\alpha_{\mathcal{V},\mathcal{U}}$ between antennas $A_{\mathcal{U}}$ and $A_{\mathcal{V}}$ for maximum ratio transmission (MRT) precoding can be acquired through A_m as

$$\alpha_{\mathcal{U},\mathcal{V}} = \frac{(t_{\mathcal{U}}\, Z_C\, r_m)(t_m\, Z_C\, r_{\mathcal{U}})^*}{(t_{\mathcal{V}}\, Z_C\, r_m)(t_m\, Z_C\, r_{\mathcal{V}})^*} = \frac{t_{\mathcal{U}}\, r_{\mathcal{U}}^*}{t_{\mathcal{V}}\, r_{\mathcal{V}}^*} \tag{10.108}$$

$$\alpha_{\mathcal{V},\mathcal{U}} = \frac{(t_{\mathcal{V}}\, Z_C\, r_m)(t_m\, Z_C\, r_{\mathcal{V}})^*}{(t_{\mathcal{U}}\, Z_C\, r_m)(t_m\, Z_C\, r_{\mathcal{U}})^*} = \frac{t_{\mathcal{V}}\, r_{\mathcal{V}}^*}{t_{\mathcal{U}}\, r_{\mathcal{U}}^*} \tag{10.109}$$

When there are antennas adjacent to antennas $A_{\mathcal{U}}$ and $A_{\mathcal{V}}$ other than A_m, the multiple calibration coefficients between $A_{\mathcal{U}}$ and $A_{\mathcal{V}}$ can be averaged to improve the accuracy of the calibration coefficients in (10.108) and (10.109). A simple algorithm to compute the calibration coefficients for all antennas using iterative method algorithm is as follows [23]:

1. Let $\boldsymbol{\Lambda}_0 \in \mathbb{C}^{\mathrm{M}\times\mathrm{M}}$ the initial matrix of the calibration coefficient, where

$$\boldsymbol{\Lambda}_0 = \begin{bmatrix} 1 & \alpha_{1.2}^{(0)} & - & - & - & - & - & \alpha_{1,\mathrm{M}}^{(0)} \\ \alpha_{2,1}^{(0)} & 1 & - & - & - & - & - & \alpha_{2,\mathrm{M}}^{(0)} \\ - & - & - & - & - & - & - & - \\ - & - & - & - & - & - & - & - \\ \alpha_{m,1}^{(0)} & \alpha_{m,2}^{(0)} & - & - & 1 & - & - & \alpha_{m,\mathrm{M}}^{(0)} \\ - & - & - & - & - & - & - & - \\ - & - & - & - & - & - & - & - \\ \alpha_{\mathrm{M},1}^{(0)} & \alpha_{\mathrm{M},2}^{(0)} & - & - & - & - & - & 1 \end{bmatrix} \tag{10.110}$$

The diagonal elements of matrix Λ_0 are 1 and antennas are sorted by taking the matrix rows as a priority. When pair of antennas is not adjacent, their calibration coefficients are zero, i.e.

$$\alpha_{1,\mathrm{M}}^{(0)} = 0 \text{ and } \alpha_{\mathrm{M},1}^{(0)} = 0. \tag{10.111}$$

2. Let r^{th} antenna be the reference antenna and i be the iteration index.
3. The iteration commences after deciding a reference antenna. In the ith iteration, given Λ_{i-1} the elements of Λ_i are obtained as

$$\alpha^{(i)}_{\mathcal{U},\mathcal{V}} = \frac{1}{\eta^{(i)}_{\mathcal{U},\mathcal{V}}} \sum_{m=1}^{M} \alpha^{(i-1)}_{\mathcal{U},m}\, \alpha^{(i-1)}_{m,\mathcal{V}} \tag{10.112}$$

Let $\xi^{(i)}_{\mathcal{U},\mathcal{V},m} = \alpha^{(i-1)}_{\mathcal{U},m}\, \alpha^{(i-1)}_{m,\mathcal{V}}$, then (10.112) becomes

$$\alpha^{(i)}_{\mathcal{U},\mathcal{V}} = \frac{1}{\eta^{(i)}_{\mathcal{U},\mathcal{V}}} \sum_{m=1}^{M} \xi^{(i)}_{\mathcal{U},\mathcal{V},m} \tag{10.113}$$

Let $\xi^{(i)}_{\mathcal{U},\mathcal{V}} = \left[\xi^{(i)}_{\mathcal{U},\mathcal{V},1} \; - - - \; \xi^{(i)}_{\mathcal{U},\mathcal{V},m} \; - - - \; \xi^{(i)}_{\mathcal{U},\mathcal{V},\mathrm{M}}\right]$ and $\eta^{(i)}_{\mathcal{U},\mathcal{V}}$ is the number of non-zero elements in $\xi^{(i)}_{\mathcal{U},\mathcal{V}}$. The calibration coefficients in (10.112) are averaged over $\eta^{(i)}_{\mathcal{U},\mathcal{V}}$ the number of non-zero elements in the vector $\xi^{(i)}_{\mathcal{U},\mathcal{V}}$. The iteration ends when all elements of calibration matrix (10.112) are nonzero.
4. At the end of the iteration, the rth row in the $\boldsymbol{\Lambda}_i$ generates the calibration matrix as

$$C_{cal} = \mathrm{diag}\left(\alpha_{\mathrm{r},1} \;\; \alpha_{\mathrm{r},2} \; - - - - - \; \alpha_{\mathrm{r,M}}\right) \tag{10.114}$$

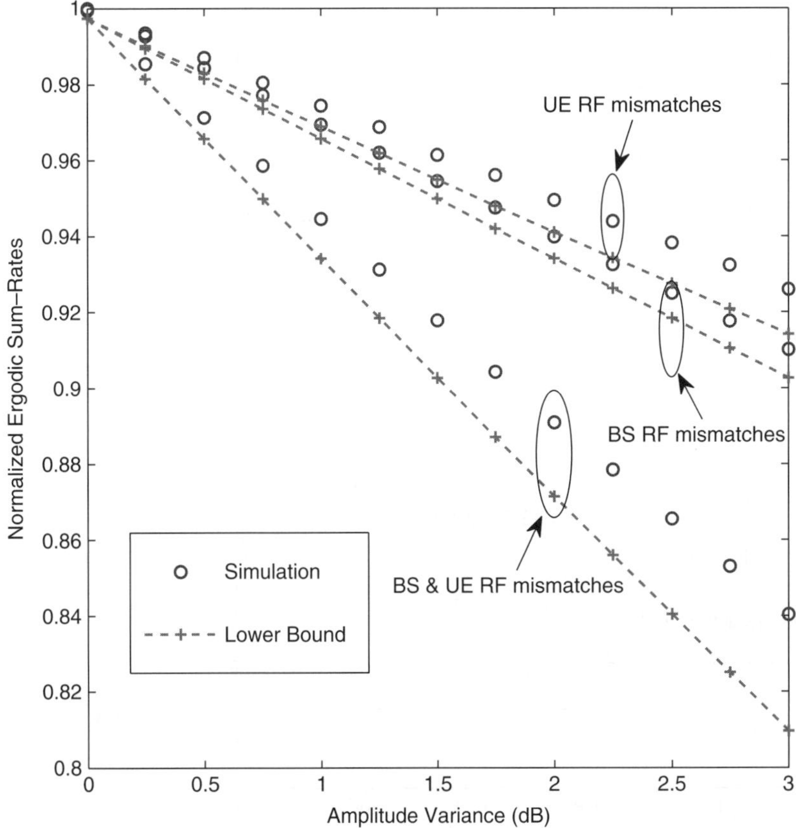

Figure 10.15 Normalized ergodic sum-rates without calibration vs. the RF hardware amplitude variance mismatch [25]

The iteration time is of the order of $\vartheta[\log(\sqrt{M})]$ so for M = 64 the iteration time is roughly two seconds. It is helpful to note that Argos computes the calibration coefficients without iterations so it is comparatively of low complexity. Nonetheless, Argos calibration coefficients of antennas far away from the reference antenna are likely to be very small due to the weak mutual coupling, which produces ill-behaved estimation that degrades the system performance. The instability of Argos calibration subdues in the mutual coupling method with a small increase in the complexity.

In the initial stage, only the calibration coefficients of adjacent antennas are computed because the mutual coupling is near to peak and the rest of the coefficients are obtained through iteration and averaging since each antenna can establish a relation with the reference through several different paths. The calibration coefficients for ZF precoding cannot be simply used for MRT precoding. The latter only maximizes the signal gain at the intended user and does not eliminate the mismatches of the RF gains and results in the performance degradation. Therefore, it is very important to design the calibration method for a specific precoding scheme.

Figures 10.15 and 10.16 depict the association between the normalized (with respect to ideal case) ergodic sum rates of the system without calibration and the amplitude and phase of the mismatch of RF gain respectively. Figure 10.15 shows the relation between sum rates and the amplitude variance of RF mismatch up to 3 dB and system SNR 15 dB. It can be observed that the sum rates decrease linearly with increasing amplitude variance at both BS and UEs. However, the effect of amplitude mismatches at BS is greater than that at the UEs.

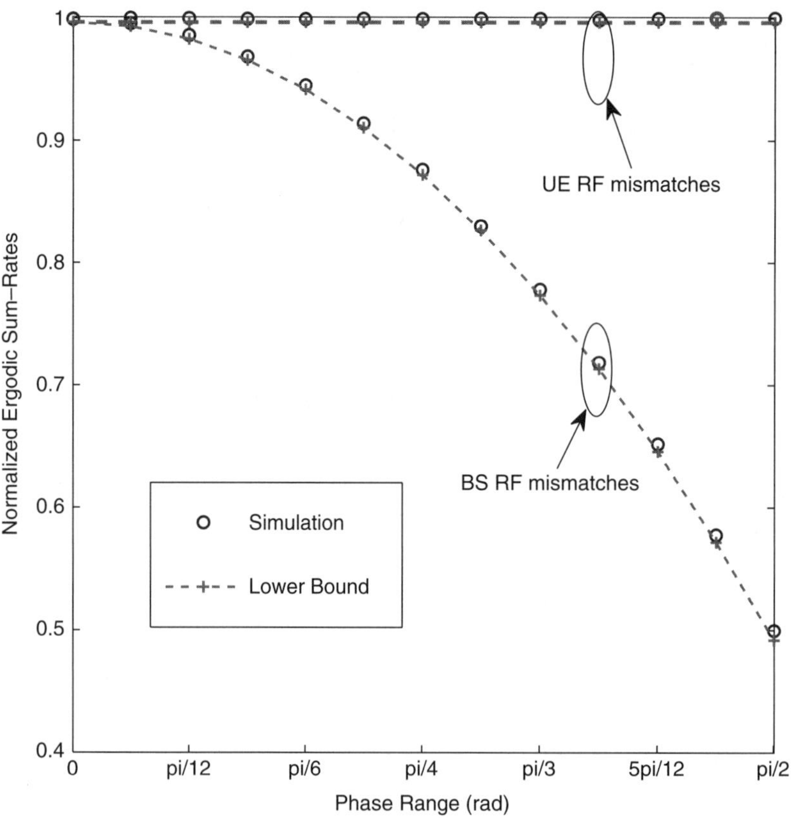

Figure 10.16 Normalized ergodic sum-rates without calibration vs. the phase of the RF mismatch [25]

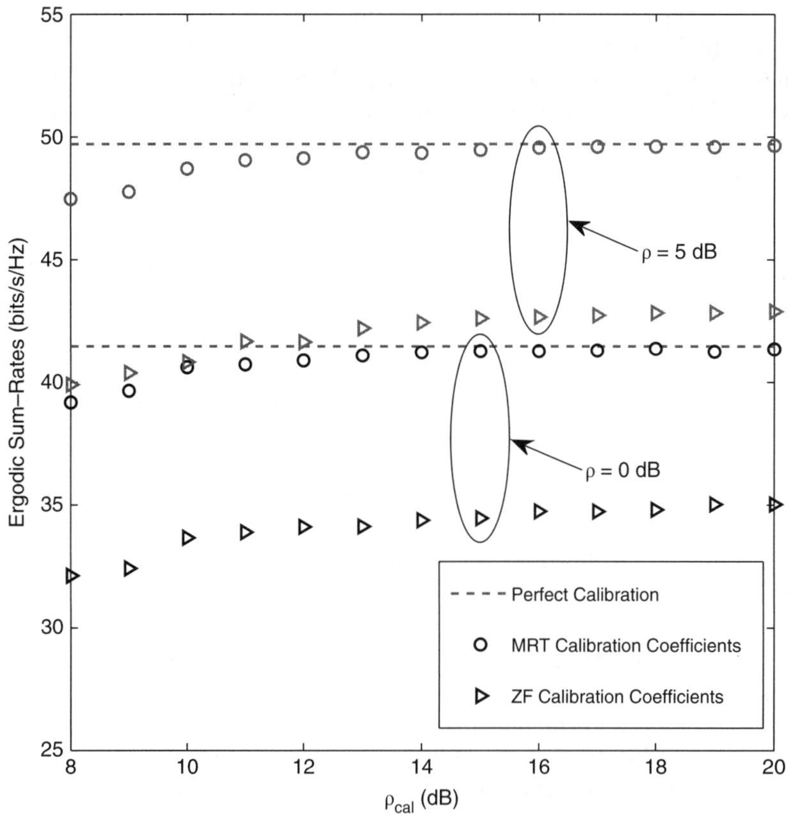

Figure 10.17 Ergodic sum-rates vs. calibration SNR (ρ_{cal}) with the MRT calibration coefficients and ZF calibration coefficients, respectively [25].

Figure 10.16 depicts the relation between normalized ergodic sum rates for system SNR 15 dB without calibration and the phase of RF mismatch in the range 0 to $\frac{\pi}{2}$. It can be seen that the phase range at the UEs has no impact of the system performance. Nonetheless, an increase of phase at the BS forces the sum rates at the BS to decrease strongly. When the phase range at $\frac{\pi}{2}$, the system sum rates decreases by 50%.

Figure 10.17 shows the impact of the calibration SNR (ρ_{cal}) on the system sum rates of precoded calibration when amplitude variance is 2 dB and the phase range $\theta_t^{BS} = \theta_r^{BS} = \frac{\pi}{4}$. We assume perfect RF gain match at the terminals and focus on the RF mismatches at the BS to evaluate the impact of calibration error. For MRT precoding, the ergodic sum rates increase with increasing ρ_{cal} and tends to a limit coinciding with the idea case (i.e. no calibration error). For ZF precoding, the ergodic sum rates tend to a limit much lower than the one with MRT precoding.

10.9 Pre-precoding/Post-precoding Channel Calibration

The hardware RF mismatch between the BS and users' devices causes DL performance degradation. Although the hardware RF mismatch generally does not affect the UL performance, it generates inaccuracies in the reciprocity estimation, which cause the DL channel precoding to be inadequate to create independent transmission streams. One solution to this problem is a robust transceiver design; however, such a solution introduces extra cost and complexity, particularly when applied to massive MIMO systems. The preferred solution is based on the estimation of the hardware mismatch elements and array antennas self/relative calibration, as we discussed above.

Once the hardware parameters are reliably found, the channel calibration is carried out to establish the reciprocity between UL/DL. Following from previous discussion, hardware calibration can be grouped into two categories based on whether the users are involved or not. Hardware estimation and calibration carried out at the BS are only referred to as *partial calibration.* This scenario cannot guarantee the QoS at the user end of the link. Hardware calibration carried out with both BS and users involved is called *full calibration.*

The hardware calibration can be accomplished either before or after the precoding. These two scenarios are named as pre-precoding calibration (pre-Cal) and post-precoding calibration (post-Cal). A single-cell multi-user massive MIMO system with M transmit antennas at BS and K single antenna users and the system operates in TDD mode is considered in [28] as illustrated in Figure 10.18. In this contribution, the pre-Cal schemes and the post-Cal schemes are investigated.

Equivalent transmit and receive circuit gains of the ith antenna at the BS are (t_i, r_i) and (t'_k, r'_k) are the equivalent transmit and receive circuit gains of the k^{th}user. The magnitudes of (t_i, r_i) and (t'_k, r'_k) are assumed identically and uniformly distributed in the amplitudes range i.e. $(1 - 0.5\delta, 1 + 0.5\delta)$. In addition, it is also assumed that the phase of the hardware circuit gains is compensated in the simulations.

The performance of the four calibration methods, namely: partial pre-Cal (p-pre-Cal), partial post-Cal (p-post-Cal), full pre-Cal (f-pre-Cal), and full Post-Cal (f-post-Cal) are compared in Figure 10.19. The empirical (verified by observation rather theory) probability density function (PDF) is plotted versus the achievable sum rates (bits/channel use) with ZF precoding when $M = 100$, $K = 40$, $0 \leq \delta \leq 1.0$, and transmit power equal 40 dB.

The PDF curves of the full calibration are sharper than the partial calibration, meaning the full calibration provides more stable performance than the partial calibration. The curves also show the performance of the post precoding calibration suffers some loss compared to its counterpart the pre-precoding calibration. This loss may be traced back to the precoding, which decreases the received SNR causing the performance degradation.

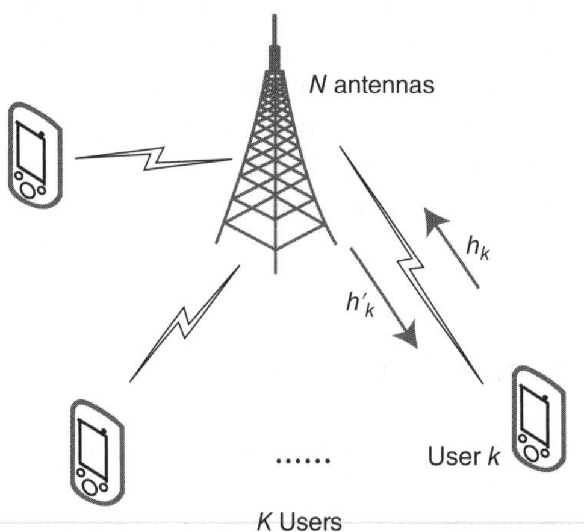

Figure 10.18 Single cell MU large-scale antenna systems [27].

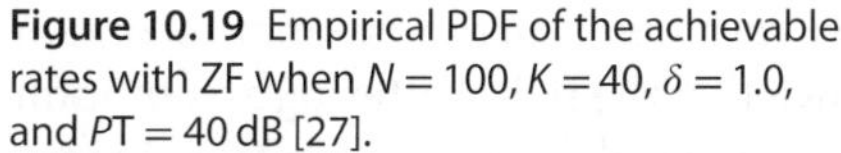
Figure 10.19 Empirical PDF of the achievable rates with ZF when $N = 100$, $K = 40$, $\delta = 1.0$, and $PT = 40$ dB [27].

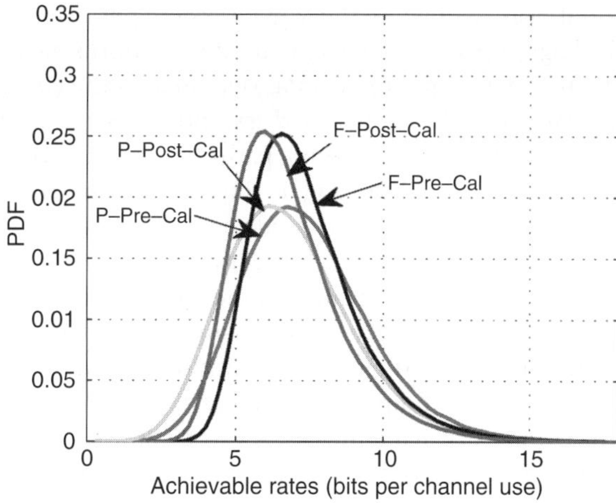

10.10 Summary

The MIMO cellular MAC and BC channels were introduced in Section 10.1. In Section 10.2, we introduced the MIMO channels definitions followed by the definition of the massive MIMO channels. In the deliberation, the transceivers hardware mismatches are not accounted for (i.e. definitions based on propagation channels). The TDD transmission protocol was clarified in Section 10.3. We established the facts about the basic concept of the reciprocity. The corresponding DL vector estimates were expressed as conjugate transpose of the UL estimates vector within the coherence interval of the channel.

Massive MIMO channel estimation when noncooperative networks operating on TDD protocol was investigated in Section 10.4. We examined the aligned pilot scheme when the TDD networks are noncooperating and derived the SINR for the UL and DL data transmission, followed by the time-shifted pilot scheme where users are arranged into groups. Users in all cells belong to a specific group and transmit their pilot sequences simultaneously over UL.

Comparing the time-shifted pilot scheme SINRs with the respective SINRs for the aligned pilot scheme, we concluded that the time-shifted pilot scheme guarantees that only cells in the same group interfere with each other and that the scheme completely nulls interference from adjacent group cells, provided the pilots do not overlap in time. The time-shifted pilot scheme promises significant gains in data transmission rates. Furthermore, using time-shifted pilot scheme and ensuring orthogonal pilot sequences guarantees cancellation of pilot contamination interference.

In Section 10.5, we introduce a new pilot transmission scheme based on a coordinated pilot protocol.

The duty of the coordination is to optimize the channel covariance matrices so that ideally the desired and the interference covariance matrices occupy different spaces. We considered estimators for the UL for all users and for the desired channel with full pilot reuse. The DL can be determined using the reciprocity property. In Section 10.6, we analysed the performance of the estimators derived in Section 10.5 for a system of large antenna number M (i.e. massive MIMO system) at the BS. We examined a coordinated pilot allocation scheme that can be applied in cellular systems. We considered scenarios such that the multipath of the desired user transmitted is restricted to a space where interfering paths are highly unlikely to exist and considered the structure that enables such assumption to be viable.

Initial assessment of the MIMO system performance was evaluated on the assumption of rich Rayleigh scattering environment. Recent works have shown that the rich scattering assumption is often inappropriate. In addition, much of the contemporary available experimental evidence appeared to indicate that *i. i. d* assumption is unrealistic in practice. These facts are possibly due to dominant LOS non-fading paths between the transmitter and receiver. In such situations, a spatially correlated Rician (i.e. Rayleigh + dominant LOS) channel model is more appropriate. In Section 10.7, we considered the MMSE estimation of the Rician channel matrix from the observations taken at the receiver during training transmission.

In Section 10.8, we examined the antenna calibration to eliminate the nonreciprocity feature provoked by the RF hardware mismatches at the BS and user device. We considered two calibration methods: the Argos calibration scheme and the antennas mutual coupling calibration method. Argos computes the calibration coefficients without averaging so it is of comparatively low complexity. Nonetheless, Argos calibration coefficients of antennas far away from the reference antenna are likely to be very small due to the weak mutual coupling, which produces ill-behaved estimation that degrades the system performance. So, the Argos method is very sensitive to the placement of the reference antenna, and the system performance is unstable unless all the other antennas have very good SNR to the reference antenna.

Mutual coupling analysis method has a matrix-based approach, which is appropriately applied to massive MIMO unlike the analysis of the Argos method, which is element based. In the initial iteration, only the calibration coefficients of adjacent antennas are computed because the mutual coupling is near to its peak and the rest of the coefficients are obtained through iteration and averaging, since each antenna can establish a relation with the reference through several different paths. Nonetheless, the instability of Argos calibration subdues in the mutual coupling method with small increase in the complexity. Section 10.9 compared different hardware mismatching calibration schemes and showed that pre-Cal outperforms post-Cal schemes.

10.A Noncooperative TDD Networks: Derivation of the Asymptotic Normalization Factor Equation

Here we derive the asymptotic normalization factor that appears in equation (10.11).

We use Lemma-1 [7]. Let $\mathbf{x}$, $\mathbf{y} \in \mathbb{C}^{M\times 1}$ be two independent vectors with distribution $\mathbb{C}\mathcal{N}\,(\mathbf{0}, c\,\mathbf{I})$. Using Lema-1 in [7], we have the following two expressions:

$$\lim_{M\to\infty} \frac{\mathbf{x}^H\mathbf{y}}{M} \approx 0$$

$$\lim_{M\to\infty} \frac{\mathbf{x}^H\mathbf{x}}{M} \approx c$$

The asymptotic normalization factor is given in (10.10b) as $\alpha_{k'i} = \frac{\|\widehat{\mathbf{h}}_{ik'i}\|}{\sqrt{M}}$. Accordingly:

$$\lim_{M\to\infty} \alpha_{k'i}^2 = \lim_{M\to\infty} \frac{[\|\widehat{\mathbf{h}}_{ik'i}\|^2]}{M} = \lim_{M\to\infty} \frac{1}{M}[(\widehat{\mathbf{h}}_{ik'i})^H(\widehat{\mathbf{h}}_{ik'i})] \tag{10.A.1}$$

The channel vectors $\widehat{\mathbf{h}}_{ik'i}$ for MS located in the ith BS is given in (10.8) as

$$\widehat{\mathbf{h}}_{ik'i} = \sum_{l=1}^{N_c} \sqrt{\rho_{k'l}\beta_{ik'l}}\ \mathbf{g}_{ik'l} + \mathbf{n}'_i \tag{10.A.2}$$

where $\mathbf{n}_i' \sim \mathbb{C}\mathcal{N}\left(0, \frac{1}{K}\mathbf{I}_\mathrm{M}\right)$. Using (10.A.2) to express $\|\widehat{\mathbf{h}}_{ik'i}\|^2$ as

$$\|\widehat{\mathbf{h}}_{ik'i}\|^2 = \left(\sum_{l=1}^{N_c} \sqrt{\rho_{k'l}\beta_{ik'l}}\ (\mathbf{g}_{ik'l})^H + (\mathbf{n}_i')^H\right)\left(\sum_{l=1}^{N_c} \sqrt{\rho_{k'l}\beta_{ik'l}}\ \mathbf{g}_{ik'l} + \mathbf{n}_i'\right) \tag{10.A.3}$$

Multiplying the terms in (10.A.3) we get

$$\begin{aligned}\|\widehat{\mathbf{h}}_{ik'i}\|^2 = \Bigg(&\sum_{l=1}^{N_c} \rho_{k'l}\beta_{ik'l}\,\|\mathbf{g}_{ik'l}\|^2 + \sum_{l_1=1}^{N_c}\ \sum_{l_2=1,l_2\neq l_1}^{N_c} \sqrt{\rho_{k'l_1}\beta_{ik'l_1}\ \rho_{k'l_2}\beta_{ik'l_2}}\ (\mathbf{g}_{ik'l_1})^H\ \mathbf{g}_{ik'l_2} + \|\mathbf{n}_i'\|^2\\ &+ \sum_{l=1}^{N_c} \sqrt{\rho_{k'l}\beta_{ik'l}}\ (\mathbf{g}_{ik'l})^H\ \mathbf{n}_i' + (\mathbf{n}_i')^H \sum_{l=1}^{N_c} \sqrt{\rho_{k'l}\beta_{ik'l}}\ \mathbf{g}_{ik'l}\Bigg)\end{aligned} \tag{10.A.4}$$

Substitute (10.A.4) in (10.A.1), and simplify each term:

$$\lim_{\mathrm{M}\to\infty} \frac{1}{\mathrm{M}}\left(\sum_{l=1}^{N_c} \rho_{k'l}\beta_{ik'l}\,\|\mathbf{g}_{ik'l}\|^2\right) = \lim_{\mathrm{M}\to\infty} \frac{1}{\mathrm{M}} \sum_{l=1}^{N_c} \rho_{k'l}\beta_{ik'l} \tag{10.A.5}$$

where $\mathrm{M}\to\infty$, we assumed the average of $\|\mathbf{g}_{ik'l}\|^2 = 1$. Furthermore, the channel vectors of different users are independent, so that as M tends to ∞ we get

$$\lim_{\mathrm{M}\to\infty} \frac{1}{\mathrm{M}} \sum_{l_1=1}^{N_c}\ \sum_{l_2=1,l_2\neq l_1}^{N_c} \sqrt{\rho_{k'l_1}\beta_{ik'l_1}\ \rho_{k'l_2}\beta_{ik'l_2}}\ (\mathbf{g}_{ik'l_1})^H\ \mathbf{g}_{ik'l_2} \xrightarrow{yields} 0 \tag{10.A.6}$$

In addition, as M tends to ∞, we have

$$\lim_{\mathrm{M}\to\infty} \frac{1}{\mathrm{M}} \sum_{l=1}^{N_c} \sqrt{\rho_{k'l}\beta_{ik'l}}\ (\mathbf{g}_{ik'l})^H\ \mathbf{n}_i' + (\mathbf{n}_i')^H \sum_{l=1}^{N_c} \sqrt{\rho_{k'l}\beta_{ik'l}}\ \mathbf{g}_{ik'l} \xrightarrow{yields} 0 \tag{10.A.7}$$

And the last term is

$$\lim_{\mathrm{M}\to\infty} \frac{1}{\mathrm{M}}\|\mathbf{n}_i'\|^2 = \frac{1}{K} \tag{10.A.8}$$

Therefore, combining (10.A.5), (10.A.6), (10.A.7), and (10.A.8) we get

$$\lim_{\mathrm{M}\to\infty} \alpha_{k'i}^2 = \sum_{l=1}^{N_c} \rho_{k'l}\beta_{ik'l} + \frac{1}{K} \tag{10.A.9}$$

10.B Beamforming Vectors for Time-Shifted Pilot Scheme

We apply similar definitions used in 10.A. The asymptotic normalization factor α_{ki} changes with signal $\boldsymbol{s}_{kl}$, beamforming vectors $\mathbf{w}_{kl}$, and transmit power of neighbouring BSs. Nonetheless, α_{ki} value will be established once $\widehat{\mathbf{h}}_{ik'i}$ estimates are computed. It is worth noting that the inner product of signal vector $\boldsymbol{s}_{kl}$ and pilot sequence $\boldsymbol{\psi}_k$ is, in general, random variable. Let us assume the signal $\boldsymbol{s}_{kl}$ is PSK modulated such that

$$|\boldsymbol{s}_{kl}\boldsymbol{\psi}_k^H|^2 \leq K \tag{10.B.1}$$

The product of channel matrix $\mathbf{h}_{il}$ and beamforming can be constraint as

$$\|\ \mathbf{h}_{il}\ \boldsymbol{w}_{kl}\|^2 \leq |\lambda_{\max}(\mathbf{h}_{il})|^2 \tag{10.B.2}$$

where $\lambda_{max}(\boldsymbol{h}_{il})$ is the largest eigenvalue of $\mathbf{h}_{il}$.

We assume there is no line of sight (LOS) propagation path between BSs, and the entries to matrices $\mathbf{h}_{il}$ are independent and Gaussian distributed. Since the largest singular value of $\mathbf{h}_{il}$ is $\lambda_{max}(\mathbf{h}_{il})$, which varies between [+1, − 1] as M → ∞. Meanwhile, we assume there is no LOS energy propagation path between BSs, and we assume the entries to matrices $\mathbf{h}_{il}$ are independent with Gaussian distribution. Since the largest singular value of $\mathbf{h}_{il}$ is $\lambda_{max}(\mathbf{h}_{il})$, which varies between [+1, − 1] as M → ∞ then, according to [30, 31]:

$$\lim_{M\to\infty} \frac{|\lambda_{\max}(\mathbf{h}_{il})|^2}{M} = 4 \tag{10.B.3}$$

We assume the beam vector $\boldsymbol{w}_{kl}$ will coincide with the eigenvector associated with the largest eigenvalue of the channel between the BSs (i.e. $\mathbf{h}_{il}^H\, \mathbf{h}_{il}$). This assumption is only true for the Rayleigh fading channel model. A LOS component together with a Ralyeigh channel (i.e. Rician) model in the propagation path between BSs is more likely in actual scenario, and this possibility raises the issue that there may be a few eigenvectors and in such a scenario $\mathbf{h}_{il}^H\, \mathbf{h}_{il}$, which would be likely to grow as M^2. The worst case is when the beamforming vector $\boldsymbol{w}_{kl}$aligned with the largest eigenvector with the possibility of a higher interference from adjacent BSs and producing poor channel vector estimates. When LOS exists between BSs, beamforming vector $\boldsymbol{w}_{kl}$ must be modified to avert to keep the asymptotic scenario unchanged. In deriving the modification, we consider the extreme case where the $\mathbf{h}_{il}$ formed only by LOS component and it has rank 1. To eliminate the interference, from neighbouring BSs are ordered to project all of its beamforming vectors onto the $\mathbf{h}_{il}$ null space thus forcing $\mathbf{h}_{il}\, \boldsymbol{w}_{kl} = 0$.

The projection aligns the beamforming vectors in M − 1 dimensions, and when M is large, the effect is minimal.

Now we consider the channel matrix between the ith and the lth BSs $\mathbf{h}_{il}$ is composed of LOS and non-LOS (NLOS) components:

$$\mathbf{h}_{il} = \mathbf{h}_{il}^{LOS} + \mathbf{h}_{il}^{NLOS} \tag{10.B.4}$$

where $\mathbf{h}_{il}^{LOS} \in \mathbb{C}^{M \times i^{th} M}$ deterministic matrix whose rank does not grow as M grows and $\mathbf{h}_{il}^{NLOS}$ is a complex Gaussian full-rank matrix.

The modified beamforming vector $\overline{\boldsymbol{w}}_{kl}$ is

$$\overline{\boldsymbol{w}}_{kl} = \frac{\overline{\mathbf{h}}_{il}^{LOS}\, \hat{\mathbf{h}}_{lkl}}{\|\overline{\mathbf{h}}_{il}^{LOS}\, \hat{\mathbf{h}}_{lkl}\|} \tag{10.B.5}$$

where $\overline{\mathbf{h}}_{il}^{LOS}$ is given as

$$\overline{\mathbf{h}}_{il}^{LOS} = \mathbf{I} - (\mathbf{h}_{il}^{LOS})^{\dagger}\, \mathbf{h}_{il}^{LOS} \tag{10.B.6}$$

and $(\mathbf{h}_{il}^{LOS})^{\dagger}$ denotes Moore-Penrose Pseudo-inverse. The projection will cancel the LOS and hence removing its interference on the channel estimates does not alter the asymptotic performance except that the receiver power would grow as $(M - m)^2$ rather than M^2 where m identifies the LOS path. The procedure removes the LOS component and considers only NLOS component of channel matrices as the interference is removed.

10.C Derivation of equations (10.48b) and (10.49b)

Considering the estimated channel expression in (10.43) and as we proved in Section 11.5.1 that the Bayesian and the MMSE channel estimates are identical, it may be easier to analyse the

MMSE estimate expressed in (10.43) as

$$\widehat{\mathbf{h}} = \mathbf{R}\,\widetilde{\mathbf{S}}^H(\sigma_n^2\,\mathbf{I}_{\tau\mathrm{M}} + \widetilde{\mathbf{S}}\mathbf{R}\,\widetilde{\mathbf{S}}^H\,)^{-1}\;\mathbf{y}$$

Define a matrix $\mathbf{G}$ as

$$\mathbf{G} \triangleq \;\mathbf{R}\,\widetilde{\mathbf{S}}^H(\sigma_n^2\,\mathbf{I}_{\tau M} + \widetilde{\mathbf{S}}\mathbf{R}\,\widetilde{\mathbf{S}}^H\,)^{-1} \tag{10.C.1}$$

Therefore,

$$\widehat{\mathbf{h}}_{\mathrm{MMSE}} = \mathbf{G}\,\mathbf{y} \tag{10.C.2}$$

The MSE $\mathcal{M}$ for the channel estimate is defined as

$$\mathcal{M} \triangleq \mathbb{E}\{\|\widehat{\mathbf{h}} - \mathbf{h}\|_F^2\} \tag{10.C.3}$$

Substituting (11.D.2) into (11.D.3), we get

$$\mathcal{M} \triangleq \mathbb{E}\{\|\mathbf{G}\,\mathbf{y} - \mathbf{h}\|_F^2\}$$

The square Frobenius norm for a matrix $\mathbf{A}$ is defined by

$$\|\mathbf{A}\|_F^2 \;:= \mathrm{tr}(\mathbf{A}\,\mathbf{A}^H\,) \tag{10.C.4}$$

Let set $\mathbf{A} = \mathbf{G}\,\mathbf{y} - \mathbf{h}$, and substitute for $\mathbf{A}$ in (10.C.4). Then we get

$$\mathcal{M} = \mathrm{tr}[\mathbb{E}\{(\mathbf{G}\,\mathbf{y} - \mathbf{h})(\mathbf{G}\,\mathbf{y} - \mathbf{h})^H\}] \tag{10.C.5a}$$

Simplifying (11.D.5), we have

$$\mathcal{M} = \mathrm{tr}\{\;\mathbb{E}[\mathbf{G}\,\mathbf{y}\,\mathbf{y}^H\,\mathbf{G}^H \;- \mathbf{G}\,\mathbf{y}\,\mathbf{h}^H \;- \mathbf{h}\,\mathbf{y}^H\mathbf{G}^H + \mathbf{h}\mathbf{h}^H]\;\} \tag{10.C.5b}$$

$$\mathcal{M} = \mathrm{tr}\{\;[\mathbf{G}\,\mathbb{E}\{\mathbf{y}\,\mathbf{y}^H\}\,\mathbf{G}^H \;- \mathbf{G}\,\mathbb{E}\{\mathbf{y}\,\mathbf{h}^H\} - \mathbb{E}\{\mathbf{h}\,\mathbf{y}^H\}\mathbf{G}^H + \mathbb{E}\{\mathbf{h}\mathbf{h}^H\}]\;\} \tag{10.C.5c}$$

We have the covariance $\mathbf{R} \triangleq \mathbb{E}\{\mathbf{h}\,\mathbf{h}^H\}$ so (10.C.5c) simplifies to

$$\mathcal{M} = \mathrm{tr}\{\;[\mathbf{G}\,\mathbb{E}\{\mathbf{y}\,\mathbf{y}^H\}\,\mathbf{G}^H \;- \mathbf{G}\,\mathbb{E}\{\mathbf{y}\,\mathbf{h}^H\} - \mathbb{E}\{\mathbf{h}\,\mathbf{y}^H\}\mathbf{G}^H + \mathbf{R}]\;\} \tag{10.C.5d}$$

We also have $\mathbf{y} = \widetilde{\mathbf{S}}\mathbf{h} + \mathbf{n}$

It can easily be shown that

$$\mathbb{E}\{\mathbf{y}\,\mathbf{y}^H\} = \widetilde{\mathbf{S}}\,\mathbf{R}\,\widetilde{\mathbf{S}}^H + \sigma_{\mathbf{n}}^2\,\mathbf{I}_{M\tau}$$

$$\mathbb{E}\{\mathbf{y}\,\mathbf{h}^H\} = \widetilde{\mathbf{S}}\,\mathbf{R}$$

$$\mathbb{E}\{\mathbf{h}\,\mathbf{y}^H\} = \mathbf{R}\widetilde{\mathbf{S}}^H$$

Substituting the above $\mathbf{G}$ expression into (10.C.5d) and simplifying, we get

$$\mathcal{M} = \mathrm{tr}\{\mathbf{R} - \mathbf{G}\,\widetilde{\mathbf{S}}\mathbf{R}\} \tag{10.C.6a}$$

Substituting for $\mathbf{G}$ in (10.C.6a), we obtain

$$\mathcal{M} = \mathrm{tr}\{\mathbf{R} - \mathbf{R}\,\widetilde{\mathbf{S}}^H(\sigma_n^2\,\mathbf{I}_{\tau M} + \widetilde{\mathbf{S}}\mathbf{R}\,\widetilde{\mathbf{S}}^H)^{-1}\,\widetilde{\mathbf{S}}\mathbf{R}\}$$

Expression (10.C.6b) can be rewritten as

$$\mathcal{M} = \mathrm{tr}\left\{\mathbf{R} - \mathbf{R}\,\widetilde{\mathbf{S}}^H\left(\mathbf{I}_{\tau M} + \widetilde{\mathbf{S}}\frac{\mathbf{R}}{\sigma_n^2}\,\widetilde{\mathbf{S}}^H\right)^{-1}\widetilde{\mathbf{S}}\,\frac{\mathbf{R}}{\sigma_n^2}\right\}$$

We operate the matrix inversion to simplify the second term in (10.C.6c):

$$(\mathbf{I} + \mathbf{AB})^{-1}\mathbf{A} = \mathbf{A}(\mathbf{I} + \mathbf{BA})^{-1}$$

$$\left(\mathbf{I}_{\tau M} + \widetilde{\mathbf{S}}\frac{\mathbf{R}}{\sigma_n^2}\widetilde{\mathbf{S}}^H\right)^{-1}\widetilde{\mathbf{S}}\frac{\mathbf{R}}{\sigma_n^2} = \widetilde{\mathbf{S}}\frac{\mathbf{R}}{\sigma_n^2}\left(\mathbf{I} + \widetilde{\mathbf{S}}^H\widetilde{\mathbf{S}}\frac{\mathbf{R}}{\sigma_n^2}\right)^{-1}$$

Using (10.C.d), we can reduce (10.C.6c) to

$$\mathcal{M} = \mathrm{tr}\left\{\mathbf{R} - \mathbf{R}\widetilde{\mathbf{S}}^H\widetilde{\mathbf{S}}\frac{\mathbf{R}}{\sigma_n^2}\left(\mathbf{I}_{N_c M} + \widetilde{\mathbf{S}}^H\widetilde{\mathbf{S}}\frac{\mathbf{R}}{\sigma_n^2}\right)^{-1}\right\} \tag{10.C.6b}$$

$$\mathcal{M} = \mathrm{tr}\left\{\mathbf{R} - \mathbf{R}\left(\mathbf{I}_{N_c M} + \widetilde{\mathbf{S}}^H\widetilde{\mathbf{S}}\frac{\mathbf{R}}{\sigma_n^2}\right)^{-1}\widetilde{\mathbf{S}}^H\widetilde{\mathbf{S}}\frac{\mathbf{R}}{\sigma_n^2}\right\}$$

$$\mathcal{M} = \mathrm{tr}\left\{\mathbf{R}\left(\mathbf{I}_{N_c M} - \left(\mathbf{I}_{N_c M} + \widetilde{\mathbf{S}}^H\widetilde{\mathbf{S}}\frac{\mathbf{R}}{\sigma_n^2}\right)^{-1}\widetilde{\mathbf{S}}^H\widetilde{\mathbf{S}}\frac{\mathbf{R}}{\sigma_n^2}\right)\right\} \tag{10.C.6c}$$

We use the matrix inversion identity below to reduce (10.C.6c),

$$\mathbf{I} - (\mathbf{I} + \mathbf{A})^{-1}\mathbf{A} = (\mathbf{I} + \mathbf{A})^{-1}$$

Therefore, we have (11.D.6e)

$$\mathcal{M} = \mathrm{tr}\left\{\mathbf{R}\left(\mathbf{I}_{N_c M} + \widetilde{\mathbf{S}}^H\widetilde{\mathbf{S}}\frac{\mathbf{R}}{\sigma_n^2}\right)^{-1}\right\} \tag{10.C.7}$$

Further, from (10.C.6b) we have,

$$\mathcal{M}_1 = \mathrm{tr}\left\{\mathbf{R}_1 - \mathbf{R}_1^2\left(\frac{\sigma_n^2}{\tau}\mathbf{I}_M + \sum_{l=1}^{N_c}\mathbf{R}_l\right)^{-1}\right\} \tag{10.C.8}$$

References

1 Marzetta, T.L. (2010). Noncooperative cellular wireless with unlimited numbers of base station antennas. *IEEE Transaction on Wireless Communications* 9 (11): 3590–3600.

2 Yin, H., Gesbert, D., Filippou, M., and Liu, Y. (2013). A coordinated approach to channel estimation in large-scale multiple-antenna systems. *IEEE Journal on Selected Areas in Communications* 31 (2): 264–273.

3 Marzetta, T.L. (2006) 'How much training is required for multiuser MIMO?', Asilomar Conference on Signals, Systems and Computers, 359 - 363.

4 Lee, H., Park, S., and Bahk, S. (2016) 'Enhancing spectral efficiency using aged CSI in massive MIMO systems', IEEE Global Communications Conference (GLOBECOM), 1-6.

5 Zheng, X., Zhang, H., Xu, W., and You, X. (2014) 'Semi-orthogonal pilot design for massive MIMO systems using successive interference cancellation', IEEE Global Communications Conference, 3719 - 3724.

6 Elijah, O., Leow, C.Y., Abdul Rahman, T. et al. (2016). A comprehensive survey of pilot contamination in massive MIMO—5G system. *IEEE Communications Surveys & Tutorials* 18 (2): 905–923.

7 Fernandes, F., Ashikhmin, A., and Marzetta, T.L. (2013). Inter-cell interference in non-cooperative TDD large scale antenna systems. *IEEE Journal on selected Areas in Communications* 31 (2): 192–201.

8 Ali, S., Chen, Z., and Yin, F. (2017). Eradication of pilot contamination and zero forcing precoding in the multi-cell TDD massive MIMO systems. *IET Journal on Communications* 11 (13): 2027–2034.

9 Kay, S.M. (1993). *Fundamentals of Statistical Signal Processing: Estimation Theory*. Englewood Cliffs, NJ: Prentice Hall.

10 Björnson, E. and Ottersten, B. (2010). A framework for training-based estimation in arbitrarily correlated Rician MIMO channels with Rician disturbance. *IEEE Transactions on Signal Processing* 58 (3): 1807–1820.

11 McKay, M.R. and Collings, I.B. (2005). General capacity bounds for spatially correlated Rician MIMO channels. *IEEE Transactions on Information Theory* 51 (9): 3121–3145.

12 Vu, M., and Paulraj, A. (2005) 'Capacity optimization for Rician correlated MIMO wireless channels', Proceedings of Asilomar Conference on Signals, Systems, and Computers, 133-138.

13 Dumont, J., Loubaton, P., and Lasaulce, S. (2006). On the capacity achieving transmit covariance matrices of MIMO correlated Ricean channels: a large system approach. *IEEE Globecom* 1–6.

14 Shepard, C., Yu, H., Anand, N., LI, L. E., Marzetta, T., Yang, R., and Zhong, L. (2012) 'Argos: practical many-antenna base stations', 18th Annual IEEE International Conference on Mobile Computing Networking, 53-64

15 Guillaud, M. and Kaltenberger, F. (2013). 'Towards Practical Channel reciprocity Exploitation: Relative Calibration in the Presence of Frequency Offset'. In: *IEEE Wireless Communications and Networking Conference*, 2525–2530.

16 Vieira, J., Rusek, F., Tufvesson, F. (2014) 'Reciprocity calibration methods for Massive MIMO based on antenna coupling', IEEE Globecom - Wireless Communications Symposium, 3708-3712

17 Larsson, E.G., Edfors, O., Tufvesson, F., and Marzetta, T.L. (2014). Massive MIMO for next generation wireless systems. *IEEE Communications Magazine* 52 (92): 186–195.

18 Rice University **W**ireless Open **A**ccess **R**esearch **P**latform. http://warp.rice.edu.

19 Nishimori, K., Cho, K., Takatori, Y., and Hori, T. (2001). Automatic calibration method using transmitting signals of an adaptive array for TDD systems. *IEEE Transactions on Vehicular Technology* 50 (6): 1636–1640.

20 Liu, J., Vandersteen, G., Craninckx, J., Libois, M., Wouters, M., Petre, F., and Barel, A. (2006) 'A novel and low-cost analog front-end mismatch calibration scheme for MIMO-OFDM WLANs', IEEE Radio Wireless Symposium, 219-222

21 Kaltenberger, F., Jiang, H., Maxime, G.M., and Knopp, R. (2010). Relative channel reciprocity calibration in MIMO/TDD systems. In: *IEEE Future Network and Mobile Summit Conference*, 1–10.

22 Su, L., Yang, C., Wang, G., and Lei, M. (2014). Retrieving channel reciprocity for coordinated multi-point transmission with joint processing. *IEEE Transactions on Communications* 62 (5): 1541–1553.

23 Shi, J., Luo, Q. and You, M. (2011) 'An efficient method for enhancing TDD over the air reciprocity calibration', IEEE Wireless Communications and Networking Conference, pp. 339–344

24 Wei, H., Wang, D., Zhu, H. et al. (2016). Mutual coupling calibration for multiuser massive MIMO systems. *IEEE Transactions on wireless communications* 15 (1): 606–619.

25 Durgin, G.D. and Rappaport, T.S. (1999) 'Effects of multipath angular spread on received voltage envelopes the spatial cross-correlation of received voltage envelopes', IEEE Vehicular Technology Conference, 996-1000.

26 Zelst, A.V. and Hammerschmidt, J.S. (2002) 'A single coefficient spatial correlation model for multiple-input multiple-output (MIMO) radio channels', *27th General Assembly of the International Union of Radio Science (URSI)*, Maastricht, the Netherlands.

27 Lee, J.-H. and Cheng, C.-C. (2012). Spatial correlation of multiple antenna arrays in wireless communication systems. *Progress In Electromagnetics Research* 132: 347–368.

28 Zhang, W., Ren, H., Pan, C. et al. (2015). Large-scale antenna systems with UL/DL hardware mismatch: achievable rates analysis and calibration. *IEEE Transactions on Communications* 63 (4): 1216–1229.

29 Dette, H. (2002). Strong approximation of eigenvalues of large dimensional Wishart matrices by roots of generalized languere polynomials. *Journal of Approximation Theory,* 118: 290–304. Elsevier Science (USA).

30 Tulino, A.M and Verdu, S. (2004) 'Random Matrix Theory and Wireless Communications', Foundations and trends in communications and information theory, Now Publishers.

31 Foschini, G.J. (1996). Layered space-time architecture for wireless communication in a fading environment when using multi-element antennas. *Bell Labs Technical Journal* 41–59.

32 Foschini, G.J., Golden, G.D., Valenzuela, R.A., and Wolniansky, P.W. (1999). Simplified Processing for High Spectral Efficiency Wireless Communication Employing Multi-Element Arrays. *IEEE Journal on Selected Areas in Communications* 17 (11): 1841–1852.

33 Telatar, I.E. (1999). Capacity of Multiple-antenna Gaussian Channels. *European Transactions on Telecommunications* 10: 585–595.

11

Linear Precoding Strategies for Multi-User Massive MIMO Systems

11.1 Introduction

There had been intensive research activities in the last decade aimed at improving the throughput of interference-limited systems. Precoding has attracted the interest of many researchers in this field. Generally, precoding is the design of the signal at the transmitter prior to transmission to provide effective delivery of the information to the intended user over a multi-user multiple-input, multiple-output (MIMO) system. Early works on precoding examined single-cell, multi-user scenarios where the main throughput limitation is due to the intra-cell interference. Subsequently, the works in the precoding field extended to cover inter-cell interference in cellular and heterogeneous networks. Today, precoding has found applications in many systems such as satellite communications, power line communications, and optical communications, to name but a few.

Precoding arises by two fundamentally different types: nonlinear precoding and linear precoding. The former is also known as Costa coding or dirty paper coding (DPC) and is designed when the interference is known to the transmitter. DPC precoding at the transmitter cancels the interference and optimally achieves the capacity region of the MIMO broadcast channel (BC). However, application of the DPC calls for considerable extra computational complexity at both the transmitter and receiver. In contrast, linear precoding is a low complexity that is able to transmit the same number of data streams as a DPC–based system and thus achieves equal multiplexing gain. In linear precoding, each antenna transmits a data stream that is the linear combination of all data streams with beamforming weights, which allows all transceivers to share a common data downlink (DL).

A key challenge for transmission over multi-user MIMO channels is the segregation and equalization of the parallel data streams. In order to achieve the capacity and performance gains expected from the MIMO system, we have to deal with channel distortion and interference. These two problems can be resolved using a well-designed receive equalizer for the former to remove channel corruptions and use a transmit precoder for the later to null out the inter-users interference. The precoding-equalizing can be attained at the transmitter or at the receiver, or jointly optimized at the transmitter and the receiver. The Mobile stations (MSs) mostly have limited processing power, so they can only execute simple algorithms with low computational complexity.

An example of the DL transmission with the base station (BS) multiple antennas transmit to multiple mobile users, each user has access only to its received signal and cannot cooperate with the other users. Thus, receive processing is impractical and the system must resort to precoding. Precoding requires the availability of channel state information (CSI) at the transmitter, which, in time-division duplex (TDD) transmission protocol, can be estimated using pilot training to estimate the uplink (UL) CSI, and the DL exploiting the reciprocity concept. The CSI estimation

5G Physical Layer Technologies, First Edition. Mosa Ali Abu-Rgheff.

under fast-varying wireless channels is a complicated scheme. Another difficulty is that the transmitter must find a set of the best channels in each transmission interval that have the most-orthogonal channel vectors.

The system model we use in the chapter comprises a BS with M antennas and the number of users is denoted K, where each user is provided by a single or multiple antenna N. The channel matrix is denoted by $\mathbf{H} \in \mathbb{C}^{N \times M}$ and the additive Gaussian noise $\mathbf{n} \in \mathbb{C}^{N \times 1}$. Furthermore, a detailed mathematical derivation for most of the formulae we introduced is also provided in 11.A–11.F.

11.2 Group-Level and Symbol-Level Precoding

The multi-user DL precoding schemes can be classified into two classes [1, 2]: multi-group multicasting precoding *employs group-level precoding* to transmit multiple precoded messages simultaneously and each precoded message is intended for a group of users. The precoder design is conditional on the channels in each group.

Group-level precoding is also employed when multiple precoded data streams are transmitted simultaneously; each precoded stream is intended for a single user. This precoding class is the most common and is referred to as MIMO BC precoding. The BC precoding is dependent on the channels of the individual users.

Symbol-level precoding is employed when multiple precoded symbols are transmitted simultaneously and each symbol is sent to a single user. The precoding design is dependent on both the channels and symbols of the users. Precoding at the symbol level is capable of providing substantial benefits due to manipulating the multi-user interference (MUI) vector, such that its direction moves away in any random direction from the desired signal direction and its power is added to the system power. However, these benefits come at the cost of much higher computing complexity in the precoder and system design.

A linear precoder complements the signal at one side of the channel to the other side by splitting the transmit signal into orthogonal eigenbeams and allocates higher power level along the beams where the channel is strong and lower levels or no power along the weak channels. Further, precoding design changes depend on the performance criteria and the type of channel state information at the transmitter (CSIT). A flat-fading channel capacity can be achieved by CSIT-independent coding, together with CSIT-dependent linear precoding. A linear precoding allocates power over both space and time. However, power allocation over time marginally improves the capacity of a fading channel at low signal power to noise power ratios (SNRs). On the other hand, allocation over space can meaningfully increase the capacity at all SNRs. This feature drives precoding techniques to exploit spatial CSIT.

A conventional precoding system comprises a channel encoder and a multi-user precoder, as illustrated in Figure 11.1. The encoder encompasses a channel encoder, interleaver, and symbol

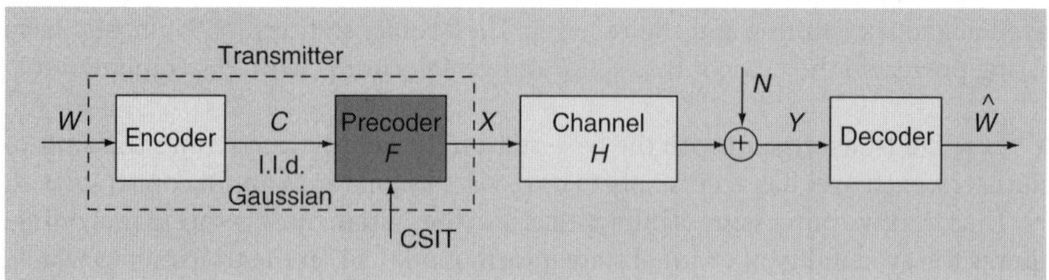

Figure 11.1 An optimal configuration for exploiting CSIT [3].

mapping. Denote the encoded input signal as matrix **C** normalised to transmit power and let its covariance be $\mathbf{Q} = \mathbb{E}[\mathbf{C}\,\mathbf{C}^H]$. The precoder processes the combined multi-user signals before transmission. At the receiver, the noise-corrupted received signal is decoded to recover the data bits using an effective channel comprising the precoder and the actual channel.

11.3 Linear Precoding Schemes

A linear precoder comprises a general structure illustrated in Figure 11.1 with an input shaper and a multimode beamformer per beam power allocation. Consider the precoder matrix **F** and let us apply the singular value decomposition (SVD) on the precoding matrix **F** as

$$\mathbf{F} = \mathbf{U}_{\mathrm{F}}\mathbf{D}\mathbf{V}_{\mathrm{F}}^{\mathrm{H}} \tag{11.1}$$

The precoder has essentially two roles: decoupling the input data into orthogonal streams in the form of eigenbeams, and allocating power to these beams. If the precoder orthogonal beams match the channel eigen-directions, there will be no interference between the transmitted data streams. This depicts the precoder role of the decoupling effects. In addition, the precoder allocates power to the beams, and if the beams have equal power, then the overall radiation pattern will be omni-directional. If the beams have different levels, the overall pattern is directional.

The orthogonal beam directions are given by the vectors $\mathbf{U}_{\mathrm{F}}$ where each column describes a beam direction. The beam power allocations are the square of the diagonal elements of the matrix **D**. The vectors $\mathbf{V}_{\mathrm{F}}$ combine the input symbols into each beam and accordingly are called the input-shaping matrix. The linear precoding structure is depicted in Figure 11.2 where d_1, d_2 are elements of the diagonal matrix **D**. The precoder design imposes certain transmit power constraint to satisfy the transmit power level:

$$\mathrm{tr}(\mathbf{F}\,\mathbf{F}^{\mathrm{H}}) = 1 \tag{11.2}$$

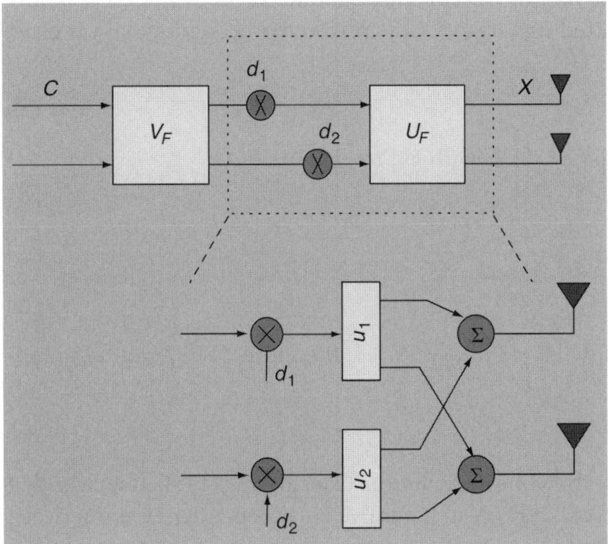

Figure 11.2 A linear precoder structure as a multimode beamformer [3].

11.4 SU-MIMO Model

Consider a point-to-point link with M, N antennas at the transmitter and at the receiver, respectively. The single user (SU) transmits encoded signal matrix **C**. The transmitted signal propagates along L multiple paths developed by scattering and reflection from physical obstacles in the terrain, such as buildings, trees, etc. The presence of the multipath enhances the capacity of MIMO channel in such a way that it grows as long as the SNR increases. Furthermore, *SVD precoding* offers the optimal precoding solution for single-user MIMO systems [4, 5]. Denote the received SU signal vector as **y**:

$$\mathbf{y} = \mathbf{HC} + \mathbf{n} \tag{11.3}$$

where $\mathbf{H} \in \mathbb{C}^{N \times M}$ channel matrix with i.i.d complex Gaussian entries and $\mathbf{n} \in \mathbb{C}^{N \times 1}$ is the uncorrelated additive white Gaussian noise (AWGN). SU signal encoded matrix **C** is independent of noise **n**. The receiver knows earlier the precoding matrix **F** and handles **HF** as effective channel. We consider a simple transmission scheme employing the SVD precoding. The SU-MIMO channel matrix is decomposed into a number non-equal gain parallel single-input single output (SISO) subchannels. We will assume that the channel is known at the transmitter. For example, using the reciprocal TDD system, the DL channel may be determined by the estimated UL, as we explored in Chapter 10.

The SVD of the channel matrix **H** is given as

$$\mathbf{H} = \mathbf{UDV}^{H} \tag{11.4}$$

where **U** is $N \times N$ complex unitary matrix, **V** is $M \times M$ complex unitary matrix, and **D** is $M \times N$ diagonal matrix with non-negative real elements (representing the singular values) on the diagonal. The ith singular values of **H** is denoted by λ_i and are organized in descending order (i.e. $\lambda_1 \geq \lambda_2 \geq \dots \dots \dots \geq \lambda_{\min(M,N)}$). The number of the singular values in **H** are given by the rank of **H**, which is given by the min(M, N). Let us assume the encoded data matrix covariance **Q** is square and we can eigevalue decomposed as $\mathbf{Q} = \mathbf{U_Q}\, \mathbf{\Lambda}_Q\, \mathbf{U}_Q^H$. According to [3], we set $\mathbf{V_F} = \mathbf{U_Q}$.

The precoding is initiated by multiplying the SU encoded data matrix **C** by matrix **V** before transmission, so the precoded transmitted signal is **VC**. The received precoded signal $\boldsymbol{y}$ is given as

$$\boldsymbol{y} = \mathbf{HVC} + \mathbf{n} \tag{11.5}$$

At the receiver, we multiply both sides of (11.5) with $\mathbf{U}^H$ to give $\tilde{\boldsymbol{y}}$ as

$$\tilde{\boldsymbol{y}} = \mathbf{U}^H\mathbf{HVC} + \mathbf{U}^H\,\mathbf{n} \tag{11.6}$$

Substitute (11.4) for **H** into (11.6):

$$\tilde{\boldsymbol{y}} = \mathbf{U}^H\mathbf{UDV}^H\mathbf{VC} + \mathbf{U}^H\,\mathbf{n} \tag{11.7}$$

Since **U** and **V** are unitary matrices, $\mathbf{U}^H\mathbf{U} = \mathbf{I}$ and $\mathbf{V}^H\mathbf{V} = \mathbf{I}$, therefore, (11.7) can be simplified to

$$\tilde{\boldsymbol{y}} = \mathbf{DC} + \mathbf{U}^H\,\mathbf{n} \tag{11.8a}$$

Denote $\tilde{\mathbf{n}} = \mathbf{U}^H\,\mathbf{n}$ where $\tilde{\mathbf{n}}$ and **n** have the same distribution. Considering (11.8), it is clear that the multiple streams precoding using SVD of **H** created min(M, N) independent links between the transmitter and the receiver such that

$$\tilde{y}_i = \lambda_i\mathbf{C} + \tilde{\mathbf{n}}, \tag{11.8b}$$

where $i = 1, 2, \ldots\ \ldots\ \ldots\ \ldots\ \ldots\ \ldots, \min(\mathrm{M}, \mathrm{N})$. Apparently, the quality of the parallel streams is degraded by increasing noise level (i.e. $\tilde{\mathbf{n}} \neq \mathbf{n}$). In practical systems, performance degradation can be alleviated using appropriate power allocation (i.e. water-filling) scheme that optimizes the power allocation to the ith stream p_i such that

$$\sum_{i=1}^{\min(\mathrm{M},\mathrm{N})} p_i = \mathrm{P_T} \tag{11.9}$$

where $\mathrm{P_T}$ is total transmit power.

11.5 Multi-User MIMO Precoding System Model

11.5.1 Broadcast Channel (BC) System Model

We consider a single cell supported by single BS transmitter with M antennas as depicted in Figure 11.3, and examine the BC in a multi-user MIMO system. Assume $K \leq \mathrm{M}$, the kth user is provided with N_k antennas when communicates with the system. The kth user channel matrix is denoted by matrix $\mathbf{H}_k \in \mathbb{C}^{\mathrm{N}_k \times \mathrm{M}}$.

Furthermore, we assume that $\mathrm{M} \geq \mathrm{N}$ where $\mathrm{N} = \sum_{k=1}^{K} \mathrm{N}_k$ and the BC CSI is known at the BS. It is important to be clear about the difference between BC and DL and multiple access channel (MAC) and UL. In simple words, DL refers to a single radio link for signal transmission from BS to a MS, and UL refers to a single radio link for signal transmission from MS to the BS. So both DL and UL are used by a single link. On the other hand, the BC comprises many DL links, and the MAC consists of multiple UL links.

We assume the channel coefficients are independent random variables with zero mean circularly symmetric complex Gaussian distributed (i.e. Rayleigh fading). The multi-user transmitted symbols $\boldsymbol{S}$ are expressed as

$$\mathbf{S} = [\mathbf{s}_1, \ldots\ldots\ldots\ldots\ldots, \mathbf{s}_K] \tag{11.10}$$

where $\boldsymbol{s}_k$ is the symbol vector destined to the kth MS expressed as

$$\boldsymbol{s}_k^T = [s_{k,1}, \ldots\ldots\ldots\ldots\ldots, s_{k,\mathrm{M}}] \tag{11.11}$$

The multi-user data matrix $\boldsymbol{S} \in \mathbb{C}^{\mathrm{M} \times K}$ has zero mean and covariance defined by $\mathrm{M} \times \mathrm{M}$ matrix, the diagonal elements corresponding to user power constrained to fix transmit power:

$$\mathbb{E}[\boldsymbol{s}_k\, \boldsymbol{s}_k^H] = \mathrm{tr}(\mathbf{R}_s^k) = p_k \tag{11.12a}$$

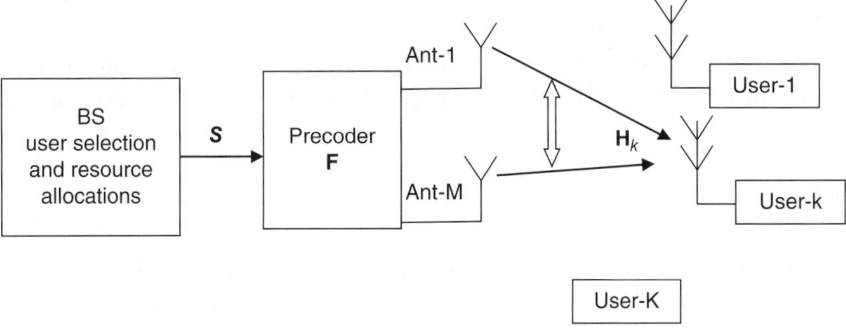

Figure 11.3 Multi-user MIMO precoding system model.

where $\mathbf{R}_s^k$ is M × M covariance matrix and p_k is the allocated power to k^{th} user. Denote the total transmit power to be $\mathrm{P_T}$; then we get

$$\sum_{k=1}^{K} p_k \leq \mathrm{P_T} \tag{11.12b}$$

Define the precoding matrix as

$$\mathbf{F} := [\boldsymbol{f}_1 \boldsymbol{f}_2 \ldots \ldots \ldots \ldots ., \boldsymbol{f}_K] \tag{11.13}$$

The received signal $\boldsymbol{y}_k$ at the kth MS is

$$\boldsymbol{y}_k \in \mathbb{C}^{\mathrm{N}_k \times 1} = \mathbf{H}_k\, \mathbf{F}\, \boldsymbol{s}_k + \mathbf{n}_k \tag{11.14}$$

where $\mathbf{n}_k$ is Gaussian noise at the kth MS expressed as

$$\boldsymbol{n}_k = [n_{k,1}, \ldots \ldots \ldots ., n_{k,\mathrm{N}_k}]^T \tag{11.15}$$

where $\mathrm{n}_k \sim \mathbb{C}\mathcal{N}(0, 1)$ and $\mathbf{H}_k^H \in \mathbb{C}^{\mathrm{N}_k \times \mathrm{M}}$.

Conventionally, the precoders are designed subject to a total power constraint $\mathrm{P_T}$ as,

$$\mathbb{E}[||S||^2] = \mathrm{tr}\{\mathbf{F}\mathbf{R}_s\mathbf{F}^H\} \leq \mathrm{P_T} \tag{11.16}$$

where $\mathrm{P_T} > 0$, so the individual per antenna power is constrained $\leq \frac{\mathrm{P_T}}{\mathrm{M}}$.

11.5.2 Multiple Access Channels (MAC) System Model with Non-Equal Antennas at Each User

In the majority of massive MIMO literature, it is assumed that a single MS is endowed with a single antenna. However, in recent years, there has been a growing interest in the multiple antennas MSs in many practical applications. Accordingly, it would be of theoretical and practical benefit to examine the massive MIMO system with distributed antenna sets (ASs).

Consider a frequency-flat fading MIMO MAC channel with one BS serving K MSs [6]. The antennas at the BS are arranged in L distributed antenna sets (ASs). The lth antenna set consisting of M_l antennas. As in the BC model, the kth MS is equipped with N_k antennas. Additionally, we define N and M such that

$$\mathrm{N} := \sum_{k=1}^{K} \mathrm{N}_k \ \text{ and } \ \mathrm{M} := \sum_{l=1}^{L} \mathrm{M}_l. \tag{11.17}$$

Let us denote the $\mathrm{N}_k \times 1$ transmitted vector of the kth MS as $\mathbf{x}_k$. The covariance matrix of $\mathbf{x}_k$ is given by

$$\mathbb{E}\{\mathbf{x}_k\, \mathbf{x}_{k'}^H\} = \left\{ \begin{matrix} \frac{\mathrm{P}_k}{\mathrm{N}_k}\boldsymbol{Q}_k & \text{if } k = k' \\ \mathbf{0} & \text{otherwise} \end{matrix} \right\} \tag{11.18}$$

where P_k is the total transmitted power of the kth MS, and $\boldsymbol{Q}_k$ is an $\mathrm{N}_k \times \mathrm{N}_k$ positive semi definite matrix with the constraint $\mathrm{tr}(\boldsymbol{Q}_k) \leq \mathrm{N}_k$. The signal $\boldsymbol{y}$ received during a symbol interval can be written as

$$\boldsymbol{y} = \sum_{k=1}^{K} \mathbf{H}_k\, \mathbf{x}_k + \mathbf{n} \tag{11.19}$$

where $\mathbf{H}_k \in \mathbb{C}^{\mathrm{M} \times \mathrm{N}_k}$ channel matrix between k^{th} MS and the BS, and $\mathbf{n} \sim \mathcal{CN}(0, \sigma_n^2\, \mathbf{I}_\mathrm{M})$ is a complex Gaussian noise. The normalized covariance channel matrices of $\mathbf{H}_k$ are given by

$$\mathbb{E}\{\mathrm{tr}(\mathbf{H}_k\, \mathbf{H}_k^H)\} = \frac{\mathrm{MN}_k}{\mathrm{N}} \tag{11.20}$$

Additionally, the channel matrix $\mathbf{H}_k$ possesses the following structure:

$$\mathbf{H}_k = \overline{\mathbf{H}}_k + \tilde{\mathbf{H}}_k, \tag{11.21}$$

where the elements of $\mathbf{H}_k$ are defined as

$$\overline{\mathbf{H}}_k := (\overline{\mathbf{H}}_{1k}^T \ \overline{\mathbf{H}}_{2k}^T \ \ldots\ldots\ldots\ldots \overline{\mathbf{H}}_{Lk}^T)^T \tag{11.22a}$$

$$\tilde{\mathbf{H}}_k := (\tilde{\mathbf{H}}_{1k}^T \ \tilde{\mathbf{H}}_{2k}^T \ \ldots\ldots\ldots\ldots \tilde{\mathbf{H}}_{Lk}^T)^T \tag{11.22b}$$

where $\overline{\mathbf{H}}_{lk}$ is an $\mathrm{M}_l \times \mathrm{N}_k$ deterministic matrix, and $\tilde{\mathbf{H}}_{lk}$ is a jointly correlated matrix defined as [4]

$$\tilde{\mathbf{H}}_{lk} = \mathbf{U}_{lk}\,(\mathbf{M}_{lk} \circ \mathbf{W}_{lk})\,\mathbf{V}_{lk}^H \tag{11.23}$$

We use $\mathbf{M}_{lk} \circ \mathbf{W}_{lk}$ to denote the Hadamard product (element-wise multiplication of matrices) and $\mathbf{U}_{lk}$ and $\mathbf{V}_{lk}$ are *unitary matrices,* $\mathbf{M}_{lk}$ is an $\mathrm{M}_l \times \mathrm{N}_k$ deterministic matrix with non-negative elements, and $\mathbf{W}_{lk}$ is a complex Gaussian random matrix with independent and identically distributed (i.i.d.), zero mean and unit variance entries.

The jointly correlated channel model accounts for the correlation at both link ends and characterises their mutual dependence. It is a realistic model for massive MIMO channels since the effectiveness of the Kronecker model decreases as the number of antennas increases [7, 8]. We assume that the channel matrices of different links are independent.

11.5.3 Linear Precoding for Massive MIMO MAC with Equal Antennas at Each User

So far, we have considered channels in flat or slowly fading environment in TDD mode systems, with average pedestrians' mobility. We assumed the channel coefficients are independent random variables with zero mean circularly symmetric complex Gaussian distributed. In most scenarios that are studied, single antenna MS is assumed, i.e. no mutual coupling at the transmit link end, and the Kronecker fading channel model is adopted. Accordingly, in TDD transmission, the instantaneous CSI can be estimated accurately at the transmitter using UL pilot training and the instantaneous CSI of the DL can be obtained by applying the reciprocity concept. The CSI estimates are needed at the transmitter for multi-user precoding on the BC.

Consider a single cell MIMO MAC system with a BS fitted with M antennas with K independent users, each has N transmit antennas, and we assume in this case that all users have same number of transmit antennas. The MIMO receive signal $\mathbf{y} \in \mathbb{C}^{\mathrm{M}\times 1}$ is given as

$$y = \sum_{k=1}^{K} \mathbf{H}_k\,\mathbf{x}_k + \mathbf{n} \tag{11.24}$$

where $\mathbf{s}_k \in \mathbb{C}^{\mathrm{N}\times 1}$, $\mathbf{F}_k \in \mathbb{C}^{\mathrm{N}\times \mathrm{N}}$, $\mathbf{H}_k \in \mathbb{C}^{\mathrm{M}\times \mathrm{N}}$ denotes the transmitted data vector, linear precoding matrix, and the channel matrix of the kth user, respectively, $\mathbf{n} \in \mathbb{C}^{\mathrm{M}\times 1}$ is a zero-mean complex Gaussian noise vector with covariance matrix $\mathbf{I}_{\mathrm{M}}$, i.e. each receiver has unit variance. In addition, we assume the transmitter knows the statistical CSI for all users. The transmitted signal vector $\mathbf{x}_k$ is expressed as

$$\mathbf{x}_k = \mathbf{F}_k\,\mathbf{s}_k \tag{11.25}$$

The transmit signal $\mathbf{x}_k$ complies with the power constraint:

$$\mathbb{E}\,[\mathbf{x}_k^H\ \mathbf{x}_k] = \mathrm{tr}(\mathbf{F}_k\ \mathbf{F}_k^H) \le P_k,\ \ k = 1, 2, \ldots .., K \tag{11.26}$$

Assuming a jointly correlated fading MIMO channel, we implement the Weichselberger's model [7] derived in 8.6. This model jointly characterizes the correlation at the transmitter and receiver sides. The kth user $\mathbf{H}_k$ is modelled as [9]:

$$\mathbf{H}_k = \mathbf{U}_{\mathrm{R}_k} (\tilde{\mathbf{G}}_k \mathrm{o} \mathbf{W}_k) \mathbf{U}_{\mathrm{T}_k}^H, \tag{11.27a}$$

where

$$\mathbf{U}_{\mathrm{R}_k} = [\mathbf{u}_{\mathrm{R}_k,1}, \mathbf{u}_{\mathrm{R}_k,2}, \dots\dots\dots., \mathbf{u}_{R_k,\mathrm{M}}] \in \mathbb{C}^{\mathrm{M}\times\mathrm{M}} \tag{11.27b}$$

$$\mathbf{U}_{\mathrm{T}_k} = [\mathbf{u}_{\mathrm{T}_k,1}, \mathbf{u}_{\mathrm{T}_k,2}, \dots\dots\dots., \mathbf{u}_{\mathrm{T}_k,N_t}] \in \mathbb{C}^{\mathrm{N}\times\mathrm{N}} \tag{11.27c}$$

are deterministic unitary matrices, respectively, $\tilde{\mathbf{G}}_k \in \mathbb{C}^{\mathrm{M}\times\mathrm{N}}$ is a deterministic matrix with real-value non-negative elements, and $\mathbf{W}_k \in \mathbb{C}^{\mathrm{M}\times\mathrm{N}}$ is a random matrix with independent identically distributed (i.i.d.) Gaussian elements with zero-mean and unit variance.

Define $\mathbf{G}_k := \tilde{\mathbf{G}}_k \, o \, \tilde{\mathbf{G}}_k$ and let $g_{k,n,m}$ denote the $(n, m)^{th}$ element of matrix $\mathbf{G}_k$. Here, $\mathbf{G}_k$ is the *coupling matrix* and $g_{k,n,m}$ corresponds to the average coupling energy between $\mathbf{u}_{R_k,n}$ and $\mathbf{u}_{\mathrm{T}_k,\mathrm{m}}$. The transmit and receive correlation matrices of the kth *user* are given as

$$\mathbf{R}_{t,k} = \mathbb{E}_{\mathbf{H}_k}[\mathbf{H}_k^H \mathbf{H}_k] = \mathbf{U}_{\mathrm{T}_k} \mathbf{\Gamma}_{\mathrm{T}_k} \mathbf{U}_{\mathrm{T}_k}^H \tag{11.28}$$

$$\mathbf{R}_{r,k} = \mathbb{E}_{\mathbf{H}_k}[\mathbf{H}_k \mathbf{H}_k^H] = \mathbf{U}_{\mathrm{R}_k} \mathbf{\Gamma}_{\mathrm{R}_k} \mathbf{U}_{\mathrm{R}_k}^H \tag{11.29}$$

where $\mathbf{\Gamma}_{\mathrm{T}_k}$ and $\mathbf{\Gamma}_{\mathrm{R}_k}$ are diagonal matrices with main diagonal elements

$$[\mathbf{\Gamma}_{\mathrm{T}_k}]_{nn} = \sum_{n=1}^{\mathrm{N}} g_{k,n,m}, \quad n = 1, 2, \dots\dots, \mathrm{N} \tag{11.30}$$

and

$$[\mathbf{\Gamma}_{R_k}]_{m\,m} = \sum_{n=1}^{N_t} g_{k,n,m}, \quad m = 1, 2, \dots\dots, \mathrm{M} \tag{11.31}$$

11.6 Linear Multi-User Transmit Channel Inversion Precoding for BC

The basic idea of channel inversion precoding (CIP) is a simple linear algebra arrangement to separate the multi-user data transmission into multiple parallel streams of data. In other words, we zero out the interference between streams of transmitted links. We assume that CSI is available at the transmitter. The channel inversion (CI)-based transmit precoding can be performed at the transmitter.

Denote the signal transmitted vector as $\mathbf{x}_{BC} \in \mathbb{C}^{\mathrm{M}\times 1}$ and the precoded vector as $\tilde{\mathbf{x}}_{BC}$. To explain the CI-based precoding, let us consider a MIMO flat-fading channel with M transmit antennas and N receive antennas. The received vector $\mathbf{y}$ without precoding is given as

$$\mathbf{y} \in \mathbb{C}^{\mathrm{N}\times 1} = \mathbf{H}_{BC}\,\mathbf{x}_{BC} + \mathbf{n} \tag{11.32}$$

where matrix $\mathbf{H}$ is $\mathrm{N}\times\mathrm{M}$. Before transmission, the precoding at the transmitter transforms vector $\mathbf{x}_{BC}$ into $\tilde{\mathbf{x}}_{BC}$ where

$$\tilde{\mathbf{x}}_{BC} = \mathbf{H}_{BC}^{\dagger}\,\mathbf{x}_{BC} \tag{11.33}$$

where $\mathbf{H}_{BC}^{\dagger}$ denotes Moore-Penrose inverse (MPI) of $\mathbf{H}_{BC}$. Note the channel matrix in most cases is not a square matrix, so we use the MPI. The CI equalized the SNR in all streams of transmission and removed the interference between streams. At the receive end, the data are recovered from the received signal:

$$\tilde{\mathbf{y}} = \mathbf{H}_{BC}\,\tilde{\mathbf{x}}_{BC} + \mathbf{n} \tag{11.34}$$

Substituting (11.33) into (11.34), we get

$$\tilde{\mathbf{y}} = \mathbf{H}_{BC}\,\mathbf{H}_{BC}^{\dagger}\,\mathbf{x}_{BC} + \mathbf{n} = \mathbf{x}_{BC} + \mathbf{n} \tag{11.35}$$

We can conclude that the CI preprocessing transforms the multiple data transmission channel into independent streams transmitted over AWGN channel [10]. Furthermore, the CI processing cannot be initiated without channel reciprocity, which enables the transmitter to obtain the CSI from the uplink prior to BC transmission.

The size of the **H** matrix to be inverted depends on the number of transmit and receive antennas and the length of the channel impulse response. Although CI transmit precoding is simple to implement, CI precoding is not popular with a system that has a *large number of antennas* because of high computational cost to formulate the MPI of the channel. In addition, the sum rate for the CIP scheme is poor. In fact, it exhibits a large capacity loss in more extreme fading transmission environments and approaches zero capacity in severe Rayleigh fading [11]. The channel inversion is related to the zero-forcing (ZF), which we examine in the next section.

11.7 Zero-Forcing Precoding using the Wiesel et al. Method

11.7.1 Multi-User Linear Zero-Forcing (ZF) Precoding for BC

The conventional channel inversion, presented in the preceding section aims at removing the interference generated from other users sharing the same channel. Such a precoding scheme is optimal when the estimated channel of the uplink is perfect. However, due to channel estimation errors, the CI precoding performs poorly at low signal-to-noise ratios (SNRs) and for any number of users in multi-user systems.

Another precoding scheme that achieves the same goal as the CIP is the ZF precoding. The ZF precoder aims at zeroing the inter-user interference and achieves close to the sum capacity even when the number of users tends to infinity (i.e. it achieves the ergodic sum capacity of BC channel when the number of users is large) or the system is interference-limited (the interference is more dominant than noise). In other words, the ZF precoding ensures removing the MUI by projecting the transmission of the desired data stream into the null-space of all other (interfering) users' transmission.

Consider a simple model comprising a single cell with BS with M antennas serving K single antenna users. We formulate the BC between the BS and the kth user $\mathbf{h}_k \in \mathbb{C}^{1\times \mathrm{M}}$ and the BC channel matrix expressed as $\mathbf{H} \in \mathbb{C}^{K\times \mathrm{M}} = [\mathbf{h}_1 \ldots \ldots \ldots \mathbf{h}_K]^T$. The transmitted signal vector $\mathbf{x} \in \mathbb{C}^{\mathrm{M}\times 1} = \mathbf{Fs}$. The received signal vector $\mathbf{y} \in \mathbb{C}^{K\times 1} = [y_1, \ldots \ldots \ldots, y_K]^T$ and the receiver noise $\mathbf{n} \in \mathbb{C}^{\mathrm{K}\times 1} = [n_1, \ldots \ldots \ldots, n_1]^T$ is assumed to be circularly symmetric Gaussian noise where $n_k \sim \mathcal{N}(0, 1)$. The received signal at the kth user is given by

$$y_k = \mathbf{h}_k^H\,\mathbf{x} + n_k \;\; for\, k = 1, \ldots \ldots \ldots, K, \tag{11.36}$$

where the entries of $\mathbf{h}_k$ are zero mean i.i.d complex Gaussian variables with unit variance.

The multi-user broadcast signal is

$$\mathbf{y} = \mathbf{H}\,\mathbf{x} + \mathbf{n}$$

where vector **s** satisfies

$$\mathbb{E}[\mathbf{s}_k \mathbf{s}_k^H] = \mathbf{I}_K \tag{11.37}$$

and $\mathbf{n}_k$ is additive Gaussian noise vector such that $\mathbb{E}[\mathbf{n}_k\, \mathbf{n}_k^H] = \sigma^2\, \mathbf{I}_\mathrm{N}$. Since the precoding matrix **F** is used to remove inter-user interference, for this to occur we have

$$(\mathbf{HF})_{k,j} = 0 \quad \mathit{if}\ k \neq j \qquad k = 1, \ldots \ldots \ldots \ldots, K \tag{11.38}$$

We assume, without loss of generality, that $\mathcal{R}e\{[(\mathbf{HF})_{k,k}]\} \geq 0$ and $\mathrm{Im}\{[(\mathbf{HF})_{k,\,k}]\} = 0$. The kth user desired signal power is given as

$$|[\mathbf{H}\,\mathbf{F}]_{k,k}|^2 \tag{11.39a}$$

The multi-user interference power is given as

$$\sum_{j=1, j\neq k}^{K} |[\mathbf{HF}]_{k,j}|^2 \tag{11.39b}$$

The ratio of received desired signal power to the interference plus noise power γ_k is given as

$$\gamma_k = \frac{|(\mathbf{HF})_{k,k}|^2}{\sum_{\substack{i=1\\ i\neq k}}^{K} |(\mathbf{HF})_{k,i}|^2 + \sigma^2} \tag{11.40a}$$

Since **F** is designed to attain zero interference, then (11.40a) becomes

$$\gamma_k = p_k = |(\mathbf{HF})_{k,k}|^2, \tag{11.40b}$$

when $[\mathbf{n}_k\, \mathbf{n}_k^H] = \mathbf{I}_\mathrm{N}$.

Furthermore, γ_k is SNR at the transmitter and p_k is the transmit power for the kth user terminal and are identical. In addition, ZF precoding provides a tradeoff between performance and design complexity. The ZF precoding is comprehensively researched in the literature, and several design versions are well-known for the optimization under various constraints.

11.7.2 ZF Precoder Design with Total Transmit Power Constraint

ZF precoding implemented at the BS transmitter to remove the interference at the multi-user receivers satisfying (11.38). Moreover, we assume that $\mathcal{R}e\{[(\mathbf{HF})_{k,k}]\} \geq 0$ and $\mathrm{Im}\{[(\mathbf{HF})_{k,\,k}]\} = 0$ for $=1, \ldots\ \ldots\ \ldots$, K. According to (11.40b), the ZF in matrix articulation is given by

$$\mathbf{HF}_{ZF} = \mathrm{diag}\{\sqrt{\mathbf{P}}\} \tag{11.41}$$

where $\sqrt{\mathbf{P}}$ is a vector with real non-negative elements i.e. $\sqrt{\mathbf{P}} = [\sqrt{p_1}, \ldots \ldots \ldots, \sqrt{p_K}]^\mathrm{T}$. Accordingly, the ZF precoding breaks up the BC into K SISO channels. The precoder total transmit power $\mathrm{P_T}$ is constraint as

$$\mathbb{E}[\mathbf{x}\,\mathbf{x}^H] = \mathbb{E}[\mathbf{Fs}\,\mathbf{s}^H\,\mathbf{F}^H] = \mathbb{E}[\mathbf{F}\,\mathbf{F}^H] = \mathrm{P_T} \tag{11.42}$$

So, per-antenna total power is restricted to $\leq \frac{\mathrm{P_T}}{\mathrm{M}}$. The BC channel matrix **H** is a $K \times \mathrm{M}$ matrix. Denote the inverse of matrix **H** as $\mathrm{M} \times \mathrm{K}$ matrix; according to [12], the pseudo-inverse $\mathbf{H}^\dagger$ is optimal in the precoder design with respect to the total power constraint in most practical scenarios.

Let us introduce $f(\mathbf{P})$ as an arbitrary cost function of $\mathbf{P}$:

$$\max_{\mathbf{P}\geq\mathbf{0},\mathbf{F}} f(\mathbf{P}) \tag{11.43a}$$

$$\text{s.t } \mathbf{HF}_{ZF} = \text{diag}\{\sqrt{\mathbf{P}}\}; \tag{11.43b}$$

$$\text{tr}\{\mathbf{F}_{\text{ZF}}\,\mathbf{F}_{ZF}^{H}\} \leq \text{P}_{\text{T}} \tag{11.43c}$$

The optimal solution is given by

$$\mathbf{F}_{\text{ZF}}^{\text{opt}} = \mathbf{H}^{\dagger}\text{diag}\{\sqrt{\mathbf{P}^{\text{opt}}}\}, \tag{11.44}$$

where $\mathbf{P}^{\text{opt}}$ is the optimization can be expressed in problem (11.45):

$$\max_{\mathbf{P}\geq\mathbf{0}} f(\mathbf{P}) \tag{11.45a}$$

$$\text{S.t.} \sum_k p_k\,[(\mathbf{H}^{\dagger})^{H}\,\mathbf{H}^{\dagger}]_{k,k} \leq \text{P}_{\text{T}} \tag{11.45b}$$

Assuming that $\mathbf{P} \geq \mathbf{0}$, then the problem (11.45) is concave maximization, which can be solved using the water-filling method.

11.7.3 Optimal ZF Precoding with per-Antenna Power Constraint

We wish to find the optimal ZF precoder under individual per-antenna power constraint. With such a scenario that maximizes an arbitrary function, $f(\mathbf{P})$ can be articulated as the following optimization problem:

$$f(\mathbf{P}^{\text{pot}}) = \left\{\begin{array}{c} \max_{\mathbf{P}\geq\mathbf{0}} f(\mathbf{P}) \\ \text{s.t } \mathbf{HF} = \text{diag}[\sqrt{\mathbf{p}}] \\ [\mathbf{F}\,\mathbf{F}^{H}]_{m,m} \leq \dfrac{\text{P}_{\text{T}}}{\text{M}} \text{ for all } m \end{array}\right\} \tag{11.46}$$

The optimization described in (11.46) is a nonconvex problem and its closed-form solution (if any) is difficult to find. The easier approach is to derive bounds to the optimal solution between upper U and lower L limits:

$$\text{L} \leq f(\mathbf{P}^{\text{pot}}) \leq \text{U}, \tag{11.47}$$

where,

$$\text{L} = \left\{\begin{array}{c} \max_{\mathbf{P}\geq\mathbf{0}} f(\mathbf{P}) \\ \text{s.t } \sum_k p_k\,|\mathbf{H}^{\dagger}{}_{m,k}|^2 \leq \frac{\text{P}_{\text{T}}}{\text{M}} \end{array}\right\} \text{ for all m} \tag{11.48a}$$

$$\text{U} = \left\{\begin{array}{c} \max_{\mathbf{P}\geq\mathbf{0}} f(\mathbf{P}) \\ \text{s.t } \sum_k p_k\,[(\mathbf{H}^{\dagger})^{H}\,\mathbf{H}^{\dagger}]_{k,k} \leq \text{P}_{\text{T}} \end{array}\right\} \text{ for all } k \tag{11.48b}$$

The optimization depends on the system parameters and can be solved using optimization tools that are available in software platforms on the market.

Fairness in wireless systems is simply a metric used to determine whether users/applications are receiving a fair share of the system resources. Fairness can be defined by quantitative measures and qualitative measures. Two of such measures are most commonly used in wireless are max-min fairness and proportional fairness. Max-min fairness can be achieved by resource allocation if it is feasible. Increasing the resource allocation of any user can result in the decrease in the resource allocation to the others. The max-min is based on maximising the minimum of power allocation for users. The fairness criterion that we want to optimize can be expressed as follows:

$$\max_{\mathbf{P}\geq\mathbf{0},\mathbf{F}} \ \min_k p_k \tag{11.49a}$$

$$\text{s.t} \quad \mathbf{HF}_{\text{ZF}} = \text{diag}\{\sqrt{\mathbf{P}}\}\ ; \tag{11.49b}$$

$$\{\mathbf{F}_{\text{ZF}}\,\mathbf{F}^H_{ZF}\}_{m,m} \leq \frac{\text{P}_\text{T}}{\text{M}} \quad \text{for all } m. \tag{11.49c}$$

The network throughput implies the average amount or average rate of data transported by reliable communications, which is defined by Shannon data rate. The throughput objective function is given as

$$\max_{\mathbf{P}\geq\mathbf{0},\mathbf{F}} \sum_k \log_2[1 + p_k] \tag{11.50a}$$

$$\text{s.t} \qquad \mathbf{HF} = \text{diag}(\sqrt{\mathbf{P}}) \tag{11.50b}$$

$$[\mathbf{F}\,\mathbf{F}^H]_{m,m} \leq \frac{\text{P}_\text{T}}{\text{M}} \text{ for all } m \tag{11.50c}$$

The two optimization problems are nonconcave and are too complicated to solve by analysis. Closed-form solutions to the optimization problems in (11.49a) and (11.49b) are difficult to formulate by analysis. However, the two problems can be solved adequately by standard optimization tools using a high-level software platform.

11.8 The Outage Probability

Consider channel matrix $\mathbf{H} \in \mathbb{C}^{\text{K}\times\text{M}}$ with entries that are independent random variables with zero mean and unit variance and assume an accurate CSI is known at both transmitter and the receiver. Since the channel entries are random variables, then the instantaneous capacity C is a random variable as well. Outage probability is widely accepted as a suitable behaviour metric to evaluate the random fluctuations of the instantaneous capacity [11]. Outage probability $\text{pr}_{out}(R)$ is defined as the probability that a given target capacity R cannot be met by the instantaneous capacity C using the random channel matrix.

In this section, we explore various attributes of the outage probability in wireless communications. Let us examine the MIMO DL system and denote the kth user signal power to interference power plus noise power ratio (SINR) as SINR_k and let $F_{\text{SINR}_k}(.)$ be the cumulative distribution function (CDF) of SINR_k. Then according to the channel conditions to have an acceptable outage probability of $\epsilon > 0$ implies that

$$\Pr[\log_2(1 + \text{SINR}_k) < R_{k,out}] = F_{\text{SINR}_k}(2^{R_{k,out}} - 1) = \epsilon \tag{11.51}$$

To meet the probability constraint in (11.51), we have to transmit to the kth user with the ϵ – outage rate denoted as $R_{k,\,out}$ given by

$$R_{k,out} = \log_2(1 + F^{-1}_{\text{SINR}_k}(\epsilon)) \tag{11.52}$$

The ϵ – outage sum rate is given by [13]:

$$R_{\text{sum}} = \sum_k R_{k,out} = \sum_k \log_2(1 + F^{-1}_{\text{SINR}_k}(\epsilon)) \tag{11.53}$$

Assuming an equal power allocation at the transmitter, the instantaneous capacity C *is* expected to fall within a symmetrical interval A given in [14] as

$$\mathbb{E}[C] - A < C < \mathbb{E}[C] + A \tag{11.54a}$$

where $\mathbb{E}[C]$ is the mean capacity known. The interval A is given by

$$A = \frac{qL\sqrt{\rho}}{\mathrm{M}^{\frac{1}{4}}} \tag{11.54b}$$

where ρ is the SNR, $L = 1.44$ and $q > 0$ is a real factor for adjusting the capacity interval. The symmetrical interval is $[\mathbb{E}[C] \pm A]$ occurs with probability at worst equal to

$$1 - 2\exp\left[-\frac{q^2\sqrt{\mathrm{M}}}{2}\right] \tag{11.55}$$

Let us employ an appropriate outage probability ϵ so the probability of outage is given by

$$\Pr_{out}(R) = \Pr(C < R) \leq \epsilon \tag{11.56}$$

Since C is Gaussian distributed random variable, the probability of the symmetrical interval is given by

$$\Pr(R < C < R + 2\,\omega) \geq 1 - 2\epsilon, \tag{11.57a}$$

where

$$\omega = \mathbb{E}[C] - R \tag{11.57b}$$

The symmetrical interval characteristic of Gaussian distributed C is depicted in Figure 11.4. In massive MIMO systems, it is a difficult undertaking to determine $\mathbb{E}[C]$ accurately. The appropriate minimum number of antennas $\mathrm{M}_{\min}$ that satisfies the outage probability ϵ in massive antennas MIMO systems is derived as

$$\mathrm{M}_{\min} = \arg_{\mathrm{M}}[\mathrm{f}(\mathrm{M}) = 0], \tag{11.58a}$$

where

$$\mathrm{f}(\mathrm{M}) \triangleq \exp\left(-\frac{(q^*)^2\,\mathrm{M}^{\frac{1}{4}}}{2}\right) - \epsilon, \tag{11.58b}$$

The adjusting scalar factor q^* is given by

$$q^* = \frac{(\log_2\mathbb{E}[\exp(\,C)] - \log_2(\omega - R)\,\mathrm{M}^{\frac{1}{4}}}{L\sqrt{\rho}} \tag{11.59}$$

Concurrently, authors of [15] were investigating a similar problem under various channel fading conditions for MIMO systems with large but finite number of antennas. The authors obtained closed-form formulae for the outage factor defined as the negative slope of the outage probability versus the number of antennas. Further they confirmed that diverse QoS demands can be assured with rather fewer transmit/receive antennas. In addition, they explored the consequences of temporal/spatial correlation on the minimum number of antennas and discovered

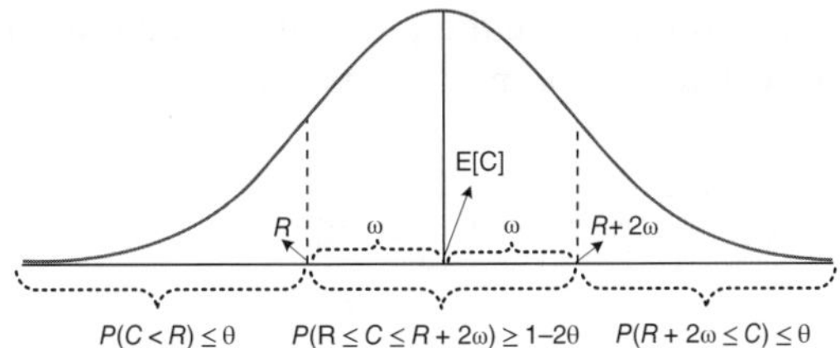

Figure 11.4 Symmetrical characteristic of Gaussian distributed C [14].

that for weak/non-excessive correlation, the correlation effect is negligible. However, correlation effect increases in highly correlation scenarios.

System outage can be attributed to the antennas themselves. Malfunctioning antennas reduce the survival of the system performance. The survival probability in the transmitter and receiver antennas in military MIMO systems is associated with the outage [16]. The survival probability affirms information about the working state of each transmit and receive antenna to protect the system performance. Denote the transmitting and the receiving survival probabilities vectors $\mathbf{p}_t$ and $\mathbf{p}_r$ that are considered to be binomial distributed real random vectors as

$$\mathbf{p}_t = [\mathrm{p}_t(1), \ldots \ldots, \mathrm{p}_t(\mathrm{M})]^T \tag{11.60a}$$

$$\mathbf{p}_r = [\mathrm{p}_r(1), \ldots \ldots, \mathrm{p}_r(\mathrm{N})]^T \tag{11.60b}$$

Consequently, the matrix of survival probabilities $\mathbf{P}$ is given by

$$\mathbf{P} = \mathbf{p}_r \otimes \mathbf{p}_t, \tag{11.60c}$$

where $\otimes$ is Kronecker product. So as to accurately define the propagation of the survival channels, we articulate the channel matrix by a product of the survival probability matrix and the channel coefficient matrix $\mathbf{H}$ weighing each channel coefficient $\mathrm{h}_{n,m}$ with $p_{n,m}$ using Hadamard product. Consequently, the channel matrix model is $\mathbf{P} \circ \mathbf{H}$. The capacity of the survival system mode is given by

$$C = \log_2 \det(\mathbf{I}_\mathrm{N} + \mu(\mathbf{P} \circ \mathbf{H})(\mathbf{P} \circ \mathbf{H})^H) \tag{11.61}$$

where o is Hadamard (element-wise) product, and $\mu \triangleq \frac{\mathrm{P_T}}{\sigma_\mathrm{n}^2}\mathrm{M}$, and $\frac{\mathrm{P_T}}{\sigma_\mathrm{n}^2}$ is the transmit SNR. Let the target data rate of the transmitter be R. The outage probability $\mathrm{Pr}_{\mathrm{out}}(R)$ is defined as the probability that random variable C falls below R. Accordingly, the outage probability of the survival system is given by

$$\mathrm{Pr}_{\mathrm{out}}(R) = \mathrm{pr}[C \leq R] = \lim_{c \to R} F_C(c), \tag{11.62}$$

where $F_C(c)$ is the CDF of the random variable capacity C and c is the limit of $F_C(c)$ at the point $c = R$.

11.9 Precoding for MIMO Channels with Johan et al. Method

The Johan et al. method for precoding MIMO channels is defined as a filter matrix at the transmitter and filter matrix at the receiver working jointly in the precoding process. The method is used to optimise the filter matrices. In addition, the power allocation defined by diag($\sqrt{\mathbf{P}}$) in Wiesel et al. method is replaced by a new factor β described in terms of the channel matrix and the total transmit power. β can be considered as a power adjustment factor.

11.9.1 Introduction

We now formulate the precoding using the MIMO system model illustrated in Figure 11.5 to articulate the optimization problem.

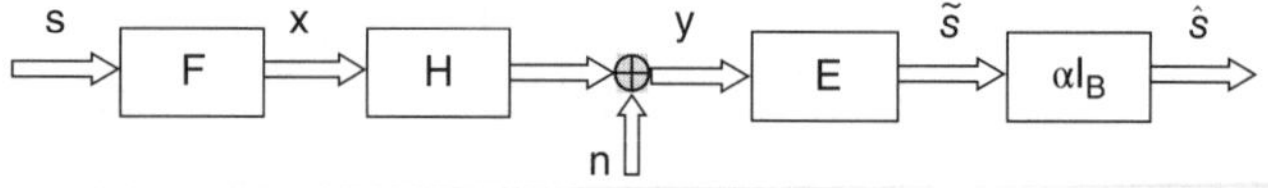

Figure 11.5 MIMO system with linear precoding filter **F** and linear equalizer filter **E** [17].

We consider a single cell MIMO system with a BS provided with M antennas and a user terminal with N antennas. The model comprises the transmit filter $\mathbf{F} \in \mathbb{C}^{M\times B}$ for precoding, the channel $\mathbf{H} \in \mathbb{C}^{N\times M}$, the equalizer filter at the receiver $\mathbf{E} \in \mathbb{C}^{B\times N}$ where the number of transmitted information symbols are $B \leq \min(M, N)$. The transmitted vector $\mathbf{x} \in \mathbb{C}^{M\times 1}$ is the desired symbols $\mathbf{s} \in \mathbb{C}^{B\times 1}$ and $\mathbf{x} \in \mathbb{C}^{M\times 1} = \mathbf{Fs}$ is transmitted signal. So, the average transmit power ρ is given as

$$\mathbb{E}\|\mathbf{Fs}\|_2^2 = \text{tr}(\mathbf{FR_s}\,\mathbf{F}_{FZ}^H) = \rho \tag{11.63}$$

where $\mathbf{R_s} = \mathbb{E}[\mathbf{s}\,\mathbf{s}^H]$. The transmitted signal is the desired $\mathbf{s}$ transformed by the transmit filter $\mathbf{F}$. After transmission over the channel $\mathbf{H}$, the received signal is degraded by the noise $\mathbf{n}$ and filtered through the receive filter $\mathbf{E}$ to output as the estimate $\tilde{\mathbf{s}}$:

$$\tilde{\mathbf{s}} = \mathbf{E}(\mathbf{HFs} + \mathbf{n}) \in \mathbb{C}^B \tag{11.64}$$

where $\mathbf{n} \in \mathbb{C}^{N\times 1}$ is the interference including Gaussian noise vector. We compare different transmit and receive filters using the mean square error (MSE) ε given as

$$\varepsilon = \mathbb{E}[\|\mathbf{s} - \hat{\mathbf{s}}\|_2^2] = \mathbb{E}[\|\mathbf{s} - \alpha\tilde{\mathbf{s}}\|_2^2], \tag{11.65}$$

The scalar constant α at the receiver can be used as an automatic gain control (AGC) in the MIMO system to minimize the MSE. It is worth noting that the MSE with the scalar constant (i.e. α) is for comparing different precoding methods. We derive the MSE to find the transmit filter and the receive filter matrices.

The SNR γ is given by the transmit power per data stream $\frac{\rho}{B}$ divided by the noise power per receive antenna $(\frac{\text{tr}(\mathbf{R_n})}{N})$, so the SNR is given by

$$\gamma = \frac{\frac{\rho}{B}}{\frac{\text{tr}(\mathbf{R_n})}{N}}, \tag{11.66}$$

where the noise matrix $\mathbf{R_n} = \sigma_\mathbf{n}^2\,\mathbf{I}_N$

11.9.2 ZF Transmit Filter F Matrix

The linear equalizing filter $\mathbf{E}$ processes the signal and forces $\tilde{\mathbf{s}}$ to be interference free estimate of $\mathbf{s}$. Accordingly, we get

$$\tilde{\mathbf{s}}|_{\mathbf{n}=0} = \mathbf{EHFs} \to \mathbf{s} \tag{11.67a}$$

Equation (11.67a) compels $\mathbf{F}$, $\mathbf{H}$, and $\mathbf{E}$ to be identity mapping:

$$\mathbf{EHF} = \mathbf{I}_B \tag{11.67b}$$

where the mapping by filter $\mathbf{E}$ of $\tilde{\mathbf{s}}$ to $\mathbf{s}$ gives rise to identity matrix $\mathbf{I}_B$.

Since the mapping at the receive filter removes all the interference caused at the transmitter and since the transmitter has no restraint on the noise, this is the only action the transmitter could do to minimize the transmit power at the transmit filter $\mathbf{F}$ using the next minimization problem:

$$\mathbf{F}_{ZF} = \arg\min_{\mathbf{F}} \;\; \mathbb{E}[\|\mathbf{F}s\|_2^2] \tag{11.68a}$$

$$s.t \quad \mathbf{EHF} = \mathbf{I}_B, \tag{11.68b}$$

Equation (11.68a) determines the value of $\mathbf{F}$ for, which $\mathbb{E}[.]$ achieves its minimum value. Using the Lagrange multipliers in an appropriate optimization method that satisfies the constraint in (11.68b), we get the transmit precoding filter $\mathbf{F} \in \mathbb{C}^{M\times B}$ as

$$\mathbf{F}_{ZF} \in \mathbb{C}^{M\times B} = (\mathbf{EH})^H\,[\mathbf{EH}\,(\mathbf{EH})^H]^{-1} \tag{11.69a}$$

It is interesting to compare the transmit precoding filter when the receiver equalizing filter **E** is included in the modelling with the transmit filter $\mathbf{F}_{ZF}$ derived in [18] as

$$\mathbf{F_{ZF}} = \mathbf{H^H}\,(\mathbf{HH^H})^{-1} \tag{11.69b}$$

We can understand that (11.69a) and (11.69b) become identical if we let channel **H** in (11.69b) to be replaced by **EH** in (11.69a). In other words, the transmit filter in (11.69a) observes a composite channel **EH** rather than channel **H**. In deriving (11.69a) we assumed $\mathbf{R_s} = 1$. So if $\mathbf{R_s} \neq 1$ and since the transmit power changes with the channel, we use an experiential approach to introduce a scaling factor β_{ZF} to gauge the transmit power to a defined value at the precoding filter, i.e. $\beta_{\mathrm{ZF}}\,\mathbf{F}_{ZF}$. Equation (11.69a) is then modified to

$$\mathbf{F}_{ZF} \in \mathbb{C}^{\mathrm{M\times B}} = \beta_{\mathrm{ZF}}(\mathbf{EH})^H\,[\mathbf{EH}\,(\mathbf{EH})^H]^{-1} \tag{11.70}$$

Denote the SNR at the transmitter ρ_{ZF} to represent the available transmit power when noise variance (power) is a normalized one, then the transmit power ρ_{ZF} is given as

$$\beta^2_{\mathrm{ZF}}\,\mathrm{tr}(\mathbf{F}_{\mathrm{ZF}}\,\mathbf{R_s}\mathbf{F}^{\mathrm{H}}_{\mathrm{ZF}}) \equiv \rho_{ZF}, \tag{11.71}$$

where $\mathbf{R_s} = \sigma^2_{\mathbf{s}}\,\mathbf{I}_{\mathrm{B}}$. Therefore, from (11.70) and (11.71) the scaling factor β is derived in Appendix 11.A as

$$\beta_{ZF} = \sqrt{\frac{\rho_{ZF}}{\mathrm{tr}(\mathbf{EH}\,(\mathbf{EH})^{\mathrm{H}})^{-1}\,\mathbf{R_s}}} \tag{11.72}$$

Optimizing the transmit filter **F** can be obtained using an alternative design based on maximization of the SINR replacing the MSE minimization as described in detail in [18].

When equal transmit power is allocated to each stream and assume received noise per each stream is unity, then using (11.72), the received SINR per stream-γ_k^{ZFP} is

$$\gamma_k^{ZFP} = \frac{\beta^2_{ZF}}{interference} = \frac{\rho_{ZF}}{\mathrm{tr}[(\mathbf{EH}\,\mathbf{R_s}\,(\mathbf{EH})^{\mathrm{H}})^{-1}\,]} \tag{11.73}$$

11.9.3 ZF Receive Filter E Matrix

Next, we derive the receive ZF filter matrix **E**, assuming that the constraint that $\tilde{\mathbf{s}}$ is an interference free estimate of **s** is still valid:

$$\tilde{\mathbf{s}}|_{\mathbf{n=0}} = \mathbf{EHFs} \tag{11.74a}$$

However, **s** is unknown to the receiver but the receive filter must affirm that the **F** transmit filter, the channel **H** together with receive filter **E**, have to achieve an identity mapping as expressed by (11.68b). The constraint in (11.68b) can be accomplished at the receiver by minimising the MSE as

$$\mathbb{E}[\|\mathbf{s} - \tilde{\mathbf{s}}\|_2^2] = \mathbb{E}[\|\mathbf{En}\|_2^2] \tag{11.74b}$$

Accordingly, **E** minimises (11.74b) and removes the interference to realize a receive filter optimization problem as

$$\mathbf{E}_{ZF} = \arg\min_{\mathbf{E}}\mathbb{E}[\|\mathbf{En}\|_2^2] \tag{11.75a}$$

$$\text{s.t } \mathbf{EHF} = \mathbf{I}_{\mathrm{B}} \tag{11.75b}$$

Using the Lagrange multipliers in an appropriate optimisation method that satisfies the constraint in (11.75b), we get the receive precoding filter $\mathbf{E} \in \mathbb{C}^{\mathrm{B\times N}}$ in the next equation:

$$\mathbf{E}_{ZF} \in \mathbb{C}^{\mathrm{B\times N}} = [(\mathbf{HF})^H\,\mathbf{R_n^{-1}}\,\mathbf{HF}]^{-1}\,(\mathbf{HF})^H\,\mathbf{R_n^{-1}} \tag{11.76}$$

It is worth noting that (11.76) is given by rotate left of (11.69a) when **F** is replaced with **E** and, of course, $\mathbf{R_n} = \sigma_{\mathbf{n}}^2 \mathbf{I}_N$ at the transmitter.

11.9.4 ZF Outage Probability for Minimum Transmit Power

We define the outage Pr_{out} as the probability that the sum rate for all streams is less than or equal to the predefined rate R such that

$$\mathrm{Pr}_{out} := \Pr(\mathrm{N}\log_2(1+\gamma_k) \leq R \tag{11.77}$$

Substituting (11.73) into (11.77), we get

$$\mathrm{Pr}_{out} = \Pr\left(\mathrm{N}\log_2\left(1+\frac{\rho_{ZF}}{\mathrm{tr}[(\mathbf{EHR_s}\,(\mathbf{EH})^{\mathbf{H}})^{-1}\,]}\right) \leq R\right) \tag{11.78}$$

Since $\mathbf{R_s}$ is constant covariance of the transmitted symbols **s**, Substitute the eigenvalues $\{\lambda_k\}$ of $\mathbf{EHR_s}\,(\mathbf{EH})^{\mathbf{H}}$ in (11.78) as follows:

$$\mathrm{tr}(\,[(\mathbf{EHR_s}\,(\mathbf{EH})^{\mathbf{H}})^{-1}]) = \sum_{k=1}^{\mathrm{N}} \frac{1}{\lambda_k} \tag{11.79}$$

The outage Pr_{out} simplified to

$$\mathrm{Pr}_{out} = \Pr\left(\mathrm{B}\log_2\left(1+\frac{\rho_{ZF}}{\sum_{k=1}^{\mathrm{N}}\frac{1}{\lambda_k}}\right) \leq R\right) \tag{11.80}$$

11.9.5 ZF Precoder Design to Allocate Unequal Power

ZF precoding design allocates unequal transmit power across the transmit antennas to optimise the performance while satisfying the total transmit power constraint. The optimization problem of the system throughput can be expressed mathematically as

$$\max_{p_{k\geq 0},F} \sum_k \log_2(1+\gamma_k^{ZFP}) \tag{11.81a}$$

$$\text{s.t:}\quad \mathbf{HF}_{ZF} = \mathrm{diag}\{\sqrt{p_1},\dots\dots\dots,\sqrt{p_{\mathrm{N}}}\} \tag{11.81b}$$

$$\mathbb{E}(\|\mathbf{Fs}\|^2) \leq \rho \tag{11.81c}$$

where p_k is the transmit power for k^{th} stream. From (11.50b), we have the constraint can be simplified to the following expression:

$$\mathbf{F}_{ZF} = (\mathbf{H}^{\dagger})\,\mathrm{diag}(\sqrt{\mathbf{P}}) \tag{11.82a}$$

Substituting pseudo-inverse channel $\mathbf{H}^{\dagger}$ into (11.82a), we get

$$\mathbf{F}_{ZF} = \mathbf{H}^{\mathrm{H}}(\mathbf{HH}^{\mathrm{H}})^{-1}\,\mathrm{diag}(\sqrt{\mathbf{P}}) \tag{11.82b}$$

where $\sqrt{\mathbf{P}} = (\sqrt{p_1},\dots\dots\dots,\sqrt{p_k},\dots\dots\dots\dots,\sqrt{p_{\mathrm{B}}})$

Consider the transmit precoding filter **F** and the receive equalizing filter **E** to modify (11.82b) to

$$\mathbf{F}_{ZF} = (\mathbf{EH})^{\mathrm{H}}[(\mathbf{EH})(\mathbf{EH})^{\mathrm{H}}]^{-1}\,\mathrm{diag}(\sqrt{\mathbf{P}}) \tag{11.83}$$

$$\mathrm{tr}(\mathbf{FF}^{\mathrm{H}}) = \mathrm{tr}[[(\mathbf{EH})(\mathbf{EH})^{\mathrm{H}}]^{-1}]\mathrm{diag}(\mathbf{P}) \tag{11.84}$$

where we assume $\mathbb{E}(\boldsymbol{ss}^H) = \mathbf{I}$. Let p_k be the transmit power for the kth stream. Then constraint (11.81c) using (11.84) becomes

$$\sum_{k=1}^{\mathrm{M}} \mathrm{p}_k[\mathbf{H}^{\dagger}(\mathbf{H}^{\dagger})^{\mathrm{H}}]_{kk} = \sum_{k=1}^{\mathrm{M}} \mathrm{p}_k[(\mathbf{EH})(\mathbf{EH})^{\mathrm{H}}]_{kk}^{-1} \leq \rho \tag{11.85}$$

The optimal p_k is the solution to the following optimization problem:

$$\max_{p_k} \sum_{k} \log_2(1 + \gamma_k^{ZFP}) \tag{11.86a}$$

$$\text{s.t}: \quad \sum_{\mathrm{k}} \mathrm{p}_k[(\mathbf{EH})(\mathbf{EH})^{\mathrm{H}}]_{kk}^{-1} \leq \rho \tag{11.86b}$$

We use γ_k^{ZFP} is the SINR at the kth receiver output where transmitted power $p_k$is cleared out interference and channel is equalized. If we assumed unit noise power then $\gamma_k^{ZFP} = p_k$. This optimization problem we are considering is a logarithmic function, similar to the familiar optimization problem of a water-filling power allocation. Logarithmic optimization may drive some instantaneous p_k to zero depending on the ρ and values of $(\mathbf{EH})^H$. The optimal solution is derived in Appendix 11.B as

$$p_k = \frac{\rho + \sum_k [\mathbf{EH}(\mathbf{EH})^H]_{kk}^{-1}}{\mathrm{N}[\mathbf{EH}(\mathbf{EH})^H]_{kk}^{-1}} - 1 \quad \mathrm{k} = 1, \ldots \ldots .., \mathrm{M} \tag{11.87}$$

11.9.6 ZF Outage Probability for Unequal Power Allocation across Transmit Antennas

Since water-filling may derive some of p_k to zero, so depending on the realization of $\mathbf{EH}(\mathbf{EH})^H$ it may also happen that all optimal p_k are positive. The set of $\mathbf{EH}(\mathbf{EH})^H$ realizations that satisfies the optimal p_k to be positive are collected into an event denoted $\mathcal{P}$ then conditioned on $\mathcal{P}$ occurs with probability $\Pr(\mathcal{P})$ the outage probability is given as [18]

$$\Pr_{out} = \Pr\left(\sum_{k=1}^{\mathrm{N}} \log_2(1 + p_k) < R|\mathcal{P}\right)\Pr(\mathcal{P}) + \Pr\left(\sum_{k=1}^{\mathrm{N}} \log_2(1 + p_k) < R|\overline{\mathcal{P}}\right)\Pr(\overline{\mathcal{P}}) \tag{11.88}$$

where event $\overline{\mathcal{P}}$ is occurs with probability

$$\Pr(\overline{\mathcal{P}}) = 1 - \Pr(\mathcal{P})$$

Substituting (11.87) into (11.88), we get

$$\begin{aligned}\Pr_{out} = {} & \Pr\left(\sum_{k=1}^{\mathrm{N}} \log_2\left(\frac{\rho + \sum_k [\mathbf{EH}\,(\mathbf{EH})^H]_{kk}^{-1}}{\mathrm{M}[\mathbf{EH}\,(\mathbf{EH})^H]_{kk}^{-1}}\right) \leq R|\mathcal{P}\right)\Pr(\mathcal{P}) \\ & + \Pr\left(\sum_{k=1}^{\mathrm{N}} \log_2\left(\frac{\rho + \sum_k [\mathbf{EH}\,(\mathbf{EH})^H]_{kk}^{-1}}{\mathrm{M}[\mathbf{EH}\,(\mathbf{EH})^H]_{kk}^{-1}}\right) \leq R|\overline{\mathcal{P}}\right)\Pr(\overline{\mathcal{P}})\end{aligned} \tag{11.89}$$

Assuming equal share of events $\mathcal{P}$ and $\overline{\mathcal{P}}$ (i.e. 50 : 50), we can bound the outage probability in (11.89) as

$$\Pr_{out} \geq \Pr\left(\sum_{k=1}^{\mathrm{N}} \log_2\left(\frac{\rho + \sum_k [\mathbf{EH}\,(\mathbf{EH})^H]_{kk}^{-1}}{\mathrm{M}[\mathbf{EH}\,(\mathbf{EH})^H]_{kk}^{-1}}\right) \leq R\right) \tag{11.90}$$

11.10 Matched Filter (MF) Precoding

11.10.1 Transmit MF F Matrix

In this subsection we derive the transmit matched filter (MF) matrix **F** assuming the transmitter is a priori knows the constant receive filter **E** in addition to the instantaneous channel matrix **H**. We use the precoding model illustrated in Figure 11.5. Conventionally, the MF filter is located at the detection part of the communication system to manage the noise and maximize SNR at the receiver output. It was proposed in [19, 20] to move the MF filter from receive to the transmit part of the system to provide pre-RAKE diversity and improve the performance of SISO in the multiple paths environment signal. The concept is developed and adapted by [17] for applications as the multi-user MIMO channel precoding, since the MS receiver poses a limited processing capability. Unlike the receive MF, the transmit MF maximizes the SNR at the MS receiver by maximizing the user transmit power. Usually, the optimization process entails using Lagrangian multipliers and Karush–Kuhn–Tucker (KKT) conditions, instead, we maximize the received desired signal and introduce a constraint, which sets the transmit power of every transmitted signal to a MS specific value. The desired signal in the received signal $\tilde{\mathbf{s}}$ is selected by correlating with $\mathbf{s}^H$:

$$\mathbf{F}_{\mathrm{MF}} = \arg\max_{\mathbf{F}} \frac{|\mathbb{E}[\mathbf{s}^H\,\tilde{\mathbf{s}}]|^2}{\mathbb{E}[\|\mathbf{En}\|_2^2]} \tag{11.91a}$$

$$s.t \quad \mathbb{E}[\|\mathbf{Fs}\,\|_2^2 \leq \rho \tag{11.91b}$$

where **En** is the interference including Gaussian noise, $\tilde{\mathbf{s}}$ is the noiseless received signal that is

$$\tilde{\mathbf{s}} = \mathbf{EHF}\boldsymbol{s} \tag{11.91c}$$

Substitute (11.91c) in (11.91a) we get

$$\mathbf{F}_{\mathrm{MF}} = \arg\max_{\mathbf{F}} \frac{\mathbb{E}[\|\mathbf{s}^H\mathbf{EHFs}\,\|_2^2]}{\mathbb{E}[\|\mathbf{En}\|_2^2]} \tag{11.92}$$

The solution to (11.92) is derived in 11.C as

$$\mathbf{F}_{\mathrm{MF}} \in \mathbb{C}^{\mathrm{M\,x\,B}} = \beta_{\mathrm{MF}}\,(\mathbf{EH})^H \tag{11.93a}$$

where

$$\beta_{MF} = \sqrt{\frac{\rho}{\mathrm{tr}((\mathbf{EH})^H\mathbf{R_s}\,(\mathbf{EH})}} \tag{11.93b}$$

11.10.2 Receive MF E Matrix

The MF receiver does not consider the interference but maximizes the SNR at the filter **E** output. In deriving the matrix fraction of the received signal **E**, we segregate the desired signal in the estimate ($\tilde{\mathbf{s}}$) by correlation as we did in the transmit MF filter but now we do not impose any constraint on the transmit power as it is already satisfied at the transmit filter that is

$$\mathbf{E}_{\mathrm{MF}} = \arg\max_{\mathbf{E}} \frac{|\mathbb{E}[\mathbf{s}^H\,\tilde{\mathbf{s}}]|^2}{\mathbb{E}[\|\mathbf{En}\|_2^2]} \tag{11.94}$$

The solution to the optimization problem (11.94) can be attained by setting the derivation of the cost function $f(\mathbf{E})$ with respect to **E** to zero:

$$|\mathbb{E}[\mathbf{s}^H\tilde{\mathbf{s}}]|^2 = |\mathbb{E}[\mathbf{s}^H\mathbf{EHFs}\,]|^2 = |\,\mathbf{EHFR_s}|^2 \tag{11.95}$$

$$f(\mathbf{E}) = \frac{|\,\mathbf{EHFR_s}|^2}{\mathbf{E}^H\mathbf{E}\,\mathbf{R_n}} \tag{11.96}$$

The solution of the optimization in (11.94) is given by

$$\mathbf{E}_{\mathrm{MF}} \in \mathbb{C}^{\mathrm{B}\times\mathrm{N}} = \mathbf{R_s}\,(\mathbf{HF})^H\,\mathbf{R_n^{-1}} \tag{11.97}$$

The received signal is

$$\mathbf{y} = \mathbf{E}_{\mathrm{MF}}(\mathbf{HF}_{\mathrm{MF}}\,\mathbf{s} + \mathbf{n}) = \beta_{\mathrm{MF}}\mathbf{E}_{\mathrm{MF}}\mathbf{H}\,(\mathbf{EH})^H\mathbf{s} + \mathbf{E}_{\mathrm{MF}}\mathbf{n}) \tag{11.98a}$$

$$\mathbf{y} = \beta_{MF}\mathbf{E}_{\mathrm{MF}}\mathbf{H}\,\mathbf{H}^H\mathbf{E}_{\mathrm{MF}}^H\mathbf{s} + \mathbf{E}_{\mathrm{MF}}\mathbf{n} \tag{11.98b}$$

where $\mathbf{E}_{\mathrm{MF}}\mathbf{H} \in \mathbb{C}^{\mathrm{B}\times\mathrm{M}}$ and $\mathrm{M} \geq \mathrm{N}$ for massive MIMO and since $\mathrm{B} = \min(\mathrm{M}, \mathrm{N})$ then we assume $\mathrm{B} = \mathrm{N}$. Accordingly, $\mathbf{E}_{\mathrm{MF}}\mathbf{H} \in \mathbb{C}^{\mathrm{N}\times\mathrm{M}}$ is not a square matrix and we apply SVD to get

$$\mathbf{E}_{\mathrm{MF}}\mathbf{H} = \mathbf{U}\,\boldsymbol{\Gamma}\,\mathbf{V}^H \tag{11.99a}$$

$$\mathbf{E}_{\mathrm{MF}}\mathbf{H}\,(\mathbf{E}_{\mathrm{MF}}\mathbf{H})^H = \mathbf{U}\,\boldsymbol{\Gamma}\,\mathbf{V}^H\,\mathbf{V}\boldsymbol{\Gamma}^T\,\mathbf{U}^H = \mathbf{U}\boldsymbol{\Lambda}\,\mathbf{U}^H \tag{11.99b}$$

where $\mathbf{U} \in \mathbb{C}^{\mathrm{N}\times\mathrm{N}}$ square complex unitary matrix, $\boldsymbol{\Gamma} \in \mathbb{R}^{\mathrm{N}\times\mathrm{M}}$ rectangular real matrix with non-negative elements, $\mathbf{V} \in \mathbb{C}^{\mathrm{M}\times\mathrm{M}}$ square complex unitary matrix and $\boldsymbol{\Lambda} \in \mathbb{R}^{\mathrm{N}\times\mathrm{N}} = \boldsymbol{\Gamma}\,\boldsymbol{\Gamma}^T$ is a real matrix whose elements are the eigenvalues of $\mathbf{EH}\,(\mathbf{EH})^H$, which a square matrix. Then (11.98b) simplifies to Eq. (11.100):

$$\mathbf{y} = \beta_{MF}\mathbf{U}\,\boldsymbol{\Lambda}\,\mathbf{U}^H\mathbf{s} + \mathbf{E}_{\mathrm{MF}}\mathbf{n} \tag{11.100}$$

where $\mathbf{E}_{\mathrm{MF}} \in \mathbb{C}^{\mathrm{B}\times\mathrm{N}}$ complex rectangular matrix is given by

$$\mathbf{E}_{\mathrm{MF}} = \begin{bmatrix} e_{1,1} & - & \mathrm{e}_{1,\mathrm{k}} & - & - & \mathrm{e}_{1,\mathrm{N}} \\ - & - & - & - & - & - \\ - & - & - & - & - & - \\ - & - & - & - & - & - \\ e_{\mathrm{B},1} & - & \mathrm{e}_{\mathrm{B},\mathrm{k}} & - & - & \mathrm{e}_{\mathrm{B},\mathrm{N}} \end{bmatrix} \tag{11.101}$$

The received signal at the kth antenna is

$$\mathrm{y_k} = \beta_{\mathrm{MF}}\left(\sum_{l=1}^{\mathrm{N}} \lambda_l\,|u_{kl}|^2\right)\mathrm{s_k} + \beta_{\mathrm{MF}}\sum_{i=1,i\neq k}^{\mathrm{N}}\left(\sum_{l=1}^{\mathrm{N}} \lambda_{\mathrm{k}}\,u_{kl}\,u_{il}\right)\mathrm{s}_i + \mathrm{n}_k \tag{11.102}$$

where $\mathrm{n}_k = [\mathbf{E}_{\mathrm{MF}}\,\mathbf{n}]_k$ and $\sigma_{\mathrm{k}}^2 = \mathbb{E}[\,\mathrm{n}_k\,\mathrm{n}_k^H]$. The SINR γ_k at the kth antenna is

$$\gamma_k = \frac{\beta_{\mathrm{MF}}^2\,\frac{\rho}{\mathrm{N}}\left(\sum_{l=1}^{\mathrm{N}} \lambda_l\,|u_{kl}|^2\right)^2}{\beta_{\mathrm{MF}}^2\,\frac{\rho}{\mathrm{N}}\sum_{i=1,i\neq k}^{\mathrm{N}}\left|\sum_{l=1}^{\mathrm{N}} \lambda_{\mathrm{k}}\,u_{kl}\,u_{il}^*\right|^2 + \sigma_{\mathrm{k}}^2} \tag{11.103a}$$

We can simplify (11.103a) to

$$\gamma_k = \frac{\left(\sum_{l=1}^{\mathrm{N}} \lambda_l\,|u_{kl}|^2\right)^2}{\sum_{i=1,i\neq k}^{\mathrm{N}}\left|\sum_{l=1}^{\mathrm{N}} \lambda_{\mathrm{k}}\,u_{kl}\,u_{il}^*\right|^2 + \sigma_{\mathrm{k}}^2\,\rho^{-1}\,\mathrm{N}\,\beta_{\mathrm{MF}}^{-2}}, \tag{11.103b}$$

where β_{MF} is given in (11.93b) as

$$\beta_{\mathrm{MF}} = \sqrt{\frac{\rho}{\mathrm{tr}((\mathbf{EH})^H\mathbf{R_s}\,(\mathbf{EH})}}$$

The outage probability pr_{out} for the sum rate for all streams is less or equal as predefined rate R is given by

$$\mathrm{pr}_{out} = \Pr\left(\sum_{k=1}^{N} \log_2(1+\gamma_k) \le R\right) \tag{11.104}$$

where γ_k is given in (11.103b).

11.11 Wiener Filter (WF) Precoding

11.11.1 Transmit WF F Matrix

A plot of the MSE vs. SNR for the transmit MF and transmit ZF filter are depicted in Figure 11.6. Under high SNR, typically SNR ≥ 30 dB, the MSE of the transmit MF filter tends to a limiting value greater than that akin to the transmit ZF filter. However, at low SNR, 0 ≤ SNR ≤ 30 dB, the transmit MF filter MSE is much lower than that corresponding to the transmit ZF. Similar dependence on the SNR can be seen at the respective receive precoding filters. The receive Wiener filter (WF) finds the optimum balance between the signal maximization of the receive MF and the interference eradication of the receive ZF respective filters because the MSE of the receive WF is always smaller than the MSEs of the both receive filters.

However, unlike the transmit MF and the transmit ZF, the noise covariance $\mathbf{R_n}$ in addition to the matrices of $\mathbf{H}$, $\mathbf{E}$, and $\mathbf{R_s}$ have to be known prior to transmitting for the design of the transmit WF. The transmit WF precoding is derived from the optimisation problem that minimises the MSE between the information transmitted $\boldsymbol{s}$ and its received estimate that is

$$\mathbf{F}_{\mathrm{WF},\beta} = \mathrm{argmin}_{\mathbf{F}_{\mathrm{WF},\beta}} \mathbb{E}(\|\boldsymbol{s} - \beta^{-1}\tilde{\mathbf{s}}\|_2^2) \tag{11.105a}$$

$$\text{s.t} \qquad \mathbb{E}[\|\mathbf{Fs}\|_2^2] = \rho \tag{11.105b}$$

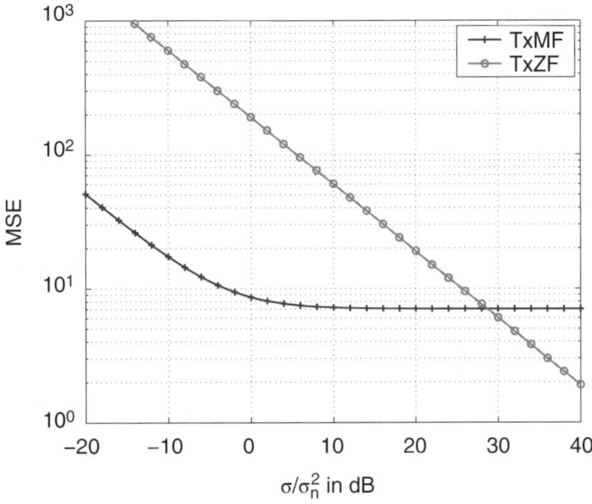

Figure 11.6 MSE of transmit MF and transmit ZF vs. SNR [19].

The scaling factor β has to compose the amplitude of the desired fraction of the received signal attainable to encounter the ramification of the noise. The solution to the optimisation problem expressed in (11.105) can be determined by formulating the Lagrangian cost function L(**F**, β, λ) as described in the next articulation:

$$L(\mathbf{F},\beta,\lambda) = \mathbb{E}(\|s - \beta^{-1}\tilde{s}\|_2^2) + \lambda\{(\mathrm{tr}(\mathbf{F}\,\mathbf{R}_s\,\mathbf{F}^H) - \rho\} \tag{11.106}$$

We differentiate the Lagrangian function in (11.106) with respect **F**, β, λ and comparing the derivation to zero as described in 11.D to get

$$\mathbf{F}_{\mathrm{WF}} \in \mathbb{C}^{\mathrm{M}\times\mathrm{B}} = \beta_{\mathrm{WF}}\mathbf{F}^{-1}(\mathbf{EH})^H \tag{11.107a}$$

where

$$\mathbf{F} \in \mathbb{C}^{\mathrm{M}\times\mathrm{M}} = \left((\mathbf{EH})^H\,\mathbf{EH} + \frac{\mathrm{tr}(\mathbf{E}\,\mathbf{R}_\mathbf{n}\,\mathbf{E}^H)}{\rho}\,\mathbf{I}_\mathrm{M}\right) \tag{11.107b}$$

$$\beta_{WF} = \sqrt{\frac{\rho}{\mathrm{tr}[\mathbf{F}_{\mathrm{WF}}^{-2}\,(\mathbf{EH})^H\,\mathbf{R}_s\,(\mathbf{EH})]}} \tag{11.107c}$$

11.11.2 Receive WF Matrix

The receive WF minimizes the MSE at the receive WF filter matrix **E** without any constraint, as described in the next formulation of the optimization problem:

$$\mathbf{E}_{\mathrm{WF}} = \mathrm{argmin}_{\mathbf{E}_{\mathrm{WF}}}\,\mathbb{E}(\|s - \tilde{s}\|_2^2) \tag{11.108}$$

The solution to the optimization problem expressed in (11.108) can be found by differentiating (11.108) with respect to $\mathbf{E}_{\mathrm{WF}}$ and equating the derivative to zero to get

$$\mathbf{E}_{\mathrm{WF}} = \{[(\mathbf{HF})^H\,\mathbf{R}_\mathbf{n}^{-1}\,(\mathbf{HF}) + \mathbf{R}_\mathbf{s}^{-1}]^{-1}\,(\mathbf{HF})^H\,\mathbf{R}_\mathbf{n}^{-1}\} \tag{11.109}$$

Equation (11.109) reflects the interdependence of the receive $\mathbf{E}_{\mathrm{WF}}$ on the $\mathbf{F}_{\mathrm{WF}}$, as highlighted in [17]. As the SNR deceases below a certain value, the first term in (11.109) (the inverse term) decreases compared to the second term and the receive WF converges to the receive MF. On the contrary, as SNR increases above the SNR threshold, the second term can be neglected and the receive WF converges to the receive ZF that is $\mathbf{E}_{\mathrm{WF}} \to \mathbf{E}_{\mathrm{MF}}$ as SNR decreases and $\mathbf{E}_{\mathrm{WF}} \to \mathbf{E}_{\mathrm{ZF}}$ as SNR increases above the threshold. The received Wiener precoded signal is

$$\mathbf{y} = \mathbf{E}_{\mathrm{WF}}\{\mathbf{F}_{\mathrm{WF}}\,\mathbf{H}\,\mathbf{s} + \mathbf{n}\} \tag{11.110}$$

$$\mathbf{y} = \mathbf{E}_{\mathrm{WF}}\{\beta_{\mathrm{WF}}\,\mathbf{F}_{WF}\,\mathbf{H}\,\mathbf{s} + \mathbf{n}\}$$

$$\mathbf{y} = \mathbf{E}_{\mathrm{WF}}\beta_{\mathrm{WF}}\mathbf{F}^{-1}(\mathbf{EH})^H\,\mathbf{H}\,\mathbf{s} + \mathbf{E}_{\mathrm{WF}}\mathbf{n}$$

The SINR of the transmit WF precoding at the kth receive antenna is

$$\gamma_k = \frac{\beta_{\mathrm{WF}}^2\,|(\mathbf{F}_{\mathrm{WF}}\,\mathbf{H})_{kk}|^2}{\beta_{\mathrm{WF}}^2\,\sum_{i=1,i\neq k}^{\mathrm{N}}|(\mathbf{E}_{\mathrm{WF}}\mathbf{F}^{-1}(\mathbf{EH})^H\,\mathbf{H})_{ki}|^2 + \sigma_\mathrm{k}^2} \tag{11.111a}$$

where $\mathrm{n}_k = [\mathbf{E}_{\mathrm{WF}}\,\mathbf{n}]_k$ and $\sigma_\mathrm{k}^2 = \mathbb{E}[(\mathbf{E}_{\mathrm{WF}}\,\mathbf{n})\,(\mathbf{E}_{\mathrm{WF}}\,\mathbf{n})^H]_k$

$$\gamma_k = \frac{|(\mathbf{F}_{WF}\,\mathbf{H})_{kk}|^2}{\sum_{i=1,i\neq k}^{\mathrm{N}}|(\mathbf{F}_{\mathrm{WF}}\,\mathbf{H})_{ki}|^2 + \sigma_\mathrm{k}^2\,\beta_{\mathrm{WF}}^{-2}} \tag{11.111b}$$

where

$$\beta_{\mathrm{WF}}^{-2} = \frac{1}{\rho}\,\mathrm{tr}[\mathbf{F}_{\mathrm{WF}}^{-2}\,(\mathbf{EH})^H\,\mathbf{R}_s\,(\mathbf{EH})] \tag{11.111c}$$

The outage pr_{out} probability that the sum rate for all streams is less or equal the predefined rate R such that

$$pr_{out} = pr\left(\sum_{k=1}^{N} \log_2(1+\gamma_k) \le R\right) \tag{11.112}$$

where γ_k is given in (11.117b).

11.12 Regularized Zero-Forcing (RZF) Precoding

When channel matrix is known at the transmitter, we have shown that simple inversion of the matrix channel avows the received signal to be separated into independent user streams. When the channel matrix is known at the transmitter, ZF forces the interference at the receiver to null. We have ascertained that ZF performs inadequately for any number of users except in scenarios with high SNRs.

Authors in [21, 22] proposed regularising the ZF precoding at the transmitter by a scaled identity matrix before transmission is attained. The authors conducted a thorough analysis to find the regularisation parameter that allows a certain amount of interference remained alive to maximize the SINR at each user receive end. Accordingly, the regularised ZF can accomplish considerable improvement in performance and achieves near sum capacity performance. The regularisation concept introduced a new precoding that can be considered as a modification of ZF precoding. Such a simple adjustment to the ZF helps to achieve a sum rate that develops linearly with the minimum number of antennas contradictory to the sum rate produced by the channel inversion. Further the regularized form of ZF improves performance particularly at low SNR. The numerical evaluation of the regularized ZF carried out by [21] showed as a number of antennas and user number increased simultaneously, the ergodic sum capacity of the DL grows linearly with the min (M, K), the sum rate of ZF tends to $\frac{\log_2 e}{\sigma^2}$, and sum rate of RZF increases linearly.

One method that can be used to regularise the channel inversion is to introduce a multiplier of the identity matrix. Towards a further understanding of the regularisation concept, we articulate the regularisation precoding process without including the receive side processing. Let us consider a simple system model comprising a single cell with a BS supported by M antennas and K single antenna users served at the same data rate and assume that $\mathrm{M} = K$. Define the transmitted vector $\mathbf{x} \in \mathbb{C}^{\mathrm{M}\times 1}$ and the users set of symbols $\mathbf{s} \in \mathbb{C}^{\mathrm{M}\times 1}$. Denote the regularised ZF precoding filter matrix $\mathbf{F}_{\mathrm{RZF}}$ given by modifying (11.69b):

$$\mathbf{F}_{\mathrm{RZF}} \in \mathbb{C}^{\mathrm{M}\times \mathrm{B}} = \mathbf{H}^{H}\,[\mathbf{H}\mathbf{H}^{H} + \alpha\mathbf{I}_K]^{-1} \tag{11.113}$$

where α is a constant added to retain some of the multi-user interference to improve the received signal. The transmitted signal through the channel becomes

$$\mathbf{H}\mathbf{F}_{\mathrm{RZF}}\mathbf{s} = \mathbf{H}\,\mathbf{H}^{H}\,[\mathbf{H}\mathbf{H}^{H} + \alpha\mathbf{I}_K]^{-1}\,\mathbf{s} \tag{11.114}$$

Next we assess the amount of the desired signal and the interference powers that are generated by using the decomposition. The SVD of $\mathbf{H}$ is given as

$$\mathbf{H} = \mathbf{U}\boldsymbol{\Gamma}\,\mathbf{V}^{H}$$

then it can be shown that

$$\mathbf{H}\,\mathbf{H}^{H} = \mathbf{U}\,\boldsymbol{\Lambda}\,\mathbf{U}^{H} \tag{11.115}$$

where $\mathbf{U} \in \mathbb{C}^{K\times K}$, $V \in \mathbb{C}^{\mathrm{M}\times \mathrm{M}}$ are unitary matrices (that is $\mathbf{U}^{H} = \mathbf{U}^{-1}, \mathbf{V}^{H} = \mathbf{V}^{-1}$):

$\mathbf{\Gamma} \in \mathbb{R}^{K\times M}$ and $\mathbf{\Lambda} = \mathbf{\Gamma}\,\mathbf{\Gamma}^T \in \mathbb{R}^{K\times K}$ diagonal matrix with elements being eigenvalues of $\mathbf{H}^H$. Using the composition in (11.115), the transmited signal in (11.114) becomes

$$\begin{aligned}\mathbf{HF}_{\text{RZF}}\mathbf{s} &= \mathbf{U}\,\mathbf{\Lambda}\,\mathbf{U}^H\ (\mathbf{U}\,\mathbf{\Lambda}\,\mathbf{U}^H + \alpha\mathbf{I}_K)^{-1}\boldsymbol{s}\\ &= \mathbf{U}\frac{\mathbf{\Lambda}}{(\mathbf{\Lambda}+\alpha\mathbf{I}_K)}\mathbf{U}^H\boldsymbol{s}\ (\mathbf{\Lambda}+\alpha\mathbf{I}_K)^{-1}\mathbf{U}^H\boldsymbol{s}\\ &= \mathbf{U}\frac{\mathbf{\Lambda}}{(\mathbf{\Lambda}+\alpha\mathbf{I}_K)}\mathbf{U}^H\boldsymbol{s}\end{aligned} \tag{11.116}$$

where $\mathbf{\Lambda}$ diagonal matrix with element are eigenvalues of $\mathbf{H}\,\mathbf{H}^H$.

$$\mathbf{\Lambda} = \begin{pmatrix} \lambda_1 & 0 & - & - & - & 0\\ 0 & \lambda_2 & - & - & - & 0\\ - & - & - & - & - & -\\ - & - & - & - & - & -\\ - & - & - & - & - & -\\ 0 & - & - & - & - & \lambda_K \end{pmatrix} \tag{11.117}$$

The (unnormalised) signal and interference received by k^{th} user is the kth entry of (11.116) given as

$$[\mathbf{HF}_{\text{RZF}}\mathbf{s}\,]_k = \left[u_{k,1}\frac{\lambda_1}{\lambda_1+\alpha}\ \ - - -\ \ u_{k,K}\frac{\lambda_K}{\lambda_K+\alpha}\right]\begin{bmatrix} u^*_{1,1} & - & u^*_{k,1} & - & - & u^*_{K,1}\\ - & - & - & - & - & -\\ - & - & - & - & - & -\\ - & - & - & - & - & -\\ - & - & - & - & - & -\\ u^*_{1,K} & - & u^*_{k,K} & - & - & u^*_{K,K} \end{bmatrix}\begin{bmatrix} \mathrm{s}_1\\ -\\ \mathrm{s}_k\\ -\\ -\\ \mathrm{s}_K \end{bmatrix} \tag{11.118}$$

where $u_{l,\,\mathrm{k}}$ is the $(l,\,\mathrm{k})^{th}$ element of the matrix $\mathbf{U}$.

The (un-normalised) k^{th} user desired signal given by (11.118) as

$$\left(\sum_{l=1}^{K}\frac{\lambda_l}{\lambda_l+c}\,|u_{k,l}|^2\right)\mathrm{s}_\mathrm{k} \tag{11.119}$$

All of the remaining term in (11.119) including $(l \neq k)$ is interference.

The kth user normalised received signal is

$$y_k = \frac{1}{\sqrt{\rho}}\left(\sum_{l=1}^{K}\frac{\lambda_l}{\lambda_l+c}\,|u_{k,l}|^2\right)\mathrm{s}_\mathrm{k} + n'$$

where n' intermixes interference and Gaussian noise. The average normalisation factor ρ for the precoded signal is derived now. From (11.114) we have

$$\mathbb{E}[\rho] = \mathbb{E}\|\mathbf{F}_{RZF}\,\mathbf{s}\|^2 = \mathbb{E}\|\mathbf{F}_{RZF}\,\|^2\,\mathbb{E}\|\,\mathbf{s}\|^2 = \mathbb{E}\|\mathbf{F}_{RZF}\,\|^2 \tag{11.120a}$$

Since we assumed $\mathbb{E}(s_k s_k^H) = \mathbf{I}_K$, we get

$$\mathbb{E}[\rho] = \mathbb{E}[(\mathbf{H}^H\,[\mathbf{HH}^H + \alpha\mathbf{I}_K]^{-1})(\mathbf{H}^H\,[\mathbf{HH}^H + \alpha\mathbf{I}_K]^{-1})^H]$$

Substituting expression (11.115) for $\mathbf{HH}^H$, it can easily be shown that

$$\mathbb{E}[\rho] = \mathrm{tr}\left[\frac{\mathbf{\Lambda}}{(\mathbf{\Lambda}+\alpha\,\mathbf{I}_K)}\right] = \sum_{l=1}^{K}\frac{\lambda_l}{(\lambda_l+\alpha)^2} \tag{11.120b}$$

Accordingly, the scale factor β_{RZF} is given by,

$$\beta_{RZF} = \frac{1}{\sqrt{\rho}} = \frac{1}{\sqrt{\sum_{l=1}^{K} \frac{\lambda_l}{(\lambda_l+\alpha)^2}}} \tag{11.121}$$

It can be shown that the expected total power is

$$\mathbb{E}\|\mathbf{HF}_{\mathrm{RZF}}\mathbf{s}\,\|^2 = \mathbb{E}\left[\mathbf{U}\left(\frac{\boldsymbol{\Lambda}}{(\boldsymbol{\Lambda}+\alpha\mathbf{I}_K)}\right)^2\mathbf{U}^H\right] \tag{11.122}$$

$$\mathbb{E}\|\mathbf{HF}_{\mathrm{RZF}}\mathbf{s}\,\|^2 = \sum_{k=1}^{K} \frac{\lambda_k^2}{(\lambda_k+\alpha\,)^2} \tag{11.123}$$

The kth user desired power can be derived from (11.121) as

$$\mathbb{E}\left[\left(\sum_{l=1}^{K} \frac{\lambda_l}{\lambda_l+\alpha}\,|u_{k,l}|^2\right)^2\right] \tag{11.124a}$$

Expression (11.124a) can be simplified [7] to,

$$= \beta^2\,\frac{1}{\mathrm{K(K+1)}}\left[\left[\left(\sum_{l=1}^{K}\frac{\lambda_l}{\lambda_l+\alpha}\right)^2\right] + \frac{1}{\mathrm{k}}\left[\sum_{l=1}^{K}\left(\frac{\lambda_l}{\lambda_l+\alpha}\right)^2\right]\right] \tag{11.124b}$$

Denote the SINR for the kth user as γ_k. Thus

$$\gamma_k = \frac{\text{Desired power}}{\text{Interference} + \sigma_k^2} \tag{11.125}$$

The kth user interference power can be expressed by the expected power – the desired power,

$$\text{The interference power} = \sum_{k=1}^{K}\frac{\lambda_k^2}{(\lambda_k+\alpha\,)^2} - \frac{1}{\mathrm{K(K+1)}}\left[\left[\left(\sum_{l=1}^{K}\frac{\lambda_l}{\lambda_l+\alpha}\right)^2\right] + \frac{1}{\mathrm{k}}\left[\sum_{l=1}^{K}\left(\frac{\lambda_l}{\lambda_l+\alpha}\right)^2\right]\right] \tag{11.126a}$$

The normalised interference power is obtained by multiplying by β_{RZF}^2,

$$\beta_{RZF}^2\sum_{k=1}^{K}\frac{\lambda_k^2}{(\lambda_k+\alpha\,)^2} - \frac{1}{\mathrm{K(K+1)}}\left[\left[\left(\sum_{l=1}^{K}\frac{\lambda_l}{\lambda_l+\alpha}\right)^2\right] + \frac{1}{\mathrm{k}}\left[\sum_{l=1}^{K}\left(\frac{\lambda_l}{\lambda_l+\alpha}\right)^2\right]\right] \tag{11.126b}$$

Substituting (11.124a) and (11.126b) in (11.125) we get,

$$\gamma_k = \frac{\beta_{RZF}^2\left[\frac{1}{\mathrm{K(K+1)}}\left[\left[\left(\sum_{l=1}^{K}\frac{\lambda_l}{\lambda_l+\alpha}\right)^2\right] + \frac{1}{\mathrm{k}}\left[\sum_{l=1}^{K}\left(\frac{\lambda_l}{\lambda_l+\alpha}\right)^2\right]\right]\right]}{\beta_{RZF}^2\sum_{k=1}^{K}\frac{\lambda_k^2}{(\lambda_k+\alpha\,)^2} - \frac{\beta_{RZF}^2}{\mathrm{K(K+1)}}\left[\left[\left(\sum_{l=1}^{K}\frac{\lambda_l}{\lambda_l+\alpha}\right)^2\right] + \frac{1}{\mathrm{k}}\left[\sum_{l=1}^{K}\left(\frac{\lambda_l}{\lambda_l+\alpha}\right)^2\right]\right] + \sigma_k^2}$$

which can be approximated and simplified to

$$\gamma_k \approx \frac{\left(\sum_{l=1}^{K} \frac{\lambda_l}{\lambda_l+\alpha}\right)^2}{(\mathrm{K}+1)\sum_{k=1}^{K} \frac{\lambda_k^2}{(\lambda_k+\alpha)^2} - \left(\sum_{l=1}^{K} \frac{\lambda_l}{\lambda_l+\alpha}\right)^2 + (\mathrm{K}+1)\sum_{l=1}^{K}\left(\frac{\lambda_l}{\lambda_l+\alpha}\right)^2 + \mathrm{K}(\mathrm{K}+1)\beta_{RZF}^{-2}\,\sigma_k^2} \tag{11.127}$$

And the outage probability for RZF is given as

$$\mathrm{pr}_{out} = \mathrm{pr}\left(\sum_{k=1}^{K} \log_2(1+\gamma_k) \le R\right) \tag{11.128}$$

where γ_k is given in (11.127)

11.13 Block Diagonalization (BD)

11.13.1 Multi-User BD Precoding

We start with a system model for the analysis of the block diagonalization (BD) precoding comprising a single cell supported by single BS transmitter with M antennas serving K users simultaneously and each user is equipped with multiple antennas. Denote the number of receive antennas used by the users as $(\mathrm{N}_1, \mathrm{N}_2, \ldots, \mathrm{N}_k \ldots \mathrm{N}_K)$ such that the kth user receives antennas is denoted by N_k. A block diagram of the system model is shown in Figure 11.7.

The BD is a generalization of channel inversion for situations when *multiple antennas are used by individual users.* The BD method either maximizes the throughput or optimizes power allocation.

The BD is examined by a number of researchers. In 2004 the authors of [24] proposed the BD as a generalization of channel inversion when there are multiple antennas at each receiver and are considered the optimization of the new precoding for maximum transmission rate and minimum power at high SNR. Later in 2008, authors of [23] proposed a BD precoding technique to be used with arbitrary number of antennas at the user terminals. The precoding matrix is designed in two steps: the first step is to remove the MUI and the second is to optimize the system performance and aver to reach the maximum sum rate capacity when the number of antennas at the user terminals is less or equal to the number of antennas at the BS. In 2011,

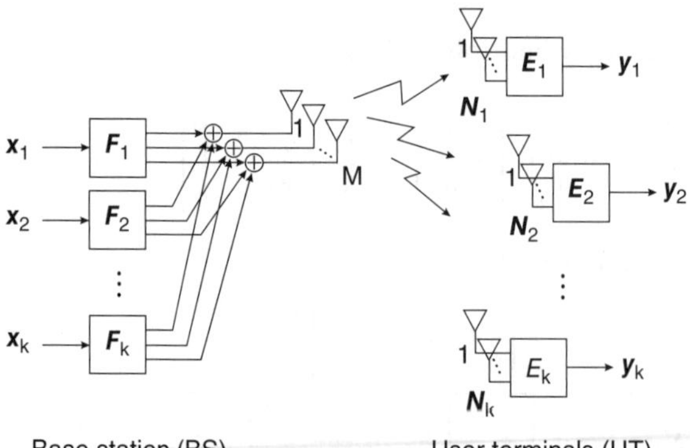

Figure 11.7 Block diagram of a multi-user MIMO downlink system [23].

two publications [25, 26] appeared in the same year. In [25], the authors examined the computation efficiency of the channel SVD based BD and proposed a BD method and argued that it offers a significantly lower complexity compared to the SVD-BD scheme with evidence on the performance improvement. In [26], the authors study the noise enhancement caused by the BD method, which degrades the system performance at a low and medium SNR. The authors proposed to extend $\overline{\mathbf{H}}_k$ in order to minimize the interference and noise to improve the system performance. The proposal model is simulated for each user with only two antennas and the BS antennas equal twice the number of users. The simulation results showed improved performance compared to the conventional BD method. In [27], the authors derived the precoding matrix at the transmitter, which is composed of two matrices and the receiver matrix. Authors of [28] proposed QR method to attain for MIMO BC systems and the computation complexity is less than that of the conventional BD.

11.13.2 BD Transmit Filter and Receive Filter Matrices

Consider the MU-MIMO DL channel in a system of a single cell with BS is supported with M antennas serving K users, each with multiple antennas. The MIMO composite channel matrix $\mathbf{H}$ is given by

$$\mathbf{H} \in \mathbb{C}^{\mathrm{N}\times\mathrm{M}} = [\mathbf{H}_1^T \ \mathbf{H}_2^T \ldots\ldots.\mathbf{H}_K^T]^T \tag{11.129}$$

where $\mathbf{H}_k \in \mathbb{C}^{\mathrm{N}_k\times\mathrm{M}}$ is the kth user's channel matrix. The MIMO channel to each user is assumed to be flat fading with Gaussian distribution $\mathcal{CN}(0, \mathbf{I})$. The total number of antennas for all users is given as $\mathrm{N} = \sum_{k=1}^{K} \mathrm{N}_k$. To affirm that all streams are detectable, we choose $\sum_{k=1}^{K} \mathrm{N}_k \leq \mathrm{M}$ and assign $\mathbf{H}$ to be row full rank, otherwise the number of supported simultaneous streams is less than N.

Denote the data vector transmitted on the BC for the kth as $\mathbf{s}_k \in \mathbb{C}^{\mathrm{N}_k\times 1}$, $k = 1, \ldots., K$ and for all users the individual vectors are stacked to form a single vector $\mathbf{s} \in \mathbb{C}^{\mathrm{N}\times 1} = [\mathbf{s}_1^T, \ldots\ldots., \mathbf{s}_K^T]^T$. The conjoint transmit filter precoding and receive filter equalizing are denoted as $\mathbf{F}$ and $\mathbf{E}$, respectively, as used in 11.9 analysis. The transmit precoding matrix $\mathbf{F}$ is defined as

$$\mathbf{F} \in \mathbb{C}^{\mathrm{M}\times\mathrm{N}} = [\mathbf{F}_1, \ldots\ldots\ldots., \mathbf{F}_K] \tag{11.130}$$

where $\mathbf{F}_k \in \mathbb{C}^{\mathrm{M}\times\mathrm{N}_k}$ is the kth user precoding matrix. The received precoded vector at the input of the receive filter is given by

$$\mathbf{y} \in \mathbb{C}^{\mathrm{N}\times 1} = \mathbf{E}\{\mathbf{HFs} + \mathbf{n}\} \tag{11.131}$$

where $\mathbf{F}$ and $\mathbf{E}$ are the joint transmit filter matrix and receive filter matrix. The receive vector is given by

$$\mathbf{y} \in \mathbb{C}^{\mathrm{N}\times 1} = [\mathbf{y}_1^T, \ldots\ldots\ldots.., \mathbf{y}_K^T]^T \tag{11.132}$$

The kth user's receiver vector at the input of the receive filter is given by

$$\mathbf{y_k} \in \mathbb{C}^{\mathrm{N}_k\times 1} = \left[\mathbf{H}_k\mathbf{F}_k\, s_k + \mathbf{H}_k \sum_{i=1, i\neq k}^{K} \mathbf{F}_i\, s_i + \mathbf{n}_k\right], \quad k = 1, \ldots\ldots\ldots\ldots., K \tag{11.133}$$

and

$$\mathbf{n} \in \mathbb{C}^{\mathrm{N}\times 1} = [\mathbf{n}_1^T, \ldots\ldots.., \mathbf{n}_K^T]^T \tag{11.134}$$

where $\mathbf{n}_\mathrm{k} \sim \mathcal{N}(\mathbf{0}, \boldsymbol{\sigma}_\mathbf{n}^2)$ is the kth user's Gaussian noise defined as

$$\mathbf{n}_\mathrm{k} = [n_{k,1}, n_{k,2}, \ldots., n_{k,l}, \ldots., n_{k,\mathrm{N}_k}]^T \tag{11.135}$$

The complex Gaussian noise $n_{k,l}$ at the lth receive antenna of the kth user is Gaussian distributed as $n_{k,l} \sim \mathcal{CN}(0, \sigma^2_{k,l})$. The noise components in (11.131) are independent of each other and each element is independent of the kth user transmitted signal $\mathbf{s}_k$.

The receive filter matrix **E** can be composed as

$$\mathbf{E} \in \mathbb{C}^{N_K \times N} = \begin{bmatrix} \mathbf{E}_1 & - & - & - & - & \mathbf{0} \\ - & - & - & - & - & - \\ - & - & - & - & - & - \\ - & - & - & - & - & - \\ \mathbf{0} & - & - & - & - & \mathbf{E}_K \end{bmatrix} \tag{11.136}$$

where $\mathbf{E}_k \in \mathbb{C}^{N_k \times N_k}$ is the kth user receive filter matrix. The kth user received signal at the output of the receive filter is given by $\mathbf{E}_k\, \mathbf{y}_k$.

The key feature of the BD concept is to ascertain the precoding matrix **F** derives the multi-user interference zero. The regularised block diagonalization (RBD) precoding at the transmitter is implemented to achieve two objectives. The first is to completely remove the MUI or at least balance the MUI with noise, and the second is to optimise the system performance. The first objective can be accomplished reducing the overlap of the effective channels of different users and establish the parallel SISO channels. The second step is achieved by optimizing the performance of the parallel SISO streams. The precoding matrix **F** in (11.130) is revised to

$$\mathbf{F} = \beta\, \mathbf{F}_a \cdot \mathbf{F}_b \tag{11.137}$$

where $\mathbf{F}_a$ is employed to remove the interference and is defined as

$$\mathbf{F}_a \in \mathbb{C}^{M \times N} = [\mathbf{F}_{a_1}, \mathbf{F}_{a_2}, \ldots \ldots \mathbf{F}_{a_K}] \tag{11.138a}$$

where $\mathbf{F}_{a_k} \in \mathbb{C}^{M \times N_k}$ is derived in the following analysis and $\mathbf{F}_b$ is exploited to optimize the system performance according to a specific criterion considering that the MU MIMO channel has been transformed into a set of parallel SU MIMO streams given by

$$\mathbf{F}_b = \begin{bmatrix} \mathbf{F}_{b_1} & \mathbf{0} & - & - & - & \mathbf{0} \\ \mathbf{0} & \mathbf{F}_{b_2} & - & - & - & - \\ - & - & - & - & - & - \\ - & - & - & - & - & - \\ - & - & - & - & - & - \\ \mathbf{0} & \mathbf{0} & - & - & - & \mathbf{F}_{b_K} \end{bmatrix} \tag{11.138b}$$

where $\mathbf{F}_{b_k} \in \mathbb{C}^{N_k \times N_k}$.

Let us define $\overline{\mathbf{H}}_k$ as

$$\overline{\mathbf{H}}_k \in \mathbb{C}^{\overline{N} \times M} = \begin{bmatrix} \mathbf{H}_1 \\ \\ \\ \mathbf{H}_{k-1} \\ \mathbf{H}_{k=1} \\ \\ \\ \mathbf{H}_K \end{bmatrix} \tag{11.139}$$

where $\overline{\mathrm{N}} = \mathrm{N} - \mathrm{N}_k$ The composite channel matrix $\mathbf{H}$ of all users after precoding with matrix $\mathbf{F}_a$ is given by

$$\mathbf{H}\,\mathbf{F}_a \in \mathbb{C}^{\mathrm{N}\times K\mathrm{M}} = \begin{bmatrix} \mathbf{H}_1 \\ \mathbf{H}_2 \\ \\ \\ \mathbf{H}_K \end{bmatrix} [[\mathbf{F}_{a_1}, \mathbf{F}_{a_2}, \ldots\ldots\ldots \mathbf{F}_{a_K}]] \tag{11.140}$$

The desired k^{th} user effective channel matrix is $\mathbf{H}_k\mathbf{F}_{a_k}$ and the interference is generated by the other users is due to $\overline{\mathbf{H}}_k\mathbf{F}_{a_k}$. So to eliminate the MUI we enforce the constraint that

$$\overline{\mathbf{H}}_k\mathbf{F}_{a_k} = \mathbf{0} \text{ for all } K-1 \text{ interfering users s.t. } \mathbb{E}\|\mathbf{F}_a\mathbf{F}_b\,\mathbf{s}\|_k^2 = \frac{\mathrm{P}_T}{\mathrm{M}} \tag{11.141}$$

Equation (11.141) defines the BD constraint, which is similar to ZF constraint. The $\overline{\mathbf{H}}_k\mathbf{F}_{a_k}$ matrix is related to the overlap of effective channels of different users $\mathbf{H}_k\mathbf{F}_{a_k}$. We design the precoding matrix $\mathbf{F}_a$ and the scaling factor β using the minimum mean square error (MMSE) criterion. Let us define the BD precoding matrix $\beta\mathbf{F}_a$ in terms of the MMSE as in the following optimization problem:

$$\mathbf{F}_a = \min_{\mathbf{F}_a} \mathbb{E}\left[\left\|\mathbf{s} - \frac{(\mathbf{H}\mathbf{F}_a\mathbf{s} + \mathbf{n})}{\beta}\right\|_2^2\right] \tag{11.142}$$

where β in (11.142) is a scaling factor to satisfy the transmit power constraint. Matrix $\mathbf{F}_a$ is chosen such that the off-diagonal block matrices of $\mathbf{H}\mathbf{F}_a$ is minimized under high SNRs. The main diagonal block matrices are optimized by the design of the matrix $\mathbf{F}_b$. The power transmission of desired k^{th} user is $\|\mathbf{H}_k\mathbf{F}_{a_k}\|^2$ is subjected to interference power generated by the other users, which is equal to $\|\overline{\mathbf{H}}_k\mathbf{F}_{a_k}\|^2$. The $\mathbf{F}_a$ matrix is designed to minimize the interference plus noise power. Optimization of the minimum interference power can be expressed by modifying the MMSE-based optimization [23] to

$$\mathbf{F}_a = \min_{\mathbf{F}_a} \mathbb{E}\left[\sum_{k=1}^{K} \|\overline{\mathbf{H}}_k\mathbf{F}_{a_k}\|^2 + \frac{\|\mathbf{n}\|^2}{\beta^2}\right] \tag{11.143a}$$

Equation (11.143a) can be simplified to

$$\mathbf{F}_a = \min_{\mathbf{F}_a} \left[\sum_{k=1}^{K} \{(\overline{\mathbf{H}}_k\mathbf{F}_{a_k})(\overline{\mathbf{H}}_k\mathbf{F}_{a_k})^H\} + \frac{\mathrm{N}\,\sigma_{\mathbf{n}_k}^2}{\beta^2}\right] \tag{11.143b}$$

where the receiver noise vector $\mathbf{n} \in \mathbb{C}^{\mathrm{N}\times 1}$ has a zero mean and variance over all individual receivers is $\sigma_{\mathbf{n}_k}^2$. The total transmit power-scaling factor β is selected to satisfy the transmit power P_T constraint:

$$\beta^2\,\|\mathbf{F}_a\mathbf{F}_b\,\mathbf{s}\|^2 \le \mathrm{P}_T \tag{11.144a}$$

where the left side of (11.144a) can be converted to a summation as

$$\beta^2\,\|\mathbf{F}_a\mathbf{F}_b\,\mathbf{s}\|^2 = \beta^2\mathbb{E}\left\{\sum_{k=1}^{K} \mathrm{tr}(\mathbf{F}_{ak}\mathbf{F}_{bk})(\mathbf{s}_k\,\mathbf{s}_k^H)(\mathbf{F}_{ak}\mathbf{F}_{bk})^H\right\} \tag{11.144b}$$

where the users' data vector $\mathbf{s}$ is given by

$$\mathbb{E}[\mathbf{s}_k\,\mathbf{s}_k^H] = \mathbf{I}_\mathrm{N}$$

The derivation of suggested that $\mathbf{F}_{bk}$ is a unitary matrix, so $\mathbf{F}_{bk}(\mathbf{F}_{bk})^H = \mathbf{I}_M$ and (11.144a) is simplified to

$$\beta^2 \left\{ \sum_{k=1}^{K} \text{tr}(\mathbf{F}_{ak}(\mathbf{F}_{ak})^H) \right\} \leq \text{P}_T \tag{11.145a}$$

Accordingly,

$$\beta^2 = \frac{\text{P}_T}{\sum_{k=1}^{K} \text{tr}(\mathbf{F}_{ak}(\mathbf{F}_{ak})^H)} \tag{11.145b}$$

We substitute (11.145b) in (11.143b) and simplify:

$$\mathbf{F}_a = \min_{\mathbf{F}_a} \left[\sum_{k=1}^{K} \left\{ \text{tr}(\overline{\mathbf{H}}_k \mathbf{F}_{a_k})(\overline{\mathbf{H}}_k \mathbf{F}_{a_k})^H + \frac{\text{N}\,\sigma^2_{\mathbf{n}_k}}{\text{P}_T} \text{tr}(\mathbf{F}_{ak}(\mathbf{F}_{ak})^H) \right\} \right] \tag{11.146a}$$

Using the cyclic–permutations property of the trace, we can now rewrite (11.146a):

$$\mathbf{F}_a = \min_{\mathbf{F}_a} \left[\sum_{k=1}^{K} \left\{ \text{tr}(\overline{\mathbf{H}}_k^H\ \overline{\mathbf{H}}_k \mathbf{F}_{a_k} \mathbf{F}_{a_k}^H) + \frac{\text{N}\,\sigma^2_{\mathbf{n}_k}}{\text{P}_T} \text{tr}(\mathbf{F}_{ak}(\mathbf{F}_{ak})^H) \right\} \right] \tag{11.146b}$$

$$\mathbf{F}_a = \min_{\mathbf{F}_a} \left[\sum_{k=1}^{K} \left\{ \text{tr} \left[\left(\overline{\mathbf{H}}_k^H \overline{\mathbf{H}}_k + \frac{\text{N}\,\sigma^2_{\mathbf{n}_k}}{\text{P}_T} \mathbf{I}_\text{M} \right) \mathbf{F}_{ak} \mathbf{F}_{a_k}^H \right] \right\} \right] \tag{11.146c}$$

$$\mathbf{F}_a = \min_{\mathbf{F}_a} \left[\sum_{k=1}^{K} \left\{ \text{tr} \left[\mathbf{F}_{a_k}^H \left(\overline{\mathbf{H}}_k^H \overline{\mathbf{H}}_k + \frac{\text{N}\,\sigma^2_{\mathbf{n}_k}}{\text{P}_T} \mathbf{I}_\text{M} \right) \mathbf{F}_{ak} \right] \right\} \right] \tag{11.146d}$$

Let us apply SVD to $\overline{\mathbf{H}}_k$ to get

$$\overline{\mathbf{H}}_k \in \mathbb{C}^{\overline{\text{N}} \times \text{M}} = \overline{\mathbf{U}}_k \overline{\mathbf{\Sigma}}_k \overline{\mathbf{V}}_k^H \tag{11.147}$$

where $\overline{\mathbf{\Sigma}}_k \in \mathbb{C}^{\overline{\text{N}}_k \times \text{M}}$ is a diagonal matrix, whose elements are the singular values of the matrix $\overline{\mathbf{H}}_k$, $\overline{\mathbf{U}}_k \in \mathbb{C}^{\overline{\text{N}}_k \times \overline{\text{N}}_k}$ and $\overline{\mathbf{V}}_k \in \mathbb{C}^{\text{M} \times \text{M}}$ are complex unitary matrices. Substituting (11.147) in (11.146d), we get

$$\mathbf{F}_a = \min_{\mathbf{F}_a} \left[\sum_{k=1}^{K} \left\{ \text{tr} \left[\mathbf{F}_{a_k}^H \left(\overline{\mathbf{V}}_k \overline{\mathbf{\Sigma}}_k^T \overline{\mathbf{\Sigma}}_k \overline{\mathbf{V}}_k^H + \frac{\text{N}\,\sigma^2_{\mathbf{n}_k}}{\text{P}_T} \mathbf{I}_\text{M} \right) \mathbf{F}_{a_k} \right] \right\} \right] \tag{11.148a}$$

Both $\overline{\mathbf{\Sigma}}_k^T \overline{\mathbf{\Sigma}}_k$ and $\mathbf{I}_\text{M}$ are real diagonal matrices and we can rewrite (11.148a) as

$$\mathbf{F}_a = \min_{\mathbf{F}_a} \left[\sum_{k=1}^{K} \left\{ \text{tr} \left[\mathbf{F}_{a_k}^H \overline{\mathbf{V}}_k \left(\overline{\mathbf{\Sigma}}_k^T \overline{\mathbf{\Sigma}}_k + \frac{\text{N}\,\sigma^2_{\mathbf{n}_k}}{\text{P}_T} \mathbf{I}_\text{M} \right) \overline{\mathbf{V}}_k^H \mathbf{F}_{ak} \right] \right\} \right] \tag{11.148b}$$

Optimisation of $\mathbf{F}_a$ should satisfy the ZF constraint in (11.141), we denote the ZF constraint as $\mathbf{F}_a^{(\text{BD})}$ and the regularised ZF constraint in (11.148b) as $\mathbf{F}_a^{(\text{RBD})}$. We derive the ZF constrain solution first.

Denote $\overline{\text{L}}_\text{k}$ as rank($\overline{\mathbf{H}}_k$) and examine the SVD to $\overline{\mathbf{H}}_k$ to get

$$\overline{\mathbf{H}}_k \in \mathbb{C}^{\overline{\text{N}} \times \text{M}} = \overline{\mathbf{U}}_k \overline{\mathbf{\Sigma}}_k \overline{\mathbf{V}}_k^H = \overline{\mathbf{U}}_k \overline{\mathbf{\Sigma}}_k \left[\overline{\mathbf{V}}_k^{(1)}\ \overline{\mathbf{V}}_k^{(0)} \right]^H \tag{11.149}$$

where we have factorised $\overline{\mathbf{V}}_k$ into two parts, $\overline{\mathbf{V}}_k^{(1)} \in \mathbb{C}^{\mathrm{M}\times\overline{\mathrm{L}}_k}$ contains the first $\overline{\mathrm{L}}_k$ nonzero singular vectors and $\overline{\mathbf{V}}_k^{(0)} \in \mathbb{C}^{\mathrm{M}\times(\mathrm{M}-\overline{\mathrm{L}}_k)}$ holds the last $\mathrm{M}-\overline{\mathrm{L}}_k$ zero singular vectors. Consequently $\overline{\mathbf{V}}_k^{(0)}$ forms the orthogonal basis for the null space of $\overline{\mathbf{H}}_k$. The solution to the constraint (11.141) is

$$\mathbf{F}_{a_k}^{(\mathrm{BD})} = \overline{\mathbf{V}}_k^{(0)} \tag{11.150a}$$

Optimisation solution to (11.148b) is given by [27] as

$$\mathbf{F}_{a_k}^{(\mathrm{RBD})} = \overline{\mathbf{V}}_k \left\{ \overline{\boldsymbol{\Sigma}}_k^T \overline{\boldsymbol{\Sigma}}_k + \frac{\mathrm{N}\,\sigma_{\mathbf{n}_k}^2}{\mathrm{P}_T} \mathbf{I}_\mathrm{M} \right\}^{-\frac{1}{2}} \tag{11.150b}$$

Consider the effective channel $\mathbf{H}_{\mathrm{eff}_k} = \mathbf{H}_k\,\mathbf{F}_{a_k}$ with $\mathrm{L}_k = \mathrm{rank}(\mathbf{H}_{\mathrm{eff}_k})$. Consider the second SVD on the effective channel matrix,

$$\mathbf{H}_{\mathrm{eff}_k} = \mathbf{U}_k\boldsymbol{\Sigma}_k\mathbf{V}_k^H = \mathbf{U}_k \begin{bmatrix} \boldsymbol{\Sigma}_k & \mathbf{0} \\ \mathbf{0} & \mathbf{0} \end{bmatrix} \begin{bmatrix} \mathbf{V}_k^{(1)} & \mathbf{V}_k^{(0)} \end{bmatrix}^H \tag{11.151}$$

The diagonal elements of $\boldsymbol{\Sigma}_k$ are the singular values and can be used to compute the power loading $\boldsymbol{\Lambda}_k$, $k = 1, \ldots, K$ that satisfies the total power constraint using the water-filling method. Loading matrix $\boldsymbol{\Lambda}$ is a diagonal matrix with diagonal elements $\boldsymbol{\Lambda}_1, \ldots \ldots \ldots, \boldsymbol{\Lambda}_K$.

The $\mathbf{F}_{\mathrm{b}_k}$ is given by

$$\mathbf{F}_{\mathrm{b}_k}^{(\mathrm{BD})} = \mathbf{V}_k^{(1)}\,\boldsymbol{\Lambda}^{(\mathrm{BD})} \tag{11.152a}$$

$$\mathbf{F}_{\mathrm{b}_k}^{(\mathrm{RBD})} = \mathbf{V}_k\,\boldsymbol{\Lambda}^{(\mathrm{RBD})} \tag{11.152b}$$

The kth user's receive filter matrix $\mathbf{E}_k$ has to be known by each receiver, and is given by [27] as

$$\mathbf{E}_k = \mathbf{U}_k^H \tag{11.153}$$

The BD method enforces a constraint on the link between the number of transmit and receive antennas. In addition, BD method provokes the noise enhancement due to excluding noise's strength in the consideration. Thus, at low and medium signal-to-noise ratio (SNR) system, the performance of the BD scheme is poor. Authors in [27] proposed a method that achieves extra performance improvement to the conventional BD scheme with lower complexity.

11.14 Transmit MF Precoding Filters and MMSE Receive Filters in MIMO Broadcast Channel

The design of an appropriate precoding is essential for the massive MIMO systems. Simple linear signal processing such as MF and MMSE precoding/detection can be used in single cell or multi-cell coordinated processing to achieve the performance gains expected in the massive MIMO technology.

In moderate to high SNR scenario, MF receivers suffer from an interference floor. Therefore, receive schemes for massive MIMO systems, rather than simple MF, is required. In large scale antennas at BS and user terminals, the MMSE receiver outperforms the MF under high SNR regime. In addition, the complexity of the matrix inversion used by the MMSE at the receiver can be reduced drastically using truncated polynomial expansion technology (TPE). TPE is dealt with in Section 11.15, which makes the MMSE receivers a practical candidate for massive MIMO systems.

The optimal MMSE filter matrix, $\mathbf{W}_{\text{MMSE}}$ derived in 11.E as

$$\mathbf{W}_{\text{MMSE}} = \left[\mathbf{H}^H\,\mathbf{H} + \frac{\sigma^2}{\rho}\,\mathbf{I}_{\text{M}}\right]^{-1}\mathbf{H}^H \tag{11.154a}$$

Equation (11.154a) can also be written as

$$\mathbf{W}_{\text{MMSE}} = \mathbf{H}^H\left[\mathbf{H}\mathbf{H}^H + \frac{\sigma^2}{\rho}\,\mathbf{I}_{\text{M}}\right]^{-1} \tag{11.154b}$$

The transmitted signal MF precoded is

$$\mathbf{HFs} = \beta_{\text{MF}}\mathbf{H}\mathbf{H}^H\,\mathbf{s} \tag{11.155}$$

The MF precoded received signal $\boldsymbol{y}$ at the input of the MMSE receive filter is

$$\boldsymbol{y} = \beta_{MF}\,\mathbf{H}\mathbf{H}^H\boldsymbol{s} + \mathbf{n} = \mathbb{H}\boldsymbol{s} + \mathbf{n} \tag{11.156a}$$

The receive filter output signal $\tilde{\boldsymbol{y}}$ is

$$\tilde{\mathbf{y}} = \mathbf{W}_{\text{MMSE}}\,\mathbf{y} = \mathbf{W}_{\text{MMSE}}\,\mathbb{H}\,\boldsymbol{s} + \mathbf{W}_{\text{MMSE}}\,\mathbf{n} \tag{11.156b}$$

The kth spatial stream equalized received signal y_k is given as

$$\mathbf{y}_k = [\mathbf{W}_{MMSE,k}]\,\boldsymbol{y} = [\mathbf{W}_{MMSE,k}][\mathbb{H}\,\boldsymbol{s} + \mathbf{n}] \tag{11.156c}$$

This formula is simplified in 11.F to give SINR_k as the kth element of diagonal matrix $[.]_{,k}$ in terms of the system composite channel $\mathbb{H}$ as in [29, 30]:

$$(\mathbf{SINR})_k = \frac{1}{\sigma^2\left[[\mathbb{H}^H\,\mathbb{H}] + \frac{\sigma^2}{\mathbf{R}_s}\,\mathbf{I}_{\text{M}}\right]^{-1}_{k,k}} - 1 \tag{11.157}$$

11.15 Linear Precoding Based on Truncated Polynomial Expansion

11.15.1 Introduction

With the aim of avoiding matrix inversion, Taylor approximation is applied in a technique commonly known as the polynomial expansion scheme. The inversion concept can simply be expressed using Taylor approximation such as

$$(1 \pm x)^{-1} = \sum_{l=0}^{\infty}(\pm x)^l,\ |x| < 1 \tag{11.158a}$$

In most applications, the approximation is limited to a small order J apptroximation, i.e.

$$(1 \pm x)^{-1} = \sum_{l=0}^{J}(\pm x)^l \tag{11.158b}$$

The polynomial expansion concept is used by Muller [31] for approximating a large antenna arrays matrix inversion for designing a communication system equaliser. Kohno et al. [32] used a linear algebraic approach for co-channel interference cancellation. In [33], Muller et al. observed that the truncated polynomial of L-order is identical to L-stage linear interference cancellation for vector channels when applied to linear multi-user detection. In [34], polynomial expansion is used in multi-user detection to reduce the computational complexity for synchronous DS-CDMA systems. Matrix polynomial instead of matrix inversion is used to provide low complexity precoding in DL large scale MIMO system and achieved near MMSE capacity performance in [35]. Recently, there have been a number of publications on the application of polynomial expansion in the multi-user cellular communications.

In [36] a large-scale multi-user multi-cell MIMO system is examined employing TPE precoding rather than a large matrix inversion. The network throughput of the TPE precoding is comparable to that of the RZF precoding system of similar size and resources. Furthermore, the TPE precoding is capable of achieving higher throughput using TPE order such as five. In [37] the work is extended, but unlike that in [38] where total transmit power is assumed to increase with the number of user terminals, in [37] the total transmit power is fixed. Extension of the work includes the TPE precoding analysis and the optimization of the throughput.

Polynomial expansion is based on the Cayley-Hamilton theorem, which states that the inverse of a matrix of dimension M can be formulated as a weighted sum of its first M powers. Mathematically, this implies that the inverse of matrix **A** can be expressed as

$$\mathbf{A}^{-1} = \frac{(-1)^{\mathrm{M}-1}}{\det(\mathbf{A})} \sum_{l=1}^{\mathrm{M}} \alpha_l\, \mathbf{A}^{l-1} \tag{11.159}$$

where α_l are the coefficients of the characteristic polynomial.

The key idea of the TPE is to approximate the matrix inverse by a polynomial with J terms, where J needs not be to scale with the system dimensions to maintain the desired output SINR. Research results in [35] show that only a few terms of the inverted matrix polynomials are sufficient to achieve almost the sum rate of the massive MIMO downlink MMSE precoding performance. A new precoding design using the truncated polynomial is explored in the next subsection.

11.15.2 Modelling the TPE Precoding for BC

To demonstrate the principles of TPE precoding we consider a simple system model comprising a single cell with a BS supported by M antennas and K single antenna users uniformly distributed within the cell and served simultaneously at the same data rate. The system operates TDD protocol synchronized BS acquires instantaneous user-link channel and uses it for the downlink transmission implementing the channel reciprocity. The data symbol s_k of the kth user and the K users data form a set of independent and identically distributed (i.i.d.) zero mean complex Gaussian symbols expressed by the symbol vector $\boldsymbol{s} \sim \mathcal{CN}(\mathbf{0}_{\mathrm{K}\times 1}, \mathbf{I}_K)$ destined to *all users* in the cell as

$$\boldsymbol{S} \in \mathbb{C}^{\mathrm{K\,x\,M}} = [\boldsymbol{s}_1^T, \ldots \ldots .., \boldsymbol{s}_K^T]^T \tag{11.160}$$

where $\mathbb{E}[\boldsymbol{s}\,\boldsymbol{s}^H] = \mathbf{I}_K$. The data vector $\boldsymbol{s}$ is precoded by matrix $\mathbf{F}_{RZF}$ given by

$$\mathbf{F} \in \mathbb{C}^{\mathrm{M}\times K} = [\mathbf{f}_1, \ldots \ldots .., \mathbf{f}_K] \tag{11.161}$$

where $\mathbf{f}_k \in \mathbb{C}^{\mathrm{M}\times 1}$ the precoding vector of the kth for the user symbol s_k.

The BS is subjected to a total power constraint:

$$\mathrm{tr}[\mathbf{F}\mathbf{F}^H] = P_T \tag{11.162}$$

where P_t is the average transmit power per user.

An important aspect of massive MIMO systems is that the effective useful precoded channel of a user converges to its average value when BS antennas grow large. Consequently, it is adequate to apply only statistical CSI for each user when computing the achievable information rate. In addition, the performance loss vanishes when BS antennas tend to be massive. This characteristic is known in the literature as channel hardening [39].

The computational complexity of the multi-users precoding grows with the system dimension. For example, the near to optimal RZF precoding is considered to be excessively complicated to be implemented in a practical system. It required fast inversions of large matrices in every coherence interval. The complexity can be abated by using a TPE. TPE precoding hardware implementation is complexity friendly in the commencement of the transmission of the first symbol. The TPE precoding enables the system performance intermediate between MF precoding and RZF and most of such a gap is associated with the small scalar coefficients that are used. The TPE precoding can be customized to the cell size, performance, and hardware requirements. Consequently, the cells SINR are functions of the TPE coefficients in the all cells. The TPE order J denotes the number of the matrix polynomial terms that are used in the precoding. Denote the estimated channel between the BS and all users in the cell, $\widehat{\mathbf{H}} \in \mathbb{C}^{M \times K}$ given by

$$\widehat{\mathbf{H}} = [\widehat{\mathbf{h}}_1, \ldots \ldots \ldots \ldots, \widehat{\mathbf{h}}_K] \tag{11.163}$$

Define the scalar factor $\tau \in [0, 1]$ as an indicator for the quality of the channel estimates: $\tau = 0$ for perfect estimate and $\tau = 1$ represents statistical knowledge only. We introduce another formation of $\mathbf{A}^{-1}$ based on standard Taylor Series expansion of the function f(x) as

$$\mathrm{f(x)} = (1 - \mathrm{x})^{-1} = \sum_{l}^{\infty} \mathrm{x}^l \text{ for } |\mathrm{x}| < 1 \tag{11.164a}$$

We also have the inverse of matrix $\mathbf{X}$ can be expressed as

$$\mathbf{X}^{-1} = \kappa[\mathbf{I} - (\mathbf{I} - \kappa\mathbf{X})]^{-1} \tag{11.164b}$$

Comparing (11.164a) with (11.164b) to get

$$\mathbf{X}^{-1} = \kappa \sum_{l}^{\infty} (\mathbf{I} - \kappa\mathbf{X})^l \tag{11.164c}$$

Equation (11.164c) holds if the parameter κ such that

$$0 < \kappa < \frac{2}{\max_n \lambda_n(\mathbf{X})}, \tag{11.165}$$

where λ_n is matrix $\mathbf{X}$ eigenvalues. Since the eigenvalues of $(\mathbf{I} - \kappa\mathbf{X})^l$ tend to zero as l grows to infinity, the low order terms are the most important ones. In other words, only the first few terms in the summation in (11.164c) are the most significant. So, with negligible errors, we can shorten (truncate) the polynomial to a few terms. Let us assume the TPE applies the first J terms. This is known TPE precoding using polynomial expansion truncated J order.

We now explore the TPE scheme and compare its performance with that of RFZ precoding of matrix $\mathbf{F}_{\mathrm{RZF}}$ derived in Section 11.12 and given by (11.113) as

$$\mathbf{F}_{\mathrm{RZF}} = \beta_{RZF}\mathbf{H}^H [\mathbf{H}\mathbf{H}^H + \alpha\mathbf{I}_{\mathrm{K}}]^{-1}$$

where the scaling factor β_{RZF} is given by setting the power constraint P_T to satisfy (11.163), and α is the regularisation parameter used in the cell. Considering (11.164), we use the precoding matrix $\mathbf{F}_{\mathrm{RZF}}$ in our analysis i.e.

$$\mathbf{F} = \beta\, [\widehat{\mathbf{H}}\widehat{\mathbf{H}}^H + \alpha\mathbf{I}_{\mathrm{K}}]^{-1}\widehat{\mathbf{H}} \tag{11.166a}$$

Accordingly, we get [38]

$$\beta[\widehat{\mathbf{H}}\widehat{\mathbf{H}}^H + \alpha\mathbf{I}_M]^{-1}\widehat{\mathbf{H}} \approx \sum_{l=0}^{J-1} \omega_l \, [\widehat{\mathbf{H}}\widehat{\mathbf{H}}^H]^l\widehat{\mathbf{H}} \tag{11.166b}$$

where $\omega_j = \omega_0, \ldots \ldots, \omega_{J-1}$ are scaling factors and TPE order $J = \min(M, K)$. Let us examine the SINR at the kth user's receiver given by (11.166b) and let us denote $\mathbf{e}_k$ for the kth column of the identity matrix $\mathbf{I}_K$. Applying the definition to $\mathbf{F}$, we get

$$\mathbf{f}_k = \mathbf{F}\,\mathbf{e}_k \tag{11.167}$$

Further, we transform the following expression to

$$[\mathbf{h}_k^H\,\overline{\mathbf{F}}][\mathbf{h}_k^H\,\overline{\mathbf{F}}]^H = [\mathbf{h}_k^H\,\mathbf{F}][\mathbf{h}_k^H\,\mathbf{F}]^H - [\mathbf{h}_k^H\mathbf{f}_k][[\mathbf{h}_k^H\mathbf{f}_k]^H \tag{11.168a}$$

$$[\mathbf{h}_k^H\,\overline{\mathbf{F}}][\mathbf{h}_k^H\,\overline{\mathbf{F}}]^H = [\mathbf{h}_k^H\,\mathbf{F}][\mathbf{h}_k^H\,\mathbf{F}]^H - [\mathbf{h}_k^H\mathbf{F}\,\mathbf{e}_k][[\mathbf{h}_k^H\mathbf{F}\,\mathbf{e}_k]^H \tag{11.168b}$$

The transmit truncated polynomial expansion (TPE) precoded signal $\mathbf{X}$ is given by

$$\mathbf{x} = \mathbf{F}_{\text{TPE}}\,\mathbf{s} = \sum_{l}^{J-1} \omega_l\,\tilde{\mathbf{x}}_l \tag{11.169a}$$

where $\tilde{\mathbf{x}}_j$ arises from approximating the TPE, $\mathbf{s}$ data symbol vector and

$$\tilde{\mathbf{x}}_j = \left\{ \begin{array}{cc} \widehat{\mathbf{H}}\,\mathbf{s}, & l = 0 \\ \widehat{\mathbf{H}}[\widehat{\mathbf{H}}^H\tilde{\mathbf{x}}_{l-1}], & 1 \le l \le J-1 \end{array} \right\} \tag{11.169b}$$

We next derive the equivalent TPE precoding for $\text{SINR}_k$of the RZF-based in (11.172b).

$$[\mathbf{h}_k^H\mathbf{f}_k][\mathbf{h}_k^H\mathbf{f}_k]^H = [\mathbf{h}_k^H\mathbf{F}\,\mathbf{e}_k]\,[\mathbf{h}_k^H\mathbf{F}\,\mathbf{e}_k]^H \tag{11.170a}$$

We exchange RZF with TPE in (11.170a) to get

$$[\mathbf{h}_k^H\mathbf{f}_k][\mathbf{h}_k^H\mathbf{f}_k]^H \to [\mathbf{h}_k^H\mathbf{F}_{\text{TPE}}\,\mathbf{e}_k][\mathbf{h}_k^H\mathbf{F}_{\text{TPE}}\,\mathbf{e}_k]^H \tag{11.170b}$$

Substitute (11.166b) for $\mathbf{F}_{\text{TPE}}$ in (11.170a) to get

$$\to \left[\mathbf{h}_k^H \sum_{l=0}^{J-1} \omega_l[\mathbf{H}\mathbf{H}^H]^l\mathbf{H}\,\mathbf{e}_k\right]\left[\mathbf{h}_k^H \sum_{m=0}^{J-1} \omega_m[\mathbf{H}\mathbf{H}^H]^m\mathbf{H}\,\mathbf{e}_k\right]^H \tag{11.170c}$$

Let us define $\mathbf{A}_k$ as

$$\mathbf{A}_k := \left[\mathbf{h}_k^H \sum_{l=0}^{J-1} [\mathbf{H}\mathbf{H}^H]^l\mathbf{H}\,\mathbf{e}_k\right]\left[\mathbf{h}_k^H \sum_{m=0}^{J-1} [\mathbf{H}\mathbf{H}^H]^m\mathbf{H}\,\mathbf{e}_k\right]^H \tag{11.171a}$$

where

$$[\mathbf{A}_k]_{l,m} = \mathbf{h}_k^H[\mathbf{H}\mathbf{H}^H]^l\mathbf{H}\mathbf{e}_k\mathbf{e}_k^H\,\widehat{\mathbf{H}}^H\,[\widehat{\mathbf{H}}\widehat{\mathbf{H}}^H]^m\,\mathbf{h}_k \tag{11.171b}$$

$$\mathbf{h}_k^H\mathbf{f}_k(\mathbf{h}_k^H\mathbf{f}_k)^H = \boldsymbol{\omega}^H\mathbf{A}_k\,\boldsymbol{\omega} \tag{11.172}$$

Let us define $[\mathbf{B}_k]_{l,m}$ as

$$[\mathbf{B}_k]_{l,m} := \mathbf{h}_k^H[\mathbf{H}\mathbf{H}^H]^{l+m+1}\,\mathbf{h}_k \tag{11.173a}$$

where = 0, ..., $J-1$, $m = 0, \ldots \ldots, J-1$.

$$[\mathbf{h}_k^H \overline{\mathbf{F}}][\mathbf{h}_k^H \overline{\mathbf{F}}]^H = [\mathbf{h}_k^H \mathbf{F}][\mathbf{h}_k^H \mathbf{F}]^H - [\mathbf{h}_k^H \mathbf{f}_k][[\mathbf{h}_k^H \mathbf{f}_k]^H$$

$$[\mathbf{h}_k^H \mathbf{F}][\mathbf{h}_k^H \mathbf{F}]^H \rightarrow \mathbf{h}_k^H \mathbf{F}_{\text{TPE}} \mathbf{F}_{\text{PE}}^H \mathbf{h}_k$$

$$\rightarrow \mathbf{h}_k^H \rightarrow \mathbf{h}_k^H \sum_{l=0}^{J-1} \omega_l \{[\widehat{\mathbf{H}}\widehat{\mathbf{H}}^H]^l \widehat{\mathbf{H}}\}\{\widehat{\mathbf{H}}^H [\widehat{\mathbf{H}}\widehat{\mathbf{H}}^H]^m\} \mathbf{h}_k \omega_m^H \tag{11.173b}$$

$$[\mathbf{B}_k] = \mathbf{h}_k^H \{[\widehat{\mathbf{H}}\widehat{\mathbf{H}}^H]\widehat{\mathbf{H}}\{\widehat{\mathbf{H}}^H[\widehat{\mathbf{H}}\widehat{\mathbf{H}}^H]\}\mathbf{h}_k \tag{11.173c}$$

$$[\mathbf{B}_k]_{l,m} = \mathbf{h}_k^H [\widehat{\mathbf{H}}\widehat{\mathbf{H}}^H]^l \widehat{\mathbf{H}}\,\widehat{\mathbf{H}}^H [\widehat{\mathbf{H}}\widehat{\mathbf{H}}^H]^m \mathbf{h}_k \tag{11.173d}$$

Hence, we have proved that

$$[\mathbf{B}_k]_{l,m} = \mathbf{h}_k^H [\widehat{\mathbf{H}}\widehat{\mathbf{H}}^H]^{l+m+1} \mathbf{h}_k$$

Accordingly, we have

$$[\mathbf{h}_k^H \overline{\mathbf{F}}][\mathbf{h}_k^H \overline{\mathbf{F}}]^H = \boldsymbol{\omega}^H \mathbf{B}_k \boldsymbol{\omega} - \boldsymbol{\omega}^H \mathbf{A}_k \boldsymbol{\omega}$$

Consequently, the SINR for the TPE precoder is given by

$$\text{SINR}_k = \frac{\boldsymbol{\omega}^H \mathbf{A}_k \boldsymbol{\omega}}{\boldsymbol{\omega}^H \mathbf{B}_k \boldsymbol{\omega} - \boldsymbol{\omega}^H \mathbf{A}_k \boldsymbol{\omega} + \sigma^2}, \tag{11.174a}$$

and $\boldsymbol{\omega}$ is defined as

$$\boldsymbol{\omega} := [\omega_0, \ldots \ldots \ldots \ldots, \omega_{J-1}]^T \tag{11.174b}$$

TPE precoding gives many benefits to the cellular massive MIMO systems. Regardless of the fact that TPE coefficients are bounded by the instantaneous channel coefficients, the TPE precoded system SINRs converge to deterministic level which enables TPE coefficients to be computed using channel statistics. Furthermore, different TPE orders can be employed in different cells and in cell-specific power constraints. More importantly the TPE order can be adjusted to suit the available hardware. As a rule, polynomial complexity is between linear and cubic in the parameters. However, this is not restricting the TPE benefits, since the TPE coefficients can be computed offline using the channel statistics.

Even if the channel would change, its statistics change at a slow rate and the impact will be negligible compared to the RZF precoding.

Comparison of average per UT rate versus transmitted power to noise power ratio between the conventional RZF precoding and its equivalent TPE precoding was validated by computer simulations by the authors of [36, 38]. The purpose of the validation was to compare the system throughput of the conventional RZF precoding and its equivalent TPE precoding. A single cell with BS M = 128 serving 32 users for varying channel estimates error $\tau = 0.1, 0.4, 0.7$ was simulated for SNR up to 20 dB employing RZF and its protocol testing environment (PTE) equivalent with $J = 3$. The results of the simulation are depicted in Figure 11.8. Comparing these results shows equal performance for the two precoding under low SNR (SNR $\leq$ 5 dB) disregarding the CSI error. As SNR increased with low CSI error, the RZF outperformed the PTE but as the CSI error increased to 0.7, their average per user rate in bit/sec/Hz performance tends to be almost equal.

The effect of increasing the J order of the PTE is studied considering a similar system model but using larger BS antennas M = 512 serving 1282 users for varying channel estimates error $\tau = 0.1$. The system performance is expressed on average per user rate bit/sec/Hz illustrated

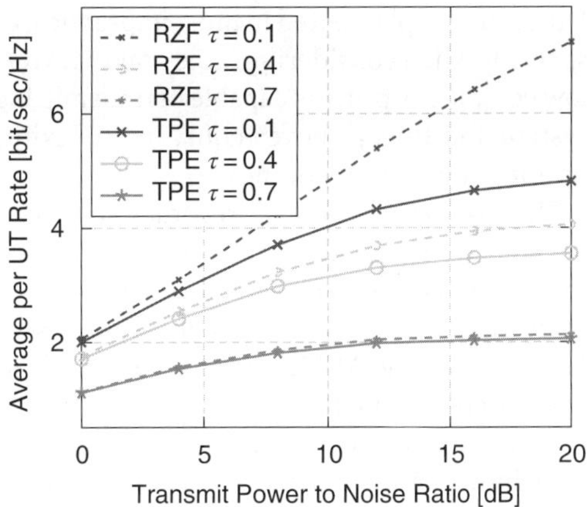

Figure 11.8 Comparison of average per UT rate vs. transmit power to noise ratio between conventional RZF precoding and the proposed TPE precoding with different orders *J* [38].

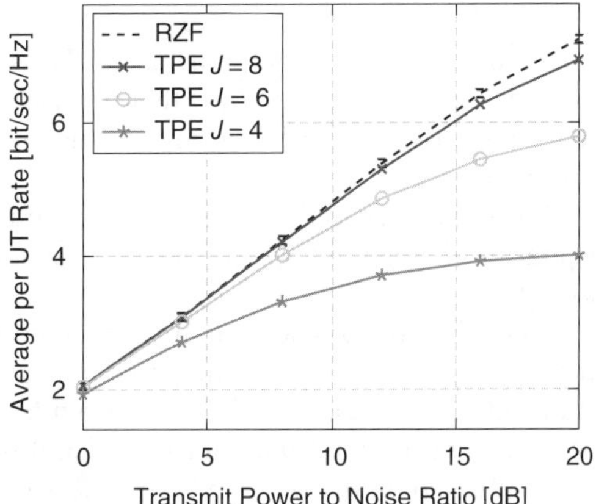

Figure 11.9 Average UT rate vs. transmit power to noise ratio for different orders J in the TPE precoding [38].

in Figure 11.9. As the TPE *J* increased from two to four and with increasing SNR, the performance of the two precoders becomes more imminent. It is reasonable to conclude that the performance of the PTE cannot exceed that of the RZF, since PTE is an approximation to the RZF precoder.

11.16 Summary

A key challenge for transmission over multi-user MIMO channels is the separation and equalization of the parallel data streams. In an effort to achieve the expected capacity and performance gains in large scale MIMO system we have to manage the channel distortion and the interference. These two issues can be resolved using a well-designed receive equalizer for the former to remove channel corruptions and use a transmit precoder for the latter to null out the inter-users inter-cell interference. The precoding-equalizing can be attained at the transmitter or at the receiver, or jointly optimized at the transmitter and the receiver. The MSs broadly have limited processing power, so they can only execute simple algorithms that have low computational complexity.

Section 11.1 introduced the precoding strategies including nonlinear precoding known as DPC or Costa precoding. The latter requires the knowledge of interference at the transmitter.

DPC is not implemented in most impractical cases because of its high computation requirement specially when considering large scale MIMO systems. On the other hand, linear precoding is a low complexity process capable of transmitting the same number of data streams as DPC–based system and thus achieves equal multiplexing gain. The following sections examined various options of linear precoding.

The basic generic linear precoding structure and the articulation of the precoding matrix was developed in Section 11.2. In Section 11.3, a single user link with M, N antennas at the transmitter and at the receiver, respectively, was analysed with fixed total transmit power in terms of the SVD of the channel matrix. Multi-user MIMO precoding was examined in Section 11.4. First, the BC and MAC systems were modelled and then linear precoding for massive BC and MAC were considered.

Linear CIP at the transmitter of multi-user was conferred in Section 11.5, which began by explaining the basic concept of channel inversion, followed by a detailed analysis of BC channel inversion model.

Specific precoding strategies were defined, analysed, and evaluated in the Sections 11.6–11.13.

Section 11.6 considered multi-user ZF precoding used in BC, starting with the formulation, analysis, and evaluation of the SINR per user's terminal, followed by the derivation of the optimal ZF matrix with transmit power constraint along with optimal ZF matrix for per antenna power constraint.

Section 11.7 focused on the definition and fairness, maximizing the throughput, and managing the channel outage. Details of these metrics for ZF precoding design were articulated as well.

Unlike the previous section, in 11.8 we derived the precoding–equalization matrices when they are operating jointly. The precoding and the equalization matrices are specified as the transmit filter matrix **F** and the receive filter matrix **E**. In Section 11.8.1, we constructed the MIMO model for multi-antenna BS and multi-antenna user terminals. The optimization of both filters is based on the MSE and the SNR per stream. We started with ZF precoding in Section 11.8.2 and derived $\mathbf{F}_{ZF}$ and $\mathbf{E}_{ZF}$ in Section 11.8.3, while the outage probability was analysed in Section 11.8.4. Section 11.8.5 developed the ZF design with unequal power allocation and the outage rate for unequal power allocation was analysed in 11.8.6.

Section 11.9 addressed the matched filter precoding. The transmit filter matrix was obtained in Section 11.9.1 and the receive filter matrix in 11.9.2. WF precoding was taken care of in Section 11.10. The WF transmit and receive matrices were formulated in Sections 11.10.1 and 11.10.2.

Regularizing the ZF precoding at the transmitter allows a certain amount of interference to be alive, so the desired signal can maximize the SINR at each user receive end and achieves near sum capacity performance. The RZF is considered in Section 11.11. The BD precoding is presented in Section 11.12 and the multi-user BD precoding is modelled in Section 11.12.1. The articulation of the BD transmit filter and receive filter matrices are dealt with in Section 11.12.2.

In Section 11.13 we presented a new type of precoders as an alternative to the RZF/ZF precoders. These precoders are based on a nearly perfect approximation to the RZF using a TPE. The new (TPE) precoders have a low computation complexity but retain the achievable rate attained using RZF precoders.

We analysed and compared the performance of the TPE precoders. An essential parameter in the TPE precoder is the TPE order, which is the number of truncated polynomial coefficients used that are sufficient to reach the performance similar to that of the RZF counterpart. The PTE order is a comparative indication of how low the complexity of the TPE precoding is. As expected, the effective SINRs are functions of the TPE coefficients used. Furthermore, the TPE precoding order that can be adapted to match different cells depends on factors such as cell size, performance requirements, and hardware resources, so different TPE orders are required in different cells.

It is well known that the RZF precoding is the best possible performance next to the optimal scenario.

However, RZF precoding performance depends greatly on the choice of regularization coefficients. Indeed, the PTE precoding is capable of achieving higher throughput at lower complexity.

11.A Derivation of the Scaling Factor β_{ZF}

The transmit ZF precoding matrix is given in Eq. (11.70) as

$$\mathbf{F}_{\mathrm{ZF}} = (\mathbf{EH})^{\mathrm{H}}\,[\mathbf{EH}(\mathbf{EH})^{\mathrm{H}}]^{-1}$$

The transmit power ρ_{ZF} is given by Eq. (11.71) as

$$\beta_{\mathrm{ZF}}^2\,\mathrm{tr}(\mathbf{F}_{\mathrm{ZF}}\,\mathbf{R}_{\mathbf{s}}\mathbf{F}_{\mathrm{ZF}}^{\mathrm{H}}) \equiv \rho_{\mathrm{ZF}} \tag{11.A.1}$$

where $\mathbf{R}_{\mathbf{s}}$ is the covariance of the desired signal. Substituting for $\mathbf{F}$ in (11.A.1) we get

$$\beta_{\mathrm{ZF}}^2\,\mathrm{tr}[\{(\mathbf{EH})^{\mathrm{H}}\,[\mathbf{EH}(\mathbf{EH})^{\mathrm{H}}]^{-1}\}\mathbf{R}_{\mathbf{s}}\,\{(\mathbf{EH})^{\mathrm{H}}\,[\mathbf{EH}(\mathbf{EH})^{\mathrm{H}}]^{-1}\}^{H}] \tag{11.A.2}$$

Simplifying (11.A.2) it can easily be shown that

$$\beta_{\mathrm{ZF}}^2\,\mathrm{tr}[\,[\mathbf{EH}(\mathbf{EH})^{\mathrm{H}}]^{-1}\}\mathbf{R}_{\mathbf{s}}\,] = \rho_{ZF}$$

Accordingly, we have

$$\beta_{ZF} = \sqrt{\frac{\rho_{ZF}}{\mathrm{tr}\{(\mathbf{EH}\,(\mathbf{EH})^{\mathrm{H}})^{-1}\,\mathbf{R}_{\mathbf{s}}\}}} \tag{11.A.3}$$

11.B ZF Precoder Design Optimum User Power in Unequal Power Allocation

Recall Eqs. (11.87a) and (11.91); we deduce that

$$\max_{p_k \geq 0} f(\mathbf{p}) = \sum_{k=1}^{\mathrm{M}} \log_2(1 + p_k) \tag{11.B.1a}$$

$$\text{s.t} \sum_{k=1}^{\mathrm{M}} p_k\,[\mathbf{EH}\,(\mathbf{EH})^{H}]_{kk}^{-1} \leq \rho \tag{11.B.1b}$$

We assumed a Gaussian noise of zero mean and unit variance, $\mathbf{p}$ denotes the variable transmit powers $(p_1, p_2, \ldots\ \ldots\ \ldots .., p_{\mathrm{M}})$, and $f(\mathbf{p})$ is a scalar function of M variables. The optimum solution is given by the gradient of f with respect to each of the variables. Furthermore, if we differentiate f with respect of (say p_i) then all the terms in f involving p_j with $j \neq i$ disappear. The gradient of f can be expressed as

$$\frac{\partial f}{\partial p_k} = \frac{1}{\log_2(1 + p_k)} \tag{11.B.2}$$

Even though there are M variables, the summation in (11.B.1b) adds up all of these terms in a series, and the result has to be $\leq \rho$. This implies that we need just one Lagrange multiplier λ. Let us denote g(**p**) to include the constraint as

$$g(\mathbf{p}) = \sum_{k=1}^{M} p_k\, [\mathbf{EH}\,(\mathbf{EH}^H]_{kk}^{-1} - \rho \tag{11.B.3}$$

The Lagrangian equation is

$$L(\mathbf{P}, \lambda) = f(\mathbf{p}) - \lambda g(\mathbf{p}) \tag{11.B.4}$$

Substitute (11.B.1a) and (11.B.3) in (11.B.4) we get

$$L(\mathbf{P}, \lambda) = \sum_{k=1}^{M} \log_2(1 + p_k) - \lambda \left(\sum_{k=1}^{M} p_k\, [\mathbf{EH}\,(\mathbf{EH})^H]_{kk}^{-1} - \rho \right) \tag{11.B.5}$$

We differentiate $L(\mathbf{P}, \lambda)$ with respect to p_k and λ. We transform $\log_2(.)$ to natural log

$(i.\text{e.} \log(.) \text{ or } \log_e(.))$:

$$\log_2(1 + p_k) = \frac{1}{\log 2} \log(1 + p_k)$$

We can rewrite (11.B.5) as

$$L(\mathbf{P}, \lambda) = \frac{1}{\log 2} \sum_{k=1}^{M} \log((1 + p_k)) - \lambda \left(\sum_{k=1}^{K} p_k\, [\mathbf{EH}\,(\mathbf{EH})^H]_{kk}^{-1} - \rho \right) \tag{11.B.6}$$

Differentiate (11.B.6) to get the following two equations:

$$\frac{\partial L}{\partial p_k} = \frac{1}{\log_2} \frac{1}{(1 + p_k)} - \lambda\, [\mathbf{EH}\,(\mathbf{EH})^H]_{kk}^{-1} \tag{11.B.7a}$$

$$\frac{\partial L}{\partial \lambda} = -\left(\sum_{k=1}^{M} p_k\, [\mathbf{EH}\,(\mathbf{EH})^H]_{kk}^{-1} - \rho \right) \tag{11.B.7b}$$

We set (11.B.7) and (11.B.8) to zero:

$$\frac{1}{\log 2} \frac{1}{1 + p_k} - \lambda\, [\mathbf{EH}\,(\mathbf{EH})^H]_{kk}^{-1} = 0 \tag{11.B.8a}$$

$$\sum_{k=1}^{M} p_k\, [\mathbf{EH}\,(\mathbf{EH})^H]_{kk}^{-1} - \rho = 0 \tag{11.B.8b}$$

From (11.B.8a) we have

$$\frac{1}{\lambda \log 2} = [1 + p_k]\ \ [\mathbf{EH}\,(\mathbf{EH})^H]_{kk}^{-1} \tag{11.B.9a}$$

$$\lambda \log 2\, [\mathbf{EH}\,(\mathbf{EH})^H]_{kk}^{-1} [1 + p_k] = 1 \tag{11.B.9b}$$

And rearrange to get

$$\lambda \log 2\, [\mathbf{EH}\,(\mathbf{EH})^H]_{kk}^{-1} + p_k \lambda \log 2\, [\mathbf{EH}\,(\mathbf{EH})^H]_{kk}^{-1} = 1 \tag{11.B.10}$$

Sum (11.B.10) over all k to get

$$\lambda \log 2 \left[\sum_{k=1}^{M} [\mathbf{EH}\,(\mathbf{EH})^H]_{kk}^{-1} \right] + \lambda \log 2 \left[\sum_{k=1}^{M} p_k [\mathbf{EH}\,(\mathbf{EH})^H]_{kk}^{-1} \right] = \mathrm{M} \tag{11.B.11}$$

where $\sum_{k=1}^{M} 1 = 1, 1, 1 \ldots \ldots \ldots$,. Substituting (11.B.8b) for the second term in the LHS of (11.B.11) to get

$$\lambda \log 2 \left[\sum_{k=1}^{M} [\mathbf{EH}\,(\mathbf{EH})^H]_{kk}^{-1}\right] + \lambda\rho \log 2 = \mathrm{M} \tag{11.B.12}$$

Now simplify (11.B.12):

$$\lambda \log 2 \left[\left[\sum_{k=1}^{M} [\mathbf{EH}\,(\mathbf{EH})^H]_{kk}^{-1}\right] + \rho\right] = \mathrm{M} \tag{11.B.13}$$

So

$$\lambda = \frac{1}{\log 2} \frac{\mathrm{M}}{\left[\rho + \left[\sum_{k=1}^{M} [\mathbf{EH}\,(\mathbf{EH})^H]_{kk}^{-1}\right]\right]} \tag{11.B.14}$$

Substitute for λ in (11.B.14) and use the expression in (11.B.9b) to get

$$p_k = \frac{\rho + \left[\sum_{k=1}^{M} [\mathbf{EH}\,(\mathbf{EH})^H]_{kk}^{-1}\right]}{\mathrm{M}\,[\mathbf{EH}\,(\mathbf{EH})^H]_{kk}^{-1}} - 1 \tag{11.B.15}$$

11.C Transmit Matched Filter (MF) Precoding

In designing the transmit filter **F** precoding, we assume the transmitter has a priori knowledge of a constant receive filter **E** and $\mathbf{R_s}$ in addition to the instantaneous channel matrix. Denote the desired information $\mathbf{s} \in \mathbb{C}^{M\times 1}$ precoded by a transmit MF **F** and transmitted as **Fs**. Further, we denote $\mathbb{E}[\|\mathbf{Fs}\|_F^2] = \rho$ and $\mathbb{E}[\mathbf{ss}^H] = \mathbf{R_s}$ where ρ denotes average transmit power. The received noiseless signal $\tilde{\mathbf{s}} = \mathbf{EHFs}$. The optimisation problem can be formulated as

$$\mathbf{F} = \mathrm{argmax}_{\mathbf{F}} \frac{|\mathbb{E}[\mathbf{s}^H\tilde{\mathbf{s}}]|^2}{\mathbb{E}[\|\mathbf{En}\|_F^2]} \tag{11.C.1a}$$

$$\text{s.t.}\quad \mathbb{E}[\|\mathbf{Fs}\,\|_2^2] = \rho \tag{11.C.1b}$$

Furthermore, we have $[\mathbf{ss}^H] = \mathbf{R_s}$. We may simplify (11.C.1b) as follows:

$$\mathbb{E}[\|\mathbf{Fs}\|_2^2] = \mathbb{E}[[\mathbf{Fs}]^H\,\mathbf{Fs}] = \mathbb{E}[s^H\mathbf{F}^H\mathbf{Fs}] = \mathbf{F}^H\mathbf{F}\mathbb{E}[ss^H] = \mathrm{tr}[\mathbf{FF}^H\,\mathbf{R_s}] = \mathrm{tr}[\mathbf{F}^H\,\mathbf{R_s}\,\mathbf{F}] = \tag{11.C.1c}$$

The transmit power constraint becomes

$$s.t.\quad \mathrm{tr}[\mathbf{F}^H\,\mathbf{R_s}\mathbf{F}] \le \rho \tag{11.C.2}$$

The correlation of the desired signal with information signal transmitted is

$$|\mathbb{E}[\mathbf{s}^H\tilde{\mathbf{s}}]|^2 = |\mathbb{E}[\mathbf{s}^H\mathbf{EHFs}\,]|^2 = |\mathbf{R_s}\,\mathbf{EHF}|^2 \tag{11.C.3}$$

$$\mathbf{F} = \mathrm{argmax}_{\mathbf{F}} \frac{|\mathbf{R_s}\,\mathbf{EHF}|^2}{\mathbb{E}[\|\mathbf{En}\|_F^2]} \tag{11.C.4}$$

The transmitter has no influence on the receiver noise power $\mathbb{E}[\|\mathbf{En}\|_2^2]$ and can be thought of as a constant scaling factor. The transmit filter maximizes the desired signal portion at the receiver.

We formulate the Lagrangian multipliers as

$$L = |\mathbf{R_s}\ \mathbf{EHF}|^2 - \lambda(\mathrm{tr}[\mathbf{F}^H\ \mathbf{R_s F}] - \rho) \quad (11.C.5)$$

$$\frac{\partial L}{\partial F} = 2\mathbf{R_s}(\mathbf{EH})^H - 2\lambda\,\mathbf{R_s F} \quad (11.C.6a)$$

$$\frac{\partial L}{\partial \lambda} = -(\mathrm{tr}[\mathbf{F}^H\ \mathbf{R_s F}] - \rho) \quad (11.C.6b)$$

We set (11.C.6a) and (11.C.6b) to zero and get

$$\mathbf{F} = \frac{(\mathbf{EH})^H}{\lambda} \quad (11.C.7)$$

$$\mathrm{tr}[\mathbf{F}\,\mathbf{R_s}\,\mathbf{F}^H\ ] = \rho \quad (11.C.8)$$

Substitute for $\mathbf{F}$ in (11.C.8) using (11.C.7) we get

$$\mathrm{tr}\left[\frac{(\mathbf{EH})^H\ \ \mathbf{R_s}\,(\mathbf{EH})}{\rho}\right] = \lambda^2 \quad (11.C.9)$$

$$\lambda = \sqrt{\frac{\mathrm{tr}[(\mathbf{EH})^H\ \ \mathbf{R_s}\,(\mathbf{EH})]}{\rho}}$$

Substitute for λ from (11.C.9) in (11.C.7):

$$\mathbf{F} = \frac{(\mathbf{EH})^H}{\lambda} = \sqrt{\frac{\rho}{\mathrm{tr}[(\mathbf{EH})^H\,\mathbf{R_s}\,(\mathbf{EH})]}}\,(\mathbf{EH})^H \quad (11.C.10)$$

$$\mathbf{F} = \beta\,(\mathbf{EH})^H \in \mathbb{C}^M \quad (11.C.11a)$$

where

$$\beta = \sqrt{\frac{\rho}{\mathrm{tr}[(\mathbf{EH})^H\ \ \mathbf{R_s}\,(\mathbf{EH})]}} \quad (11.C.11b)$$

11.D Wiener Filter (WF) Precoding

The optimum $\mathbf{F}$ is given by the solution to the following optimisation problem:

$$\{\mathbf{F}_{\mathrm{WF}}, \beta_{\mathrm{WF}}\ \} = \mathrm{argmin}_{\mathbf{F},\beta}\ \mathbb{E}(\|\boldsymbol{s} - \beta^{-1}\tilde{\mathbf{s}}\|_2^2) \quad (11.D.1a)$$

$$\text{s.t. } \mathbb{E}[\|\mathbf{Fs}\|_2^2] = \rho \quad (11.D.1b)$$

where ρ is the average transmit power, which is fixed. We can simplify $\mathbb{E}\|\mathbf{Fs}\|^2$ as

$$\mathbb{E}[\|\mathbf{Fs}\|_2^2] = \mathrm{tr}[\mathbf{F}\,\mathbf{R_s}\,\mathbf{F}^H] = \rho \quad (11.D.2)$$

The error vector $\boldsymbol{e}$ is defined as

$$\boldsymbol{e} \triangleq \mathbf{s} - \beta^{-1}\,\tilde{\mathbf{s}} \quad (11.D.3)$$

where $\tilde{\mathbf{s}}$ is given by

$$\tilde{\mathbf{s}} = \mathbf{E}(\mathbf{HF}\,\boldsymbol{s} + \boldsymbol{n})$$

The MSE matrix $\mathbb{E}\|\boldsymbol{e}(\,\mathbf{F})\|^2$ is given by

$$\boldsymbol{e} \triangleq \mathbf{s} - \beta^{-1}\,\tilde{\mathbf{s}}.$$

The MSE is given by the MSE matrix $\mathbb{E}\{\|\boldsymbol{e}(\mathbf{F})\|^2\}$ as

$$\mathrm{MSE}(\mathbf{F},\beta) = \mathbb{E}[\|\boldsymbol{e}(\mathbf{F})\|_2^2] \triangleq \mathbb{E}[(\mathbf{s}-\beta^{-1}\tilde{\mathbf{s}})(\mathbf{s}-\beta^{-1}\tilde{\mathbf{s}})^H] \tag{11.D.4a}$$

$$= \mathbb{E}[(\mathbf{s}\mathbf{s}^H + \beta^{-2}\tilde{\mathbf{s}}\tilde{\mathbf{s}}^H - \beta^{-1}\tilde{\mathbf{s}}\,\mathbf{s}^H - \beta^{-1}\mathbf{s}\tilde{\mathbf{s}}^H)]$$

$$\mathrm{MSE}(\mathbf{F},\beta) = \mathrm{tr}[\mathbf{R}_\mathbf{s}] + \mathrm{tr}[\beta^{-2}\mathbf{R}_{\tilde{\mathbf{s}}\tilde{\mathbf{s}}^H} - \beta^{-1}\mathbf{R}_{\tilde{\mathbf{s}}\mathbf{s}^H} - \beta^{-1}\mathbf{R}_{s\tilde{\mathbf{s}}^H}] \tag{11.D.4b}$$

$$\mathbf{R}_{\tilde{\mathbf{s}}\tilde{\mathbf{s}}^H} = \mathbf{EHF}\,\mathbf{R}_\mathbf{s}\mathbf{F}^H\mathbf{H}^H + \mathbf{R}_\mathbf{n} \tag{11.D.5a}$$

$$\mathbf{R}_{\tilde{\mathbf{s}}\mathbf{s}^H} = \mathbf{EHF}\,\mathbf{R}_\mathbf{s} \tag{11.D.5b}$$

$$\mathbf{R}_{s\tilde{\mathbf{s}}^H} = \mathbf{R}_\mathbf{s}\,[\mathbf{EHF}]^H \tag{11.D.5c}$$

$$\mathrm{MSE}(\mathbf{F},\beta) = \mathrm{tr}[\mathbf{R}_\mathbf{s}] + \beta^{-2}\,\mathrm{tr}[\mathbf{EHF}\,\mathbf{R}_\mathbf{s}(\mathbf{EHF})^H + \mathbf{E}\mathbf{R}_\mathbf{n}\mathbf{E}^H] - \beta^{-1}\mathrm{tr}[\,\mathbf{HF}\,\mathbf{R}_\mathbf{s} + \mathbf{R}_\mathbf{s}\,[\mathbf{HF}]^H] \tag{11.D.6a}$$

$$\mathrm{MSE}(\mathbf{F},\beta) = \mathrm{tr}[\mathbf{R}_\mathbf{s}] + \beta^{-2}\,\mathrm{tr}[(\mathbf{EHF})\,\mathbf{R}_\mathbf{s}(\mathbf{EHF})^H + \mathbf{E}\mathbf{R}_\mathbf{n}\mathbf{E}^H] - 2\beta^{-1}\,\{\mathrm{tr}[\,\mathcal{R}e(\mathbf{EHF}\,\mathbf{R}_\mathbf{s})]\} \tag{11.D.6b}$$

We can find the WF precoding matrix $\mathbf{F}$ and a formula for the scaling factor β by formulating the Lagrangian function $\mathrm{L}(\mathbf{F},\beta,\lambda)$.

$$\mathrm{L}(\mathbf{F},\beta,\lambda) = \mathrm{MSE}(\mathbf{F},\beta) + \lambda[\mathrm{tr}[\mathbf{F}\,\mathbf{R}_\mathbf{s}\,\mathbf{F}^H] - \rho] \tag{11.D.7}$$

Differentiate Eq. (11.D.7) with respect to, β, λ,

$$\frac{\partial L}{\partial \mathbf{F}} = \beta^{-2}\,(\mathbf{EH})^T(\mathbf{EH})^*\mathbf{F}^*\mathbf{R}_\mathbf{s}^T - \beta^{-1}(\mathbf{EH})^T\mathbf{R}_\mathbf{s}^T + \lambda\mathbf{F}^*\mathbf{R}_\mathbf{s}^T = \mathbf{0}_{\mathrm{M}\times\mathrm{B}} \tag{11.D.8}$$

$$\frac{\partial L}{\partial \beta} = \mathrm{tr}\left\{\frac{2}{\beta^2}\mathcal{R}e(\mathbf{EHF}\,\mathbf{R}_\mathbf{s})] + \frac{2}{\beta^3}\,[(\mathrm{tr}(-\mathbf{E}[\mathbf{HF}\,\mathbf{R}_\mathbf{s}(\mathbf{HF})^H + \mathbf{R}_\mathbf{n})\mathbf{E}^H]\right\} = 0 \tag{11.D.9}$$

$$\frac{\partial L}{\partial \lambda} = \mathrm{tr}[\mathbf{F}\mathbf{R}_\mathbf{s}\mathbf{F}^H] - \rho = 0 \tag{11.D.10}$$

$\mathbf{R}_\mathbf{s}^T$ Multiply both sides of (11.D.8) by (β^2) and $(\mathbf{R}_\mathbf{s}^T)^{-1}$ to get

$$(\mathbf{EH})^T(\mathbf{EH})^*\mathbf{F}^* = \beta(\mathbf{EH})^T - \beta^2\,\lambda\mathbf{F}^* \tag{11.D.11a}$$

$$\beta(\mathbf{EH})^T = [\mathbf{EH})^T(\mathbf{EH})^* + \beta^2\,\lambda\mathbf{I}_\mathrm{M}]\mathbf{F}^*$$

$$\mathbf{F}(\beta^2\,\lambda) = [\mathbf{EH})^H(\mathbf{EH}) + \beta^2\,\lambda\mathbf{I}_\mathrm{M}]^{-1}\beta(\mathbf{EH})^H \tag{11.D.11b}$$

Let us define $\mathbf{F}(\beta^2\,\lambda)$ in (11.D.11b) as

$$\mathbf{F}_{\mathrm{WN}}(\beta^2\,\lambda) = \beta\,\tilde{\mathbf{F}}(\beta^2\,\lambda) \tag{11.D.12a}$$

where

$$\tilde{\mathbf{F}}(\beta^2\,\lambda) = [\mathbf{EH})^H(\mathbf{EH}) + \beta^2\,\lambda\mathbf{I}_\mathrm{M}]^{-1}\,(\mathbf{EH})^H \tag{11.D.12b}$$

Using (11.D.12a) with the scaling factor β we have

$$\beta^2\mathrm{tr}[\tilde{\mathbf{F}}(\lambda\beta^2)\mathbf{R}_\mathbf{s}\tilde{\mathbf{F}}^H(\lambda\beta^2)] = \rho \tag{11.D.13a}$$

Therefore, we get

$$\beta = \sqrt{\frac{\rho}{\mathrm{tr}[\tilde{\mathbf{F}}(\lambda\beta^2)\mathbf{R}_\mathbf{s}\tilde{\mathbf{F}}^H(\lambda\beta^2)]}} \tag{11.D.13b}$$

We introduce $\xi = \lambda\beta^2 \in \mathbb{R}$ and manipulate the derivative w.r.t β in (11.D.9) as follows:

$$\mathrm{tr}\,\{2\beta\mathcal{R}e(\mathbf{EHF}\,\mathbf{R}_\mathbf{s})] + 2\,[(\mathrm{tr}(-\mathbf{E}[\mathbf{HF}\,\mathbf{R}_\mathbf{s}(\mathbf{HF})^H + \mathbf{R}_\mathbf{n})\mathbf{E}^H]\} = 0 \tag{11.D.14a}$$

Substitute (11.D.12a) for $\mathbf{F}$ in (11.D.14a) to get

$$\beta^2\mathrm{tr}[\mathcal{R}e(\mathbf{EH}\tilde{\mathbf{F}}(\xi)\,\mathbf{R}_\mathbf{s}] - \beta^2\,\mathrm{tr}[\mathbf{EH}\tilde{\mathbf{F}}(\xi)\mathbf{R}_\mathbf{s}(\mathbf{H}\tilde{\mathbf{F}}(\xi))^H\mathbf{E}^H] - \mathrm{tr}[\mathbf{E}\mathbf{R}_\mathbf{n}\mathbf{E}^H] = 0 \tag{11.D.14b}$$

Now from [17] we have

$$\mathrm{tr}[\mathcal{Re}(\mathbf{EH\tilde{F}}(\xi)\mathbf{R_s})] = \mathrm{tr}[(\mathbf{EH\tilde{F}}(\xi)\mathbf{R_s})] \tag{11.D.15}$$

Using (11.D.12), we get

$$(\mathbf{EH}) = \mathbf{\tilde{F}}^H(\xi)\,[\mathbf{EH})^H(\mathbf{EH}) + \beta^2\,\lambda\mathbf{I}_\mathrm{M}] \tag{11.D.16}$$

Substitute (11.D.16) in (11.D.15):

$$\begin{aligned}\mathrm{tr}[(\mathbf{EH\tilde{F}}(\xi)\mathbf{R_s})] &= \mathrm{tr}[(\mathbf{\tilde{F}}^H(\xi)\,[\mathbf{EH})^H(\mathbf{EH}) + \beta^2\,\lambda\mathbf{I}_\mathrm{M}]\mathbf{\tilde{F}}(\xi)\mathbf{R_s})]\\ \mathrm{tr}[(\mathbf{EH\tilde{F}}(\xi)\mathbf{R_s})] &= \mathrm{tr}[(\,[\mathbf{EH})^H(\mathbf{EH}) + \xi\mathbf{I}_\mathrm{M}]\mathbf{\tilde{F}}(\xi)\mathbf{R_s})\mathbf{\tilde{F}}^H(\xi)]\end{aligned} \tag{11.D.17}$$

Substitute (11.D.17) in (11.D.14):

$$\beta^2\mathrm{tr}[(\,[\mathbf{EH})^H(\mathbf{EH}) + \xi\mathbf{I}_\mathrm{M}]\mathbf{\tilde{F}}(\xi)\mathbf{R_s})\mathbf{\tilde{F}}^H(\xi)] - \beta^2\,\mathrm{tr}[\mathbf{EH\tilde{F}}(\xi)\mathbf{R_s}(\mathbf{H\tilde{F}}(\xi))^H\mathbf{E}^H] - \mathrm{tr}[\mathbf{ER_nE}^H] = 0 \tag{11.D.18}$$

We show that (11.D.18) can be simplified to

$$\begin{aligned}&\beta^2\mathrm{tr}[\xi\mathbf{\tilde{F}}(\xi)\mathbf{R_s}\mathbf{\tilde{F}}^H(\xi)] - \mathrm{tr}[\mathbf{ER_nE}^H] = 0\\ &\xi\beta^2\mathrm{tr}[\xi\mathbf{\tilde{F}}(\xi)\mathbf{R_s}\mathbf{\tilde{F}}^H(\xi)] - \mathrm{tr}[\mathbf{ER_nE}^H] = 0\end{aligned} \tag{11.D.19}$$

Use (11.D.13) to get

$$\rho\xi = \mathrm{tr}[\mathbf{ER_nE}^H]$$

Therefore, we have

$$\xi = \frac{\mathrm{tr}[\mathbf{ER_nE}^H]}{\rho} \tag{11.D.20}$$

Substitute for ξ in (11.D.20) in (11.D.12b) to get

$$\mathbf{\tilde{F}}(\beta^2\,\lambda) = \left[(\mathbf{EH})^H(\mathbf{EH}) + \frac{\mathrm{tr}[\mathbf{ER_nE}^H]}{\rho}\mathbf{I}_\mathrm{M}\right]^{-1}(\mathbf{EH})^H \tag{11.D.21}$$

Define $\mathbf{F}$ as

$$\mathbf{F} \in \mathbb{C}^{\mathrm{M}\times\mathrm{M}} := \mathbf{EH})^H(\mathbf{EH}) + \frac{\mathrm{tr}[\mathbf{ER_nE}^H]}{\rho}\mathbf{I}_\mathrm{M} \tag{11.D.22}$$

Then we get the WN filter matrix as

$$\mathbf{F}_\mathrm{WN} = \beta_\mathrm{WN}\,\mathbf{F}^{-1}(\mathbf{EH})^H \tag{11.D.23a}$$

$$\beta = \sqrt{\frac{\rho}{\mathrm{tr}[\mathbf{F}^{-2}\,(\mathbf{EH})^H\mathbf{R_s}\,(\mathbf{EH})\,]}} \tag{11.D.23b}$$

11.E MMSE Matrix

Consider a spatial multiplexing system with M_t transmit antennas, M_r receive antennas. The total power transmitted on M antennas at one transmission interval is ρ. The corresponding received signal is given by

$$\mathbf{y} = \mathbf{Hs} + \mathbf{n} \tag{11.E.1}$$

where the elements of the noise vector **n** are i.i.d., $\mathbf{n} \sim \mathbb{CN}(\mathbf{0},\ \sigma^2\mathbf{I}_N)$. An equalizer with **W** is applied to **y** to obtain an estimate of **s** as follows:

$$\tilde{\mathbf{s}} = \mathbf{W}\,\mathbf{y} = \mathbf{W}\mathbf{H}\mathbf{s} + \mathbf{W}\mathbf{n} \tag{11.E.2}$$

where the matrices **W** and **H** are expanded in (11.E.3) and (11.E.4):

$$\mathbf{W} = \begin{bmatrix} \mathbf{w}_1 \\ - \\ \mathbf{w}_k \\ - \\ \mathbf{w}_N \end{bmatrix} = \begin{bmatrix} w_{1,1} & - & - & - & - & - & w_{1,N} \\ - & - & - & - & - & - & - \\ w_{k,1} & - & - & - & - & - & w_{k,N} \\ - & - & - & - & - & - & - \\ w_{M,1} & - & - & - & - & - & w_{M,N} \end{bmatrix} \tag{11.E.3}$$

$$\mathbf{H} = \begin{bmatrix} \mathbf{h}_1 \\ - \\ \mathbf{h}_k \\ - \\ \mathbf{h}_N \end{bmatrix} = \begin{bmatrix} \mathrm{h}_{1,1} & - & - & \mathrm{h}_{1,k} & - & - & \mathrm{h}_{1,M} \\ - & - & - & - & - & - & - \\ \mathrm{h}_{k,1} & - & - & \mathrm{h}_{k,k} & - & - & \mathrm{h}_{k,M} \\ - & - & - & - & - & - & - \\ \mathrm{h}_{N,1} & - & - & \mathrm{h}_{N,k} & - & - & \mathrm{w}_{N,M} \end{bmatrix} \tag{11.E.4}$$

The MSE matrix is defined by the covariance matrix of the error vector $\boldsymbol{e}$ as $\mathbb{E}\|\boldsymbol{e}(\mathbf{W})\|^2$ where $\boldsymbol{e} \triangleq \mathbf{s} - \tilde{\mathbf{s}}$. We have the MSE, which is expressed as

$$\begin{aligned} \text{MSE} &= \mathbb{E}\|\boldsymbol{e}(\mathbf{W})\|^2 \triangleq \mathbb{E}[(\mathbf{s} - \tilde{\mathbf{s}})(\mathbf{s} - \tilde{\mathbf{s}})^H] = \mathbb{E}[(\mathbf{W}\mathbf{y} - s)(\mathbf{W}\,\mathbf{y} - s)^H] \\ &= \mathbb{E}[(\mathbf{W}\mathbf{y}\,\mathbf{y}^T\mathbf{W}^{\mathrm{H}} - \mathbf{W}\,\mathbf{y}\,s^H - s\mathbf{y}^H\mathbf{W}^H + ss^H)] \\ &= \mathbf{W}\mathbb{E}[\,\mathbf{y}\,\mathbf{y}^H]\mathbf{W}^{\mathrm{H}} - \mathbf{W}\,\mathbb{E}[\mathbf{y}\,s^H] - \mathbb{E}[s\mathbf{y}^H]\mathbf{W}^H + \mathbb{E}[ss^H] \end{aligned}$$

$$\text{MSE} = \mathbf{W}\mathbf{R}_{yy^H}\mathbf{W}^{\mathrm{H}} - \mathbf{W}\,\mathbf{R}_{ys^H} - \mathbf{R}_{sy^H}\,\mathbf{W}^{\mathrm{H}} + \mathbf{R}_s \tag{11.E.5}$$

We now take the partial derivative of (11.E.5) with respect to **W** and set it to zero to find the optimum matrix **W** for MMSE as follows:

$$\frac{\partial MSE(\mathbf{W})}{\partial \mathbf{W}} = 2\mathbf{R}_{yy^H}\mathbf{W}^H - \mathbf{R}_{ys^H} - \mathbf{R}^H_{sy^H} = 0 \tag{11.E.6}$$

From (11.E.6) we get

$$2\mathbf{R}_{yy^H}\mathbf{W}^H = \mathbf{R}_{ys^H} + \mathbf{R}^H_{sy^H}$$

$$\begin{aligned} \mathbf{R}_{yy^H} &\triangleq \mathbb{E}[y\,y^H] = \mathbb{E}[(\mathbf{H}\,s + \mathbf{n})\,[\,(\mathbf{H}\,\mathbf{s} + \mathbf{n})]^H] \\ &= \mathbb{E}[(\mathbf{H}\mathbf{s}\,\mathbf{s}^H\mathbf{H}^H]) + \mathbb{E}[(\mathbf{n}\,\mathbf{n}^H)\,] \end{aligned} \tag{11.E.7a}$$

$$\mathbb{E}[\mathbf{n}\,\mathbf{n}^H] = \sigma^2\mathbf{I}_N \tag{11.E.7b}$$

$$\mathbf{R}_{yy^H} = \mathbf{R}_s\mathbf{H}\mathbf{H}^H + \sigma^2\mathbf{I}_N \tag{11.E.7c}$$

$$\mathbf{R}_{ss} = \mathbb{E}[s\,s^H] = \mathbf{R}_s \tag{11.E.8}$$

$$\mathbf{R}_{ys^H} = \mathbb{E}[\mathbf{y}\,s^H] = \mathbb{E}[(\,\mathbf{H}\,s + \mathbf{n})\,s^H] = \mathbf{R}_s\mathbf{H} \tag{11.E.9}$$

$$\mathbf{R}_{sy^H} = \mathbb{E}[s\,(\,\mathbf{H}\,s + \mathbf{n})^H\,] = \mathbf{R}_s\mathbf{H}^H \tag{11.E.10}$$

Substitute (11.E.7c), (11.E.8), (11.E.9), (11.E.10) into (11.E.6):

$$2[\mathbf{R}_s\mathbf{H}\mathbf{H}^H + \sigma^2\mathbf{I}_N]\mathbf{W}^H - \mathbf{R}_s\mathbf{H} - \mathbf{R}_s\mathbf{H} = 0 \tag{11.E.11}$$

$$2[\mathbf{R}_s\mathbf{H}\mathbf{H}^H + \sigma^2\mathbf{I}_N]\mathbf{W}^H - 2\,\mathbf{R}_s\mathbf{H} = 0$$

Divide both sides of (11.E.11) by $\mathbf{R}_s$ we get

$$\left[\mathbf{H}\mathbf{H}^H + \frac{\sigma^2}{\mathbf{R}_s}\mathbf{I}_N\right]\mathbf{W}^H = \mathbf{H}$$

$$\mathbf{W}^H \left[\mathbf{H}^H\mathbf{H} + \frac{\sigma^2}{\mathbf{R}_s}\mathbf{I}_N\right] = \mathbf{H}$$

$$\mathbf{W} = \left[\mathbf{H}^H\,\mathbf{H} + \frac{\sigma^2}{\mathbf{R}_s}\,\mathbf{I}_N\right]^{-1}\,\mathbf{H}^H \tag{11.E.12b}$$

11.F SINR for MMSE Receiver for MF the Transmit Precoding

We use the following equation we derived in Appendix E:

$$\text{MSE} = \mathbf{W}\mathbf{R}_{yy^H}\,\mathbf{W}^H - \mathbf{W}\,\mathbf{R}_{ys^H} - \mathbf{R}_{sy^H}\,\mathbf{W}^H + \mathbf{R}_s \tag{11.F.1a}$$

Substitute (11.E.7b), (11.E.7c), and 1(11.E.8)–(11.E.10) in the MSE:

$$\text{MSE} = \mathbf{W}[\mathbf{R}_s\mathbf{H}\mathbf{H}^H + \sigma^2\mathbf{I}_N]\mathbf{W}^H - \mathbf{R}_s\,[\mathbf{W}\,\mathbf{H} + \mathbf{H}^H\,\mathbf{W}^H] + \mathbf{R}_s \tag{11.F.1b}$$

$$\text{MSE} = \mathbf{R}_s\left\{\mathbf{W}\left[\mathbf{H}\mathbf{H}^H + \frac{\sigma^2}{\mathbf{R}_s}\mathbf{I}_N\right]\mathbf{W}^H - [\mathbf{W}\,\mathbf{H} + \mathbf{H}^H\,\mathbf{W}^H] + \mathbf{I}_M\right\} \tag{11.F.1c}$$

Furthermore, we have shown in 11.E that the matrix **W** that minimises the MSE is given by

$$\mathbf{W} = \left[\mathbf{H}^H\,\mathbf{H} + \frac{\sigma^2}{\mathbf{R}_s}\,\mathbf{I}_M\right]^{-1}\,\mathbf{H}^H \tag{11.F.2a}$$

Equation (11.F.2a) can also be written as

$$\mathbf{W} = \mathbf{H}^H\left[\mathbf{H}\mathbf{H}^H + \frac{\sigma^2}{\mathbf{R}_s}\,\mathbf{I}_M\right]^{-1} \tag{11.F.2b}$$

In order to express the MMSE, we need to plug matrix **W** in (11.F.2b) in each term of (11.F.1c) that comprises the matrix. We start with

$$\mathbf{W}\,\mathbf{H} + \mathbf{H}^H\,\mathbf{W}^H = \mathbf{H}^H\mathbf{H}\left[\mathbf{H}^H\mathbf{H} + \frac{\sigma^2}{\mathbf{R}_s}\,\mathbf{I}_M\right]^{-1} + \mathbf{H}^H\mathbf{H}\left[\mathbf{H}^H\mathbf{H} + \frac{\sigma^2}{\mathbf{R}_s}\,\mathbf{I}_M\right]^{-1} \tag{11.F.3a}$$

$$[\mathbf{W}\,\mathbf{H} + \mathbf{H}^H\,\mathbf{W}^H\,] = 2[\mathbf{H}\mathbf{H}^H]\left[\mathbf{H}^H\mathbf{H} + \frac{\sigma^2}{\mathbf{R}_s}\,\mathbf{I}_M\right]^{-1} \tag{11.F.3b}$$

Accordingly, the MMSE is

$$\text{MMSE} = \mathbf{R}_s\left\{\mathbf{W}\left[\mathbf{H}\mathbf{H}^H + \frac{\sigma^2}{\mathbf{R}_s}\mathbf{I}_N\right]\mathbf{W}^H - [\mathbf{W}\,\mathbf{H} + \mathbf{H}^H\,\mathbf{W}^H] + \mathbf{I}_M\right\} \tag{11.F.4}$$

We can simplify (11.F.4) to

$$\text{MMSE} = \mathbf{R}_s\left\{\mathbf{I}_M - \left[\mathbf{H}^H\mathbf{H} + \frac{\sigma^2}{\mathbf{R}_s}\,\mathbf{I}_M\right]^{-1}\,[\mathbf{H}\mathbf{H}^H]\right\} \tag{11.F.5}$$

Use the Kailath variant the Woodbury to simply the inverted matrix in (11.F.5):

$$\text{MMSE} = \sigma^2\left[[\mathbf{H}^H\,\mathbf{H}] + \frac{\sigma^2}{\mathbf{R}_s}\,\mathbf{I}\right]^{-1} \tag{11.F.6}$$

It is well-known fact as the MMSE decreases, the value of the SINR increases reciprocally. The exact relation is usually expressed as

$$1 + \mathbf{SINR} = \frac{1}{\text{MMSE}} \tag{11.F.7}$$

Consequentially, substituting (11.F.6) in (11.F.7), we get

$$\mathbf{SINR} = \frac{1}{\sigma^2\left[[\mathbf{H}^H\,\mathbf{H}] + \frac{\sigma^2}{\mathbf{R}_s}\,\mathbf{I}_\mathrm{M}\right]^{-1}} - 1 \tag{11.F.8}$$

where the inverted matrix is assumed to be nonsingular and diagonal. The diagonal elements are the individual stream SINR. For example, the kth stream SINR is given by the kth row and the kth column of the diagonal matrix. For MF transmit precoding matrix, (11.F.8) changed to

$$(\mathbf{SINR})_k = \frac{1}{\sigma^2\left[[\mathbb{H}^H\,\mathbb{H}] + \frac{\sigma^2}{\mathbf{R}_s}\,\mathbf{I}_\mathrm{M}\right]^{-1}_{k,k}} - 1 \tag{11.F.9}$$

where $\mathbb{H} = \mathbf{HH}^H$ is the composite channel of the signal received by MMSE receiver.

References

1 Alodeh, M., Chatzinotas, S., and Ottersten, B. (2017). Symbol-level multi-user MISO precoding for multi-level adaptive modulation. *IEEE Transactions on Wireless Communications* 16 (8): 5511–5524.

2 Alodeh, M., Spano, D., Kalantari, S. et al. (2018). Symbol-level and multicast precoding for multiuser multiantenna downlink: A state-of-art, classification and challenges. *IEEE Communications Surveys & Tutorials* 20 (3): 1733–1757.

3 Vu, M. and Paulraj, A. (2007). MIMO wireless linear precoding. *IEEE Signal Processing Magazine* 24 (5): 86–105.

4 Mukherjee, A., and Kwon, H.M. (2007) 'Enhanced SVD-based transmit pre-processing for single-user and multi-user MIMO wireless systems with imperfect CSIT Asilomar Conference on Signals, Systems and Computers, 1659–1663.

5 Raleigh, G.G. and Cioffi, J.M. (1998). Spatio-temporal coding for wireless communications. *IEEE Transactions on Communications* 46 (3): 357–366.

6 Lu, A.-A., Gao, X., and Xiao, C. (2016). Free deterministic equivalents for the analysis of MIMO multiple access channel. *IEEE Transactions on Information Theory* 62 (8): 4604–4629.

7 Weichselberger, W., Herdin, M., Ozcelik, H., and Bonek, E. (2006). A stochastic MIMO channel model with joint correlation of both link ends. *IEEE Transactions on Wireless Communications* 5 (1): 90–100.

8 Oestges, C. (2006) 'Validity of the Kronecker model for MIMO correlated channels', IEEE Vehicular Technology Conference, 6, 2818–2822.

9 Wu, Y., Wen, C.-K., Xiao, C. et al. (2015). Linear precoding for the MIMO multiple access channel with finite alphabet inputs and statistical CSI. *IEEE Transactions on Wireless Communications* 14 (2): 983–997.

10 Haustein, T., von Helmolt, C., Jorswieck, E., et al.(2002) 'Performance of MIMO systems with channel inversion', IEEE Vehicular Technology Conference, 1, 35–39.

11 Goldsmith, A.J. and Varaiya, P.P. (1997). Capacity of fading channels with channel side information. *IEEE Transactions on Information Theory* 43 (6): 1986–1992.

12 Wiesel, A., Eldar, Y.C., and Shamai, S. (2008). Zero-forcing precoding and generalized inverses. *IEEE Transactions on Signal Processing* 56 (9): 4409–4418.

13 Bjornson, E., Ntontin, K., and Ottersten, B. (2011) 'Channel quantization design in multi-user MIMO systems: asymmetric verses practical conclusions', IEEE International Conference on Acoustics, Speech and Signal Processing, 3072–3075.

14 Long, Y., Chen, Z., and fang, J. (2016). Minimum number of antennas required to satisfy outage probability in massive MIMO systems. *IEEE Wireless Communications Letters* 5 (4): 348–351.

15 Makki, B., Svensson, T., Eriksson, T., and Alouini, M.-S. (2016). On the required number of antennas in a point to point large but finite MIMO system: outage –limited scenario. *IEEE Transactions on Communications* 64 (5): 1968–1982.

16 Luo, H., Xu, D., and Bao, J. (2018). Outage capacity analysis of MIMO system with survival probability. *IEEE Communications Letters* 22 (6): 1132–1135.

17 Joham, M., Utschick, W., and Nossek, J. (2005). Linear transmit processing in MIMO communications systems. *IEEE Transactions on Signal Processing* 53 (8): 2700–2712.

18 Mehana, A.H. and Nosratinia, A. (2014). Diversity of MIMO linear precoding. *IEEE Transactions on Information Theory* 60 (2): 1019–1038.

19 Joham, M., Utschick, W. and Nossek,J.A. (2001) 'On the equivalence of prerake and transmit matched filter', *Proc. ASST*, 313–318.

20 Esmailzadeh, R. and Nakagawa, M. (1993) 'Pre-RAKE diversity combination for direct sequence spread spectrum communications systems IEEE International Communications Conference (ICC), 1, 463–467.

21 Peel, C.B., Hochwald, B.M., and Swindlehurst, L. (2005). A vector-perturbation technique for near-capacity multiantenna multiuser communication—part I: channel inversion and regularization. *IEEE Transactions on Communications* 53 (1): 195–202.

22 Hochwald, B.M., Peel, C.B., and Swindlehurst, L. (2005). A vector-perturbation technique for near-capacity multiantenna multiuser communication—part II: perturbation. *IEEE Transactions on Communications* 53 (3): 537–544.

23 Stankovic, V. and Haardt, M. (2008). Generalized design of multi-user precoding matrices. *IEEE Transactions on Wireless Communications* 7 (3): 953–961.

24 Spencer, Q.H., Swindlehurst, A.L., and Haardt, M. (2004). Zero-forcing methods for downlink spatial multiplexing in multiuser MIMO channels. *IEEE Transactions on Signal Processing* 52 (2): 461–471.

25 Li, W. and Latva-aho, M. (2011). An efficient channel block diagonalization method for generalized zero forcing assisted MIMO broadcasting systems. *IEEE Transactions on Wireless Communications* 10 (3): 739–744.

26 Wang, H., Li, L., Wang, J., et al. (2011) 'A transmit precoding scheme for downlink multi-user MIMO systems IEEE Vehicular Technology Conference, 1–5

27 Zu, K., de Lamare, R.C., and Haardt, M. (2013). Generalized design of low-complexity block diagonalization type precoding algorithms for multiuser MIMO systems. *IEEE Transactions on Communications* 61 (10): 4232–4242.

28 Khan, M.H.A., Cho, K.M., Lee, M.H., and Chung, J.-G. (2014). A simple block diagonal for multi-user MIMO broadcast channels. *EURASIP Journal on Wireless Communications and Networking* 95: 1–8.

29 Heath, R.W., Sandhu, S., and Paulraj, A. (2001). Antenna selection for spatial multiplexing systems with linear receivers. *IEEE Communications Letters* 5 (4): 142–144.

30 Agnihotri, A. and Gupta, B. (2015) 'Performance evaluation of linear, non-linear precoding schemes for downlink multi-user MIMO systems', IEEE International Conference on Industrial Instrumentation and Control (ICIC), 484–489.

31 Muller R.R. (2001) 'Polynomial expansion equalizers for communication via large MIMO arrays', European Personal Mobile Communications Conference. 1–7.

32 Kohno, R., Hatori, M., and Imai, H. (2001). Cancellation techniques of co-channel interference in asynchronous spread spectrum multiple access systems. *Electronics and Communications in Japan* 66A (5): 20–29.

33 Muller, R.R. and Verdu, S. (2001). Design and analysis of low-complexity interference mitigation on vector channels. *IEEE Journal on Selected Areas in Communications* 19 (8): 1429–1441.

34 Zhang, J., Wu, Y., Gu, J., et al. (2005) 'Polynomial expansion based fast iterative multiuser detection algorithm for synchronous DS-CDMA systems', IEEE Vehicular Technolog Conference, 2, 988–991.

35 Zarei, S., Gerstacker, W., Muller, R.R., and Schober, R. (2013) 'Low-complexity linear precoding for downlink large-scale MIMO systems', IEEE International Symposium on Personal, Indoor, and Radio Communications: Fundamentals and PHY Track', 1119–1124.

36 Kammoun, A., Muller, A., Bjorson, E., and Debbah, M. (2014). Linear precoding based on polynomial expansion: large-scale multi-cell MIMO systems. *IEEE Journal of Selected Topics in Signal Processing* 8 (5): 861–871.

37 Muller, A., Kammoun, A., Bjorson, E., and Debbah, M. (2016). Linear precoding based on polynomial expansion: reducing-complexity in massive MIMO. *EURASIP Journal on Wireless Communications and Networking* 63: 1–22.

38 Muller, A., Kammoun, A., Bjorson, E., and Debbah, M. (2014) 'Efficient linear precoding for massive MIMO systems using truncated polynomial expansion', IEEE Sensor Array and Multichannel Signal Processing Workshop, 273–276.

39 Narasimhan, T.L. and Chockalingam, A. (2014). Channel hardening-exploiting message passing (CHEMP) receiver in large-scale MIMO systems. *IEEE Journal of Selected Topics 1946 in Signal Processing* 8 (5): 847–860.

Index

5G Physical Layer Technologies, First Edition. Mosa Ali Abu-Rgheff.